TOWARDS PROLONGATION OF THE HEALTHY LIFE SPAN

Practical Approaches to Intervention

ANNALS OF THE NEW YORK ACADEMY OF SCIENCES
Volume 854

TOWARDS PROLONGATION OF THE HEALTHY LIFE SPAN
Practical Approaches to Intervention

Edited by Denham Harman, Robin Holliday, and Mohsen Meydani

New York Academy of Sciences
New York, New York
1998

Softcover art: Four generations of the Bohall family (right to left), Roger, Robert, Steven, and Jeremy.

Library of Congress Cataloging-in-Publication Data

Towards prolongation of the healthy life span : practical approaches
   to intervention / edited by Denham Harman, Robin Holliday, and
   Mohsen Meydani.
      p.   cm. — (Annals of the New York Academy of Sciences ; v. 854)
      Includes bibliographical references and index.
      ISBN 1-57331-108-1 (cloth : alk. paper). —ISBN 1-57331-109-X
(pbk. : alk. paper)
      1. Longevity—Congresses. 2. Aging—Molecular aspects—
Congresses. 3. Aging—Physiological aspects—Congresses. I. Harman,
Denham. II. Holliday, R. (Robin), 1932- .
   III. Meydani, Mohsen. IV. Series.
   Q11.N5 vol. 854
   [QP85]
   500 s—dc21
   [612.6'8]   98-31475
                                   CIP

E-Med/B-M
Printed in the United States of America
ISBN 1-57331-108-1 (cloth)
ISBN 1-57331-109-X (paper)

ANNALS OF THE NEW YORK ACADEMY OF SCIENCES

Volume 854
November 20, 1998

TOWARDS PROLONGATION OF THE HEALTHY LIFE SPAN
Practical Approaches to Intervention[a]

Editors
DENHAM HARMAN, ROBIN HOLLIDAY, AND MOHSEN MEYDANI

Conference Organizers
JOSEPH CANNON, DENHAM HARMAN, ROBIN HOLLIDAY, DONALD K. INGRAM,
MOHSEN MEYDANI, AND SANG CHUL PARK

Advisory Board
KENICHI KITANI, IMRE ZS.-NAGY, EDWARDO PORTA, AND BYUNG P. YU

CONTENTS

Part I. Molecular and Cellular Changes with Age

---

[a]This volume is the result of a conference, entitled **Towards Prolongation of the Healthy Life Span: Practical Approaches to Intervention**, held in Adelaide, Australia on August 15–18, 1997, by the Seventh Congress of the International Association of Biomedical Gerontology.

Part VII. Poster Papers

Financial assistance was received from:
- JEAN CARPER
- GATORADE SPORTS SCIENCE INSTITUTE OF THE QUAKER OATS COMPANY
- HENKEL CORPORATION
- LIFE EXTENSION FOUNDATION
- NIKKEN FOODS COMPANY LTD.
- NOW FOODS
- NUTRITION 21
- ROCHE VITAMINS & FINE CHEMICALS
- ROSS PRODUCTS DIVISION: ABBOTT LABORATORIES
- VITAMIN RESEARCH PRODUCTS, INC.

Preface

DENHAM HARMAN

University of Nebraska College of Medicine, Department of Medicine, 600 South 42nd Street, Omaha, Nebraska 68198-4635, USA

The International Association of Biomedical Gerontology (IABG) was organized in 1985. Beginning in 1985 in New York City, the IABG has held a congress every two years; with the exception of the first and third congresses, the proceedings have been published (see below). When the meeting year of the IABG coincides with that of the International Association of Gerontology (IAG), the IABG meeting is held just prior to that of the IAG. The purpose of the IABG is to (1) make the general public more aware of the potential of biomedical aging research to increase the span of healthy productive life and to decrease the social and economic problems of age; and to (2) promote greater communication among the worldwide community of individuals engaged in biomedical aging research.

The 7th Congress of the IABG, held on August 15–18, 1997 at the Hilton Hotel in Adelaide, Australia, was entitled "Towards Prolongation of the Healthy Life Span: Practical Approaches to Intervention," indicating that emphasis would be placed on practical measures to increase the functional life span. I deeply appreciate the financial support from many companies and individuals that made this meeting possible, particularly the significant, unsolicited donation from Ms. Jean Carper, author of *Stop Aging Now.* I also am indebted to the speakers and to the audience for making the congress successful. Discussions were frequent and animated throughout this excellent four-day meeting.

This volume contains chapters by all but four of the invited speakers, as well as abstracts from all the speakers and from those who presented posters. The following are two of the missing lectures with references to the same topics: S. Christen and B. N. Ames, Electrophile-trapping by gamma-tocopherol: A new physiological function for vitamin E?[1] and C. J. C. Hsia, Nitroxides as multifunctional antioxidant enzyme mimics.[2,3]

This meeting, as in past congresses, focused on major topics of current biomedical aging research. As usual, mirroring the many different points of view in this area of research, there was no general agreement on the cause(s) of aging. There is growing evidence, however, that efforts over the past 40 years, based on one postulated cause of aging, are now being reflected in increases in the span of healthy productive life, that is, the functional life span. Thus, in the United States, the average life expectancy at birth (ALE-B) rose from 69.7 years in 1960 to 75.4 years in 1990 and to 75.7 years in 1994. Increases in ALE-B were associated with relative increases in the size of the older population. Between 1960 and 1990 the number of individuals 65 years of age or older (the elderly) grew by 89%, those age 85 and older increased by 232%, whereas the total population grew by 39 percent.

Accompanying increases in the elderly fraction of the population have been declines, at least from 1982 to 1994, in the proportion of elderly who are chronically disabled, that is, the functional life span is increasing. Further, although the average age of the population is increasing, the overall age-adjusted cancer mortality rate has been declining since 1991, and the cardiovascular death rate continues to decline.

The above data are in accord with beneficial effects expected from the growing use of antioxidant supplements since the 1960s to decrease disease and enhance life span and the widespread publicity about the ability of fruits and vegetables to decrease disease by depressing free radical reaction damage. Since the 1960s the percentage of the population of the United States taking antioxidant supplements has increased to a value of 40–50% today. Most take supplements on an irregular basis: on a daily basis about 8% take ascorbic acid, and around 4% take vitamin E.

It is reasonable to expect, on the basis of animal and epidemiological studies, that efforts over the past 40 years to decrease free radical damage have helped to increase the functional life span by contributing significantly to the decline in such "free radical" diseases as cancer and atherosclerosis, increases in the fraction of the elderly in the population, and the decline in chronic disability in this group. It is also reasonable to expect that a consensus on the basic cause(s) of aging will arise eventually from accumulating basic biomedical aging knowledge. This will expedite further productive efforts to emulate the ideal life envisioned by Oliver Wendell Holmes in his poem, "The Deacon's Masterpiece: The Wonderful One-Hoss Shay," a long functional life that ends quickly and painlessly. Finally, I hope that this volume will help to increase the number of scientists involved in the interesting, important problem of extending the healthy, productive life span.

REFERENCES

1. Christen, S., A.A. Woodall, M.K. Shigenaga, P.T. Southwell-Keely, M.W. Duncan & B.N. Ames. 1997. γ-Tocopherol traps mutagenic electrophiles such as NO(x) and complements alpha-tocopherol: Physiological implications. Proc. Natl. Acad. Sci. USA **94:** 3217–3222.
2. Kuppusamy, P., P. Wang, J.L. Zweier, M.C. Krishna, J.B. Mitchell, L. Ma, C.E. Trimple & C.J. Hsia. 1996. Electron paramagnetic resonance imaging of rat heart with nitroxide and polynitoxyl-albumiun. Biochemistry **35:** 7051–7057.
3. Zhang, R., E. Shohami, E. Bett-Yannai, R. Bass, V. Trembovler & A. Sumuni. 1998. Mechanism of brain protection by nitroxide radicals in experimental model of closed-head injury. Free Radical Biol. Med. **24:** 332–340.

Past Published Proceedings

2nd Congress: Steinhagen-Thiessen, E. & D.L. Knook, Eds. 1988. Trends in Biomedical Gerontology. TNO Institute for Experimental Gerontology, Rijswijk.
4th Congress: Fabris, N., D. Harman, D. L. Knook, E. Steinhagen-Thiessen & I. Zs.-Nagy, Eds. 1992. Physiopathological Processes of Aging: Towards a Multicausal Interpretation. Annals of the New York Academy of Sciences. Vol. 673.
5th Congress: Zs.-Nagy, I., D. Harman & K. Kitani, Eds. 1994. Pharmacology of Aging Processes: Methods of Assessment and Potential Interventions. Annals of the New York Academy of Sciences. Vol. 717.
6th Congress: Kitani, K., A. Aoba & S. Goto, Eds. 1996. Pharmacological Intervention in Aging and Age-Associated Disorders. Annals of the New York Academy of Sciences. Vol. 786.

TOWARDS PROLONGATION OF THE HEALTHY LIFE SPAN

Practical Approaches to Intervention

Aging: Phenomena and Theories

DENHAM HARMAN[a]

*University of Nebraska College of Medicine, Department of Medicine,
600 South 42nd Street, Omaha, Nebraska 68198-4635, USA*

ABSTRACT: Aging is the accumulation of diverse adverse changes that increase the risk of death. These changes can be attributed to development, genetic defects, the environment, disease, and the inborn aging process. The chance of death at a given age serves as a measure of the number of accumulated aging changes, that is, of physiologic age, and the rate of change of this measure, as the rate of aging. As living conditions in a population approach optimum, the curve of the logarithm of the chance of death versus age shifts towards a limit determined by the sum of (1) the irreducible contributions to the chance of death by aging changes that can be prevented to varying degrees, and (2) those due to the intrinsic aging process. In the developed countries living conditions are now near optimum, and the ALE-Bs are about 6–9 years less than the potential maximum of around 85 years. The inborn aging process is now the major risk factor for disease and death after about age 28. By age 28 only 1 to 2% of a cohort is dead, the remaining 98 to 99% die at an exponentially increasing rate determined by the aging process. This process ensures that few reach 100 years and none exceed about 122 years. Many theories have been advanced to account for the aging process. No single theory is generally accepted. Theories that can contribute to the important practical goal of increasing the healthy, useful span of humans will endure.

AGING

Definition

Aging is the accumulation of diverse adverse changes[1,2] that increase the risk of death.[3] These aging changes are responsible for both the commonly recognized sequential alterations that accompany advancing age beyond the early period of life and the progressive increases in the chance of disease and death associated with them. Aging changes can be attributed to development, genetic defects, the environment, disease, and an inborn factor, the aging process: these categories may not be independent.

The chance of death of an individual of a given age in a population, readily available from vital statistics data, serves as a measure of the average number of aging changes accumulated by persons of that age, that is, of that physiologic age, and the rate of change of the chance of death with time at that age, as the average rate of aging.

The chances for death in a population determine the average life expectancy at birth (ALE-B) ALE-B serves as a rough measure of the span of healthy, productive life, that is, the functional life span.

Effect of Improved Living Conditions

Conventional means of increasing the ALE-B of a population by decreasing the chances for death through improvements in general living conditions are becoming

[a]Tel: 402/559-4416; fax: 402/559-7330; e-mail: dharman@unmc.edu

increasingly futile. This is illustrated in FIGURE 1 by the curves of the logarithm of the chance of death versus age for Swedish females for various periods from 1751 to 1992;[4] a straight line represents exponential increases with age.

The chance of death[1-7] drops precipitously for a short period after birth (largely due to a declining death rate related to birth and early development), to reach a minimum around puberty. It then increases with age as deleterious effects of other contributors (for example, genetic defects, environment, and disease) to the chance of death increase. These adverse effects eventually give rise to the sequential changes associated with aging and exponential increases in the chance of disease and death.

Improvements in living conditions in Sweden, including better nutrition, housing, medical care, public health facilities, and accident prevention, decreased the chances for death towards limiting values. The decreases were greater in the young than in the old, whereas the age beyond which the chance of death rises linearly started at progressively lower ages, reaching an age of about 28 in 1992; today only 1.1% of female cohorts die before age 28.[4]

The chances for death for both sexes in Sweden, as well as in the other developed countries, are slowly declining.[4,8-10] As a result, average life expectancies at birth have been

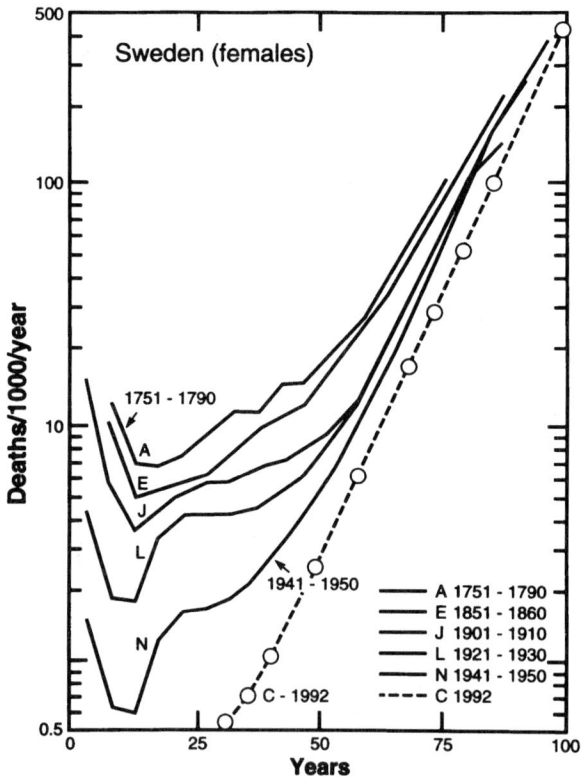

FIGURE 1. Age-specific death rates of Swedish females in various periods from 1751 to 1950 (adapted from H. R. Jones[5]) and for 1992.[4]

TABLE 1. Average Life Expectancy of Male and Females, and for the Population as a Whole, in Four Developed Coutries in 1992

	Males	Females	Whole Population
United States	73.2	79.7	75.7
Switzerland	74.3	81.2	77.8
Sweden	75.4	80.8	78.1
Japan	76.1	82.2	79.2

gradually increasing over the past four decades at a rate of one to two years per decade, except in Japan where the rate has been over three years. This is illustrated in the 1950–1992 data for Sweden[4] (FIG. 2), Switzerland[8] (FIG. 3), the United States[9] (FIG. 4), and Japan[10] (FIG. 5). ALE-Bs in 1992 for males and females, and for the population as a whole, in these countries are shown in TABLE 1. In Japan[10] life expectancies rose rapidly from the values in 1950 (males, 58.0 years; females 61.5 years), so that by 1987 they had become the longest living population.

ALE-Bs in the developed countries are now about 6–9 years less than the potential maximum of about 85 years.[11–13] Significant increases in ALE-Bs beyond 85 years can be achieved only by slowing the rate of aging from the inborn aging process. In the absence of efforts to slow this process, ALE-Bs may reach plateau values of about 80–82 years in 60–80 years.

Thus, as living conditions in a population approach the optimum, and premature deaths approach a minimum, the curve of the logarithm of the chance of death versus age shifts towards a limit determined by the sum of (1) the irreducible contributions to the chance of

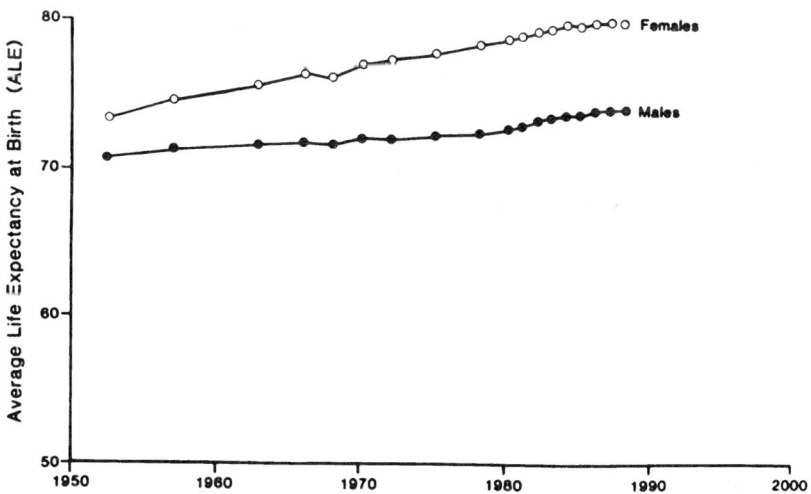

FIGURE 2. Average life expectancy at birth in Sweden, 1950–1992.

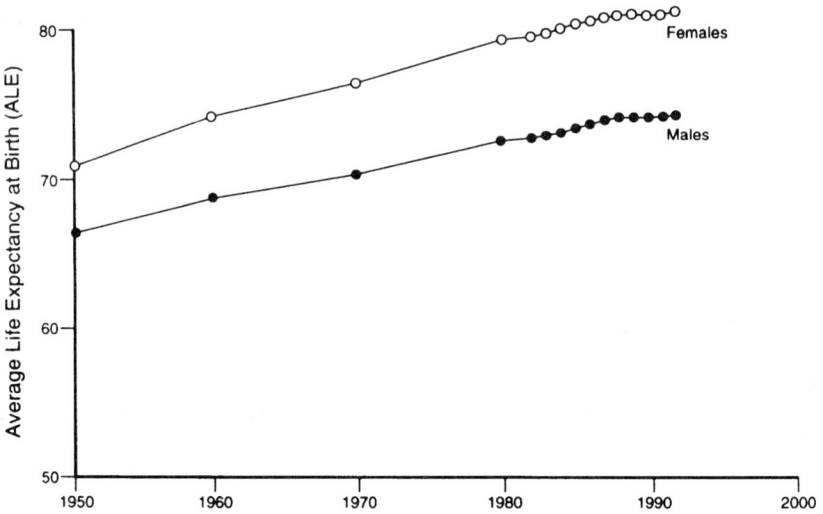

FIGURE 3. Average life expectancy at birth in Switzerland, 1950–1992.

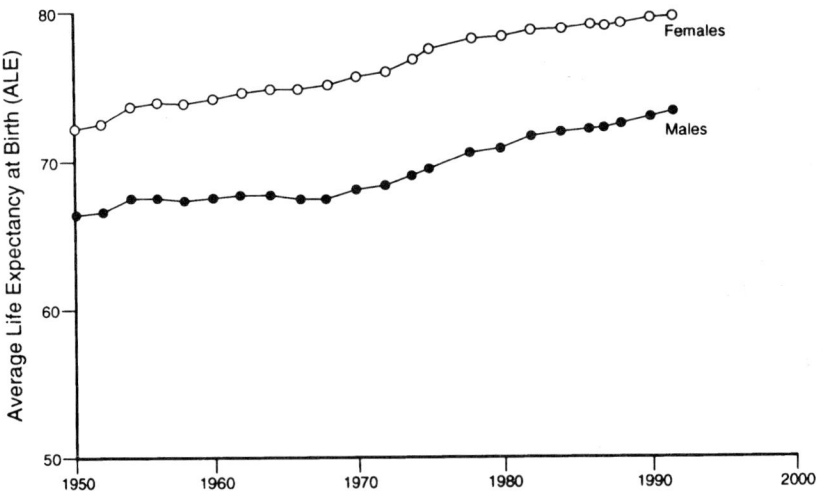

FIGURE 4. Average life expectancy at birth in the United States, 1950–1992.

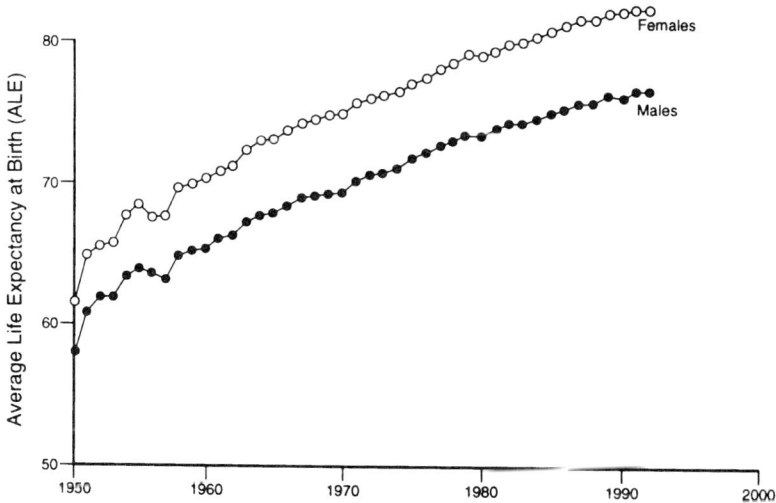

FIGURE 5. Average life expectancy at birth in Japan, 1950–1992.

death of aging changes that can be prevented to varying degrees, for example, those due to development, genetic defects, the environment, and disease, and (2) those due to an intrinsic process—the aging process. The contributions of the aging process are produced by chemical reactions that arise in the course of normal metabolism, which, collectively, produce aging changes that exponentially increase the chance of death with advancing age even under optimal living conditions.

The rate of aging, that is, the rate of production of aging changes, should vary from individual to individual, due to differences in genetic and environmental factors that modulate production of aging changes and thus contribute to differences in the age of death and of the onset of disease.

THE AGING PROCESS

Nature

The aging process produces aging changes at an unalterable exponentially increasing rate with advancing age so that few reach 100 years[14] and none live beyond about 122 years.[15] The longest lived person whose date of birth can be confirmed was Jeanne L. Calment, born on February 21, 1875 in France; she died on August 4, 1997 at age 122 in a nursing home in Arles, France.[15] The rate of aging is low early in life but rapidly increases with age due to the exponential nature of the process, illustrated in FIGURE 6 by a plot of the chances for death in 1985 for the entire population of the United States as a function of age.[6]

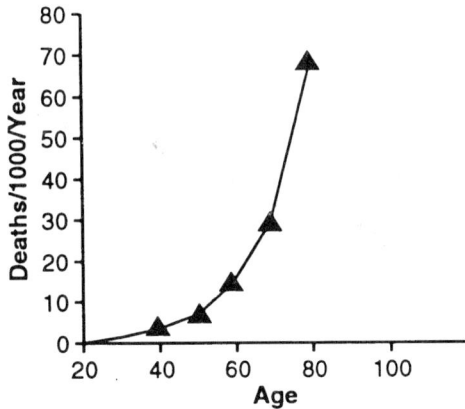

FIGURE 6. The chance of dying in 1985 as a function of age for the total population of the United States.

The relative contribution of the aging process to the chance of death increases as living conditions improve, and with advancing age. Today the chance of death in the developed countries after age 28 is essentially due to the aging process. Inasmuch as death can usually be ascribed to some disease (for example, the two major causes of death, cancer and cardiovascular disorders, or a neurodegenerative disease, such as senile dementia of the Alzheimer's type), the foregoing implies that the aging process is now the major risk factor for disease after age 28 in the developed countries. The importance of this process to our health and well-being is obscured by the protean nature of its contributions[2] to nonspecific change and to disease pathogenesis.

The aging process may be common to all living things because the phenomena of aging and death is universal. It may also be the major determinant of the manifestations of aging, for these occur even under optimal living conditions. The aging process is under genetic control; life span and manifestations of aging differ among species and individual members of a species. This process is also subject to environmental influences like other chemical reactions.

Theories

Many theories have been advanced to account for the aging process.[16-20] For example, the aging process has been attributed to molecular cross-linking,[21] changes in immunologic function,[22] damage by free-radical reactions,[23] senescence genes in the DNA,[24] and most recently, to telomere shortening.[25] No single theory is generally accepted: "This remarkable process remains a mystery,"[26] and "It is doubtful that a single theory will explain all the mechanisms of aging."[27,28]

The importance attached to increasing the healthy, useful life span of humans beyond the 3–5 years that may still be achieved by conventional measures, dictates that the aging process hypothesis be explored for practical means of achieving this goal. Those hypotheses that can contribute to this important practical problem will endure.

REFERENCES

1. KOHN, R.R. 1985. Aging and age-related diseases: Normal processes. *In* Relation between Normal Aging and Disease. H. A. Johnson, Ed.: 1–44. Raven Press. New York.
2. UPTON, A.C. 1977. Pathology. *In* The Biology of Aging, C.E. Finch & L. Hayflick, Eds.: 513–535. Van Nostrand. New York.
3. HARMAN, D. 1994. Aging· Prospects for further increases in the functional life span. Age **17**: 119–146.
4. SVERIGES OFFICIELLA STATISTIK. 1988. Befolknings-forandringar. 1987. Statistiska centralbyran: 114–115. Stockholm, Sweden.
5. JONES, H.R. 1955. The relation of human health to age, place and time. *In* Handbook of Aging and the Individual. J.E. Birren, Ed.: 333–363. Chicago University Press. Chicago, Ill.
6. NATIONAL CENTER FOR HEALTH STATISTICS. 1988. Vital Statistics of the United States. 1985. Life Tables, Vol. 2 (6). Hyattsville, MD (U.S. Dept. Health & Human Serv.), PHS Publ. No. 88-1104: 9.
7. DUBLIN, L.I., A.J. LOTKA & M. SPIEGELMAN. 1949. The contribution of medical and sanitary science to health and longevity. *In* Length of Life. 141–166. Ronald Press. New York.
8. OFFICE FEDERAL DE LA STATISTIQUE. 1993. Suisse—Table de Mortalite 1950–1992. Swiss Government. Berne, Switzerland.
9. NATIONAL CENTER FOR HEALTH STATISTICS. 1993. Annual Summary of Births, Marriages, Divorces, and Deaths United States. 1992. Monthly Vital Statistics 41: (13). Hyattsville, MD (U.S. Dept. Health Human Ser.) PHS Publ. No. 93-1120.
10. STATISTICS AND INFORMATION DEPARTMENT. 1992. Abridged Life Tables for Japan, Average Life Expectancy 1945–1992. Minister's Secretariat, Ministry Health Welfare: Japanese Government, Tokyo 162, Japan.
11. WOODHALL, B. & S. JOBLON. 1957. Prospects for future increases in average longevity. Geriatrics **12**: 586–591.
12. FRIES, J.F. 1988. Aging, natural death, and the compression of morbidity. N. Engl. J. Med. **303**: 130–135.
13. OLSHANSKY, S.J., B.A. CARNES & C. CASSEL. 1990. In search of Methuselah: Estimating the upper limits to human longevity. Science **250**: 634–640.
14. COMFORT, A. 1979. The Biology of Senescence, 3rd edit.: 81–86. Elsevier. New York.
15. YOUNG, M.C., Ed. 1997. The Guinness Book of Records: 11. Bantam Books. New York.
16. ROCKSTEIN, M., M.L. SUSSAMAN & J. CHESKY, Eds. 1974. Theoretical Aspects of Aging. Academic Press. New York.
17. HAYFLICK, L. 1985. Theories of biological aging. Exp. Gerontol. **20**: 145–159.
18. WARNER, H.R., R.N. BUTLER, R.L. SPROTT & E.L. SCHNEIDER, Eds. 1987. Modern Biological Theories of Aging. Raven Press. New York.
19. MEDVEDEV, Z.A. 1990. An attempt at a rational classification of theories of aging. Biol. Rev. Camb. Philos. Soc. **65**: 375–398.
20. HOLLIDAY, R. 1995. Understanding Ageing. Cambridge University Press. New York.
21. BJORKSTEN, J. 1968. The crosslinkage theory of aging. J. Am. Geriat. Soc. **16**: 408–427.
22. WALFORD, R.L. 1969. The immunologic theory of aging. Munksgaard. Copenhagen, Denmark.
23. HARMAN, D. 1993. Free radical involvement in aging: Pathophysiology and therapeutic implications. Drugs & Aging **3**: 60–80.
24. HAYFLICK, L. 1987. Origins of longevity. *In* Modern Biological Theories of Aging. H.R. Warner, R.N. Butler, R.L. Sprott & E.L. Schneider, Eds.: 21–34. Raven Press. New York.
25. KRUK, P.A., N.J. RAMPINO & V. A. BOHR. 1995. DNA damage and repair in telomeres: Relation to aging. Proc. Natl. Acad. Sci. USA **92**: 258–262.
26. ROTHSTEIN, M. 1986. Biochemical studies of aging. Chem. & Eng. News **64** (No. 32): 26.
27. SCHNEIDER, E.L. 1987. Theories of aging: A perspective. *In* Modern Biologic Theories of Aging. H.E. Warner, R.N. Butler, R.L. Sprott & E.L. Schneider, Eds.: 1–4. Raven Press. New York.
28. VIJG, J. 1990. Searching for the molecular basis of aging: The need for life extension models. Aging (Milano) **2**: 227–229.

Genes Involved in the Control of Cellular Proliferative Potential[a]

ROGER R. REDDEL[b]

*Children's Medical Research Institute, 214 Hawkesbury Road,
Westmead, Sydney, NSW 2145, Australia*

ABSTRACT: Evidence that control of cellular proliferative potential may be linked to telomere length, along with data indicating that other factors may also be involved, will be reviewed. According to the telomere hypothesis of senescence, the sequential loss of telomeric repeat DNA that occurs during the replication of normal somatic cells eventually dictates the onset of the permanently nonreplicative state known as senescence. Many immortalized cells express telomerase, a ribonucleoprotein enzyme that replaces the telomeric DNA that would otherwise be lost due to replication. However, some immortalized human cells may avoid telomeric shortening without using telomerase. The mechanism involved is currently unknown, but other eukaryotes are able to replace telomeric DNA through (1) recombination and copy switching or (2) retrotransposition. Human fibroblasts that lose p53 function proliferate a limited number of times beyond the population-doubling level at which their normal counterparts become senescent. Lack of functional retinoblastoma (Rb) protein (or equivalent events, such as loss of p16INK4 function, resulting in abrogation of Rb regulatory activity) also permits a temporary extension of proliferative potential. The p53 and pRb effects are additive, indicating that they exert their control on proliferative potential separately. The temporary life span extension associated with loss of p53 and/or Rb pathway function is accompanied by continued telomere shortening. The proliferation arrest that eventually ensues in p53-minus cells or in p53-minus/Rb-minus cells may be regarded as terminal proliferation arrest states serving as a backup to senescence. p53-minus/Rb-minus cells cannot proliferate further unless they acquire the ability to prevent telomeric shortening. Somatic cell hybridization and microcell-mediated chromosome transfer experiments indicate that immortalization involves the loss of function of other, as yet unidentified, genes; some of these may normally repress telomerase expression in somatic cells.

TERMINAL PROLIFERATION ARREST STATES

The studies of Hayflick and colleagues showed that normal diploid human cells proliferate only a limited number of times *in vitro* before they enter a nondividing state, referred to as senescence.[34] Although they are permanently mitotically sterile, senescent cells may remain metabolically active for a long period of time (reviewed in refs. 14, 15, and 26). Senescence is therefore a viable terminal proliferation arrest (TPA) state distinct not only from the temporary growth arrest occasioned by adverse conditions, such as nutrient deprivation or minor DNA damage on the one hand, but also from the various forms of cell death, such as apoptosis, that are permanent but result in rapid loss of viability.

[a]Work in this laboratory has been supported by the Carcinogenesis Fellowship of the NSW Cancer Council and project Grants from the National Health and Medical Research Council of Australia.
[b]Address for correspondence: Roger R. Reddel, Children's Medical Research Institute, Locked Bag 23, Wentworthville, NSW 2145, Australia. Tel.: +61-2-9687-2800; fax: +61-2-9687-2120; e-mail: rreddel@cmri.usyd.edu.au

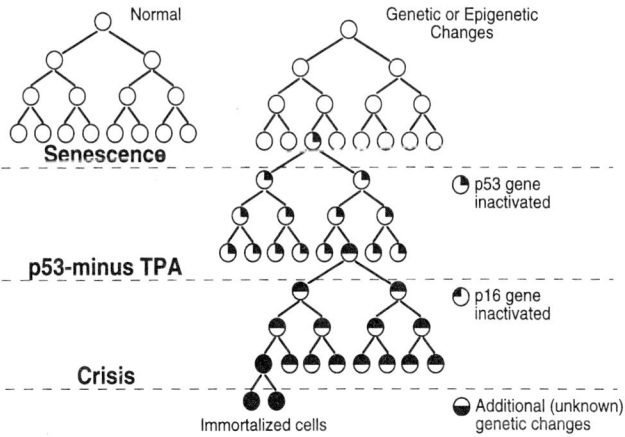

FIGURE 1. Schematic representation of viable terminal proliferation arrest (TPA) states in human fibroblasts. Normal cells permanently cease dividing in a state referred to as senescence. Other TPA states act as barriers to continued proliferation following various genetic or epigenetic changes. The genetic changes responsible for immortalization are essentially unknown, but they are associated with the activation of a telomere maintenance process such as telomerase.

Studies of virally transformed human cells revealed the existence of another viable TPA state referred to as crisis.[25] Human fibroblasts infected with simian virus 40 (SV40) or transfected with SV40 early-region DNA proliferate beyond the population doubling (PD) level at which their normal counterparts undergo senescence, but eventually the cells enter crisis and cell numbers stop increasing (reviewed in ref. 12). Alternative names were subsequently suggested for senescence and crisis, *viz.* mortality states 1 and 2 (M1 and M2).[81]

Recently it has been found that loss of functional p53 gene activity may result in fibroblasts escaping temporarily from senescence before they undergo TPA at a PD level that is considerably lower than the PD level characteristic of crisis. We refer to this as the p53-minus TPA state. Fibroblasts infected with a recombinant retrovirus carrying a dominant negative p53 gene proliferated for 17 PDs more than their normal counterparts.[7] Li-Fraumeni syndrome (LFS) fibroblasts (that were heterozygous wild-type [wt]/mutant at the p53 locus) became senescent at the expected PD level but proliferated for an additional 33 PDs following spontaneous loss of the wt p53 allele.[69] By contrast, when fibroblasts from the same LFS individual were transfected with SV40 or human papillomavirus (HPV) strain 16 genes, they proliferated for an average of 67 PDs before entering crisis.[31,61]

These data indicate the existence of at least three TPA states, also referred to as life span checkpoints.[82] It is not yet known whether there are other TPA states, or to what extent these findings will apply to cells from nonfibroblastic lineages.

This article briefly reviews what is known of the genetic basis of proliferative life span control. Because abrogation of wt p53 function allows escape from senescence,[7,69] it is likely that p53 plays a major role in this TPA state. Normal cells transduced with viral oncoproteins that bind and inactivate the protein products of both the p53 and retinoblastoma (Rb) genes proliferate until crisis (reviewed in ref. 12). This suggests that the retinoblastoma protein (pRb) plays a role in the p53-minus TPA state. There is evidence, summarized below, that normally functioning p16INK4 is required for pRb to limit proliferative capacity and that loss of p16INK4 is functionally equivalent to loss of pRb in permitting escape from the normal limits on proliferation. The genes involved in crisis are essentially unknown, but escape from crisis is associated with the activation of a telomere maintenance mechanism such as telomerase. The p53-minus TPA and crisis may be regarded as "backup" TPAs to senescence. Each of these proliferation control barriers must be breached in order for immortalization (unlimited proliferative capacity) to ensue (FIG. 1).

THE ROLE OF p53 IN PROLIFERATIVE LIFE-SPAN CONTROL

p53 is better known for its role in inducing either (1) temporary proliferation arrest following DNA damage or nutrient deprivation, or (2) apoptotic cell death (reviewed in ref. 40). Because a temporary arrest to allow DNA repair may protect the integrity of the genome, and apoptosis may rid tissues of cells with a load of DNA damage, p53 has been called "guardian of the genome"[47] and "guardian of the tissues" for these respective roles. More recently evidence has been obtained that p53 protects against teratogenesis (reviewed in ref. 29).

It is clear, however, that p53 also has a role in senescence (FIG. 2). Although there appears to be some controversy as to whether p53 levels actually increase,[1,2,46,76] there is evidence for an increase in p53 transcriptional activity and DNA binding activity as cells approach senescence.[2,6] p53 transactivates expression of p21,[sdi1/waf1/cip1] a protein that was originally identified as an inhibitor of DNA synthesis in senescent cells,[62] and removal of p21 from human fibroblasts by gene targeting technology resulted in temporary escape from senescence.[9] These observations, together with the evidence referred to above that loss of normal p53 function through expression of a dominant negative p53[7] or through spontaneous loss of the wt p53 allele in LFS cells[69] results in temporary escape from senescence, represent strong circumstantial evidence for the involvement of p53 in senescence.

The trigger for this p53 activity, however, is unknown. According to the telomere hypothesis of senescence, the telomeric attrition that occurs during replication of normal

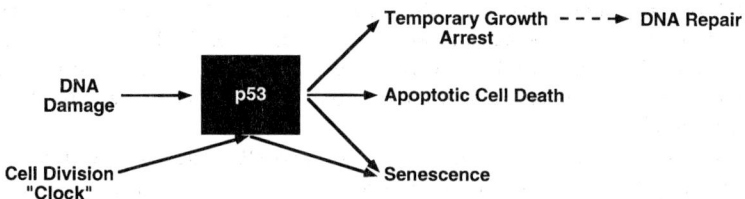

FIGURE 2. Proliferation arrest states involving p53. DNA damage and other stimuli may result in activation of p53 and the induction of either a temporary growth arrest or cell death. p53 is also involved in a viable terminal proliferation arrest state, senescence, which may be triggered by a cell proliferation–counting mechanism (telomeric shortening?) or possibly by DNA damage.

somatic cells eventually acts as the signal for the onset of senescence.[66,67] Although there is direct experimental evidence that progressive telomere shortening does indeed occur during the replicative history of normal cells, a mechanistic link between this and p53 activation or other aspects of the senescence phenotype has yet to be established. Following loss of p53 function, telomere shortening can continue to occur in cells that escape from senescence.[69] This raises the question as to whether (1) something other than telomere shortening normally triggers senescence, or (2) whether telomere shortening is the normal signal for senescence but that absence of p53 allows this to be bypassed.

As described above, p53 is involved in growth arrest states other than senescence (FIG. 2), which poses questions as to how the specificity of the response is achieved, that is, how does activation of p53 result in temporary growth arrest in one context, apoptosis in another, and senescence in yet another context? Does posttranslational modification of p53 result in a family of p53 molecules that specify different growth arrest states? Or do changes in expression of other genes specify the precise nature of the cellular response to p53 activation?

THE ROLE OF pRb, p16INK4, AND OTHER FUNCTIONALLY RELATED PROTEINS

Inactivation of both p53 and pRb by DNA tumor virus oncoproteins permits cells to proliferate until crisis,[12] that is, beyond the PD level characteristic of the p53-minus TPA state. It seems likely, therefore, that pRb may be involved in inducing growth arrest in the abnormal situation where p53 is not functional (i.e., the p53-minus TPA). In the normal situation, it is possible that pRb may cooperate with p53 in imposing senescence. pRb plays a key role in governing cell cycle progression at the restriction point, R, through its interactions with members of the E2F transcription family and other factors (FIG. 3). In its hypophosphorylated form pRb inhibits progression past R, but this inhibition may be relieved by phosphorylation (reviewed in ref. 3). A complex containing cyclin-dependent kinase (cdk) 4 or cdk6 and a member of the D-type cyclin family, such as cyclin D1, catalyzes the hyperphosphorylation of pRb, and this activity is inhibited by cdk inhibitors including p16INK4[72] (FIG. 3).

In view of these observations regarding the mechanism of pRb inactivation by phosphorylation, it may be expected that the increased proliferative potential associated with inactivation of pRb would also be seen when p16INK4 is inactivated by mutation, gene deletion, or methylation and when cdk4 or cyclin D1 is overexpressed. In cells that do not contain viral oncoproteins, increased proliferative potential may be associated with loss of p16INK4 function.[61] Further, in immortalized cells either pRb or p16INK4 is usually inactivated,[65,80] and in tumors inactivation of the "Rb pathway" by functional loss of pRb or p16INK4 or by overexpression of cdk4 or cyclin D1 is an almost universal finding. The involvement of p16INK4 in permanent growth arrest is also suggested by the observation that levels of this protein rise as cells become senescent.[30,68]

THE TELOMERE HYPOTHESIS OF CRISIS

A number of studies (e.g., ref. 31) have confirmed Olovnikov's hypothesis that telomere shortening accompanies cellular proliferation.[67] There appear to be at least two factors

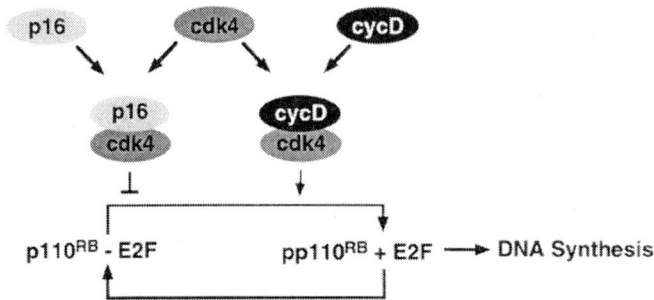

FIGURE 3. Rb phosphorylation cycle. Hypophosphorylated pRb binds members of the E2F transcription factor family (and other proteins not shown). Phosphorylation is catalyzed by an enzyme complex containing cyclin-dependent kinase 4 (cdk4) or cdk6 and a cyclin D, and inhibited by p16INK4. Hyperphosphorylated ppRb dissociates from E2F transcription factors permitting synthesis of proteins required for S-phase.

that contribute to telomere shortening (FIG. 4). The first is the "end-replication problem"[48] that results from the nature of DNA replication on the lagging strand. DNA synthesis is primed by short RNA primers, and when these are degraded, the resulting internal gaps are filled but the terminal deficit remains. Recent evidence suggests that this overhang is enlarged to 130–210 nucleotides by the action of a putative 5′-3′ exonuclease that also creates an overhang of similar length at the leading strand end.[52]

Although a connection between shortening of telomeres and senescence has yet to be formally proven, there is very strong circumstantial evidence that telomeric shortening is

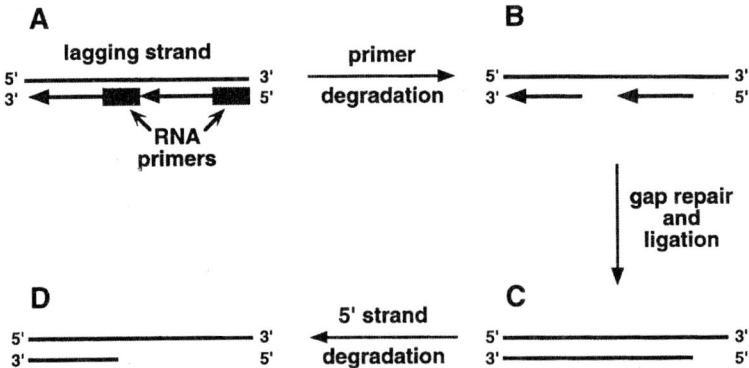

FIGURE 4. Shortening of telomeres during DNA replication in somatic cells. Lagging strand synthesis generates (**A**) a series of Okazaki fragments attached to RNA primers. **B:** Degradation of the RNA primers is followed by (**C**) filling of the internal gaps and ligation of the fragments, leaving an unfilled gap at the 5′ terminus. **D:** A putative 5′-to-3′ exonuclease degrades an additional 130–210 nucleotides, generating a 3′ G-rich tail.[52] Leading strand synthesis results in complete replication of telomeric DNA, but 5′ degradation generates a 3′ G-rich tail here also.

involved in the crisis TPA state. Thus a "telomere hypothesis of crisis" appears to be well founded. Telomeric shortening continues in cells that have escaped from senescence due to the effects of viral oncoproteins,[18,19,44] the spontaneous loss of p53 function,[69] or the loss of p21 function.[9] Crisis is associated with an increase in chromosome end-to-end fusions (resulting in dicentric and ring chromosome formation) and cell death, suggesting that the telomere length of cells in crisis is incompatible with further proliferation. Cells that escape from crisis, however, all exhibit a telomere maintenance mechanism,[11] implying that shortened telomeres are a key factor in triggering crisis. In the majority of cases telomere maintenance is achieved by activation of telomerase, but in some cell lines this occurs through an alternative mechanism.

TELOMERASE

Most immortalized human cell lines, and most human tumors, express telomerase activity.[18,42,56] Although it was initially thought that normal somatic cells did not possess telomerase activity, recent analyses have shown detectable telomerase activity in normal cells, including lymphocytes, keratinocytes in the basal layer of the epidermis, hair follicle cells, colonic crypt cells, and endothelial cells.[8,20,36 38,83] It appears that the level of telomerase activity in normal cells is not sufficient to completely compensate for the telomeric shortening that accompanies proliferation,[8] although it is possible that partial compensation may occur in normal cells. By contrast, the telomerase levels found in immortalized human cell lines and tumors result in stabilization of telomere length.[18]

Telomerase is a ribonucleoprotein complex that is still incompletely characterized. It was first detected in the unicellular ciliate *Tetrahymena*,[27] and components of the enzyme complex have subsequently been cloned from various species. These include a gene encoding the RNA molecule that acts as template for synthesis of the DNA repeats found in telomeres.[5,22,28] In *Tetrahymena*, genes encoding two putative protein subunits of molecular weight 80 and 95 kDa have been identified,[16] and an 80 kDa homologue, TP1/TLP1 (subsequently renamed TEP1), has been identified in mammalian species.[32,59] The telomerase catalytic subunit has been identified in several species.[13a,15a,17,32a,41,49,54a,58] It is not clear at this stage how many subunits are present in telomerase and whether there is more than one form of telomerase complex.

ALTERNATIVE LENGTHENING OF TELOMERES

Some human cell lines immortalized *in vitro* have no detectable telomerase activity.[11,42,57] This cannot be explained by the presence of telomerase inhibitors.[11] These telomerase-negative immortalized cells have telomeres with very heterogeneous length, ranging from short to extremely long (>50 kb).[11] The telomere lengths of one such telomerase-negative cell line were analyzed over the course of 650 PDs, and during this time there was an overall increase in the amount of telomeric DNA (ref. 69 and unpublished data). For a number of telomerase-negative cell lines, cryopreserved cells were available before and after crisis, and in all cases the telomere length was greater after crisis.[11] Taken together, these data indicate the existence of a telomerase-independent, alternative mechanism for lengthening of telomeres (ALT).[13]

Most of the ALT cell lines identified to date are SV40-immortalized fibroblasts, but ALT occurs in fibroblasts immortalized by other means, including through HPV genes, chemical carcinogen treatment, and spontaneously in the case of LFS fibroblasts, and also occurs in cell lines of epithelial origin.[11] Overall, about 25% of *in vitro*–immortalized cell lines have ALT,[13] but there are some cell strains that seem to have a predilection for expression of ALT. For example, 10/10 cell lines derived from fibroblasts from an LFS individual had ALT.[11] This contrasts with a neonatal foreskin fibroblast cell strain that gave rise to 19 SV40-immortalized cell lines, all of which were telomerase positive.[55] The reasons for these differences among cell strains are not clear. In addition, evidence for ALT has recently been found in tumor-derived cell lines and tumors.[10]

The molecular details of ALT are currently being investigated. Although ALT cells all appear to have similar telomere length patterns, it cannot be assumed at this stage that the ALT mechanism will be the same in each cell line. In eukaryotes there are two telomere maintenance mechanisms other than telomerase that may be used. In yeast species that have been rendered telomerase negative by the introduction of mutations into the telomerase RNA gene or other genetic manipulations, telomeric shortening followed by a senescence-like state occurs, but survivors arise at a fairly high frequency.[50,54] The survivors maintain their telomeres by a recombination mechanism that is dependent on RAD52.[54] *Drosophila* and related species do not have telomerase activity but maintain their telomeres by a retrotransposon-mediated mechanism: shortened telomeres are lengthened by HeT-A and TART retrotransposons.[4,53]

The nature of ALT is of intrinsic biological interest and may have important practical implications. The existence of ALT in immortalized cells raises questions as to the nature of the normal biological processes from which this is derived. Do these processes normally contribute to chromosomal maintenance and stability, to events in normal meiosis, or to chromosomal healing after damage? An understanding of the nature of ALT may be important in view of efforts to develop anticancer therapeutics that inhibit telomerase activity. It may be predicted that effective inhibition of telomerase activity in telomerase-positive tumors will apply a potent selective pressure for the emergence of drug-resistant clones using ALT.

OTHER GENES

Studies in which microcells were used to transfer normal human chromosomes into immortalized cell lines have shown that a number of chromosomes contain genes that are capable of imposing a TPA state on the cells. In these studies it has not been possible to determine whether the cells arrest in senescence or in another TPA state. Chromosomes 1q, 2, 3, 4, 6q, 7, 11, 18, and X all have this property.[35,43,45,60,63,64,70,71,75,79] It is possible that at least some of the relevant loci on these chromosomes will be found to encode repressors of the genes responsible for telomerase[64] or ALT activity.

A number of other interesting genes have been identified as having relevance to senescence and immortalization, although in most cases their role is not yet clear. A small sample includes the following. The mortalins mot-1 and mot-2 are heat-shock protein 70-related proteins with differing intracellular distributions whose expression and distribution patterns alter following immortalization.[77,78] The DNA repair enzyme O[6]-methylguanine-DNA methyltransferase is frequently absent from immortal cell lines.[33] Activity of the cytosine methylation enzyme, DNA methyltransferase, declines as cells approach senescence, but its expression is increased in some immortal cells.[21,39] Genes that are upregulated as cells approach senescence include the recently identified hic-5[73,74] and ING1[23,24] genes. Further analyses of these and other genes are required to determine which genes have a causal role in TPA states and which genes are affected secondarily.

CONCLUSION

Genes involved in controlling proliferative potential have been identified through analysis of the genetic alterations in cells that have escaped from senescence, or from the other TPA states, p53-minus TPA and crisis. The changes that result in escape from the TPA states (p53 mutations, abrogation of the pRb control mechanism by such changes as inactivation of p16INK4, and activation of a telomere maintenance mechanism such as telomerase or ALT) are the most commonly identified changes in cancer cells. Although it is very likely that telomeric shortening triggers crisis, the triggers for senescence and the p53-minus TPA state are yet to be identified, and the detailed molecular mechanisms whereby the various TPA states are imposed have yet to be defined. Analysis of these processes will extend our understanding of cellular proliferative life-span control and of the immortal phenotype and cancer.

ACKNOWLEDGMENTS

I regret that space constraints have made it impossible to cite all of the relevant publications. I thank the past and present members of my laboratory and collaborators who have contributed to the work described here, and Lindy Hodgkin for assistance with the manuscript.

REFERENCES

1. AFSHARI, C.A., P.J. VOJTA, L.A. ANNAB, P.A. FUTREAL, T.B. WILLARD & J.C. BARRETT. 1993. Investigation of the role of G_1/S cell cycle mediators in cellular senescence. Exp. Cell Res. **209:** 231–237.
2. ATADJA, P., H. WONG, I. GARKAVTSEV, C. VEILLETTE & K. RIABOWOL. 1995. Increased activity of p53 in senescing fibroblasts. Proc. Natl. Acad. Sci. USA **92:** 8348–8352.
3. BARTEK, J., J. BARTKOVA & J. LUKAS. 1996. The retinoblastoma protein pathway and the restriction point. Curr. Opin. Cell Biol. **8:** 805–814.
4. BIESSMANN, H., J.M. MASON, K. FERRY, M. D'HULST, K. VALGEIRSDOTTIR, K.L. TRAVERSE & M.-L. PARDUE. 1990. Addition of telomere-associated HeT DNA sequences "heals" broken chromosome ends in *Drosophila*. Cell **61:** 663–673.
5. BLASCO, M.A., W. FUNK, B. VILLEPONTEAU & C.W. GREIDER. 1995. Functional characterization and developmental regulation of mouse telomerase RNA. Science **269:** 1267–1270.
6. BOND, J., M. HAUGHTON, J. BLAYDES, V. GIRE, D. WYNFORD-THOMAS & F. WYLLIE. 1996. Evidence that transcriptional activation by p53 plays a direct role in the induction of cellular senescence. Oncogene **13:** 2097–2104.
7. BOND, J.A., F.S. WYLLIE & D. WYNFORD-THOMAS. 1994. Escape from senescence in human diploid fibroblasts induced directly by mutant p53. Oncogene **9:** 1885–1889.
8. BROCCOLI, D., J.W. YOUNG & T. DE LANGE. 1995. Telomerase activity in normal and malignant hematopoietic cells. Proc. Natl. Acad. Sci. USA **92:** 9082–9086.

9. BROWN, J.P., W. WEI & J.M. SEDIVY. 1997. Bypass of senescence after disruption of p21$^{CIP1/WAF1}$ gene in normal diploid human fibroblasts. Science **277:** 831–834.

10. BRYAN, T.M., A. ENGLEZOU, L. DALLA-POZZA, M.A. DUNHAM & R.R. REDDEL. 1997. Evidence for an alternative mechanism for maintaining telomere length in human tumors and tumor-derived cell lines. Nature Med. **3:** 1271–1274.

11. BRYAN, T.M., A. ENGLEZOU, J. GUPTA, S. BACCHETTI & R.R. REDDEL. 1995. Telomere elongation in immortal human cells without detectable telomerase activity. EMBO J. **14:** 4240–4248.

12. BRYAN, T.M. & R.R. REDDEL. 1994. SV40-induced immortalization of human cells. Crit. Rev. Oncog. **5:** 331–357.

13. BRYAN, T.M. & R.R. REDDEL. 1997. Telomere dynamics and telomerase activity in *in vitro* immortalised human cells. Eur. J. Cancer **33A:** 767–773.

13a. BRYAN, T.M., J.M. SPERGER, K.B. Chapman & T.R. CECH. 1998. Teomerase reverse transcriptase genes identified in *Tetrahymena thermophila* and *Oxytricha trifallax*. Proc. Natl. Acad. Sci. USA **95:** 8479–8484.

14. CAMPISI, J. 1996. Replicative senescence: An old lives' tale? Cell **84:** 497–500.

15. CAMPISI, J., G.P. DIMRI, J.O. NEHLIN, A. TESTORI & K. YOSHIMOTO. 1996. Coming of age in culture. Exp. Gerontol. **31:** 7–12.

15a. COLLINS, K. & L. GANDHI. 1998. The reverse transcriptase component of the *Tetrahymena* telomerase ribonucleoprotein complex. Proc. Natl. Acad. Sci. USA **95:** 8485–8490.

16. COLLINS, K., R. KOBAYASHI & C.W. GREIDER. 1995. Purification of *Tetrahymena* telomerase and cloning of genes encoding the two protein components of the enzyme. Cell **81:** 677–686.

17. COUNTER, C.M., M. MEYERSON, E.N. EATON & R.A. WEINBERG. 1997. The catalytic subunit of yeast telomerase. Proc. Natl. Acad. Sci. USA **94:** 9202–9207.

18. COUNTER, C.M., A.A. AVILION, C.E. LEFEUVRE, N.G. STEWART, C.W. GREIDER, C.B. HARLEY & S. BACCHETTI. 1992. Telomere shortening associated with chromosome instability is arrested in immortal cells which express telomerase activity. EMBO J. **11:** 1921–1929.

19. COUNTER, C.M., F.M. BOTELHO, P. WANG, C.B. HARLEY & S. BACCHETTI. 1994. Stabilization of short telomeres and telomerase activity accompany immortalization of Epstein-Barr virus-transformed human B lymphocytes. J. Virol. **68:** 3410–3414.

20. COUNTER, C.M., J. GUPTA, C.B. HARLEY, B. LEBER & S. BACCHETTI. 1995. Telomerase activity in normal leukocytes and in hematologic malignancies. Blood **85:** 2315–2320.

21. EL-DEIRY, W.S., B.D. NELKIN, P. CELANO, R.-W.C. YEN, J.P. FALCO, S.R. HAMILTON & S.B. BAYLIN. 1991. High expression of the DNA methyltransferase gene characterizes human neoplastic cells and progression stages of colon cancer. Proc. Natl. Acad. Sci. USA **88:** 3470–3474.

22. FENG, J., W.D. FUNK, S.-S. WANG, S.L. WEINRICH, A.A. AVILION, C.-P. CHIU, R.R. ADAMS, E. CHANG, R.C. ALLSOPP, J.H. YU, S.Y. LE, M.D. WEST, C.B. HARLEY, W.H. ANDREWS, C.W. GREIDER & B. VILLEPONTEAU. 1995. The RNA component of human telomerase. Science **269:** 1236–1241.

23. GARKAVTSEV, I., A. KAZAROV, A. GUDKOV & K. RIABOWOL. 1996. Suppression of the novel growth inhibitor p33^{ING1} promotes neoplastic transformation. Nat. Genet. **14:** 415–420.

24. GARKAVTSEV, I. & K. RIABOWOL. 1997. Extension of the replicative life span of human diploid fibroblasts by inhibition of the p33^{ING1} candidate tumor suppressor. Mol. Cell. Biol. **17:** 2014–2019.

25. GIRARDI, A.J., F.C. JENSEN & H. KOPROWSKI. 1965. SV40-induced transformation of human diploid cells: Crisis and recovery. J. Cell. Comp. Physiol. **65:** 69–84.

26. GOLDSTEIN, S. 1990. Replicative senescence: The human fibroblast comes of age. Science **249:** 1129–1133.

27. GREIDER, C.W. & E.H. BLACKBURN. 1985. Identification of a specific telomere terminal transferase activity in *Tetrahymena* extracts. Cell **43:** 405–413.

28. GREIDER, C.W. & E.H. BLACKBURN. 1989. A telomeric sequence in the RNA of *Tetrahymena* telomerase required for telomere repeat synthesis. Nature **337:** 331–337.

29. HALL, P.A. & D.P. LANE. 1997. Tumour suppressors: A developing role for p53? Curr. Biol. **7:** R144–R147.

30. HARA, E., R. SMITH, D. PARRY, H. TAHARA, S. STONE & G. PETERS. 1996. Regulation of p16^{CDKN2} expression and its implications for cell immortalization and senescence. Mol. Cell. Biol. **16:** 859–867.

31. HARLEY, C.B., A.B. FUTCHER & C.W. GREIDER. 1990. Telomeres shorten during ageing of human fibroblasts. Nature **345:** 458–460.
32. HARRINGTON, L., T. MCPHAIL, V. MAR, W. ZHOU, R. OULTON, AMGEN EST PROGRAM, M.B. BASS, I. ARRUDA & M.O. ROBINSON. 1997. A mammalian telomerase-associated protein. Science **275:** 973–977.
32a. HARRINGTON, L., W. ZHOU, T. MCPHAIL, R. OULTON, D.S.K. YEUNG, V. MAR, M.B. BASS & M.O. ROBINSON. 1997. Human telomerase contains evolutionarily conserved catalytic and structural subunits. Genes & Dev. **11:** 3109–3115.
33. HARRIS, L.C., M.A. VON WRONSKI, C.C. VENABLE, J.S. REMACK, S.R. HOWELL & T.P. BRENT. 1996. Changes in O^6-methylguanine-DNA methyltransferase expression during immortalization of cloned human fibroblasts. Carcinogenesis **17:** 219–224.
34. HAYFLICK, L. & P.S. MOORHEAD. 1961. The serial cultivation of human diploid cell strains. Exp. Cell Res. **25:** 585–621.
35. HENSLER, P.J., L.A. ANNAB, J.C. BARRETT & O.M. PEREIRA-SMITH. 1994. A gene involved in control of human cellular senescence on human chromosome 1q. Mol. Cell. Biol. **14:** 2291–2297.
36. HIYAMA, K., Y. HIRAI, S. KYOIZUMI, M. AKIYAMA, E. HIYAMA, M.A. PIATYSZEK, J.W. SHAY, S. ISHIOKA & M. YAMAKIDO. 1995. Activation of telomerase in human lymphocytes and hematopoietic progenitor cells. J. Immunol. **155:** 3711–3715.
37. HSIAO, R., H.W. SHARMA, S. RAMAKRISHNAN, E. KEITH & R. NARAYANAN. 1997. Telomerase activity in normal human endothelial cells. Anticancer Res. **17:** 827–832.
38. HÄRLE-BACHOR, C. & P. BOUKAMP. 1996. Telomerase activity in the regenerative basal layer of the epidermis in human skin and in immortal and carcinoma-derived skin keratinocytes. Proc. Natl. Acad. Sci. USA **93:** 6476–6481.
39. ISSA, J.-P.J., P.M. VERTINO, J. WU, S. SAZAWAL, P. CELANO, B.D. NELKIN, S.R. HAMILTON & S.B. BAYLIN. 1993. Increased cytosine DNA-methyltransferase activity during colon cancer progression. J. Natl. Cancer Inst. **85:** 1235–1240.
40. KASTAN, M.B. 1996. Signalling to p53: Where does it all start? BioEssays **18:** 617–619.
41. KILIAN, A., D.D.L. BOWTELL, H.E. ABUD, G.R. HIME, D.J. VENTER, P.K. KEESE, E.L. DUNCAN, R.R. REDDEL & R.A. JEFFERSON. 1997. Isolation of a candidate human telemeraze catalytic subunit gene, which reveals complex splicing patterns in different cell types. Hum. Mol. Genet. **6:** 2011–2019.
42. KIM, N.W., M.A. PIATYSZEK, K.R. PROWSE, C.B. HARLEY, M.D. WEST, P.L.C. HO, G.M. COVIELLO, W.E. WRIGHT, S.L. WEINRICH & J.W. SHAY. 1994. Specific association of human telomerase activity with immortal cells and cancer. Science **266:** 2011–2015.
43. KLEIN, C.B., K. CONWAY, X.W. WANG, R.K. BHAMRA, X. LIN, M.D. COHEN, L. ANNAB, J.C. BARRETT & M. COSTA. 1991. Senescence of nickel-transformed cells by an X chromosome: Possible epigenetic control. Science **251:** 796–799.
44. KLINGELHUTZ, A.J., S.A. BARBER, P.P. SMITH, K. DYER & J.K. MCDOUGALL. 1994. Restoration of telomeres in human papillomavirus-immortalized human anogenital epithelial cells. Mol. Cell. Biol. **14:** 961–969.
45. KOI, M., L.A. JOHNSON, L.M. KALIKIN, P.F.R. LITTLE, Y. NAKAMURA & A.P. FEINBERG. 1993. Tumor cell growth arrest caused by subchromosomal transferable DNA fragments from chromosome 11. Science **260:** 361–364.
46. KULJU, K.S. & J.M. LEHMAN. 1995. Increased p53 protein associated with aging in human diploid fibroblasts. Exp. Cell Res. **217:** 336–345.
47. LANE, D.P. 1992. Cancer: p53, guardian of the genome. Nature **358:** 15–16.
48. LEVY, M.Z., R.C. ALLSOPP, A.B. FUTCHER, C.W. GREIDER & C.B. HARLEY. 1992. Telomere end-replication problem and cell aging. J. Mol. Biol. **225:** 951–960.
49. LINGNER, J., T.R. HUGHES, A. SHEVCHENKO, M. MANN, V. LUNDBLAD & T.R. CECH. 1997. Reverse transcriptase motifs in the catalytic subunit of telomerase. Science **276:** 561–567.
50. LUNDBLAD, V. & E.H. BLACKBURN. 1993. An alternative pathway for yeast telomere maintenance rescues *est1⁻* senescence. Cell **73:** 347–360.
51. MACLEAN, K., E.M. ROGAN, N.J. WHITAKER, A.C.-M. CHANG, P.B. ROWE, L. DALLA-POZZA, G. SYMONDS & R.R. REDDEL. 1994. *In vitro* transformation of Li-Fraumeni syndrome fibroblasts by SV40 large T antigen mutants. Oncogene **9:** 719–725.
52. MAKAROV, V.L., Y. HIROSE & J.P. LANGMORE. 1997. Long G tails at both ends of human chromosomes suggest a C strand degradation mechanism for telomere shortening. Cell **88:** 657–666.

53. MASON, J.M. & H. BIESSMANN. 1995. The unusual telomeres of *Drosophila*. Trends Genet. **11:** 58–62.
54. MCEACHERN, M.J. & E.H. BLACKBURN. 1996. Cap-prevented recombination between terminal telomeric repeat arrays (telomere CPR) maintains telomeres in *Kluyveromyces lactis* lacking telomerase. Genes & Dev. **10:** 1822–1834.
54a. MEYERSON, M., C.M. COUNTER, E.N. EATON, L.W. ELLISEN, P. STEINER, S. DICKINSON CADDLE, L. ZIAUGRA, R.L. BEIJERSBERGEN, M.J. DAVIDOFF, Q. LIU, S. BACCHETTI, D.A. HABER & R.A. WEINBERG. 1997. *hEST2*, the putative human telomerase catalytic subunit gene, is up-regulated in tumor cells and during immortalization. Cell **90:** 785–795.
55. MONTALTO, M.C. & F.A. RAY. 1996. Telomerase activation during the linear evolution of human fibroblasts to tumorigenicity in nude mice. Carcinogenesis **17:** 2631–2634.
56. MORIN, G.B. 1989. The human telomere terminal transferase enzyme is a ribonucleoprotein that synthesizes TTAGGG repeats. Cell **59:** 521–529.
57. MURNANE, J.P., L. SABATIER, B.A. MARDER & W.F. MORGAN. 1994. Telomere dynamics in an immortal human cell line. EMBO J. **13:** 4953–4962.
58. NAKAMURA, T.M., G.B. MORIN, K.B. CHAPMAN, S.L. WEINRICH, W.H. ANDREWS, J. LINGNER, C.B. HARLEY & T.R. CECH. 1997. Telomerase catalytic subunit homologs from fission yeast and human. Science **277:** 955–959.
59. NAKAYAMA, J.I., M. SAITO, H. NAKAMURA, A. MATSUURA & F. ISHIKAWA. 1997. TLP1: A gene encoding a protein component of mammalian telomerase is a novel member of WD repeats family. Cell **88:** 875–884.
60. NING, Y. & O.M. PEREIRA-SMITH. 1991. Molecular genetic approaches to the study of cellular senescence. Mutat. Res. **256:** 303–310.
61. NOBLE, J.R., E.M. ROGAN, A.A. NEUMANN, K. MACLEAN, T.M. BRYAN & R.R. REDDEL. 1996. Association of extended *in vitro* proliferative potential with loss of p16[INK4] expression. Oncogene **13:** 1259–1268.
62. NODA, A., Y. NING, S.F. VENABLE, O.M. PEREIRA-SMITH & J.R. SMITH. 1994. Cloning of senescent cell-derived inhibitors of DNA synthesis using an expression screen. Exp. Cell Res. **211:** 90–98.
63. OGATA, T., D. AYUSAWA, M. NAMBA, E. TAKAHASHI, M. OSHIMURA & M. OISHI. 1993. Chromosome 7 suppresses indefinite division of nontumorigenic immortalized human fibroblast cell lines KMST-6 and SUSM-1. Mol. Cell. Biol. **13:** 6036–6043.
64. OHMURA, H., H. TAHARA, M. SUZUKI, T. IDE, M. SHIMIZU, M.A. YOSHIDA, E. TAHARA, J.W. SHAY, J.C. BARRETT & M. OSHIMURA. 1995. Restoration of the cellular senescence program and repression of telomerase by human chromosome 3. Jpn. J. Cancer Res. **86:** 899–904.
65. OKAMOTO, A., D.J. DEMETRICK, E.A. SPILLARE, K. HAGIWARA, S.P. HUSSAIN, W.P. BENNETT, K. FORRESTER, B. GERWIN, M. SERRANO, D.H. BEACH & C.C. HARRIS. 1994. Mutations and altered expression of p16[INK4] in human cancer. Proc. Natl. Acad. Sci. USA **91:** 11045–11049.
66. OLOVNIKOV, A.M. 1971. Principle of marginotomy in template synthesis of polynucleotides. Dokl. Akad. Nauk SSSR **201:** 1496–1499.
67. OLOVNIKOV, A.M. 1973. A theory of marginotomy. J. Theor. Biol. **41:** 181–190.
68. REZNIKOFF, C.A., T.R. YEAGER, C.D. BELAIR, E. SAVELIEVA, J.A. PUTHENVEETTIL & W.M. STADLER. 1996. Elevated p16 at senescence and loss of p16 at immortalization in human papillomavirus 16 E6, but not E7, transformed human uroepithelial cells. Cancer Res. **56:** 2886–2890.
69. ROGAN, E.M., T.M. BRYAN, B. HUKKU, K. MACLEAN, A.C.-M. CHANG, E.L. MOY, A. ENGLEZOU, S.G. WARNEFORD, L. DALLA-POZZA & R.R. REDDEL. 1995. Alterations in p53 and p16[INK4] expression and telomere length during spontaneous immortalization of Li-Fraumeni syndrome fibroblasts. Mol. Cell. Biol. **15:** 4745–4753.
70. SANDHU, A.K., K. HUBBARD, G.P. KAUR, K.K. JHA, H.L. OZER & R.S. ATHWAL. 1994. Senescence of immortal human fibroblasts by the introduction of normal human chromosome 6. Proc. Natl. Acad. Sci. USA **91:** 5498–5502.
71. SASAKI, M., T. HONDA, H. YAMADA, N. WAKE, J.C. BARRETT & M. OSHIMURA. 1994. Evidence for multiple pathways to cellular senescence. Cancer Res. **54:** 6090–6093.
72. SERRANO, M., G.J. HANNON & D. BEACH. 1993. A new regulatory motif in cell-cycle control causing specific inhibition of cyclin D/CDK4. Nature **366:** 704–707.

73. SHIBANUMA, M., J. MASHIMO, T. KUROKI & K. NOSE. 1994. Characterization of the TGFβ1-inducible *hic-5* gene that encodes a putative novel zinc finger protein and its possible involvement in cellular senescence. J. Biol. Chem. **269:** 26767–26774.
74. SHIBANUMA, M., E. MOCHIZUKI, R. MANIWA, J.-I. MASHIMO, N. NISHIYA, S.-I. IMAI, T. TAKANO, M. OSHIMURA & K. NOSE. 1997. Induction of senescence-like phenotypes by forced expression of *hic-5*, which encodes a novel LIM motif protein, in immortalized human fibroblasts. Mol. Cell. Biol. **17:** 1224–1235.
75. UEJIMA, H., K. MITSUYA, H. KUGOH, I. HORIKAWA & M. OSHIMURA. 1995. Normal human chromosome 2 induces cellular senescence in the human cervical carcinoma cell line SiHa. Genes Chromosomes & Cancer **14:** 120–127.
76. VAZIRI, H. & S. BENCHIMOL. 1996. From telomere loss to p53 induction and activation of a DNA-damage pathway at senescence: The telomere loss/DNA damage model of cell aging. Exp. Gerontol. **31:** 295–301.
77. WADHWA, R., S. AKIYAMA, T. SUGIHARA, R.R. REDDEL, Y. MITSUI & S.C. KAUL. 1996. Genetic differences between the pancytosolic and perinuclear forms of murine mortalin. Exp. Cell Res. **226:** 381–386.
78. WADHWA, R., S.C. KAUL & Y. MITSUI. 1994. Cellular mortality to immortalization: Mortalin. Cell Struct. Function **19:** 1–10.
79. WANG, X.W., X. LIN, C.B. KLEIN, R.K. BHAMRA, Y.-W. LEE & M. COSTA. 1992. A conserved region in human and Chinese hamster X chromosomes can induce cellular senescence of nickel-transformed Chinese hamster cell lines. Carcinogenesis **13:** 555–561.
80. WHITAKER, N.J., T.M. BRYAN, P. BONNEFIN, A.C.-M. CHANG, E.A. MUSGROVE, A.W. BRAITHWAITE & R.R. REDDEL. 1995. Involvement of RB 1, p53, p16^INK4 and telomerase in immortalization of human cells. Oncogene **11:** 971–976.
81. WRIGHT, W.E., O.M. PEREIRA-SMITH & J.W. SHAY. 1989. Reversible cellular senescence: Implications for immortalization of normal human diploid fibroblasts. Mol. Cell. Biol. **9:** 3088–3092.
82. WYNFORD-THOMAS, D. 1997. Proliferative lifespan checkpoints: Cell-type specificity and influence on tumour biology. Eur. J. Cancer **33A:** 716–726.
83. YASUMOTO, S., C. KUNIMURA, K. KIKUCHI, H. TAHARA, H. OHJI, H. YAMAMOTO, T. IDE & T. UTAKOJI. 1996. Telomerase activity in normal human epithelial cells. Oncogene **13:** 433–439.

Somatic Mutation and Aging

ALEC MORLEY[a]

*Department of Haematology, Flinders University of South Australia
and Flinders Medical Centre, Bedford Park, South Australia 5042, Australia*

ABSTRACT: A key prediction of the somatic mutation theory of aging is that there is an invariant relationship between life span and the number of random mutations. A number of studies at a number of gene loci have shown that somatic mutations of a variety of types accumulate with age. Dietary restriction, which prolongs life span, results in slowed accumulation of HPRT mutants in mice. Conversely, senescence-accelerated mice, which have been bred to have a shortened life span, show accelerated accumulation of somatic mutations.

The somatic mutation theory of aging postulates that aging results from the accumulation of mutations in somatic cells that results in failure of the cells either to survive, to proliferate, or to function at complete efficiency.[1-5] The accumulation of mutations, and thus aging, is therefore a stochastic process, although it can obviously be influenced by genetic or environmental factors that influence the rate of accumulation. The theory also conceives that aging is a secondary but inevitable consequence of the more fundamental biological process of mutagenesis.

There are three broad classes of predictions of the somatic mutation theory of aging. These predictions are all testable, and their power of discrimination increases progressively. (1) Mutation frequency increases with age. (2) Experimental manipulation of the life span of laboratory animals will be associated with an appropriate alteration in mutation frequency. (3) Experimental manipulation of mutation frequency in laboratory animals will result in appropriate alteration of life span. In our laboratory, we have studied nuclear mutations in the lymphocytes of humans and mice at the hypoxanthine phosphoribosyl transferase locus and of humans at the HLA-A locus.[6-8] At both loci mutations increase by a factor of 5–10 over life for both species. Given the much shorter life span of the mouse, the data imply that the rate of accumulation of mutations in mice is 50- to 100-fold more rapid than it is in humans, which is in agreement with the much shorter life span of the mouse.

Molecular analysis of HLA-A mutations in human lymphocytes has shown that the molecular mechanisms are heterogeneous, and the data suggest that most, if not all, of the various types of mutation accumulate with age. Approximately one third of mutations result from mitotic recombination,[9,10] and the frequency of this type of mutation increases by a factor of approximately five over one life span.[8] Recent unpublished data suggest that oxidizing free radicals are potent producers of mitotic recombination.[11] Approximately two thirds of HLA-A mutations result from point mutations or small intragenic deletions, and these also appear to increase with age.[8,12]

The length of life span is subject to environmental and genetic influences. It has been known for many years that life span can be substantially prolonged in rodents by dietary restriction. Dempsey *et al.*[13] studied mutation frequency in mice who had a dietary intake

[a]Tel: +61-8-82044431; fax: +61-8-82045450; e-mail: alec.morley@flinders.edu.au

60% that of controls and found there was virtually no increase in mutations between 6–12 months of age in such animals, as compared to a fourfold increase in controls. Their data support the somatic mutation theory of aging and also suggest that the majority of endogenous mutations are somehow related to diet. An obvious mechanism might be that dietary intake directly determines the cellular flux of oxidizing free radicals and/or determines intracellular levels of glucose or other DNA-damaging metabolites.

Life span can also be influenced by genetic factors. Odagiri *et al.* have studied the rate of accumulation in the senescence-accelerated mouse strain, a strain of mouse that has been produced by genetic inbreeding to manifest a substantially shortened life span.[14] Preliminary data suggest that the rate of accumulation of mutations in this strain is substantially increased as compared with controls.[15]

The most discriminating prediction of the somatic mutation theory of aging is that manipulations designed to increase or decrease the mutational load will have as a secondary effect a corresponding shortening or lengthening of life span. The best evidence of this type comes from the studies of Orr and Sohal who observed prolongation of life span in *Drosophila* transgenically engineered to express increased levels of superoxide dismutase and catalase.[16] There are also data showing that exposure of animals to mutagens, such as X-irradiation[17] or the alkylating agent, busulfan,[18] will result in shortening of life span and/or features such as greying of hair and cataracts, which are typical of aging.

Thus, increasing experimental evidence supports the somatic mutation theory of aging, but whether the mutations of importance are nuclear, mitochondrial, or both, remains to be determined. The evidence linking mutations and aging is reviewed in more detail in reference 18, and that linking mitochondrial mutations and aging is discussed elsewhere in this volume. An important task for the future is to determine the nature of the physical and chemical influences responsible for mutagenesis, a task that is of both theoretical and practical importance.

REFERENCES

1. Szilard, L. 1959. On the nature of aging process. Proc. Natl. Acad. Sci. USA **45**: 35–45.
2. Curtis, H.J. 1966. Biological Mechanisms of Aging. Thomas. Springfield, IL.
3. Orgel, L.E. 1973. Aging of clones of mammalian cells. Nature (London) **243**: 441–445.
4. Burnet, F.M. 1974. Intrinsic Mutagenesis: A Genetic Approach to Aging. Medical and Technical Publishing. Lancaster.
5. Morley, A.A. 1982. Is aging the result of dominant or co-dominant mutations? J. Theor. Biol. **98**: 469–474.
6. Morley, A.A., S. Cox & R. Holliday. 1982. Human lymphocytes resistant to 6-thioguanine increase with age. Mech. Aging Dev. **19**: 21–26.
7. Trainor, K.J., D.J. Wigmore, A. Chrysostomou, J. Dempsey, R. Seshadri & A.A. Morley. 1984. Mutation frequency in human lymphocytes increases with age. Mech. Aging Dev. **27**: 83–86.
8. Grist, S.A., M. McCarron, A. Kutlaca, D.R. Turner & A.A. Morley. 1992. *In vivo* human somatic mutation: Frequency and spectrum with age. Mutat. Res. **266**: 189–196.
9. Turner, D.R., S.A. Grist, M. Janatipour & A.A. Morley. 1998. Mutations in human lymphocytes commonly involve gene duplication and resemble those seen in cancer cells. Proc. Natl. Acad. Sci. USA **85**: 3189–3192.
10. Morley, A.A., S.J. Grist, D.R. Turner, A. Kutlaca & G. Bennett. 1990. The molecular nature of *in vivo* mutations in human cells at the autosomal HLA-A locus. Cancer Res. **50**: 4584–4587.
11. Turner, D.R., D.A. Holt & A.A. Morley. In preparation.
12. Male, D. Unpublished results.

13. DEMPSEY, J.L., A.A. MORLEY & M. PFEIFFER. 1993. Effect of dietary restriction on *in vivo* somatic mutation. Mutation Res. **291**: 141–145.
14. TAKEDA, T. *et al.* 1981. A new murine model of accelerated senescence. Mech. Aging Dev. **17**: 183–194.
15. ODAGIRI, Y., H. UCHIDA, M. HOSOKAWA, K. TAKEMOTO, A.A. MORLEY & T. TAKEDA. 1998. Accelerated accumulation of somatic mutations in the senescence-accelerated mouse. Nat. Genet. **19**: 116–117.
16. ORR, W.C. & R.S. SOHAL. 1994. Extension of life-span by overexpression of superoxide dismutase and catalase in *Drosophila melanogaster*. Science **263**: 1128–1130.
17. LINDOP, P.J. & R. ROTBLAT. 1961. Long-term effects of a single whole-body exposure of mice to ionizing radiations. Proc. R. Soc. Lon. B. **154**: 332–349.
18. MORLEY, A.A. 1995. The somatic mutation theory of aging. Mutation Res. **338**: 19–23.

Chromosomal Damage Rate, Aging, and Diet[a]

MICHAEL FENECH[b]

*CSIRO Division of Human Nutrition, P.O. Box 10041, Gouger Street,
Adelaide SA, Australia 5000*

ABSTRACT: Chromosomal damage as measured by frequency of translocations, acentric fragments, telomere shortening, nondisjunction, chromosome loss, aneuploidy, and micronucleus formation has been shown to increase progressively with age. Using the cytokinesis-block micronucleus technique, which provides an efficient measure of chromosomal breakage and loss, we have been able to show that aging can explain at least 25% of the variation in chromosomal damage rate in lymphocytes from both males and females. We have also performed cross-sectional and placebo-controlled intervention studies to determine the relationship between the micronucleus (MN) frequency in lymphocytes and diet, and blood status for vitamins C, E, B_{12}, and folic acid. Our studies have shown that MN frequency in the 41- to 60-year age group is significantly lower in vegetarians when compared to nonvegetarians, but the reverse was true in males aged between 20 and 40 years. This was accounted for by a deficient/low B_{12} status in vegetarian males; there was no difference in the MN frequency of vegetarian and nonvegetarian subjects aged between 61 and 90 years. Results from this study also showed significant negative correlations of MN frequency with folic acid and vitamin B_{12} but not with vitamin C or vitamin E. In separate studies on healthy men aged 50–70, we have verified the significant negative correlation between vitamin B_{12} status in plasma and MN frequency ($r = -0.315$, $p = 0.013$) in subjects who were not vitamin B_{12} deficient and observed a significant positive correlation between MN frequency and homocysteine status ($r = 0.414$, $p = 0.0086$) in those men who were not vitamin B_{12} and/or folate deficient. These data suggest that MN frequency is minimized when plasma B_{12} is above 300 pmol/L and plasma homocysteine is below 7.5 μmol/L. Double-blind placebo-controlled intervention studies conducted over four months have shown that above RDI intake of vitamin E ($30 \times$ RDI) or folic acid ($10 \times$ RDI) did not produce a significant reduction in MN frequency in men aged 50–70 years. In the latter case plasma homocysteine was reduced from a mean value of 9.33 μmol/L to 8.51 μmol/L, a level that does not correspond with minimization of MN frequency. We have also tested the hypothesis that moderate wine drinking can protect against the DNA-damaging effect of hydrogen peroxide and found that there was a strong *ex vivo* inhibition (> 70%) of hydrogen peroxide–induced MN frequency by plasma samples from blood collected one hour after consumption of red or white wine, as compared to plasma samples collected immediately before wine consumption ($p = 0.0008$). However, only samples following red wine consumption produced a significant reduction in baseline MN frequency. The above results suggest that chromosome damage can be modulated, under selected circumstances, by diverse dietary factors.

There is increasing evidence that an elevated chromosomal damage rate is a valid marker and possibly a cause of accelerated aging and carcinogenesis.[1-5] Several studies investigating various indices of chromosomal instability, such as balanced translocations, chromosome fragments, telomere shortening, nondisjunction, chromosomal loss, aneuploidy, and micronucleus (MN) formation have consistently shown that chromosomal damage increases pro-

[a]The studies were supported by Kellogg Aust. Pty. Ltd., Meat Research Corporation (Australia), Australian Wine Research Institute, Herb Valley Pty. Ltd (Australia), and Blackmores Ltd. (Australia).
[b]Tel: 618 83038880; fax: 618 83038899; e-mail: michael.fenech@dhn.csiro.au

gressively with age.[2,4,6,7] The precise causes for the observed increase of chromosomal damage with age remain unclear; however, there is a definite interest in determining whether chromosomal damage rate can be reduced by alterations of lifestyle and diet. By understanding the contribution of environmental exposure to genotoxins, diet, lifestyle, and genetic background to spontaneous chromosomal damage rate, it will presumably be possible to define more precisely the relative importance of these factors, and to identify which factors should and can be modified to minimize chromosomal damage rates.

To contribute to this field of research, we have undertaken cross-sectional studies aimed at defining the baseline chromosomal damage rate in a large population of healthy individuals and its relationship with age and diet.[5,8–10] Furthermore we have also performed a series of intervention trials to establish whether certain vitamin supplements and moderate wine intake could alter the chromosomal damage rate.[11–13] The technique we have employed is the cytokinesis-block MN assay in human lymphocytes, which provides a reliable and sensitive measure of chromosomal breakage and loss in a relevant population of somatic cells.[14,15] The aim of this paper is to review the results of our studies together with those from other laboratories who used the same or similar techniques. The material and methods for the cytokinesis-block MN assay and the various biochemical measurements referred to in this paper are detailed in our published communications.[5,8–15]

MICRONUCLEUS FREQUENCY IN RELATION TO AGING

The MN frequency in cytokinesis-blocked lymphocytes of 152 females and 113 males aged between 20 and 89 years (minimum of 15 subjects per sex per decade) was compared[9] (FIG. 1A and B). Marked differences in the MN frequency of males and females was observed: (1) there was a greater dispersion in the results for females when compared to males in all age groups older than 40 years; (2) there was a significant positive correlation between MN frequency and age in both sexes ($p < 0.0001$) but the slope of the linear regression line was steeper in females (slope = 0.499 micronuclei/year) compared to males (slope = 0.289 micronuclei/year) ($p < 0.0045$); (3) the MN frequency in females was significantly greater than the MN frequency in males ($p < 0.05$) across all decades examined (range of incremental factor was 1.47–1.65, with an average of 1.53 ± 0.03 (mean ± 1 SEM)). These results suggested that an added mechanism, possibly the loss of X chromosomes, may contribute to the MN frequency in females.

A number of studies have been performed to investigate the contribution of sex chromosomal loss at anaphase to the observed MN frequency in males and females. Using X chromosome–specific centromeric probes Hando et al.[16] demonstrated that up to 72% of the micronuclei scored in females may contain an X chromosome; however, the studies of Catalan et al.[17] and Surrales et al.[18] reported lower frequencies of 24% and 42%, respectively. Using a Y chromosome–specific probe, Nath et al.[19] demonstrated that 13.5% of total micronuclei in males contained a Y chromosome. A large proportion of micronuclei containing sex chromosomes were also shown to be kinetochore negative, which suggests that damage to the kinetochore assembly may be a key mechanism in MN formation. Using acetylated histone as a marker for activation, Surrales et al.[20] have also investigated the possibility that the inactive X chromosome may be primarily responsible for X chromosome-positive micronuclei in females and found that the inactive as well as the active X chromosome are equally represented within micronuclei. The MN data clearly show

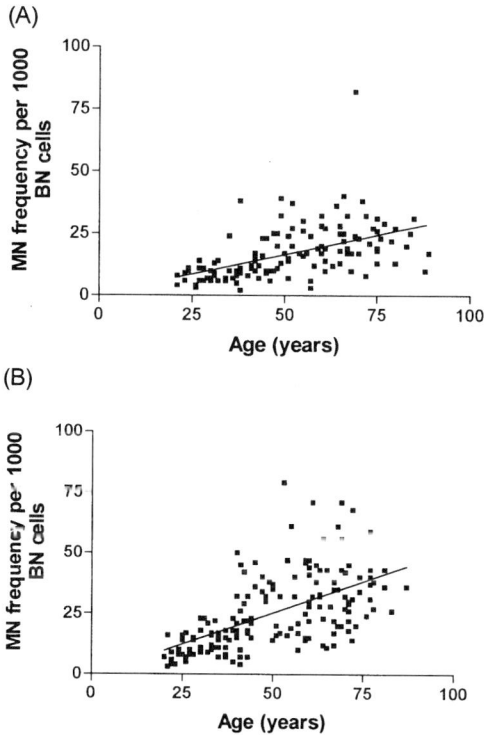

FIGURE 1. The relationship between MN frequency and age in (A) males and (B) females. Spearman's r = 0.6252; p < 0.0001; N = 126 in A; Spearman's r = 0.6510; p < 0.0001; N = 166 in B. BN = cytokinesis-blocked binucleated.

that both kinetochore negative and kinetochore positive micronuclei increase with age, suggesting that both chromosomal loss and chromosome breakage events are important events in aging, although the proportion of kinetochore positive micronuclei appears to be more prevalent in the older age groups.[16,21]

LYMPHOCYTE MICRONUCLEI IN
VEGETARIANS AND NONVEGETARIANS

Having established that the MN index increases with age and that there was a wide variation in the MN index within each age group, it was important to determine whether dietary differences could impact on the genetic damage rate. For this purpose we chose to compare MN frequencies in healthy vegetarians and nonvegetarians as one could expect marked differences in vitamin intake particularly with respect to vitamin C, folic acid, and vitamin B_{12}, all of which are strongly suspected of influencing genetic stability.[1,22] We therefore performed a cytogenetic and biochemical epidemiological study to establish if there are significant differences between vegetarians (V) and nonvegetarians (NV) in their blood vitamin status and their MN frequency in peripheral blood lymphocytes.[8] The levels of plasma vitamin C (VIT-C), vitamin E (VIT-E), vitamin B_{12} (B_{12}), and folic acid were also analyzed to assess if differences in chromosomal damage rates were associated with these potentially antimutagenic micronutrients. Volunteers were classified as either "vegetarian" if they had abstained from eating any flesh foods for at least three years prior to the study or "nonvegetarian" if they consumed meat or meat products at least five days per week for at least three years before participation in the study. The volunteers in the study consisted of 47 male and 79 female V and 66 male and 72 female NV, all of whom were nonsmokers for at least three years prior to the study. The age of the volunteers varied between 20 and 89 years. There was no significant difference in the slope of the age-related increase in MN index of V and NV for either sex. However, the MN index was significantly lower in NV males in the 20- to 40-year age group and significantly lower for V males in the 41- to 60-year age group when compared to their corresponding male counterparts (FIG. 2A). No significant differences between the MN index of V and NV males in the oldest age group was detectable, and there also was no difference in the MN index of V and NV females across all age groups (FIGS. 2A and B). V were generally found to have significantly higher plasma levels of VIT-C and folic acid, significantly lower levels of B_{12}, and similar levels of VIT-E when compared to NV. VIT-C correlated significantly and positively with the MN index in young males, but the reverse was true for B_{12} (FIG. 3A). In young females, folate and vitamin B_{12} appeared to correlate negatively with the MN index

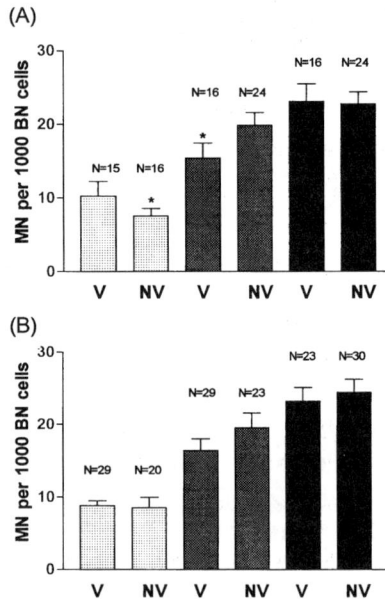

FIGURE 2. Comparison of MN frequency in (A) vegetarian and nonvegetarian males and (B) vegetarian and nonvegetarian females. To facilitate comparison of data between sexes, data for females were sex adjusted to male values using the 1.53 factor, which is the mean ratio of the MN frequency for males and females for each age group. *p < 0.05. ☐, 20–40 years; ▨, 41–60 years; ■, 61–90 years.

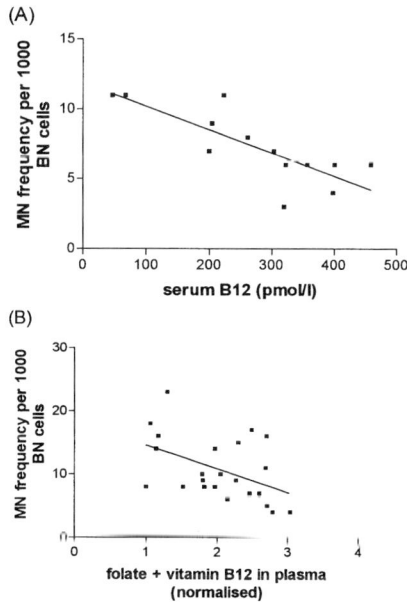

FIGURE 3. Correlations between MN frequency and folate and vitamin B_{12} status in young males and females aged between 20 and 30 years. (A) Relationship between plasma vitamin B_{12} and MN frequency in males. Spearman's r = −0.808; p < 0.0008; N = 13. (B) Relationship between MN frequency in lymphocytes of females and the combined normalized values for plasma folate and plasma B_{12}. Spearman's r = 0.4426; p < 0.0345; N = 23. Folate and B_{12} values were normalized by dividing the actual assay value by the mean value for the group and combined by adding the respective normalized folate and normalized B_{12} data.

(FIG. 3B). VIT-E had no apparent impact on the MN index. These data suggest that the level of folic acid and B_{12} may be more important than VIT-C and VIT-E in minimizing chromosomal damage rates in human lymphocytes, and the chromosomal damage rates of healthy V and healthy NV are similar.

FOLATE, VITAMIN B_{12}, HOMOCYSTEINE STATUS, AND MICRONUCLEUS FREQUENCY IN LYMPHOCYTES OF OLDER MEN

Deficient levels of folic acid and vitamin B_{12} are associated with an elevated chromosomal damage rate and high concentrations of homocysteine in the blood.[1,10,23,24] We have therefore performed a study to determine the prevalence of folate deficiency, vitamin B_{12} deficiency, and hyperhomocysteinemia in 64 healthy men aged between 50 and 70 years and to evaluate the relationship of these micronutrient levels in the blood with the MN frequency in peripheral blood lymphocytes.[12] We also performed a placebo-controlled, double-blind, intervention study to determine whether supplementation of the diet with a daily dose of 0.7 mg (as a supplement in cereal) or 2.0 mg (in a tablet) over a period of four months resulted in a significant alteration of folate status, homocysteine

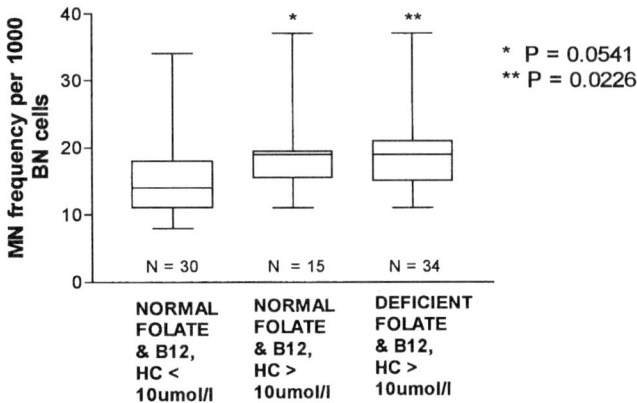

FIGURE 4. MN frequency in men (aged 50–70 years) who (1) were deficient in serum folate (< 6.0 nmol/L), and/or red blood cell folate (< 317 nmol/L), and/or serum B_{12} (< 150 pmol/L), and/or had a plasma homocysteine (HC) value > 10 μmol/L, N = 34; (2) had normal values for folate and vitamin B_{12} but had a plasma HC > 10 μmol/L, N = 15; and (3) had normal values for folate and vitamin B_{12} and a plasma HC < 10 μmol/L, N = 30. p values refer to a two-tailed Mann-Whitney U test. [*]p = 0.0541; [**]p = 0.0226.

status, and the MN index.[12] Twenty-three percent of the men were serum folate deficient (< 6.8 nmol/L), 16% were red blood cell folate deficient (< 317 nmol/L), 4.7% were vitamin B_{12} deficient (< 150 pmol/L), and 37% had plasma homocysteine levels greater than 10 μmol/L. In total 56% of the men had one or more abnormal blood values for folate, vitamin B_{12}, or homocysteine. The MN index of these men (N = 34) in cytokinesis-blocked binucleated cells (19.2 ± 1.1) was significantly elevated (p = 0.02) when compared to the MN index of the rest of the men who had normal levels of folate, vitamin B_{12}, and homocysteine (16.3 ± 1.3, N = 30) (FIG. 4). Interestingly, the MN index in men with normal folate and vitamin B_{12} but homocysteine levels greater than 10 μmol/L (19.4 ± 1.7, N = 15) was also significantly higher (p = 0.05) when compared to those with normal folate, vitamin B_{12}, and homocysteine (FIG. 4). This novel result was also supported by the observation that the MN index and plasma homocysteine were significantly (p = 0.0086) and positively correlated (r^2 = 0.172) in those subjects who were not deficient in folate or vitamin B_{12} (FIG. 5A). The MN index was not significantly correlated with folate indices, but there was a significant (p = 0.013) negative correlation with serum vitamin B_{12} (r^2 = 0.099) (FIG. 5B). Daily supplementation of the diet with 0.7 mg free folic acid in cereal for two months followed by 2.0 mg free folic acid via a tablet produced a fourfold increase in plasma folate, a 2.6-fold increase in red blood cell folate, and an 11% reduction in plasma homocysteine. However, these changes were not accompanied with a reduction in the MN index. In conclusion it is apparent that elevated homocysteine status, in the absence of vitamin deficiency, and low, but not deficient, vitamin B_{12} status are important risk factors for increased chromosomal damage in lymphocytes.

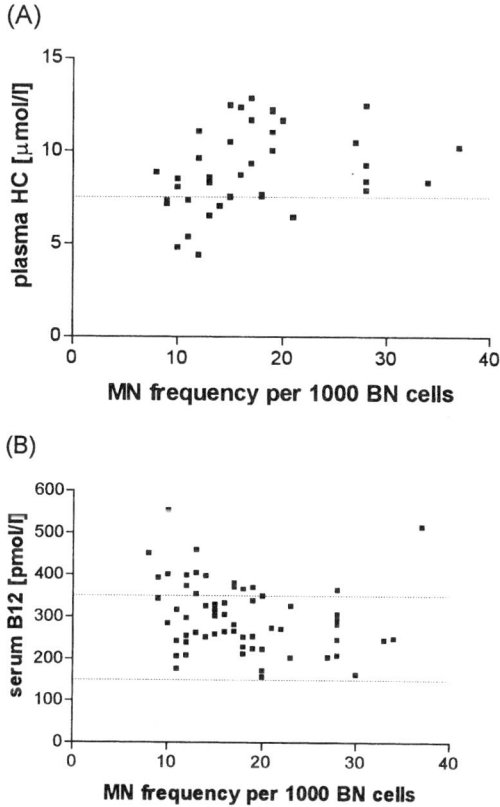

FIGURE 5. (A) Relationship between plasma homocysteine and MN frequency in subjects (50- to 70-year-old men) with normal levels of serum B_{12}, serum folate, and red blood cell folate. Spearman's r = 0.4148; p = 0.0086; N = 39. (B) Relationship between serum B_{12} and MN frequency for subjects (50- to 70-year-old men) whose serum B_{12} level was not in the deficient range. Spearman's r = −0.3418; p = 0.0127; N = 62. The results shown are for men aged between 50 and 70 years.

VITAMIN E SUPPLEMENTS AND THEIR EFFECT ON VITAMIN E STATUS IN BLOOD AND MICRONUCLEUS FREQUENCY IN LYMPHOCYTES OF OLDER MEN

A placebo-controlled double-blind intervention trial was performed with 60 male volunteers aged between 50 and 70 years to test the hypothesis that intake of *d*-alpha-tocopherol (VITE) above the recommended dietary intake (RDI) level (10.0 mg or 14.9 I.U. VITE) can protect against DNA damage in human peripheral blood lymphocytes.[11] The intervention consisted of two phases, each of eight weeks duration: during the initial phase the VITE supplement was 5 × RDI (provided in cereal), and during the second phase the

VITE supplement was 30 × RDI (provided in capsules). Blood samples were collected before the initial phase, between phases, and at the end of the second phase. The level of VITE was measured in plasma using HPLC, and the genetic damage rate in peripheral blood lymphocytes was measured using the cytokinesis-block MN assay. The study has shown that cereal supplementation is an effective route for an intake of VITE that is above the RDI; a 5 × RDI supplement resulted in a 22% elevation in the plasma VITE status. By comparison the use of 30 × RDI supplementation with capsules as the route of delivery resulted in an 89% increment in plasma VITE status. The increased VITE status during the intervention, however, had no significant impact on the spontaneous genetic damage rate in human lymphocytes. There was also no correlation between baseline genetic damage frequency and VITE status. However, a 32% ($p < 0.007$) decrease in the MN index was recorded in both the control and VITE-supplemented groups during the course of the study, which could have been due either to seasonal effects or other common components in the diet such as the carrier used for the VITE in the cereal and the capsules. The study has identified a small proportion (3.4%) of apparently healthy individuals who are abnormally sensitive to oxidative stress by hydrogen peroxide and who demonstrated that VITE supplementation did not attenuate the impact of the oxidative challenge on genetic damage rate. The above data suggest that supplements in cereal are a viable route for delivering VITE and that supplementation with VITE is unlikely to affect chromosomal damage occurring spontaneously or as a result of exposure to oxidative radicals.

MODERATE WINE CONSUMPTION PROTECTS AGAINST SPONTANEOUS AND HYDROGEN PEROXIDE–INDUCED DNA DAMAGE

We have tested the hypothesis that moderate wine drinking can protect somatic cells against the DNA-damaging effect of hydrogen peroxide, which is an endogenous source of reactive oxygen metabolites.[13] In this preliminary investigation, four male volunteers were placed on a plant-polyphenol free (PPF) diet to ensure that the wine provided was the only main source of plant phenolic compounds. After 48 hours on the PPF diet, the volunteers were required to consume 300 mL of red or white wine, and blood samples were collected 1, 3, 8, and 24 hours postconsumption while still on a PPF diet. Plasma was isolated from the blood samples and stored frozen for subsequent assays. In the subsequent assays, fresh lymphocytes from each donor were incubated in their corresponding plasma from the various intervention time points for 30 minutes. The capacity of the plasma to prevent damage to DNA in lymphocytes by hydrogen peroxide was assessed using the cytokinesis-block MN technique. The data from this preliminary investigation indicated that there was a strong inhibition (> 70%) of hydrogen peroxide–induced micronucleated (MNed) cells by the plasma samples from the blood collected one hour after consumption of wine, as compared to plasma samples from blood immediately before the consumption of wine. This protective effect was apparent for both red and white wine, although statistical significance ($p = 0.0068$) was achieved only in the white wine intervention. A higher degree of statistical significance ($p = 0.0008$) was achieved when the data for samples following the consumption of red and white wine were combined (FIG. 6). There was no difference in the hydrogen peroxide–induced MN cell frequency when comparing results immediately before starting on the PPF diet, before consumption of wine, and 8 hours after or 24 hours after wine consumption. The hydrogen peroxide–induced MN cell frequency in cells incu-

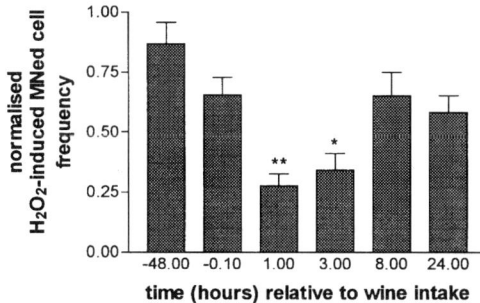

FIGURE 6. Effect of plasma collected before and after drinking wine on the MNed cell frequency induced by hydrogen peroxide. The results represent the combined data for the red and white wine intervention, for each of four male individuals aged between 20 and 40 years. The data shown have been normalized by dividing with the highest value for induced MNed cell frequency for each individual in their respective trial. The p values refer to comparisons with the results at −48 hours and were obtained using Dunn's multiple comparison test. P(ANOVA) was estimated using the Kruskal-Wallis test. p(ANOVA) = 0.0002; *p < 0.01; **p < 0.001; N = 8.

bated with plasma from blood collected 3 hours after wine consumption was intermediate to that observed for plasma collected 1 hour and 8 hours after wine intake.

The protective effect of plasma against hydrogen peroxide–induced DNA damage cannot be readily explained by the red wine content of phenolic compounds because results for red wine were similar to those for white wine, even though white wine had a much lower level of total polyphenols. A possible explanation could be that alcohol, glycerol, and ascorbate in wine, together with specific wine phenolic compounds that are also equally present in red and white wine (*e.g.*, hydroxycinnamates), may have contributed to the observed protection of nuclear material from hydrogen peroxide–derived reactive oxygen metabolites. This explanation is supported by data from *in vitro* experiments showing that incubation of lymphocytes either with alcohol or wine stripped of phenolic compounds resulted in a statistically significant (p < 0.05), dose-related reduction (up to 87% reduction) in hydrogen peroxide–induced MN cell frequency.

The effect of plasma from the intervention on baseline MNed cell frequency was assessed in the control cultures that were not exposed to hydrogen peroxide. The data obtained for samples after the consumption of white wine do not suggest any effect on baseline MNed cell frequency. By contrast, the ANOVA analysis for red wine indicates that the overall observed variation in MNed cell frequency, including the reduction in the mean value of the baseline MNed cell frequency during the first three hours after red wine intake, was statistically significant (p = 0.0371). On combining the data for red and white wine, one can discern an apparent 20% reduction in the baseline frequency of MNed cells one hour after wine consumption; however, the observed variation in the mean value for this index with time relative to the consumption of wine did not achieve statistical significance; marginal statistical significance was only achieved when the result at one hour after wine consumption was compared to the result for the −48 hour time point using a two-tailed Wilcoxon's matched pairs test (p = 0.068).

DISCUSSION

Chromosomal breakage and loss is a common event induced by carcinogens and reactive oxygen metabolites;[21] furthermore cancer cells also exhibit a high rate of spontaneous chromosomal breakage and loss.[3] The evidence that chromosomal aberration rate is positively correlated with cancer risk has been further consolidated by epidemiological data from the Italian and the Nordic cohort studies, which showed that overall cancer rates are increased approximately twofold in those with the highest tertile of chromosomal damage relative to those in the lowest tertile.[25,26] The extent to which diet, genetic background, and exposure to environmental genotoxins contributes to the observed rate of chromosomal damage remains unclear, and an understanding of the potential contribution of each of these factors is crucial in evaluating the scope for intervention. The main assumption of intervention strategies is that chromosomal damage rate is the underlying cause of aging[27] and cancer[25,26] and that a minimization of chromosomal damage rate should result in a reduced risk for cancer and a restraint of the aging process.

What level in reduction of chromosomal damage can be achieved? What level of reduction would be significant? One of the most practical ways of determining the potential for reducing chromosomal damage rate is to examine the available database for baseline or spontaneous MN frequency in a healthy population that is not exposed to abnormal levels of genotoxins through the diet or ambient environment (e.g. FIG. 1). A wide range of MN frequency values has been recorded for each age group. It would be reasonable to expect that there would be some scope for a reduction in the MN frequency of those with high values to the lowest levels observed in the same age group. The age-related increase in MN frequency clearly indicates that there is a baseline minimum of genetic damage that can be expected even in the youngest age groups with frequencies in the range of 1 to 5 micronuclei (MNi) per 1000 cytokinesis-blocked binucleated (BN) cells. One may therefore expect a reduction to such a level would be the lowest practical level achievable and that such a reduction could be significant in ultimate health terms. Another approach is to consider the MN frequency in terms of the level that could be induced by a significant dose of ionizing radiation, that is, the rad-equivalent concept;[28] the dose-response relationship of induced MN frequency and X-ray dose is very well established for human lymphocytes.[29] Using such a dose response we can estimate that an increment of 16.2 MN per 1000 BN cells is equivalent to a dose of 0.40 Gy of X rays, which is considered to be a significant dose with respect to cancer and premature aging risk.[30,31] The use of a rad-equivalent approach is supported by experimental evidence showing that the shortening of life produced by whole body gamma irradiation is due not to the induction of specific diseases but the bringing forward in time of all causes of death without any intrinsic change in their relative probabilities.[30,31] Consequently one could argue that reduction of the MN frequency by 16.2 MNi per 1000 BN cells may be considered to be significant in health terms. Another approach would be to consider the MN index as a measure of biological age and that the health significance of the MN frequency could be interpreted in terms of the risks associated with such a biological age.

Which vitamins are most likely to help reduce chromosomal damage? Will combinations of vitamins work better than single vitamins? In the past decade several studies have been performed to investigate the impact of antioxidant (redox) and other vitamin supplements on chromosomal damage and MN rates in human subjects. By and large it appears that such vitamins as beta-carotene, alpha-tocopherol, and folic acid can reduce chromosomal damage rate in situations when cells are exposed to abnormal doses of genotoxins[32-35] or when the vitamins are deficient;[36,37] however, it is still not clear whether vitamin supplements produce a net benefit in healthy subjects who are not vitamin deficient and/or abnormally exposed to DNA-damaging agents. Our cross-sectional studies on healthy subjects[8,10] do not show a negative correlation between the antioxidant vitamin C and E and MN frequency; this is also underscored by the lack of a reduction in the MN frequency following supplementation with 10–30 times the RDI of VIT-E for four months.[11] By contrast, a consistent and significant negative correlation between vitamin B_{12} and MN frequency has been observed in both young and older men.[8,10,12] In the latter case the data are significantly correlated even though none of the men were in the deficient range (as defined by clinical parameters for anemia), which suggests that the optimum for minimizing chromosomal damage in men does not correspond to the current standard of sufficiency. Our correlation data indicate that a plasma B_{12} level above 300 pmol/L may be required to minimize the MN index in older men. Also interesting is the observation that the MN frequency is positively and significantly correlated with plasma homocysteine status in men who are not folate or vitamin B_{12} deficient, suggesting that a homocysteine plasma status below 7.5 µmol/L may be required to minimize chromosomal damage rate in older men. The intervention with folate did not produce a significant change in the MN index. This could be explained by the fact that folate, unlike vitamin B_{12}, does not correlate significantly with MN frequency in men who are not folate deficient and that the homocysteine status was not sufficiently reduced by the folate supplement given: homocysteine was reduced by 11% (p = 0.002) from a mean value of 9.33 µmol/L to 8.51 µmol/L. A combination of folate and vitamin B_{12} supplementation over a longer time frame might have produced more adequate changes in the B_{12} and homocysteine status to alter chromosomal damage rate.

Other intervention studies with human volunteers have shown that (a) beta-carotene but not ascorbate supplementation produced a significant *ex vivo* protection against ionizing radiation[38] and (b) supplementation with a mixture of vitamins A, C, E; beta-carotene; folic acid; and rutin produced a 25% reduction in baseline MN frequency in older subjects and a greater than 50% reduction in gamma-ray-induced MN frequencies in both older and younger men.[39] A systematic study of combination supplements is required to identify the optimal set for reducing both spontaneous and induced MN formation.

Which diets can best contribute to lower chromosomal damage rates? Which combination of fruits and vegetables is best? There is also increased interest in identifying specific diets (*e.g.* Mediterranean diet, vegan, or vegetarian diets) or food combinations that might lower the rate of spontaneous chromosomal damage. The only studies published are those we have described comparing vegetarians and nonvegetarians and those of Verhagen *et al.*[40] who compared the MN frequency in female vegans and female omnivores and found no difference between the groups. Our data and those of Verhagen *et al.*[40] are in agreement in that both show no remarkable difference in the MN index of vegans or vegetarians in relation to omnivores. These studies, however, were cross-sectional studies and therefore limited by the variation in the spectrum of foods eaten among individuals. Intervention studies will be required to compare the impact of specific classes of fruits or vegetables in relation to reduction of chromosomal damage rates. With regard to beverages, it appears from our preliminary data that red wine flavonoids may have a protective effect against spontaneous DNA damage and that wine (red or white) can produce a marked protective effect against hydrogen peroxide. In the latter case it appears that alcohol in combination with other wine components may be the main protective agent. The role of flavonoids in reducing MN formation is also supported by two recent studies showing (a) that tea drinking is significantly correlated with reduced MN formation in humans[41] and (b) that a variety of natural flavonoids have been shown to produce significant attenuation of MN formation following *in vivo* X-irradiation in mice.[42]

CONCLUDING REMARKS

There is increasing evidence that diet may have a significant influence on chromosomal damage rates and MN frequency. The optimal intakes of specific vitamins and other micronutrients have yet to be determined but cross-sectional data in nondeficient males suggest that the current cutoff for vitamin B_{12} adequacy does not appear to be optimal for minimizing the MN index in lymphocytes of older men. Furthermore it is evident in males that an optimal homocysteine status (less than 7.5 µmol/L) may be required to minimize chromosomal damage rates. Not enough is known about the impact of specific diets on chromosomal damage rates, but it is evident that vegetarians and vegans do not have a lower MN index when compared to omnivores, and, in fact, vegetarians or vegans may be at greater risk for chromosomal damage if they become vitamin B_{12} deficient. More needs to be known about the specific impact of specific classes of fruits and vegetables on chromosomal stability. Based on knowledge to date it seems that modulation of oxidative stress by diet and optimization of B vitamin status may be the most promising avenues of research in the prevention of chromosomal damage. Such research could be of even more practical value if it were targeted to individuals with common genetic traits that predispose to oxidatively induced DNA damage or limit the appropriate utilization of folic acid or vitamin B_{12}.

ACKNOWLEDGMENTS

The author thanks J. Rinaldi, C. Aitken, S. Neville for their technical support, Dr. Ivor Dreosti and Ms. Creina Stockley for their advice, and the volunteers for participating in the various studies.

REFERENCES

1. AMES, B.N., L.S. GOLD & W.C. WILLETT. 1995. The causes and prevention of cancer. Proc. Natl. Acad. Sci. USA **92:** 5258–5265.
2. BOLOGNESI, C., A. ABBONDANDOLO, R. BARALE, R. CASALONE, L. DALPRA, M. DE FERRARI, F. DEGRASSI, A. FORNI, L. LAMBERTI, C. LANDO, L. MIGLIORE, P. PADOVANI, R. PASQUINI, R. PUNTONI, I. SBRANA, M. STELLA & S. BONASSI. 1997. Age-related increase of base-line frequencies of sister chromatid exchanges, chromosomal aberrations and micronuclei in human lymphocytes. Cancer Epid. Biomarkers & Prev. **6:** 249–256.
3. SOLOMON, E., J. BORROW & A.D. GODDARD. 1991. Chromosomal aberrations and cancer. Science **254:** 1153–1160.
4. WEIRICH-SCHWAIGER, H., H.G. WEIRICH, B. GRUBER, M. SCHWEIGER, & M. HIRSCH-KAUFFMANN. 1994. Correlation between senescence and DNA repair in cells from young and old individuals and in premature aging syndromes. Mutat. Res. **316**(1): 37–48.

5. FENECH, M. & A.A. MORLEY. 1986. Cytokinesis-block micronucleus method in human lymphocytes: Effect of *in vivo* aging and low dose X-irradiation. Mutat. Res. **161:** 193–198.
6. RAMSEY, M.J., D.H. MOORE, J.F. BRINER, D.A. LEE, L. OLSEN, J.R. SENFT & J.D. TUCKER. 1995. The effects of age and life-style factors on the accumulation of cytogenetic damage as measured by chromosomal painting. Mutat. Res. **338:** 95–106.
7. HASTIE, N., M. DEMPSTER, M.G. DUNLOP, A.M. THOMPSON, D.K. GREEN & R.C. ALLSHIRE. 1990. Telomere reduction in human colorectal carcinoma and with aging. Nature **346:** 866–868.
8. FENECH, M. & J. RINALDI. 1995. A comparison of lymphocyte micronuclei and plasma micronutrients in vegetarians and non-vegetarians. Carcinogenesis **16**(2): 223–230.
9. FENECH, M., S. NEVILLE & J. RINALDI. 1994. Sex is an important variable affecting spontaneous micronucleus frequency in cytokinesis-blocked lymphocytes. Mutat. Res. **313:** 203–207.
10. FENECH, M. & J. RINALDI. 1994. The relationship between micronuclei in human lymphocytes and plasma levels of vitamin C, vitamin E, vitamin B_{12} and folic acid. Carcinogenesis **15**(7): 1405–1411.
11. FENECH, M., I.E. DREOSTI & C. AITKEN. 1997. Vitamin-E supplements and their effects on vitamin-E status in blood and genetic damage rate in peripheral blood lymphocytes. Carcinogenesis **18**(2): 359–364.
12. FENECH, M., I.E. DREOSTI & J.R. RINALDI. 1997. Folate, vitamin B_{12}, homocysteine status and chromosomal damage rate in lymphocytes of older men. Carcinogenesis **18**(8). In press.
13. FENECH, M., C. STOCKLEY & C. AITKEN. 1997. Moderate wine consumption protects against hydrogen peroxide-induced DNA damage. Mutagenesis **12**(4): 289–296.
14. FENECH, M. 1993. The cytokinesis-block micronucleus technique: A detailed description of the method and its application to genotoxicity studies in human populations. Mutat. Res. **285:** 35–44.
15. FENECH, M. 1996. The cytokinesis-block micronucleus technique. *In* Technologies for Detection of DNA Damage and Mutations. G.P. Pfeiffer, Ed.: 25–34. Plenum Press. New York and London.
16. HANDO, J.C., J. NATH & J.D. TUCKER. 1994. Sex chromosomes, micronuclei and aging in women. Chromosoma **103:** 186–192.
17. CATALAN, J., K. AUTIO, M. WESSMAN, C. LINDHOLM, S. KNUUTILA, M. SORSA & H. NORPPA. 1995. Age-associated micronuclei containing centromeres and the X chromosomal in lymphocytes of women. Cytogenet. Cell Genet. **68**(1–2): 11–16.
18. SURRALES, J., G. FALCK & H. NORRPA. 1996. *In vivo* cytogenetic damage revealed by FISH analysis of micronuclei in uncultured human T-lymphocytes. Cytogenet. Cell Genet. **75**(2–3): 151–154.
19. NATH, J., J.D. TUCKER & J.C. HANDO. 1995. Y chromosomal aneuploidy, micronuclei, kinetochores and aging in men. Chromosoma **103**(10): 725–731.
20. SURRALES, J., P. JEPPESEN, H. MORRISON & A.T. NATARAJAN. 1996. Analysis of loss of inactive X chromosome in interphase cells. Am. J. Hum. Genet. **59:** 1091–1096.
21. FENECH, M. & A.A. MORLEY. 1989. Kinetochore detection in micronuclei: An alternative method for measuring chromosome loss. Mutagenesis **4**(2): 98–104.
22. WILLETT, W.C. 1994. Diet and health: What should we eat? Science **264:** 532–537.
23. BLOUNT, B.C. & B.N. AMES. 1995. DNA damage in folate deficiency. Bailliere's Clin. Haematology **8**(3): 461–478.
24. WAGNER, C. 1995. Biochemical role of folate in cellular metabolism. *In* Folate in health and disease. Lynn. B. Bailey, Ed.. 23–42. Marcell Dekker Inc. New York.
25. HAGMAR, L., A. BROGGER, I. HANSTEEN. S. HEIM, B. HOGSTEDT, L. KNUDSEN, B. LAMBERT, K. LINNAINMAA, F. MITELMAN, I. NORDENSON, C. REUTERWALL, S. SALOMAA, S. SKERFVING & M. SORSA. 1994. Cancer risk in humans predicted by increased levels of chromosomal aberrations in lymphocytes: Nordic Study Group on the health risk of chromosome damage. Cancer Res. **54:** 2919–2922.
26. BONASSI, S., A. ABBONDANDOLO, L. CAMURRI, L. DAL PRA, M. DE FERRARI, F. DEGRASSI, A. FORNI, L. LAMBERTI, C. LANDO, P. PADOVANI, I. SBRANA, D. VECCHIO & R. PUNTONI. 1995. Are chromosome aberrations in circulating lymphocytes predictive of future cancer onset in humans? Cancer Genet. Cytogenet. **79:** 133–135.
27. GUARENTE, L. 1996. Do changes in chromosomes cause aging? Cell **86:** 9–12.

28. TORNQVIST, M. & S. OSTERMAN-GOLKAR. 1991. Monitoring of *in vivo* dose by macromolecular adducts: Usefulness in risk estimation. *In* J.D. Groopman & P.L. Skipper, Eds.: 89–99. Molecular Dosimetry and Human Cancer. CRC Press. Boston.
29. FENECH, M. & A.A. MORLEY. 1986. Cytokinesis-block micronucleus method in human lymphocytes: Effect of *in vivo* aging and low dose X-irradiation. Mutation Res. **161:** 193–198.
30. NEARY, G.J. 1960. Aging and radiation. Nature **187:** 10–18.
31. LINDOP, P.J. & J. ROTBLAT. 1961. Shortening of life and causes of death in mice exposed to a single whole-body dose of radiation. Nature **189:** 645–648.
32. BENNER, S.E., S.M. LIPPMAN, M.J. WARGOVICH, M.J. VELASCO, M. PETERS, E.J. MORICE & W.K. HONG. 1992. Micronuclei in bronchial biopsy specimens from heavy smokers: Characterization of an intermediate biomarker of lung carcinogenesis. Int. J. Cancer **52:** 44–47.
33. BENNER, S.E., M.J. WARGOVICH, S.M. LIPPMAN, R. FISHER, M.J. VELASCO, R.J. WINN & W.K. HONG. 1994. Reduction in oral mucosa micronuclei frequency following alpha-tocopherol treatment of oral leukoplakia. Cancer Epidemiol. Biomarkers Prev. **3:** 73–76.
34. STICH, H.F., M.P. ROSIN, A.P. HORNBY, B. MATHEW, R. SANKARANARAYAN & M.K. NAIR. 1988. Remission of oral leukoplakia and micronuclei in tobacco-betel quid chewers treated with beta-carotene and with beta-carotene plus vitamin A. Int. J. Cancer **42:** 195–199.
35. BENNER, S.E., S.M. LIPPMAN, M.J. WARGOVICH, J.J. LEE, M. VEALSCO, J.W. MARTIN, B.B. TOTH, W.K. HONG. 1994. Micronuclei, a biomarker for chemoprevention trials: Results of a randomized study in oral pre-malignancy. Int. J. Cancer **59:** 457–459.
36. EVERSON, R.B., C.M. WEHR, G.L. EREXSON, & J.T. MACGREGOR. 1988. Association of marginal folate depletion with human chromosomal damage *in vivo*: Demonstration by analysis of micronucleated erythrocytes. J. Natl. Cancer Inst. **880:** 525–529.
37. BLOUNT, B.C., M.M. MACK, C.M. WEHR, J.T. MACGREGOR, R.A. HIATT, G. WANG, S.N. WICKRAMASINGHE, R.B. EVERSON & B.N. AMES. 1997. Folate deficiency causes uracil misincorporation into human DNA and chromosomal breakage: Implications for cancer and neuronal damage. Proc. Natl. Acad. Sci. USA **94:** 3290–3295.
38. UMEGAKI, K., S. IKEGAMI, K. INOUE, T. ISCHIKAWA, S. KOBAYASHI, N. SOENO & K. TOMABECHI. 1994. Beta-carotene prevents X-ray induction of micronuclei in human lymphocytes. Am. J. Clin. Nutr. **59:** 409–412.
39. GAZIEV A.I., G.R. SOLOGUB, L.A. FOMENKO, S.I. ZAICHKINA, N.I. KOSYAKOVA & R.J. BRADBURY. 1996. Effects of vitamin-antioxidant micronutrients on the frequency of spontaneous and *in vitro* gamma-ray-induced micronuclei in lymphocytes of donors: The age factor. Carcinogenesis **17**(3): 493–499.
40. VERHAGEN H., A.L. RAUMA, R. TORRONEN, N. de VOGEL, G.C. BRUIJNTJES-ROZIER, M.A. DREVO, J.J. BOGAARDS & H. MYKKANEN. 1996. Effect of a vegan diet on biomarkers of chemoprevention in females. Hum. & Exp. Toxicol. **15**(10): 821–825.
41. XUE, K., S. WANG, P. ZHOU, P. WU, R. ZHANG, Z. XU, W. CHEN & Y. WANG. 1992. Micronucleus formation in peripheral blood lymphocytes from smokers and the influence of alcohol- and tea-drinking habits. Int. J. Cancer **50:** 702–705.
42. SHIMOI, K., S. MASUDA, M. FURUGORI, S. ESAKI & N. KINAE. 1994. Radioprotective effects of antioxidative flavonoids in gamma-ray-irradiated mice. Carcinogenesis **15:** 2669–2672.

Pluripotent Protective Effects of Carnosine, a Naturally Occurring Dipeptide[a]

A.R. HIPKISS,[b,e,g] J.E. PRESTON,[c] D.T.M. HIMSWORTH,[b] V.C. WORTHINGTON,[b] M. KEOWN,[b] J. MICHAELIS,[f] J. LAWRENCE,[b] A. MATEEN,[b] L. ALLENDE,[b] P.A.M. EAGLES,[b] AND N. JOAN ABBOTT[d]

[b]Molecular Biology and Biophysics Group, [c]Institute of Gerontology, and [d]Department of Physiology, King's College London, Strand, London WC2R 2LS, United Kingdom
[e]Division of Biomolecular Engineering, CSIRO North Ryde, Sydney, NSW 2112, Australia
[f]Peptide Technology Ltd., Dee Why, Sydney, NSW 2099, Australia

ABSTRACT: Carnosine is a naturally occurring dipeptide (β-alanyl-L-histidine) found in brain, innervated tissues, and the lens at concentrations up to 20 mM in humans. In 1994 it was shown that carnosine could delay senescence of cultured human fibroblasts. Evidence will be presented to suggest that carnosine, in addition to antioxidant and oxygen free-radical scavenging activities, also reacts with deleterious aldehydes to protect susceptible macromolecules. Our studies show that, *in vitro*, carnosine inhibits nonenzymic glycosylation and cross-linking of proteins induced by reactive aldehydes (aldose and ketose sugars, certain triose glycolytic intermediates and malondialdehyde (MDA), a lipid peroxidation product). Additionally we show that carnosine inhibits formation of MDA-induced protein-associated advanced glycosylation end products (AGEs) and formation of DNA-protein cross-links induced by acetaldehyde and formaldehyde. At the cellular level 20 mM carnosine protected cultured human fibroblasts and lymphocytes, CHO cells, and cultured rat brain endothelial cells against the toxic effects of formaldehyde, acetaldehyde and MDA, and AGEs formed by a lysine/deoxyribose mixture. Interestingly, carnosine protected cultured rat brain endothelial cells against amyloid peptide toxicity. We propose that carnosine (which is remarkably nontoxic) or related structures should be explored for possible intervention in pathologies that involve deleterious aldehydes, for example, secondary diabetic complications, inflammatory phenomena, alcoholic liver disease, and possibly Alzheimer's disease.

Carnosine (β-alanyl-L-histidine) is found in brain, innervated muscle, and other tissues sometimes at concentrations as high as 20 mM. Many functions of carnosine have been proposed, including neurotransmitter, intracellular buffer, immunomodulator, antiradiation agent, metal ion chelator, wound healing agent, and antioxidant and free-radical scavenger.[1,2] More recently, carnosine has been shown to delay senescence in cultured human fibroblasts and increase the so-called Hayflick limit by up to 10 cell doublings,[3] although it is not known how the apparent antiaging effects are achieved. It is possible that these effects could be explained by carnosine's previously described antioxidant and oxygen free-radical scavenging activities,[1,2] its reaction with hypochlorite anions and ability to chelate divalent metal ions.[1,2] However, we have recently shown that carnosine can react preferentially with deleterious aldehydes and thereby protect proteins against cross-linking induced by reducing sugars.[4] We have therefore suggested that the dipeptide might be

[a]This work was supported by the World Cancer Research Fund.
[g]Tel: 44. 0171-873-2490; fax: 44. 0171-873-2285; e-mail: alan.hipkiss@kcl.ac.uk

37

a naturally occurring antiglycating agent[4–6] as its structure resembles some preferred sites in protein for nonenzymic glycosylation (lysine-histidine dipeptide sequences).

Aging is multifactorial, involving the deleterious consequences of many agents that compromise macromolecular integrity and physiological homeostasis; hence any effective antiaging agent is likely to be pluripotent in its actions. The results presented here suggest that carnosine is pluripotent. We show that it can protect proteins, DNA, and cells against deleterious aldehydes, including malondialdehyde (MDA), the highly reactive lipid peroxidation product, and we also show that carnosine can protect cultured cells against advanced glycosylation end products and a highly toxic amyloid peptide.

MATERIAL AND METHODS

Synthetic amyloid peptide βA4 (residues 25–35) was obtained from Bachem., Saffron Waldon, Essex, United Kingdom. All other proteins, 2,4-dinitrophenylhydrazine (2,4-DNPH), and sodium hypochlorite were obtained from Sigma Chemical Co., Poole, Dorset, United Kingdom. All other chemicals were obtained from BDH Ltd., Poole, Dorset, United Kingdom. Carnosine was a gift from Peptide Technology Ltd., Dee Why, Sydney, Australia.

Treatment of Proteins with Hypochlorite

Proteins, normally 1 mg/mL, were incubated in 100 mM potassium phosphate, pH 7, at 37°C, and 10 μL of reagent grade hypochlorite was added to 1 mL of protein solution unless otherwise stated.

Treatment of Proteins with Malondialdehyde

MDA was prepared by acid hydrolysis of its methyl acetal as described by Libondi *et al.*[7] Briefly, 15 mmoles of 1,1,3,3,-tetramethoxypropan was added to 1.35 mL of 1 M HCl and allowed to stand in the dark at 37°C with occasional shaking for 30 minutes prior to use. Bovine serum albumin (BSA), ovalbumin, and alpha-crystallin (1 mg/mL unless otherwise stated) were incubated at 37°C in 2 mL of 100 mM potassium phosphate buffer (containing 20 μL sodium azide) in sealed microcentrifuge tubes in the dark with 100 mM MDA for up to 18 hours, or as indicated in the text. Where employed, carnosine was added 30 minutes prior to addition of MDA.

Analysis of Proteins

Sodium dodecylsulphate polyacrylamide gel electrophoresis (SDS PAGE) was employed for protein separation and was carried out on 4–15% "ReadyGels" obtained from Bio-Rad Laboratories (Hercules, CA, USA) according to manufacturer's instructions. Gels were stained with Coomassie blue (R-250). Size-exclusion chromatography was performed on a Sephadex G100 (Pharmacia) column (85 cm × 1.9 cm) using 0.1 M ammonium bicarbonate as eluant at 4°C. Determination of protein carbonyl groups was carried out essentially as described by Hazell *et al.*[8] Proteins exposed to hypochlorite were precipitated with 5% trichloroacetic acid (TCA) after treatment. Following centrifugation the supernatant fraction was discarded; the precipitate was washed twice (with centrifugation) with 5% TCA, redissolved in 2 M HCl prior to addition of the 500 µL of 10 mM 2,4-DNPH, and incubated for two hours at 37°C, after which an equal volume of 2 M NaOH was added. The presence of carbonyl groups in the proteins were shown by an increase in absorbance at 450 nm. Proteins exposed to MDA were precipitated with 5% TCA and centrifuged at 11000 g for five minutes. The supernatant fraction was discarded, and the precipitate was washed with 5% TCA to remove free reagent and then redissolved in 500 µL of 10 mM 2,4-DNPH in 2 M HCl and allowed to stand at room temperature for one hour, vortexing every 15 minutes; 750 µL of 2 M NaOH was added, and the absorbance at 450 nm was determined.

DNA-protein Cross-linking

DNA-protein cross-links were induced by incubating calf thymus DNA (1 µg) with histone H2 (4 µg) in the presence of either formaldehyde or acetaldehyde (1–100 mM) (total volume 400 µL) for one hour at 37°C. When present, carnosine (0–200 mM) was added prior to addition of the aldehyde, and the mixture was kept on ice for 15 minutes. DNA cross-linked to protein was determined essentially as described by Zhitkovich and Costa,[9] which involved the selective precipitation of DNA cross-linked to protein by precipitation with potassium-SDS, digestion of protein with proteinase K, and fluorometric measurement (excitation at 360 nm and emission at 450 nm) of the precipitated DNA using the Hoechst reagent. Results (mean of quadruplicate determinations) are expressed as arbitrary fluorescence units.

Cell Culture

Cultured rat brain endothelial cells (RBE4) were obtained from Dr. P-O. Couraud, INSERM France, and cultured as described.[10] Cells at passages 52–58 were removed from storage in liquid N_2, plated onto collagen-coated 24-well plates at a seeding density of 2×10^6 cells/cm^{-2}, and grown to confluence in three days in 1 mL culture medium per well. The culture medium consisted of αMEM:Ham's F10 1:1 (Gibco), supplemented with 5% fetal bovine serum, 0.5 ng/mL basic fibroblast growth factor (Boehringer), 100 U/mL penicillin, and 100 µg/mL streptomycin and was changed every three days. Cells were washed three times in Hank's balanced salt solution between changes. Chinese hamster ovary (CHO) cells were cultured as described by Holliday and McFarland.[11]

Treatment with Amyloid Peptide

Synthetic Aβ(25–35), that is, β/A4 amyloid peptide fragment residues 25–35, was dissolved in double-distilled sterile water at 6 mg/mL and used immediately. Cells were treated with peptide at 50, 100, 200, 300 or 400 µg/mL for three days. Morphological changes in the cells were monitored daily using a phase contrast microscope. Cell damage was assessed at the end of incubation with peptide by measuring lactate dehydrogenase (LDH) release into the medium, glucose consumption, and mitochondrial dehydrogenase activity. LDH release was determined by measuring the rate of disappearance of NADH (after addition of pyruvate and NADH) monitored spectrophotometrically at 340 nm.[12] Glucose consumption was calculated from the loss of glucose from the medium after incubation and was measured using the hexokinase, glucose-6-phosphate dehydrogenase method.[13] Measurements of mitochondrial dehydrogenase activity employed the MTT (3-[4,5-dimethylthiazol-2-yl]-2,5-diphenyltetrazolium bromide) reduction assay, which uses cleavage of the tetrazolium salt to yield a purple dye with absorbance read at 570 nm.[14] Results are expressed as percentage of control absorbance.

RESULTS

Carnosine Can Protect Proteins against Some Age-associated Modifications

Protein cross-linking increases upon aging and may be brought about through reducing sugars or other deleterious aldehydes. *In vivo* this is clearly shown in uncontrolled diabetes where elevated levels of serum glucose interact with polypeptide amino groups to form Schiff's bases, then Amadori products and finally advanced glycosylation end products (AGEs) on the proteins of the circulatory system, the eye lens, kidney, and peripheral neurons. We have previously shown that carnosine can react preferentially with triose and pentose aldose and ketose sugars and thereby prevent cross-linking of proteins.[5,6] FIGURE 1a shows that MDA, the lipid peroxidation end product, is a very effective protein cross-linking agent and that carnosine can inhibit cross-linking is a concentration-dependent manner (FIG. 1b). Exposure to MDA also induces protein oxidation as illustrated by an increase in polypeptide carbonyl groups. FIGURE 2 shows that carnosine inhibited this process too with both crystallin and BSA. It may not be coincidental that amino groups of lysine residues and the imidazole groups of histidine residues are the groups most prone to age-associated oxidation[15] and that carnosine possesses both these functions. We also found that prolonged incubation of protein with MDA produced AGEs, as detected by anti-AGE antibodies, and that carnosine prevented AGE development (TABLE 1).

Hypochlorite anions are produced during the inflammatory response through the action of myeloperoxidase activity on hydrogen peroxide and chloride ions. Hypochlorite induces protein cross-linking and oxidation.[8] FIGURES 3 and 4 show that carnosine inhibited both these hypochlorite-mediated modifications probably because of the dipeptide's ability to react directly with hypochlorite to form a chloramine derivative.[1,2]

Cross-linking between DNA and proteins is another macromolecular modification that accumulates with age and that may result from the effects of ionizing or ultraviolet irradiation, oxygen free radicals, and/or deleterious aldehydes.[16] TABLE 2 shows that, in a model system, carnosine inhibited DNA-protein cross-linking induced by formaldehyde or acetaldehyde; the presence of carnosine in the incubation mixture inhibited cross-link generation between DNA and histone and in a concentration-dependent manner.

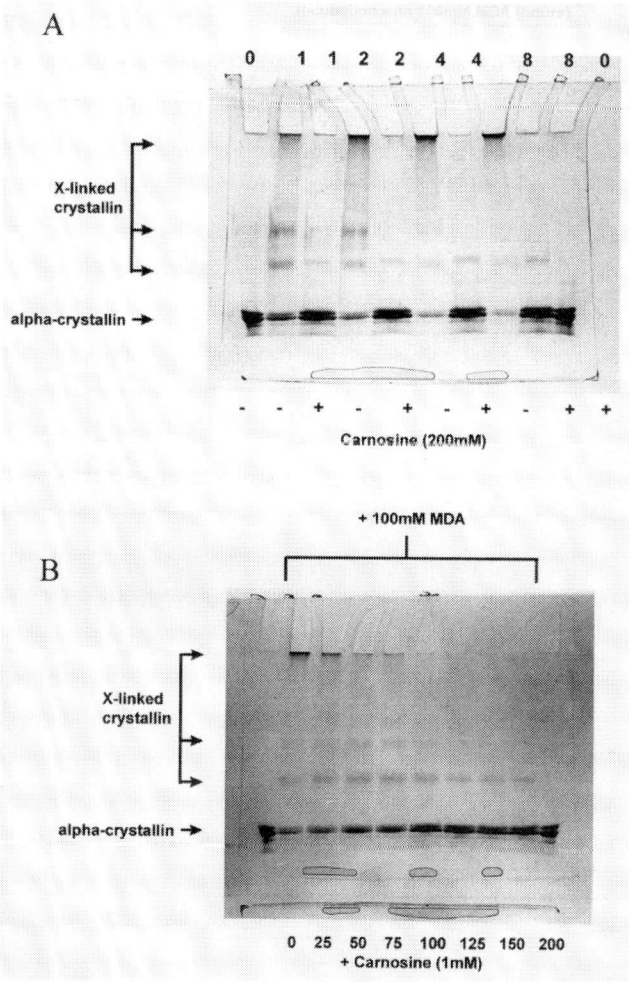

FIGURE 1A. Effects of carnosine on malondialdehyde-induced protein cross-linking. α-Crystallin (1 mg/mL) was incubated with 100 mM MDA for various times in the presence or absence of carnosine (200 mM) and then analyzed by SDS PAGE. The figures at the top of each lane indicate the length in time (in hours) the protein was incubated with MDA, and the figures at the bottom of each lane show the presence or absence of carnosine (200 mM). **B**: Effects of varying concentrations of carnosine on MDA-induced protein cross-linking. α-Crystallin (1 mg/mL) was incubated with 100 mM MDA and carnosine as described in legend to FIG. 5 and then analyzed by SDS PAGE. The figures at the bottom of each lane indicate the carnosine concentration used. All except the extreme left- and right-hand lanes were incubated with 100 mM MDA.

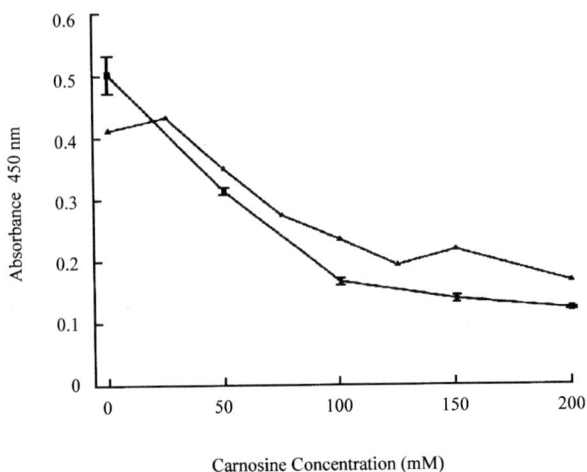

FIGURE 2. Effect of carnosine on malondialdehyde-induced formation of protein carbonyl groups. Bovine serum albumin (▲) and α-crystallin (■) (1 mg/mL) were incubated with 100 mM MDA at 37°C for 45 and 120 minutes, respectively, with carnosine (0–200 mM) prior to assay for carbonyl group formation (absorbance at 450 nm). Where present carnosine was added 30 minutes prior to addition of the MDA.

TABLE 1. The Effects of Carnosine on MDA-induced Generation of Epitopes Reactive Towards anti-AGE Antibody[a]

Incubation	Reactivity towards anti-AGE antibody					
	Day 0	Day 1	Day 3	Day 4	Day 6	Day 8
BSA only	–	–	–	–	–	–
BSA + MDA (40 mM)	–	–	(+)	(+)	+	++
BSA + MDA (40 mM) + carnosine (400 mM)	–	–	–	–	–	–

[a]Bovine serum albumin was incubated with malondialdehyde (40 mM) at 37°C for 0–8 days then assayed through dot blotting using rabbit anti-AGE antibody (glycated albumin used as original antigen). *NB*: Following Sephadex G100 column chromatography (gel filtration), reactive epitopes eluted as high molecular weight material in the void volume region.

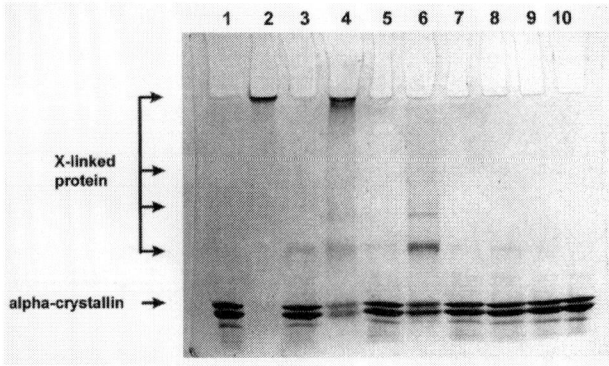

FIGURE 3. Inhibition of hypochlorite-mediated protein cross-linking by carnosine. α-Crystallin was incubated with hypochlorite (reagent grade) for 2 minutes at 37°C in the presence (lanes 3, 5, 7, 9, and 10) or absence (lanes 1, 2, 4, 6, and 8) of 50 mM carnosine (lanes 1 and 10 were controls). Hypochlorite added: 10 μL (lanes 2 and 3); 5 μL (lanes 4 and 5); 2.5 μL (lanes 6 and 7); 0.5 μL (lanes 8 and 9).

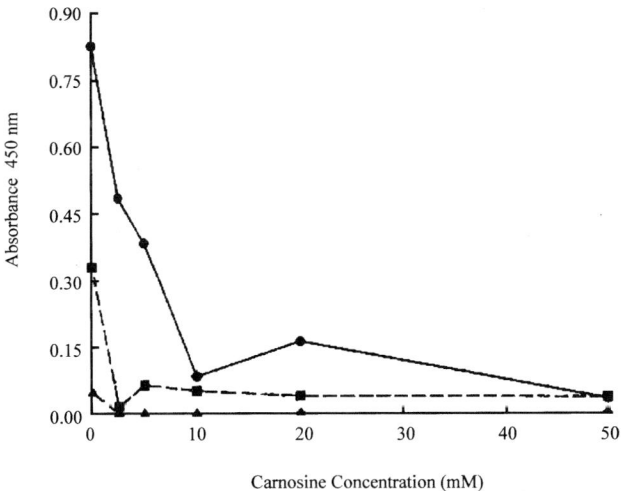

FIGURE 4. Effects of carnosine on hypochlorite-induced formation of protein carbonyl groups. Bovine serum albumin (1 mg/mL) was incubated with hypochlorite (40 and 10 μL/mL) for 5 minutes in the presence of carnosine (0 to 50 mM). Protein carbonyl groups were estimated (absorbance at 450 nm) as described in the methods section. ●–●, 40 μL/mL hypochlorite; ■–■, 10 μL/mL hypochlorite; ▲–▲, no hypochlorite added.

TABLE 2. Effects of Carnosine on Formation of DNA-protein Cross-links Induced by Formaldehyde and Acetaldehyde[a]

Carnosine concentration (mM)	DNA cross-linked to protein (measured fluorometrically, arbitrary units)	
	Acetaldehyde (1 mM)	Formaldehyde (100 mM)
0	161 ± 16	79 ± 9
50	51 ± 5	32 ± 12
200	4 ± 3	0

[a]Calf thymus DNA (1 μg) and histone H2 (4 μg) were incubated with either formaldehyde (100 mM) or acetaldehyde (1 mM) for 60 min at 37°C, in the presence/absence of carnosine (50 or 200 mM). DNA cross-linked to protein was determined as described in the methods section and measured fluorometrically. Results (mean of quadruplicate measurements) are expressed in arbitrary fluorescence units after subtraction of baseline value (no cross-linker added).

Carnosine Is Protective to Cultured Cells

We cannot be sure that any of the above processes occur at the cellular level. However, we observed that carnosine protected cells against the toxicity of acetaldehyde[6] and MDA. FIGURES 5 and 6 (a and b) show that MDA was toxic towards cultured RBE4, as indicated by decreased glucose utilization, cell-associated LDH activity, and mitochondrial dehydrogenase activity (MTT assay). By all three determinants carnosine was protective in a concentration-dependent manner. We assume that protection is mediated by reaction of the dipeptide with the aldehyde prior to any deleterious interaction with the cells.

Although AGE proteins are pathologically deleterious, AGE peptides may be even more reactive. Incubation of the amino acid lysine with deoxyribose for 14 days at pH 7

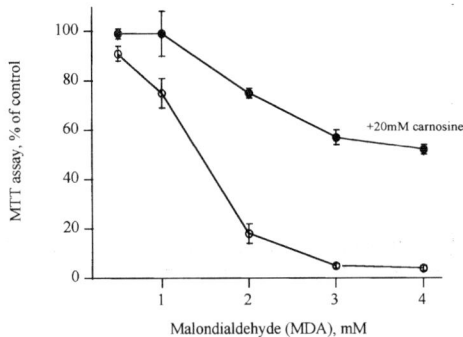

FIGURE 5. Effects of MDA (0–4 mM) on cultured rat brain endothelial cell viability as measured by mitochondrial activity (MTT assay): effect of 20 mM carnosine. Cells were cultured and assayed for mitochondrial function after 24 hours in the presence of 0–4 mM extracellular MDA. Carnosine (20 mM) was added to parallel wells 2 hours prior to addition of MDA (0–4 mM). Values are ± SEM (n = 3–15).

a)

b)

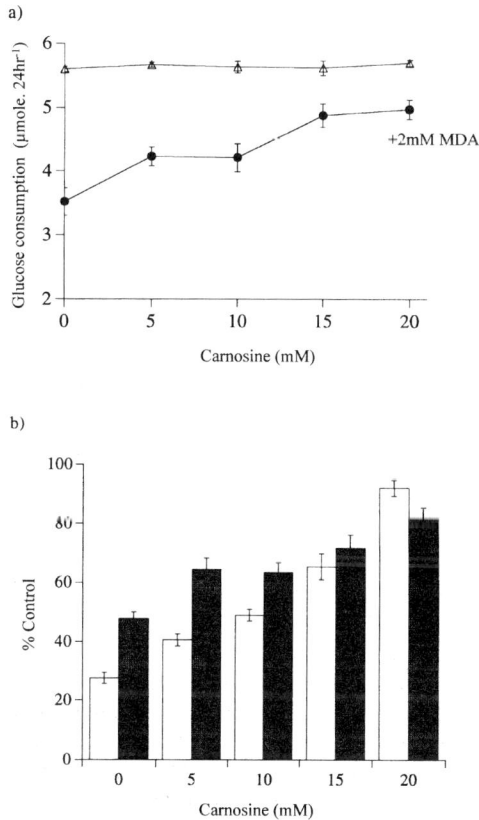

FIGURE 6. Effects of carnosine on 2 mM MDA toxicity in cultured rat brain endothelial cells as measured by (**a**) glucose utilization and (**b**) cell-associated LDH and mitochondrial activity (MTT assay). Cells were cultured as described in the text. Carnosine (0–20 mM) was added to the cells. After 2 hours MDA at 2 mM was added to the cells; 24 hours later the cells were assayed for glucose utilization, lactate dehydrogenase activity, and mitochondrial activity as described in the text. Values are mean ± SEM (n = 11–32). □, MTT reduction; ■, LDH activity.

and 37°C produced a brown solution that was toxic to cultured human fibroblasts (MRC-5) and CHO cells. TABLE 3 shows that toxicity of the lysine/deoxyribose mixture to CHO cells was both concentration- and time-dependent. Exposure of the cells to the toxic mixture for as little as two hours was sufficient to cause a decline in the number of viable cells when the culture was trypsinized and when the cells were counted and plated for viability the next day; doubling the exposure time to four hours resulted in a further reduction in viability. Addition of carnosine to the cells one hour before addition of the lysine/deoxyribose mixture was protective against the effects on viability. We do not know how protection is mediated, but possibilities include direct interaction with the toxic agent(s), prevention of binding (nonspecific or receptor mediated) of the toxin(s) to the cells, and

TABLE 3. Effect of Exposure of CHO Cells (2 or 4 hours) to Lysine/Deoxyribose Incubate on the Yield of Viable Cells: Protection by Carnosine[a]

Lysine/deoxyribose incubate concentration (mM)	Yield of viable cells (percent control)	
	minus carnosine	plus carnosine (20 mM)
(2 hours of treatment)		
10	22	38
20	8	52
(4 hours of treatment)		
10	4	53
20	0	27

[a]Lysine and deoxyribose (each at 100 mM) were incubated together at 37°C, pH 7, for 14 days and then added (at either 10 or 20 mM) to subconfluent cultures of CHO for either 2 or 4 hours in the presence or absence of carnosine (20 mM), after which time the medium was removed and replaced with normal growth medium. After 24 hours growth the cells were released by trypsinization, counted, and plated for viability. After 7 days colonies were counted.

interfering with signal transduction and/or modulation of the hyperoxic response induced by AGE-like substances.

The toxicity of the amyloid peptide that accumulates in Alzheimer's disease is thought to involve induced generation of reactive oxygen species (ROS),[17–19] although other explanations are not excluded. We found that addition of a highly toxic β/A4-amyloid peptide fragment (residues 25–35) damaged cultured RBE4, as measured by glucose uptake from the medium, release of LDH into the medium, and mitochondrial dehydrogenase activity (MTT assay) (FIG. 7). Pretreatment of the cells with 20 mM carnosine protected them against amyloid-induced toxicity (FIG. 7). The related molecule homocarnosine was as protective as carnosine, whereas the methylated derivative anserine produced little protection but was somewhat inhibitory even in the absence of the amyloid fragment (FIG. 8). The components of carnosine, β-alanine, and histidine promoted opposite effects; β-alanine was as protective as carnosine, whereas histidine was not and was inhibitory even in the absence of the fragment (FIG. 8). We do not know how carnosine-mediated protection occurs, but because superoxide dismutase (SOD, an antioxidant) and aminoguanidine (an antiglycator) were both somewhat protective (FIG. 9), involvement of oxidative and glycatory phenomena are likely. Alternative/additional explanations include interaction with the amyloid fragment or cellular receptor (e.g., the receptor for advanced glycosylation end

FIGURE 7. Effects of β-amyloid peptide on cultured rat brain endothelial cells and their modulation by carnosine. Rat brain endothelial cells (RBE4) were incubated with 200–400 μg/mL amyloid β-peptide fragment (residues 25–35) in the presence or absence of carnosine (added 2 hours previously) for 3 days. Wells were assayed for (**a**) LDH release, (μmole/mL/mg protein); (**b**) glucose consumption (μmoles/24 hours); and (**c**) mitochondrial dehydrogenase activity (MTT assay) (percent of control values were incubated in absence of amyloid peptide.) Values are mean ± SEM. n = 8–20. *p < 0.05, difference between carnosine absence and presence, paired *t*-test. ▨, alone; ■, plus 20 mM carnosine.

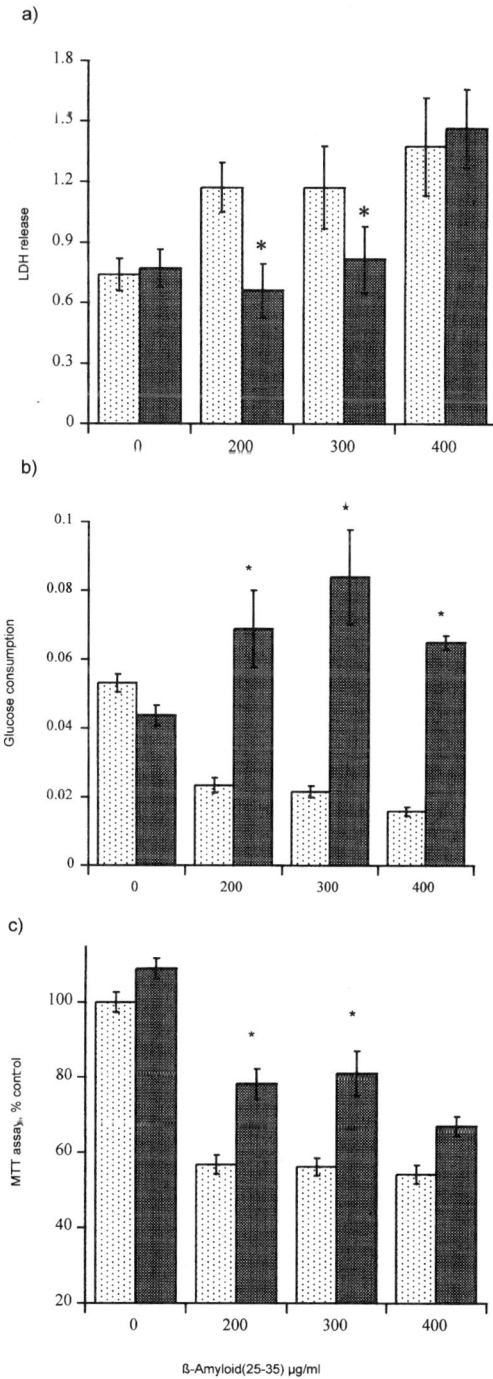

a)

b)

c)

ß-Amyloid(25-35) µg/ml

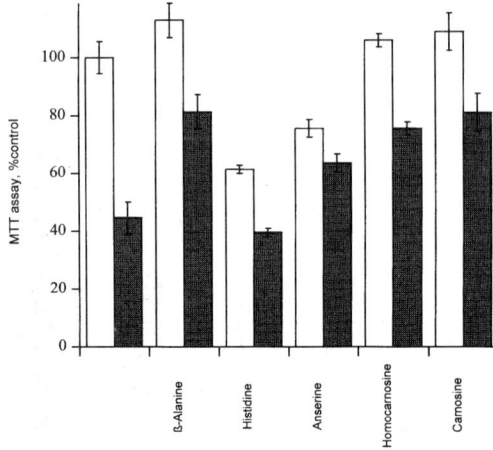

FIGURE 8. Effects of carnosine, related peptides, and amino acids, on the toxicity of amyloid β-peptide fragment (residues 25–35) to rat brain endothelial cells. Rat brain endothelial cells (RBE4) were incubated with carnosine, anserine, homocarnosine, β-alanine, and histidine, all at 20 mM. After 2 hours amyloid β-peptide fragment (residues 25–35) (300 μg/mL) was added. After 3 days cells were assayed for mitochondrial dehydrogenase activity (MTT assay). Values (percent control activity, *i.e.*, cells incubated in absence of both amyloid peptide and carnosine, etc.) are mean ± SEM. n = 6–16. □, 20 mM amino acid/peptide; ■, plus 300 μg/mL β-amyloid (25–35).

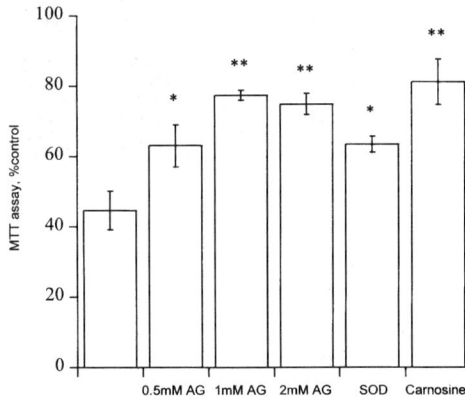

FIGURE 9. Effects of carnosine, superoxide dismutase, and aminoguanidine, on the toxicity of amyloid β-peptide fragment (residues 25–35) to rat brain endothelial cells. Rat brain endothelial cells (RBE4) were incubated with carnosine (20 mM), aminoguanidine (AG) (0.5–2 mM), or superoxide dismutase (SOD) (300 units/mL). After 2 hours amyloid β-peptide fragment (residues 25–35) (300 μg/mL) was added. After 3 days cells were assayed for mitochondrial dehydrogenase activity (MTT assay). Values (percent control activity, *i.e.*, cells incubated in absence of both amyloid peptide and carnosine, etc.) are mean ± SEM. n = 6–12. $^*p < 0.05$; $^{**}p < 0.01$, difference from control (amyloid β-peptide fragment alone) using ANOVA single factor.

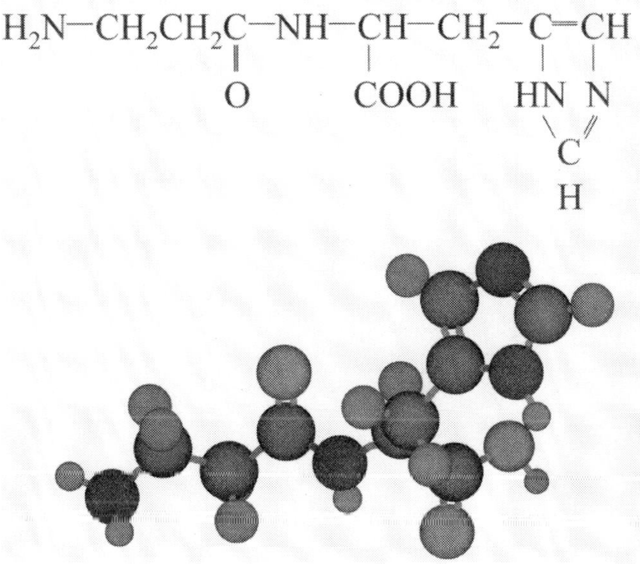

$$H_2N-CH_2CH_2C-NH-CH-CH_2-C-CH$$

FIGURE 10. Structure of carnosine (β-alanyl-L-histidine).

products (RAGE)), interference with signal transduction, and modulation of ROS generation, including reaction with MDA (see above). Another possibility is that carnosine stimulates proteolytic destruction of the amyloid fragment. It is of interest to note that glucose utilization was markedly increased in cells exposed to amyloid together with carnosine, which possibly implies an increased energy demand (*e.g.*, for ATP-dependent proteolysis?) under these circumstances. We have observed that carnosine both stimulates and inhibits degradation of aberrant polypeptides in rabbit reticulocyte cell-free extracts, and the effect was dependent on the nature of the abnormal protein: proteolysis of prematurely terminated globin chains (puromycin peptides) was stimulated by carnosine, whereas catabolism of polypeptides of normal chain length but containing a lysine analogue (aminoethylcysteine) was inhibited by the dipeptide.[20] We do not know whether carnosine affects proteolysis in the cultured RBE4, although the peptide has been observed to stimulate breakdown of some proteins in mature cultured human fibroblasts.[6]

DISCUSSION

The present results suggest that carnosine may be a sacrificial and nontoxic site for reaction with deleterious aldehydes, in general, as well as reactive oxygen species such as MDA and hypochlorite. It may not be coincidental that carnosine possesses the amino acid side chains (*i.e.*, those of lysine and histidine) most readily oxidized during protein aging[15] and structurally resembling sites of preferential glycation in proteins (a target amino group with proximal imidazole and carboxyl groups) (FIG. 10). Formaldehyde reacts with histidine to form a stable product; therefore it is possible that formaldehyde and other aldehydes react with carnosine in an analogous manner. Formation of a stable product with aldose sugars, for example, could inhibit the subsequent Amadori rearrangement necessary prior to cross-linking and formation of AGEs. It should be noted, however, that high molecular weight colored products are generated following the reaction of carnosine with many aldehydes (except formaldehyde), which suggests that this explanation of the dipeptide's inhibitory properties is at least incomplete. Additionally/alternatively, carnosine could react with the carbonyl group generated as a result of the Amadori rearrangement step to prevent cross-linking of a glycated polypeptide to another protein amino group.

Many long-lived tissues (*e.g.*, brain and muscle) form increasing amounts of lipofuscin, the age pigment, upon aging. It has been suggested that lipofuscin is formed from the interaction of MDA with peptide amino groups.[21] Brain and muscle are also rich in carnosine. We therefore suggest that lipofuscin might be derived at least in part from the reaction of carnosine with deleterious aldehydes, such as MDA, as outlined above.

In model systems carnosine prevented aldehyde-induced formation of protein carbonyl groups, cross-links, and AGEs, as well as DNA/protein cross-linking, all of which normally increase with age.[7] Interestingly, an increase in macromolecular-based carbonyl groups in hippocampal tissue from an Alzheimer's-diseased brain has been reported relative to age-matched controls.[18] We do not know whether any of the protective phenomena described above occur *in vivo*, but carnosine can protect cultured cells against two highly deleterious aldehydes (acetaldehyde[6] and malondialdehyde). Additionally we show that carnosine can protect cultured cells against AGEs and an amyloid peptide fragment whose mechanisms of toxicity are thought to involve the generation of ROS species, possibly including MDA. Although the mechanisms by which protection is mediated remain to be elucidated, direct reaction of carnosine with MDA is possible as well as the previously suggested hydroxyl radical scavenging action. SOD and aminoguanidine are antioxidant and antiglycating agents, respectively, and were both protective, which suggests that both processes are involved. Additionally we tentatively suggest that the dipeptide could conceivably modulate proteolysis depending on the substrate involved. Decreases in certain proteolytic activities have been shown to accompany aging and may partly explain the age-associated accumulation of aberrant polypeptides.[22]

It is not known whether tissue levels of carnosine generally decline with age or whether any pathologies result from lowered carnosine levels. However, it is known that the related homocarnosine concentration declines in the cerebrospinal fluid with age,[23] and that increasingly severe cataracts (aggregated, insoluble protein) are associated with progressively decreasing levels of lenticular carnosine.[1,2] Interestingly a correlation between muscle carnosine levels and species longevity has also been suggested in mammals.[24]

Carnosine inhibits sugar-induced aggregation of a β-A4 amyloidogenic peptide *in vitro*.[25] The toxic effects of the β-amyloid peptide frequently require some sort of prior "aging" phenomenon. Amyloid from an Alzheimer's-diseased brain contains β-A4 amyloid peptides associated as antiparallel β-pleated sheets. Amyloid is also enriched in isoaspartates in which the β-methylene (CH_2) group of the amino acid side chain becomes incorporated into the polypeptide backbone usually following the spontaneous deamidation of asparagine residues to L- or D-aspartic or isoaspartic acid residues.[26] (This modification is associated with general protein aging too[22]). The presence of an isoaspartate creates a β-peptide bond in the polypeptide backbone, which reduces susceptibility to proteolytic attack and inhibits α-helical structures by disrupting hydrogen bonding regularity.

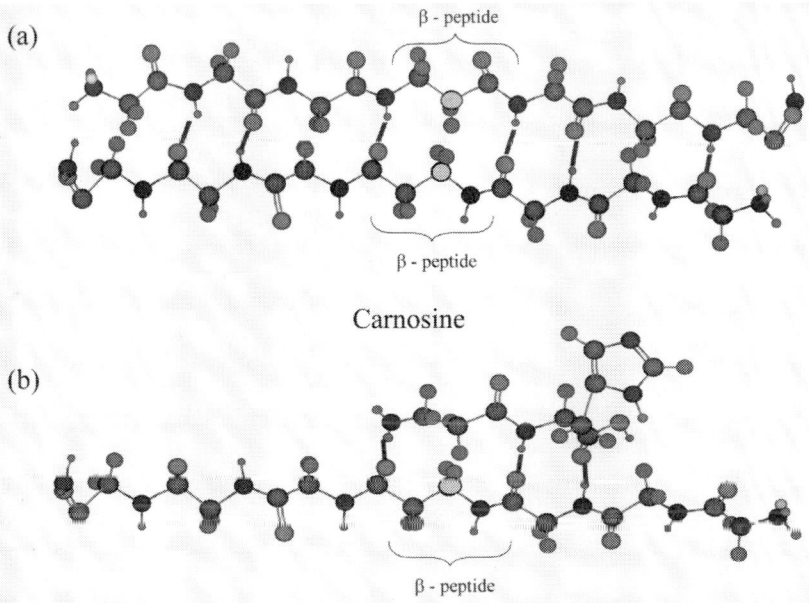

FIGURE 11 (a). Structure of two polypeptide chains each containing a β-peptide bond associated together as an antiparallel pleated sheet. The thick black lines between the two polypeptide chains represent hydrogen bonds between peptide carbonyl and imino groups. (b). Hypothetical structure of a polypeptide chain containing a β-peptide bond hydrogen bonded to carnosine. The thick black lines represent hydrogen bonds or ionic bonds between carnosine and the polypeptide chain.

Formation of the alternative secondary structure, the antiparallel β-pleated sheet, would still be possible especially with another "aged" peptide chain possessing an isoaspartate that restored maximal hydrogen bonding. This would promote self-selecting aggregation of "aged" peptides that contained β-peptide bonds (FIG. 11a). Could a carnosine (also a β-peptide) hydrogen bond to an iso-aspartate residue, disrupt interaction between polypeptide chains, and thereby prevent fragment aggregation? Model structures show that this speculation is possible (FIG. 11b). Carnosine's β-amino group could form a hydrogen bond/salt bridge to the β-carbonyl group of the iso-aspartate residue, the dipeptide's peptide bond imino group could form a hydrogen bond to the carbonyl group of the prior peptide bond, and the dipeptide's terminal carboxyl group could form a hydrogen bond/salt bridge to the imino group of the second peptide bond before the iso-aspartate residue. Little is known about any age-related change in carnosine concentrations in either a normal brain or one from Alzheimer disease; it may or may not be significant that the carnosine content of the olfactory lobe is normally particularly high and that a loss of the sense of smell has been taken by some as an early indication of Alzheimer's disease.

That carnosine is reactive towards deleterious aldehydes that can modify proteins suggests that the dipeptide could be explored therapeutically. Possibilities include controlling the secondary consequence of diabetes (production of protein cross-linking and AGEs) by its ability to react with sugar aldehydes (*e.g.*, glucose and fructose) and possibly AGEs.

Additionally, atherosclerotic phenomena associated with the generation of MDA and hypochlorite anions and their reaction with LDL amino groups might also be investigated because of carnosine's protective abilities demonstrated here. We need investigation also with arthritis, which also involves generation of aldehydes and hypochlorite. Finally the results with the amyloid peptide fragment suggest that a potential for carnosine to control Alzheimer's disease should be considered. Possible approaches could include increasing dietary carnosine intake, modulating carnosine's synthesis (via carnosine synthetase) or degradation (via serum or cellular carnosinases), or using carcinine that is not attacked by carnosinase.

ACKNOWLEDGMENTS

We thank Dr. Robin Holliday for his support and hospitality and Dr. Geoffrey Grigg for his continued interest. We thank Mr. Tommy Yan and Mr. Ehsen Khan for their help with the computer-generated peptide structures.

REFERENCES

1. BOLDYREV, A.A., V.E. FORMAZYUK & V.I. SERGIENKO. 1994. Biological significance of histidine containing dipeptides with special reference to carnosine: chemistry, distribution, metabolism and medical application. Sov. Sci. Rev. D. Physicochem. Biol. **13:** 1–60.
2. QUINN, P.R., A.A. BOLYREV & V.E. FORMAZUYK. 1992. Carnosine: its properties, functions and potential therapeutic applications. Mol. Aspects Med. **13:** 379–444.
3. McFARLAND, G.A. & R. HOLLIDAY. 1994. Retardation of the senescence of cultured human fibroblasts by carnosine. Exp. Cell Res. **212:** 167–175.
4. HIPKISS, A.R., R. HOLLIDAY, G.A. McFARLAND & J. MICHAELIS. 1993. Carnosine and senescence. Lifespan **4:** 1–3.
5. HIPKISS, A.R., J. MICHAELIS & P. SYRRIS. 1995a. Non-enzymic glycosylation of the dipeptide L-carnosine, a potential anti-protein-cross-linking agent. FEBS Lett. **371:** 81–85.
6. HIPKISS, A.R., J. MICHAELIS, P. SYRRIS & M. DREIMANIS. 1995b. Strategies for the extension of human life span. Perspect. Hum. Biol. **1:** 59–70.
7. LIBONDI, T., R. RAGONE, D. VINCENTI, P. STIUSO, G. AURICCHIO & G. COLLONA. 1994. *In vitro* cross-linking of calf lens alpha-crystallin by malondialdehyde. Int. J. Pept. Protein Res. **44:** 342–347.
8. HAZELL, L.J., J.J.M. VAN DEN BERG & R. STOCKER. 1994. Oxidation of low-density lipoprotein by hypochlorite causes aggregation that is mediated by modification of lysine residues rather than lipid oxidation. Biochem. J. **302:** 297–304.
9. ZHITKOVICH, A. & M. COSTA. 1992. A simple, sensitive assay to detect DNA-protein crosslinks in intact cells and *in vivo*. Carcinogenesis **13:** 1485–1489.
10. ROUX, F., A. DURIEU-TRAUTMANN, N. CHAVEROT, M. CLAIRE, P. MAILL, J.-M. BOURRE, A.D. STROSBERG & P.-O. COUROUD. 1994. Regulation of gamma-glutamyl transpeptidase and alkaline phosphatase activities in immortalised rat brain microvessel endothelial cells. J. Cell. Physiol. **159:** 101–113.
11. HOLLIDAY, R. & G.A. McFARLAND. 1996. Inhibition of the growth of transformed and neoplastic cells by the dipeptide carnosine. Br. J. Cancer **73:** 966–971.
12. VASSAULT, A. 1983. Lactate dehydrogenase: UV-method with pyruvate and NADH. *In* Methods of Enzymatic Analysis. H.U. Bergmeyer, J. Bergmeyer & M. Grassi, Eds. Vol. III: 118–126. Verlag Chemie. Academic Press. New York.
13. BERGMEYER, H.U., E. BERNT, F. SCHMIDT & H. SORK. 1974. D-glucose: Determination with hexokinase and glucose-6-phosphate dehydrogenase. *In* Methods of Enzymatic Analysis. H.U. Bergmeyer, Ed.: 1196–1201. Verlag Chemie. Academic Press. New York.

14. CARMICHAEL, J., W.G. DEGRAFF, A.F. GAZDER, J.D. MINNA & J.B. MITCHELL. 1987. Evaluation of a tetrazolium based semi-automatic colorimetric assay: Assessment of radiosensitivity. Cancer Res. **47**: 943–946.

15. SANTA MARIA, C., E. REVILLA, A. AYALA, C.P. DE LA CRUZ & A. MACHADO. 1995. Changes in histidine residues in Cu/Zn superoxide dismutase during aging. FEBS Lett. **374**: 85–88.

16. KUYKENDALL, J.R & M.S. BOGDANFFY. 1992. Reaction kinetics of DNA-histone crosslinking by vinyl acetate and acetaldehyde. Carcinogenesis **13**. 2095–2100.

17. SCHUBERT, D., C. BEHL, R. LESLEY, A. BRACK, R. DARGUSCH, Y. SAGARA & H. KIMUA. 1995. Amyloid peptides are toxic via a common oxidative mechanism. Proc. Natl. Acad. Sci. USA **92**: 1989–1993.

18. SMITH, M.A., G. PERRY, P.L. RICHEY, L.M. SAYRE, V.E. ANDERSON, M.F. BEAL & N. KOWALL. 1996. Oxidative damage in Alzheimer's. Nature **382**: 120–121.

19. IVERSON, L.L., R.J. MORTISHIRE-SMITH, S.J. POLLOCK & M.S. SHEARMAN. 1995. The toxicity *in vitro* of the β-amyloid peptide. Biochem. J. **311**: 1–16.

20. HIPKISS, A.R., J. MICHAELIS & J. LAURENCE. Unpublished observation.

21. BRUNK, U.T., C.B. JONES & R.S. SOHAL. 1992. A novel hypothesis of lipofuscinogenesis and cellular aging based on interactions between oxidative stress and autophagocytosis. Mutat. Res. **275**: 395–404.

22. HIPKISS, A.R. 1996. Proteins and aging. Rev. Clin. Gerontol. **6**: 3–6.

23. PERRY, T.L., S. HANSEN, D. STEDMAN & D. LOVE. 1968. Homocarnosine in human cerebrospinal fluid: An age-dependent phenomenon. J.Neurochem. **15**: 1203–1206.

24. HOLLIDAY, R. Personal communication.

25. MUNCH, G., S MAYER, J. MICHAELIS, A.R. HIPKISS, P. REIDERER, K. MULLER, A. NEUMANN, R. SCHINZEL & A. CUNNINGHAM. 1997. Influence of advanced glycation end-products and AGE-inhibitors on nucleation-dependent polymerization of β-amyloid peptide. Biochim. Biophys. Acta **1360**: 17–29.

26. SZENDREI, G.I., K.V. PRAMMER, M. VACKO, V.M.-Y. LEE & L. OTVOS, Jr. 1996. The effects of aspartic acid-bond isomerization on *in vitro* properties of the amyloid β-peptide as modeled with *N*-terminal decapeptide fragments. Int. J. Pept. Protein Res. **47**: 289–296.

The Nature of Gerontogenes and Vitagenes

Antiaging Effects of Repeated Heat Shock on Human Fibroblasts

SURESH I. S. RATTAN[a]

Laboratory of Cellular Aging, Danish Centre for Molecular Gerontology,
Department of Molecular and Structural Biology,
University of Aarhus, DK-8000 Aarhus-C, Denmark

ABSTRACT: Our survival and the physical quality of life depends upon an efficient functioning of various maintenance and repair processes. This complex network of the so-called longevity assurance processes is composed of several genes, termed vitagenes. The homeodynamic property of living systems is a function of such a vitagene network. Because aging is characterized by the failure of homeodynamics, a decreased efficiency and accuracy of the vitagene network can transmutate it into a gerontogene network. It is not clear how various components of the vitagene network operate and influence each other in a concordant or a discordant manner. Experimental strategies through which this transmutation of vitagenes into virtual gerontogenes may be elucidated include induction of molecular damage, antisense intervention, and genetic screening for varied efficiencies of the members of the vitagene family. A reversal of this approach by maintaining or recovering the activity of vitagenes will lead to a delay of aging, a decreased occurrence of age-related diseases, and a prolongation of a healthy life span.

Our survival and the physical quality of life depends upon an efficient functioning of various maintenance and repair processes. This complex network of the so-called longevity assurance processes is composed of several genes, which may be called *vitagenes*. The homeodynamic property of living systems is a function of such a vitagene network. Some of the main longevity assurance processes that constitute the vitagene network are listed in TABLE 1. These processes have to work in the presence of extrinsic and intrinsic sources of damage, such as environmental and nutritional agents, spontaneous errors of macromolecular synthesis, postsynthetic modifications of macromolecules making them inactive or abnormal, and other defects occurring during the course of normal metabolism. Apparently, this homeodynamic network of vitagenes appears to work optimally during the major part of life, allowing growth, development, differentiation, and maturation to occur. There are no obvious reasons why these maintenance, defense, and repair mechanisms could not operate forever and make an organism immortal. Yet, a progressive failure of maintenance underlines and typifies the process of aging.

Aging has many facets, and almost all the experimental data suggest that aging is an emergent, epigenetic, and meta-phenomenon, which is not controlled by a single mechanism.[1,2] Individually no tissue, organ, or system becomes functionally exhausted, even in very old organisms; yet it is their combined interaction and interdependence that determines the survival of the whole. Because it is necessary to look for genes as the ultimate controllers of all biological processes, the term *gerontogenes* has been suggested to refer

[a]Tel: +4589425034; fax: +4586123178 or +4586201222; e-mail: rattan@imsb.au.dk

TABLE 1. Major Components of the Homeodynamic Vitagene Network

Molecular level	Tissue and organ level
Maintenance and repair of the genome.	Neutralizing and removing toxic chemicals.
Fidelity of genetic information transfer.	Tissue regeneration and wound healing.
Turnover of macromolecules.	Cell death and cell replacement.
Stress-protein synthesis.	
Scavenging of free radicals.	
Cellular level	Physiological level
Maintaining the differentiated state.	Neuronal response.
Regulation of cell proliferation.	Hormonal response.
Stability of the cellular milieu (viscosity, ion	Immune response.
balance, pH).	Stress response.
Stability of the cell membranes.	Thermoregulation.

to any genetic elements that are involved in aging.[3,4] However, there is still no agreement as regards the nature of the gerontogene network and the number of genes involved in it

GERONTOGENES—REAL OR VIRTUAL?

Evolutionary theories of aging and longevity discount the notion of the adaptive nature of aging. This is because Darwinian evolution, which works primarily through the process of natural selection for reproductive success, does not leave any margin for the selection of aging and death as an advantageous trait for the individual.[5–7] Evolutionary forces work only on life processes and do not select for death. Natural selection of what is recently termed a "selecton,"[8] is in the form of an efficient vitagene network to assure its longevity for fulfilling its purpose of life, that is, reproduction and continuation of generations. The so-called programmed death or apoptotic mechanisms are essential parts of the developmental processes (or may even be a component of the homeodynamic vitagene network) necessary for the making of a reproductively successful individual

Once the Darwinian purpose of life is fulfilled, there is no reason (or selection pressure) left either for the maintenance of the body or for its destruction. Thus aging, as a failure of maintenance, occurs without a cause or a purpose. It just happens. Any search for genes that were selected specifically to cause aging is misdirected and ill-informed. Furthermore, the diversity of the forms and variations in which age-related alterations are manifested indicate that the progression of aging is not deterministic but stochastic in nature. An age-related increase in variability among individuals, in terms of any physiological, cellular, or biochemical parameter studied, is a reflection of the stochastic nature of aging. In the words of an anonymous poet, *We are born as copies, but we die as originals.*

Yet, aging appears to have a genetic component of some kind. The role of genes in aging is evident from an apparent practical limit to maximum life span within a species,[9,10] along with the evidence from studies on twins,[11,12] from the human genetic mutants of premature aging,[13] and from genetic linkage studies for the inheritance of life span and for genetic markers of exceptional longevity.[14] Thus, aging appears to be genetically regulated without involving any genes that may be held responsible as its cause. This paradoxical situation of the genetic aspects of aging and longevity on the one hand and the stochastic nature of the progression of the aging phenotype can be resolved by developing radically novel views about the nature of gerontogenes.

If aging is seen as the failure of the homeodynamic vitagene network due to stochastic causes of damage and perturbations, the term gerontogenes refers to the functional trans-mutation of the vitagene network, due to progressive accumulation of random damage. Therefore, the term gerontogenes does not refer to a tangible physical reality of real genes for aging but refers to an altered state of vitagenes, which gives the appearance of being the genes for aging. Thus, gerontogenes are not real; they are virtual.[4] More importantly, the idea of virtual gerontogenes is in line with the evolutionary explanation of the aging process as being an emergent phenomenon caused by the absence of eternal maintenance and repair instead of being an active process caused by evolutionary adaptation.

Obviously, not every gene is potentially a virtual gerontogene, although, of about 100,000 genes in the human genome, potentially every gene can affect the survival of an organism. Therefore, a distinction must be made between immediate survival or death on the one hand and the process of aging on the other. The inactivation of any essential gene involved in fundamental metabolic processes will result in the death of an organism with-out having anything to do with the process of aging. This situation is similar to the toxic effects of various chemicals that may result in the immediate or somewhat delayed death of an organism without affecting its rate of aging, as determined by well-established parameters of survival and mortality kinetics.[9,15] Similarly, a single gene mutation may give rise to an apparently accelerated-aging phenotype without qualifying as being the gene that causes aging.[13,16] Therefore, the set of possible virtual gerontogenes can be nar-rowed down to those genes that fit those criteria of survival kinetics. In this sense, the sets of vitagenes involved in the maintenance and repair of the cellular and subcellular compo-nents best qualify as the candidates to become virtual gerontogenes.

Evidence for the hypothesis that candidate virtual gerontogenes operate through the vitagene network comes from experiments performed to retard aging and to increase the life span of organisms. For example, antiaging and life-prolonging effects of calorie restriction are seen to be accompanied by the stimulation of various maintenance mecha-nisms. These include increased efficiency of DNA repair, increased fidelity of genetic information transfer, more efficient protein synthesis, more efficient protein degradation, more effective cell replacement and regeneration, improved cellular responsiveness, forti-fication of the immune system, and enhanced protection from free radical– and oxidation-induced damage.[17,18] Similarly, antiaging effects of a wide variety of hormones, vitamins, a dipeptide carnosine,[19] modified amino acids, and nucleic acid bases on human cells, rodents, and insects[2,20–22] are also mainly due to the effects of these chemicals on main-taining or recovering the efficiency of vitagenes.

Genetic selection of *Drosophila* for longer life span also appears to work mainly through an increase in the efficiency of maintenance mechanisms, such as antioxidation potential.[5,23] An increase in life span of transgenic *Drosophila* containing extra copies of Cu-Zn superoxide dismutase (SOD) and catalase genes is due primarily to enhanced defenses against oxidative damage.[24] The identification of long-lived mutants of the nem-atode *Caenorhabditis elegans*, involving various genes, may provide other examples of virtual gerontogenes because in almost all these cases increased life span is accompanied by an increased resistance to oxidative damage, an increase in the activities of SOD and catalase enzymes, and an increase in thermotolerance.[25–28]

MODULATING THE NETWORK

Estimates of the number of genes that might qualify as being a part of the vitagene/gerontogene network of mammals run up to a few hundred out of about one hundred thousand genes, and their allelic variants. Direct gene therapy directed towards the overall aging process seems to hold little promise. This is because gene therapy for aging will require methods to improve upon the "genetic hand of cards" that determines the aging and longevity of a "genetically normal" individual.

The question of total gene therapy for aging is linked with the issue of defining what is a normal combination of genes in a so-called normal and healthy individual. It is in the very nature of genetic polymorphism and the interactive nature of the genome that each individual is unique and that the term "normal" is extremely wide ranging. Therefore, improving upon an already normal situation with respect to the genetic constitution related to total life span is a no-win situation.

Assuming that there are only 50 longevity-assuring vitagenes that constitute the network in which they interact with each other, this gives rise to 2^{50} or 10^{15} (a million billion) possibilities of their interacting and influencing each other. Not considering billions of cells in an adult, even at the level of a single cell zygote, interfering with such a complex network and improving upon what is already a normal combination for that particular individual (in the absence of any obvious genetic diseases) is a mission impossible.[29] What is more likely to be achieved in the not-so-distant future is that experimental manipulation of certain genes will fine-tune or retune the network and will prevent the onset of various age-related diseases and impairments by maintaining the efficiency of homeostatic processes.

The following lines of research can form the basis of a promising strategy to understand and modulate the aging process: (1) studying the extent of maintenance and repair of the genes involved in maintaining the stability of the nuclear and mitochondrial genome; (2) studying the efficiency of transcription of various vitagenes and posttranscriptional processing of their transcripts; (3) studying the accuracy and efficiency of translation of vitagenes and analyzing the specificity, stability, and turnover of vitagene products (vitaprins?), including their posttranslational modifications; (4) searching for natural or induced mutants (including transgenic and knockout organisms) with altered levels of maintenance and repair of the crucial vitagenes; (5) searching for age-specific and age-related disease-specific biomarkers for diagnostic purposes and for monitoring the effects of potential therapeutic agents; and (6) experimental modulation of various types of maintenance mechanisms and studying its effects on other levels, such as gene stability, gene product synthesis, and turnover. Some of the main defense processes that may provide a relatively easy access to experimentation include responsiveness to stress, efficiency of signal transduction pathways, and regulators of cell-cycle progression.

EXPERIMENTAL STUDIES ON UPREGULATING
THE STRESS RESPONSE IN HUMAN FIBROBLASTS

An organism's ability to respond to stress is a major component of its homeodynamic vitagene network, and altered responsiveness is one of the most significant features of aging.[2,30] The so-called heat-shock response as an important mechanism of cellular defense is very well established.[31–33] It is also known that the extent of heat-shock response decreases during aging.[32] Therefore, it has been hypothesized that if organisms are exposed to brief thermal treatment so that their stress response–induced gene expression is upregulated and this particular pathway of maintenance and repair is stimulated, one should observe antiaging and longevity-promoting effects. Such a phenomenon in which stimulatory responses to low doses of otherwise toxic substances improve health and enhance life span is known as hormesis.[34,35] Recently, antiaging and life-prolonging effects of heat shock have been reported for *Drosophila*[36] and the nematodes.[27,28]

I have tested the effects of mild but repetitive heat shock on various cellular and biochemical characteristics of human skin fibroblasts undergoing aging *in vitro*. I undertook a series of pilot experiments to determine suitable temperature conditions that fulfilled the following criteria: (1) the thermal treatment had no effects on immediate survival of the cells, as checked by a trypan blue exclusion test; (2) the cells responded to the thermal treatment by inducing the synthesis of major heat-shock proteins (hsp), as detected by metabolic labeling of cells with radioactive amino acids followed by SDS-polyacrylamide gel electrophoresis (PAGE); and (3) thermally treated cells could be subcultured normally without any effect on their attachment frequency.

All experiments were performed on a normal human adult female skin fibroblast line, designated ASS, which has been used previously to test for the antiaging effects of the cytokinin hormone kinetin.[20] At least three parallel cultures of control (series A1 to A3) and heat-shocked cells (series H1 to H3) were serially passaged at 1:4 or 1:2 split ratio until the end of their proliferative life span, using normal culture conditions of medium-containing antibiotics, 10% fetal calf serum, and incubation at 37°C with 95% humidity. The H-series cells were given a 30-min heat shock at 41°C by immersing the culture flasks in a fine-regulated water bath. The cultures were kept at 37°C for 60 min before changing the medium. Heat-shock treatment was repeated twice a week with following restrictions: cultures were not subcultivated within 24 h of heat shock, and heat shock was not given to newly subcultivated cultures for at least 24 hours. Growth rates, population doubling (PD) rates, cell yield, and cumulative population doubling levels (CPDL) achieved *in vitro* were determined. In addition, morphological characteristics, actin filament organization, and the senescence-specific β-galactosidase staining pattern of normal and heat-shocked cells were compared. The extent of heat-shock response in terms of hsp synthesis was also checked at various PD levels, by SDS-PAGE.

The present series of experiments have demonstrated that repetitive and mild heat shock has several positive and antiaging effects on human cells in culture. Briefly the results are summarized as follows. Human fibroblasts could be exposed to mild heat shock at 41°C repeatedly during their limited proliferative life span *in vitro* without any apparent negative effects on survival, attachment frequency, PD rates, and CPDL potential. Although the temperatures higher than this (up to 43°C) could stimulate a more intense heat-shock response, in terms of hsp synthesis, cells could not survive more than seven repeated thermal treatments. Continuous survival of human fibroblasts for 140 days, during which time they underwent about 30 PDs and received 35 repeated heat shocks at 41°C, is a novel effect not observed before (details are published elsewhere[37]).

Although there was no prolongation of the proliferative life span of human fibroblasts after repeated heat-shock treatment, several other antiaging effects were observed. Most dramatically, the age-related alteration in the morphology of cells, which is one of the most obvious changes during cellular aging, was significantly slowed down in heat-shocked cells. The control cultures showed the typical age-related increase in cell size, flattened appearance, increased morphological heterogeneity, loss of arrayed arrangement, increased number of lysosomal residual bodies, increased number of actin filaments, and increased proportion of multinuclear cells during serial passaging. However, the heat-shocked cultures showed a highly reduced rate of these age-related alterations and maintained a relatively young morphology even at the end of their proliferative life span. These cells did not undergo significant enlargement, maintained to a large extent their spindle-shape and arrayed arrangement, did not accumulate many residual bodies, did not show many rod-like actin filaments, and had an almost complete absence of multinucleate cells. A reduced rate of cell enlargement was also evident from the analysis of cell yield per cm^2 of cell culture flasks, which was reduced from about 4×10^4 cells in young cultures to 1×10^4 cells in senescent-controlled cultures, but was maintained at a level two to three times higher (between 2.5 and 3×10^4 cells) in repeatedly heat-shocked cultures. Maintenance of young morphology and reduced cell size is a strong indication of antiaging effects of heat shock, as also observed for such antiaging treatments as carnosine and kinetin.

That there may be several other antiaging or hormesis-like effects of repeated mild heat shock on human cells is also evident from a comparison of the proportion of β-galactosidase positive cells in the culture. Recently, this marker has been suggested to be a good indicator of senescent cells in culture. Whereas more than 95% of cells in a high-passage control culture at the end of its proliferative life span were β-galactosidae positive, less than 5% of cells in heat-shocked cultures were detectable by this marker. We are now investigating what other cellular, physiological, biochemical, and molecular effects of repeated heat shock occur in human cells. Some of the characteristics to be tested are the rates and extent of transcription and translation of various genes, the extent of gene-specific and total DNA repair, including telomere length, the extent and rates of protein synthesis and degradation, and the accumulation of molecular damage, such as oxidative damage in DNA, lipid-peroxidation products, and abnormal proteins.

Most importantly, the above experiments show that it is possible to retune the vitagene network in such a way that its transmutation to a virtual gerontogene network is slowed down. Due to the highly complex nature of interactions within the vitagene network, it is most unlikely that a complete transmutation of vitagenes into gerontogenes can ever be prevented. Furthermore, retuning one or more vitagenes may or may not have significant effects on the final outcome in terms of reducing the rates of aging and increasing the life span. However, what can definitely be achieved by this approach is that increased levels and/or efficiency of one or more maintenance and repair pathways will have positive effects in terms of improving the physical quality of life and increasing its chances of survival, and ultimately, achieving a healthy old age.

REFERENCES

1. HOLLIDAY, R. 1995. Understanding Ageing. Cambridge University Press. Cambridge.
2. RATTAN, S.I.S. 1995. Ageing—a biological perspective. Mol. Aspects Med. **16:** 439–508.
3. RATTAN, S.I.S. 1985. Beyond the present crisis in gerontology. BioEssays **2:** 226–228.
4. RATTAN, S.I.S. 1995. Gerontogenes: Real or virtual? FASEB J. **9:** 284–286.
5. ROSE, M.R. 1991. Evolutionary Biology of Aging. Oxford University Press. New York.
6. KIRKWOOD, T.B.L. 1992. Biological origins of ageing. In Oxford Textbook of Geriatric Medicine. J.G. Evans & T.F. Williams, Eds.: 35–40. Oxford University Press. Oxford.
7. CHARLESWORTH, B. & K.A. HUGHES. 1996. Age-specific inbreeding depression and components of genetic variance in relation to the evolution of senescence. Proc. Natl. Acad. Sci. USA **93:** 6140–6145.
8. MAYR, E. 1997. The object of selection. Proc. Natl. Acad. Sci. USA **94:** 2091–2094.
9. FINCH, C.E. & M.C. PIKE. 1996. Maximum life span predictions from the Gompertz mortality model. J. Gerontol. Biol. Sci. **51A:** B183–B194.

10. JOHNSON, T.E. 1997. Genetic influences on aging. Exp. Gerontol. **32:** 11–22.
11. MCGUE, M., J.W. VAUPEL, N. HOLM & B. HARVALD. 1993. Longevity is moderately heritable in a sample of Danish twins born 1870–1880. J. Gerontol. **48:** B237–B244.
12. YASHIN, A.I. & I. IACHINE. 1995. How long can humans live? Lower bound for biological limit of human longevity calculated from Danish twin data using correlated frailty model. Mech. Ageing Dev. **80:** 147–169.
13. MARTIN, G.M. 1985. Genetics and aging: The Werner syndrome as a segmented progeroid syndrome. Adv. Exp. Med. Biol. **190:** 161–170.
14. JAZWINSKI, S.M. 1996. Longevity, genes, and aging. Science **273:** 54–59.
15. PIANTANELLI, L., A. BASSO & G. ROSSOLINI. 1994. Modelling the link between aging rate and mortality rate. Ann. N.Y. Acad. Sci. **719:** 136–145.
16. YU, C.-E. *et al.* 1996. Positional cloning of the Werner's syndrome gene. Science **272:** 258–262.
17. MASORO, E.J. 1995. Dietary restriction. Exp. Gerontol. **30:** 291–298.
18. MASORO, E.J. & S.N. AUSTAD. 1996. The evolution of the antiaging action of dietary restriction: A hypothesis. J. Gerontol. Biol. Sci. **51A:** B387–B391.
19. MCFARLAND, G.A. & R. HOLLIDAY. 1994. Retardation of the senescence of cultured human diploid fibroblasts by carnosine. Exp. Cell Res. **212:** 167–175.
20. RATTAN, S.I.S. & B.F.C. CLARK. 1994. Kinetin delays the onset of ageing characteristics in human fibroblasts. Biochem. Biophys. Res. Commun. **201:** 665–672.
21. SHARMA, S.P., P. KAUR & S.I.S. RATTAN. 1995. Plant growth hormone kinetin delays ageing, prolongs the life span and slows down development of the fruitfly *Zaprionus paravittiger*. Biochem. Biophys. Res. Commun. **216:** 1067–1071.
22. SHARMA, S.P., J. KAUR & S.I.S. RATTAN. 1997. Increased longevity of kinetin-fed *Zaprionus* fruitflies is accompanied by their reduced fecundity and enhanced catalase activity. Biochem. Mol. Biol. Int. **41:** 869–875.
23. LUCKINBILL, L.S. 1993. Prospective and retrospective tests of evolutionary theories of senescence. Arch. Gerontol. Geriatr. **16:** 17–32.
24. ORR, W.C. & R.S. SOHAL. 1994. Extension of life-span by overexpression of superoxide dismutase and catalase in *Drosophila melanogaster*. Science **263:** 1128–1130.
25. LAKOWSKI, B. & S. HEKIMI. 1996. Determination of life-span in *Caenorhabditis elegans* by four clock genes. Science **272:** 1010–1013.
26. LARSEN, P.L. 1993. Aging and resistance to oxidative damage in *Caenorhabditis elegans*. Proc. Natl. Acad. Sci. USA **90:** 8905–8909.
27. LITHGOW, G.J., T.M. WHITE, S. MELOV & T.E. JOHNSON. 1995. Thermotolerance and extended life-span conferred by single-gene mutations and induced by thermal stress. Proc. Natl. Acad. Sci. USA **92:** 7540–7544.
28. LITHGOW, G.J. 1996. Invertebrate gerontology: The age mutations of *Caenorhabditis elegans*. BioEssays **18:** 809–815.
29. RATTAN, S.I.S. 1997. Gene therapy for ageing: Mission impossible? Eur. J. Genet. Soc. **3.** 27–29.
30. RATTAN, S.I.S. & A. DERVENTZI. 1991. Altered cellular responsiveness during ageing. BioEssays **13:** 601–606.
31. HAYES, S.A. & J.F. DICE. 1996. Roles of molecular chaperones in protein degradation. J. Cell Biol. **132:** 255–258.
32. HOLBROOK, N.J. & R. UDELSMAN. 1994. Heat shock protein gene expression in response to physiologic stress and aging. *In* The Biology of Heat Shock Proteins and Molecular Chaperones. R.I. Morimoto, A. Tissières & C. Georgopoulos, Eds.: 577–593. Cold Spring Harbor Laboratory Press. Cold Spring Harbor, NY.
33. JINDAL, S. 1996. Heat shock proteins: Applications in health and disease. Trends Biotechnol. **14:** 17–20.
34. NEAFSEY, P.J. 1990. Longevity hormesis: A review. Mech. Ageing Dev. **51:** 1–31.
35. POLLYCOVE, M. 1995. The issue of the decade: Hormesis. Eur. J. Nucl. Med. **22:** 399–401.
36. KHAZAELI, A.A., M. TATAR, S.D. PLETCHER & J.W. CURTSINGER. 1997. Heat-induced longevity extension in *Drosophila*. I. Heat treatment, mortality, and thermotolerance. J. Gerontol. Biol. Sci. **52A:** B48–B52.
37. RATTAN, S.I.S. 1998. Repeated mild heat shock delays ageing in cultured human skin fibroblasts. Biochem. Mol. Biol. Int. **45:** 753–759.

Causes of Aging

ROBIN HOLLIDAY[a]

CSIRO Molecular Science, Sydney Laboratory, P.O. Box 184,
North Ryde, NSW 2113, Australia

ABSTRACT: A broad biological approach makes it possible to understand why aging exists and also why different mammalian species have very different maximum longevities. The adult organism is maintained in a functional state by at least ten major mechanisms, which together constitute a substantial proportion of all biological processes. These maintenance mechanisms eventually fail, because the evolved physiological and anatomical design of higher animals is incompatible with continual survival. The life span of each mammalian species depends on the efficiency of maintenance of their cells, tissues, and organs, and there is much evidence that such maintenance is more effective in long-lived species, such as humans, than in short-lived small mammals. It is also evident that there is an inverse relationship between reproductive potential and longevity, which would be expected if available metabolic resources are shared between investment in reproduction and investment in the preservation of the adult body. It is proposed that the eventual failure of maintenance leads to the pathological changes seen in age-associated disease. Although we now have a biological understanding of the aging process, much future research will be needed to uncover the cellular and molecular changes that give rise to age-associated diseases. The major aim of such research is to devise procedures to delay or prevent the onset of these diseases.

Those who maintain that aging is an unsolved problem in biology tend to take a narrow view, believing that a single cause of aging exists, or that it is controlled by a few "gerontogenes." A broad view, which encompasses a considerable proportion of the whole of biological knowledge, makes it clear why aging exists. This knowledge provides answers to three basic questions, at least at the biological level: Why do we age? Why do we live as long as we do? and Why do different mammalian species have very different maximum life spans? In answering these questions, a great deal is revealed about the mechanisms that underpin eventual senescence and death. Almost all the material in the following discussion will be found in my book *Understanding Ageing*,[1] which is fully referenced, and other recent reviews are available.[2–4]

EARLY EVOLUTIONARY ORIGINS OF AGING

The aging of somatic cells must have occurred quite early in the evolution of multicellular animals. Initially, primitive animals probably had considerable powers of regeneration and renewal, as do the coelenterates and flatworms today. Such organisms may be potentially immortal, although in natural environments their life would be ended by one of many environmental hazards. As more complex animals evolved the distinction between the germline and the soma, or body, became much more clear-cut, and in particular organisms evolved where all the cells of the body are postmitotic, except the germline cells.

[a]Tel: 61 2 9490 5156; fax: 61 2 9490 5010; e-mail: randl.holliday@bigpond.com

This is the case in nematodes and many insects. Such animals, when kept under good environmental conditions, clearly have a finite life span. This can be attributed to the fact that nondividing cells, active in metabolism, cannot be expected to survive indefinitely.

At first sight, it seems that aging is nonadaptive, inasmuch as an organism that can survive and reproduce indefinitely is fitter in Darwinian terms than one that reproduces for a given period of time and then dies. Why then did aging evolve in the first place? The answer to this question, oddly enough, lies in the Darwinian realization that organisms normally produce far more offspring than can possibly survive and reproduce themselves. The environment is hostile, and individuals are competing for limited resources. This competition results in the natural selection of the fittest. In these circumstances the probability of an organism surviving and reproducing for a long period become very small, so potential immortality confers very little, if any, adaptive advantage. In other words, such organisms are not necessarily the fittest because resources are used to maintain the soma for a long period of time. It is a better strategy for the survival of an organism's lineage to invest resources into growth to adulthood and reproduction, rather than in long-term maintenance of the soma. Thus, the organism that evolves a soma with a limited survival time is at an advantage over one that attempts to maintain the soma indefinitely. This disposable soma theory neatly explains the early origins of aging in animals.[5,6]

Subsequently, as evolution proceeded, there arose many variations in the pattern of aging. Many adult vertebrates grew continuously, and these tended to have very long life spans. Although the signs of senescence were less obvious than in species that had constant adult body size, their survival for a century or so was still a minute fraction of evolutionary time. Life span variability is seen particularly in fish, where small species may survive for a year and very large ones for several decades.[2,7]

Mammals and birds clearly evolved from cold blooded vertebrates, which had a finite life span, so in a sense mammals and birds merely inherited the life style strategy that included senescence and aging. The next section briefly reviews some features of the mammalian body plan that make aging inevitable.

THE EVOLVED DESIGN OF MAMMALS

A vast amount of information is available about the cells, tissues, and organs of mammals. Much of this comes from the biomedical investigations of the human body encompassing many disciplines. These show that many organ systems have very limited capacity for regeneration and renewal, and it is these features of our anatomy that make senescence and aging inevitable. The neurons of the brain are postmitotic and very active in metabolism. Although DNA can be repaired and proteins turned over, cells that are lost cannot be replaced. There are many reasons why one individual cell cannot survive indefinitely. Some DNA lesions are not repaired, and some altered or abnormal proteins cannot be degraded by proteases and therefore accumulate. The brain is very definitely a nonrenewable structure. The same applies to the retina (an extension of the brain). The rods and cones continually synthesize photoreceptors, and the oldest are removed. This process does not achieve a steady state, and remnants of partially degraded photoreceptor elements accumulate in the cells themselves, or in the underlying epithelial layer. Eventually the degenerative process of retinopathy occurs. The crystallin proteins of the lens of the eye are laid down at an early stage and cannot be replaced. Lens transparency depends on their molecular homogeneity. Unfortunately proteins are subject to many chemical changes, including the processes of oxidation, glycation, racemization of amino acids, and deamidation. Because these cannot be prevented or reversed, the molecules gradually lose their initial properties, and cataracts may occur.

Collagen and elastin are also very long-lived proteins that are subject to chemical change. It is well established that collagen becomes progressively cross-linked with age, thereby losing its initial elasticity. The heart is a highly efficient pump, but like the brain, it has very limited capacity for repair or renewal. The muscle cells are postmitotic and unlike most skeletal muscle they cannot be replaced by the division of myoblasts. The anatomy of the major blood vessels is also incompatible with efficient repair. The cross-linking of elastin and collagen results in hardening of arteries, and the inner wall is subject to damage, including the buildup of atherosclerotic plaques. The basic anatomical problem is that there is only one vascular system, and it cannot be shut down for repair. It is, in fact, very difficult to repair a machine while it is operating, and the same is true of the vascular system. A potentially immortal organism would need to have two vascular systems, one of which could be shut down and repaired, while the other kept operating. We did not evolve in that way.

Teeth provide an instructive example of the way components of the body have evolved "to last a lifetime." Clearly the shape and size of adult teeth are genetically determined, but they are also subject to wear and tear, as well as decay. This is one of many examples that demonstrate the artificiality of the distinction that is often made between "wear and tear" theories of aging (or the stochastic accumulation of various defects) and the "program" theories. Both, in fact, are interrelated and important. Some herbivores that continually crop plants have incisors that keep growing at the base, which is clearly a secondary adaptation to produce "immortal" teeth. Many other herbivores, however, do not have this ability, and it is well known that an estimate of a horse's age can be made by examination of the wear on its teeth.

MAINTENANCE OF THE ORGANISM

Although the evolved design of many body components is incompatible with indefinite survival, this does not mean that maintenance mechanisms are unsuccessful. The life history of a mammalian organism comprises development and growth to the adult, a fairly long period of reproduction, followed by loss of fertility, and the senescence, and death. Maintenance of cell, tissue, and organ function is essential during development and reproduction. The total resources available to a mammalian organism are allocated to three major functions: first, ongoing metabolism, second, all aspects of reproduction, and third, a set of maintenance mechanisms. These three functions consume all available metabolic energy, and although there may be some overlap between them, it is possible to itemize their main features, as shown in TABLE 1. The major maintenance mechanisms are as follows.

Wound Healing

Damage to skin and muscle can be effectively repaired, and broken bones can rejoin. Loss of blood is prevented by clotting, and the smaller arteries and veins can be replaced.

TABLE 1. The Allocation of All Available Energy Resources in Mammals

Normal functions	Reproduction	Maintenance
Biochemical synthesis	Gonads, gametes, and sex	Wound healing
Metabolism	Development	Immunity
Respiration	Gestation	Protein turnover
Cell turnover	Suckling	Defense against free radicals
Movement	Care of offspring	Proofreading
Feeding and digestion	Growth to adult	DNA repair
Excretion		Detoxification
		Epigenetic stability
		Apoptosis
		Fat storage
		Homeostasis

Nevertheless, mammals do not have strong regenerative capacity. Severed limbs or digits are not replaced, and major nerves that are cut cannot be rejoined. In this respect, some lower vertebrates have greater regenerative capacity, inasmuch as lost limbs can be regrown.

Immunity

All organisms are subject to attack by pathogens and parasites, and a complex immune system has evolved to protect the organism. Immunology is, of course, a science in its own right, and leaving aside the "immunologic" theory of aging,[8] the immune response is not thought to have any relationship to the study of longevity or aging. Nevertheless, it is a vital maintenance mechanism, and without it an organism does not survive very long.

DNA Repair

Although DNA is a stable molecule, it is continually subject to intrinsic and extrinsic damage. It is highly likely that oxygen free radicals are an important source of damage.[4] A battery of repair enzymes exists that continually monitors DNA for abnormalities in structure, excises or removes such damage, and then fills in any gaps by repair synthesis and rejoining. One of the commonest defects in DNA is the loss of purine residues. Indeed, it has been estimated that up to 10,000 of these lesions occur in each cell per day.[9] All this damage is effectively repaired. There may be lesions, however, that are not repaired, perhaps because they are less common, and the necessary enzymes have never evolved to deal with them.[10] Also, there may be adjacent lesions on both strands of a DNA molecule that are difficult to repair and that can lead to chromosome breaks.

Synthesis of Macromolecules

DNA repair overlaps with mechanisms to ensure that DNA is synthesized with extreme accuracy. The insertion of an incorrect base by the replicating polymerase is usually corrected by an editing excision/replacement mechanism. However, if this fails, there is a backup mismatch repair system. The removal of errors in DNA synthesis depends on many enzymes and accessory proteins. RNA and proteins are made with less accuracy; nevertheless it would be wasteful, as well as harmful, to synthesize defective molecules, so it is not surprising that proofreading mechanisms exist to detect and remove errors. All these proofreading mechanisms consume energy. The question of the optimum accuracy of synthesis of macromolecules is an interesting one. In general, it seems to be the case that rapid synthesis results in more errors, and slower synthesis allows time for more efficient editing. There must be an optimum or some balance between the two, which may well not be the same for all mammalian species (see below).

Protein Turnover

As has been mentioned, protein molecules are subject to many postsynthetic modifications. Some modifications are, of course, a normal part of the maturation of proteins and play essential roles in their function, but there are many others that are abnormal, with potentially deleterious effects on the cell. These molecules are usually recognized and removed by proteases and the proteosome. This is a very important ongoing process, essential for the normal function of cells. Amino acids that cannot be reused are broken down and the nitrogen excreted in the form of urea. The removal of abnormal proteins is not completely successful, particularly if the protein is inaccessible or is part of a nonreplaceable structure (such as the walls of major arteries). Also, altered proteins may form high molecular weight aggregates that are resistant to proteolytic digestion such as AGEs (advanced glycation end products) or the amyloid plaques in Alzheimer's disease. The gradual accumulation of these high molecular weight protein or peptide aggregates are an important part of the aging process.

Detoxification

Animals have complex diets, and toxic chemicals are a common component. In particular, plants often defend themselves against animals by synthesizing such compounds. In response, mammals have evolved a large set of detoxifying enzymes, collectively known as the P450 cytochromes. These comprise a very complex family of enzymes located in the liver, but also in other tissues that can degrade a very wide range of chemicals. Nowadays, these include many man-made chemicals that would never have been encountered during evolution. Thus, the detoxification system has an inbuilt "overkill" capacity to deal with any new chemicals that may arise in the diet or environment.

Defences against Free Radicals

Oxygen free radicals are continually generated by respiration and some other metabolic processes. Although very short lived, they are highly reactive and can damage DNA, proteins, and membranes. Organisms have developed major defences against free radical attack. There are enzymes that break down free radicals, such as superoxide dismutase, catalase, glutathione peroxidase, and reductase. Metabolites exist that react with free radicals, acting as free-radical "sinks," such as carotenoids or other antioxidants. It is likely that the evolution of the respiratory organelle, the mitochondrion, protects the chromosomal DNA in the nucleus from free-radical attack. It is well known that the small mitochondria DNA genome mutates at a much higher rate than chromosomal DNA.

Epigenetic Controls

Differentiated cells stably maintain their particular biochemical and morphological characteristics. This depends on the activities of genes responsible for the cells' specialized functions, together with the inactivity of all the genes needed for all other specialized cells. These controls of gene activity are generally referred to as epigenetic, and they are superimposed on the information in DNA, which is present in all cells. Many believe that epigenetic controls are entirely due to proteins that bind to specific DNA sequences, but there are now many indications that chemical modification of DNA is an essential component. The major modified base in mammals is 5-methylcytosine, and it is known that the pattern of this methylation is inherited through mitotic division, and therefore stably maintained in those specialized cells that are capable of division, as well as in postmitotic cells. Obviously, it is extremely important to maintain epigenetic controls, because if normal regulation is lost, then a cell can adopt an abnormal phenotype and become, for example, a neoplastic cell. This can occur through mutation, but epigenetic defects are also likely to be involved.[11]

Apoptosis

The suicide mechanism known as apoptosis is triggered in a variety of contexts. It removes unwanted cells during development, or in the immune system, but it also comes into play when damaged or abnormal cells arise. Otherwise such cells would have harmful effects on the organism. Although it has been suggested that aging and apoptosis may be linked, the relationship is not at all simple. Apoptosis is, at least in part, a maintenance process to prevent deleterious changes. If apoptosis does not come into play, for whatever reason, then an abnormal cell will survive, and this may contribute to senescence and aging.

Homeostatic Mechanisms

These comprise a large set of physiological or regulatory processes that maintain cells, tissues, and organs in a normal functional state. The most important homeostatic mechanism in mammals and birds is the control of body temperature. This produces a much more uniform internal environment, with less dependence on the external one. Therefore, mammals and birds can colonize a wider range of environments than cold-blooded vertebrates. It also allows many biochemical processes to be optimized, with the activity of many proteins adapted to body temperature. Many other homeostatic mechanisms depend on hormones or growth factors, which ensure that potential variables (such as blood-sugar levels) are controlled. Too many examples exist to review here, but it is worth mentioning that a stress response, such as the heat-shock response, can be regarded as a cellular homeostatic maintenance mechanism, which protects cells from an even greater rise in temperature.

Fat Storage

In natural environments it is common for the availability of food to vary considerably. Thus, periods of glut may alternate with periods of scarcity. To ensure survival in the absence of food, mammals have evolved an efficient energy storage mechanism that can tide them over periods in a harsh environment. The laying down and reuse of fat can therefore be regarded as a maintenance mechanism.

There are several features of the totality of maintenance mechanisms that should be emphasized. First, the study of all these processes comprises a major part of all biological research. Inasmuch as aging and death are ultimately the result of failure of maintenance, it is not unreasonable to propose that all this research is in one way or another related to the study of aging itself. Second, it is fashionable to invoke specific "gerontogenes" that in some way control longevity and aging, but there are innumerable genes that specify the components of all maintenance mechanisms. All of these relate in one way or another to the efficiency and also the eventual failure of maintenance. We know of many examples of single gene mutations that have pleiotropic effects on the phenotype, and some of these clearly relate to aging. Third, the various theories of aging that have been proposed usually relate quite closely to failure of maintenance. Thus, the oxygen free radical theory of aging is directly related to the failure to nullify their dangerous effects. This, in turn, overlaps with the somatic mutation theory, which is clearly related to the failure of DNA repair. The protein error theory proposes that abnormal proteins can cause escalating damage by reducing the accuracy of synthesis. Clearly this is related to the failure to remove abnormal molecules by proteolysis. This failure also results in the accumulation of abnormal protein molecules, which constitutes another theory of aging. The immunologic theory of aging suggests that the immune system eventually loses its ability to distinguish self from nonself antigens and therefore inflicts pathological damage on cells and tissues. The dys-differentiation theory of aging proposes that ectopic protein synthesis (*i.e.*, the synthesis of a specialized protein in an inappropriate cell) is an important feature of senescence. This is related to the loss of epigenetic controls. It is very likely that there is some truth in all of these theories of aging, because aging is multicausal.[1]

Finally, new information about the importance and complexity of maintenance is continually being obtained, and a recent example is the discovery of peptide antibiotics in human skin.[12] Clearly this is an important defense mechanism against bacterial infection, which is rather distinct from the more familiar immune responses to infection.

REPRODUCTION, MAINTENANCE, AND LONGEVITY

The "disposable soma" theory of the evolution of aging and longevity predicts that there should be some trade-off between resources invested in rapid growth and reproduction and resources invested in maintenance of the soma. The balance between reproduction and maintenance depends on the level of environmental hazard. In a high-risk environment, where annual mortality is very high, it would then be expected that development and reproduction would be rapid; fewer resources would be devoted to maintenance, and therefore a shorter life span would be seen. In a low-risk environment, the annual mortality is much lower, so we would expect the evolution of slow-breeding long-lived species. In the adaptive radiation of mammals, there have been evolutionary trends that increase reproduction and reduce longevity. The carnivores provide one example, because the highly specialized stoats and weasels need a continual supply of food (*i.e.*, they live in a high-risk environment), produce large litters that mature rapidly, and have a short life span in captivity. The other trend is an increase in longevity, exemplified by the primates. Small monkeys reproduce rapidly and have short life spans; larger monkeys, small apes, the great apes, and humans have progressively longer life spans. This is associated with fewer offspring and a much lower annual mortality. If lower mortality is regarded as a "successful" adaptation to the environment, then it is also clearly associated with natural selection for longer life spans.

The foregoing use of the terms life span and longevity refers to the documented length of life of mammalian species kept in captivity, that is, the protected environment of a zoo or a laboratory cage. There is some relationship of this measured life span with the likely survival of that species in a natural environment, which is usually hard to determine. However, if it is assumed that the population size is constant, it is possible to calculate the average expectation of life at birth, provided the various reproductive parameters are known. These are the gestation period, the litter size, the time to develop to a fertile adult, and the interlitter interval. For early human-hunter gatherers, assuming a constant population size, the expectation of life at birth was only about 16 years; for females that reached reproductive age, it was about 28 years.[13]

The best available reproductive and life span-in-captivity data for 47 mammalian species demonstrates a clear inverse relationship between maximum life span and reproduc-

TABLE 2. Correlation between Maintenance Parameters and Maximum Life Span of Mammalian Species[a]

Positive Correlations
    Longevity of fibroblasts *in vitro*
    Longevity of erythrocytes *in vivo*
    DNA repair
    Poly-ADP ribose polymerase
    γ-Ray-induced ADP-ribose transferase
    Carotenoids in serum

Negative Correlations
    Cross-linking of collagen
    Production of oxygen free radicals
    Auto-oxidation of tissues
    Metabolic rate and oxidized DNA bases
    DNA methylation decline
    Carcinogen binding to DNA
    Mutagenicity of activated carcinogen
    Incidence of cancer

[a]For sources, see refs. 1, 11, and 15.

tive potential.[1,14] The fecundity/life span ratio is highest in small living rodents and rabbits, then decreases through small carnivores, small primates, large carnivores, larger herbivores, pachyderms, the great apes, and humans. Many attempts have been made over the years to relate maximum life span to metabolic rate, weight, and brain size, or any combination of these in mammalian species. It is often found that bats (*Chiroptera*) with a high metabolic rate provide an exception to any general rule. It is striking that bats have long life spans and low rates of reproduction, as expected from their low-risk lifestyle. The analysis of reproductive potential and maximum life span strongly confirms a prediction of the disposable soma theory.

Another prediction is that the efficiency of maintenance should relate to maximum longevity. A number of comparative studies have been carried out, although more are needed. In almost every case there is the expected relationship between efficiency of the maintenance parameter studied and the maximum life span of the species. In other cases, the relationship is inverse, but this is also in the expected direction. The studies that have been published are listed in TABLE 2 (for sources, see refs. 1, 11, and 15).

AGE-RELATED DISEASES

It is commonly assumed by clinicians that the pathological conditions commonly seen in the elderly are distinct from "natural aging." Dementia in the sixth and seventh decade is Alzheimer's disease, but dementia in the tenth decade, or thereafter, is natural aging. Also, it is said that an age-related disease should be considered as such, because many elderly people never exhibit the symptoms. Thus, many centenarians have no obvious sign of heart disease.

From a biological point of view, what should we expect? It is clear that aging comprises a deterioration of many organ systems that are not obviously related to each other. The loss of neurons in the brain is not obviously related to the accumulation of atherosclerotic plaques in aortas or the loss of transparency of the lens. The cross-linking of collagen and skin wrinkling are not obviously related to retinopathy, and so on. There is a degree of synchrony in all known age-related changes, but the synchrony is certainly not exact. We would therefore expect that in a given individual, one tissue or organ system deteriorates in advance of others. This is then usually diagnosed as a disease or pathological state. In another individual a different degenerative condition may appear. Although aging itself cannot be regarded as a disease, the various diseases associated with old age can certainly be regarded as part of the overall processes of aging. This applies to dementia, cardiovascular and cerebrovascular disease, osteoarthritis, osteoporosis, late-onset diabetes, renal failure, loss of sight or hearing, carcinomas, as well as many other less well-known conditions. These diseases can be broadly related to the failure of various maintenance mechanisms, as shown in TABLE 3.

It cannot be overemphasized that all the research devoted to the study of age-related diseases is in fact related to gerontology itself. This research has three aims: (1) better treatment of the disease in question, (2) the elucidation of the cause, or causes, of the disease, and (3) the development of procedures to postpone or prevent the onset of the disease. Aims two and three can certainly be regarded as being within the province of gerontology, and therefore more research on aging itself is very likely to throw much light on the origins and development of age-related disease. This was clearly recognized by the

TABLE 3. General Relationships between Cell or Tissue Maintenance and Some Major Human Age-associated Diseases

Failure of maintenance	Major pathologies
Neurons	Dementias
Retina, lens	Blindness
Insulin metabolism	Type II diabetes
Blood vessels and heart	Cardiovascular and cerebrovascular disease
Bone structure	Osteoporosis
Immune system	Autoimmune disorders
Epigenetic controls	Cancer
Joints	Osteoarthritis
Glomeruli	Renal failure

geriatrician Steiglitz more than 55 years ago in his article "The social urgency of research on aging."[16] Unfortunately, much persuasion is necessary to convince the present community of clinical and biomedical research scientists.

CONCLUSIONS

We now have answers to the three basic questions posed at the beginning of this article. We age because we evolved from organisms that also age. We age because our evolved body structure is incompatible with continual survival. We age because our various maintenance mechanisms eventually fail to preserve the normal structure and function of cells and tissues. We live as long as we do because we have evolved a lifestyle with low annual mortality.[13] This has allowed for more resources to be invested in maintenance and less in reproduction. By contrast, species that live in a high-risk environment can only survive by investing much more heavily in reproduction, with correspondingly fewer resources allocated to maintenance. Thus, the adaptive radiation of mammals to many ecological niches has also resulted in the evolution of longevities over an approximately 50-fold range. Thus, when considered at the level of the organism, aging is no longer an unsolved problem in biology.

Nevertheless, at the level of fine detail, the actual molecular and cellular changes that produce the aging phenotype, there is a great deal to learn. An understanding of these changes will come from further studies of maintenance mechanisms, and more important, the reasons why maintenance eventually fails. This new knowledge will greatly increase our understanding of the origins of age-associated disease and will concomitantly make it possible to prevent or delay the onset of these diseases. The aim of all this research is not to increase the overall life span but to significantly extend the "health span" so that the quality of the elderly is greatly improved and the costs of health care for the aged are greatly reduced.

REFERENCES

1. HOLLIDAY, R. 1995. Understanding Ageing. Cambridge University Press. Cambridge.
2. FINCH, C.E. 1990. Longevity, Senescence and the Genome. University Press. Chicago.
3. GRIMLEY EVANS, J. & T. FRANKLIN WILLIAMS, Eds. 1992. Oxford Textbook of Geriatric Medicine. Oxford University Press. Oxford.
4. MARTIN, G.M., S.N. AUSTAD & T.K. JOHNSON. 1996. Genetic analysis of aging: Role of oxidative damage and environmental stress. Nat. Genet. **13:** 25–34.
5. KIRKWOOD, T.B.L. & R. HOLLIDAY. 1979. The evolution of ageing and longevity. Proc. R. Soc. Lond. B **205:** 531–546.
6. KIRKWOOD, T.B.L. 1985. Comparative and evolutionary aspects of longevity. *In* Handbook of the Biology of Aging. C.E. Finch & E.L. Schneider, Eds.: 27–44. Van Nostrand Reinhold. New York.
7. COMFORT, A. 1979. The Biology of Senescence, 3rd Ed. Churchill Livingstone. London.
8. WALFORD, R.L. 1969. The Immunologic Theory of Ageing. Munksgaard. Copenhagen.
9. LINDAHL, T. 1979. DNA glycosylases, endonucleases for apurinic/pyrimidinic sites and base excision-repair. Prog. Nucleic Acid Res. Mol. Biol. **22:** 135–192.
10. LINDAHL, T. 1993. Instability and decay of the primary structure of DNA. Nature **362:** 709–715.
11. HOLLIDAY, R. 1996a. Neoplastic transformation: The contrasting stability of human and mouse cells. *In* Genetic Instability in Cancer. T. Lindahl, Ed.: 103–115. Cancer Surveys **18:** Cold Spring Harbor Laboratory Press. New York.
12. HARDER, J., J. BARTELS, E. CHRISTOPHERS & J.M. SCHRODER. 1997. A peptide antibiotic from human skin. Nature **387:** 861.
13. HOLLIDAY, R. 1996. The evolution of human longevity. Perspect. Biol. Med. **40:** 100–107.
14. HOLLIDAY, R. 1994. Longevity and fecundity in eutherian mammals. *In* Genetics and Evolution of Aging. M.R. Rose & C.E. Finch, Eds.: 217–225. Kluwar Academic Publishers. The Netherlands.
15. HOLLIDAY, R. 1997. Understanding Aging. Philos. Trans. R. Soc. Lond. B **352:** 1793–1797.
16. STEIGLITZ, E.J. 1942. The social urgency of research in ageing. *In* Problems of Ageing: Biological & Medical Aspects, 2nd Ed. E.V. Cowdray, Ed.: 890–907. Williams & Williams. Baltimore.

Intrinsic and Extrinsic Factors in Muscle Aging

JOSEPH G. CANNON[a]

Noll Physiological Research Center and Department of Kinesiology, Pennsylvania State University, University Park, Pennsylvania 16802-6900, USA

ABSTRACT: The regenerative potential of skeletal muscle, and overall muscle mass, decline with age. This regenerative potential may be influenced by autocrine growth factors intrinsic to the muscle itself. Extrinsic host factors that may influence muscle regeneration include hormones, growth factors secreted in a paracrine manner by accessory cells, innervation, and antioxidant mechanisms. Unaccustomed exercise, which involves mechanical overload of myofibers, provides a convenient method for studying muscle regeneration in both humans and animal models. An inflammatory response ensues in which distinctive populations of macrophages infiltrate the affected tissue: some of these macrophages are involved in phagocytosis of damaged fibers; other macrophages arriving at later times may deliver growth factors or cytokines that promote regeneration. These include fibroblast growth factor and insulin-like growth factor, which are important regulators of muscle precursor cell growth and differentiation, as well as nerve growth factor, which is essential for maintenance or reestablishment of neuronal contact. Other cytokines, including interleukin-1, tumor necrosis factor, interleukin-15, and ciliary neurotrophic factor, have a strong influence on the balance between muscle protein synthesis and breakdown. The functional activity of invading macrophages can be influenced by age, by factors in myofibers and extracellular matrix, and can be influenced systemically by the antioxidant status of the host.

Aging is associated with a progressive loss of muscle mass. In healthy humans, the decline in mass begins in the fourth decade of life and the rate of loss is approximately 8% per decade thereafter.[1] Although some of this loss may be related to decreased physical activity, those who maintain a high activity level through life still exhibit age-related losses. The mechanisms that may be responsible for the decline are the focus of widespread investigation. A number of factors extrinsic to the muscle itself are likely to be important for skeletal muscle viability. Age-associated alterations in anabolic steroid metabolism may play a role. Changes in pituitary function, especially growth hormone secretion, have also been implicated. Although many factors are likely to be important, this review will focus on the potential role of cytokines, especially those delivered in a paracrine manner by infiltrating inflammatory cells.

INTRINSIC VERSUS EXTRINSIC FACTORS

In 1989, Carlson and Faulkner sought to determine whether the regenerative properties of skeletal muscle depended upon the age of the muscle tissue itself or age-related changes in the supporting milieu.[2] Extensor digitorum longus (EDL) muscles from young rats (2–3 months of age) were transplanted into old rats (24 months) and vice versa. After 60 days of recovery, the transplanted muscles were removed and assessed for mass and force pro-

[a]Tel: 814/865-0322; fax: 814/865-4602; e-mail: jgc2@psu.edu

duction. For both outcome variables, the age of the transplanted muscle made no difference, but the age of the recipient was critical. Both young and old muscle transplanted into
young recipients exhibited similar characteristics and were approximately double those of
muscle (of either age) transplanted into old rats. Clearly, extrinsic host factors were important for muscle viability.

EXPERIMENTAL APPROACHES

The above study by Carlson and Faulkner investigated characteristics of regeneration
following the perturbations brought out by transplantation. Other interventions that induce
regeneration *in vivo* include crush injury,[3] ablation of synergistic muscle that results in
increased chronic loading of the experimental muscle,[4] intramuscular injection of toxins
that cause localized fiber necrosis,[5] and rodent hindlimb suspension followed by resump-
tion of weight bearing that results in overloading of atrophied muscle.[6] Another common
experimental approach involves forcing muscle to lengthen as active tension develops.
This occurs in human quadriceps muscle when a subject runs downhill. In anesthetized
rodents, controlled lengthening contractions can be carried out by extension of a limb with
a servomotor and simultaneous activation of associated flexor muscles via electrical stim
ulation of the motor neuron.[7] The muscle damage associated with this activity is thought
to be caused by recruitment of an insufficient number of myofibers to bear the applied
load.

RESPONSES TO MYOFIBER OVERLOAD

The responses following at least two types of overloading protocols (hindlimb suspen-
sion/reloading and lengthening contractions) seem to express similar patterns of recovery.
In the first 1–3 days following the intervention, myofiber necrosis is evident with extensive infiltration of myofibers with macrophages. In rats, these macrophages express the
phenotypic marker ED1;[6] in mice the macrophages express ERMP20 and F4/80.[8] It is generally assumed that these are phagocytic macrophages responsible for clearing out damaged cellular material. Over the next 4–7 days, myofiber infiltration subsides, but
macrophages expressing different phenotypic markers accumulate in the interstitial spaces
between fibers. In rats, these macrophages express ED2;[6] in mice, the interstitial macrophages express ERBMDM-1.[8] The function of these macrophages is less clear, but their
appearance is temporally associated with histological evidence of myofiber regeneration,
that is, centrally located nuclei in the myofibers and expression of developmental isoforms
of myosin heavy chain protein. The muscle-specific actions of macrophage-derived cytokines is summarized in TABLE 1.

INFLUENCE OF MACROPHAGES ON MYOFIBERS

During regeneration of myofibers, whether in response to acute injury or simply as part
of the constant remodeling that occurs throughout life, macrophages represent potential
sources of growth factors and chemoattractants. Mature myofibers do not undergo cell

TABLE 1. Influence of Cytokines[a] on Skeletal Muscle Cells

	Myoblasts		Mature Fibers	
Cytokine	proliferation	differentiation	proteolysis	protein synthesis
IL-1		inhib[c]	stim	inhib
TNF			stim	
IGF	stim[b]	stim		
FGF	stim			
IL-6 superfamily				
LIF				
CNTF			stim	stim
IL-15				stim
TGF superfamily				
TGFβ		inhib		
myostatin				inhib

[a]IL-1: interleukin-1; IGF: insulin-like growth factor; LIF: leukemia inhibitory factor; TGFβ: transforming growth factor-beta; TNF: tumor necrosis factor; FGF: fibroblast growth factor; CNTF: ciliary neurotrophic factor.
[b]Stimulates.
[c]Inhibits.

division; instead, dormant "satellite cells" begin to proliferate, differentiate, and fuse into myotubes or existing fibers.[9] *In vitro* studies with muscle cell lines and primary myoblast cultures indicate that several cytokines, including IGF, FGF, IL-1, and TNF, stimulate satellite cell proliferation[10] (all cytokine acronyms defined in TABLE 1 legend). However, these same cytokines (with the exception of IGF) also inhibit differentiation. IGF is unique in its ability to stimulate differentiation. The extent to which each of these cytokines represents extrinsic versus intrinsic influence on muscle viability is currently under study. IGF and FGF are produced in an autocrine manner by regenerating myofibers,[10] but all of the cytokines mentioned can be secreted by macrophages as well.[11] Cocultures of satellite cells and macrophages exhibited greater muscle cell proliferation and differentiation than homogeneous satellite cell cultures.[12] In addition, secretory products of macrophages (including PDGF, TGFβ, FGF, and LIF) appear to act as chemoattractants for satellite cells[13] and thus may be critical for recruiting quiescent muscle precursor cells to the focus of myofiber injury.

INFLUENCE OF MYOFIBERS ON MACROPHAGES

Signaling between muscle tissue and macrophages is not unidirectional. The proteoglycan composition of the extracellular matrix can influence the functional characteristics of macrophages.[14] Furthermore, release of hyaluronic acid from disrupted matrix stimulates macrophages via specific binding to CD44 receptors.[15] IGF delays apoptosis of myeloid precursor cells;[16] thus it is possible that growth factors produced by regenerating myofibers may influence the life cycle or function of infiltrating macrophages.

NEURONAL INTERACTIONS WITH MACROPHAGES AND MYOCYTES

Innervation has long been recognized as an essential element of muscle viability. One mechanism may involve stimulation of growth factor expression. Chronic motor nerve stimulation was shown to increase concentrations of FGF in rabbit EDL muscle.[17] Macrophages, in turn, may be important for neuronal viability. Macrophage-derived IL-1 stimulates nerve growth factor (NGF) expression in nerve tissue.[18] NGF, in turn, enhances production of the neuronal-cell adhesion molecule (N-CAM) that may be critical for establishment of neuronal-myocyte connections.[19]

Schwann cells of peripheral nerves produce a cytokine related to interleukin-6 known as ciliary neurotrophic factor (CNTF). This cytokine inhibits motor neuron degeneration in certain neuropathies and induces motor nerve terminal sprouting in healthy animals. However, systemic administration of CNTF causes net skeletal muscle loss.[20] Although CNTF stimulates protein synthesis, this is overwhelmed by a much greater stimulation of protein breakdown and a mild anorexia. These effects are not observed in isolated muscle incubated with recombinant CNTF *in vitro*, indicating that intermediates or cofactors are necessary for CNTF actions on protein synthesis (as seems to be true for IL-1 and TNF).[20]

INTRINSIC AND EXTRINSIC CONTROL OF MUSCLE MASS

During acute infection, skeletal muscle protein is broken down, and the amino acids are resynthesized into protective plasma proteins with antimicrobial and antioxidant properties.[21] In addition, glutamine liberated from skeletal muscle is used as metabolic fuel by activated leukocytes. *In vivo* experiments have shown that IL-1 and TNF are two primary stimuli for these events. In chronic inflammation, such as rheumatoid arthritis, patient lean body mass was inversely related to TNF secretion by circulating mononuclear cells.[22] In rodent models of sepsis, inhibition of IL-1 activity with IL-1 receptor antagonist demonstrated that IL-1 inhibited protein synthesis as well as stimulated protein breakdown.[23] However, IL-1 and TNF have no direct effect on muscle tissue *in vitro*, indicating that intermediate factors or cofactors are necessary for their proteolytic action.

Recently, two cytokines expressed by skeletal muscle itself have been shown to influence muscle protein synthesis. IL-15, originally identified as a T-cell growth factor, was shown to stimulate myosin heavy chain accumulation in a myoblast cell line as well as in primary myotube cultures.[24] The stimulation was additive with maximally effective doses of IGF-I, suggesting that these growth factors acted through independent mechanisms. Another muscle-specific cytokine related to TGFβ, myostatin, was shown to have an inhibitory influence on muscle mass.[25] Mutant mice unable to express the gene for this cytokine exhibited skeletal muscle hyperplasia and hypertrophy. The balance of these recently discovered cytokines, in concert with previously identified growth factors may be critical to the long-term viability of skeletal muscle tissue. The stimuli for expression of these factors has yet to be identified.

MACROPHAGES AND AGING: POTENTIAL IMPACT ON SKELETAL MUSCLE

Several functional characteristics of macrophages may change with age. There is evidence that phagocytosis,[26] chemotaxis to some stimuli,[26] and tumoricidal activity[27] are all depressed with advancing age. On the other hand, *in vivo* concentrations of IL-1[28] and IL-6,[29] and mononuclear cell secretions of several cytokines in response to mitogenic stimuli *in vitro* may increase with advancing age.[30]

A decline in phagocytic activity may lead to an inability to clear damaged or dysfunctional tissue in an effective manner. Furthermore, any subtle shift in the balance toward catabolic cytokine expression could, over many years, account for the losses of muscle mass associated with aging.

REFERENCES

1. GRIMBY, G. & B. SALTIN. 1983. The aging muscle. Clin. Physiol. **3:** 209–218.
2. CARLSON, B.M. & J.A. FAULKNER. 1989. Muscle transplantation between young and old rats: Age of host determines recovery. Am. J. Physiol. **256:** C1262–C1266.
3. SCHULTZ, E., D.L. JARYSZAK & C.R. VALIERE. 1985. Response of satellite cells to focal skeletal muscle injury. Muscle & Nerve **8:** 217–222.
4. DEVOL, D.L., P. ROTWEIN, J.L. SADOW, J. NOVAKOFSKI & P.J. BECHTEL. 1990. Activation of insulin-like growth factor gene expression during work-induced skeletal muscle growth. Am. J. Physiol. **259:** E89–E95.
5. JENNISCHE, E. & G.L. ANDERSSON. 1987. Regenerating skeletal muscle cells express insulin-like growth factor I. Acta Physiol. Scand. **130:** 327–332.
6. ST. PIERRE, B.A. & J.G. TIDBALL. 1994. Differential response of macrophage subpopulations to soleus muscle reloading after rat hindlimb suspension. J. Appl. Physiol. **77:** 290–297.
7. ASHTON-MILLER, J.A., Y. HE, V.A. KADHIRESAN, D.A. McCUBBREY & J.A. FAULKNER. 1992. An apparatus to measure *in vivo* biomechanical behavior of dorsi- and plantarflexors of mouse ankle. J. Appl. Physiol. **72:** 1205–1211.
8. ST. PIERRE, B.A., J.H. FLASKERUD & J.G. CANNON. 1996. Macrophage dynamics during myofiber necrosis and regeneration (Abstract). Physiologist **39:** A92.
9. BISCHOFF, R. 1994. The satellite cell and muscle regeneration. *In* Myology. A.G. ENGEL & C. FRANZINI-ARMSTRONG, Ed.: 97–118. McGraw-Hill. New York.
10. FLORINI, J.R., D.Z. EWTON & K.A. MAGRI. 1991. Hormones, growth factors, and myogenic differentiation. Annu. Rev. Physiol. **53:** 201–216.
11. WERB, Z., J.L. UNDERWOOD & D.A. RAPPOLEE. 1992. The role of macrophage-derived growth factors in tissue repair. *In* Mononuclear Phagocytes. R.V. FURTH, Ed.: 404–409. Kluwer Academic Publishers. Amsterdam.
12. CANTINI, M., M.L. MASSIMINO, A. BRUSON, C. CANTINI, L.D. LIBERA & U. CARRARO. 1994. Macrophages regulate proliferation and differentiation of satellite cells. Biochem. Biophys. Res. Commun. **202:** 1688–1696.
13. ROBERTSON, T.A., M.A.L. MALEY, M.D. GROUNDS & J.M. PAPADIMITRIOU. 1993. The role of macrophages in skeletal muscle regeneration with particular reference to chemotaxis. Exp. Cell Res. **207:** 321–331.
14. POSTLETHWAITE, A.E. & A.H. KANG. 1992. Fibroblasts and matrix proteins. *In* Inflammation: Basic Principles and Clinical Correlates. J.I. GALLIN, I.M. GOLDSTEIN & R. SNYDERMAN, Ed.: 747–773. Raven Press, Ltd. New York.
15. NOBLE, P.W., F.R. LAKE, P.M. HENSON & D.W.H. RICHES. 1993. Hyaluronate activation of CD44 induces insulin-like growth factor-1 expression by a tumor necrosis factor-α-dependent mechanism in murine macrophages. J. Clin. Invest. **91:** 2368–2377.
16. MINSHALL, C., S. ARKINS, J. STRAZA, J. CONNERS, R. DANTZER, G.G. FREUND & K.W. KELLEY. 1997. IL-4 and insulin-like growth factor-I inhibit the decline in Bcl-2 and promote the survival of IL-3-deprived myeloid progenitors. J. Immunol. **159:** 1225–1232.

17. MORROW, N.G., W.E. KRAUS, J.W. MOORE, R.S. WILLIAMS & J.L. SWAIN. 1990. Increased expression of fibroblast growth factors in a rabbit skeletal muscle model of exercise conditioning. J. Clin. Invest. **85:** 1816–1820.
18. LINDHOLM, D., R. HEUMANN, M. MEYER & H. THOENEN. 1987. Interleukin-1 regulates synthesis of nerve growth factor in non-neuronal cells of rat sciatic nerve. Nature **330:** 658–943.
19. DANILOFF, J.K., G. LEVI, M. GRUMET, F. RIEGER & G.M. EDELMAN. 1986. Altered expression of neuronal cell adhesion molecules induced by nerve injury and repair. J. Cell. Biol. **103:** 929–945.
20. ESPAT, N.J., T. AUFFENBERG, J.J. ROSENBERG, M. ROGY, D. MARTIN, C.H. FANG, P.O. HASSELGREN, E.M. COPELAND & L.L. MOLDAWER. 1996. Ciliary neurotrophic factor is catabolic and shares with IL-6 the capacity to induce an acute phase response. Am. J. Physiol. **271:** R185–R190.
21. FISCHER, J.E. & P.-O. HASSELGREN. 1991. Cytokines and glucocorticoids in the regulation of the hepato-skeletal muscle axis in sepsis. Am. J. Surg. **161:** 266–271.
22. ROUBENOFF, R., R.A. ROUBENOFF, J.G. CANNON, J.J. KEHAYIAS, H. ZHUANG, B. DAWSON-HUGHES, C.A. DINARELLO & I.R. ROSENBERG. 1994. Rheumatoid cachexia: Cytokine-driven hypermetabolism accompanying reduced body cell mass in chronic inflammation. J. Clin. Invest. **93:** 2379–2386.
23. COONEY, R., E. OWENS, C. JURASINSKI, K. GRAY, J. VANNICE & T. VARY. 1994. Interleukin-1 receptor antagonist prevents sepsis-induced inhibition of protein synthesis. Am. J. Physiol. **267:** E636–E641.
24. QUINN, L.S., K.L. HAUGK & K.H. GRABSTEIN. 1995. Interleukin-15: A novel anabolic cytokine for skeletal muscle. Endocrinology **136:** 3669–3672.
25. MCPHERRON, A.C., A.M. LAWLER & S.-J. LEE. 1997. Regulation of skeletal muscle mass in mice by a new TGFβ superfamily member. Nature **387:** 83–90.
26. ANTONACI, S., E. JIRILLO, M.T. VENTURA, A.R. GAROFALO & L. BONOMO. 1984. Non-specific immunity in aging: Deficiency of monocyte and polymorphonuclear cell-mediated functions. Mech. Ageing Dev. **24:** 367–375.
27. MCLACHLAN, J.A., C.D. SERKIN, K.M. MORREY & O. BAKOUCHE. 1995. Antitumoral properties of aged human monocytes. J. Immunol. **154:** 832–843.
28. LIAO, Z., J.H. TU, C.B. SMALL, S.M. SCHNIPPER & D.L. ROSENSTREICH. 1993. Increased urine interleukin-1 levels in aging. Gerontology **39:** 19–27.
29. DAYNES, R.A., B.A. ARANEO, W.B. ERSHLER, C. MALONEY, G.-Z. LI & S.-Y. RYU. 1993. Altered regulation of IL-6 production with normal aging. J. Immunol. **150:** 5219–5230.
30. FAGIOLO, U., A. COSSARIZZA, E. SCALA, E. FANALES-BELASIO, C. ORTOLANI, E. COZZI, D. MONTI, C. FANCESCHI & R. PAGANELLI. 1993. Increased cytokine production in mononuclear cells of healthy elderly people. Eur. J. Immunol. **23:** 2375–2378.

Age-associated Changes in the Response of Skeletal Muscle Cells to Exercise and Regeneration[a]

MIRANDA D. GROUNDS[b]

Department of Anatomy and Human Biology, The University of Western Australia, Nedlands 6907, Western Australia

ABSTRACT: This paper looks at the effects of aging on the response of skeletal muscle to exercise from the perspective of the behavior of muscle precursor cells (widely termed satellite cells or myoblasts) and regeneration. The paper starts by outlining the ways in which skeletal muscle can respond to damage resulting from exercise or other trauma. The age-related changes within skeletal muscle tissue and the host environment that may affect the proliferation and fusion of myoblasts in response to injury in old animals are explored. Finally, *in vivo* and *in vitro* data concerning the wide range of signaling molecules that stimulate satellite cells and other aspects of regeneration are discussed with respect to aging. Emphasis is placed on the important role of the host environment, inflammatory cells, growth factors and their receptors (particularly for FGF-2), and the extracellular matrix.

EXERCISE AND REGENERATION

There are at least three possible cellular responses that can occur in muscles subjected to exercise as outlined below.

Low-level "Sublethal" Damage Insufficient to Provoke Regeneration

Specific forms of exercise, in particular lengthening contractions (eccentric muscle actions, *e.g.*, those that occur when descending any incline), can result in disruption of the myofibrillar structure, especially that of the Z-bands,[1] and also in minor membrane damage, resulting in "leakiness" of the sarcolemma. It seems that such minor or sublethal injury to myofibers[2] can be repaired locally by rapid restoration of the wounded sarcolemnal membrane[3] so that cellular breakdown is limited and focal necrosis does not occur.[1,4] The extent to which such sublethal damage occurs after exercise is unknown.

Minor damage does occur in response to modifications in muscle loading causing alterations to myofibrils and nascent sarcomere formation in myotendinous regions associated with an accumulation of macrophages.[5] The macrophages may be specifically attracted to factors and then secrete other factors that assist in remodeling in this area, and yet this inflammatory activity at the myotendinous junction occurs without any associated myofiber necrosis and regeneration.[5] In such situations of sublethal damage there should be no need for satellite cells (myoblasts) to undergo replication. However, Darr and Shultz[4] sug-

[a]Research support over many years from the National Health and Medical Research Council of Australia is gratefully acknowledged.
[b]Tel: 618 9380 3486; fax: 618 9380 1051; e-mail: mgrounds@anhb.uwa.edu.au

gest that satellite cells may become activated and replicate even on fibers where overt necrosis is not detectable at the light microscopic level, although it is not known whether these proliferating myoblasts actually fuse under conditions of minor damage (see also discussion below under *Hyperlasia versus Hypertrophy*). It is worth noting that, in denervated muscle, although there is an increase in satellite cell proliferation above the basal rate, these labeled myoblasts do not fuse but instead subsequently "disappear" from the muscle, either by emigration or cell death.[6,7]

Necrosis and Regeneration

Where muscle damage is more severe, this will precipitate an influx of calcium ions that results in focal necrosis of myofibers.[2,8–10] Such necrosis undoubtedly occurs after intense or unaccustomed physical exercise.[10,11] In this situation, the damaged area of the myofiber is rapidly sealed off from the remainder of the myofiber by new sarcolemmal formation, observed ultrastructurally at 12 hours after injury[12] and demonstrated at 8 hours by the exclusion of horseradish peroxidase [9] There is an associated rapid influx of inflammatory cells, followed by satellite cell proliferation and fusion to repair the damaged segment of the myofiber. This is a classical regeneration response.[13,14] In a quantitative study of muscle injury after exercise, a low level of regeneration was also reported in normal unexercised control adult rat muscle.[11] Although there is good evidence that old muscle is more susceptible than young and adult muscle to injury after exercise,[15,16] this is not addressed in the present paper.

Hyperplasia versus Hypertrophy

The extent to which particular kinds of exercise result in hyperplasia (an increase in the number of muscle nuclei due to satellite cell proliferation and fusion) or in hypertrophy (the classical increase in muscle fiber size due to new protein synthesis) is widely debated. Hyperplasia could result from the fusion of satellite cells with stretched, hypertrophying myofibers,[17] similar to the situation seen with growing muscles during development, in order to maintain some optimal nuclear/cytoplasmic ratio as the fiber increases in size. Although there is very strong evidence of satellite cell proliferation in rat muscle hypertrophying in response to overloading,[18–20] fusion with the parent myofiber is not always confirmed, and there is also evidence that the satellite cells can form new myofibers (see below).[21] The extent to which such satellite cell proliferation and fusion with hypertrophying parent myofibers does occur within exercised adult muscles is unclear. Much data support the idea that exercise training results in injury that is sufficient to provoke a regenerative response.[11,22]

Hyperplasia traditionally arises from regeneration (in response to necrosis as outlined above) where satellite cells proliferate and fuse to repair segments of damaged myofibers; in some instances this can result in split or branched myofibers that have the appearance of new myofibers although they are actually continuous with a parent myofiber.[23,24] There is evidence that genuine new myofibers may be formed *de novo* in interstitial connective tissue between the existing myofibers, from experiments with chicken muscle (hypertrophying in response to weights attached to a wing)[17,25] and young rat muscle (hypertrophying in response to ablation of synergistic muscles).[21] However, the discontinuity of such nascent myofibers can only be proved by serially sectioning the tissue. It is not known if such new myofiber formation occurs in response to exercise. These three situations of hyperplasia (fusion with hypertrophying myofibers, muscle regeneration, and new myofiber formation) require the activation of quiescent satellite cells and their resultant proliferation and fusion. What are the signals that stimulate these events? Are different signals required for these three different situations? Is the availability of the signals or the response of the myogenic cells to these signals affected by age? Before attempting to address some of these questions, the age-related changes within skeletal muscle and the host environment will be examined, because these factors may well influence the behavior of the muscle cells.

CHANGES WITH AGE

Age-related Changes within Skeletal Muscle Tissue

Innervation

Old age is associated with a progressive loss of muscle mass due to atrophy of individual myofibers, as a result of denervation combined with a reduction in the number of myofibers.[26,27] Old age is also associated with a decrease in force and power due to a loss and change in contractile properties of the motor units in muscle.[16,28] Because the early events of regeneration (up until fusion) are unaffected by innervation,[6,27,29] these age-related changes should have little impact on the capacity for muscle repair after exercise-induced injury. However, other age-related changes within skeletal muscle will have an impact on the regenerative response.

Extracellular matrix

Extracellular matrix (ECM) in skeletal muscle includes both interstitial connective tissue and the external (basal) lamina, which is in intimate contact with satellite cells and myofibers.[30] A general increase in interstitial fibrous connective tissue is associated with aging: the amount of endomysial collagen doubles between 3 and 26 weeks of age in mice,[31] and it is well documented that increasing fibrosis occurs in regenerating muscles of older animals.[32,33] An increase in fibrous connective tissue and "rigidity" will also affect the response of the muscle to exercise. An age-related increase in fibrosis and fibroblastic activity may account for the increased myofiber branching seen in regenerating muscles of old rats.[24,34] There might also be associated age-related changes in fibroblast-derived soluble growth factors that have been shown to play a paracrine role in myoblast proliferation.[35] Apart from the interstitial connective tissue, an increase in the external lamina encircling satellite cells has been reported with age.[36,37] In Duchenne's muscular dystrophy and the animal dystrophies there is a marked increase in extracellular collagen and altered forms of collagen with time[31](see also ref. 38). Age-related changes in the amount and composition of the ECM components, particularly of the external lamina, could adversely affect the efficiency of muscle regeneration as discussed later under THE SIGNALS.

Vasculature

Other age-related changes within skeletal muscle relate to the vasculature,[39] with a reduced blood supply,[40] decreased capillary density,[41] and changes in vascular pathology[42] being reported in older subjects. Vascularity and revascularization are affected by many factors, including exercise,[43] and exercise is usually decreased in older subjects. Rapid revascularization can be a major factor in efficient muscle repair, particularly after large injuries, and a decrease in vascularity could have an adverse effect on muscle regeneration, as it could reduce the effciency of inflammatory cell infiltration, the importance of which is discussed later.

Age-related Changes in the Systemic Host Environment

In addition to such local changes within the muscle tissue itself, there are systemic changes in the complex endocrine system with age.[33,44,45] Reduced serum levels of growth hormone and insulin-like growth factor-I (IGF-I) in old humans and rats[33] may have a direct effect on the proliferation and fusion of satellite cells. Other changes in blood-borne factors influence the immune system, and this is of critical importance due to the central role of inflammatory cells during muscle regeneration.[46] Although it was reported that macrophage function and hence muscle regeneration is severely impaired in old compared with young SJL/J mice,[47] this dramatic effect of host age on macrophage function does not occur in other strains of mice[44] and is clearly a function of the hormonal status, because it is seen only in old male (but not female) SJL/J host mice and is ablated by orchidectomy.[44] Thus age-related changes in hormonal status can impact on the efficiency of the inflammatory cell response of the host. Delayed macrophage infiltration and associated regeneration after muscle injury was also reported in old (24 month) relative to young rats.[34] Studies in humans show that macrophage activity declines[48] and the extent of mobilization of polymorphonuclear leukocytes (PML) is decreased in older subjects;[49] it is also widely recognized that there is an age-associated decline in T-cell mediated immune parameters[50] and that this can also be affected by certain types of exercise.[35,40,51] These influences of the host environment would seem to be of considerable importance in determining the efficiency of muscle regeneration in old animals as discussed below.

Age-related Decrease in Muscle Regeneration

There is clear evidence that muscle can regenerate well in old hosts. However, muscle regeneration in old hosts is generally less successful than in young hosts with respect to both morphological[34] and functional properties[15] (reviewed by Carlson,[27]) although these differences can be subtle in certain situations.[44,52] Although a similar capacity for myoblast proliferation was demonstrated in autoradiographic studies in muscles regenerating after crush injury in old (40 week) compared with young (4 week) host mice, myoblast replication was retarded in the old hosts.[53] Classical cross-transplantation studies of whole extensor digitorum longus muscles between old (24 month) and young (4 month) rats, and examined at 60 days, showed that it was the age of the host (rather than the muscle graft) that determined the success of regeneration in muscles examined at 60 days.[32] The age-related success in these long-term grafts was attributed to the capacity for axonal regeneration and hence functional reinnervation of the graft.[27] However, experiments in our laboratory suggest that the status of inflammatory cells is another crucial variable in the host environment that influences the success of muscle regeneration.[44,54] When whole muscle grafts were cross-transplanted between two strains of mice with strikingly different regenerative capacity (SJL/J have superior regeneration compared with BALB/c mice[55]) the pattern of regeneration reflected the strain of the host, again showing that the host environment (rather than the muscle itself) can determine the efficiency of muscle regeneration.[54] Because this is not accounted for by genetic differences between the bone marrow–derived cells from the two strains,[56] it seems most likely that some factor (possibly blood borne) in the SJL/J host mice affects leukocytes so that they are in a "more activated" state. Earlier experiments with minced muscle autografts in young and old mice showed that impaired macrophage function in old (compared with young) hosts prevented the removal of necrotic tissue and hence new muscle formation, and this inflammatory cell defect was clearly linked to the hormonal status of the host.[44] With respect to the related situation in old and young hosts, it seems likely that inflammatory cells might generally be "less active" in the old host environment, compared with young animals. This idea is supported by strong evidence that an age-related decline in macrophage activity contributes to the slower healing of wounds in old mice.[48] There are a wide range of potential factors that might account for the variation in the "state of activation" of circulating leukocytes: these include hormones and cytokines and the capacity of the cells to respond to them.[14,44,45,50] Before discussing these signals it is pertinent to review the question of whether the number and proliferative potential of satellite cells show any decrease with age.

Age-related Decline in Satellite Cells?

Although the relative and absolute proportions of satellite cells to muscle nuclei are affected by innervation[57] and decrease from birth to maturity in rodents, there is little further decrease between muscles of adult and old animals (see ref. 53). The question then arises as to whether the satellite cells lose their capacity to proliferate in old animals? The answer to this question relies in part upon knowing the extent of satellite proliferation in normal uninjured adult and aging muscle. Measurements of the proliferative capacity of human satellite cells indicate that the population of satellite cells undergoes considerable proliferation during the first two decades of life when muscles are growing, but that after this time the population is constant with little or no replication into old age (86 years).[58] Thus it appears that a similar proliferative capacity might be expected for satellite cells from adult and old muscles in response to damage.

Information about the turnover of myonuclei would also indirectly provide information about the proliferation of satellite cells throughout the life of a myofiber. Until recently it was not possible to determine whether there was any turnover of myonuclei within undamaged adult muscle fibers, or if the same myonuclei persisted throughout the life of an individual. This can now be investigated by measuring the length of telomeres

(TTAGGG repeats located at the ends of eukaryotic chromosomes), which are known to decrease with proliferation and are used as an indicator of cellular aging.[59] It has been shown that 86 bp of telomeric DNA is lost with each round of human satellite cell replication in culture.[60] A comparison of telomere restriction fragment (TRF) lengths of myonuclei from young, adult, and old human muscle (9 months to 86 years) showed no significant decrease in the mean TRF length from birth until old age,[61] indicating a tremendous stability of these myonuclei over time, which, in turn, reinforces the idea that satellite cells are essentially quiescent and have minimal turnover in normal uninjured adult muscles. However, comparison of the minimal values of TRF identified a very small increase of 13 bp per year, showing that there is actually a very small turnover of muscle nuclei throughout the life of the myofibers.[61] These elegant studies therefore indicate that there must be some proliferation and fusion of satellite cells throughout life, albeit at an extremely low rate. The simplest explanation is that in adult muscle this occurs sporadically in response to hypertrophy or accidental muscle damage,[11] although it might possibly reflect an extremely low endogenous level of myonuclei turnover. By contrast, a dramatic decrease in mean TRF length is seen in muscle from patients with Duchenne's muscular dystrophy where the muscle is subjected to repeated cycles of necrosis and regeneration.[62]

Tissue culture studies confirm the *in vivo* observations that satellite cells from old muscle have the capacity to replicate[63–65] and that the rate of proliferation is not decreased with age.[66,67] However, tissue cultured muscle cells from old rats consistently have an increased "lag phase" before the onset of replication.[64,66,67] In conclusion, it appears that the number and proliferative capacity of satellite cells is not impaired in old (compared with adult) muscle, although the response of these cells may be slower for a range of reasons.

It has often been proposed that the replicative potential of satellite cells may be exhausted by repeated cycles of regeneration in diseases such as muscular dystrophy,[65,68,69] and a similar situation might arise theoretically after repeated bouts of extreme exercise over a lifetime.[4] Although *in vivo* studies indicate that dystrophic muscles retain the capacity to regenerate after experimental injury,[70,71] tissue culture studies generally support the idea of a loss of proliferative potential of satellite cells from old dystrophic muscles,[68,69](for more refs., see refs. 38 and 62). Some caution must be exercised in the interpretation of tissue culture studies of cells from old and young muscles, as there may be difficulties in extracting satellite cells from muscles of different ages and pathologies;[38] for example, it has been estimated that less than 0.01% of myogenic cells are normally extracted from adult muscle under standard procedures.[72] Furthermore the environment from which the cells are extracted may affect their ability to respond under standard tissue culture conditions optimized for growth of myogenic cells from young or adult muscle.[38,73]

Genetics

The influence of genetics adds another layer of complexity to the effects of aging. For example, new muscle formation in minced muscle grafts is far more effective in SJL/J than in BALB/c mice,[44] and, in crush-injured muscles, superior regeneration is seen in SJL/J mice and is associated with twice the number of inflammatory cells at three days.[55,74] Inherent differences in myogenicity between these two strains is also demonstrated in tissue culture, where myoblasts from SJL/J mice show an earlier onset of expression of MyoD and myogenin,[75] larger and more frequent myotubes, and a lower dependency on the ECM substrate[76] in comparison with BALB/c myoblasts. So genetics are yet another factor that must be taken into consideration from the perspective of both the host and muscle-related factors.

THE SIGNALS

Vast numbers of growth factors (*e.g.*, fibroblast growth factors (FGF), platelet-derived growth factor-BB (PDGF-BB), IGF, transforming growth factor-β (TGF-β)) have been shown to affect the proliferation and fusion of myoblasts under tissue culture conditions (reviewed in refs. 14, 38, 45, 77–79) and the situation for FGF-2 (widely referred to as basic FGF) is discussed in detail below. Some ECM molecules also have a direct effect on the movement, attachment, proliferation, and fusion of myoblasts[14,38,78]: these include the laminins that are associated with the external lamina,[27,35,76,80,81] specific proteoglycans that are essential for the binding of growth factors like FGF to their receptors (see below), and proteolytic fragments of fibronectin and laminin, which are important chemotactic signals.[82] Recently, there has been additional interest in such factors as hepatocyte growth factor that appears to be an early mitogen for myoblasts,[77, 83] cytokines like interferon-α,[84] interleukin-6,[45,85,86] and leukemia inhibitory factor (LIF).[86,87] A factor produced by crush-injured muscle is of considerable interest, because it has been shown to be a specific and potent mitogen for myoblasts,[67,79,88,89] and recent evidence indicates that this active factor may be hepatocyte growth factor.[83] Bischoff[88] concluded that although this factor can activate satellite cells, a serum factor is required for cells to move through the cell cycle and replicate.

In exercised muscle, a great deal of attention has been focused on IGF-I through its effects on increasing protein (sarcomere) formation. An increase in IGF-I is seen in muscle after acute eccentric exercise,[90] in compensatory hypertrophy,[19,35,56] and both IGF-I and another isoform IGF-Ieb increase in response to stretching within two hours.[91] Local production of IGF-I undoubtedly stimulates the growth of postnatal muscle and an increase in muscle mass.[91] Although mRNA for IGF-I is seen in myoblasts and myotubes *in vivo* in injured muscle, and the pattern corresponds closely to that for myoblast proliferation,[19,92] it is not clear what role IGF-I actually plays as a mitogen during muscle regeneration in response to exercise, compared with its effects on differentiation and protein production.[93,94]

Studies comparing the response to mitogens of primary muscle cultures from old and young rodents consistently show some decrease in the response of old satellite cells. Mezzogiorno and colleagues[73] investigated the response of old (26 month) mouse muscles to a range of growth factors (FGF, PDGF-BB, IGF-II, ACTH, and LIF) and concluded that there was a generalized reduction in the response to all mitogens tested. Of particular interest was the demonstration that the production of paracrine factors was very different between old and young muscle cells, and they proposed that this led to differences in the local environment *in vivo* that probably played a major role in the response of young compared to old muscle cells.[73] Age-related differences were also seen among cultured satellite cells from 3- , 12- , and 24-month-old rats, with respect to the number and affinity of receptors for IGF-II associated with a delayed onset of proliferation in old cells (although the proliferation rates were similar).[66] A delayed response was similarly seen to mitogens from crushed muscle and to FGF-2.[67] The delayed response to FGF-2 with aging satellite cells[67,95] is discussed in more detail below.

All of these growth factors and ECM molecules must interact in the complex *in vivo* environment. Many growth factors such as the FGF and TGF-β also stimulate angiogenesis. Many others, including proteolytic fragments of ECM molecules, stimulate the chemotaxis of inflammatory cells and myoblasts.[79,82] These growth factors and ECM molecules are produced by a wide variety of cells, including myoblasts, fibroblasts, endothelial cells, resident macrophages, dendritic cells,[96] and infiltrating leukocytes.[45,46] Of the infiltrating leukocytes, it is widely recognized that macrophages play a particularly important role in muscle regeneration.[46,96,97]

Role of Leukocytes

When muscle is damaged, PML accumulate very rapidly (within minutes) at the injury site; they predominate initially but are largely replaced by macrophages by 24 hours after crush injury.[14,16,89,99] Rapid evascularization of PML in response to chemokines produced by tissue damage has been widely studied and is a very important event in general tissue repair. Tissue culture studies of chemotaxis in muscle[98] and other tissues show that the PML produce soluble factors that chemoattract macrophages to the damage site (see ref. 82). However, the soluble factors produced by PML do not chemoattract myoblasts. Large numbers of platelets may also be present after severe trauma, and they produce many factors (*e.g.*, PDGFs) that facilitate wound repair.

Macrophages, which predominate during skeletal muscle regeneration, are essential for the effective removal of necrotic tissue and produce a vast array of growth factors and enzymes that influence many aspects of the regenerative process, including angiogenesis; the ECM environment; and the chemotaxis, proliferation, and differentiation of myoblasts.[14,79,82,98] There is also evidence that damaged myofibers themselves (in the absence of circulating leukocytes) produce chemotactic signals that attract both macrophages and myoblasts to the injury site.[82]

Thus, in order to evaluate the real importance of various growth factors and ECM molecules during myogenesis, it is essential to assess the effects of them in the complex *in vivo* environment. There have been remarkably few instances of such *in vivo* studies in postnatal regenerating skeletal muscles.

Fibroblast Growth Factors

It is well documented from tissue culture studies that FGF-2 (previously known as bFGF) is one of the most potent mitogens for myoblasts, and it would appear to play a critical role during myogenesis in developing muscles.[100] On the basis of these data, and the correlation *in vivo* between immunohistochemical studies showing high FGF-2 expression in situations of good muscle regeneration,[74,101] it was considered that exogenous administration of FGF-2 might enhance new muscle formation, particularly in BALB/c mice where regeneration is usually poor.[55,74] However, FGF-2 administered *in vivo* by various regimes (by injection ± heparin, in hydron or elvax implants) to experimentally injured, denervated, or dystrophic muscle had no effect on myoblast proliferation or the overall histological appearance.[102] The failure of exogenous FGF-2 to enhance the regenerative response indicates that availability of FGF-2 may not normally be the limiting factor *in vivo;* instead, the cellular responses may be determined by the expression of specific FGF-2 receptors and associated heparan sulphate proteoglycans that regulate the binding of FGF-2.[103]

It is now recognized that proteoglycans sequester heparin-binding growth factors close to cell surfaces and are able to protect them from proteolytic degradation. They are an essential prerequisite for the binding of such factors to their high-affinity cell surface and signal-transducing receptors. Integral membrane species of heparan sulphate molecules, which regulate FGF activity, are known as syndecans.[104] Of particular interest to muscle repair after injury is the report that cellular infiltrates in wounds release a peptide that induces mammalian cells to express syndecans as part of the repair process.[105] If such cellular infiltrates are reduced in old hosts, this could affect the production of syndecans, the speed of response to FGF-2, and hence the onset of satellite cell proliferation and regeneration.

Tissue culture studies on satellite cells show binding of FGF-2 at 18 h (the earliest time examined) and at 42 h postplating in primary cultures from 4-week- and 9-month-old rats, respectively.[95] This correlates with the delayed entry into the cell cycle and the delayed response to FGF-2 seen in satellite cells from old rats.[67] The results suggest that expression of functional FGF receptors on satellite cells may represent an important step in the activation of quiescent satellite cells. An earlier study, which examined the response of young and old muscle in tissue culture to FGF-2, reported no differences in the pattern of myoblast proliferation.[73] However, this study did not look at the onset of the response, and this is probably a critical factor *in vivo.* Although other factors have been tested and no age-related differences have been observed (reviewed by ref. 45), it is probable that the precise timing of the onset of myoblast activation and replication was not the focus of these studies. Unfortunately, in tissue culture studies it is not possible to study cells prior to about 12 h postplating, as the cells have not fully attached; thus observations on quiescent (time 0) satellite cells and the early phases of activation are not possible.

Other in Vivo *Studies*

Earlier studies with daily intramuscular injections of the synthetic corticosteroid dexamethasone (1 µg/Kg to 100 µg/Kg) showed no improvement in muscle regeneration after crush injury in BALB/c mice,[106] although tissue culture studies report a stimulation of myoblast proliferation at these low doses. The effect of the cytokine, interferon-α (IFN-α) was also studied *in vivo,* as interferons are well-known regulators of cellular events and there is conflicting evidence regarding the effects on stimulating myoblast proliferation and fusion in culture: daily intramuscular injections of IFN-α (2.25×10^3 IU/dose) showed impaired regeneration in SJL/J mice with persisting necrotic tissue, reduced myotube formation, and increased fibrosis at 10 days after crush injury.[84] In contrast with these studies, *in vivo* administration of LIF is reported to enhance muscle regeneration.[87,107] The addition of extra macrophages also improves muscle regeneration *in vivo,*[108] as does the addition of extract from crushed muscles,[109] supporting the idea that regeneration can be assisted by the exogenous administration of various factors. Anabolic effects of exogenous IGF-I have been demonstrated in dystrophic muscles, although this is probably due largely to a reduction in protein degradation.[93] A further example of exogenous administration of a factor is studies with the thyroid hormone, triiodothyronine, which increased the severity of the dystrophy particularly in younger mdx mice,[110] probably due to metabolic effects and a modulation of myosin synthesis. Mouse strains with inherited defects in genes for specific growth factors and ECM molecules, and engineered "null mutant mice" that lack selected genes, provide many ready opportunities to assess the importance of such factors on exercise, regeneration, and aging *in vivo*.

CONCLUSIONS

Older muscle generally has a very good capacity for myoblast proliferation and fusion, and hence new muscle formation, although this is slightly less efficient than in younger hosts. It seems likely that optimal cytokine and hormonal production declines with age (and this is also affected by exercise, diet, immune status, and genetics), that such systemic blood-borne factors in the host environment are particularly critical for determining the efficiency of muscle repair in old animals, and that this may be mainly by an effect on the immune response. This is good news. If the factors involved can be identified, this should enable systemic manipulation of the host, possibly by administration of exogenous factors, to enhance muscle repair in old subjects. (The problem of effective reinnervation of regenerated muscle in old hosts is another issue). Clearly other factors intrinsic to the skeletal muscle itself, including changes in the ECM, vascularity, and the expression of growth factors and particularly their receptors by satellite cells, can also contribute to the less efficient regeneration generally seen in old hosts. However, these intrinsic parameters are less readily manipulated.

ACKNOWLEDGMENTS

I thank Peter Hamer for his critical reading of the manuscript and Marilyn Davies for her generous assistance.

REFERENCES

1. FRIDEN, F., M. SJOSTROM & B. EKBLOM. 1983. Myofibrillar damage following intense eccentric exercise in man. Int. J. Sports Med. **4:** 170–176.
2. CULLEN, M.J., S.T. APPLEYARD & L. BINDOFF. 1979. Morphologic aspects of muscle breakdown and lysosomal activation. Ann. N. Y. Acad. Sci. **317:** 440–464.
3. MCNEIL, P.L. & R. KHARKEE. 1992. Disruption of muscle fiber plasma membrane: Role in exercise-induced damage. Am. J. Anat. **140:** 1097–1109.
4. DARR, K.C. & E. SCHULTZ. 1987. Exercise-induced satellite cell activation in growing and mature skeletal muscle. J. Appl. Physiol. **63:** 1816–1821.
5. ST PIERRE, B.A. & J.G. TIDBALL. 1994. Macrophage activation and remodeling at myotendinous junctions after modifications in muscle loading. Am. J. Pathol. **145:** 1463–1471.
6. MCGEACHIE, J.K. & M.D. GROUNDS. 1989. The onset of myogenesis in denervated mouse skeletal muscle regenerating after injury. Neuroscience. **28:** 509–514.

7. VIGUIE, C.A. & B.M. CARLSON. 1994. Nuclear numbers in long-term denervated rat EDL muscle fibers. FASEB J. **8:** A60.

8. CULLEN, M.J. & J.J. FULTHORPE. 1975. Stages in fiber breakdown in Duchenne Muscular Dystrophy: An electron microscope study. J. Neurol. Sci. **24:** 179–200.

9. GROUNDS, M.D. *et al.* 1992. Necrosis and regeneration in dystrophic and normal skeletal muscles. *In* Duchenne Muscular Dystrophy: Animal Models and Genetic Manipulation. B.A. Kakulas, J. Mc Howell,& A Roses, Eds.: Chapter 12:141–153 Raven Press. New York.

10. ARMSTRONG, R.B. 1990. Initial events in exercise-induced muscle injury. Med. Sci. Sports Exercise. **22:** 429–435.

11. SMITH, H.K. *et al.* 1997. Skeletal muscle damage and repair in the rat hindlimb following single or repeated daily bouts of downhill exercise. Int. J. Sports Med. **18:** 94–100.

12. PAPADIMITRIOU, J.M. *et al.* 1990. The process of new plasmalemma formation in focally injured skeletal muscle fibers. J. Struct. Biol. **103:** 124–134.

13. MAZANET, R. & C. FRANZINI-ARMSTRONG 1986. The satellite cell. *In* Myology. A.G. Engel & B.Q. Banker, Eds.: 285–307. McGraw-Hill. New York.

14. GROUNDS, M.D. 1991. Towards understanding skeletal muscle regeneration. Pathol. Res. Pract. **187:** 1–22.

15. BROOKS, S.V. & J.A. FAULKNER. 1996. The magnitude of the initial injury induced by stretches of maximally activated muscle fibers of mice and rats increases with old age. J. Physiol. **497:** 573–580.

16. FAULKNER, J.A., S.V. BROOKS & E. ZERBA. 1995. Muscle atrophy and weakness with aging: Contraction-induced injury as an underlying mechanism. J. Gerontol. **50A:** 124–129.

17. KENNEDY, J.M., L.J. SWEENEY & L.Z. GAO. 1989. Ventricular myosin expression in developing and regenerating muscle, cultured myoblasts, and nascent myofibers of overloaded muscle in the chicken. Med. Sci. Sports Exercise **21:** S187–S197.

18. SCHIAFFINO, S., S.P. BORMIIOLLI & M. ALOISI. 1972. Cell proliferation in rat skeletal muscle during early stages of compensatory hypertrophy. Virchows Arch. Abt. B Zellpathol. **11:** 268–273.

19. ADAMS, G.R. & F. HADDAD. 1996. The relationships among IGF-1, DNA content and protein accumulation during skeletal muscle hypertrophy. J. Appl. Physiol. **81:** 2509–2516.

20. ROSENBLATT, J.D., D. YONG & D.J. PARRY. 1994. Satellite cell activity is required for hypertrophy of overloaded rat muscle. Muscle & Nerve **17:** 608–613.

21. SALLEO, A. *et al.* 1980. New muscle fiber production during compensatory hypertrophy. Med. Sci. Sports Exercise. **12:** 268–273.

22. MCCORMICK, K.M. & D.P. THOMAS. 1992. Exercise-induced satellite cell activation in senescent soleus muscle. J. Appl. Physiol. **72:** 888–893.

23. ONTELL, M. & K.C. FENG. 1981. The three-dimensional architecture and pattern of motor innervation of branched striated myotubes. Anat. Rec. **200:** 11–31.

24. BALAIVAS, M. & B.M. CARLSON. 1991. Muscle fiber branching: Differences between grafts in young and old rats. Mech. Ageing Dev. **60:** 43–53.

25. MCCORMICK, K.M. & E. SCHULTZ. 1992. Mechanism of nascent fiber formation during avian skeletal muscle hypertrophy. Dev. Biol. **150:** 319–334.

26. CARTEE, G.D. 1995. What insights into age-related changes in skeletal muscle are provided by animal models? J. Gerontol. **50A:** 137–141.

27. CARLSON, B.M. 1995. Factors influencing the repair and adaption of muscles in aged individuals: Satellite cells and innervation. J. Gerontol. **50A:** 96V100.

28. KADHIRESAN, V.A., C.A. HASSETT & J.A. FAULKNER. 1996. Properties of single motor units in medial gastrocnemius muscles of adult and old rats. J. Physiol. **493:** 543–552.

29. SCHMALBRUCH, H. & D.M. LEWIS. 1994. A comparison of the morphology of denervated with aneurally regenerated soleus muscle of rat. J. Muscle Res. Cell Motil. **15:** 256–266.

30. SANES, J.R. 1986. The extracellular matrix. *In* Myology. A.G. Engel & B.Q. Banker, Eds.: 155–175. McGraw-Hill. New York.

31. MARSHALL, P.A., W.P.E. & G. GOLDSPINK. 1989. Accumulation of collagen and altered fiber-type ratios as indicators of abnormal muscle gene expression in the mdx dystrophic mouse. Muscle & Nerve **12:** 528–537.

32. CARLSON, B.M. & J.A. FAULKNER. 1989. Muscle transplantation between young and old rats: Age of host determines recovery. Am. J. Physiol. **256:** 1262–1266.

33. ULLMAN, M. *et al*. 1990. Effects of growth hormones on muscle regeneration and IGF-1 concentration in old rats. Acta Physiol. Scand. **140:** 521–525.

34. SADEH, M. 1988. Effects of aging on skeletal muscle regeneration. J. Neurol. Sci. **87:** 67-74.

35. QUINN, L.S., L.D. ONG & R.A. ROEDER. 1990. Paracrine control of myoblast proliferation and differentiation by fibroblasts. Dev. Biol. **140:** 8–19.

36. SNOW, M. H. 1977. The effects of aging on satellite cells in skeletal muscles of mice and rats. Cell Tissue Res. **185:** 399 408.

37. FRANZINI-ARMSTRONG, C. 1979. Satellite and invasive cells in frog sartorius. *In* Muscle Regeneration. A. Mauro *et al*, Eds.: 133–138. Raven Press. New York.

38. GROUNDS, M.D. & Z. YABLONKA-REUVENI. 1993. Molecular and cell biology of skeletal muscle regeneration. *In* Molecular and Cell Biology of Muscular Dystrophy. T.A. Partridge, Ed.: **4:** 210–256. Chapman & Hall. London.

39. JERUSALEM, F. 1986. The microcirculation of muscle. *In* Myology. A.G. Engel & B.Q. Banker, Eds.: 343–356. McGraw-Hill. New York.

40. MCCULLY, K.K. & J.E. POSNER. 1995. The application of blood flow measurements to the study of aging muscle. J. Gerontol. **50A:** 130–136.

41. COGGAN, A.R. *et al*. 1992. Histochemical and enzymatic comparison of the gastrocnemius of young and elderly men and women. J. Gerontol. Biol. Sci. **47:** B71–B76.

42. COOPER, L.T., J.P. COOKE & V. J. DZAU. 1995. The vascular pathology of aging. J. Gerontol. Biol. Sci. **49:** B191–B195.

43. ROBERTS, P. & J.K. MCGEACHIE. 1992. The influence of revascularization, vasoactive drugs and exercise on the regeneration of skeletal muscle, with particular reference to muscle transplantation. Basic Appl. Myology 2: 5–16.

44. GROUNDS, M.D. 1987. Phagocytosis of necrotic muscle in muscle isografts is influenced by the strain, age and sex of host mice. J. Pathol. **153:** 71–82.

45. CANNON, J.G. 1995. Cytokines in aging and muscle homeostasis. J Gerontol. **50A:** 120–123.

46. TIDBALL, J.G. 1995. Inflammatory cell response to acute muscle injury. Med. Sci. Sports Exercise **27:** 1022–1023.

47. ZACKS, S.I. & M.F. SHEFF. 1982. Age-related impeded regeneration of mouse minced anterior tibial muscle. Muscle & Nerve **5:** 152–161.

48. DANON, D., M.A. KOWATCH & G.S. ROTH. 1989. Promotion of wound repair in old mice by local injection of macrophages. Proc. Natl. Acad. Sci. USA **86:** 2018–2020.

49. CANNON, J.G. *et al*. 1994. Aging and stress-induced changes in complement activation and neutrophil mobilization. J. Appl. Physiol. **76:** 2616–2620.

50. RALL, L.C. *et al*. 1996. Effects of progressive resistance training on immune response in aging and chronic inflammation. Med. Sci. Sports Exercise **28:** 1356–1365.

51. SHEPARD, R.J. & P.N. SHEK. 1994. Potential impact of physical activity and sport on the immune system—a brief review. Br. J. Sports Med. **28:** 247–255.

52. GUTMANN, E. & B.M. CARLSON. 1976. Regeneration and transplantation of muscles in old rats and between young and old rats. Life Sci. **18:** 109–114.

53. MCGEACHIE, J.K. & M.D. GROUNDS. 1995. Retarded myogenic cell replication in regenerating skeletal muscles of old mice: An autoradiographic study in young and old BALB/c and SJL/J mice. Cell Tissue Res. **280:** 277–282.

54. ROBERTS, P., J.K. MCGEACHIE & M.D. GROUNDS. 1997. The host environment determines strain specific differences in the timing of skeletal muscle regeneration: Cross-transplantation studies between SJL/J and BALB/c mice. J. Anat. **191:** 585–594.

55. MITCHELL, C.A., J.K. MCGEACHIE & M.D. GROUNDS. 1992. Cellular differences in the regeneration of murine skeletal muscle—a quantitative histological study in SJL/J and BALB/c mice. Cell Tissue Res. **269:** 159–166.

56. MITCHELL, C.A., J.M. PAPADIMITRIOU & M.D. GROUNDS. 1995. The genotype of bone-marrow derived inflammatory cells does not account for differences in skeletal muscle regeneration between SJL/J and BALB/c mice. Cell Tissue Res. **208:** 407–413.

57. SCHULTZ, E. 1984. A quantitative study of satellite cells in regenerated soleus and extensor digitorum longus muscles. Anat. Rec. **208:** 501–506.

58. MOULY, V. *et al*. 1997. Satellite cell proliferation: Starting points and key steps to regeneration and cell mediated therapy. Basic Appl. Myology **7.** In press.

59. SHAY, J.W., H. WERBIN & W.E. WRIGHT. 1994. Telomere shortening may contribute to aging and cancer: A perspective. Mol. Cell. Differ. **2**: 1–21.

60. DECARY, S., V. MOULY & G.S. BUTLER-BROWNE. 1996. Telomere length as a tool to monitor satellite cell amplification for cell-mediated gene therapy. Hum. Gene Ther. **7**: 1347-1350–1350.

61. DECARY, S. *et al.* 1997. Replicative potential and telomere length in human skeletal muscle: Implications for satellite cell mediated gene therapy. Hum. Gene Ther. **8**. In press.

62. BUTLER-BROWNE, G.S., S. DECARY & V. MOULY. 1997. Premature telomere shortening in diseased skeletal muscle. Keystone Symposium on Molecular Biology of Muscle Development **1-6**: 25.

63. ALLEN, R.E., P.K. MCALLISTER & K. C. MASAK. 1980. Myogenic potential of satellite cells in skeletal muscle of old rats. A brief note. Mech. Ageing Dev. **13**: 105–109.

64. SCHULTZ, E. & B.H. LIPTON. 1982. Skeletal muscle satellite cells: Changes in proliferative potential as a function of age. Mech. Ageing Dev. **20**: 377–383.

65. WRIGHT, W.E. 1985. Myoblast senescence in muscular dystrophy. Exp. Cell Res. **157**: 343–354.

66. DODSON, M.V. & R.E. ALLEN. 1987. Interaction of multiplication stimulating activity/rat insulin-like growth factor II with skeletal muscle satellite cells during aging. Mech. Ageing Dev. **39**: 121–128.

67. JOHNSON, S.E. & R.E. ALLEN. 1993. Proliferating cell nuclear antigen (PCNA) is expressed in activated rat skeletal muscle satellite cells. J. Cell. Physiol. **154**: 39–43.

68. BLAU, H.M., C. Webster & G.K. PAVLATH. 1983. Defective myoblasts identified in Duchenne muscular dystrophy. Proc. Natl. Acad. Sci. USA **80**: 4856–4860.

69. ONTELL, M.P. *et al.* 1992. Transient neonatal denervation alters the proliferative capacity of myosatellite cells in dystrophic (129REJdy/dy) muscle. J. Neurobiol. **23**: 407–419.

70. BOURKE, D.L., M. ONTELL & F. TAYLOR. 1988. Spontaneous regeneration of older dystrophic muscle does not reflect its regenerative capacity. Am. J. Anat. **181**: 1–11.

71. GROUNDS, M.D. & J.K. MCGEACHIE. 1992. Skeletal muscle regeneration after crush injury in dystrophic MDX mice: An autoradiographic study. Muscle & Nerve **15**: 580–586.

72. ROSENBLATT, J.D. *et al.* 1995. Culturing satellite cells from living single muscle fiber explants. *In Vitro* Cell Dev. Biol. **31**: 773–779.

73. MEZZOGIORNO, A. *et al.* 1993. Paracrine stimulation of senescent satellite cell proliferation by factors released by muscle or myotubes from young mice. Mech. Ageing Dev. **70**: 35–44.

74. ANDERSON, J.E. *et al.* 1995. The time course of bFGF expression in crush injured skeletal muscles of SJL/J and BALB/c mice. Exp. Cell Res. **216**: 325–334.

75. MALEY, M.A. L. *et al.* 1994. Intrinsic differences in MyoD and myogenin expression between primary cultures of SJL/J and BALB/c skeletal muscle. Exp. Cell Res. **211**: 99–107.

76. MALEY, M.A., M.J. DAVIES & M.D. GROUNDS. 1995. Extracellular matrix, growth factors, genetics: Their influence on cell proliferation and myotube formation in primary cultures of adult mouse skeletal muscle. Exp. Cell Res. **219**: 169–179.

77. ALLEN, R.E. *et al.* 1995. Hepatocyte growth factor activates quiescent skeletal muscle satellite cells *in vitro.* J. Cell Physiol. **165**: 307–312.

78. DODSON, M.V. *et al.* 1996. Extrinsic regulation of domestic animal-derived satellite cells. Domest. Anim. Endocrinol. **13**: 107–126.

79. BISCHOFF, R. 1997. Chemotaxis of skeletal muscle satellite cells. Dev. Dyn. **208**: 505–515.

80. SCHULER, F. & L.M. SOROKIN. 1995. Expression of laminin isoforms in mouse myogenic cells *in vitro* and *in vivo*. J. Cell Sci. **108**: 3795–3805.

81. GROUNDS, M.D. *et al.* 1998. The expression of extracellular matrix during adult skeletal muscle regeneration: How the basement membrane, interstitium, and myogenic cells collaborate. Basic Appl. Myology **8**(2): 129–142.

82. GROUNDS, M.D. & M.J. DAVIES. 1996. Chemotaxs in myogenesis. Basic Appl. Myology **6**: 469–483.

83. ALLEN, R.E., R. TATSUMI & S.M. SHEEHAN. 1997. HGF/SF is the activating factor in crushed muscle extract and is an autocrine growth factor for satellite cells. Keystone Symposium on Molecular Biology of Muscle Development **1–6**: p. 43.

84. GARRETT, K. *et al.* 1992. Interferon inhibits myogenesis *in vitro* and *in vivo*. Basic Appl. Myology **2**: 291–298.

85. CANTINI, M. & U. CARRARO. 1996. Control of cell proliferation by macrophage-myoblast inter-actions. Basic Appl. Myology **6:** 485–489.
86. AUSTIN, L. *et al.* 1992. The effects of leukaemia inhibitory factor and other cytokines on murine and human myoblast prolieration. J. Neurol. Sci. **112:** 185–191.
87. KUREK, J.B. *et al.* 1996. Leukaemia inhibitory factor (LIF) treatment stimulates muscle regen-eration in the mdx mouse. Neurosci. Lett. **212:** 167–170.
88. BISCHOFF, R. 1990. Control of satellite cell proliferation. *In* Myoblast Transfer Therapy. R. Griggs & G. Karpati, Eds.: **280:** 147–158. Plenium Press. New York.
89. HAUGK, K.L. *et al.* 1996. Crushed muscle extracts: A model system to investigate growth fac-tor regulation of satellite cell activities in meat animals. Basic Appl. Myology **6:** 163–173.
90. YAN, Z., R.B. BIGGS & F.W. BOOTH. 1993. Insulin-like growth factor immunoreactivity increases in muscle after acute eccentric contraction. J. Appl. Physiol. **74:** 410–414.
91. YANG, S. *et al.* 1996. Cloning and characterization of an IGF-1 isoform expressed in skeletal muscle subjected to stretch. J. Muscle Res. Cell Motil. **17:** 487–495.
92. EDWALL, D. *et al.* 1989. Induction of insulin-like growth factor 1 messenger ribonucleic acid during regeneration of rat skeletal muscle. Endocrinology **124:** 820–825.
93. ZDANOWICZ, M.D. *et al.* 1995. Effect of insulin-like growth factor 1 in murine muscular dystro-phy. Endocrinology **136:** 4880–4886.
94. MACGREGOR, J. & W.S. PARKHOUSE. 1996. The potential role of insulin-like growth factors in skeletal muscle regeneration. Can. J. Appl. Physiol. **21:** 236–250.
95. JOHNSON, S.E. & R.E. ALLEN. 1995. Activation of skeletal muscle satellite cells and the role of fibroblast growth factor receptors. Exp. Cell Res. **219:** 449–453.
96. PIMORADY-ESFAHANI, A., M.D. GROUNDS & P.G. MCMENAMIN. 1997. Macrophages and den-dritic cells in normal and regenerating murine skeletal muscle. Muscle & Nerve **20:** 158–166.
97. ROBERTSON, T.A. *et al.* 1990. Fusion between myogenic cells *in vivo*: An ultrastructural study in regenerating murine skeletal muscle. J. Struct. Biol. **105:** 170–182
98. ROBERTSON, T.A. *et al.* 1993. The role of macrophages in skeletal muscle regeneration with particular reference to chemotaxis. Exp. Cell Res. **207:** 321–331.
99. FIELDING, R.A. *et al.* 1993. Acute phase response to exercise III Neutrophil and IL-1beta accu-mulation in skeletal muscle. Am. J. Physiol. **265:** R166–R172.
100. OLWIN, B.B., K. HANNON & A.J. KUDLA. 1994. Are fibroblast growth factors regulators of myogenesis *in vivo*? Prog. Growth Factor Res. **5:** 145–158.
101. ANDERSON, J.E. *et al.* 1993. Comparison of basic fibroblast growth factor in X-linked dystro-phin-deficient myopathies of human, dog and mouse. Growth Factors **9:** 107–121.
102. MITCHELL, C.A., J.M. MCGEACHIE & M.D. GROUNDS. 1996. Exogenous administration of basic FGF does not enhance the regeneration of murine skeletal muscle. Growth Factors **13:** 1–19.
103. BRICKMAN, Y. *et al.* 1995. Heparan sulphates mediate the binding of basic fibroblast growth factor to a specific receptor on neural precursor cells. J. Biol. Chem. **270:** 1–8.
104. BERNFIELD, M. & K.C. HOOPER. 1991. Possible regulation of FGF activity by syndecan, an inte-gral membrane heparan sulphate proteoglycan. Ann. N.Y. Acad. Sci. **638:** 182–194.
105. GALLO, R.L. *et al.* 1994. Syndecans, cell surface heparan sulphate proteoglycans, are induced by a proline-rich antimicrobial peptide from wounds. Proc. Natl. Acad. Sci. USA **9:** 11035–11039.
106. MITCHELL, C.A., M.D. GROUNDS & J.K. MCGEACHIE. 1991. The effect of low dose Dexametha-sone on skeletal muscle regeneration *in vivo*. Basic Appl. Myology. **1:** 139–144.
107. BARNARD, W. *et al.* 1994. Leukemic inhibitory factor (LIF) infusion stimulates skeletal muscle regeneration after injury: Injured muscle expresses LIF mRNA. J. Neurol. Sci. **123:** 108–113.
108. YAROM, R. *et al.* 1982. Enhancement of human muscle growth in diffusion chambers by bone marrow cells. Virchows Archiv. B. Cell Pathol. **41:** 171–180.
109. BISCHOFF, R. & C. HEINTZ. 1994. Enhancement of skeletal muscle regeneration. Dev. Dyn. **201:** 41–54.
110. ANDERSON, J.E. & E. KARDAMI. 1994. The effects of hyperthyroidism on muscular dystrophy in the mdx mouse: Greater dystrophy in cardiac and soleus muscles. Muscle & Nerve **17:** 64–73.

Age-associated Changes in the Innervation of Muscle Fibers and Changes in the Mechanical Properties of Motor Units

ANTHONY R. LUFF[a]

Department of Physiology, Monash University,
Clayton, Victoria 3168, Australia

ABSTRACT: In both humans and animals there is a progressive loss of muscle strength with age. Tests of handgrip and knee extension in men show that some decline in strength is evident by the age of 55 years and is pronounced by the age of 65, compared with the 25- to 35-year period when strength is at a maximum. A comparable age-related decline in peak force development has also been shown in hindlimb muscles of aged rats. Motoneurons and consequently motor units are lost with age, and this is apparent in man after the age of 60. Again, a comparable decline has been demonstrated in the motoneuron population of hindlimb muscles of rats aged 20–24 months. Loss of motoneurons in young adults (through either injury or disease) results in the remaining intact motoneurons sprouting to innervate the denervated fibers. This capacity for sprouting has been shown to be seriously impaired in the hindlimb muscles of aged rats. Furthermore, the well-established relationship between motor unit size and fatigability (smaller units tend to be more fatigue resistant) also tends to break down, with large units just as likely to be fatigable as fatigue resistant. The normally large, fatigable motor units also appear to be reduced in size in the aged muscles. The age-related loss of motoneurons and associated loss of muscle fibers accounts in part for the reduced functional capacity of muscle with age. The reason for the impairment of the aged motoneuron remains to be investigated, but it may relate to the integrity of the oxidative metabolic pathways within the cell, given that mitochondrial respiratory chain function is known to be reduced with age.

One of the more obvious consequences of aging for both humans and animals is the loss of muscle mass, particularly the more distal muscles, and the associated loss of muscle strength. Both physical capacity and quality of life are impaired as a result. This is one of the major issues faced by an increasingly aged population, by those who care for them, and by the broader community. The reasons for the loss of muscle mass are complex. This and the influence that the nerves innervating the muscles have in this process will be considered in the paper. Several excellent reviews in this area have appeared in recent years, and reference should be made to these for more detailed information.[1–5]

Age-related loss of muscle mass has two primary causes. First, psychosocial factors result in *disuse atrophy*, resulting from a person simply being less active. This relates to a lack of inclination for physical activity, the (misplaced) expectations of society that aged people should be less active, and the consequences of debilitating musculoskeletal and cardiovascular diseases. Second, there are intrinsic physiological factors, and it is these that will be discussed below.

[a]Tel: +613 9905 2507; fax: +613 9905 2547; e-mail: tony.luff@med.monash.edu.au

MUSCLE MASS AND MUSCLE FORCE

In general terms, muscle mass and muscle strength start to decline from about the age of 50 in humans.[6] Initially, the rate of loss is only about 15% per decade from 50 to 70 years. This accelerates with increasing age, rising to a further 30% loss from 70 to 80 years of age,[7] although Lexell *et al.*[8] have shown a loss of muscle mass in humans of about 10% from age 24 to 50 years, with a further 30% loss for 50 to 80 years. Changes in muscle mass and muscle force may not necessarily change in a strictly proportional manner. For example, ultrasound measurements of the cross-sectional area of the quadriceps showed a 25% decrease for people in their 80s compared with people in their 20s but there was a 39% reduction in the maximal voluntary contraction (MVC) developed by the quadriceps.[9] A similar change in the MVC was found in a study of the elbow flexors, where the average force in a group of aged persons (60 to 81 years) was 65% that of a group of younger (22 to 38 years) people.[10] There is also evidence to indicate that the muscles in the arms and legs are affected disproportionately and indeed that there are distinct differences in the extent to which specific muscles are affected. Loss of muscle strength in leg muscles was about 40% compared with about 30% in arm muscles between 30 and 80+ years of age.[11]

Results from studies on aged animals generally confirms the findings in humans with reductions in muscle mass of 32% in gastrocnemius, 26% in quadriceps, and 11% in soleus in aged (27–28 months) rats compared with adult rats aged 9–10 months.[12] Kanda and Hashizume[13] noted a 24% loss in medial gastrocnemius (MG) muscle mass of aged rats, and Einsiedel and Luff[14] found a 32% decrease in the maximum isometric force developed by the rat MG. Brooks and Faulkner[15] found comparable changes in the muscles of aged mice. The results of Holloszy *et al.*[12] are also consistent with the observation that different muscles suffer atrophy to a different extent; this may relate to their muscle fiber type composition (see below). The results also indicated that weight-bearing muscles showed a greater loss than non-weight-bearing muscles.

Loss of muscle mass may also relate to the age-related decline in whole-body maximal oxygen consumption.[16] However, the time course of the loss of muscle mass is very different from the decline in maximal oxygen consumption, which starts at the age of 25 years. Furthermore, Pollack *et al.*[17] found that aerobic capacity was maintained while estimated muscle mass declined in 50- to 82-year-old masters athletes over a 10-year training period.

There is, therefore, a well-documented decline in muscle mass of 20–40% in humans between 30 and 80 years of age and a comparable decline between the ages of 12 and 28 months in rodents. Whereas some of this decline in all species can be attributed to disuse atrophy, it is widely believed that there is also a specific physiological basis. The possible reasons for a loss of muscle mass and muscle force can be summarized as follows: changes in the intrinsic force-generating capacity of muscle, reduction in size of muscle fibers (fiber atrophy), reduction in the number of muscle fibers, and some combination of the above.

CHANGES IN THE INTRINSIC FORCE-GENERATING CAPACITY OF MUSCLE

Muscle force is proportional to the cross-sectional area of the muscle fibers and can be expressed as force per unit cross-sectional area or specific force. Using permeabilized fibers from the extensor digitorum longus (EDL) of adult (12 months old) and aged (27 months old) mice, Brooks and Faulkner[18] found no difference in specific force, although the values they obtained were relatively low (110–118 mN mm^{-2}). Earlier studies involving whole intact muscles[15,19] had shown that specific force was reduced in hindlimb muscles of aged animals. However, estimates of specific force in single motor units where total fiber area is known (from glycogen depletion of the fibers) showed that there was no age-related change in specific force.[13,20] Results from aged[9,21,22] human muscle had also shown an age-related reduction in specific force.

The basis of this change in specific force is difficult to understand, particularly in the light of the recent work on permeabilized fibers.[18] Specific force depends on the myofibrillar lattice and on the proportion of nonmyofibrillar material in the fiber cross-section. It seems unlikely that either would change sufficiently to account for the differences observed in intact muscle.

MUSCLE FIBER LOSS

One of the core arguments about the loss of muscle mass with age is the extent to which it can be attributed to loss of muscle fibers or atrophy of fibers. The next two sections will examine this issue in more detail. It also needs to be kept in mind that any fiber loss can only be compensated for by hypertrophy of the remaining fibers. New fibers are not normally produced, and if they are it is probably only to a very limited extent. It is in some ways surprising, but the determination of total muscle fiber number is difficult and prone to significant methodological errors. One method is to count all the fibers appearing in a single cross-section through the belly of a muscle. This is only an acceptable method for a simple fusiform muscle where it can be reasonably assumed that all the fibers lie within that section. An alternative, though somewhat heroic method, is to lightly digest the muscle and then tease out and count all the fibers. Methods involving estimating fiber number from the mean cross-sectional area (CSA) of the fibers and the CSA of the muscle would have to be considered less reliable.

Probably one of the key studies in humans is that of Lexell et al.[23] involving examination of cross-sections of the quadriceps muscle taken at autopsy from males aged from 20 to 80 years of age. Fiber number was found to decline by 39%, which approximately paralleled the loss of muscle force. Interestingly, the decline seemed to be progressive, commencing at about the age of 30. A loss of muscle fibers is consistent with other human-based investigations,[24] including the rectus abdominis.[25]

Conversely, much of the more recent animal-based data indicates no loss of fibers with age. This is the case for soleus and EDL muscles for rats between 9 and 27–30 months of age.[26–28] Consistent with such results was the finding of Holloszy et al.[12] in the rat plantaris that the reduction in muscle mass was consistent with a reduction in the CSA of the fibers, suggesting that there was no change in fiber number. It has also been suggested that there is a relative loss of type II (fast-twitch) fibers in advanced age.[29,30]

The results from the human and animal-based studies are equivocal. In general the human data indicate that there is a loss of muscle fibers; conversely the rodent data suggest there is no loss of muscle fibers. It has to be acknowledged that the rodent data were obtained from simple fusiform hindlimb muscles, whereas the human data were obtained from more complex pennate muscles. Despite this, several reviews maintain that there is a loss of fibers with age.[4,5,31] This is a conclusion that is consistent with findings on innervation and motoneuron changes (see below).

MUSCLE FIBER SIZE

Virtually all studies, whether in humans or animals, and for nearly all muscles studied, have shown a reduction in muscle fiber size—as a reduction in CSA with increasing age. The only differences seem to be in the degree of reduction in the different muscle fiber types. Grimby *et al.*,[32] for example, found a selective reduction in CSA of type II fibers in vastus lateralis of humans above 70 years of age, with only slight reductions prior to the age of 60–70 years. Coggan *et al.*[33] also found no reduction in CSA of type I (slow-twitch) fibers but a 13–30% reduction in type II fibers in the aged gastrocnemius muscle.

Comparable changes have been observed in animal studies. Alnaqeeb and Goldspink[34] found a significant decline in the CSA of type II fibers in soleus between rats aged 10 months and 24 months, but no reduction in size of type I fibers. In the rat plantaris the CSA of type II fibers declined 37% and the type I fibers by only 21% between the ages of 9–10 months and 28–30 months.[12]

It is perhaps not surprising that most reduction in muscle fiber CSA is seen in type II fibers, particularly the IIB fibers (fast-twitch, predominantly glycolytic and readily fatigable). Order of recruitment of motor units[35] and therefore muscle fibers dictates that type I fibers will remain in relatively regular use, even in aged subjects, whereas the type II fibers and particularly the IIB fibers will rarely be recruited and therefore subject to disuse atrophy.

AGE-RELATED CHANGES AT THE NEUROMUSCULAR JUNCTION

The neuromuscular junction (NMJ) is the crucial link between the motoneuron and the muscle. Normally, the NMJ has a "high safety factor," meaning that the arrival of an action potential at the nerve terminal almost invariably results in an action potential in the muscle fiber. Consequently, any change in the security of synaptic transmission at the NMJ may profoundly affect the extent to which individual muscle fibers are recruited for normal function. Results of both animal and human studies indicate that there is remodeling and fragmentation of the NMJ with age, which suggests a progressive degeneration of the NMJ. Considerable fragmentation in the distribution of acetylcholine receptors[36] and in acetylcholinesterase staining[37] has been demonstrated, and also an increase in the incidence of branches or boutons that are spatially separate or only connected by fine nerve filaments, suggesting fragmentation of the terminal.[38] An increase in what are termed "growth configurations," that is, nerve terminals innervated by two or more axon branches, was found in rats from the age of 21 months to 26 months.[39] Specifically, in soleus there was an increase from an already high 30% of terminals, with growth configurations to 37.5% of terminals at 26 months of age. A gradual and progressive loss of synaptic contact was found in rat soleus and diaphragm muscles aged 21 months, with entire NMJs being lost.[40] A comparison of the morphology of gastrocnemius NMJs of old (27 month old) with young (6 month old) mice showed that 85% of NMJs in young mice could be considered normal; however, this was reduced to 40% in the old mice.[41] The results suggest a significant age-related deterioration in the structure and functional capacity of the NMJ, ultimately involving outright failure or loss of the NMJ.

MOTONEURONS AND MOTOR UNITS

The concept of the "motor unit" was defined by Liddell and Sherrington[42] and consists of the motoneuron and all the muscle fibers innervated by that motoneuron. The number of fibers innervated by a particular motoneuron is referred to as the "innervation ratio," and although this can vary by up to an order of magnitude within a single muscle, it can vary by several orders of magnitude between different muscles. Essentially, if a motoneuron is lost, a motor unit is lost. The consensus view from both animal and human studies is that motor units are lost with age. Early studies of Gutmann and Hanzlikova[43] demonstrated a significant loss of motor units in aged rats. More recently, with the use of morphometric and physiological techniques, Edstrom and Larsson[44] found that the average number of motor units in soleus declined from 49 in 3- to 6-month-old rats to 29 in 20- to 24-month-old rats, a reduction of about 40 percent. Einsiedel and Luff[14] found a reduction in motor unit number in the rat MG from an average of 93 to 66. Consistent with these results, the number of motoneurons in a particular motoneuron pool, identified by retrograde labeling with horseradish peroxidase, was found to decline with age.[45,46] Recently, Hashizume and Kanda[47] found a significant decrease in the number of MG motoneurons from an average of 132 in middle-aged rats to 121 in aged rats (27 months). Interestingly, there was no change in the number of motoneurons supplying the ulnar nerve in the rat forelimb. In addition the mean soma size of both motoneuron pools was reduced with age.[47]

Estimates of motor unit numbers derived by dividing maximum whole muscle force by the average maximum force of single isolated units have consistently shown an overall reduction in motor unit numbers in aged animals. Einsiedel and Luff[14] found a decrease from an average of 93 to 66 motor units (a 29% decline) in rat MG, somewhat less than that obtained by Edstrom and Larsson[44] in the rat soleus where motor unit number declined from an average of 49 to 29, a decrease of 41 percent. Comparable results have been obtained in the mouse.[48] These indirect determinations are supported by direct determination of motoneuron numbers using retrograde labeling of motoneurons with horseradish peroxidase (HRP).[46,49]

In addition to the overall loss of motoneurons, there appears to be a preferential loss of the larger motoneurons.[45,50] More specifically, in rat hindlimb muscles, the number of IIB muscle fibers was reduced at 14 months without reduction in numbers of type I and IIA fibers, whereas by the age of 27 to 31 months the type IIA and IIB fibers had started to decline along with a preferential loss of the largest motoneurons with the lowest oxidative capacity.[46,49] This would also be consistent with the preferential loss of type IIB muscle fibers (see above), given that in general the largest motoneurons innervate the type II fibers, particularly the type IIB.

Human studies indicate a similar pattern of motoneuronal loss. Tomlinson and Irving[51] determined the number of motoneurons in the lumbrosacral region of the spinal cord from subjects aged 13 to 95 years. There appeared to be little change up to the age of 60 years, but from that age, the number of motoneurons declined linearly, such that subjects in their 90s had about 70% the number of motoneurons of subjects up to 60 years of age. A significant study by Cambell *et al.*,[52] in which motor unit number was determined indirectly in the extensor digitorum brevis (EDB, a relatively small muscle in the lower leg), used electromyographic and electrophysiological methods. This showed that motor unit number remained reasonably constant (average ~200 units) up to the age of 55 to 60 years and then declined linearly with age. From 60 to 95 years of age the number of units was highly variable from subject to subject (as it was also up to the age of 60 years), and in a few subjects aged in excess of 80 years less than ten units remained. Among the surviving motor units, the proportion of slow-twitch type units increased, and they tended to be somewhat larger. Estimated motor unit number in the human biceps-brachialis muscles declined by nearly half, from an average of 911 in subjects aged less than 60 years to 479 in subjects in excess of 60 years.[53] A comparable degree of motor unit loss has also been shown in a group of subjects aged 60 to 80 years, compared with a group aged 20 to 40 years in a very distal muscle, the thenar muscle.[54] A recent study,[55] involving the use of a "fully automated" recording and analysis system, in order to eliminate subjective factors biasing motor unit estimates, confirmed the age-related loss of motor units in EDB and the thenar muscles but found no significant reduction in biceps.

Both human and animal studies have demonstrated a significant loss of motor units with age, and this loss occurs predominantly after the age of 60 in humans. Although this is certainly true for those more distal muscles that have been investigated, the extent to which more proximal muscles exhibit this loss is less certain.

MOTONEURON SPROUTING AND MOTOR UNIT REMODELING

In young adult animals (and humans) when motoneurons are lost through injury or disease, the remaining motoneurons will sprout and reinnervate the denervated fibers. This is a powerful compensatory mechanism, and motor unit innervation ratios may increase five- to tenfold.[56,57] Sprouting occurs either from the nerve terminals (terminal sprouting) or from the first node of Ranvier (nodal sprouting) and is rapid, with significant recovery of function within about two weeks. Observations in aged muscle suggest that the capacity for motoneuron sprouting is relatively limited, indicating some impairment of motoneuron function.

In an experiment specifically designed to test the sprouting capacity of MG motoneurons in the aged rat,[20] it was found that there was no increase in size of the units in response to an extensive partial denervation of the muscle. The remodeling of the units with "clumping" or clustering of constituent muscle fibers suggests that some sprouting had occurred, but at the expense of the more distal connections. This is consistent with the finding that nerve terminals in the EDL muscle of aged rats showed a relatively limited sprouting response to partial denervation, compared with younger rats, resulting in the conclusion that, in EDL, the nerve terminals (and motoneurons) had approached the maximum capacity for sprouting.[58] Furthermore, Pestronk *et al.*,[59] using botulinum toxin to "denervate" fibers in the rat soleus muscle, found that although this resulted in significant terminal sprouting in the muscles of young adult rats, there was no significant change in aged (28 month) rats. A recent study[60] of motor unit characteristics in the human thenar muscle found that the twitch amplitude of single units was on average 34% greater in the aged group, and, in addition, the units in the aged muscle had somewhat slower contraction times. Given that these authors had previously found an approximate 50% loss of motor units in the thenar muscle,[54] it suggests that although the units retained some sprouting capacity, it was insufficient to fully compensate for the loss of motor units. Stålberg and Fawcett,[61] although making no estimate of unit number, did find that the "macro-EMG" signal increased with age in the vastus lateralis and tibialis anterior muscles (but not in the biceps brachii) of healthy human subjects, which was interpreted as an increase in size of the units.

There appears to be an age-related process of motor unit remodeling involving denervation and reinnervation of fibers. Whereas such a process probably occurs in muscles of young adult animals, it only really becomes manifest in the older animals. Considerable work in this area has been done by Larsson and his group and is well summarized in a recent review.[62] Essentially, in the soleus muscles of young animals, the fibers of a unit are arranged randomly; however, in the soleus of old animals, the arrangement of fibers is not random, with an increase in both long and short interfiber distances. The latter feature appears as a "clumping" of fibers and is strongly indicative of a denervation–reinnervation process. Significant clumping of constituent fibers of units was also found following partial denervation of the MG in aged rats.[20]

Other age-related changes in motor unit characteristics have been identified. Generally, there appears to be a slowing of contractile characteristics, although whether this is due to changes in contractile properties of both fast and slow units[44] or whether it is due to a preferential loss of fast-type units[57] remains unclear. The relationship between unit force and fatigability is well defined in the motor unit population of the MG of young adult rats, with a group of small and fatigue-resistant units, and then a continuum with the most fatigable units being relatively medium to large in size.[14] In the aged animal this well-defined relationship ceases to exist. A group of smaller and highly fatigue-resistant units persists, but there is no longer any relationship between unit size and fatigability.[14] Disproportionate changes in size of units with age has been recorded,[13] where the mean maximum force of fast units in rat MG at 27 months of age had declined to about 70% of that of 12-month-old rats, but the mean force of slow units had more than doubled, almost entirely due to an increase in the innervation ratio. It should be noted that the force produced by slow units in 12-month-old rats is only about one tenth that of fast units.

CONCLUSIONS

A significant factor in the age-related loss of muscle mass and muscle strength is the loss of muscle fibers and motoneurons and the inability of the remaining motoneurons to fully compensate for that loss. The remodeling of motor units indicates that a process of denervation and reinnervation is occurring, but that overall the functional capacity of the motoneuron is impaired. It is reasonable to presume that in a young adult the bioenergetic and biosynthetic demands on the motoneuron are well within its capacity and it has the potential to support and maintain a peripheral field several times greater than normal. With age, this potential capacity may become greatly diminished, to the point where the functional capacity of the motoneuron is only capable of maintaining a "normal-sized" peripheral field. Any significant requirement to meet an increased demand cannot be met. This raises two separate but related questions. First, is the problem an inherent failure of the motoneuron or is it that the muscle fibers are unable or incapable of supporting the motoneuron? Second, if there is impairment of motoneuron function, what is the nature and origin of that impairment?

There are several ways in which motoneuron function could be impaired; these include the bioenergetic capacity of the neuron and the influence of neurotrophic factors. Impairment of mitochondrial function will ultimately impair motoneuron function. Possible mechanisms are well known and include the increase in mitochondrial DNA mutations[63] and oxidative damage.[64] In the past few years the importance of neurotrophic factors in motoneuron survival and function has been recognized. However, it seems unlikely that a single factor is involved. For example, whereas leukemia inhibitory factor (LIF) promotes motoneuron survival and axonal regeneration,[65] the inactivation of LIF by gene targeting resulted in no apparent change in the number or structure of motoneurons and no impairment of their function.[66] Factors that are probably important for motoneurons include brain-derived neurotrophic factor, neurotrophin-3 (NT3), NT4/5, and LIF.[67] It is possible that a diminution of some combination of these factors would have a deleterious effect on motoneuron function.

REFERENCES

1. CARTEE, G.D. 1994. Aging skeletal muscle: Response to exercise. Exercise Sports Sci. Rev. 22: 91–120.
2. CARMELI, E. & A.Z. REZNICK. 1994. The physiology and biochemistry of skeletal muscle atrophy as a function of age. Proc. Soc. Exp. Biol. Med. 206: 103–113.
3. BROOKS, S.V. & J.A. FAULKNER. 1993. Skeletal muscle weakness in old age: Underlying mechanisms. Med. Sci. Sports Exercise 26: 432–439.
4. DOHERTY, T.J., A.A. VANDERVOORT & W.F. BROWN. 1993. Effects of ageing on the motor unit: A brief review. Can. J. Appl. Physiol. 18: 331–358.
5. BOOTH, F.W., S.H. WEEDEN & B.S. TSENG. 1994. Effect of aging on human skeletal muscle and motor function. Med. Sci. Sports Exercise 26: 556–560.
6. ROGERS, M.A. & W.J. EVANS. 1993. Changes in skeletal muscle with aging: Effects of exercise training. Exercise Sports Sci. Rev. 21: 65–102.
7. DANNESKIOLD-SAMSOE, B., V. KOFOD, J. MUNTER et al. 1984. Muscle strength and functional capacity in 78–81 year old men and women. Eur. J. Appl. Physiol. 52: 310–314.
8. LEXELL, J., D. DOWNHAM & M. SJOSTROM. 1986. Distribution of different fibre types in human skeletal muscles. Fibre type arrangement in m. vastus lateralis from three groups of healthy men between 15 and 83 years. J. Neurol. Sci. 72: 211–222.
9. YOUNG, A., M. STOKES & M. CROWE. 1985. The size and strength of the quadriceps muscle of young and old men. Clin. Physiol. 5: 145–154.
10. DOHERTY, T.J., A.A. VANDERVOORT, A.W. TAYLOR & W.F. BROWN. 1993. Effects of motor unit losses on strength in older men and women. J. Appl. Physiol. 74: 868–874.
11. FAULKNER, J.A., S.V. BROOKS & E. ZERBA. 1990. Skeletal muscle weakness and fatigue in old age: Underlying mechanisms. Annu. Rev. Ger. Gerontol. Geriatr. 10: 147–166.
12. HOLLOSZY, J.O., M. CHEN, G.D. CARTEE & J.C. YOUNG. 1991. Skeletal muscle atrophy in old rats: Differential changes in the three fiber types. Mech. Ageing Dev. 60: 199–213.
13. KANDA, K. & K. HASHIZUME. 1989. Changes in properties of the MG motor units in aging rats. J. Neurophysiol. 61: 737–746.
14. EINSIEDEL, L.J. & A.R. LUFF. 1992a. Alterations in the contactile properties of motor units within the ageing rat medial gastrocnemius. J. Neurol. Sci. 112: 170–177.
15. BROOKS, S.V. & J.A. FAULKNER. 1988. Contractile properties of skeletal muscle from young, adult and aged mice. J. Physiol. 404: 71–78.
16. FLEG, J.L. & E.G. LAKATTA. 1988. Role of muscle loss in the age associated reduction in VO_{2max}. J. Appl. Physiol. 65: 1147–1151.
17. POLLACK, M.L., C. FOSTER, D. KNAPP, J.L. ROD & D.H. SCHMIDT. 1987. Effect of age and training in aerobic capacity and body composition of master athletes. J. Appl. Physiol. 62: 725–731.

18. BROOKS, S.V. & J.A. FAULKNER. 1994. Isometric, shortening, and lengthening contractions of muscle fiber segments from adult and old mice. Am. J. Physiol. **267:** C507–C513.

19. PHILLIPS, S.K., R.W. WISEMAN, R.C. WOLEDGE & M.J. KUSHMERICK. 1993. Neither changes in phosphorous metabolite levels nor myosin isoforms can explain the weakness in aged mouse muscle. J. Physiol. **463:** 157–167.

20. EINSIEDEL, L.J. & A.R. LUFF. 1992b. Effect of partial denervation on motor units in the ageing rat medial gastrocnemius. J. Neurol. Sci. **112:** 178–184.

21. BRUCE, S.A., D. NEWTON & R.C. WOLEDGE. 1989. Effect of age on voluntary force and cross-sectional area of human adductor pollicis muscle. Q. J. Exp. Physiol. **74:** 359–362.

22. PHILLIPS, S.K., K.M. ROOK, N.C. SIDDLE & R.C. WOLEDGE. 1992. Muscle weakness in women occurs at an earlier age than in men, but strength is preserved by hormone replacement therapy. Clin. Sci. **84:** 95–98.

23. LEXELL, J., C.C. TAYLOR & M. SJOSTROM. 1988. What is the cause of the ageing atrophy? Total number, size and proportion of different fibre types studied in the whole vastus lateralis muscle from 15- to 83-year-old men. J. Neurol. Sci. **84:** 275–294.

24. GRIMBY, G. & B. SALTIN. 1983. The ageing muscle. Clin. Physiol. **3:** 209–218.

25. INOKUCHI, S., H. ISHIKAWA, S. IWANOTO & T. KIMURA. 1975. Age related changes in the histochemical composition of rectus abdominis muscle of the adult human. Hum. Biol. **47:** 231–349.

26. BROWN, M. 1987. Change in fiber size not number in aging skeletal muscle. Age Aging **16:** 244–248.

27. EDDINGER, T.J., R.L. MOSS & R.G. CASSENS. 1985. Fiber number and type composition in extensor digitorum longus, soleus, and diaphragm muscle with aging in Fisher 344 rats. J. Histochem. Cytochem. **33:** 1033–1041.

28. ALNAQEEB, M.A. & G. GOLDSPINK. 1986. Changes in fibre type, number and diameter in developing and ageing skeletal muscle. J. Anat. **153:** 31–45.

29. LARSSON, L., B. SJÖDIN & J. KARLSSON. 1978. Histochemical and biochemical changes in skeletal muscle with age in sedentary males, age 22–65 years. Acta Physiol. Scand. **102:** 31–39.

30. JAKOBSSON, F., K. BORG & L. EDSTROM. 1990. Fiber type, composition, structure and cytoskeletal protein location of fibers in anterior tibialis muscle: Comparison between young adults and physically active aged humans. Acta Neuropathol. **80:** 459–468.

31. FAULKNER, J.A. & S.V. BROOKS. 1995. Muscle fatigue in old animals. Unique aspects of fatigue in elderly humans. *In* Fatigue. S.C. Gandevia *et al.*, Ed.: 471–480. Plenum Press. New York.

32. GRIMBY, G., B. DANNESKIOLD-SAMSOE, K. HVID & B. SALTIN. 1982. Morphology and enzymatic capacity in arm and leg muscles in 78–81 year old men and women. Acta Physiol. Scand. **115:** 125–134.

33. COGGAN, A.R., R.J. SPINA, D.S. KING *et al.* 1992. Histochemical and enzymatic comparison of the gastrocnemius muscle of young and elderly men and women. J. Gerontol. Sci. **46:** B71–B76.

34. ALNAQEEB, M.A. & G. GOLDSPINK. 1987. Changes in fibre type, number and diameter in developing and ageing skeletal muscle. J. Anat. **153:** 31–45.

35. HENNEMAN, E. & L.M. MENDELL. 1981. Functional organization of motoneuron pool and its inputs. *In* Handbook of Physiology. The Nervous System. Motor Control. Am. Physiol. Soc. sect. 1, vol. 11, pt. 1, chapter 11: 423–507. Bethesda, MD.

36. GUTMANN, E. & V. HANZLIKOVA. 1965. Age changes of motor endplates in muscle fibers of the rat. Gerontologia **11:** 12–24.

37. ODA, K. 1989. Age changes in motor innervation and acetyl choline receptor distribution on human skeletal muscle fibers. J. Neurol. Sci. **66:** 327–338.

38. ROBBINS, N. & M.A. FAHIM. 1985. Progression of age changes in mature mouse motor nerve terminals and its relation to locomotor activity. J. Neurocytol. **14:** 1019–1036.

39. STEBBINS, C.L., E. SCHULTZ, R.T. SMITH & E.L. SMITH. 1985. Effects of chronic exercise during aging on muscle and end-plate morphology in rats. J. Appl. Physiol. **58:** 45–51.

40. CORDOSIS, C.A. & D.M. LAFONTAINE. 1987. Aging rat neuromuscular functions: A morphometric study of cholinesterase-stained whole mounts and ultrastructure. Muscle & Nerve **10:** 100–213.

41. LUDATSCHER, R.M., M. SILBERMANN, D. GERSHON & A.Z. REZNICK. 1985. Evidence of Schwann cell degeneration in the aging mouse motor end-plate region. Exp. Gerontol. **20:** 80–91.

42. LIDDELL, E.G.T. & C.S. SHERRINGTON. 1925. Recruitment and some other features of reflex inhibition. Proc. R. Soc. **97:** 488–518.
43. GUTMANN, E. & V. HANZLIKOVA. 1966. Motor units in old age. Nature **209:** 921–922.
44. EDSTRÖM, L. & L. LARSSON. 1987. Effects of age on contractile and enzyme-histochemical properties of fast- and slow-twitch single motor units in the rat. J. Physiol. **392:** 129–145.
45. HASHIZUME, K., K. KANDA & R.E. BURKE. 1988. Medial gastrocnemius motor nucleus in the rat: Age-related changes in the numbers and size of motoneurons. J. Comp. Neurol. **269:** 425–430.
46. ISHIHARA, A. & H. ARAKI. 1988. Effects of age on the number and histochemical properties of muscle fibers and motoneurons in the rat extensor digitorum longus muscle. Mech. Ageing Dev. **45:** 213–221.
47. HASHIZUME, K. & K. KANDA. 1995. Differential effects of aging on motoneurons and peripheral nerves innervating the hindlimb and forelimb muscles of rats. Neurosci. Res. **22:** 189–196.
48. CACCIA, M.R., J.B. HARRIS & M.A. JOHNSON. 1979. Morphology and physiology of skeletal muscle in aging rodents. Muscle & Nerve **2:** 202–212.
49. ISHIHARA, A., H. NAITOH & S. KATSUTA. 1987. Effects of ageing on the total number of muscle fibers and motoneurons of the tibialis anterior and soleus muscles in the rat. Brain Res. **435:** 355–358.
50. ANSVED, T. & L. LARSSON. 1990. Quantitative and qualitative morphological properties of the soleus motor nerve and the L5 ventral root in young and old rats. J. Neurol. Sci. **96:** 269–282.
51. TOMLINSON, B.E. & D. IRVING. 1977. The numbers of limb motor neurons in the human lumbosacral cord throughout life. J. Neurol. Sci. **34:** 213–219.
52. CAMPBELL, M.J., A.J. MCCOMAS & F. PETITO. 1973. Physiological changes in ageing muscles. J. Neurol. Neurosurg. Psychiatry **36:** 174–182.
53. BROWN, W.F., M.J. STRONG & R.S. SNOW. 1988. Methods for estimating numbers of motor units in biceps-brachialis muscles and losses of motor units with aging. Muscle & Nerve **11:** 423–432.
54. DOHERTY, T.J. & W.F. BROWN. 1993. The estimated numbers and relative sizes of thenar motor units as selected by multiple point stimulation in young and older adults. Muscle & Nerve **16:** 355–366.
55. GALEA, V. 1996. Changes in motor unit estimates with aging. J. Clin. Neurophysiol. **13:** 253–260.
56. BROWN, M.C., R.L. HOLLAND & W.G. HOPKINS. 1981. Motor nerve sprouting. Annu. Rev. Neurosci. **4:** 17–42.
57. LUFF, A.R. & L.J. EINSIEDEL. 1992. Motoneurone sprouting: A mechanism for recovery of muscle function. Proc. Aust. Physiol. Pharmacol. Soc. **23:** 40–50.
58. ROSENHEIMER, J.L. 1990. Ultraterminal sprouting in innervated and partially denervated adult and aged rat muscle. Neuroscience **38:** 763–770.
59. PESTRONK, A., D.B. DRACHMAN & J.W. GRIFFIN. 1980. Effects of aging on nerve sprouting and regeneration. Exp. Neurol. **70:** 65–82.
60. DOHERTY, T.J. & W.F. BROWN. 1997. Age-related changes in the twitch contractile properties of human thenar motor units. J. Appl. Physiol. **82:** 93–101.
61. STÅLBERG, E. & P.R.W. FAWCETT. 1982. Macro EMG in healthy subjects of different ages. J. Neurol. Neurosurg. Psychiatry **45:** 870–878.
62. LARSSON, L. 1995. Motor units: Remodeling in aged animals. J. Gerontol. (Series A) **50A:** 91–95.
63. LAWEN, A., R.D. MARTINUS, G.L. MCMULLEN et al. 1994. The universality of bioenergetic disease: The role of mitochondrial mutation and the putative interrelationship between mitochondria and plasma membrane NADH oxidoreductase. Mol. Aspects Med. **15:** S13–27.
64. AMES, B.N., M.K. SHINEGANA & T.M. HAGAN. 1995. Mitochondrial decay in aging. Biochim. Biophys. Res. Acta **1271:** 165–170.
65. CHEEMA, S.S., L.J. RICHARDS, M. MURPHY & P.F. BARTLETT. 1994. Leukaemia inhibitory factor rescues motoneurones from axotomy-induced cell death. Neuroreport **5:** 989–992.
66. SENDTNER, M., R. GOTZ, B. HOLTMANN et al. 1996. Cryptic physiological support of motoneurons by LIF revealed by double gene targeting of CNTF and LIF. Curr. Biol. **6:** 686–694.
67. SENDTNER, M., B. HOLTMANN & R.A. HUGHES. 1996. The response of motoneurons to neurotrophins. Neurochem. Res. **21:** 831–841.

Oxidative Stress and Aging

Role of Exercise and Its Influences on Antioxidant Systems

LI LI JI,[a] CHRIS LEEUWENBURGH, STEVE LEICHTWEIS, MITCH GORE, RUSSEL FIEBIG, JOHN HOLLANDER, AND JEFFERY BEJMA

Department of Kinesiology and Nutritional Sciences,
and Institute on Aging, University of Wisconsin-Madison,
2000 Observatory Drive, Madison, Wisconsin 53706, USA

ABSTRACT: Strenuous exercise is characterized by an increased oxygen consumption and disturbance of intracellular prooxidant–antioxidant homeostasis. At least three biochemical pathways, that is, mitochondrial electron transport chain, xanthine oxidase, and polymorphoneutrophil have been identified as potential sources of intracellular free radical generation during exercise. These deleterious reactive oxygen species pose a serious threat to the cellular antioxidant defense system, such as diminished reserve of antioxidant vitamins and glutathione, and have been shown to cause oxidative damage in exercising and/or exercised muscle and other tissues. However, enzymatic and nonenzymatic antioxidants have demonstrated great versatility and adaptability in response to acute and chronic exercise. The delicate balance between prooxidants and antioxidants during exercise may be altered with aging. Study of the complicated interaction between aging and exercise under the influence of reactive oxygen species would provide more definitive information as to how much aged individuals should be involved in physical activity and whether supplementation of nutritional antioxidants would be desirable.

Among the various theories attempting to explain the aging process, the free radical theory of aging has received increasing recognition during the past two decades.[1,2] A basic tenet of this theory is that reactive oxygen species (ROS) are produced as a normal by-product of aerobic life and that accumulation of oxidative damage caused by ROS underlies the fundamental changes found in senescence. At least three lines of evidence support the theory: (1) aging has been shown to correlate with production of ROS and adequacy of cellular antioxidant defense systems;[2–4] (2) an increasing number of age-related and degenerative diseases have been found to have an etiological component associated with ROS generation;[3] and (3) strategies that are effective to reduce oxidative stress are also found to affect aging. A clear example is dietary or caloric restriction in rodents.[2]

Physical exercise is an intimate part of the life cycle, as organisms need the mobility to pursue food, escape predators, and ensure reproduction. The most prominent biological change occurring during exercise is the increased metabolic rate, matched by an enhanced rate of mitochondrial respiration and oxidative phosphorylation in the eukaryotes. It is estimated that during maximal muscular contraction in men oxygen consumption at the local muscle fibers can reach as high as 100-fold of the resting levels, while the whole body oxygen consumption increases by ~20-fold.[5] Such a high rate of oxygen flux may provoke increased electron "leakage" found even at the resting condition and impose an oxidative stress to the vicinity of the mitochondria and other critical organelles essential for cell life.[6] Here an interesting and fundamental question arises, as to whether the

[a]Tel: 608/262-7250; fax: 608/262-1656; e-mail: ji@soemadison.wisc.edu

increased ROS production and oxidative stress associated with physical exercise can be a contributing factor to aging during the course of evolution, provided that the free radical theory of aging holds true. It is obviously premature to address this intriguing question at present; however, it would be beneficial for the gerontologists and exercise physiologists to integrate their thinking and research with a common theme wherein the ROS plays a central role.

In the present review the authors choose to focus on skeletal muscle for the following three reasons. (1) The topic of aging and oxidative stress has been reviewed by many experts in the field previously[2–4] and by the excellent articles in this series. However, relatively few reviews are devoted to skeletal muscle oxidative stress and aging. (2) Deterioration of skeletal muscle function is an important issue in medical gerontology because of the critical role of muscle for mobility and normal life. This statement is made without even considering the diaphragm, a vital skeletal muscle for respiration. (3) Skeletal muscle has displayed some unique characteristics during aging both in terms of free radical production and antioxidant systems. It is important to keep in mind that the extent of cell oxidative damage is determined by the rate of not only ROS production, but also ROS removal, provided by the antioxidant defense systems (including the capacity to repair the damage). Thus, age-related changes in muscle antioxidant capacity and possible influences of physical exercise will be emphasized in the present review.

ROS GENERATION IN SKELETAL MUSCLE

The free radical theory of aging has been supported by strong evidence that senescent organisms produce ROS at a higher rate than their young counterparts.[1–3] There are several well-documented sources of ROS production in the cell, among which the mitochondria have received particular attention because of their central role in oxidative phosphorylation.[7] The electron transport chain consumes ~85% of all the O_2 used in the cell, and 1–5% of the oxygen becomes various ROS as by-products.[7,8] Concentration of $O_2^{\bullet-}$ in the mitochondrial inner membrane is maintained to be 8×10^{-12} M,[4] whereas the steady state H_2O_2 production of the heart mitochondria has been estimated to be 0.3–0.6 nmol/min/mg protein.[8] Aging apparently increases ROS production in the mitochondria. Nohl and Hegner[9] showed that mitochondrial production of $O_2^{\bullet-}$ in rat heart was ~40% higher, comparing 24-month-old rats with 3-month-old rats. The majority of the $O_2^{\bullet-}$ produced is dismutated to H_2O_2 by mitochondrial superoxide dismutase (Mn SOD). Sohal and Sohal[10] demonstrated that H_2O_2 release from the mitochondria of aged flies was doubled and that the main site of $O_2^{\bullet-}$ and H_2O_2 production lies between complexes I and III. Using dichlorofluorescein as a probe, Kim *et al.*[11] reported an age-associated increase in ROS production in the mitochondria of rat liver, and the age differences widened dramatically when the mitochondria were incubated with NADPH, ADP, and Fe^{3+}.

The reason for the enhanced ROS production as organisms get old is not entirely clear, although several scenarios have been postulated. A major paradigm focuses on the finding that aged mitochondria have a lower level of cytochrome aa_3 content in proportion to other electron carriers, posing a potential danger that electrons may be transferred out of sequence, thereby forming $O_2^{\bullet-}$ at the inner membrane.[12] Another hypothesis was based on the age-related increase in the hydrophillic property of the mitochondrial inner membrane, wherein polyunsaturated fatty acids undergo an increasing rate of peroxidative modification with age.[13] This change of membrane profile, reflected by its fluidity, can attenuate its protection against the autoxidation of electron-transferring ubisemiquinone, resulting in increased $O_2^{\bullet-}$ formation.[12] These theories require further experimental approval.

In addition to mitochondria, several other cellular sources of ROS production have been identified, including (1) microsomes and cytochrome P450 complex that have special detoxification function against xenobiotics, (2) peroxisomes that produce H_2O_2 as a by-product of amino acid oxidation, and (3) activated neutrophils and other phagocytes that actively produce ROS to destroy damaged cell components and invading organisms.[3,4] However, the first two pathways are of relatively lesser importance in skeletal muscle that contains low microsome and peroxisome content. On the other hand, xanthine oxidase has been hypothesized to play a major role in $O_2^{\bullet-}$ production in skeletal muscle during intermittent hypoxia and reoxygenation, which could occur when muscles undergo ischemic contraction.[14] Each of the three ROS production pathways in the muscle, that is, mitochondria, xanthine oxidase, and neutrophils, may be activated under different physiological conditions.[15]

If the aforementioned ROS generative mechanisms are operational during normal cell function throughout the life span, then intense physical activity is expected to further exacerbate their production for the following reasons. (1) Oxygen consumption in skeletal muscle and myocardium is dramatically increased during exercise, exposing the mitochondria to a higher level of oxygen flux and electron leakage. (2) Strenuous exercise may inflict physical and mechanical damage to tissue, thereby stimulating neutrophil invasion and activating the phagocytosis pathway.[16] (3) Exercise results in a redistribution of blood flow to the various organs, rendering some tissues to hypoxia (ischemia) and reoxygenation (reperfusion), which provides an additional pathway for ROS production.[15]

Experimental data reveal that strenuous exercise indeed elicits an increased ROS production in aerobic tissues. Using the electron paramagnetic resonance (EPR) spectroscopy method, Davies et al.[17] demonstrated that free radical signals were intensified in rat hindlimb muscle and liver after an acute bout of exhaustive running. The main free radical species was identified as semiquinone, consistent with the mitochondrial theory mentioned above. Jackson et al.[18] reported a 70% increase in the EPR signals in electrically stimulated contracting muscle compared to the resting controls. Again using EPR, Kumur et al.[19] showed an increased generation of free radical signals in the myocardium after an acute bout of exhaustive exercise in rats and that vitamin E supplementation attenuated the free radical production. These studies, although in only limited numbers, provided unequivocal evidence that physical exercise promotes increased ROS generation. Because of the limited availability of EPR and the associated technical difficulty, alternative methods have been developed to measure ROS production during physical work. Reid et al.[20] demonstrated that the oxidation of DCF, a synthetic intracellular ROS probe, was increased in isolated contracting diaphragm muscle. The fact that SOD could attenuate DCF oxidation implies that $O_2^{\bullet-}$ was produced under the experimental conditions. O'Neill et al.[21] recently showed in perfused feline muscle that $^{\bullet}OH$ radical production was significantly increased during and after contraction. Using a modified DCF assay, we have recently found that rat deep vastus lateralis (DVL) muscle, immediately excised and homogenized after an exhaustive bout of running, produced ROS at a higher rate than the rested control muscles (FIG. 1). Our data also reveal that exercise provoked a greater effect on old muscle than young muscle working at a comparable intensity.

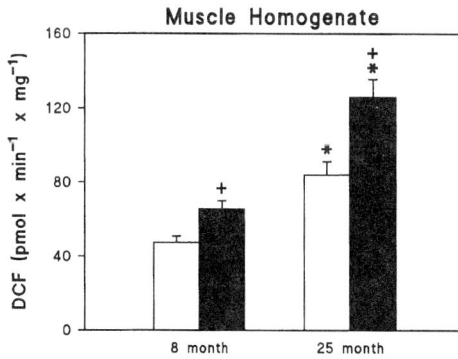

FIGURE 1. Formation of dichlorofluorescein (DCF) in vastus lateralis muscle homogenates of 8- and 25-month-old Fischer 344 rats with or without a single bout of exercise to exhaustion. Exercise intensity for young rats was 25 m/min, 5% grade; for old rats, 15 m/min, 5% grade. Exhaustion time for young and old rats was both approximately 1 hour. Incubation medium contained 20 mM Tris-HCl, 130 mM KCl, 5 mM $MgCl_2$, 20 mM NaH_2PO_4, 30 mM glucose, and 5 μM reduced DCF (pH 7.4). * $p < 0.05$, 25 month vs. 8 month. + $p < 0.05$, exercised, ■ vs. rested, □.

The above research findings lead us to believe that enhanced free radical generation at the cellular and tissue levels during physical work is a normal part of life, probably in most animal species. Because the majority of animal species (laboratory rats and humans living in the industrialized countries are exceptions) live a fairly active life, it is not unreasonable to hypothesize that the enhanced ROS production during physical activity plays an important part in aging. This hypothesis seems to be consistent with the rate of living theory, which postulates that the maximal life span of a species is inversely related to its metabolic rate, that is, oxygen consumption (VO_2) per unit of body weight. Thus, smaller mammals and birds having a higher metabolic rate due to a greater body surface area:body weight ratio, and hence heat loss, would have a shorter life span, whereas larger animals live relatively longer because of the lower VO_2 and possibly less ROS exposure.[2,23]

ANTIOXIDANT RESPONSE AND ADAPTATION

Aerobic organisms would not have survived during evolution if they had not developed a highly efficient and adaptive antioxidant defense system.[23] Antioxidants can generally be divided into two categories. The enzymatic antioxidants include superoxide dismutase (SOD), catalase (CAT), and glutathione peroxidase (GPX), supported by such other auxiliary enzymes as glutathione reductase (GR), glucose 6 phosphate dehydrogenase (G6PDH), and glutathione sulfur-transferase (GST). The nonenzymatic antioxidants include antioxidant vitamins (α-tocopherol, ascorbic acid, and β-carotene), thiols (mainly glutathione, GSH), and a variety of such low molecular weight compounds as lipoic acid, uric acid, and ubiquinone. Each of the antioxidants is located in specific cellular sites and is specialized in removing certain ROS, although considerable overlap and cooperation are demonstrated among antioxidants.[23] An important feature of antioxidants is that their cellular concentrations are heavily influenced by nutritional factors. Some are mandatory in the diet, and others require special amino acids or trace elements for biosynthesis.[4,24] The specific function and regulatory mechanism of each individual antioxidant have been studied thoroughly in the past two decades.[4,24] However, several critical questions still remain unclear. For example, biological factors contributing to the tissue-specific changes in these enzymatic and nonenzymatic antioxidants during aging and/or in response to exercise are poorly understood. Furthermore, it is still a matter of debate whether there is a need for exogenous antioxidant supplementation in aged and/or physically active individuals.

Age-related Changes in Antioxidants

Profound changes in antioxidant content and activity have been observed in various tissues and organs across all species.[22] According to the free radical theory of aging one might expect to see a general decline of cellular antioxidant defense capacity at old age. Available data (mostly from rodents) suggest that these changes are not uniform (for a detailed review, see ref. 22). Two possible mechanisms may explain this paradox. First, aging is associated with a deterioration of protein synthesis and cell differentiation capacity in most tissues, particularly in such postmitotic tissues as heart and eye lens. Therefore, antioxidant consumption and degradation probably are not adequately replenished at old age. Second, antioxidants have demonstrated considerable adaptability in response to prooxidant exposure. Localized oxidative stress in specific organs, tissues, and organelles may stimulate cellular uptake and synthesis of certain antioxidants under complicated genetic, hormonal, and nutritional regulation.[24] A well-documented example is skeletal muscle, which exhibits marked increases in antioxidant enzyme activity with aging.[25–29] Ji et al.[25] showed that activities of all major antioxidant enzymes (i.e., SOD, CAT, and GPX), as well as GST, GR, and G6PDH activities, were significantly higher in the DVL (a muscle actively recruited during endurance exercise) of old versus young rats. These changes occurred despite a general age-related decline of mitochondrial oxidative capacity. The work of Leeuwenburgh et al.[27] confirmed these findings in both DVL and soleus muscles in rats. In addition, γ-glutamyl transpeptidase (GGT) activity was significantly elevated, indicating aged muscle has a greater potential to take up GSH. Age-adaptation of antioxidant enzymes appears to be muscle fiber specific, with the most prominent increases found in such type 1 (slow-twitch oxidative) muscles as soleus, followed by such type 2a (fast-twitch oxidative) muscles as DVL, whereas type 2b muscles showed little effect.[27,29] Luhtala et al.[28] reported that elevation of muscle antioxidant enzymes during aging was markedly affected by dietary restriction in Fischer 344 rats. The progressive increases in CAT and GPX activities from 11 to 34 months of age were prevented by a 30% reduction of food intake, while an age-related increase in Mn SOD was also attenuated.

The mechanism responsible for the increased antioxidant enzyme activities in aging skeletal muscle is still elusive. One possibility is that mitochondria from aged muscles produce more ROS that may stimulate antioxidant enzyme gene expression. This scenario is consistent with the finding that mitochondrial fractions of antioxidant enzyme activity showed a greater increase in the senescent skeletal muscle[25,28] and myocardium.[30] Aged muscles are more susceptible to injury, which can provoke an acute phase response of the immune system, causing further ROS production.[5] Thus, aged muscles may be in a chronic inflammatory state, wherein steady state ROS production is elevated.[16] Gene regulation of antioxidant enzymes in aging muscle has been seldom studied. Oh-ishi *et al.*[29] reported that 24-month-old rats had higher activities of CuZn SOD, Mn SOD, GPX, and CAT in soleus muscle compared to 4-month-old rats. CuZn SOD protein content was also elevated in the aged muscle. However, no significant change was noticed in the relative abundance of mRNA for either SOD isozyme. Preliminary data from our laboratory also reveal no significant age differences in mRNA levels for CuZn SOD, Mn SOD, and GPX in DVL muscle despite prominent increases in their activities with old age (unpublished data). Only CAT mRNA showed a significant increase with age in soleus muscle (unpublished), which coincided with a large (2- to 3-fold) increase in CAT activity.[27] Thus, gene regulation of muscle antioxidant enzymes in old age may be subjected to complicated regulation involving both transcriptional and posttranscriptional mechanisms.

Aging is associated with a decline of cellular thiol reserve in most tissues.[22] However, data from our laboratory suggest that skeletal muscle and heart may be spared of this effect. Leeuwenburgh *et al.*[27] showed that aging caused no significant alteration of GSH content or GSH/GSSG (glutathione disulfide) ratio in rat DVL muscle, whereas in soleus there was a 37% increase in GSH content in old rats along with a higher GSH/GSSG ratio. Fiebig *et al.*[31] showed a significant increase in total glutathione content (GSH+GSSG) in the heart of 27- versus 5-month-old rats. The elevated myocardial GSH content was coupled with a twofold increase in GGT activity, suggesting a greater potential of the γ-glutamyl cycle. As for antioxidant vitamins, the limited available data reveal that aging does not provoke a significant change in α-tocopherol levels in rat skeletal muscle.[32] Interestingly, vitamin E concentration seems to be increased in a number of tissues, including the heart in senescence.[33] Vitamin C levels showed a general decline at older age.[22]

Antioxidant Response and Adaptation to Exercise

An acute bout of strenuous exercise has been shown to alter cellular and tissue antioxidant status profoundly.[34] This is primarily caused by the high levels of ROS production during exercise, but such other factors as altered blood flow, energy status, and availability of reducing power may also influence antioxidant function in the various tissues.[15] A good example is the effect of acute exercise on GSH homeostasis. As the most abundant nonprotein thiol source, GSH concentration in the cell is remarkably high (in the millimolar range). GSH not only serves as a substrate for GPX and GST to reduce hydrogen and organic peroxides, but GSH also scavenges singlet oxygen and ·OH.[35] Recent evidence shows that GSH may reduce tocopherol radicals and the semidehydroascorbate radical, thereby preventing lipid peroxidation.[36] GSH can be recycled by GR at the expense of NADPH, a relatively abundant source of cellular reducing power. Liver is the primary organ for *de novo* GSH synthesis and supplies 90% of circulatory GSH to maintain a whole-body homeostasis. Liver GSH content has been shown to decline dramatically (30–50%) after an acute bout of prolonged exercise.[37–39] This decrease is caused by three possible factors. First, hepatic GSH is oxidized to GSSG by oxygen free radicals and locally produced oxidants at a higher rate than the GR-catalyzed reduction rate. Some of the GSSG formed may be exported causing a net loss of GSH.[40] Second, hepatic synthesis of GSH is limited by γ-glutamylcysteine synthetase activity and requires ATP and cysteine. Strenuous exercise may reduce the availability of these compounds in the liver due to competition from other metabolic pathways.[41] Third, and probably the most important, there is an enhanced GSH output into the plasma stimulated by glucagon, catecholamines, and vasopression,[40] all of which increase their releases during prolonged exercise. Such extrahepatic tissues as kidney, heart, skeletal muscle, and erythrocytes presumably take up circulatory GSH via the γ-glutamyl cycle during exercise to ensure adequate GSH supply.[41] However, tissue GSH uptake requires translocation of its ingredient amino acids across the cell membrane, which is limited by GGT, an enzyme having low activity in heart and skeletal muscle.[39,42] Thus, during an acute exercise bout of relatively short duration (up to ~1 h), GSH oxidation to GSSG in these tissues may be matched by both a GR-catalyzed reduction and by GSH uptake from the circulation. Therefore, no substantial disturbance of GSH status is observed even at a high workload.[43,44] However, during prolonged exercise, as liver GSH reserve and output diminishes, tissue GSH uptake can no longer sustain the enhanced GSH oxidation by ROS, resulting in a decrease of GSH:GSSG ratio and a net GSH deficit.[37–39,45] It is conceivable that exogenous GSH supplementation would be beneficial to preserve GSH homeostasis thereby reducing an exercise-induced oxidative stress.

It has long been recognized that in mammals and birds antioxidant enzyme activity in skeletal muscle is higher in the wild species compared to their domestic counterparts.[46] Within the body, such tissues that have a higher oxygen consumption rate, as liver, heart, and brain, or those chronically exposed to ROS, such as eye lens, have greater antioxidant protection than those with lower oxygen or ROS exposure.[23] Antioxidant enzyme activity and GSH content vary widely in skeletal muscle depending on the metabolic characteristics of muscle types. The more aerobic type 1 muscle possesses greater antioxidant potential than type 2a and type 2b muscle.[44] Whereas these interspecies and interorgan/tissue differences in antioxidant potential may reflect long-term adaptation during evolution, a relatively short period of oxidative stress has also been shown to induce certain antioxidants.[23,24] In humans, a higher antioxidant enzyme activity has been reported to correlate with maximal aerobic capacity (VO_{2max}), and trained athletes have greater SOD and CAT activities in skeletal muscle than untrained people.[6] Mena et al.[47] reported that amateur and professional cyclists had higher erythrocyte SOD activity than sedentary subjects and that professional cyclists had higher CAT and GPX activities than the amateur cyclists.

Numerous studies have shown that antioxidant enzyme activities are elevated in skeletal muscle after endurance training involving repeated bouts of prolonged exercise.[34] It is believed that the increased muscle ROS production constitutes the underlying reason for this training adaptation, although the regulatory mechanism is still poorly understood. Several authors have provided detailed reviews on this subject recently.[5,6,34] Therefore only a few highlights will be mentioned in the present review. (1) Training adaptation of antioxidant enzymes is highly tissue- and muscle fiber–specific. In general, only the oxidative types of muscle that are selectively recruited during endurance exercise (including diaphragm) show an increase in antioxidant activity.[48–50] (2) The occurrence and magnitude of training adaptation in antioxidant enzymes depend largely on the volume of training, particularly the duration of each training session.[49] This is consistent with the notion that ROS production is roughly proportional to the level of oxygen consumption during exercise. (3) Among the various antioxidant enzymes, GPX has demonstrated the most consistent and prominent training adaptation. It may reflect the fact that GPX is the terminal enzyme to remove H_2O_2 and organic peroxides in the cell.[8] It may also be explained in part by its relatively small protective margin, compared to SOD, in the various tissues and muscle fibers.[34] (4) In addition to antioxidant enzymes, GSH has displayed a training adaptation in selective muscle fibers.[50–52] This may be explained in part by a higher GGT activity in these muscles, facilitating GSH uptake from the plasma during and after exercise.[50] (5) Except for a few species, antioxidant vitamins cannot be synthesized in most mammals. Therefore, endurance training can theoretically reduce tissue levels of these vital compounds that are actively consumed during exercise, unless they are supplemented in the diet. Vitamin E concentration has been shown to decrease in a number of tissues, including skeletal muscle, after endurance training.[53–55] The reduction of vitamin E concentration in the inner mitochondrial membrane can render the organelles more susceptible to oxidative damage.

The mechanism by which antioxidant enzymes can be upregulated within a relatively short period of time in response to physical exercise is largely unknown. In the prokaryotes, two ROS-responsive genes have been identified, that is, oxyR and soxR, which control the expression of catalase and SOD, respectively.[56] Antioxidant enzyme regulation in eukaryotic cells has just begun to be understood in recent years. Available data suggest that training adaptation of antioxidant enzymes is accomplished both by transcriptional and posttranscriptional mechanisms. Oh-ishi et al.[57] showed that increased Mn SOD activity in trained rat soleus muscle was accompanied by elevated mRNA levels and immunoreactive Mn SOD protein content, suggesting a possible transcriptional control. By contrast, CuZn SOD demonstrated no alteration in either mRNA abundance or protein levels despite a prominent increase in activity after training, suggesting a posttranscriptional control. Recently, there has been an extensive search for the signaling pathways of antioxidant enzyme gene expression in mammalian tissues. Two universally expressed transcription factors, AP-1 and NF-κB, have been shown to play an important role.[58,59] NF-κB is activated by a variety of prooxidants, including ROS. Upon stimulation, including oxidative stress, the inhibitory subunit IκB, previously bound to the NF-κB complex, dissociates from the two main subunits, p50 and p65, allowing them to translocate into the nucleus and serve as potential inducers of antioxidant enzymes.[60] The Mn SOD gene has been shown to contain an AP-1 binding motif, which could be a potential regulatory site.[61] We have recently reported a rapid activation of AP-1 binding using nuclear extracts from rat skeletal muscle in response to an acute bout of prolonged exercise.[62] Maximal binding was found to occur within 1 h after exercise ceased, followed by a slow return to resting levels at ~10 hours. By contrast, levels of NF-κB binding showed a latent rise at 2 h after the ceasation of exercise and remained elevated up to 10 hours. These preliminary findings suggest that increased ROS during exercise might have stimulated AP-1 and NF-κB binding to nuclear proteins of contracting muscle, which could potentially modify gene expression of the antioxidant enzymes.

Although evidence for training inducon of antioxidant enzymes in skeletal muscle is abundant, the functional implication of such adaptation is far from resolved. A logical question is whether the trained muscles are more protected from exercise-induced oxidative damage. Davies et al.[17] proposed that endurance training helps reduce oxidative damage by increasing mitochondrial oxidative enzymes and consequently decreasing oxygen flux in each respiratory chain. Higuchi et al.,[63] however, argued that the increase in mitochondrial Mn SOD with training is relatively small compared to increases in other mitochondrial enzymes and is therefore unlikely to offer improved protection. Ji et al.[64] showed that trained rats displayed a lesser degree of sulfhydryl oxidation and downregulation of oxidative enzymes in muscle mitochondria after an acute bout of exhaustive exercise, compared to untrained rats. They attributed this to an increased mitochondrial GPX activity after training. To investigate the functional implication of antioxidant enzyme training adaptation, we isolated muscle mitochondria from trained and untrained rats that were subsequently exposed to exogenous $O_2^{\cdot-}$. Although mitochondria respiratory capacity was suppressed by these oxidants in both groups, mitochondria from trained rats showed significantly less inhibition of state 3 respiration rate (FIG. 2a) and respiratory control index (FIG. 2b) than their sedentary counterparts when equal doses of $O_2^{\cdot-}$ were included in the media. Similar results were obtained when the mitochondria were subjected to inhibition by H_2O_2 exposure. These results imply that, when working at the same workload and hence exposed to similar levels of ROS, mitochondria from trained muscle are capable of maintaining a higher rate of oxidative phosphorylation and ATP production.

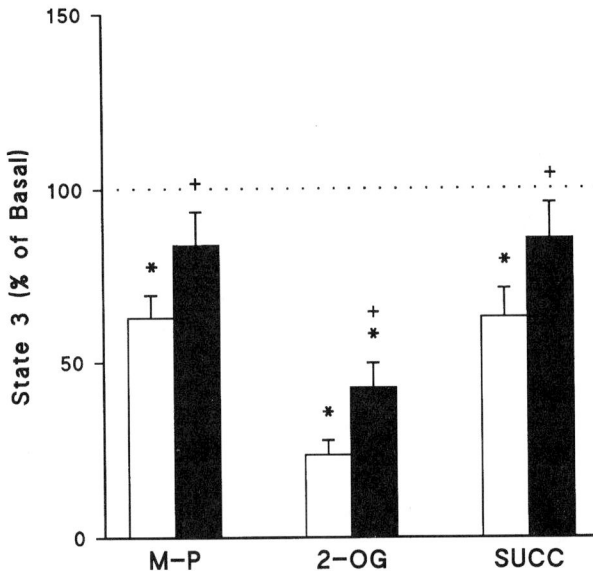

FIGURE 2A. Inhibition of muscle mitochondrial state 3 respiration rate (**A**) and respiratory control index (RCI) (**B**) by $O_2^{\cdot-}$ in trained and untrained Fischer 344 rats (age 4.5 mo). Mitochondria were isolated from mixed vastus lateralis muscle of rats killed 48 hours after the last training session at the resting state. Respiration medium contained 20 mM Tris-HCl, 130 mM KCl, 5 mM MgCl$_2$, 20 mM NaH$_2$PO$_4$, and 30 mM glucose (pH 7.4). Respiration substrates were 2 mM malate + 2 mM pyruvate (M-P), 4 mM 2-oxoglutarate (2-OG), or 4 mM succinate with 2.4 µM rotenone (SUCC). State 3 respiration was initiated by adding 750 nmol ADP after state 4 respiration was established. $O_2^{\cdot-}$ was generated by including 22 µM hypoxanthine and 0.13 units/mL xanthine oxidase. * p < 0.05, compared to values without $O_2^{\cdot-}$ treatment. + p < 0.05, trained, ■ vs. control, □.

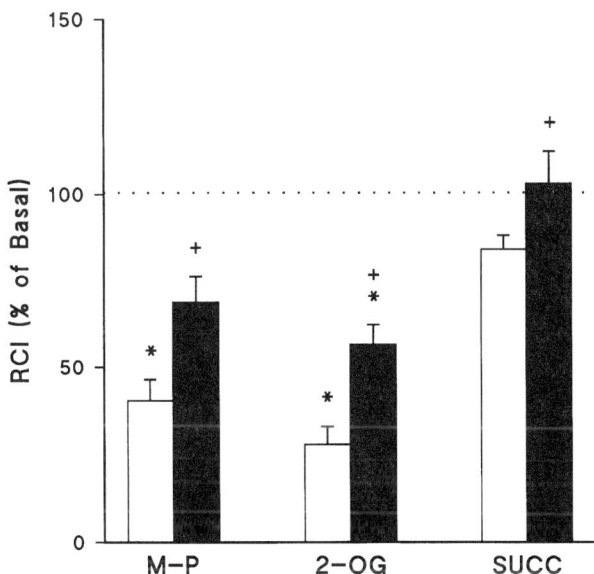

FIGURE 2B. See legend on p. 110.

EXERCISE AND OXIDATIVE STRESS IN AGING SKELETAL MUSCLE

Despite the overwhelming evidence of cellular and tissue oxidative damage at old age, surprisingly, there is a scarcity of data concerning the age-related oxidative changes in skeletal muscle. Most of the evidence has come from measurement of lipid peroxidation products, such as accumulation of lipofuscin[65] and increased MDA formation[25,27,32] in aging muscle. Elevated antioxidant enzyme activity and GSH levels have been widely regarded as adaptive responses to an age-related oxidative stress in senescent muscle.[25–29] It is well known that muscle mitochondrial respiratory capacity and oxidative potential decline with age, which results in functional difficulty for the aged skeletal muscle to perform mechanical tasks.[66] Senescent muscles also lose cross-section area and muscle mass, known as sarcopenia. It has recently been proposed that oxidative stress plays a major role in the etiology of sarcopenia.[67] However, experimental data linking the loss of muscle functional capacity with oxidative stress are yet to be provided. With regard to the role of oxidative stress in skeletal muscle aging, the following three questions are highly relevant. Is aged muscle more susceptible to exercise-induced oxidative stress? Does exercise training induce antioxidant enzymes and other antioxidants in aged muscle as found in young muscle? Does training increase the resistance of aged muscle to oxidative damage?

Susceptibility of Aged Muscles to Exercise-induced Oxidative Stress

Although mitochondria from aged muscles presumably have a greater potential of generating ROS due to the subtle changes in the structure of the electron transport chain,[12] several studies failed to detect an exacerbated lipid oxidation in aging muscle in response to an acute bout of exercise compared to young muscle.[25,68] This is probably because aged animals cannot reach a high workload that elicits oxidative damage found in the young muscles, and aged muscles have a higher antioxidant defense capacity to remove the increased ROS.[25,28] However, Meldani et al.[69] showed that supplementation of vitamin E offered a greater protection in reducing urinary lipid peroxide secretion in the older human subjects than in the young ones. Zerba et al.[70] observed that old mouse muscle was more susceptible to oxidative injury caused by lengthening contraction (eccentric exercise), as a lower isotonic force could elicit the same degree of damage in old muscle as found in the young muscle. In that study, no direct measurement of the oxidative damage marker was made, but pretreatment of the older muscle with SOD showed a protective effect. Inasmuch as eccentric exercise involves mechanical injury and acute phase response of the immune system,[16] these data indicate that age effects on exercise-induced oxidative damage in muscle may be manifested in multiple pathways and that neutrophil activation may be an important mechanism for ROS production and oxidative stress in aged muscle.

Training Adaptation of Antioxidant Systems in Aging Skeletal Muscle

Senescent skeletal muscles lose considerable regenerative ability due to a decreased rate of protein turnover, satellite cell population, and proliferative capacity.[71] Increased protein breakdown, partly explained by oxidative damage and consequently selective degradation, coupled with decreased protein synthesis, results in a progressive decline in oxidative enzyme levels and energy production in aging skeletal muscle.[5,72] Endurance training has been shown to effectively restore losses of muscle protein content and mitochondrial oxidative capacity in old age.[66] However, there are conflicting data concerning training adaptation of antioxidant systems in aged skeletal muscle. Ji et al.[73] showed that along with increased activities of muscle citrate synthase, malate dehydrogenase, and lactate dehydrogenase, training also resulted in a 60% increase in GPX activity in 27.5-month-old Fischer 344 rats (FIG. 3). Hammeren et al.[74] also reported a significant increase in GPX activity with training in several skeletal muscles of old Fischer 344 rats. However, Leeuwenburgh et al.[27] found no antioxidant enzyme adaptation in either DVL or soleus muscle from old rats. The failure to induce antioxidant enzyme levels in aged muscle obviously cannot be accounted for by a reduced cell proliferative capacity, because training successfully increased mitochondrial oxidative enzyme activity and protein content.[66,73,74] A possible explanation is that antioxidant enzyme activities in the senescent muscle are already high; therefore the training threshold is raised, requiring greater intensity to provoke a training effect.

There is a general concern that aged animals are subjected to a deficiency of antioxidant nutrients.[75] Exercise training may potentially exacerbate this nutritional deficiency due to increased consumption or decreased dietary intake, or both.[55] However, a clear consensus in this important area of gerontology is lacking. Kretzchmar et al.[76] reported that trained men at age 36–57 had higher levels of GSH in the plasma than their sedentary counterparts. Increased muscle GSH output was proposed to play a role in maintaining plasma GSH homeostasis with training at old age. Leeuwenburgh et al.[27] showed no differences in muscle GSH content between trained and untrained old rats, although aging increased GSH levels in the muscle. Starnes et al.[32] compared muscle α-tocopherol levels

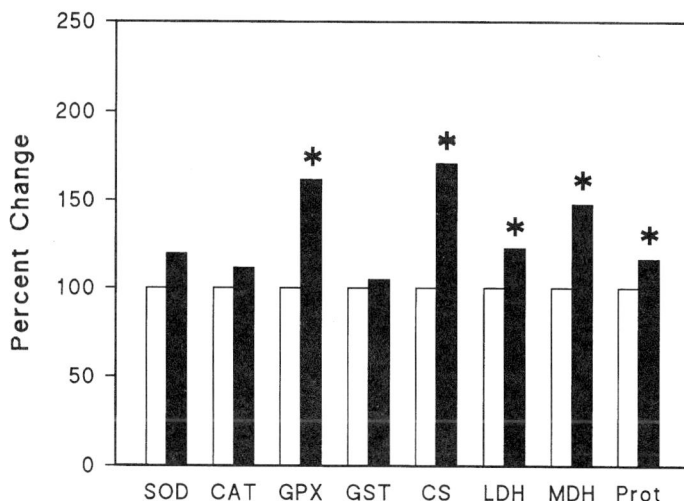

FIGURE 3. Relative changes of muscle antioxidant and metabolic enzyme activities in 27.5-month-old Fischer 344 rats after a 10-wk endurance training program. Enzyme activities were measured spectrophotometrically in the homogenate prepared from rat vastus lateralis muscle.[73] SOD, superoxide dismutase; CAT, catalase; GPX, glutathione peroxidase; GST, glutathione sulfur-transferase; CS, citrate synthase; LDH, lactate dehydrogenase; MDH, malate dyhydrogenase; Prot, muscle protein content. * $p < 0.05$, trained, ■ vs. untrained, □.

between trained and untrained 24-month-old rats and found a significant decrease in the trained muscle. Old trained rats also displayed a higher level of muscle thiobarbituric acid–reactive substance when challenged by ascorbic acid and ferrous ion. These findings are consistent with the earlier data concerning training effect on muscle vitamin E levels in young rats.[53,54] Taken together, it seems that aged skeletal muscle does not necessarily lose adaptability to training. It is still capable of upregulating antioxidant enzyme synthesis and maintaining GSH status. However, aged muscle may be susceptible to a deficit of antioxidant vitamins that the body cannot synthesize. It is conceivable that physically active elderly individuals may benefit from dietary antioxidant supplementation.

Does Training Increase the Resistance of Aged Muscle to Oxidative Damage?

When exercising at a given workload, oxygen consumption is no different between trained and untrained muscle. Therefore, endurance-trained aged muscle has the advantage of distributing the electron flux among increased mitochondrial electron transport chains as compared to untrained muscle.[17] This should theoretically result in a lesser degree of ROS production. Furthermore, trained muscles have higher levels of antioxidant enzyme activity and GSH content, providing a greater ROS removal. Thus, training at old age should reduce muscle oxidative injury during acute physical exertion. Indeed, DVL and soleus muscles from the aged rats displayed significantly lower malondialdehyde (MDA) levels than did their sedentary counterparts after endurance training.[27] The same training effect was found in the aged myocardium after endurance training.[77] However, these data should be interpreted with caution because muscle MDA levels could also be influenced by the rate of efflux from the muscle.[23] The low MDA content may result from enhanced output due to circulatory training adaptation in old animals.

The effect of endurance training on mitochondrial respiratory function and resistance to ROS has been studied in aged skeletal muscle.[78] Although training elicited no apparent improvement on the state 3 respiration rate and respiration control index (RCI), the magnitude of RCI inhibition by equal doses of $O_2^{\bullet-}$ and H_2O_2 was significantly smaller in the trained versus control rats. If these results from this *in vitro* study could be extrapolated to an *in vivo* condition, aged skeletal muscle undergoing exercise training would be more capable of maintaining adequate mitochondrial function when exposed to acute oxidative stress.

CONCLUSION

Skeletal muscle is under increasing oxidative stress during aging. This is evidenced by the higher levels of lipid peroxidation and antioxidant defense in aged muscle. However, there is a paucity of data supporting a linkage between muscle oxidative stress markers and physiological function. Physical exercise promotes the generation of ROS in the skeletal muscle primarily via the mitochondrial electron transport chain, but multiple cellular sources may contribute to the increased ROS during and after exercise. Senescent muscle appears to be more susceptible to exercise-induced oxidative damage possibly due to its greater vulnerability to mechanical injury that activates an acute phase response. Exercise training provides a clear benefit by upregulating muscle antioxidant enzymes and glutathione status, and this adaptability appears to sustain at old age. However, intensive exercise can deplete muscle antioxidant vitamin levels, which may compromise its overall antioxidant protective margin. Thus, the physically active elderly should be careful in selecting the appropriate mode and intensity of exercise and ensure sufficient dietary antioxidant intake to minimize a possible oxidative stress while enjoying the full benefits of exercise.

REFERENCES

1. HARMAN, D. 1956. Aging: A theory based on free radical and radiation chemistry. J. Gerontol. **11:** 298–300.
2. SOHAL, R.S. & R. WEINDRUCH. 1996. Oxidative stress, caloric restriction, and aging. Science **273:** 59–63.
3. AMES, B.N., M.K. SHIGENAGA & T.M. HAGEN. 1993. Oxidant, antioxidants, and degenerative diseases of aging. Proc. Natl. Acad. Sci. USA **90:** 7915–7922.
4. YU, B.P. 1994. Cellular defenses against damage from reactive oxygen species. Physiol. Rev. **74:** 139–162.
5. MEYDANI, M. & W.J. EVANS. 1993. Free radicals, exercise, and aging. *In* Free Radicals in Aging. B.P. Yu, Ed.: 183-204. CRC Press. Boca Raton, Florida.
6. JENKINS, R.R. 1993. Exercise, oxidative stress, and antioxidants: A review. Int. J. Sports Nutr. **3:** 356–375.
7. SHIGENAGA, M.K., T.M. HAGEN & B.N. AMES. 1994. Oxidative damage and mitochondrial decay in aging. Proc. Natl. Acad. Sci. USA **91:** 10771–10778.
8. CHANCE, B., H. SIES & A. BOVERIS. 1979. Hydroperoxide metabolism in mammalian organs. Physiol. Rev. **59:** 527–605.

9. Nohl, H. & D. Hegner. 1978. Do mitochondria produce oxygen radicals *in vivo*? Eur. J. Biochem. **82:** 563–547.
10. Sohal, R.S. & B.H. Sohal. 1991. Hydrogen peroxide release by mitochondria increases during aging. Mech. Ageing Dev. **57:** 187–202.
11. Kim, J.D., R.J.M. McCarter & B.P. Yu. 1996. Influence of age, exercise, and dietary restriction on oxidative stress in rats. Aging Clin. Exp. Res. **8:** 123–129.
12. Nohl, H. 1986. Oxygen radical release in mitochondria: Influence of aging. *In* Free Radicals, Aging, and Degenerative Diseases. J.E. Johnson, Ed.: 77–98. Alan R. Liss, Inc. New York.
13. Chen, J.J. & B.P. Yu. 1996. Alteration in mitochondrial membrane fluidity by lipid peroxidation products. Free Radical Biol. Med. **17:** 411–418.
14. McCord, J.M. & R.S. Roy. 1982. The pathophysiology of superoxide: Roles in inflammation and ischemia. Can. J. Physiol. Pharmacol. **60:** 1346–1352.
15. Ji, L.L. & S.L. Leichtweis. 1997. Exercise and Oxidative Stress: Source of free radicals and their impact on antioxidant systems. Age **20:** 91–106.
16. Cannon, J.G. & J.B. Blumberg. 1994. Acute phase immune responses in exercise. *In* Exercise and Oxygen Toxicity. C.K. Sen, Ed.: 447–479. Elsevier Science. New York.
17. Davies, K.J.A., A.T. Quintanilha, G.A. Brooks & L. Packer. 1982. Free radicals and tissue damage produced by exercise. Biochem. Biophys. Res. Commun. **107:** 1198-1205.
18. Jackson, M.J., R.H.T. Edwards & M.C.R. Symons. 1985. Electron spin resonance studies of intact mammalian skeletal muscle. Biochim. Biophys. Acta **847:** 185–190.
19. Kumar, C.T., V.K. Reddy, M. Prasad, K. Thyagaraju & P. Reddanna. 1992. Dietary supplementation of vitamin E protects heart tissue from exercise-induced oxidative stress. Mol. Cell. Biochem. **111:** 109–115.
20. Reid, M.B., K.E. Haack, K.M. Franchek, P.A. Valberg, L. Kobzik & M.S. West. 1992. Reactive oxygen in skeletal muscle. I. Intracellular oxidant kinetics and fatigue *in vitro*. J. Appl. Physiol. **73:** 1797–1804.
21. O'Neill, C.A., C.L. Stebbins, S. Bonigut, B. Halliwell & J.C. Longhurst. 1996. Production of hydroxyl radicals in contracting skeletal muscle of cats. J. Appl. Physiol. **81(3):** 1197–1206.
22. Matsuo, M. 1993. Age-related alterations in antioxidant defense. *In* Free Radicals in Aging. B.P. Yu, Ed.: 143–181. CRC Press. Boca Raton, Florida.
23. Halliwell, B. & J.M.C. Gutteridge. 1989. Free Radicals in Biology and Medicine, 2nd Ed. Clarendon Press. Oxford, England.
24. Harris, E.D. 1992. Regulation of antioxidant enzymes. FASEB J. **6:** 2675–2683.
25. Ji, L.L., D. Dillon & E. Wu. 1990. Alteration of antioxidant enzymes with aging in rat skeletal muscle and liver. Am. J. Physiol. **258:** R918–R923.
26. Lawler, J.M., S.K. Powers, T. Visser, H. Van Dijk, M.J. Korthuis & L.L. Ji. 1993. Acute exercise and skeletal muscle antioxidant and metabolic enzymes: Effect of fiber type and age. Am. J. Physiol. **265:** R1344–R1350.
27. Leeuwenburgh, C., R. Fiebig, R. Chandwaney & L.L. Ji. 1994. Aging and exercise training in skeletal muscle: Response of glutathione and antioxidant enzyme systems. Am. J. Physiol. **267:** R439–R445.
28. Luhtala, T.A., E.B. Roecker, T. Pugh, R.J. Feuers & R. Weindruch. 1994. Dietary restriction attenuates age-related increases in rat skeletal muscle antioxidant enzyme activities. J. Gerontol. **49:** B321–B328.
29. Oh-Ishi, S., T. Kizaki, H. Yamashita, N. Nagata, K. Suzuki, N. Taniguchi & H. Ohno. 1995. Alterations of superoxide dismutase iso-enzyme activity, content, and mRNA expression with aging in rat skeletal muscle. Mech. Ageing Dev. **84:** 65–76.
30. Ji, L.L., D. Dillon & E. Wu. 1991. Myocardial aging: Antioxidant enzyme system and related biochemical properties. Am. J. Physiol. **261:** R386–392.
31. Fiebig, R., C. Leeuwenburgh, M. Gore & L.L. Ji. 1996. The interactive effects of aging and training on myocardial antioxidant enzymes and oxidative stress. Age **19:** 83–89.
32. Starnes, J.W., G. Cantu, R.P. Farrar & J.P. Kehrer. 1989. Skeletal muscle lipid peroxidation in exercise and food restricted rats during aging. J. Appl. Physiol. **67:** 69–75.
33. Weglicki, W.B., Z. Luna & P.P. Nair. 1969. Sex and tissue specific differences in concentrations of α-tocopherol in mature and senescent rats. Nature **221:** 185.

34. JI, L.L. 1995. Exercise and free radical generation: Role of cellular antioxidant systems. *In* Exercise and Sport Science Review. J. Holloszy, Ed.: 135–166. Williams & Wilkins Co. Baltimore, Maryland.

35. MEISTER, A. & M.E. ANDERSON. 1983. Glutathione. Annu. Rev. Biochem. **52:** 711–760.

36. PACKER, L., A.L. ALMADA, L.M. ROTHFUSS & D.S. WILSON. 1989. Modulation of tissue vitamin E levels by physical activity. Ann. N.Y. Acad. Sci. **570:** 311–321.

37. LEW, H., S. PYKE & A. QUINTANILHA. 1985. Changes in glutathione status of plasma, liver, and muscle following exhaustive exercise in rats. FEBS Lett. **185:** 262–266.

38. LEEUWENBURGH, C. & L.L. JI. 1996. Glutathione regulation during exercise in unfed and refed rats. J. Nutr. **126:** 1833–1843.

39. LEEUWENBURGH, C. & L.L. JI. 1995. Glutathione depletion in rested and exercised mice: Biochemical consequence and adaptation. Arch. Biochem. Biophys. **316:** 941–949.

40. SIES, H. & P. GRAF. 1985. Hepatic thiol and glutathione efflux under the influence of vasopressin, phenylephrine and adrenaline. Biochem. J. **226:** 545–549.

41. JI, L.L. & C. LEEUWENBURGH. 1996. Glutathione and exercise. *In* Pharmacology in Exercise and Sports. S. Somani, Ed.: 97–123. CRC Press. New York.

42. LEEUWENBURGH, C., S. LEICHTWEIS, R. FIEBIG, J. HOLLANDER, M. GORE & L.L. JI. 1996. Glutathione depletion in the myocardial: Effect of acute exercise. Mol. Cell. Biochem. **156:** 17–24.

43. JI, L.L. & R.G. FU. 1992. Responses of glutathione system and antioxidant enzymes to exhaustive exercise and hydroperoxide. J. Appl. Physiol. **72:** 549–554.

44. JI, L.L., R.G. FU & E.W. MITCHELL. 1993. Glutathione and antioxidant enzyme in skeletal muscle: Effect of fiber type and exercise intensity. J. Appl. Physiol. **73:** 1854–1859.

45. SEN, C.K., E. MARIN, M. KRETZSCHMAR & O. HANNINEN. 1992. Skeletal muscle and liver glutathione homeostasis in response to training, exercise, and immobilization. J. Appl. Physiol. **73:** 1265–1272.

46. BURGE, W.E. & A.J. NEIL. 1916–1917. Comparison of the amount of catalase in the muscle of large and of small animals. Am. J. Physiol. **43:** 433–437.

47. MENA, P., M. MAYNAR, J.M. GUTIERREZ, J. MAYNAR, J. TIMON & J.E. CAMPILLO. 1991. Erythrocyte free radical scavenger enzymes in bicycle professional racers. Adaptation to training. Int. J. Sports Med. **12:** 563–566.

48. LAUGHLIN, M.H., T. SIMPSON, W.L. SEXTON, O.R. BROWN, J.K. SMITH & R.J. KORTHUIS. 1990. Skeletal muscle oxidative capacity, antioxidant enzymes, and exercise training. J. Appl. Physiol. **68:** 2337–2343.

49. POWERS, S.K., D. CRISWELL, J. LAWLER, L.L. JI, D. MARTIN, R. HERB & G. DUDLEY. 1994. Influence of exercise intensity and duration on antioxidant enzyme activity in skeletal muscle differing in fiber type. Am. J. Physiol. **266:** R375–R380.

50. LEEUWENBURGH, C., J. HOLLANDER, S. LEICHTWEIS, R. FIEBIG, M. GORE & L.L. JI. 1997. Adaptations of glutathione antioxidant system to endurance training are tissue and muscle fiber specific. Am. J. Physiol. **272:** R363–R369.

51. SEN, C.K., E. MARIN, M. KRETZSCHMAR & O. HANNINEN. 1992. Skeletal muscle and liver glutathione homeostasis in response to training, exercise and immobilization. J. Appl. Physiol. **73:** 1265–1272.

52. MARIN, E., O. HANNINEN, D. MULLER & W. KLINGER. 1990. Influence of acute physical exercise on glutathione and lipid peroxide in blood of rat and man. Acta Physiol. Hung. **76:** 71–76.

53. AIKAWA, K.M., A.T. QUINTANILHA, B.O. DELUMEN, G.A. BROOKS & L. PACKER. 1984. Exercise endurance training alters vitamin E tissue levels and red blood cell hemolysis in rodents. Biosci. Rep. **4:** 253–257.

54. GOHIL, K., L. ROTHFUSS, J. LANG & L. PACKER. 1987. Effect of exercise training on tissue vitamin E and ubiquinone content. J. Appl. Physiol. **63:** 1638–1641.

55. PACKER, L. 1986. Oxygen radicals and antioxidants in endurance exercise. *In* Biochemical Aspects of Physical Exercise. G. Benzi, L. Packer & N. Siliprandi, Eds.: 73–92. Elsevier Science Publishers. New York.

56. STORZ, G., L.A. TARTAGLIA & B.N. AMES. 1990. Transcriptional regulator of oxidative stress-inducible genes: Direct activation by oxidation. Science **248:** 189–94.

57. OH-ISHI, S., T. KIZAKI, H. YAMASHITA, H. YAMASHITA, T. IZAWA, T. KOMABAYASHI, N. NAGATA, K. SUZUKI, N. TANIGUCHI & H. OHNO. 1997. Effect of endurance training on superoxide dismutase

activity, content, and mRNA expression in rat skeletal muscle. Clin. Exp. Pharm. Physiol. **24:** 326–332.

58. DEMPLE, B. & C.F. AMABILE-CUEVAS. 1991. Redox redux: The control of oxidative stress responses. Cell **67:** 837–839.

59. SEN, C.K. & L. PACKER. 1996. Antioxidant and redox regulation of gene transcription. FASEB J. **10:** 709–720.

60. MEYER, M., R. SCHRECK & P.A. RAEUERLE. 1993. H_2O_2 and antioxidants have opposite effects on activation of NF-κB and AP-1 in intact cells: AP-1 as secondary antioxidant-responsive factor. EMBO J. **12:** 2005–2015.

61. HO, Y.S., A.J. HOWARD & J.D. CRAPO. 1991. Molecular structure of a functional rat gene for manganese-containing superoxide dismutase. Am. J. Respir. Cell. Mol. Biol. **4:** 278–286.

62. GORE, M., R. FIEBIG, J. HOLLANDER & L.L. JI. 1997. Acute exercise alters mRNA abundance of antioxidant enzyme and nuclear factor B activation in skeletal muscle, heart and liver. Med. Sci. Sports Exercise **29:** S229.

63. HIGUCHI, M., L.J. CARTIER, M. CHEN & J.O. HOLLOSZY. 1985. Superoxide dismutase and catalase in skeletal muscle: Adaptive response to exercise. J. Gerontol. **40:** 281–286.

64. JI, L.L., F.W. STRATMAN & H.A. LARDY. 1988. Enzymatic downregulation with exercise in rat skeletal muscle. Arch. Biochem. Biophys. **263:** 137–149.

65. GUTMANN, E. 1977. Muscle. *In* Handbook of the biology of aging. C.E. Finch & L. Hayflick, Eds.: 445–469. Van Nostrand Reinhold. New York.

66. FITTS, R.H., J.P. TROUP & F.A. WITZMANN. 1984. The effect of ageing and exercise on skeletal muscle function. Mech. Ageing Dev. **27:** 161–172.

67. WEINDRUCH, R. 1995. Interventions based on the possibility that oxidative stress contributes to sarcopenia. J. Gerontol. **50:** 157–161.

68. CANNON, J.G., S.F. ORENCOLE, R.A. FIELDING, M. MEYDANI, S. MEYDANI, M.A. FITARONE, J.B. BLUMBERG & W.J. EVENS. 1990. Acute phase response in exercise: Interaction of age and vitamin E on neutrophils and muscle enzyme release. Am. J. Physiol. **256:** R1214.

69. MEYDANI, M., J.G. CANNON, J. BURRILL, S.F. ORENCOLE, R.A. FIELDING, M.A. FITARONE, J.B. BLUMBERG & W.J. EVENS. 1990. Protective effect of vitamin E on exercise induced oxidative damage in young and elderly subjects. Free Radical Biol. Med. **9:** 109.

70. ZERBA, E., T.E. KOMOROWSKI & J.A. FAULKNER. 1990. Free radical injury to skeletal muscle of young, adult, and old mice. Am. J. Physiol. **258:** C429–C435.

71. CARLSON, B.M. 1995. Factors influencing the repair and adaptation of muscles in aged individuals: Satellite cells and innervation. J. Gerontol. **50:** 96–100.

72. HANSFORD, R.G. 1983. Bioenergetics in aging. Biochim. Biophys. Acta **726:** 41–80.

73. JI, L.L., E. WU & D.P. THOMAS. 1991. The effect of exercise training on metabolic and antioxidant functions in senescent rat skeletal muscle. Gerontology **37:** 317–325.

74. HAMMEREN, J., S. POWERS, J. LAWLER, D. CRISWELL, D. LOWENTHAL & M. POLLOCK. 1993. Exercise training-induced alterations in skeletal muscle oxidative and antioxidant enzyme activity in senescent rats. Int. J. Sports Med. **13:** 412–416.

75. KATZ, M.L. & W.G. ROBINSON. 1986. Nutritional influences on autoxidation, lipofuscin accumulation, and aging. *In* Free Radicals, Aging, and Degenerative Diseases. J.E. Johnson, R. Walford, D. Harman & J. Miguel, Eds.: 221–262. Alan R. Liss, Inc. New York.

76. KRETZSCHMAR, M., U. PFEIFER, G. MACHNIK & W. KLINGER. 1991. Influence of age, training and acute physical exercise on plasma glutathione and lipid peroxidation in man. Int. J. Sports Med. **12:** 218.

77. FIEBIG, R., C. LEEUWENBURGH, M. GORE & L.L. JI. 1996. The interactive effects of aging and training on myocardial antioxidant enzymes and oxidative stress. Age **19:** 8–89.

78. CHANDWANEY, R., S. LEICHTWEIS, C. LEEUWENBURGH & L.L. JI. 1998. Oxidative stress and mitochondrial function in skeletal muscle: Effects of aging and exercise training. Age **21:** 109–117.

Mitochondrial Aging: Open Questions[a]

KENNETH B. BECKMAN[b] AND BRUCE N. AMES[c]

Department of Molecular and Cell Biology, University of California,
Barker Hall, Berkeley, California 94720-3202, USA

ABSTRACT: Interest in the role of mitochondria in aging has intensified in recent years. This focus on mitochondria originated in part from the free radical theory of aging, which argues that oxidative damage plays a key role in degenerative senescence. Among the numerous mechanisms known to generate oxidants, leakage of the superoxide anion and hydrogen peroxide from the mitochondrial electron transport chain are of particular interest, due to the correlation between species-specific metabolic rate ("rate of living") and life span. Phenomenological studies of mitochondrial function long ago noted a decline in mitochondrial function with age, and on-going research continues to add to this body of knowledge. The extranuclear somatic mutation theory of aging proposes that the accumulation of mutations in the mitochondrial genome may be responsible in part for the mitochondrial phenomenology of aging. Recent studies of mitochondrial DNA (mtDNA) deletions have shown that they increase with age in humans and other mammals. Currently, there exist numerous important and fundamental questions surrounding mitochondria and aging. Among these are (1) How important are mitochondrial oxidants in determining overall cellular oxidative stress? (2) What are the mechanisms of mitochondrial oxidant generation? (3) How are lesions and mutations in mtDNA formed? (4) How important are mtDNA lesions and mutations in causing mitochondrial dysfunction? (5) How are mitochondria regulated, and how does this regulation change during aging? (6) What are the dynamics of mitochondrial turnover? (7) What is the relationship between mitochondrial damage and lipofuscinogenesis? (8) What are the relationships among mitochondria, apoptosis, and aging? and (9) How can mitochondrial function (ATP generation and the establishment of a membrane potential) and dysfunction (oxidant generation) be modulated and degenerative senescence thereby treated?

If the promulgation of a scientific hypothesis in the popular press is a measure of how well it has captivated the scientific community, then the "mitochondrial free radical theory of aging" has surely been successful. In recent years, on the pages of *Scientific American*[1,2] (as well as in scores of daily newspapers and health magazines), mitochondrial oxygen free radicals have been fingered as a major contributor to age-related degeneration. In the scientific literature, the attention being paid to mitochondria has paralleled an overall swell of interest in free radicals themselves, as can be seen in the number of recent review articles devoted to the topic.[3–14] Also, the burgeoning field of mitochondrial genetics, in particular the elucidation of mitochondrial genetic defects in a number of progressive degenerative diseases, has added to the interest in mitochondria.[9,15–20]

What is the mitochondrial free radical theory of aging? It entails the following central concepts, which have been put forth in papers from a number of researchers: (1) with increasing age, there is a loss in mitochondrial metabolic and oxidative phosphorylation

[a]This work was supported by National Cancer Institute Outstanding Investigator Grant CA39910 and National Institute of Environmental Health Sciences Center Grant ES01896 to B.N. Ames.
[b]Tel: 510/642-5163 or -9218; fax: 510/643-7935; e-mail: kbeckman@uclink4.berkeley.edu
[c]Tel: 510/642-5165; fax: 510/643-7935; e-mail: bnames@uclink4.berkeley.edu

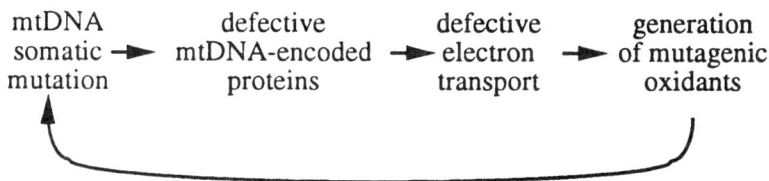

FIGURE 1. The hypothetical "vicious cycle." Mutations in mtDNA result in defective mitochondrial proteins, defective electron transport, and ultimately, the generation of mutagenic oxidants. These, in turn, may cause mtDNA damage and mutagenesis, completing the cycle.

(OXPHOS) activity, such that cells may become limited by their capacity to generate ATP; (2) concomitantly, there is an increase in mitochondrial free radical generation, due to increased "leakage" of electrons from the electron transport chain, which may damage cellular macromolecules; and (3) the loss of OXPHOS capacity may itself be caused by damage wrought by the oxidants generated by mitochondria themselves. Hypotheses surrounding the involvement of mitochondrial DNA (mtDNA) in aging comprise a corollary to the mitochondrial free radical theory of aging. In short, it has been proposed that mtDNA, which encodes a small number of critical mitochondrial genes and may be exceptionally sensitive to oxidative damage, is a weak link. Damage to mtDNA by mitochondrial oxidants, it has been suggested, may result in faulty mitochondrial proteins, decreased OXPHOS, increased mitochondrial oxidant generation, and increased mtDNA damage. This hypothetical positive feedback loop is generally referred to as the "vicious cycle"[21] (FIG. 1).

Evidence that supports the mitochondrial free radical theory, of which there is much, has been discussed in recent research reports and review articles, including a number of articles in this volume. We will not attempt to write another such review here. Rather, we have identified some of the most outstanding unanswered questions. The following is a brief (and by no means complete) list of questions, the resolution of which may push mitochondria farther into the spotlight of molecular gerontology, and will at the very least refine our knowledge of the role of mitochondria in aging.

"MITOCHONDRIAL MYSTERIES"

How Important Are Mitochondrial Oxidants Relative to Other Endogenous Sources?

In support of the mitochondrial free radical theory of aging, it is often written that 1–2% of electron flow through the mitochondrial electron transport chain (ETC) results in the univalent generation of superoxide ($O_2^{\cdot-}$). Upon what evidence are these statements based? Are mitochondria, rather than other endogenous sources, the cell's most significant source of oxidants? The initial experiments in this area were straightforward and were reviewed by Chance and co-workers.[22] First, it was shown that isolated mitochondria (from rat and pigeon tissues, generally) generate hydrogen peroxide (H_2O_2) at a rate equivalent to about 2% of total oxygen consumption. Second, isolated submitochondrial particles (from which the $O_2^{\bullet-}$-detoxifying enzyme, superoxide dismutase, has been removed) were observed to generate from 1.5 to 2.1 times as much $O_2^{\bullet-}$ as the isolated organelles generated H_2O_2, indicating near stoichiometry between the former and the latter. Since two molecules of $O_2^{\bullet-}$ ultimately dismutate—either spontaneously or with the help of superoxide dismutase—to form H_2O_2, these experiments resulted in the often-repeated mitochondrial "leakage" statistic of 1–2% of electron flow.

It is not a criticism of this pioneering work in mitochondrial oxidant generation to stress that one cannot extrapolate from such experiments to the *in vivo* situation. For one, measurements of mitochondrial H_2O_2 generation were generally carried out in air-saturated buffer, a condition of higher oxygen tension than exists *in vivo*, which may overestimate the rate of H_2O_2 generation (shown to be sensitive to the concentration of oxygen[22]). Second, mitochondrial H_2O_2 generation was routinely measured under supraphysiological conditions of such saturating substrates as succinate, a potentially confounding factor recently examined by Hansford, whose experiments suggest that the leakage of $O_2^{\bullet-}$ from mitochondria may be less dramatic *in vivo*.[23] When these workers measured H_2O_2 generation using physiological concentrations of substrates, they measured a rate of generation of H_2O_2 below 0.1% of oxygen consumption.

Therefore, little can be said with certainty about the *in vivo* generation of oxidants. To what extent does this matter? As Hansford *et al.* point out, even their lower estimates of mitochondrial H_2O_2 generation may represent a considerable oxidative stress. Although this is undoubtedly true, until robust and physiologically defensible *in vivo* estimates have been achieved, perhaps by methods that have yet to be devised, it would be wise to refrain from repeating concrete and potentially erroneous rates of endogenous mitochondrial oxidant generation. As is shown schematically in FIGURE 2, mitochondrial oxidant generation represents only one of numerous pro- and antioxidant mechanisms whose modulation could accelerate or retard aging, a point that has recently been made by others.[23] Last, despite evidence that the ubisemiquinone site of the electron transport chain *can* generate $O_2^{\bullet-}$,[23] it is not evident that this is the principle site of $O_2^{\bullet-}$ generation *in vivo*,[24] and so there exist uncertainties about both the extent *and* the mechanisms of oxidant generation by mitochondria *in vivo*.

Does Mitochondrial Generation of Oxidants Increase with Age?

In addition to these unanswered questions about the relative importance of mitochondrial versus nonmitochondrial oxidants in the overall scheme of oxidative stress, it also remains to be definitively shown that the generation of mitochondrial oxidants increases with age. Although some workers report an age-related increase,[25, 26] others have failed to repeat this work.[23] Although there are numerous studies that show an increase in oxidative stress in older cells,[27] as well as recent studies showing increases in specific oxidative damage to mitochondrial enzymes with age,[28] these phenomena could be the result of any number of physiological changes besides increased mitochondrial oxidant generation (for example, they could easily result from decreased antioxidant defenses). One approach to tackling this question has been to show that the oxidative damage of isolated mitochondria results in increased generation of oxidants,[26] an observation that supports the concept of a positive feedback loop of oxidative damage and oxidant generation in aging.

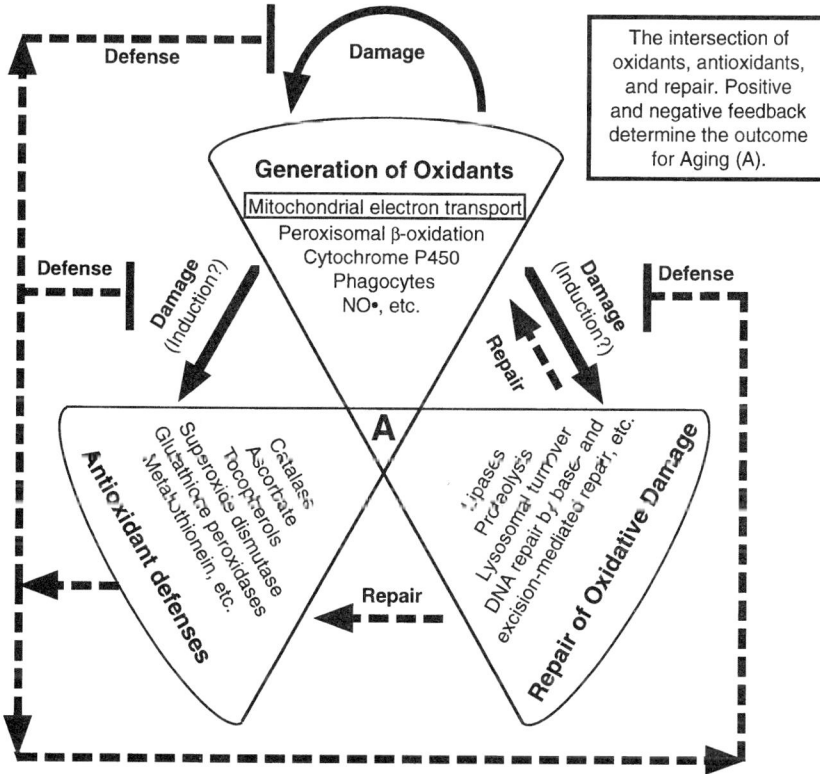

FIGURE 2. The relationship among the generation of oxidants, antioxidant defenses, and the repair of oxidative damage, in determining the overall outcome for aging. Mitochondrial oxidant generation is therefore only one of many regulated processes that may affect aging.

What Are the Most Significant Targets of Mitochondrial Oxidant Generation?

A number of related questions concern the nature of the oxidant species present in mitochondria. What are the steady-state concentrations of the oxygen-free radical $O_2^{\cdot-}$ versus its dismutation product H_2O_2 *in vivo*?[29] Is nitric oxide, which can react with $O_2^{\cdot-}$ to generate the oxidant peroxynitrite, present or generated in mitochondria and a significant factor in mitochondrial decay?[30] If "oxidative damage" increases in mitochondria with age, which terminal-oxidizing species are most likely to cause it, and which targets (proteins, lipids, DNA) are most susceptible to damage?

How Are mtDNA Lesions (Damage and Somatic Mutations) Formed?

As has been recently reviewed,[31] there is evidence from a number of different species that somatic deletions of mtDNA accumulate with age (as well as limited evidence for the accumulation of point mutations). What is still a mystery, however, is the mechanism of deletion formation. It has been proposed that mtDNA, located in close proximity to mitochondrial oxidants, is likely to suffer greater oxidative damage than nuclear DNA,[30,32–34] and that such oxidative damage may be responsible for the induction of mitochondrial deletions.[15,35] Indeed, an early estimate of oxidative damage of mtDNA suggested that it was more extensive than that of nuclear DNA,[32] and this evidence is often cited in support of the oxidative etiology of mtDNA deletions. Moreover, it has also often been written that mitochondria are relatively deficient in DNA repair activities, a claim substantiated by early studies of the repair of UV damage to mtDNA.[36] Is it true that mtDNA is more extensively damaged and less efficiently repaired than nuclear DNA?

Unfortunately this is a difficult question to answer, because when all of the studies of mtDNA oxidation are compared, the range of estimates spans almost four orders of magnitude.[37] Even when such a comparison is limited to studies in which similar techniques were employed (e.g., the study of the oxidative adduct 8-oxo-deoxyguanosine by HPLC chromatography with electrochemical detection), the highest and lowest estimates differ by more than 100-fold. The reason for these discrepancies have to do both with the inherent difficulties in accurately quantifying oxidative DNA damage,[38,39] as well as the specific difficulties introduced by analyzing low levels of DNA,[37] which is generally the case when analyzing mtDNA. In fact, the lowest values reported of mtDNA oxidation[40] represent values that are equivalent to the lowest estimates of oxidative damage to nuclear DNA (and considerably lower than the majority of nuclear DNA estimates). For this reason, as well as the fact that the relatively small number of studies that have measured oxidative mtDNA damage have used different starting materials, it cannot be said with any degree of certainty that mtDNA is more oxidized than nuclear DNA. Such an assertion awaits improvements in analytical techniques and further research. It should be noted, however, that recent research has shown that mtDNA is more extensively damaged by *extracellular* oxidants than is nuclear DNA,[41,42] a susceptibility that extends to several nonoxidative mutagens.

The assertion that mtDNA is inefficiently repaired also appears to be only partially true. Although early reports that mitochondria lack the ability to repair ultraviolet-induced DNA damage have been recently confirmed, repair of oxidative and other adducts has now been well documented,[43] and an enzyme responsible for the repair of oxidative damage has recently been purified from mitochondria,[44] although in other recent studies, mitochondrial repair was observed to proceed more slowly than nuclear repair.[41] Lastly, the assertion that mitochondria lack homologous recombination has also been recently challenged, based on the biochemical characterization of recombination activity in mitochondrial fractions.[45] If recombination is catalyzed by mitochondrial enzymes, then the accumulation of mtDNA deletions with advancing age may represent an aberrant regulation of this process rather than the result of increased oxidative mtDNA damage, as has been proposed.

How Significant Are mtDNA Lesions (Damage and Somatic Mutations)?

In addition to knowing the extent and genesis of mtDNA lesions, it will be important to answer an even more fundamental question: Do mtDNA lesions "matter?" Cells contain hundreds to thousands of copies of mtDNA; because of this genetic redundancy, it is not self-evident that mtDNA lesions represent more than a biomarker. Indeed, in many of the inherited, progressive diseases of mtDNA, in which mtDNA mutations are transmitted from mother to child and occur at proportions that are far higher than those measured during aging, physiological manifestations of the disease are generally unobservable unless the mtDNA mutations represent more than half of all mtDNA molecules.[18] It is therefore fair to ask whether the levels of mtDNA mutations observed in aging tissues, which are generally lower than 0.1% for a given mutation (and almost always lower than 1%), are likely to have an effect on mitochondrial function.

This is not a new concern, and a number of different ideas have been proposed to reconcile the low observed frequency of mtDNA mutations with a loss of mitochondrial function. One is the idea that whereas a single mtDNA deletion may not be present at a high frequency, it may merely represent the "tip of the iceberg" of mtDNA mutations. To date, however, there has appeared little evidence to substantiate this idea, and the fact that simple Southern blotting of mtDNA of old versus young animals does not reveal a large population of deleted molecules argues against it. An alternative suggestion is based on the observation that mtDNA deletions are not distributed homogeneously, but rather are concentrated in a small number of cells.[20,46] If this population of deletion-containing cells is likely to undergo cell death and disappear from the pool of living cells, it may be that a low measured frequency of mtDNA deletions masks an ongoing process of cell death.[71,35] In other words, the fraction of mtDNA deletion-containing cells may represent a dynamically changing population of cells in the process of dying.

How Are Mitochondria Regulated during Aging?

As has been reviewed elsewhere, a host of age-related mitochondrial alterations have been documented, including decreases in OXPHOS activity.[31] What is less clear from such studies is that a drop in mitochondrial activity is evidence of pathology, rather than the outcome of a physiological alteration. What are the effects of changes in endocrine function on mitochondria during aging? Work has recently focused on the effect of dietary restriction (a procedure well known to retard aging), on mitochondrial function and has shown that it delays some of the decline in mitochondrial performance that is observed in control animals.[47] Does this represent the slowing of a pathological process of mitochondrial damage or merely the attenuating of an age-related change in endocrine control of metabolism? The fact that many molecular gerontologists have approached the mitochondrial free radical theory of aging from the perspective of biochemistry and genetics has meant that physiological mechanisms have been less well examined. A cell biological approach to mitochondrial aging may also be warranted. An important aspect of mitochondrial function is their turnover, as some mitochondria are destroyed (digested by phagolysosomes) and others divide (by fission). How is mitochondrial turnover affected by aging?

How Heterogeneous Is the Population of Mitochondria?

As we discuss at length elsewhere in this volume,[48] another important question concerns mitochondrial heterogeneity. Even within a single cell, mitochondria are not a homogeneous population with uniform characteristics, but rather show a range of sizes, ultrastructural properties, and biochemical activities. Intercellular heterogeneity in mitochondrial structure and function add another layer of complexity. Indeed, with the use of mitochondrial potential-sensitive dyes, we have recently shown that rat liver hepatocytes can be divided into subpopulations with different mitochondrial membrane potential, and that the distribution of these subpopulations changes during aging.[49] Similar studies have recently been reported using isolated mitochondria from liver and brain tissue.[50] Understanding the causes and consequences of these shifts in mitochondrial populations should provide new information on mitochondrial dynamics, as well as an avenue for therapeutic interventions.[48]

Are Mitochondria Determinants of Species Maximum Life-span Potential?

Finally, even if mitochondrial oxidants and loss of mitochondrial respiration are shown to be involved in age-related degeneration, it is still unclear whether or not the dramatic differences in life span between species are also due to differences in mitochondrial physiology. It has been noted that primates and birds, which deviate from the (otherwise quite good) interspecies correlation between specific metabolic rate and maximum life-span potential,[27] live longer than their specific metabolic rates would predict. As a result, the total lifetime oxygen consumptions of these groups are quite a bit higher than other groups.[51] However, mitochondria of different species may exhibit different rates of intrinsic oxidant generation. Are the mitochondria of long-lived species designed to minimize oxidant generation? Experiments from two groups suggest that longevity is indeed associated with a lesser capacity for mitochondrial oxidant generation.[52–55]

Isolated pigeon mitochondria consume from two- to threefold as much oxygen as isolated rat mitochondria from the same tissues. However, pigeon mitochondria generate only a third to a half as much H_2O_2 as rat mitochondria under identical conditions. As a result, the calculated percentage of O_2 converted to $O_2^{\cdot-}/H_2O_2$ by mitochondria from lung, liver, and brain is about tenfold lower in pigeons than in rats and corresponds to the roughly tenfold longer life span of pigeons.[52] A second, independent comparison of mitochondrial oxidant generation by pigeons and rats found similar results.[53] Although this work does not prove that mitochondrial oxidants determine life span, it illustrates two points. First, it suggests that mitochondrial oxidant generation may be under genetic control and subject to evolutionary pressures. Second, it suggests that comparative biochemistry may teach us as much about mitochondrial oxidative metabolism as do age-related studies of a single species or tissue.

CONCLUSIONS

The free radical theory of aging has taken center stage among the competing perspectives on aging and propelled mitochondria into the spotlight. In addition to enjoying the recognition that is finally being paid to the potential importance of free radicals in aging, it is important that scientists who are interested in mitochondrial oxidants use the renewed attention that is being paid to their field in order to reexamine old questions, in hopes of clarifying issues that have remained, despite much work, somewhat murky.

REFERENCES

1. WEINDRUCH, R. 1996. Caloric restriction and aging. Sci. Am. Jan.: 46–52.
2. WALLACE, D.C. 1997. Mitochondrial DNA in aging and disease. Sci. Am. Aug.: 40–47.
3. YU, B.P. & R. YANG. 1996. Critical evaluation of the free radical theory of aging: A proposal for the oxidative stress hypothesis. Ann. N.Y. Acad. Sci. **786:** 1–11.
4. YIN, D. 1996. Biochemical basis of lipofuscin, ceroid, and age pigment-like fluorophores. Free Radical Biol. Med. **21:** 871–88.
5. WACHSMAN, J.T. 1996. The beneficial effects of dietary restriction: Reduced oxidative damage and enhanced apoptosis. Mutat. Res. **350:** 25–34.
6. STADTMAN, E.R. 1992. Protein oxidation and aging. Science **257:** 1220–1224.
7. SOHAL, R.S. 1993. The free radical hypothesis of aging: An appraisal of the current status. Aging Clin. Exp. Res. **5:** 3–17.
8. SOHAL, R.S. & R. WEINDRUCH. 1996. Oxidative stress, caloric restriction, and aging. Science **273:** 59–63.
9. SHIGENAGA, M.K. *et al.* 1994. Oxidative damage and mitochondrial decay in aging. Proc. Natl. Acad. Sci. USA **91:** 10771–10778.
10. NOHL, H. 1993. Involvement of free radicals in ageing: A consequence or cause of senescence. Br. Med. Bull. **49:** 653–667.
11. MARTIN, G.M. et al. 1996. Genetic analysis of ageing: Role of oxidative damage and environmental stresses. Nat. Genet. **13:** 25–34.
12. KNIGHT, J.A. 1995. The process and theories of aging. Ann. Clin. Lab. Sci. **25:** 1–12.
13. KING, C.M. & Y.A. BARNETT. 1995. Oxidative stress and human ageing. Biochem. Soc. Trans. **23:** 375S.
14. AMES, B.N. *et al.* 1993. Oxidants, antioxidants, and the degenerative diseases of aging. Proc. Natl. Acad. Sci. USA **90:** 7915–7922.
15. ARNHEIM, N. & G. CORTOPASSI. 1992. Deleterious mitochondrial DNA mutations accumulate in aging human tissues. Mutat. Res. **275:** 157–167.
16. BITTLES, A.H. 1992. Evidence for and against the causal involvement of mitochondrial DNA mutation in mammalian ageing. Mutat. Res. **275:** 217–225.
17. KOWALD, A. & T.B. KIRKWOOD. 1993. Mitochondrial mutations, cellular instability and ageing: Modelling the population dynamics of mitochondria. Mutat. Res. **295:** 93–103.
18. WALLACE, D.C. *et al.* 1995. Mitochondrial DNA mutations in human degenerative diseases and aging. Biochim. Biophys. Acta **1271:** 141–151.
19. KOWALD, A. & T.B. KIRKWOOD. 1996. A network theory of ageing: The interactions of defective mitochondria, aberrant proteins, free radicals and scavengers in the ageing process. Mutat. Res. **316:** 209–236.
20. LEE, C.M. et al. 1997. Age-associated alterations of the mitochondrial genome. Free Radical Biol. Med. **22:** 1259–1269.
21. WONG, A. & G. CORTOPASSI. 1997. mtDNA mutations confer cellular sensitivity to oxidant stress that is partially rescued by calcium depletion and cyclosporin A. Biochem. Biophys. Res. Commun. **239:** 139–145.
22. CHANCE, B. *et al.* 1979. Hydroperoxide metabolism in mammalian organs. Physiol. Rev. **59:** 527–605.
23. HANSFORD, R.G. *et al.* 1997. Dependence of H_2O_2 formation by rat heart mitochondria on substrate availability and donor age. J. Bioenerg. Biomembr. **29:** 89–95.
24. NOHL, H. *et al.* 1996. Conditions allowing redox-cycling ubisemiquinone in mitochondria to establish a direct redox couple with molecular oxygen. Free Radical Biol. Med. **20:** 207–213.
25. SOHAL, R.S. & B.H. SOHAL. 1991. Hydrogen peroxide release by mitochondria increases during aging. Mech. Ageing Dev. **57:** 187–202.
26. SOHAL, R.S. et al. 1994. Oxidative damage, mitochondrial oxidant generation and antioxidant defenses during aging and in response to food restriction in the mouse. Mech. Ageing Dev. **74:** 121–133.
27. BECKMAN, K.B. & B.N. AMES. 1998. The free radical theory of aging matures. Physiol. Rev. **78:** 547–581.
28. YAN, L.-J. *et al.* 1997. Oxidative damage during aging targets mitochondrial aconitase. Proc. Natl. Acad. Sci. USA **94:** 11168–11172.

29. FORMAN, H.J. & A. AZZI. 1997. On the virtual existence of superoxide anions in mitochondria: Thoughts regarding its role in pathophysiology. FASEB J. **11:** 374-375.

30. RICHTER, C. & M. SCHWEIZER. 1997. Oxidative stress in mitochondria. *In* Oxidative Stress and the Molecular Biology of Antioxidant Defenses. J.G. Scandalios, Ed. Cold Spring Harbor Laboratory Press. Plainview, N.Y.

31. OZAWA, T. 1997. Genetic and functional changes in mitochondria associated with aging. Physiol. Rev. **77:** 425–464.

32. RICHTER, C. *et al.* 1988. Normal oxidative damage to mitochondrial and nuclear DNA is extensive. Proc. Natl. Acad. Sci. USA **85:** 6465–6467.

33. RICHTER, C. 1995. Oxidative damage to mitochondrial DNA and its relationship to ageing. Int. J. Biochem. Cell Biol. **27:** 647–653.

34. RICHTER, C. 1992. Reactive oxygen and DNA damage in mitochondria. Mutat. Res. **275:** 249–255.

35. CORTOPASSI, G. & E. WANG. 1995. Modelling the effects of age-related mtDNA mutation accumulation, complex I deficiency, superoxide and cell death. Biochim. Biophys. Acta **1271:** 171–176.

36. CLAYTON, D.A. *et al.* 1994. The absence of a pyrimidine dimer repair mechanism in mammalian mitochondria. Proc. Natl. Acad. Sci. USA **71:** 2777–2781.

37. BECKMAN, K.B. & B.N. AMES. 1996. Detection and quantification of oxidative adducts of mitochondrial DNA. Methods Enzymol. **264:** 442–453.

38. CADET, J. *et al.* 1997. Artifacts associated with the measurement of oxidized DNA bases. Environ. Health Perspect. **105:** 1034–1039.

39. HELBOCK, H.J. *et al.* 1998. DNA oxidation matters: The HPLC-electrochemical detection assay of 8-oxo-deoxyguanosine and 8-oxo-guanine. Proc. Natl. Acad. Sci. USA **95:** 288–293.

40. HIGUCHI, Y. & S. LINN. 1995. Purification of all forms of HeLa cell mitochondrial DNA and assessment of damage to it caused by hydrogen peroxide treatment of mitochondria or cells. J. Biol. Chem. **270:** 7950–7956.

41. YAKES, F.M. & B. VAN HOUTEN. 1997. Mitochondrial DNA damage is more extensive and persists longer than nuclear DNA damage in human cells following oxidative stress. Proc. Natl. Acad. Sci. **94:** 514–519.

42. SALAZAR, J.J. & B. VAN HOUTEN. 1997. Preferential mitochondrial DNA injury caused by glucose oxidase as a steady generator of hydrogen peroxide in human fibroblasts. Mutat. Res. **385:** 139–149.

43. CROTEAU, D.L. & V.A. BOHR. 1997. Repair of oxidative damage to nuclear and mitochondrial DNA in mammalian cells. J. Biol. Chem. **272:** 25409–25412.

44. CROTEAU, D.L. *et al.* 1997. An oxidative damage-specific endonuclease from rat liver mitochondria. J. Biol. Chem. **272:** 27338–27344.

45. THYAGARAJAN, B. *et al.* 1996. Mammalian mitochondria possess homologous DNA recombination activity. J. Biol. Chem. **271:** 27536–27543.

46. WANG, E. *et al.* 1997. The rate of mitochondrial mutagenesis is faster in mice than humans. Mutat. Res. **377:** 157–166.

47. KANG, C.-M. *et al.* 1998. Age-related mitochondrial DNA deletions: Effects of dietary restriction. Free Radical Biol. Med. **24:** 148–154.

48. HAGEN, T.M., C.M. WEHR, & B.N. AMES. 1998. Mitochondrial decay in aging: Reversal through supplementation of acetyl-L-carnitine and *N-tert*-butyl-α-phenyl-nitrone. Ann. N.Y. Acad. Sci. This volume.

49. HAGEN, T.M. *et al.* 1997. Mitochondrial decay in hepatocytes from old rats: Membrane potential declines, heterogeneity and oxidants increase. Proc. Natl. Acad. Sci. USA **94:** 3064–3069.

50. SASTRE, J. *et al.* 1998. A Ginko biloba extract (EGb 761) prevents mitochondrial aging by protecting against oxidative stress. Free Radical Biol. Med. **24:** 298–304.

51. PEREZ-CAMPO, R. *et al.* 1994. Longevity and antioxidant enzymes, non-enzymatic antioxidants and oxidative stress in the vertebrate lung: A comparative study. J. Comp. Physiol. [b] **163:** 682–689.

52. BARJA, G. *et al.* 1994. Low mitochondrial free radical production per unit O_2 consumption can explain the simultaneous presence of high longevity and high aerobic metabolic rate in birds. Free Radicical Res. Commun. **21:** 317–327.

53. KU, H.H. *et al.* 1993. Relationship between mitochondrial superoxide and hydrogen peroxide production and longevity of mammalian species. Free Radical Biol. Med. **15:** 621–627.
54. SOHAL, R.S. *et al.* 1993. Biochemical correlates of longevity in two closely related rodent species. Biochem. Biophys. Res. Commun. **196:** 7–11.
55. SOHAL, R.S. *et al.* 1995. Mitochondrial superoxide and hydrogen peroxide generation, protein oxidative damage, and longevity in different species of flies. Free Radical Biol. Med. **19:** 499–504.

Mitochondrial DNA Mutations and Age

TAKAYUKI OZAWA[a]

Department of Biomedical Chemistry, Faculty of Medicine,
University of Nagoya, 65 Tsuruma-cho, Showa-ku, Nagoya 466, Japan

ABSTRACT: Apoptic cell death is reported to be prominent in the stable tissues of the failing heart, in cardiomyopathies (CM), in the sinus node of complete heart block, in B cells of diabetes mellitus, and in neurodegenerative diseases. Recently, mitochondrial (mt) control of nuclear apoptosis was demonstrated in the cell-free system. The mt bioenergetic crisis induced by exogenously added factors such as respiratory inhibitors leads to the collapse of mt transmembrane potential, to the opening of the inner membrane pore, to the release of the apoptotic protease activating factors into cytosol, and subsequently to nuclear DNA fragmentation. However, the endogenous factor for the mt bioenegertic crisis in naturally occurring cell death under the physiological conditions without vascular involvement has remained unknown. Recently devised, the total detection system for deletion demonstrates the extreme fragmentation of mtDNA in the cardiac myocytes of senescence, and mt CM harboring maternally inherited point mutations in mtDNA and on the cultured cell line with or without mtDNA disclosed that mtDNA is unexpectedly fragile to hydroxyl radial damage and hence to oxygen stress. The great majority of wild-type mtDNA fragmented into over two hundreds types of deleted mtDNA related to oxidative damage, resulting in pleioplasmic defects in the mt energy transducing system. The mtDNA fragmentation to this level is demonstrated in cardiac myocytes of normal subjects over age 80, of an mtCM patient who died at age 20 and one who died at age 19, of a recipient of heart transplantation at age 7 with severe mtCM, and in mtDNA of a cultured cell line under hyperbaric oxygen stress for two days, leading a majority of cells to apoptotic death on the third day. The extreme fragility of mtDNA could be the missing link in the apoptosis cascade that is the physiological basis of aging and geriatrics of such stable tissues as nerve and muscle.

In 1949, Ephrussi and his collaborators[1] discovered a respiratory-deficient strain of yeast forming a small colony, called cytoplasmic "petite" mutants. Their studies showed that mutations resulting in the petite phenotype are inherited in a non-mendelian fashion, and they therefore postulated that the lesions are an extrachromosomal or cytoplasmic element, designated as the rho (ρ) factor. Hence, cytoplasmic petite mutants are also referred to as ρ^- mutants. Although this was not known at the time of the discovery, the ρ factor was subsequently shown to be identical to mitochondrial (mt) DNA, and ρ^- mutants arise from large deletions in mtDNA.[2]

In 1964, Schatz *et al.*[3] first detected in purified yeast, mitochondria DNA and quantitated it by biochemical procedures. By using the petite mutants, the circular genetic map of the yeast mt genome was first published in 1975 by Linnane's group.[4] In 1981, a group in Cambridge[5] reported the gross genetic structure of human mtDNA: the 16,596 base-pair (bp) molecule is numbered as a nucleotide position (np); the sequence was later referred to as the Cambridge sequence (Camb seq);[6,7] the genes located on mtDNA encode some important subunits of the mt respiratory chain and ATP synthetase (7 subunits of complex I, 3 subunits of complex IV, 2 subunits of complex V, and one subunit of complex III);

[a]Tel: 81-52-744-2066; fax: 81-52-744-2067; e-mail: ozawa@med.nagoya-u.ac.jp

each 13 structure gene and 2 ribosomal (r) RNA gene is economically tight packed, being punctuated not by an intron as in the case of nuclear genes, but by 22 transfer (t) RNA genes.

Based on the genetic information and on Northern blot analyses, our group[8,9] elucidated in 1985–1987 that the mt gene–encoded subunits, but not the nuclear gene–encoded subunits, are selectively decreased in the respiratory chain complexes of skeletal muscle from patients with mt myopathy, named after mt morphological and biochemical abnormalities.[10] Subsequently, the specific defect was demonstrated to be caused by inherited genetic abnormalities by Southern blot analyses[11] that documented heteroplasmically existing large deletions in the patients' mtDNA, namely, the existence of a ρ^- mutant in human tissue as in the case of yeast. The mtDNA mutations associated with several degenerative diseases were reported also by several other groups.[12,13] Hence, known morphological[14] and histochemical[15] abnormalities of mitochondria associated with human age could be attributed to the ρ^- mutant cells.

Expanding the findings of biochemical and molecular biological studies on human tissues as well as yeast, in 1989, we[16] proposed that the somatic accumulation of mtDNA mutations during human life is a major cause of human aging and degenerative diseases, muscle weakness of senescence, declining mental capacity, age-related progressive decline of ventricular performance, and mt diseases. This "mitochondrial theory of aging" stems from the following: the high frequency of gene mutation in mtDNA, analogous to an extremely high mutational rate in yeast; the small size of the mt genome and its known information content; the inefficient repair mechanism for mtDNA and the lack of protective histone, unlike nuclear (n) DNA; and the somatic segregation of individual mtDNA during eukaryotic cell division. Since 1989, timely invented polymerase chain reaction (PCR) technology[17] that makes it possible to amplify a small quantity of DNA was brought into practical use, and many publications have documented an extensive array of age-dependent accumulation of deleted (Δ) mtDNA, resulting in the ρ^- mutant cells among many human tissues, especially in postmitotic stable tissues, such as nerve and muscle,[17–22] where cells tend to accumulate somatic mtDNA mutations during an individual's life span.

In parallel with the survey of point mutations by the total base sequencing of entire mtDNA[23] and the total detection of somatically acquired ρ^- mutations by PCR,[24] it was disclosed that the increase in the hydroxyl radical adduct of guanine, a biomarker of oxidative DNA damage,[25] in mtDNA correlates closely with the cumulative increase in ΔmtDNA associated with aging and degenerative diseases.[22,26,27] This fact implies that the oxidative damage due to reactive oxygen species (ROS) is an important underlying cause of the somatic accumulation of ρ^- mutant cells. Cellular oxidative damage was proposed as a main cause of aging by the "free radical theory of aging."[28] Unifying both ideas of the mitochondrial theory and free radical theory of aging, the author proposed the "redox mechanism of mitochondrial aging"[29] as the molecular, genetic, and bioenergetic basis for the progressive decline of cellular activity and naturally occurring cell death (apoptosis), leading human tissues to aging. The mechanism is based on an age-related increase in somatic mtDNA mutations and oxidative damage in human tissue degeneration and atrophy;[30] in the nigrostriatum of the parkinsonian brain;[17,31] in human skeletal muscle;[22,26] and in decreased oxygen use of skeletal muscle,[32] leading to extensive tissue oxygenation.[33]

To draw the whole figure of somatic deletions, we recently devised a total detection system for mtDNA deletion (TD system).[24] The TD system documented extreme fragility of wild-type (ω) mtDNA with oxidative damage that leads the ωmtDNA molecule to disintegrate into hundreds of types of ΔmtDNA fragments associated with normal aging,[34] with Parkinson's disease,[31] and with young mt diseases patients[34–36] (FIG. 1, TABLE 2). The fragmentation of mtDNA and apoptotic cell death could be mimicked even within three days in the cultured human fibroblast cells under oxygen stress.[37] The exposure of a cultured fibroblast cell line (ρ^+) under oxygen stress, 95% oxygen for three days, led a great majority of the ρ^+ cells to an apoptotic cell death with mtDNA fragmentation (FIG. 2), whereas the derivative cells (ρ^0) without mtDNA, lacking a functional respiratory chain, were relatively immune.

Cell apoptosis was considered to be programmed in nDNA to eliminate unwanted cells and reported to be prominent in tissues during development. The recent reports on many types of adult tissues and cultured cell lines have disclosed, however, that several pathophysiological conditions predisposing cells to energy depletion lead cells to a naturally occurring death; this situation could be regarded as bioenergetic cell death. Hence, apoptosis could be an important cause of the aging/pathological process in tissue degeneration and atrophy. However, the proposed mechanism of apoptosis has remained unclear, even controversial, until recently. Some evidence has suggested nDNA decomposition into oligonucleosomes, forming a ladder pattern on electrophoresis[38] as the hallmark of apoptosis. However, a primary role of nDNA in apoptosis was denied by the apoptosis of anuclear cytoplasts.[39–41] The arrangement of recent reports[42–44] in order could outline the mainstream of the apoptosis cascade, as schematically illustrated in FIGURE 3. The active respiratory chain creates an electrochemical proton gradient ($\Delta\mu_{H^+}$) that can be in the form of an mt transmembrane potential ($\Delta\Psi_m$) or ΔpH.[45] The accumulation of ΔmtDNA and the decrease in ωmtDNA result in defective respiratory chains leading cells to the depletion of $\Delta\mu_{H^+}$ and hence to $\Delta\Psi_m$ collapse.[46] Next, the $\Delta\Psi_m$-dependent permeability transition (PT) pore is opened, releasing intra-mt apoptotic protease, CPP32, activating factors into the cytosol and nDNA digestion into oligonucleosomes (the process down from $\Delta\Psi_m$ collapse was regarded as mitochondrial control of nuclear apoptosis[43]). The age-related accumulation of mtDNA mutations predisposing ωmtDNA to fragmentation[30] could be linked with the apoptosis cascade being its main tributary upstream that merges with other tributaries, derived from exogenous factors, such as anoxia,[47,48] into one common bioenergy deficit state, leading cells to apoptosis and tissues to degeneration:[49]

mtDNA mutations → bioenergy deficit → $\Delta\Psi_m$ collapse → apoptosis → tissue degeneration

MUTATIONS IN mtDNA

ATP, an essential for cellular viability, is almost exclusively produced by mitochondria, and ATP production depends upon the proper functioning of the mt energy–transducing system, including the respiratory chain that transduces the redox energy into $\Delta\mu_{H^+}$. $\Delta\mu_{H^+}$ in the form of ΔpH drives the ATP synthesis[45] by proton motive ATP synthase (ATPase),[50] and that in the form of $\Delta\Psi_m$ sustains the so-called PT pore,[46,51] controlling intra-mt ions and solutes. The basic degenerative process in aging involves mtDNA mutations resulting

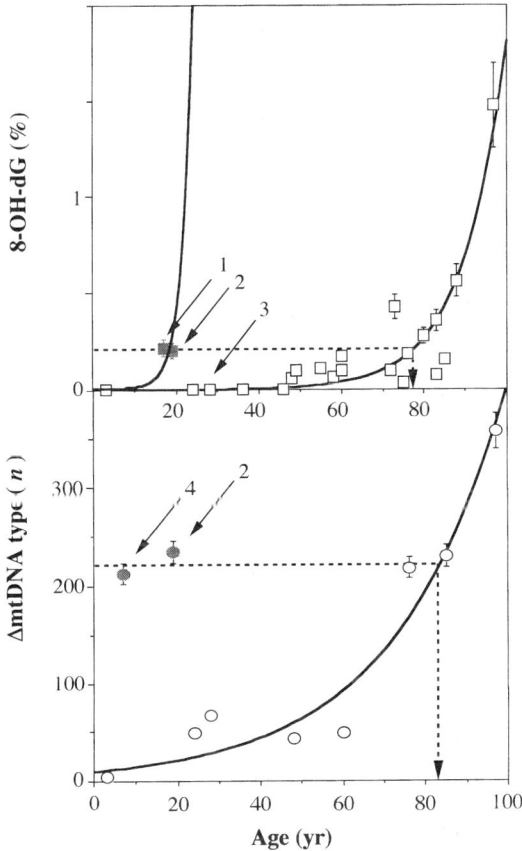

FIGURE 1. Age-associated correlative increase in the total number of ΔmtDNA types and oxygen damage. **Upper panel**: mtDNAs were extracted from autopsied cardiac muscles of 21 human subjects, 8 males and 13 females, aged 3 to 97 years, without cardiological symptoms and were obtained at random. Samples of mtDNA were enzymatically hydrolyzed into nucleosides, subjected to precolumn concentration, and analyzed by the microHPLC/MS system.[113] Both selected ion monitoring and ionization MS spectra of 8-OH-dG and dG were recorded. The percent of 8-OH-dG to the total guanine increases exponentially with the ages of subjects [$r = 0.84$]. Overlaid plots are the 8-OH-dG percent of mtCM patients who died by heart failure at age 17 (female, a closed square, arrow **1**) and at age 19 (male, MCM (mitochondrial cardiomyopathy) P 1, a closed square, arrow **2**),[35] respectively. The 8-OH-dG percent of a negative control (female, ID119, an open square, arrow **3**) harboring nonsevere base substitutions is a marginal level. The arrow with a dashed line indicates that the 8-OH-dG percent of mtCM patients is equivalent to that of the normal 78-year-old subjects. **Lower panel**: The number (n) of ΔmtDNA types was determined by the total detection system[24] on 8 mtDNA specimens (from 3 males and 5 females) of enough quantity out of 21 samples after the 8-OH-dG analyses, shown in the upper panel. ΔmtDNA type n increases exponentially with the ages of subjects, resulting in a decrease of the wild-type mtDNAs down to 11%, with a strong negative correlation with age ($r = 0.89$). Overlaid plots are ΔmtDNA type n of a female DCM patient[36] who received a heart transplant at age 7 (DCM P-3, a closed circle, arrow **4**), and that of a male mtCM patient who died at age 19 (MCM P-1, a closed circle, arrow **2**, as shown in the upper panel). Mean ΔmtDNA type n of the positive controls is equivalent to the normal 82-year-old subject.[34]

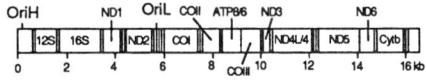

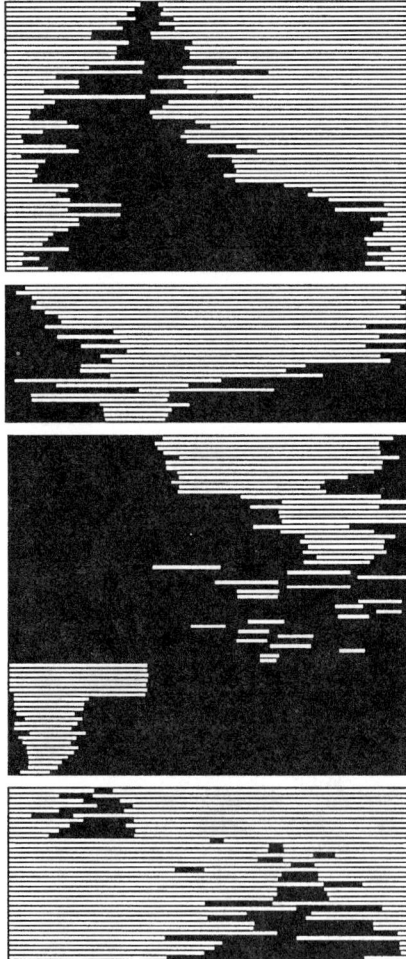

FIGURE 2. Types and size distribution of deletions detected in mtDNAs from the ρ^+ cells in the presence or absence of oxygen stress. The total-detection system[24] was applied to determine the n of ΔmtDNA types acquired during a 3-day exposure of the 701.2.8c to 95% oxygen. The deleted regions were marked with the black bars. They are arranged according to their sizes. Genomes in mtDNA are schematically illustrated between panels a and b. The ΔmtDNAs lacking either OriL (light strand) and OriH (heavy strand), or both, or having both are shown in the bulk of the bars in the respective orders. **a**: ΔmtDNAs are shown before the exposure to 95% oxygen; **b**: ΔmtDNAs are shown after the exposure for 3 days.[37]

---

in a defective energy-transducing system, leading cells to naturally occurring cell death under physiological conditions.

Many kinds of mtDNA mutations, such as point mutations and large deletions, have been reported to be the cause of degenerative diseases, named as the diseases of the mtDNA.[52] However, the cause–effect relationship between the reported mutational genotype and the clinical phenotype remained unclear in these early studies, mainly because the mutational survey was carried out within the limited region of mtDNA. Recently, accumulated data with the total base sequencing of the entire mtDNA[23] clarified the dominant feature of the mtDNA diseases. Their clinical signs and symptoms are triggered by maternally inherited[53] or somatically acquired[51] multiple point mutations located on all ωmtDNA,[55] the mutational genotypes based on the severity of point mutations and on their combination corresponding to the clinical phenotypes, ranging from the asymptomatic to incapacitating symptoms (TABLE 1). The combination of point mutations synergetically accelerates the accumulation of the somatic mutations in mtDNA, the oxidative damage and large deletions.[22] The oxidative damage disintegrates ωmtDNA into hundreds of types of ΔmtDNA fragments associated with normal aging,[34] Parkinson's disease,[31] and the young patients with mt diseases[34–36] (FIG. 1, TABLE 2).

Point Mutational Genotype and Its Phenotype

In the beginning of the point mutational survey, comparison of a sequenced mtDNA in the patients' somatic cells with the Camb seq[53] revealed various point mutations. Since then, many point mutations has been supposed to be the cause of several degenerative diseases such as Leber's hereditary optic neuropathy,[6,56] myoclonus epilepsy associated with ragged-red fibers,[7,53] mitochondrial encephalomyopathy with lactic acidosis and seizure (MELAS),[57,58] Kearns-Sayre syndrome,[59] Parkinson's disease,[60] some of the primary cardiomyopathies (CM) (later named mtCM)[61,62] Huntington's disease,[63] and diabetes mellitus (DM) with deafness.[64] However, these early studies of point mutation in mtDNA by restriction analysis or by sequence analysis within limited regions has shown that there is no obvious correlation between the point mutation and its clinical phenotype. This has resulted from lack of the real standard sequence from which point mutations diverged, lack of information to distinguish the point mutational genotypes from the entire mtDNA sequences, and lack of genetic controls to correlate the clinical phenotype. In addition, enthusiastic oversimplification of the relation raised the inconsistency; for example, the np3243 A-to-G transition in tRNA$^{Leu(UUR)}$ was first found in the MELAS patients[57,58] and cited as the MELAS mutation;[52] however, it was found in the mtCM patients without the MELAS symptoms[65] and in the patients with DM.[64,66]

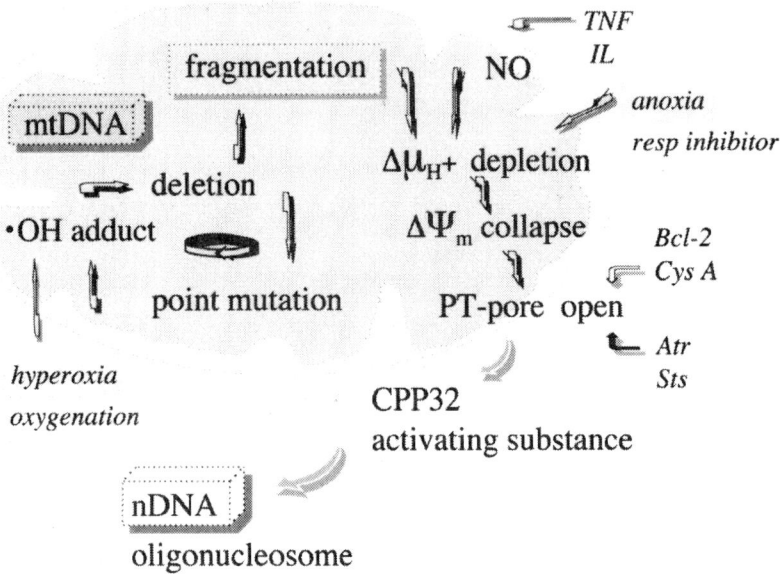

FIGURE 3. Bioenergetics and the cascade of cellular apoptosis are schematically illustrated based on the following reports. In mtDNA, each individual harbors unique base substitutions diverging from the common ancestor of the modern human, and they include point mutation(s) in the case of mt diseases.[29] A particular base substitution occurs in mtDNA of tissues of aging humans.[72] Somatic nucleotide substitutions accumulate in mtDNA of mt diseases.[54] Point mutations accelerate oxidative damage and fragmentation of mtDNA of premature aging.[35,36] Extensive tissue oxygenation, focal hyperoxia, associates with mtDNA mutations and with age.[33] Hyperoxia induces an apoptotic cell death associated with fragmentation of mtDNA.[37] Oxidative damage and deletions of mtDNA synergetically increase in human tissues with age.[22,26] Oxidative damage and extensive mtDNA fragmentation associate with age.[34] Age-associated accumulation of oxidative damage and deletion lead tissues to the loss of mt electron transport chain (ETC) activities.[114,161] NO inhibits CO activity,[162] and its synthase (NOS) is extensively inducted by such cytolytic factors as TNF[102,103] and IL-1,[104,105] leading the target cell to apoptosis. Anoxia or a respiratory inhibitor causes an acute apoptosis.[48,159] A drop in $\Delta\Psi_m$ is one of the first events in apoptosis.[163] The uncouplers of oxidative phosphorylation or divalent cations cause the collapse of $\Delta\Psi_m$ leading to apoptosis.[43] Apoptosis is prevented by Bcl-2[150] or by Cys A,[51] a specific ligand of ANT (adenine nucleotide translocator). The PT induction in response to the ANT ligand, Atr, is inhibited by a specific ANT ligand, bongkrekic acid, or by hyperexpression of Bcl-2.[43] An apoptosis-inducing reagent, Sts, results in the PT pore opening and the mt swelling eluting such mt solutes as dATP and cyt c that activate an inactive form of an ICE family protease, CPP32, to its active form.[44] CPP32 cleaves various substrates including nuclear lamin,[164] exposing nDNA to Ca^{2+}-endonuclease digestion[137] down into nucleosomes. The hatched line represents the outer or inner mt membranes. The shaded area represents the intermembrane space and the darker shaded area the matrix.

TABLE 1. Genotypes and Phenotypes of Patients with mitCM[a]

Genotypes	Patients	sexes, ages	mit⁻	mit⁻	syn⁻	symtps	finds	arrhyths	complts
I.mit⁻	HCM P-3	m, alive 25	**AP6**, CO3			CTR 47	neg T		ret pig
	HCM P-10	m, alive 41	AP6, N3, N1, N2 CO2, N3, N5			CTR 54	neg T		
	HCM P-7	m, alive 65	AP6, N5, N5, b			CTR 47	neg T		
II.mit⁻	HCM P-5	m, died 53	AP6, N3, AP6	CO1		CTR 53	EF 41	AF,CRBBB	
	DCM P-1	m, died 75	N6, b	CO1, b		cd	EF 17	AF	
	DCM P-2	m, died 45	AP6, N3	CO1		CTR 67	EF 12	AF	
III.syn⁻	HCM P-8	f, alive 18	N1, N4, N5		tRNA^Leu		hypok		
	MCM P-1	m, died 19	AP6, N3, N2,		tRNA^Asp		hypok		seiz
IV.mit⁻	HCM P-9	m, alive 21	N1	**N2**	tRNA^Thr	cd	hypok	**CLBBB,** LVFS red	ment
+ syn⁻	HCM P-4	m, died 20		b	rRNA	cd	neg T	WPW	deaf, NP, seiz ment
	DCM P-3	f, transpl 7		CO1, b	rRNA	CTR 68	EF 10	st	mitr reg, transpl

TABLE 1. Continued

Genotypes	Patients	sexes, ages	mit⁻	mit⁻⁻	syn⁻	sympts	finds	arrhyths	complts
	HCM P-6	f, died 54	AP6, N3, N2, N4	N1	tRNA^{Leu*}	cm		AV bl	DM, NP
	FICM P-1	m, died 1	AP6, N3, N2, N1, N2, N2, CO1, b	CO2, b	tRNA^{Ile}	CTR 71		pvc bradycardia	arrest seiz
	HCM P-2	m, died 45	AP6, N3, N2, AP8, N6	N5	tRNA^{Thr}	CTR 68	EF 20 hypok		NP
	HCM P-1	m, alive 21	AP6, N3, N1	AP6, N4	tRNA^{Leu*}	CTR 58	hypok, EF 45	WPW, st	DM, deaf ment

[a]Abbreviations: mit⁻ base substitution that causes replacement of a nonconserved amino acid; mit⁻⁻, replacement of an amino acid conserved among 6 known kinds of mammalian species, with block capital as replacement of the amino acid associated with change of its polarity; syn⁻, base substitution in the genes coding for components of the mt protein synthesis system that substitutes the conserved base and/or base pair among mammals; MCM, mitochondrial cardiomyopathy; CO, cytochrome oxidase; b, cytochrome b; N, NADH dehydrogenase; AP, ATPase; tRNA^{Leu*}, transfer RNA$^{Leu(UUR)}$; rRNA, 12S ribosomal RNA; sympts, symptoms; finds, findings; arrhythmias; complts, complications; cm, cardiomegaly; CTR, cardiothoracic ratio, %; neg T, negative T wave; EF, ejection fraction, %; hypok, hypokinesis; AF, atrial fibrillation; CRBBB, complete right bundle branch block; CLBBB, complete left bundle branch block; LVFS red, reduction of left ventricular fractional shortening during systole; AV bl, atrioventricular block; WPW, the Wolff-Parkinson-White syndrome; st, sinus tachycardia; pvc, premature ventricular contraction; ret pig, retinitis pigmentosa; ment, mental retardation; deaf, deafness; NP, nephropathy; seiz, convulsive seizure; mitr reg, mitral regurgitation; DM, diabetes mellitus; transpl, heart transplantation; arrest, cardiac arrest.

TABLE 2. Types of mtDNA among Subjects[a]

Subject	Sex	Age	Disease	ΔmtDNA Type (n)	Subtype of ΔmtDNA OriL+/H+	OriL⁻	OriH⁻	OriL⁻/H⁻	omtDNA (%)	8-OH-dG per 10⁴ dG
A.K.	F	3	VSD	5	4	1	0	0	>99	<1
S.T.	M	24	Accident	49	16	15	8	10	85	<1
N.N.	F	28	Pul. Emb.	67	23	14	8	22	71	<1
Y.I.	F	48	Thymoma	43	13	8	4	18	73	5.9
Y.T.	M	60	Gastric ca.	49	13	23	1	12	71	17.5
Y.Y.	F	76	SAH	218	66	68	37	47	47	18.6
K.A.	M	85	Colon ca.	230	61	64	33	72	58	15.3
H.M.	F	97	Gastric ca.	358	78	88	63	129	11	148
M.K.	F	7	DCM	212	37	58	38	79	47*	38.8*
T.K.	M	19	mtCM	235	48	59	31	97	16	20.1
rho+			normoxia	49	14	15	5	15	80*	<1*
rho+			95% O₂	187	35	55	28	69	53*	29.9*

[a]All samples of mtDNA were extracted from autopsied heart muscle, except cultured cell line. Abbreviations: ΔmtDNA, mtDNA with deletions; omtDNA, wild type mtDNA; OriL+, replication origin light strand; OriH+, replication origin, heavy strand; F, female; M, male; VSD, ventricular septal defect; Pul. Emb., pulmonary embolism; ca., cancer; SAH, subarachnoidal hemorrhage; DCM, dilated cardiomyopathy;[27] mtCM, mitochondrial cardiomyopathy;[35] *, calculated from the regression formula.[34]

Hence, the following have performed to overcome inconsistency between the point mutational genotype and its clinical phenotype.

(1) The Camb seq was mainly derived from a single placental mtDNA of a normal subject, but some regions were from a HeLa cell mtDNA. In addition, several ambiguous nucleotides were assumed to be the same as in the bovine mtDNA sequence.[5] Thus, the Camb seq could not be the real standard sequence for the genotype analysis. On the basis of restriction mapping, Wilson's group[67] concluded that all modern human mtDNA stems from one woman, mtEve, living 200,000 years ago. However, the base sequence clarified by the Wilson group is restricted to only 9% of the entire mtDNA sequence. Hence, much effort in our laboratory has been expended to sequence the entire mtDNA of modern individuals to determine the standard mtDNA sequence of our common ancestor. In the study on mtCM, we determined the entire mtDNA sequences of 65 individuals, including 32 patients with mtCM and 10 normal controls among Japanese, American, and Australians of European origin, by the automated sequencing of the entire mtDNA. The accumulated database with one million bp could infer the unique nucleotide changes of the Camb seq.[27] The total base sequence of the mtEve as the standard sequence was induced.[30] Comparing the entire mtDNA sequence of an individual with the standard sequence, base substitutions diverging from the standard were detected.[29] The genetic phylogenetic tree of the patients with mt disease was constructed,[55] and a cascade of sequentially diverged clusters of the substituted base was noted. The nucleotide substitutions unique to the individuals show often heteroplasmic coexistence with the wild type in a tissue.[68] By assuming random segregation of a selectively neutral mutant in mammals, it is estimated that it would take at least 20 generations to obtain a pure mtDNA population from a mixed population.[69]

In addition to the maternally inherited germ-line mutations, nucleotide substitution is documented to occur in a single generation of Holstein cows, probably due to a genetic "found effect" during oogenesis,[70] that is, amplification of one or a few mtDNA molecules as template will yield one predominant genotype in the mature oocyte that contains 100 to 1,000 times more mtDNA than is found in somatic cells.[71] A particular point mutation at np3243 A-to-G transition was reported in the cells of aging humans.[72] The somatically acquired point mutations were also detected in the cloned skeletal muscle mtDNA from MELAS patients (10 clones/60 clones), being significantly higher than in those from normal skeletal muscle (0/60), as well as in a normal placenta (2/60).[54]

(2) Among the mtCM patients, we detected that major nucleotide substitutions close to the standard sequence are synonymous; hence they could be regarded as a polymorphism in the human mt genome with little pathogenicity. However, some are nonsynonymous mutations that seem to have a variety of pathogenicities. With some modifications of the nomination in the field of yeast mtDNA,[2] we defined mit⁻ as a base substitution causing replacement of the amino acid that is not conserved among six known kinds of mammalian species (human, bovine, rat, mouse, seal, and whale); mit⁻⁻ as the replacement of conserved amino acid; and syn⁻ as the base substitution in the genes coding for components of the mt protein synthesis system that substitutes the conserved base and/or bp among mammalians in tRNA and the conserved base among known biological spices in the rRNA gene (rRNA).

(3) Looking over the whole mutational spectra among the patients, it was noticed that the division into four genotypes could explain the variety of seriousness in the phenotype of mtCM, as shown in TABLE 1.

A. mit⁻ Genotype

The genotype is defined as the only mit⁻ that exists in the entire mtDNA sequence. The possible pathogenicity of each individual mit⁻ could be regarded as slight. In summary, patients with the mit⁻ genotype expressed moderate cardiac hypertrophy and a negative T wave on their ECGs. However, their signs and symptoms remained stable.

B. mit⁻⁻ Genotype

The genotype is defined as mit⁻⁻ whether or not mit⁻ exists in the entire mtDNA sequence. Patients with this genotype showed a more serious clinical phenotype than those with the mit⁻ genotype. All three of the listed patients complained of dyspnea and showed atrial fibrillation, atrioventricular block, or deduced ejection fraction. The mit⁻⁻ in the cytochrome oxidase (CO) gene seems to enhance cardiac hypertrophy.

C. syn⁻ Genotype

The genotype is defined as syn⁻ whether or not the mit⁻ exists in the entire mtDNA sequence. Patients with this genotype showed different signs and symptoms from the other genotypes: little sign of cardiac hypertrophy with no arrhythmias, but diffuse hypokinesis of the left verticular wall. In yeast, syn⁻ strains harbor an impaired system of mt protein synthesis, being pleiotropically deficient in the respiratory and ATPase complexes, similar to the strains having the large deletions.[2] The mtCM patients with this genotype also show the pleiotropic symptoms. Patients with normal heart size show that the defect in protein biosynthesis by syn⁻ overcomes the hypertrophic response mediated by the growth factors. On the other hand, the pleiotropic defect in the respiratory chain could cause the hypokinesis of the ventricular wall, increasing the risk of sudden death.

D. syn⁻ + mit⁻⁻ Genotype

The genotype is defined as both syn⁻ and mit⁻⁻ whether or not mit⁻ coexists in the entire mtDNA sequence. Patients with this genotype showed severe and variegated signs and symptoms, and short longevity. Cardiomegaly with a cardiothoracic ratio (CTR) of over 60% and serious arrhythmias, such as conduction block and the WPW syndrome, were common signs among the patients. We noticed that the additional mit⁻⁻ in the patient expressed a synergistic and additive effect on the patient's clinical symptoms.

The excised heart at the time of transplantation of dilated cardiomyopathy (DCM) P 3, age 7, retained one more additional mit⁻⁻ compared with her syn⁻ positive control, hypertrophic cardiomyopathy (HCM) P-4, who died at age 20.[55] A patient with fatal infantile cardiomyopathy (FICM) P-1, harboring one syn⁻ and two mit⁻⁻, died at 12 months after delivery and showed variegated signs and symptoms (TABLE 1). The median survival time of six deceased patients with this genotype, 21 years, calculated by the method of Kaplan and Meier,[73] is more than 50 years shorter than the average life expectation of normal subjects.[74]

From accumulated data, it seems feasible that myocardial hypertrophy is triggered by the mit⁻ and more potently by the mit⁻⁻, leading cardiomyocytes to bioenergetic crisis. Katz[75] suggested that myocardial overload initiates an unnatural growth response that appears to shorten cardiac myocyte survival, possibly because the same growth factors that mediate the hypertrophic response of the adult heart can also induce apoptosis. Experimentally, cardiac hypertrophy and remodeling are initiated by a wave of apoptosis of rat cardiomyocytes.[76] The occurrence of severe arrhythmias prominent among mtCM patients with mit⁻⁻ is consistent with the finding[77] that apoptosis is a possible cause of gradual development of complete heart block and fatal arrhythmias associated with absence of the AV node, sinus node, and internodal pathways that are destroyed by a noninflammatory degeneration with no abnormal fibrosis or infiltrate. In this respect, it was noted that the mit⁻⁻ in the CO gene is associated with severe cardiomegaly, heart failure, and arrhythmias (TABLE 1). The mit⁻⁻ in the CO gene with the additional syn⁻ associated with the early onset of severe cardiomegaly of the pediatric patients, DCM P-3, and FICM P-1. These facts suggest that defected CO accelerates oxidative damage and myocardial apoptosis.

The syn⁻ with mit⁻⁻ triggers degenerative changes in other organs, especially those with postmitotic cells, *viz.*, convulsive seizure, mental retardation, deafness, nephropathy, and DM (TABLE 1). HCM P-6 and P-1, who harbor the syn⁻ at np3243 A-to-G, could be included in diabetic CM;[78] the 3243 syn⁻ has been reported to be common among the patients with DM and deafness[64] and patients with an insulin-deficient type of DM.[66]

Deletion and Fragmentation of mtDNA

Large deletions spanning over several genes could result in a more dominant phenotype, as ρ^- mutant in yeast, than point mutations. However, in the early studies on mtDNA deletion, Grivell[79] cited that, curiously, there was no obvious correlation between the severity of the clinical symptoms or biochemical abnormality and either the location of the reported deletion or the number of deleted genes. In retrospect, the curiosity is caused by the survey of deletion within a limited region of mtDNA, using a particular mtDNA probe for the Southern blot analyses or a single primer pair for PCR. The problem was solved by the total detection of mtDNA deletions using 180 primer pairs covering the all-around ωmtDNA duplex.[24,34]

With respect to somatic mutations, the proposed redox mechanism of mitochondrial aging[55] predicts that the somatic mutations satisfy the following qualifications: (1) the mutations arise afresh with each generation and accumulate age dependently; (2) the absolute level of accumulated mutations is accountable for the age-related decline of mt function and bioenergetic deficit; and (3) the mutations correlate closely with oxidative damage and cell death.

There are two types of mtDNA deletion, the Southern blot–detectable deletion and the PCR-detectable one. Soon after the practical use of PCR, multiple types of ΔmtDNA were detected in myocardium using a single PCR primer pair.[80] Hence, the author noted[81] that the PCR-detectable multiple forms of ΔmtDNA *pleioplasmically* coexist with ωmtDNA in a tissue, in contrast to the Southern blot–detectable one that is detected often among the patients with early-onset mt myopathies,[12] chronic progressive ophthalmoplegia,[11] or Pearson's syndrome.[13] That is, a single ΔmtDNA of a large quantity, 20–85% of the total mtDNA detectable by the Southern blot, *heteroplasmically* coexists with ωmtDNA, presumably originating from a clonal expansion of an initial deletion event occurring early in oogenesis.

Quantitative data on a PCR-detectable ΔmtDNA among individuals with various ages indicate that there are four orders of magnitude fewer ΔmtDNA in infancy than in old age[82,83] and that a newborn harbors only very low amounts of the 5 kilo–bp deletion that is commonly observed in different tissues of adults,[18] not being detected in the corresponding fetal tissues.[19] Hence, the PCR-detectable multiform of ΔmtDNA seems to arise afresh with each generation, satisfying the first qualification.

On the other hand, a cell harbors hundreds of mitochondria and mtDNA copies, and the fractional concentration of each ΔmtDNA detected by the conventional PCR using a single primer pair is usually 0.01–0.3% of the total mtDNA.[84–86] Based on the low absolute level of a single ΔmtDNA, a question arose whether an observed ΔmtDNA is the cause or the effect of the aging.[85,86] However, the number and the size of ΔmtDNA visualized by PCR depend on a particular primer pair used, such that the more distantly separated primers enable detection of the larger deletions.[87] Hence, a PCR-detectable ΔmtDNA was suggested to be the "tip of the iceberg" of the spectrum of somatic mutations.[88] To settle the problem, the TD system, that enables us to detect all possible ΔmtDNA over 0.5 kilo bp, was recently devised[24] and applied to mtDNA specimens from the normal hearts of various ages.[34] Surprisingly, the whole "iceberg" in a subject of age 97 is visualized in as many as 358 types of ΔmtDNA fragments, including 280 types of "minicircles" that lack either one replication origin (Ori) or both, associated with a decrease in ωmtDNA down to 11 percent. Among normal subjects, the ωmtDNA disintegration into ΔmtDNA fragments is demonstrated to increase progressively with age, correlating with the oxidative damage, as shown in FIGURE 1. Similar fragmentation, a rearrangement that is postulated to be an intermediate in the formation of deletion,[89] and depletion of ωmtDNA below the detection limit was demonstrated by using a long PCR in skeletal muscle of aged subjects.[90] Hence, PCR-detectable forms of ΔmtDNA satisfy the second qualification.

In the mtCM patients harboring severe point mutations in mtDNA,[35,36] the similar correlative oxidative damage and mtDNA fragmentation in myocardia at ages 7 to 19 are documented to be premature aging equivalent to normal subjects over age 80 (FIG. 1). The apoptotic signs in the tissue specimens and in the cultured cells under oxygen stress correlating the fragmentation of mtDNA satisfy the third qualification.

Oxidative Damage

As early as the discovery of oxygen, Priestley[91] wrote about oxygen toxicity: "oxygen might burn the candle of life too quickly, and soon exhaust the animal powers within." In the 1950s, Harman pointed out the toxicity of the oxygen free radical in relation to aging.[92] However, it took a long time to elucidate the mechanism of the toxicity related to ROS that cause the oxidative damage of mtDNA, located upstream of the cascade reactions for naturally occurring cell death (FIG. 3).

In the respiratory chain, the majority of electrons reduce molecular oxygen in the terminal oxidase CO, forming H_2O. However, part of the electrons eluted from the respiratory chain performs one-electron or two-electron reductions of oxygen-forming ROS. In 1924, Warburg[93] insisted that the oxygen molecule, activated by CO, directly accepts four-electron reduction, forming water without H_2O_2 formation. However, Michaelis[94] pro-posed that the oxidation of bivalent organic molecules proceeds in two obligatory univalent steps, the intermediate being a free radical, *viz.*, a one-electron reductant, O_2^-, and a two-electron reductant, H_2O_2. It thus had been a matter of great interest to discover the production site of such intermediates in the respiratory chain. In 1973, on the basis of optical studies on oxy- and peroxy-CO by Chance,[95] it became clear that the intermediates remain within the active site of CO until the final four-electron reduction of oxygen forming water is achieved, probably for protection against cellular intoxication. From the general properties of the mt generation of H_2O_2 and the effect of hyperbaric oxygen, it was postulated[96] that besides the well-known flavin reaction, formation of H_2O_2 may be due to interaction with an energy-dependent component of the respiratory chain at the cytochrome (cyt) *b* level. These findings indicated that the active sites of the complex IV and III, cyt *a* and *b*, play a crucial role not only for cellular energy production, but also for protection against cellular oxidative damage. Hence, the genetically defective cyt *a* and *b*, even at a low absolute level, could result in a serious outcome in cellular viability. Actually, severe DCM or HCM is expressed in the patients who harbor mit^{--} in CO and *b* and/or syn$^-$ in tRNA that affect the transcription of the cyt genes, as listed in TABLE 1.

From the above mechanism of oxygen reduction, the elution of electrons from the respiratory chain and the generation of ROS is expected not only by the genetic defect, but also by physiological changes in the cellular redox state. Hyperoxia[96] increases H_2O_2 release by lung mitochondria because of too much oxygen over the enzymic capability to dispose ROS. Skulachev[46] pointed out that mammalian uncoupled respiration or a plant's noncoupled respiration is an effective device for preventing oxidative damage and cell death by safely maintaining low levels of oxygen and its one-electron reductants.

Beside the toxicity of ROS, there are several findings suggesting some physiological role of ROS. (1) Boveris and Chance[95] demonstrated that H_2O_2 production by animal mitochondria, negligible in active respiration (state 3) or in the presence of uncouplers, becomes quite measurable in resting respiration (state 4). State 4 increases reduced electron carriers and superoxide generation in flavins, NADH oxidoreductase, CoQ, cyt *b*, and nonheme iron proteins.[97–99] State 4 mitochondria from rat liver or from pigeon heart generate about 0.3–0.6 nmol H_2O_2/min per mg of protein.[95] This H_2O_2 generation represents approximately 2% of the total oxygen use under these conditions; thus, during the life of an individual, a vast sum of the redox energy is consumed for the generation of ROS. (2) Our group[100] found that cyt *c*, an essential component of the respiratory chain, is a twenty times more efficient catalyst than the ferrous ion in promoting the formation of the most reactive oxygen radical, hydroxyl radical (•OH), from H_2O_2 by the Fenton reaction, both in NADPH-driven respiration of the leukocyte plasma membrane and in the nonenzymatic H_2O_2 solution. (3) The •OH production was demonstrated by interaction of O_2^- with nitric oxide (NO)[101] of which synthase (NOS) was extensively inducted by such cytolytic factors as tumor necrosis factor (TNF)[102,103] and interleukin-1β (IL-1),[104,105] leading the target cell to apoptosis.

With respect to the above findings, •OH could be regarded as not a mere by-product of the mt respiration, but as an active one with an important bioenergetic role to cause physiological cell death and to eliminate unwanted/transformed cells. The mtDNA is located inside the mt inner membrane where ROS continuously leak from the respiratory chain,[106] hence being directly susceptible to attack by ROS, despite the cellular defenses against damage from ROS.[107] During evolution from yeast to mammals, mtDNA has down-sized one fifth, losing its intron, in which mutations are inert. Hence, human mtDNA becomes extremely fragile, acquiring the oxidative damage and larger deletions than a single-cell organism. It seems reasonable that the genes coding the cellular energy plant have to be manageably fragile in order to be the primary target of the apoptotic process.[30]

The underlying mechanism for the large deletion is a double-stranded separation, resulting in the generation of stretches of single-stranded DNA, the breaks of which form endonuclease-sensitive sites, with a consequence of oxygen radical attack.[108] An •OH adduct of deoxyguanosine (dG), 8-hydroxy-deoxyguanosine (8-OH-dG), was noted as a hallmark of oxidative damage of DNA.[25] In the nucleus, where an efficient repair system operates, the oxidized nucleotides in nDNA are rapidly excised and excreted into urine.[109] However, in mitochondria where the system is inefficient, they accumulate in mtDNA with age,[22,26,110] leading mtDNA to the somatic point mutations,[54] the double-stranded separation,[111] to single-stranded break,[108] and to the large deletion.[22] A study[112] on the mutagenesis of 8-OH-dG in a mammalian cell clearly demonstrated that a synthetic pro-tooncogene containing 8-hydroxy-guanine induces random point mutations at the modified site and adjacent positions on the gene replication. A defective electron-transport chain encoded by the mutated mtDNA enhances the •OH formation, resulting in more accumulation of 8-OH-dG and ΔmtDNA fragments.[35,36] Such a vicious cycle of the •OH damage and mutations in mtDNA (Fig. 3) results in those changes being synergistic and exponential.[34]

In human heart mtDNA specimens, quantitative determination of 8-OH-dG using a microHPLC/mass spectrometer[113] demonstrated an age-related progressive increase in a ΔmtDNA with 7.4 kbp deletion,[22] correlating closely ($r = 0.93$) with the accumulation of 8-OH-dG up to 1.5% of the dG at age 97. In human diaphragm muscle, the accumulation of 8-OH-dG reached a level of 0.5% in an 85-year-old individual with the association of multiple deletions.[26] Consistent with heart and diaphragm, a similar accumulation in 8-OH-dG[110] was reported in human brain mtDNA up to 0.87% at age 90. The TD system revealed that a progressive age-related increase in the total number (n) of ΔmtDNA types correlates with accumulation of 8-OH-dG in heart mtDNA (TABLE 2, FIG. 1), reflecting a long-term accumulation of the oxidative damage during human life, analogous sedimentary rocks in a glacier. A remarkable mirror image of the size distribution of ΔmtDNA[34] (cf. FIG. 2) and a strong linear correlation ($r = 0.97$) between minicircles and ΔmtDNA, preserving both Oris (TABLE 2), suggest random occurrence of ΔmtDNA without a preferential site. Hence, it seems reasonable to presume random double-stranded separation by accumulated 8-OH-dG,[22,111] single-stranded breaks by •OH attacks,[108] and rejoining of mtDNA as a preferable mechanism for its fragmentation into hundreds of ΔmtDNAs that further accelerates the oxygen damage. Experimentally, these changes in mtDNA were correlated with the decline in the respiratory chain activity in laboratory animals.[114,115] An extensive oxygenation of skeletal muscle that indicates mt dysfunction causing suppressed oxygen use, hence relative tissue hyperoxia, is demonstrated noninvasively among senescent individuals and the patients with mtCM and/or myopathy harboring hazardous point mutation.[33] Similar reduced oxidative metabolism is reported in the cortex of patients with Alzheimer's dementia.[116] Therefore, the vicious cycle of progressive oxidative damage, fragmentation of ωmtDNA, defected ETC, and relative tissue hyperoxia seems to result in those changes being synergistic and exponential associated with age (FIG. 1).

The mtDNA diseases could be regarded as premature aging in tissues where the tissue oxygenation and the mtDNA fragmentation are abnormally accelerated by the mutations.

Premature aging with bioenergetic cell death could be mimicked by the cultured cell line under hyperoxia within three days.[37] The exposure of a cultured fibroblast cell line (ρ^+) under oxygen stress, 95% oxygen, for three days led the great majority of the ρ^+ cells to an apoptotic cell death with mtDNA disintegration into 187 types of ΔmtDNA fragments with a dominant mirror image of their size distribution (Fig. 2). The derivative cells (ρ^0), however, without mtDNA and lacking a functional ETC, were relatively immune; more than 80% of the ρ^0 cells survived.

These results indicate that human mtDNA is unexpectedly fragile to the •OH attack leading the cell to energy crisis and to apoptosis. (1) This oxidative damage and mtDNA fragmentation leads cells to a progressive decline of bioenergetic activities and to naturally occurring cell death during the chronic aging process. (2) Regarding mt diseases, inherited/acquired serious mutations in mtDNA could result in the biosynthesis of an abnormal subunit of the energy transducing system and/or in a pleiotropic defect of the system, leading cells to abnormal production of ROS and somatic mutations during a subacute process. The genetic analyses on myocardia of the pediatric patients have documented the abnormally increased oxidative damage and ΔmtDNA, leading to fragmentation of mtDNA into hundreds of kinds of minicircles[35,55] (TABLE 2, FIG. 1). These deleterious mutations predispose the patients to premature aging, mental retardation, heart failure, and maladaptive cell growth. (3) Fragmentation of mtDNA and acute cell death could be induced within a short period by the change of an extracellular environmental factor such as oxygen partial pressure.[37] The not yet fully elucidated mechanism/receptor seems to operate in cellular communication with such signals as TNF of IL.

MOLECULAR GENETICS AND BIOENERGETICS OF APOPTOSIS

The term "apoptosis" came from morphological observations,[117] and functionally means programmed cell death or naturally occurring cell death under physiological conditions. Apoptosis plays an important role in the elimination of unnecessary cells in human morphogenesis and such harmful cells as radical-producing or transforming cells. Recently, with such advanced techniques as nDNA nick-end labeling on microscopic specimens, cell apoptosis was reported to be prominent in bioenergy-deficit cells. Hence, as a possible pathophysiological cause of age- or disease-related tissue degeneration and atrophy, apoptosis is implicated in many types of cell death, for example, cardiac myocytes in the failing heart,[118] those in chronic heart failure,[119] those in DCM,[120] those in mtCM harboring serious point mutations, those in pressure overload–induced heart hypertrophy and remodeling in the rat,[76] sinus node cells in patients with complete heart block and fatal arrhythmias,[77] β cells in the insulin-dependent DM model,[121,122] Langerhans' islets in rats treated with interleukin-1,[105] and neuronal cells in neurodegenerative diseases.[123]

However, the proposed mechanism of apoptosis has remained unclear, even controversial, until recently. This is so because many apoptosis-inducible factors and survival factors exist and are effective at different points along the apoptosis cascade reaction, because cellular bioenergetics, leading cells to suicide, has not been fully elucidated, and because we have not known of the extreme fragility of mtDNA. Hence, an ordered arrangement of reports on apoptosis is obligatory in order to link oxidative damage and mtDNA fragmentation with the apoptosis cascade.

The most readily measurable morphological features of apoptotic cell death are nuclear, namely, chromatin condensation and endonuclease-mediated nDNA fragmentation, producing a "ladder" of oligonucleosomal-sized nDNA fragments visible by gel electrophoresis,[124] which were considered to be the hallmark of apoptosis. To the contrary, no obvious morphological change of mitochondria or of other organelles was observed until the end stage of apoptosis. Hence, it has been tempting to consider that nDNA fragmentation plays a primary and causative role in apoptotic cell death. However, no clear evidence exists to support it. Conversely, the fact that the target cell nucleus is not required for cell-mediated, granzyme-, or Fas-based cytotoxicity[41] raises the possibility that apoptotic nuclear damage may be an epiphenomenon with respect to cell death. A series of reports[39-41] indicated that anucleate cytoplasts can undergo apoptosis and that the antiapoptotic protein, Bcl-2,[125] and other extracellular survival signals can protect them, indicating that nuclear signaling is not required for apoptosis or for Bcl-2/survival factor protection.

The cell suicide program is best illustrated by genetic studies in the nematode *Caenorhabditis elegans*.[126] Two genes involved in the control of apoptosis in *C. elegans* have been well characterized. One gene, *ced-9*, encodes a protein that prevents cells from undergoing apoptosis, whereas another gene, *ced-3*, encodes a protease whose activity is required to initiate apoptosis. The *bcl-2* family of genes are the mammalian counterparts of *ced-9*.[127] The *ced-3* protein is a cysteine protease related to the IL-1β-converting enzyme (ICE) family of proteases in mammalian cells.[128-132] The closest mammalian homologue of the *ced-3* protein is CPP32 in terms of sequence identity and substrate specificity.[133] Like the *ced-3* protein, CPP32 normally exists in the cytosolic fraction as its inactive form of 32 kDa that becomes activated proteolytically to the 17/11 kDa or 20/11 kDa active form,[130,134] called apopain,[131] in cells undergoing apoptosis. The activated CPP32 cleaves several substrates, including death substrate poly(ADP-ribose) polymerase[135] and nuclear lamin,[136] exposing the nucleus to Ca^{2+}-endonuclease[137] and/or DNase-1[138] digestion. However, the intracellular factors and/or bioenergetics that initially activate CPP32 have remained unknown.

Recently, a cell-free system was established to duplicate the features of the apoptotic program, including activation of CPP32 and nDNA fragmentation.[44] The system consists of nuclei added in cytosol from normally growing cells and intact mitochondria, carefully prepared by our method.[114] The PT pore (also termed "mt megachannel" (MMC)[139,140]) of the intact mitochondria is well sustained by $\Delta\Psi_m$, hence no change was observed in the nuclei. Nuclear apoptosis was initiated by the release of apoptotic protease-activating factors from mitochondria by addition of an apoptosis-inducing reagent, staurosporine (*Sts*) or atractyloside (*Atr*), which disrupts the intact PT pore (FIG. 3). A phosphocellulose column separates the factors into two kinds, *viz.*, the flow-through factor, identified as deoxyATP (dATP), and the column-binding protein factor, as cyt *c*. Depletion of cyt *c* with anti-cyt *c* antibody loses the dATP-dependent activation of CPP32 and the nDNA fragmentation. Conversely, without dATP, cyt *c* does not activate CPP32. Nuclear apoptosis with the presence of the intact mitochondria is induced by the external addition of dATP and cyt *c* to the system; hence, both substances are required and sufficient for the activation of CPP32.[44]

The intact mitochondrial innermembrane is quasiimpermeable for small solute, allowing the creation of $\Delta\mu_H+$ in the form of $\Delta\Psi_m$, thus keeping these apoptotic protease activation factors inside the membrane and the collapse of the $\Delta\Psi_m$-dependent PT pore and mt swelling releases them into cytosol. Associated with PT, such small molecules as glutathione are rapidly effluxed before the large amplitude swelling[141] and before the apoptosis-inducing protein with mt swelling and disruption of the mt outer membrane.[142]

The precise biochemical mechanism of cyt c functioning in the activation of CPP32 remains to be determined. However, it was found that cyt c is a twenty times more efficient catalyst than ferrous ion in promoting •OH radical formation.[100] The reported molar ratio of cytosolic inorganic plus organic Fe[143] and mt cyt c[144] in rat liver cells is approximately 20:1; however, efficiency to produce •OH is 1:20. Hence, the efflux of cyt c into cytosol would be highly cytotoxic due to the •OH formation from H_2O_2. Sensible cells translocate apo-cyt c, which is translated in the cytoplasmic ribosomes, into the mt intermembrane space through a unique pathway,[145] and it is loosely attached to the surface of the inner membrane.[146] Finally heme is installed by the cyt c heme lyase, a peripheral protein of the inner membrane, and the holoprotein folds into its native structure. Hence, catalytically active cyt c is located *exclusively* in mitochondria in an intact cell, as was shown by using radiolabeled cyt c.[147]

The release of intra-mt apoptotic protease activating factors is caused by the bioenergetic crisis and the collapse of $\Delta\Psi_m$. Cyclosporin A, a specific ligand of adenine nucleotide translocator (ANT), a component of the PT pore, prevents PT and apoptosis.[46,51] Cells undergo a reduction of the $\Delta\Psi_m$ before they exhibit common signs of nuclear apoptosis.[149–153] Mitochondrial control of nuclear apoptosis is demonstrated from the fact that the mt PT constitutes a critical early event of the apoptotic process.[43] In a cell-free system combining purified mitochondria and nuclei, mitochondria undergoing PT in response to the ANT ligand *Atr* are prevented by a specific ANT ligand bongkrekic acid or by hyperexpression of Bcl-2. However, besides these exogenous ANT ligands as well as other factors, such as anoxia[47,48] and respiratory inhibitors,[141,148,149] an endogenous factor that causes the bioenergetic crisis and the collapse of $\Delta\Psi_m$ under physiological conditions remained unknown until the extreme fragility of mtDNA was recently disclosed (FIGS. 1 and 2).[24,34–37]

It was suggested that apoptosis involved the generation of ROS and that a protooncogene *bcl2* product, Bcl-2, protects against apoptosis by inhibiting the generation or action of ROS.[150] A Bcl-2 family, apoptosis-promoting gene *bax* product, Bax, counteracts Bcl-2;[151] hence it is suggested that the ratio of family members, such as Bcl-2/Bax, determines the survival or death of cells following an apoptotic stimulus.[150] Bcl-2 and a splice variant of Bcl-x[152] have been shown to heterodimerize with other members of the Bcl-2 protein family, including Bax. These oncoproteins were proposed to manipulate ROS damage and cell death.[150] Consistent with the proposal, it was shown that the antioxidant suppressed dopamine-induced apoptosis in mouse thymocytes.[153] Several studies showed that Bcl-2 can protect cellular membranes from oxidative damage[154,155] and the generation of ROS.[156] The NO-induced apoptosis is effectively protected by overexpression or transfection of *bcl-2*.[157,158] On the other hand, studies on the Bcl-2 protection of cell apoptosis under hypoxia concluded that ROS are not essential for apoptosis and that Bcl-2 protects against apoptosis in ways that do not depend on the inhibition of ROS production or activity.[47,48] We could interpret from these facts that both the hypoxia-induced apoptosis and the naturally occurring cell death with age are triggered by the bioenergetic crisis of cells, the decrease in $\Delta\Psi_m$, and the PT pore opening, which is blocked by Bcl-2.

On the basis of the above report, the apoptosis cascade based on molecular genetics and bioenergetics, is illustrated in FIGURE 3. At the point of bioenergetic crisis and $\Delta\Psi_m$ collapse in the cascade, a main function of apoptosis due to such endogenous factors as mtDNA fragmentation[34] and a defective respiratory chain, leading the tissue to oxygenation,[33] combines with other functions due to such exogenous factors as hypoxia,[47,48] the respiratory inhibitors[141,148,149] the depletion of reductants,[141,159,160] the oxidative phosphorylation uncouplers,[43] the divalent cations,[43,141] or NO.[157,158] Here in the bioenergetic apoptotic cascade, apparently contradictory apoptosis-inducible factors, that is, hyperoxia and hypoxia, are understandable.

CONCLUSION AND PERSPECTIVE

The comprehensive analyses of the entire mtDNA mutations, including the detection of point mutations, oxidative damage, and deletions, could reveal the mutational genotype unique to an individual. The analyses could disclose the types and combination of point mutations deciding the severity of somatic oxidative damage and ρ^- mutations. It was demonstrated that there is a definite correlation between the point mutational genotype and the phenotype. The practicable survey of point mutations will be useful for genetic diagnosis, predicting the patients' life spans and for the management of such patients undergoing cardiac transplantation and/or gene therapy.

The total survey of somatic oxidative damage and deletions disclosed that mt genes are unexpectedly fragile to oxygen radical damage, hence to oxygen stress, easily disintegrated into hundreds of types of ΔmtDNA fragments, resulting in a defective mt energy-transducing system and a cellular bioenergetic crisis. This could be the missing link between the mt gene mutation and the naturally occurring cell death under physiological conditions. Namely, mtDNA fragmentation due to oxidative stress leads cells to bioenergetic crisis, to the collapse of $\Delta\Psi_m$, to the release of the apoptotic protease-activating factors into cytosol, to uncontrolled cell death, and to tissue degeneration and atrophy. It was also shown that the mtDNA fragmentation largely depends on the redox state of the mt respiratory chain and hence seems to be controllable by exogenous factors. Management of this process will enable us to protect against unwanted cell death, to arrest age-related degeneration, and to accelerate harmful/transformed cell death.

REFERENCES

1. EPHRUSSI, B., H. HOTTINGUER & A.M. CHIMENES. 1949. Action de l'acriflavine sur les levures. I. La mutation "petite colonie". Ann. Inst. Pasteur **76:** 351–367.
2. TZAGOLOFF, A. 1982. Mitochondrial Genetics. *In* Mitochondria: 267–322. Plenum Press. New York and London.
3. SCHATZ, G., E. HALSBRUNNER & H. TUPPY. 1964.Deoxyribonucleic acid associated with yeast mitochondria. Biochem. Biophys. Res. Commun. **15:** 127–132.
4. MOLLOY, P.L., A.W. LINNANE & H.B. LUKINS. 1975. Biogenesis of mitochondria: Analysis of deletion of mitochondrial antibiotic resistance markers in petite mutants of *Saccharomyces cerevisiae*. J. Bacteriol. **122:** 7–18.
5. ANDERSON, S. *et al.* 1981. Sequence and organization of the human mitochondrial genome. Nature **290:** 457–465.
6. WALLACE, D.C. *et al.* 1988. Mitochondrial DNA mutation associated with Leber's hereditary optic neuropathy. Science **242:** 1427–1430.

7. YONEDA, M., Y. TANNO, S. HORAI, T. OZAWA, T. MIYATAKE & S. TSUJI. 1990. A common mito-chondrial DNA mutation in the t-RNA(Lys) of patients with myoclonus epilepsy associated with ragged-red fibers. Biochem. Int. **21:** 789–796.

8. TANAKA, M., M. NISHIKIMI, H. SUZUKI, T. OZAWA, E. OKINO & H. TAKAHASHI. 1986. Multiple cyto-chrome deficiency and deteriorated mitochondrial polypeptide composition in fatal infantile mitochondrial myopathy and renal dysfunction. Biochem. Biophys. Res. Commun. **137:** 911–916.

9. TANAKA, M., M. NISHIKIMI, T. OZAWA, S. MIYABAYASHI & K. TADA. 1987. Lack of subunit II of cytochrome c oxidase in a patient with mitochondrial myopathy. Ann. N.Y. Acad. Sci. **488:** 503–504.

10. DIMAURO, S., E. BONILLA, M. ZEVIANI, M. NAKAGAWA & D.C. DEVIVO. 1985. Mitochondrial myopathies. Ann. Neurol. **17:** 521–538.

11. OZAWA, T. et al. 1988. Maternal inheritance of deleted mitochondrial DNA in a family with mitochondrial myopathy. Biochem. & Biophys. Res. Commun. **154:** 1240–1247.

12. HOLT, I.J., A.E. HARDING & J.A. MORGAN-HUGHES. 1988. Deletions of muscle mitochondrial DNA in patients with mitochondrial myopathies. Nature **331:** 717–719.

13. ROTIG, A. et al. 1989. Mitochondrial DNA deletion in Peason's marrow/pancreas syndrome. Lancet **i:** 902–903.

14. TAUCHI, H. & T. SATO. 1968. Age changes in size and numbers of mitochondria of human hepatic cells. J. Gerontol. **23:** 454–461.

15. MULLER-HOCKER, J. 1992. Mitochondria and ageing [Review]. Brain Pathology **2:** 149–158.

16. LINNANE, A.W., S. MARZUKI, T. OZAWA & M. TANAKA. 1989. Mitochondrial DNA mutations as an important contributor to ageing and degenerative diseases. Lancet **i:** 642–645.

17. IKEBE, S. et al. 1990. Increase of deleted mitochondrial DNA in the striatum in Parkinson's dis-ease and senescence. Biochem. Biophys. Res. Commun. **170:** 1044–1048.

18. LINNANE, A.W., A. BAUMER, R.J. MAXWELL, H. PRESTON, C.F. ZHANG & S. MARZUKI. 1990. Mito-chondrial gene mutation: The ageing process and degenerative diseases. Biochem. Int. **22:** 1067–1076.

19. CORTOPASSI, G.A. & N. ARNHEIM. 1990. Detection of a specific mitochondrial DNA deletion in tissues of older humans. Nucleic Acids Res. **18:** 6927–6933.

20. HATTORI, K. et al. 1991. Age-dependent increase in deleted mitochondrial DNA in the human heart: Possible contributing factor to "presbycardia." Am. Heart J. **121:** 1735–1742.

21. CORRAL-DEBRINSKI, M., J.M. SHOFFNER, M.T. LOTT & D.C. WALLACE. 1992. Association of mito-chondrial DNA damage with aging and coronary atherosclerotic heart disease. Mutat. Res. **275:** 169–180.

22. HAYAKAWA, M., K. HATTORI, S. SUGIYAMA & T. OZAWA. 1992. Age-associated oxygen damage and mutations in mitochondrial DNA in human hearts. Biochem. Biophys. Res. Commun. **189:** 979–985.

23. TANAKA, M., M. HAYAKAWA & T. OZAWA. 1996. Automated sequencing of mitochondrial DNA. In Methods in Enzymology, 264, Mitochondrial genetics and biogenesis. G.M. Attardi & A. Chomyn, Eds.: 407–421. Academic Press. Orlando, FL.

24. HAYAKAWA, M., K. KATSUMATA, M. YONEDA, M. TANAKA, S. SUGIYAMA & T. OZAWA. 1995. Mito-chondrial DNA minicircles, lacking replication origins, exist in the cardiac muscle of a young normal subject. Biochem. Biophys. Res. Commun. **215:** 952–960.

25. RICHTER, C., J.-W. PARK & B.N. AMES. 1988. Normal oxidative damage to mitochondrial and nuclear DNA is extensive. Proc. Natl. Acad. Sci. USA **85:** 6465–6467.

26. HAYAKAWA, M., K. TORII, S. SUGIYAMA, M. TANAKA & T. OZAWA. 1991. Age-associated accumu-lation of 8-hydroxydeoxyguanosine in mitochondrial DNA of human diaphragm. Biochem. Biophys. Res. Commun. **179:** 1023–1029.

27. OZAWA, T. 1994. Mitochondrial cardiomyopathy [Review]. Herz **19:** 105–118.

28. HARMAN, D. 1960. The free radical theory of aging: The effect of age on serum mercaptan lev-els. J. Gerontol. **15:** 38–40.

29. OZAWA, T. 1995. Mechanism of somatic mitochondrial DNA mutations associated with age and diseases. [Review]. Biochim. Biophys. Acta **1271:** 177–189.

30. OZAWA, T. 1997. Genetic and functional changes in mitochondria associated with aging. Phys-iol. Rev. **77:** 425–464.

31. OZAWA, T., M. HAYAKAWA, K. KATSUMATA, M. YONEDA, S. IKEBE & Y. MIZUNO. 1997. Fragile mitochondrial DNA: The missing link in the apoptotic neuronal cell death in Parkinson's disease. Biochem. Biophys. Res. Commun. **235:** 158–161.

32. SUGIYAMA, S., K. YAMADA & T. OZAWA. 1995. Preservation of mitochondrial respiratory function by coenzyme Q10 in aged rat skeletal muscle. Biochem. Mol. Biol. Int. **37:** 1111–1120.

33. OZAWA, T., K. SAHASHI, Y. NAKASE & B. CHANCE. 1995. Extensive tissue oxygenation associated with mitochondrial DNA mutations. Biochem. Biophys. Res. Commun. **213:** 432–438.

34. HAYAKAWA, M., K. KATSUMATA, M. YONEDA, M. TANAKA, S. SUGIYAMA & T. OZAWA. 1996. Age-related extensive fragmentation of mitochondrial DNA into minicircles. Biochem. Biophys. Res. Commun. **226:** 369–377.

35. KATSUMATA, K., M. HAYAKAWA, M. TANAKA, S. SUGIYAMA & T. OZAWA. 1994. Fragmentation of human heart mitochondrial DNA associated with premature aging. Biochem. Biophys. Res. Commun. **202:** 102–110.

36. OZAWA, T. *et al.* 1995. Genotype and phenotype of severe mitochondrial cardiomyopathy: A recipient of heart transplantation and the genetic control. Biochem. Biophys. Res. Commun. **207:** 613–620.

37. YONEDA, M., K. KATSUMATA, M. HAYAKAWA, M. TANAKA & T. OZAWA. 1995. Oxygen stress induces an apoptotic cell death associated with fragmentation of mitochondrial genome. Biochem. Biophys. Res. Commun. **209:** 723–729.

38. WYLLIE, A.H., J.F.R. KERR & A.R. CURRIE. 1980. Cell death: The significance of apoptosis. Int. Rev. Cytol. **68:** 251–305.

39. JACOBSON, M.D., J.F. BURNE & M.C. RAFF. 1994. Programmed cell death and Bcl-2 protection in the absence of a nucleus. EMBO J. **13:** 1899–1910.

40. SCHULZE-OSTHOFF, K., H. WALCZAK, W. DROGE & P.H. KRAMMER. 1994. Cell nucleus and DNA fragmentation are not required for apoptosis. J. Cell Biol. **127:** 15–20.

41. NAKAJIMA, H., P. GOLSTEIN & P.A. HENKART. 1995. The target cell nucleus is not required for cell-mediated granzyme- or fas-based cytotoxicity. J. Exp. Med. **181:** 1905–1909.

42. SHIMIZU, S., Y. EGUCHI, W. KAMIIKE, H. MATSUDA & Y. TSUJIMOTO. 1996. Bcl-2 expression prevents activation of the ICE protease cascade. Oncogene **12:** 2251–2257.

43. ZAMZAMI, N. *et al.* 1996. Mitochondrial control of nuclear apoptosis. J. Exp. Med. **183:** 1553–1544.

44. LIU, X., C.N. KIM, J. YANG, R. JEMMERSON & X. WANG. 1996. Induction of apoptotic program in cell-free extracts: Requirement for dATP and cytochrome c. Cell **86:** 147–157.

45. MITCHELL, P. 1979. David Keilin's respiratory chain concept and its chemiosmotic consequences. *In* Les Prix Nobel en 1978. The Nobel Foundation. Stockholm.

46. SKULACHEV, V.P. 1996. Role of uncoupled and non-coupled oxidations in maintenance of safely low level of oxygen and its one-electron reductants [Review]. Q. Rev. Biophys. **29:** 169–202.

47. JACOBSON, M.D. & M.C. RAFF. 1995. Programmed cell death and Bcl-2 protection in very low oxygen. Nature **374:** 814–816.

48. SHIMIZU, S., Y. EGUCHI, H. KOSAKA, W. KAMIIKE, H. MATSUDA & Y. TSUJIMOTO. 1995. Prevention of hypoxia-induced cell death by Bcl-2 and Bcl-xL. Nature **374:** 811–813.

49. OZAWA, T. 1997. Oxidative damage and fragmentation of mitochondrial DNA in cellular apoptosis. Biosci. Rep. **17:** 237–250.

50. WALKER, J.E. 1995. Determination of the structures of respiratory enzyme complexes from mammalian mitochondria [Review]. Biochim. Biophys. Acta **1271:** 221–227.

51. ZORATTI, M. & I. SZABO. 1995. The mitochondrial permeability transition. Biochim. Biophys. Acta—Rev. Biomembranes **1241:** 139–176.

52. WALLACE, D.C. 1992. Diseases of the mitochondrial DNA [Review]. Annu. Rev. Biochem. **61:** 1175–1212.

53. WALLACE, D.C. *et al.* 1988. Familial mitochondrial encephalomyopathy (MERRF): Genetic, pathophysiological, and biochemical characterization of a mitochondrial DNA disease. Cell **66:** 601–610.

54. KOVALENKO, S.A., M. TANAKA, M. YONEDA, A.F. IAKOVLEV & T. OZAWA. 1996. Accumulation of somatic nucleotide substitutions in mitochondrial DNA associated with the 3243 A-to-G tRNA(leu)(UUR) mutation in encephalomyopathy and cardiomyopathy. Biochem. Biophys. Res. Commun. **222:** 201–207.

55. OZAWA, T. 1995. Mitochondrial DNA mutations associated with aging and degenerative diseases [Review]. Exp. Gerontol. **30:** 269–290.
56. YONEDA, M. *et al.* 1989. Mitochondrial DNA mutation in family with Leber's hereditary optic neuropathy [letter]. Lancet **i:** 1076–1077.
57. GOTO, Y., I. NONAKA & S. HORAI. 1990. A mutation in the transfer RNA$^{Leu(UUR)}$ gene associated with the MELAS subgroup of mitochondrial encephalomyopathy. Nature **348:** 651–653.
58. INO, H. *et al.* 1991. Mitochondrial leucine tRNA mutation in a mitochondrial encephalomyopathy [letter; comment]. Lancet **337:** 234–235.
59. MITA, S. *et al.* 1990. Recombination via flanking direct repeats is a major cause of large-scale deletions of human mitochondrial DNA. Nucleic Acids Res. **18:** 561–567.
60. IKEBE, S., M. TANAKA & T. OZAWA. 1995. Point mutations of mitochondrial genome in Parkinson's disease. Am. Heart J. **28:** 281–295.
61. OZAWA, T. *et al.* 1990. Multiple mitochondrial DNA deletions exist in cardiomyocytes of patients with hypertrophic or dilated cardiomyopathy. Biochem. Biophys. Res. Commun. **170:** 830–836.
62. OBAYASHI, T. *et al.* 1992. Point mutations in mitochondrial DNA in patients with hypertrophic cardiomyopathy. Am. Heart J. **124:** 1263–1269.
63. CHEN, X., E. BONILLA, M. SCIACCO & E.A. SCHON. 1995. Paucity of deleted mitochondrial DNAs in brain regions of Huntington's disease patients. Biochim. Biophys. Acta **1271:** 229–233.
64. van den OUWELAND, J.M.W. *et al.* 1992. Mutation in mitochondrial tRNAleu(UUR) gene in a large pedigree with maternally transmitted type II diabetes mellitus and deafness. Nat. Genet. **1:** 368–371.
65. OZAWA, T. *et al.* 1991. Patients with idiopathic cardiomyopathy belong to the same mitochondrial DNA gene family of Parkinson's disease and mitochondrial encephalomyopathy. Biochem. Biophys. Res. Commun. **177:** 518–525.
66. KADOWAKI, H. *et al.* 1993. Mitochondrial gene mutation and insulin-deficient type of diabetes mellitus [letter]. Lancet **341:** 893–894.
67. CANN, R.L., M. STONEKING & A.C. WILSON. 1987. Mitochondrial DNA and human evolution. Nature **325:** 31–36.
68. SATO, W. *et al.* 1992. Genetic analysis of three pedigrees of mitochondrial myopathy, encephalopathy, lactic acidosis, and stroke-like episodes (MELAS). Am. J. Hum. Genet. **50:** 655–657.
69. UPHOLT, W.B. & I.B. DAWID. 1977. Mapping of mitochondrial DNA of individual sheep and goats: Rapid evolution in the D loop region. Cell **11:** 571–583.
70. HAUSWIRTH, W.W. & P. LAIPIS. 1982. Mitochondrial DNA polymorphism in a maternal lineage of Holstein cows. Proc. Natl. Acad. Sci. USA **79:** 4686–4690.
71. PIKO, L. & L. MATSUMOTO. 1976. Number of mitochondria and some properties of mitochondrial DNA in the mouse egg. Dev. Biol. **19:** 1–10.
72. ZHANG, C., A. LINNANE & P. NAGLEY. 1993. Occurrence of a particular base substitution (3243 A to G) in mitochondrial DNA of tissues of ageing humans. Biochem. Biophys. Res. Commun. **195:** 1104–1110.
73. KAPLAN, E.L. & P. MEIER. 1958. Nonparametric estimation for incomplete observations. J. Am. Stat. Assoc. **52:** 457–481.
74. OZAWA, T., K. KATSUMATA, M. HAYAKAWA, M. YONEDA, M. TANAKA & S. SUGIYAMA. 1995. Mitochondrial DNA mutations and survival rate. Lancet **355:** 189.
75. KATZ, A.M. 1995. Cell death in the failing heart: Role of an unnatural growth response to overload [Review]. Clin. Cardiol. **18:** IV36–44.
76. TEIGER, E. *et al.* 1996. Apoptosis in pressure overload-induced heart hypertrophy in the rat. J. Clin. Invest. **97:** 2891–2897.
77. JAMES, T.N., E. ST. MARTIN, P.W.r. WILLIS & T.O. Lohr. 1996. Apoptosis as a possible cause of gradual development of complete heart block and fatal arrhythmias associated with absence of the AV node, sinus node, and internodal pathways. Circulation **93:** 1424–1438.
78. WILLIAMS, G.H. & E. BRAUNWALD. 1988. Endocrine and nutritional disorders and heart disease. *In* Heart Disease. E. Braunwald, Ed.: 1800–1827. W.B. Saunders. Philadelphia, London, Toronto, Montreal, Sidney, Tokyo.
79. GRIVELL, L.A. 1989. Mitochondrial DNA: Small, beautiful and essential. Nature **341:** 569–571.

80. Sato, W., M. Tanaka, K. Ohno, T. Yamamoto, G. Takada & T. Ozawa. 1989. Multiple populations of deleted mitochondrial DNA detected by a novel gene amplification method. Biochem. Biophys. Res. Commun. **162:** 664–672.

81. Ozawa, T., M. Tanaka, W. Sato, K. Ohno, M. Yoneda & T. Yamamoto. 1991. Types and mechanism of mitochondrial DNA mutations in mitochondrial myopathy and related diseases. *In* Molecular Basis of Neurological Disorders and their Treatment. J.W. Gorrod, O. Albano, E. Ferrari & S. Papa, Eds.: 171–190. Chapman & Hall. London, New York, Tokyo, Melbourne, Madras.

82. Sugiyama, S., K. Hattori, M. Hayakawa & T. Ozawa. 1991. Quantitative analysis of age-associated accumulation of mitochondrial DNA with deletion in human hearts. Biochem. Biophys. Res. Commun. **180:** 894–899.

83. Simonetti, S., X. Chen, S. DiMauro & E.A. Schon. 1992. Accumulation of deletions in human mitochondrial DNA during normal aging: Analysis by quantitative PCR. Biochim. Biophys. Acta **1180:** 113–122.

84. Ozawa, T., M. Tanaka, S. Ikebe, K. Ohno, T. Kondo & Y. Mizuno. 1990. Quantitative determination of deleted mitochondrial DNA relative to normal DNA in parkinsonian striatum by a kinetic PCR analysis. Biochem. Biophys. Res. Commun. **172:** 483–489.

85. Cooper, J.M., V.M. Mann & A.H. Schapira. 1992. Analyses of mitochondrial respiratory chain function and mitochondrial DNA deletion in human skeletal muscle: Effect of ageing. J. Neurol. Sci. **113:** 91–98.

86. Remes, A.M., I.E. Hassinen, M.J. Ikaheimo, R. Herva, J. Hirvonen & K.J. Peuhkurinen. 1994. Mitochondrial DNA deletions in dilated cardiomyopathy: a clinical study employing endomyocardial sampling. J. Am. Coll. Cardiol. **23:** 935–942.

87. Zhang, C., A. Baumer, R.J. Maxwell, A.W. Linnane & P. Nagley. 1992. Multiple mitochondrial DNA deletions in an elderly human individual. FEBS Lett. **297:** 34–38.

88. Soong, N.W., D.R. Hinton, G. Cortopassi & N. Arnheim. 1992. Mosaicism for a specific somatic mitochondrial DNA mutation in adult human brain. Nat. Genet. **2:** 318–323.

89. Poulton, J., M.E. Deadman, L. Bindoff, K. Morten, J. Land & G. Brown. 1993. Families of mtDNA rearrangements can be detected in patients with mtDNA deletions: Duplications may be a transient intermediate form. Hum. Mol. Genet. **2:** 23–30.

90. Kovalenko, S.A., J.M. Kopsidas, J.M. Kelso & A.W. Linnane. 1997. Deltoid human muscle mtDNA is extensively rearranged in old age subjects. Biochem. Biophys. Res. Commun. **232:** 147–152.

91. Priestley, J. 1775. Experiments and observations on different kinds of air. Vol. 3. 101–102. J. Johnson at St. Paul's Churchyard. London.

92. Harman, D. 1956. Aging: A theory based on free radical and radiation chemistry. J. Gerontrol. **11:** 298–300.

93. Warburg, O. 1924. Über Eisen, den sauerstoff Übertragenden Bestandteil des Atmungsferments. Biochem. Z. **152:** 479–494.

94. Michaelis, L. 1946. Fundamentals of oxidation and reduction. *In* Currents in Biochemical Research. D.E. Green, Ed.: 207–227. Interscience. New York.

95. Boveris, A. & B. Chance. 1973. The mitochondrial generation of hydrogen peroxide: General properties and effect of hyperbaric oxygen. Biochem. J. **134:** 707–716.

96. Turrens, J.F., B.A. Freeman & J.D. Crapo. 1982. Hyperoxia increases H_2O_2 release by lung mitochondria and microsomes. Arch. Biochem. Biophys. **217:** 411–421.

97. Ksenzenko, M., A.A. Konstantinov, G.B. Khomutov, A.N. Tikhonov & E.K. Ruuge. 1983. Effect of electron transfer inhibitors on superoxide generation in the cytochrome bc1 site of the mitochondrial respiratory chain. FEBS Lett. **155:** 19–24.

98. Massey, V. 1994. Activation of molecular oxygen by flavins and flavoproteins [Review]. J. Biol. Chem. **269:** 22459–22462.

99. Cross, A.R. & O.T. Jones. 1991. Enzymic mechanisms of superoxide production [Review]. Biochim. Biophys. Acta **1057:** 281–298.

100. Hayakawa, M., T. Ogawa, S. Sugiyama & T. Ozawa. 1989. Hydroxyl radical and leukotoxin biosynthesis in neutrophil plasma membrane. Biochem. Biophys. Res. Commun. **161:** 1077–1085.

101. HOGG, N., V.M. DARLEY-USMAR, M.T. WILSON & S. MONCADA. 1992. Production of hydroxyl radicals from the simultaneous generation of superoxide and nitric oxide. Biochem. J. **281**: 419–424.
102. GENG, Y.J., Q. WU, M. MUSZYNSKI, G.K. HANSSON & P. LIBBY. 1996. Apoptosis of vascular smooth muscle cells induced by *in vitro* stimulation with interferon-gamma, tumor necrosis factor-alpha, and interleukin-1 beta. Arterioscler. Thromb. Vasc. Biol. **16**: 19–27.
103. DONG, H.D., Y. KIMOTO, S. TAKAI & T. TAGUCHI. 1996. Apoptosis as a mechanism of lectin-dependent monocyte-mediated cytotoxicity. Immunol. Invest. **25**: 65–78.
104. ANKARCRONA, M., J.M. DYPBUKT, B. BRUNE & P. NICOTERA. 1994. Interleukin-1 beta-induced nitric oxide production activates apoptosis in pancreatic RINm5F cells. Exp. Cell Res. **213**: 172–177.
105. DUNGER, A., P. AUGSTEIN, S. SCHMIDT & U. FISCHER. 1996. Identification of interleukin 1-induced apoptosis in rat islets using *in situ* specific labelling of fragmented DNA. J. Autoimmunity **9**: 309–313.
106. CHANCE, B., H. SIES & A. BOVFERIES. 1979. Hydroperoxide metabolism in mammalian organs. Physiol. Rev. **59**: 527–605.
107. YU, B.P. 1994. Cellular defenses against damage from reactive oxygen species. Physiol. Rev. **74**: 139–162.
108. DENG, R.Y. & I. FRIDOVICH. 1989. Formation of endonuclease III sensitive sites as a consequence of oxygen radical attack on DNA. Free Radical Biol. Med. **6**: 123–129.
109. FRAGA, C.G., M.K. SHIGENAGA, J.W. PARK, P. DEGAN & B.N. AMES. 1990. Oxidative damage to DNA during aging: 8-hydroxy-2'-deoxyguanosine in rat organ DNA and urine. Proc. Natl. Acad. Sci. USA **87**: 4533–4537.
110. MECOCCI, P. *et al.* 1993. Oxidative damage to mitochondrial DNA shows marked age-dependent increases in human brain. Ann. Neurol. **34**: 609–616.
111. BELGUISE-VALLADIER, P. & R.P.P. FUCHS. 1991. Strong sequence-dependent polymorphism in adduct-induced DNA structure: Analysis of single *N*-2-acetyl-aminofluorene residues bound within the *Nar*I mutation hot spot. Biochemistry **30**: 10091–10100.
112. KAMIYA, H., K. MIURA, H. ISHIKAWA, H. INOUE, S. NISHIMURA & E. OHTSUKA. 1992. c-Ha-*ras* containing 8-hydroxyguanine at codon 12 induces point mutations at the modified and adjacent positions. Cancer Res. **52**: 3483–3485.
113. HAYAKAWA, M., T. OGAWA, S. SUGIYAMA, M. TANAKA & T. OZAWA. 1991. Massive conversion of guanosine to 8-hydroxy-guanosine in mouse liver mitochondrial DNA by administration of azidothymidine. Biochem. Biophys. Res. Commun. **176**: 87–93.
114. HAYAKAWA, M., S. SUGIYAMA, K. HATTORI, M. TAKASAWA & T. OZAWA. 1993. Age-associated damage in mitochondrial DNA in human hearts. Mol. Cell. Biochem. **119**: 95–103.
115. TAKASAWA, M., M. HAYAKAWA, S. SUGIYAMA, K. HATTORI, T. ITO & T. OZAWA. 1993. Age-associated damage in mitochondrial function in rat hearts. Exp. Gerontol. **28**: 269–280.
116. HOYER, S. 1986. Senile dementia and Alzheimer's disease. Brain blood flow and metabolism. Prog. Neuro-Psychopharmacol. Biol. Psychiatry **10**: 447–478.
117. KERR, J.F.R., A.H. WYLLIE & A.R. CURRIE. 1972. Apoptosis: A basic biological phenomenon with wide-ranging implications in tissue kinetics. Br. J. Cancer **26**: 239–257.
118. KATZ, A.M. 1995. The cardiomyopathy of overload: An unnatural growth response [Review]. Eur. Heart J. **16**: 110–114.
119. SHAROV, V.G., H.N. SABBAH, H. SHIMOYAMA, A.V. GOUSSEV, M. LESCH & S. GOLDSTEIN. 1996. Evidence of cardiocyte apoptosis in myocardium of dogs with chronic heart failure. Amer. J. Pathol. **148**: 141–149.
120. YAO, M., A. KEOGH, P. SPRATT, C.G. dos REMEDIOS & P.C. KIESSLING. 1996. Elevated DNase I levels in human idiopathic dilated cardiomyopathy: An indicator of apoptosis? J. Mol. Cell. Cardiol. **28**: 95–101.
121. WELSH, N. *et al.* 1995. Differences in the expression of heat-shock proteins and antioxidant enzymes between human and rodent pancreatic islets: Implications for the pathogenesis of insulin-dependent diabetes mellitus. Mol. Med. **1**: 806–820.
122. O'BRIEN, B.A., B.V. HARMON, D.P. CAMERON & D.J. ALLAN. 1996. Beta-cell apoptosis is responsible for the development of IDDM in the multiple low-dose streptozotocin model. J. Pathol. **178**: 176–181.

123. SIMONIAN, N.A. & J.T. COYLE. 1996. Oxidative stress in neurodegenerative diseases [Review]. Annu. Rev. Pharmacol. Toxicol. **36:** 83–106.

124. WYLLIE, A.H. 1980. Glucocorticoid-induced thymocyte apoptosis is associated with endogenous endonuclease activation. Nature **284:** 555–556.

125. TSUJIMOTO, Y. & C.M. CROCE. 1986. Analysis of the structure, transcripts, and protein products of bc1-2, the gene involved in human follicular lymphoma. Proc. Natl. Acad. Sci. USA **83:** 5214–5218.

126. HENGARTNER, M.O. & R.H. HORVITZ. 1994. The ins and outs of programmed cell death durlng *C. elegans* development. Philos. Trans. R. Soc. Lond B **345:** 243–246.

127. HENGARTNER, M.O. & R.H. HORVITZ. 1994. *C. elegans* cell survival gene ced-9 encodes a functional homolog of the mammalian protooncogene bc1-2. Cell **76:** 665–676.

128. YUAN, J.-Y., S. SHAHAM, S. LEDOUX, M.H. ELLIS & R.H. HORVITZ. 1993. The *C. elegans* cell death gene ced-3 encodes a protein similar to mammalian interleukin-1β converting enzyme. Cell **75:** 641–652.

129. LAZEBNIK, Y.A., S.H. KAUFMANN, S. DESNOYERS, G.G. POIRIER & W.C. EAMSHAW. 1994. Cleavage of poly(ADP-ribose) polymerase by a proteinase with properties like ICE. Nature **371:** 346–347.

130. NICHOLSON, D.W. *et al.* 1995. Identification and inhibition of the ICE/CED-3 protease necessary for mammalian apoptosis [see comments]. Nature **376:** 37–43.

131. SCHLEGEL, J. *et al.* 1996. CPP32/Apopain is a key interleukin 1β converting enzyme-like protease involved in Fas-mediated apoptosis. J. Biol. Chem. **271:** 1841–1844.

132. WANG, X., N.G. ZELENSKI, J. YANG, J. SAKAI, M.S. BROWN & J.L. GOLDSTEIN. 1996. Cleavage of sterol regulatory element binding proteins (SREBPs) by CPP32 during apoptosis. EMBO J. **15:** 1012–1020.

133. XUE, D. & H.R. HORVITZ. 1995. Inhibition of the *Caenorhabditis elegans* cell-death protease CED-3 by a CED-3 cleavage site in baculovirus p35 protein. Nature **377:** 248–251.

134. WANG, X. *et al.* 1995. Purification of an interleukin-1β converting enzyme-related cysteine protease that cleaves sterol regulatory element-binding proteins between the leucine zipper and transmembrane domains. J. Biol. Chem. **270:** 18044–18050.

135. TEWARI, M. *et al.* 1995. Yama/CPP32β, a mammalian homolog of ced-3, is a CrmA-inhibitable protease that cleaves the death substrate poly(ADP-ribose) polymerase. Cell **81:** 801–809.

136. LAZEBNIK, Y.A., S. COLE, C.A. COOKE, W.G. NELSON & W.C. EAMSHAW. 1993. Nuclear events of apoptosis *in vitro* in cell-free mitotic extracts: A model system for analysis of the active phase of apoptosis. J. Cell Biol. **123:** 7–22.

137. GAIDO, M.L. & J.A. CIDLOWSKI. 1991. Identification, purification, and characterization of a calcium-dependent endonuclease (NUC18) from apoptotic rat thymocytes. NUC18 is not histone H2B. J. Biol. Chem. **266:** 18580–18585.

138. PEITSCH, M.C. *et al.* 1993. Characterization of the endogenous deoxyribonuclease involved in nuclear DNA degradation during apoptosis (programmed cell death). EMBO J. **12:** 371–377.

139. BERNARDI, P., S. VASSANELLI, P. VERONESE, R. COLONNA, I. SZABO & M. ZORATTI. 1992. Modulation of the mitochondrial permeability transition pore: Effect of protons and divalent cations. J. Biol. Chem. **267:** 2934–2939.

140. SZABO, I., P. BERNARDI & M. ZORATTI. 1992. Modulation of the mitochondrial megachannel by divalent cations and protons. J. Biol. Chem. **267:** 2940–2946.

141. REED, D.J. & M.K. SAVAGE. 1995. Influence of metabolic inhibitors on mitochondrial permeability transition and glutathione status. Biochim. Biophys. Acta **1271:** 43–50.

142. SKULACHEV, V.P. 1996. Why are mitochondria involved in apoptosis? Permeability transition pores and apoptosis as selective mechanisms to eliminate superoxide-producing mitochondria and cell. FEBS Lett. **397:** 7–10.

143. THIERS, R.E. & B.L. VALLEE. 1957. Distribution of metals in subcellular fractions of rat. J. Biol. Chem. **226:** 911–920.

144. DAINZANI, M.U. & I. VITI. 1955. The content and distribution of cytochrome *c* in the fatty liver of rats. Biochem. J. **59:** 141–145.

145. MAYER, A., W. NEUPERT & R. LILL. 1995. Translocation of apocytochrome c across the outer membrane of mitochondria. J. Biol. Chem. **270:** 12390–12397.

146. GONZALES, D.H. & W. NEUPERT. 1990. Biogenesis of mitochondrial c-type cytochromes. J. Bioenerg. Biomembr. **22:** 753–768.

147. BEINERT, H. 1951. The extent of artificial redistribution of cytochrome c in rat liver homogenate. J. Biol. Chem. **190**: 287–292.
148. SHIMIZU, S. *et al.* 1996. Bcl-2 blocks loss of mitochondrial membrane potential while ICE inhibitors act at a different step during inhibition of death induced by respiratory chain inhibitors. Oncogene **13**: 21–29.
149. SHIMIZU, S. *et al.* 1996. Retardation of chemical hypoxia-induced necrotic cell death by Bcl-2 and ICE inhibitors: Possible involvement of common mediators in apoptotic and necrotic signal transductions. Oncogene **12**: 2045–2050.
150. KORSMEYER, S.J., X.-M. YIN, Z.N. OLTVAI, D.J. VEIS-NOVACK & G.P. LINETTE. 1995. Reactive oxygen species and the regulation of cell death by the Bcl-2 gene family. Biochim. Biophys. Acta **1271**: 63–66.
151. OLTVAI, Z.N., C.L. MILLIMAN & S.J. KORSMEYER. 1993. Bcl-2 heterodimerizes *in vivo* with a conserved homologue, Bax, that accelerates programmed cell death. Cell **74**: 609–619.
152. BOISE, L.H. *et al.* 1993. bcl-x, a bcl-2-related gene that functions as a dominant regulator of apoptotic cell death. Cell **74**: 597–608.
153. OFFEN, D., I. ZIV, S. GORODIN, A. BARZILAI, Z. MALIK & E. MELAMED. 1995. Dopamine-induced programmed cell death in mouse thymocytes. Biochim. Biophys. Acta **1268**: 171–177.
154. KANE, D. *et al.* 1993. Bcl-2 inhibition of neural death: Decreased generation of reactive oxygen species. Science **262**: 1274–1277.
155. HOCKENBERG, D.M., Z.N. OLTVAI, X.-M. YIN, C.L. MILLIMAN & S.J. KORSMEYER. 1993. Bcl-2 functions in an antioxidant pathway to prevent apoptosis. Cell **75**: 241–251.
156. ZAMZAMI, N. *et al.* 1995. Sequential reduction of mitochondrial transmembrane potential and generation of reactive oxygen species in early programmed cell death. J. Exp. Med. **182**: 367–377.
157. ALBINA, J.E., B.A. MARTIN, W. HENRY, Jr., C.A. Louis & J.S. REICHNER. 1996. B cell lymphoma-2 transfected P815 cells resist reactive nitrogen intermediate-mediated macrophage-dependent cytotoxicity. J. Immunol. **157**: 279–283.
158. MESSMER, U.K., U.K. REED & B. BRUNE. 1996. Bcl-2 protects macrophages from nitric oxide-induced apoptosis. J. Biol. Chem. **271**: 20192–20197.
159. JONES, D.E. 1995. Mitochondrial dysfunction during anoxia and acute cell injury. Biochim. Biophys. Acta **1271**: 29–33.
160. MEISTER, A. 1995. Mitochondrial changes associated with glutathione deficiency. Biochim. Biophys. Acta **1271**: 35–42.
161. SUGIYAMA, S., M. TAKASAWA, M. HAYAKAWA & T. OZAWA. 1993. Changes in skeletal muscle, heart and liver mitochondrial electron transport activities in rats and dogs of various ages. Biochem. Mol. Biol. Int. **30**: 937–944.
162. SAKAI, T. *et al.* 1996. Role of nitric oxide and superoxide anion in leukotoxin, 9,10-epoxy-12-octadecanoate-induced mitochondrial dysfunction. Free Radical Biol. Med. **20**: 607–612.
163. PETIT, P.X., H. LECOEUR, E. ZORN, C. DAUGUET, B. MIGNOTTE & M.L. GOUGEON. 1995. Alterations in mitochondrial structure and function are early events of dexamethasone-induced thymocyte apoptosis. J. Cell Biol. **130**: 157–167.
164. LAZEBNIK, Y.A. *et al.* 1995. Studies of the lamin proteinase reveal multiple parallel biochemical pathways during apoptotic execution. Proc. Natl. Acad. Sci. USA **92**: 9042–9046.

Oxidative Damage and Mutation to Mitochondrial DNA and Age-dependent Decline of Mitochondrial Respiratory Function[a]

YAU-HUEI WEI,[b] CHING-YOU LU, HSIN-CHEN LEE, CHENG-YOONG PANG, AND YI-SHING MA

Department of Biochemistry and Center for Cellular and Molecular Biology, School of Life Science, National Yang-Ming University, Taipei, Taiwan 112, Republic of China

ABSTRACT: Mitochondrial respiration and oxidative phosphorylation are gradually uncoupled, and the activities of the respiratory enzymes are concomitantly decreased in various human tissues upon aging. An immediate consequence of such gradual impairment of the respiratory function is the increase in the production of the reactive oxygen species (ROS) and free radicals in the mitochondria through the increased electron leak of the electron transport chain. Moreover, the intracellular levels of antioxidants and free radical scavenging enzymes are gradually altered. These two compounding factors lead to an age-dependent increase in the fraction of the ROS and free radical that may escape the defense mechanism and cause oxidative damage to various biomolecules in tissue cells. A growing body of evidence has established that the levels of ROS and oxidative damage to lipids, proteins, and nucleic acids are significantly increased with age in animal and human tissues. The mitochondrial DNA (mtDNA), although not protected by histones or DNA-binding proteins, is susceptible to oxidative damage by the ever-increasing levels of ROS and free radicals in the mitochondrial matrix. In the past few years, oxidative modification (formation of 8-hydroxy-2'-deoxyguanosine) and large-scale deletion and point mutation of mtDNA have been found to increase exponentially with age in various human tissues. The respiratory enzymes containing the mutant mtDNA-encoded defective protein subunits inevitably exhibit impaired respiratory function and thereby increase electron leak and ROS production, which in turn elevates the oxidative stress and oxidative damage of the mitochondria. This vicious cycle operates in different tissue cells at different rates and thereby leads to the differential accumulation of mutation and oxidative damage to mtDNA in human aging. This may also play some role in the pathogenesis of degenerative diseases and the age-dependent progression of the clinical course of mitochondrial diseases.

Human aging is a multifactorial process that leads to the gradual loss of the capability of an individual to maintain homeostasis, which is characterized by a progressive decline in function at tissue and cell levels. It causes a decrease of the individual's ability to respond

[a]The work described in this article was supported by research Grants from the National Science Council (NSC86-2314-B010-09 and NSC87-2314-B010-089) and the National Health Research Institutes (DOH87-HR-505), Executive Yuan, Republic of China. One of the authors, Yau-Huei Wei, wishes to express his appreciation to the National Science Council for its considerable support of the studies on the mitochondrial role in human aging.

[b]Tel: 886-2-28267118; fax: 886-2-28264843; e-mail: joeman@mailsrv.ym.edu.tw

to a wide range of challenges or stresses and increases the susceptibility of the subject to age-associated diseases and death.

Many theories have been formulated to explain the biochemical and molecular basis of human aging. Among them, the "free-radical theory of aging," which was first proposed more than four decades ago by Harman,[1] has received much attention of scientists in biomedical fields. He proposed that the accumulation of damage to biomolecules, caused by the attack of free radicals, plays a major role in human aging. Subsequently, he implemented this theory with the suggestion that mitochondria are the major target of free radical attack that leads to human aging.[2] In the past two decades, this theory has been widely tested and gained great support from work done in many laboratories involved in the molecular and cellular biological research of aging. Miquel and co-workers[3,4] were among the early investigators who provided substantial support to this notion by showing that mitochondrial DNA (mtDNA) damage and lipofuscin pigment formation in animal tissues are concurrently increased during aging. On the basis of the fact that mitochondria are the major intracellular source of reactive oxygen species (ROS) and that mtDNA is vulnerable to oxidative damage,[5,6] Linnane *et al.*[7] further hypothesized that accumulation of somatic mutations of mtDNA is a major contributor to human aging and degenerative diseases.

We argue that the production of ROS and free radicals in mitochondria is increased during aging as a result of the ever-increasing electron leak from the defective respiratory system and that the ensuing oxidative damage and mutation to mtDNA cause a further decline of respiratory function. In this article, we summarize recent findings of age-related oxidative damage and mutation of mtDNA and discuss their contributions to the decline of mitochondrial respiratory function in human aging.

OXIDATIVE STRESS AND AGING

Human and animal cells obtain chemical energy in the form of ATP through tight coupling of oxidative phosphorylation to respiration in mitochondria. Under normal physiological conditions, about 1–5% of the oxygen consumed by mitochondria are converted to superoxide anions, hydrogen peroxide, and other ROS.[6,8] It has long been established that several sites of the respiratory chain are involved in the generation of ROS and free radicals, such as ubisemiquinone and flavosemiquinone,[9–11] which are generated in one-electron transfer reactions and maintained at relatively high steady state levels in animal mitochondria. It was estimated that each mitochondrion in rat liver can produce about 3×10^7 superoxide anions per day.[6] Although within a certain concentration range, ROS have important physiological functions, such as regulating smooth muscle relaxation and acting as a secondary messenger to activate the transcription factors, including NF-kB and AP-1,[12–14] an excess of ROS is harmful to cells.[15] Human cells can counteract oxidative stress by expressing an array of free radical scavenging enzymes, which include manganese-superoxide dismutase (MnSOD), copper/zinc-superoxide dismutase (Cu/ZnSOD), glutathione peroxidase, and catalase. Although these and other nonenzymatic antioxidants may cope with the ROS and free radicals formed as by-products of normal aerobic metabolism, these antioxidant defense systems are not perfect at all. Thus, there is an age-dependent increase in the fraction of ROS and free radicals that may escape these cellular defense mechanisms and exert damage to cellular constituents, including DNA, RNA, proteins, and membrane lipids.[15] We have previously demonstrated that lipid peroxidation, as measured by the level of malondialdehyde, of mitochondria is enhanced in various human tissues during aging.[16] Recently, we cultured skin fibroblasts from individuals of different ages who had been proved to have none of the known mitochondrial disorders or degenerative diseases. The activities of respiratory enzymes of the fibroblasts from old donors were found to be generally lower than those from young donors (TABLE 1). This is in line with the observations that respiratory function declines with age in human tissues.[15] Moreover, we found that the enzyme activities of MnSOD and catalase of the fibroblasts changed in an age-dependent manner. This may induce an imbalance between prooxidants and antioxidants and thus elevate the oxidative stress, which can explain the observed age-dependent increase of specific contents of lipid peroxides in the skin fibroblasts (TABLE 2). It is noteworthy that the MnSOD activity was increased, but catalase activity decreased significantly in the fibroblasts of a 75-year-old woman who harbored the 4,977 bp deleted mtDNA at about 55% of the total mtDNA. The lipid peroxide content in the fibroblasts of this woman was found to be about twofold higher than that of the control. On the other hand, the intracellular content of 8-hydroxy-2′-deoxyguanosine (8-OH-dG), a specific product of oxidative damage to DNA, has been shown to increase with age in the tissues of mammals and insects.[17–19] These findings lead us to suggest that enhanced production of ROS and free radicals due to an electron leak from the impaired respiratory chain may play an important role in human aging. Additionally, Agarwal and Sohal[19] demonstrated that life expectancy of the housefly is inversely related to the 8-OH-dG content in the body tissue of the insect. Moreover, it was observed that the contents of 8-OH-dG in the body tissues of the flies overexpressing Cu/ZnSOD and catalase were much lower than those of the wild-type flies.[20] Caloric restriction, which extends the average and the maximum life span of the mammals, was found to result in a relatively slower rate of age-dependent accumulation of 8-OH-dG in various tissues of the C57BL/6 strain of mice.[21] These observations suggest that oxidative damage to cellular DNA plays an important role in aging.

TABLE 1. Comparison of Electron Transport Activities of the Respiratory Enzyme Complexes of Mitochondria of the Fibroblasts from Subjects in Different Age Groups and a 75-year-old Woman Harboring the 4,977 bp mtDNA Deletion

Age Group (year)	Enzyme Activity (nmol/min/mg)		
	NCCR[a]	SCCR	CCO
10–30	$720 \pm 84 \ (2)^b$	$221 + 29 \ (2)$	$103 \pm 24 \ (2)$
31–60	$590 \pm 52 \ (2)$	$55 \pm 23 \ (7)$	$23 \pm 12 \ (7)$
61–80	$340 \pm 152 \ (2)$	$44 \pm 19 \ (6)$	$29 \pm 6 \ (6)$
75[c]	600	48	14

[a]NCCR, NADH-cytochrome *c* reductase, SCCR, succinate-cytochrome *c* reductase; CCO, cytochrome *c* oxidase.

[b]Values in the parentheses indicate the number of subjects whose fibroblasts were used for the indicated enzyme assay.

[c]This woman harbored the 4,977 bp-deleted mtDNA at a proportion of about 55% in the sun-exposed skin on her face but had none of the known mitochondrial diseases.

TABLE 2. Comparison of RNA Levels and Enzyme Activities of Free Radical Scavenging Enzymes and Contents of Lipid Peroxides of Skin Fibroblasts from Subjects in Different Age Groups and a 75-year-old Woman Harboring the 4,977 bp mtDNA Deletion

Age Group (year)	MnSOD[a] RNA (MnSOD/β-actin)	MnSOD (μmol/min/mg)	Catalase (μmol/min/mg)	Lipid Peroxides (nmol/mg protein)
10-30	$0.15 \pm 0.07(2)^b$	$18.61 \pm 3.20(6)$	$16.40 \pm 2.06(6)$	$0.73 \pm 0.21(6)$
31-60	$0.24 \pm 0.04(2)$	$21.07 \pm 3.65(13)$	$13.30 \pm 1.57(13)$	$0.98 \pm 0.39(9)$
61-80	$0.33 \pm 0.13(4)$	$17.86 \pm 3.00(13)$	$13.80 \pm 4.25(13)$	$1.16 \pm 0.41(8)$
75[c]	0.24	23.21	7.60	1.44

[a]MnSOD, manganese-dependent superoxide dismutase.
[b]Values in the parentheses denote the number of subjects whose fibroblasts were used for the indicated assay or determination.
[c]This 75-year-old woman had the 4,977 bp-deleted mtDNA at a proportion of about 55% in her sun-exposed face skin but had none of the known mitochondrial diseases.

On the other hand, aging-associated accumulation of functionally inactive or structurally altered proteins has been demonstrated in organisms, including nematodes, flies, and humans.[22,23] It was observed that both the accumulation of protein carbonyls and the loss of glucose-6-phosphate dehydrogenase activity are increased during aging.[22] Thus, the intracellular levels of proteolytic enzymes that hydrolyze oxidatively modified proteins are insufficient for the effective disposal of the aging-associated increase of the aberrant proteins.[22] On the other hand, it was shown that the activities of proteases involved in the degradation of aberrant proteins are markedly decreased in the tissues of old animals.[24] It has long been proposed that perturbation of proteolysis is involved in the formation of the fluorescent aging pigment, lipofuscin, and possibly in the manifestation of cellular aging.[25] Lipofuscin is thought to result from cross-linking between oxidatively modified proteins and lipid peroxidation products, which are concurrently increased during aging.[3,4] In fact, lipofuscin and lipofuscin-like secondary lysosomes accumulate with age in various animal tissues.[26–28] These and related observations have led to the proposal that oxidative damage to cellular constituents, including lipids, proteins, and DNA, and their accumulation in tissue cells, are major contributors to the aging process.[15,16]

The free radical theory of aging has received great support from the observation that *Drosophila melanogaster* with homozygous mutations in either Cu/ZnSOD or catalase gene exhibits increased sensitivity to oxidative stress and has reduced viability and shorter life span.[29–31] Because glutathione peroxidase is absent in *D. melanogaster*, Cu/ZnSOD and catalase provide the main enzymatic antioxidant defenses.[32] The fruit flies that overexpress Cu/ZnSOD alone or in combination with the overexpression of catalase were found to exhibit increased resistance to oxidative stress and have significantly less oxidative damage to proteins and a longer life span.[20,33,34] Moreover, abundant experimental data have suggested that the activities and capacities of antioxidant systems of tissue cells decline with age, leading to the gradual loss of prooxidant/antioxidant balance and accumulation of oxidative damage in the aging process. These observations have lent great support to the notion that oxidative stress plays an important role in the aging process.

OXIDATIVE DAMAGE TO MITOCHONDRIA IN AGING

Mitochondria are the major intracellular producers of ROS and are thus subjected to direct attack of ROS in animal and human cells. The electron leak from the impaired respiratory chain often results in enhanced production of ROS through one-electron reduction of molecular oxygen in mitochondria.[8–10] Recently, Luo *et al.*[35] demonstrated that cultured skin fibroblasts from patients with Complex I deficiency exhibited a 2- to 10-fold increase in the production of hydroxyl radicals under basal conditions and up to a 20-fold increase after treatment with redox active agents, including menadione or doxorubicin. Moreover, they found that the products of lipid peroxidation, including hexanal, 4-hydroxynon-2-enal and malondialdehyde were significantly elevated in fibroblasts from the patients. This is the first clear demonstration of excessive production of hydroxyl radicals and lipid peroxides in human cells with a defective respiratory chain. Because respiratory function declines with age, it is thus reasonable to find that the rate of production of superoxide anions and hydrogen peroxide in mitochondria increases with age in tissue cells of the mammals[36,37] and insects.[38] Sohal and co-workers[38] showed that exposure of live flies to hyperoxia and of submitochondrial particles to X-ray irradiation increased the rate of hydrogen peroxide production in the mitochondrial membranes. The enhanced production of ROS inevitably elevates the intracellular oxidative stress and elicits oxidative damage to the cells. It was also found that the increase of hydrogen peroxide production is correlated with the oxidative damage to the inner membrane of mitochondria.[38] In another study, Sohal *et al.*[20] demonstrated that the amplitude of the age-related increase in the rate of mitochondrial generation of hydrogen peroxide in the *D. melanogaster* overexpressing Cu/ZnSOD or catalase was decreased to 60% of that of the wild-type fruit flies. Therefore, the rate of mitochondrial hydrogen peroxide release is an important factor in determining the extent of oxidative damage sustained by mitochondria. About a decade ago, Ames and co-workers[17,18] first reported that oxidative damage to mtDNA is much more extensive than that to nuclear DNA. They found that the content of 8-OH-dG in mtDNA was about 16 times higher than that in nuclear DNA in rat liver. Furthermore, the 8-OH-dG content in liver mtDNA of the 24-month-old rat was about three times higher than that of the three-month-old rat. In a recent study, Yakes and van Houten[39] demonstrated that upon treatment with 200 µM hydrogen peroxide for 15 or 60 min, human fibroblasts exhibited threefold more damage to mtDNA than that to nuclear DNA. Moreover, they found that after treatment of the fibroblasts with H_2O_2 for 1 h, the damage to nuclear DNA was completely repaired within 1.5 h but no repair of mtDNA was observed. It was thus concluded that under oxidative stress mtDNA damage is more extensive and persists longer than nuclear DNA. Recently, Ozawa and colleagues showed that oxidative modification of mtDNA, as indicated by the formation of 8-OH-dG, increases in an age-dependent manner in the human diaphragm, heart muscle, and brain tissues.[40–42] Furthermore, the levels of oxidative stress and oxidatively modified proteins and membrane lipids in mitochondria have been shown to increase during aging.[22,43–46] On the other hand, Asuncion *et al.*[47] showed that mitochondrial glutathione is markedly oxidized during aging in rodents. The ratio between oxidized glutathione (GSSG) and reduced glutathione (GSH), an index of oxidative stress, and the 8-OH-dG content of mtDNA were found to increase concurrently with age in the liver, kidney, and brain of the rat and mouse. Furthermore, they showed that oral administration of antioxidants could protect against both glutathione oxidation and mtDNA damage in rats and mice. This study established a direct relationship between mtDNA damage and glutathione oxidation in mitochondria. This is consistent with the observations of Sohal and co-workers[19,23] that protein oxidation and DNA damage in mitochondria are concurrently increased in the tissues of aging fruit flies. Taken together, these findings suggest a close relationship between oxidative stress, indicated by glutathione oxidation or depletion of other antioxidants, and oxidative damage to various biomolecules, particularly mtDNA, in mitochondria in the aging process.

AGING-ASSOCIATED HUMAN mtDNA MUTATION

Each human and animal cell contains several hundred to more than a thousand mitochondria, each carrying two to ten copies of mtDNA.[48] Human mtDNA is a 16,569 bp circular double-stranded DNA molecule.[49] This extrachromosomal genetic system contains genes coding for 13 polypeptides essential for respiration and oxidative phosphorylation, 2 rRNAs, and a set of 22 tRNAs that constitute the protein synthesis machinery in mitochondria.[50] Human mtDNA is a naked compact DNA molecule without protection of histones and replicates rapidly without proofreading and efficient DNA repair systems.[51–54] It is transiently attached to the mitochondrial inner membrane, in which a considerable amount of ROS is continually produced by the respiratory chain.[44,55] These characteristics have rendered mtDNA vulnerable to attack by ROS and free radicals that are continually generated in the inner membrane-located respiratory chain of mitochondria. The aforementioned observation that mtDNA has about a 100-fold higher content of 8-OH-dG than does nuclear DNA[17] implies that there may exist some mutations of mtDNA in aging tissues of the human and animals.

Indeed, a number of point mutations, deletions, tandem duplications, and rearrangements of mtDNA have been found in various tissues of old humans.[56–68] Mutant mtDNA(s) often coexists with the wild-type mtDNA within a cell (a condition termed heteroplasmy), and the degree of heteroplasmy often varies in different tissues of the same individual.[60,65] Furthermore, the distribution of the mutant mtDNA molecules in the body tissues of an aged individual or a patient with mitochondrial myopathy is uneven and sporadic and usually results in a kind of tissue mosaic. It has been well established that many of these mtDNA mutations accumulate in an exponential manner with age in postmitotic tissues.[63–68] Some of the aging-associated mtDNA mutations were originally observed in patients with mitochondrial diseases. The most widely-spread mtDNA mutation is the so-called "common deletion," which removes a 4,977 bp DNA segment of mtDNA by a mechanism that involves a 13 bp direct repeat, flanking the 5′- and 3′-end breakpoints at nucleotide position (np) 8470/8482 and np 13447/13459, respectively.[69] This mtDNA deletion was first observed in the muscle of patients with mitochondrial myopathies, including the Kearns-Sayre syndrome, chronic progressive external ophthalmoplegia, and Pearsons' syndrome.[70,71] Multiple large-scale deletions of mtDNA have also been found in various tissues of aged individuals.[60,65–67] Moreover, an A3243G point mutation of mtDNA, which is associated with mitochondrial myopathy, encephalopathy, lactic acidosis and stroke-like episodes (MELAS) syndrome, and another A8344G mtDNA mutation that is associated with myoclonic epilepsy and ragged-red fibers (MERRF) syndrome[72] have also been detected in the muscle of elderly subjects.[61,62] Additionally, at least five different types of tandem duplications were found in the D-loop region of mtDNA from several tissues of normal subjects, and the incidence and abundance of mtDNA with these tandem duplications are increased with age.[63,65]

It is important to note that the proportion of each of these aging-associated mtDNA mutations, including deletions, point mutations, and tandem duplications present in old human tissues, rarely exceeds 1%.[63–68] Usually, the aging-associated mtDNA deletions are only detectable by polymerase chain reaction (PCR) and are frequently quantified with a subjective choice of primers and PCR conditions by the investigators. We have observed that some deletions of mtDNA cannot be picked up by certain primer pairs but are detectable by the other primers. Recently, a more detailed detection system was designed for extensive screening of mtDNA deletions in human tissues.[73] By using 180 kinds of PCR primer pairs, this system enables one to detect all the possible deletions of mtDNA over 500 bp. Hyakawa *et al.*[73] applied this system to analyze mtDNAs from normal hearts of human subjects of various ages. They observed an extensive fragmentation of mtDNA into minicircles with various sizes of deletion, and the incidence and abundance of the mutant mtDNAs increased dramatically with age and were well correlated with oxidative damage to mtDNA.[46,73] On the other hand, the long-range PCR technique was recently adopted to amplify the entire mitochondrial genome to facilitate the detection of age-associated mtDNA deletions in old human tissues.[74–77] It is worth mentioning that the reported mutations and the aforementioned oxidative damage to mtDNA represent only the tip of the iceberg of the mtDNA damage occurring in the aging process and that multiple mutations of mtDNA often coexist in various tissues of aged individuals.

Although the proportion of the mtDNA molecules with deletion was found to correlate with the 8-OH-dG content of mtDNA,[46,47] it is poorly understood as to how oxidative stress or ROS causes mtDNA mutations. Adachi *et al.*[78] have recently demonstrated that ROS may cause large-scale deletion of mtDNA in the rodent. They detected a 4 kb deletion of mtDNA in the heart of the BALB/c mice that had received chronic intraperitoneal injection of doxorubicin, which has the side effects of inducing cardiomyopathy and eliciting profound lipid peroxidation of heart mitochondria.[78] Moreover, they found that administration of coenzyme Q_{10} (a free radical scavenger) to the mice could effectively prevent the mtDNA deletion and decreased the lipid peroxide content of the heart mitochondria. This finding provides the first direct evidence to support the notion that ROS and free radicals are involved in the formation of the deleted mtDNA. Although it remains to be demonstrated as to how mtDNA mutations are induced by ROS and free radicals, useful information for a better understanding of the possible mechanisms involved in mtDNA mutations have been gathered from several lines of research discussed below.

Analysis of the sequences of break points flanking the deletions of mtDNA revealed that they occurred frequently within the large arc between the two origins of replication (O_H and O_L) of mtDNA.[60,69] The break points of many of these mtDNA deletions are flanked by direct repeat sequences. Slipped mispairing during DNA replication between direct repeats,[69] homologous replication,[79] and topoisomerase II cleavage[80] have been suggested to be the possible mechanisms involved in the deletion of mtDNA. Mitochondrial tRNA genes have been proposed to be hot spots for point mutations, and more than 10 different point mutations have been found in the tRNA$^{Leu(UUR)}$ gene.[81] On the other hand, the insertion sites of the tandem duplications in the D-loop region of human mtDNA have been found to be localized in the regions containing either a poly C run or a direct repeat sequence.[63,82] This finding suggests that slipped replication frequently occurs at the poly C stretches in the D-loop region and thereby elicits the observed small tandem duplications. Moreover, certain regions of mtDNA have been recently demonstrated to bear bent or anti-bent DNA structure and are particularly sensitive to oxidative insult of osmium tetroxide or ROS and are thus extremely prone to mutation.[83,84] The putative hot spots in mtDNA for oxidative modification and mutation could be near or at specific regions that assume bent, anti-bent, and non-B DNA structures in mtDNA.[83] These observations suggest that the unusual structures in these regions of human mtDNA may serve as motifs recognized by some sequence-specific molecular machineries that elicit or catalyze the aging-associated mtDNA mutations. In addition, it was hypothesized that genotoxic intermediates of lipid peroxidation may have a role in eliciting age-associated mtDNA mutations.[44,55] The region of mtDNA that is attached to the ROS-generating sites in the mitochondrial inner membrane should be more susceptible to oxidative damage, strand breakage, and mutation. Furthermore, ROS-induced mutagenesis has been observed to be DNA polymerase specific.[85] Thus, it is possible that the frequency of occurrence and the type of mtDNA mutations are determined, at least in part, by the interaction between mitochondrial DNA polymerase and the DNA template with ROS-induced damage during DNA replication.

In addition, many types of mtDNA mutations have been reported to occur more frequently in sun-exposed skin at relatively high levels.[65,86,87] Recently, we found that the frequency of occurrence and abundance of mtDNA deletions in the human lung are significantly increased by cigarette smoking and aging.[88] These observations suggest that free radicals generated by such environmental insults as light irradiation and air pollutants may also play an important role in the initiation and/or promotion of mtDNA mutations during the aging process.

MITOCHONDRIAL RESPIRATORY FUNCTION DECLINES WITH AGE

It was first demonstrated in 1989 that the respiratory function of mitochondria is progressively declined in human liver[89] and skeletal muscle,[90] respectively, during human aging. The respiratory control, the efficiency of oxidative phosphorylation, the rates of resting (State 4) and ADP-stimulated (State 3) respiration, and the activities of the respiratory enzyme complexes all have been observed to decline with age in various human tissues.[89–94] These findings have led to the proposal that impairment of mitochondrial respiratory function is associated with or responsible for aging in the human. In addition, it has also been found that the number of muscle fibers that are deficient in cytochrome c oxidase increases with age in the skeletal and heart muscle.[95,96] Moreover, it was found that the extent of mtDNA mutation is strongly correlated with the amplitude of decrease in respiratory enzyme (*e.g.*, cytochrome c oxidase) activity.[92,93] Inasmuch as the age-dependent decline of the glutamate-malate-supported respiration was found to be more dramatic than that of the succinate-supported respiration,[89] we conjectured that mutation(s) to the seven genes in mtDNA that encode subunits of NADH dehydrogenase may be involved in this aging-associated respiratory function decline. We quickly confirmed this idea by showing that both the frequency of occurrence and abundance of the mtDNA with 4,977 bp deletion increase with age in liver[58,64] and many other tissues.[64,68] Because mtDNA contains no introns and has a high information density, the large-scale deletions often cause the removal or truncation of multiple structural genes and tRNA genes and thus lead to multiple deficiencies of the respiratory chain. In addition, several age-related tandem duplications in the D-loop of human mtDNA[63,67] may affect the function of replication origin O_H and two transcriptional promoters for each strand of mtDNA.[97] The D-loop is the most important control region for the regulation of the genetic function of human mtDNA. Therefore, oxidative damage or mutation of the sequence in the D-loop of mtDNA may cause alterations of mtDNA replication and transcription. In addition, oxidative modification to the nucleobases may also elicit errors in replication and gene expression of mtDNA. Therefore, accumulation of oxidatively damaged mtDNA may well be involved in the age-dependent progressive decline of respiratory functions, especially in postmitotic cells.[75,82,98] On the other hand, it has been reported that the steady-state levels of mitochondrial transcripts are significantly reduced during aging of *D. melanogaster*, and these changes correlate very well with the life span of the insect.[99] This decreased expression of mitochondrial RNAs might be caused in part by mutation and oxidative damage to the mtDNA. Interestingly, it was observed that not only cytochrome *c* oxidase, which contains three protein subunits encoded by mtDNA, but also glutamate dehydrogenase, a nuclear DNA-encoded enzyme present in the mitochondrial matrix, gradually loses enzymatic activity during the aging of *D. melanogaster*.[99] These observations suggest that although mtDNA is more vulnerable to oxidative damage, some defect(s) in the nuclear genome may also play a role in aging associated decline of the bioenergetic function of mitochondria.

However, it is noteworthy that the proportions of each of the age-related mutant mtDNAs in various human tissues are not so high as those observed in the target tissues of the patients

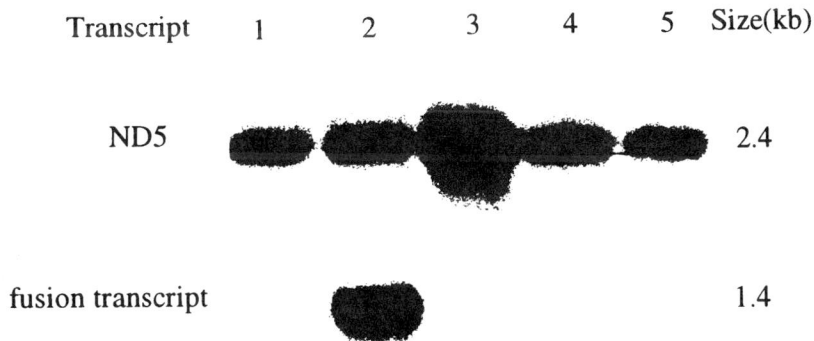

FIGURE 1. Northern hybridization of mitochondrial ND5 gene transcript and the transcript of a 4,977 bp deletion-induced fusion gene containing part of ATPase8 and ND5 genes in skin fibroblasts from subjects of different ages. Thirty micrograms of total RNA from the primary culture of fibroblasts established from each of the study subjects was subjected to electrophoresis on a 1% denaturing agarose gel at 50 volts for 8 hours. The RNA species in the gel were transferred to a Hybond-N+ nylon membrane and probed with the labeled DNA fragment of the ND5 gene (np13721-np13981) of human mtDNA. The RNA bands were then visualized on a Kodak X-ray film by autoradiography. Lanes 1 to 5 represent the RNAs from the primary cultures of skin fibroblasts established from individuals 40, 75, 42, 41, and 46 years old, respectively. The RNA pattern in lane 2 indicates that besides the normal 2.4 kb transcript, an additional 1.4 kb RNA species transcribed from the fused gene of ATPase8 and ND5, generated by the 4,977 bp deletion, was also observed in the skin fibroblasts of a 75-year-old woman harboring about 55% of the 4,977 bp deleted mtDNA.

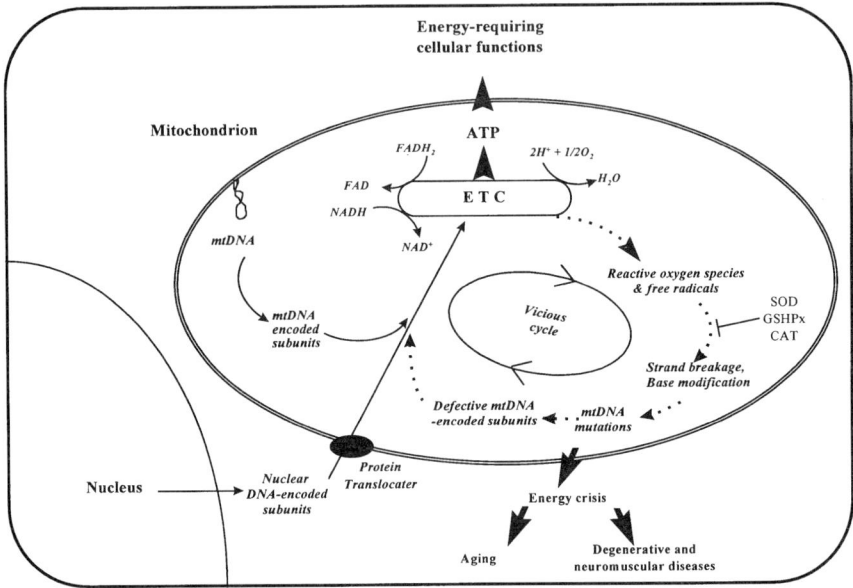

FIGURE 2. Mitochondrial theory of aging and age-related degenerative diseases. The mitochondrial electron transport system, which contains mtDNA-encoded and nuclear DNA-encoded protein subunits, is actively involved in ATP synthesis-coupled respiration that consumes about 95% of the oxygen uptake of the tissue cells. A fraction of the oxygen is incompletely reduced by one-electron transfer (*e.g.*, via ubisemiquinone) to generate the reactive oxygen species (ROS) and organic free radicals, which may cause oxidative damage and mutation of the nearby mtDNA molecules that are attached, at least transiently, to the inner membrane. The oxidatively modified and/or mutated mtDNAs are transcribed and translated to produce defective protein subunits that are assembled to form defective respiratory enzymes. The impaired electron transport chain (ETC) not only works less efficiently in ATP synthesis but also generates more ROS, which will further enhance the oxidative damage to various biomolecules in mitochondria. This vicious cycle operates in an age-dependent manner and results in the widely observed aging-associated occurrence and accumulation of oxidative damage and mutation of mtDNA, which in turn lead to a progressive decline in the bioenergetic function of tissue cells in the aging process. On the other hand, free radical scavenger systems, DNA repair systems, and mitochondrial degradation and turnover that correct or remove the oxidative damage become less efficient during aging. As a result, accumulation of oxidatively damaged and mutated mtDNA and defective mitochondria, and alteration of mitochondrial turnover, act synergistically to cause the decline of bioenergetic functions associated with the aging process of the human.

with mitochondrial myopathies. It appears difficult to comprehend how such low amounts of the mutant mtDNAs can exert significant deleterious effects on mitochondrial respiratory function. Moreover, there are 2 to 10 copies of mtDNA in each mitochondrion and thousands of mitochondria in a typical human cell, and thus complementation of the mutant mtDNA with the wild-type mtDNA may underrepresent the effect of the mutant DNA. Indeed, it has been found that one needs only 6% of the wild-type mtDNA to suppress the phenotypic expression of the mutant mtDNA in cultured cells established from the patients with the MELAS syndrome.[100–102] In addition, in most diseases associated with mtDNA deletions, the proportion of the deleted mtDNA must exceed 60% of total mtDNA before the biochemical defects or clinical manifestations can be observed.[72] We found that mtDNA molecules with the 4,977 bp deletion are transcribed, and a fusion transcript was observed in the skin fibroblasts from an elderly woman who had none of the clinical features of the known mitochondrial diseases (FIG. 1). It is possible that the fusion transcripts and expressed altered protein molecules act as a kind of stressor and thereby exert deleterious effects on the mitochondrial functions in the tissues of aged individuals.[103] The age dependent decline of mitochondrial respiratory functions may be also caused by direct ROS damage to proteins, aside from the effects of the mutated or oxidatively damaged mtDNA (FIG. 2). However, it is possible that, in a synergistic manner, all the mutations and oxidative damages to mtDNA impair the function of the electron transport chain and elicit a profound increase in the rate of ROS generation. A broad spectrum of oxidative damage and mutation of mtDNA may be effected through a recently proposed vicious cycle in the aging process.[104] However, a clear causal relationship among oxidative modification and mutation of mtDNA, mitochondrial dysfunction, and human aging remains to be established.

SUMMARY

Mitochondrial respiration and oxidative phosphorylation are gradually uncoupled from each other and the activities of the respiratory enzymes decline in various human tissues upon aging. The immediate consequence of such decline of respiratory function is the decrease of ATP synthesis and increase of the production of ROS and free radicals in mitochondria due to the increased electron leakage of the electron transport chain. It has been shown that the rate of production of superoxide anions and hydrogen peroxide in mitochondria increases with age. Moreover, the intracellular levels of antioxidants and activities of free radical scavenging enzymes are significantly altered in the aging process. These two compounding factors lead to an age-dependent increase in the fraction of the ROS and free radicals that may escape the antioxidant defense mechanism and cause an ever-increasing oxidative damage to various biomolecules in mitochondria and the cell as a whole. A large body of experimental evidence have been accumulated to establish that lipid peroxidation, protein modification, and mtDNA mutation are concurrently increased in the aging process. The mtDNA, which is exposed to ROS and is not protected by histones or DNA-binding proteins, is more susceptible than is nuclear DNA to oxidative insult during aging. In the past few years, oxidative modification and mutation of mtDNA have been found to increase exponentially with age in human and animal tissues. The respiratory enzymes containing the mutant mtDNA-encoded defective protein subunits exhibit impaired respiratory function and thereby increase the production of ROS and free radicals, which will further elevate the oxidative stress and oxidative damage to mitochondria. We suggest that this vicious cycle plays an important role in human aging and in the pathogenesis of age-related degenerative diseases.

REFERENCES

1. HARMAN, D. 1956. Ageing: Theory based on free radical and radiation chemistry. J. Gerontol. **11:** 298–300.
2. HARMAN, D. 1981. The aging process. Proc. Natl. Acad. Sci. USA **78:** 7124–7128.
3. MIQUEL, J., J.E. ECONOMOS & J.E. JOHNSON JR. 1980. Mitochondrial role in cell ageing. Exp. Gerontol. **15:** 575–591.
4. MIQUEL, J. 1991. An integrated theory of aging as the result of mitochondrial DNA mutation in differentiated cells. Arch. Gerontol. Geriatr. **12:** 99–117.

5. NOHL, H. & D. HEGNER. 1978. Do mitochondria produce oxygen radicals *in vivo*? Eur. J. Biochem. **82:** 563–567.
6. CHANCE, B., H. SIES & A. BOVERIS. 1979. Hydroperoxide metabolism in mammalian organs. Physiol. Rev. **59:** 527–605.
7. LINNANE, A.W., S. MARZUKI, T. OZAWA & M. TANAKA. 1989. Mitochondrial DNA mutations as an important contributor to ageing and degenerative disease. Lancet **i:** 642–645.
8. BOVERIS, A., N. OSHINO & B. CHANCE. 1972. The cellular production of hydrogen peroxide. Biochem. J. **128:** 617–630.
9. TURRENS, J.F. & A. BOVERIS. 1980. Generation of superoxide anion by the NADH dehydrogenase of bovine heart mitochondria. Biochem. J. **191:** 421–427.
10. BOVERIS, A. & B. CHANCE. 1973. The mitochondrial generation of hydrogen peroxide: General properties and effect of hyperbaric oxygen. Biochem. J. **134:** 707–716.
11. WEI, Y.H., C.P. SCHOLES & T.E. KING. 1981. Ubisemiquinone radicals from the cytochrome b-c_1 complex of mitochondrial electron transport chain-demonstration of QP-S radical formation. Biochem. Biophys. Res. Commun. **99:** 1411–1419.
12. PAHL, H.L. & P.A. BAEUERLE. 1994. Oxygen and the control of gene expression. BioEssays **16:** 497–502.
13. SEN, C.K. & L. PACKER. 1996. Antioxidant and redox regulation of gene transcription. FASEB J. **10:** 709–720.
14. LANDER, H.M. 1997. An essential role for free radicals and derived species in signal transduction. FASEB J. **11:** 118–124.
15. RICHTER, C., V. GOGVADZE, R. LAFFRANCHI, R. SCHLAPBACH, M. SCHNIZER, M. SUTER, P. WALTER & M. YAFFEE. 1995. Oxidants in mitochondria: From physiology to disease. Biochim. Biophys. Acta **1271:** 67–74.
16. WEI, Y-H., S-H. KAO & H-C. LEE. 1996. Simultaneous increase of mitochondrial DNA deletions and lipid peroxidation in human aging. Ann. N. Y. Acad. Sci. **786:** 24–43.
17. RICHTER, C., J.W. PARK & B.N. AMES. 1988. Normal oxidative damage to mitochondrial and nuclear DNA is extensive. Proc. Natl. Acad. Sci. USA **85:** 6465–6467.
18. FRAGA, C.G., M.K. SHIGENAGA, J.W. PARK, P. DEGAN & B.N. AMES. 1990. Oxidative damage to DNA during aging: 8-Hydroxy-2′-deoxyguanosine in rat organ DNA and urine. Proc. Natl. Acad. Sci. USA **87:** 4533–4537.
19. AGARWAL, S. & R.S. SOHAL. 1994. DNA oxidative damage and life expectancy in houseflies. Proc. Natl. Acad. Sci. USA **91:** 12332–12335.
20. SOHAL, R.S., A. AGARWAL, S. AGARWAL & W.C. ORR. 1995. Simultaneous overexpression of copper- and zinc-containing superoxide dismutase and catalase retards age-related oxidative damage and increases metabolic potential in *Drosophila melanogaster*. J. Biol. Chem. **270:** 15671–15674.
21. SOHAL, R.S., S. AGARWAL, M. CANDAS, M. FORSTER & H. LAL. 1994. Effect of age and caloric restriction on DNA oxidative damage in different tissues of C57BL/6 mice. Mech. Ageing Dev. **76:** 215–224.
22. STADTMAN, E.R. 1992. Protein oxidation and aging. Science **257:** 1220-1224.
23. SOHAL, R.S., S. AGARWAL, A. DUBEY & W.C. ORR. 1993. Protein oxidative damage is associated with life expectancy of houseflies. Proc. Natl. Acad. Sci. USA **90:** 7255–7259.
24. LAVIE, L., A.Z. REZNICK & D. GERSHON. 1992. Decreased protein and puromycinyl-peptide degradation in livers of senescent mice. Mutat. Res. **275:** 217–225.
25. IVY, G.O., F. SCHOTTLER, J. WENZEL, M. BAUDRY & G. LYNCH. 1984. Inhibitors of lysosomal enzymes: Accumulation of lipofuscin-like dense bodies in the brain. Science **226:** 985–987.
26. GRAY, J.E., R.N. WEAVER & A. PURMALIS. 1974. Ultrastructural observations of chronic progressive nephrosis in the Sprague-Dawley rat. Vet. Pathol. **11:** 153–164.
27. HAYASHIDA, M., B.P. YU, E.J. MASORO, K. IWASAKI & T. IKEDA. 1986. An electron microscopic examination of age-related changes in the rat kidney: The influence of diet. Exp. Gerontol. **21:** 535–553.
28. IVY, G.O., R. ROOPSINGH, S. KANAI, M. OHTA, Y. SATO & K. KITANI. 1996. Leupeptin causes an accumulation of lipofuscin-like substances and other signs of aging in kidneys of young rats: Further evidence for the protease inhibitor model of aging. Ann. N. Y. Acad. Sci. **786:** 12–23.
29. MACKAY, W.J. & G.C. BEWLEY. 1989. The genetics of catalase in *Drosophila melanogaster*: Isolation and characterization of acatalasemic mutants. Genetics **122:** 641–652.

30. PHILLIPS, J.P., S.D. CAMPBELL, D. MICHAUD, M. CHARBONNEAU & A.J. HILLIKER. 1989. Null mutation of copper/zinc superoxide dismutase in *Drosophila* confer hypersensitivity to paraquat and reduced longevity. Proc. Natl. Acad. Sci. USA **86:** 2761–2765.

31. GRISWOLD, C.M., A.L. MATTHEWS, K.E. BEWLEY & J.W. MAHAFFEY. 1993. Molecular characterization and rescue of acatalasemic mutants of *Drosophila melanogaster*. Genetics **134:** 781–788.

32. SOHAL, R.S., L. ARNOLD & W.C. ORR. 1990. Effect of age on superoxide dismutase, catalase, glutathione reductase, inorganic peroxides, TBA-reactive material, GSH/GSSG, NADPH/NADP⁺ and NADH/NAD⁺ in *Drosophila melanogaster*. Mech. Ageing Dev. **56:** 223–235.

33. ORR, W.C. & R.S. SOHAL. 1993. Effects of Cu-Zn superoxide dismutase overexpression on life span and resistance to oxidative stress in transgenic *Drosophila melanogaster*. Arch. Biochem. Biophys. **301:** 34–40.

34. ORR, W.C. & R.S. SOHAL. 1994. Extension of life-span by overexpression of superoxide dismutase and catalase in *Drosophila melanogaster*. Science **263:** 1128–1130.

35. LUO, X., S. PITKANEN, S. KASSOVSKA-BRATINOVA, B.H. ROBINSON & D. LEHOTAY. 1997. Excessive formation of hydroxyl radicals and aldehydic lipid peroxidation products in cultured skin fibroblasts from patients with Complex I deficiency. J. Clin. Invest. **99:** 2877–2882.

36. SOHAL, R.S. & B.H. SOHAL. 1991. Hydrogen peroxide release by mitochondria increases during aging. Mech. Ageing Dev. **57:** 187–202.

37. SOHAL, R.S., H.H. KU, S. AGARWAL, M.J. FORSTER & H. LAL. 1994. Oxidative damage, mitochondrial oxidant generation and antioxidant defenses during aging and in response to food restriction in the mouse. Mech. Ageing Dev. **74:** 121–133.

38. SOHAL, R.S. & A. DUBEY. 1994. Mitochondrial oxidative damage, hydrogen peroxide release, and aging. Free Radical Biol. Med. **16:** 621–626.

39. YAKES, F.M. & B. VAN HOUTEN. 1997. Mitochondrial DNA damage is more extensive and persists longer than nuclear DNA damage in human cells following oxidative stress. Proc. Natl. Acad. Sci. USA **94:** 514–519.

40. OZAWA, T., K. SAHASHI, Y. NAKASE & B. CHANCE. 1995. Extensive tissue oxygenation associated with mitochondrial DNA mutations. Biochem. Biophys. Res. Commun. **213:** 432–438.

41. HAYAKAWA, M., K. TORII, S. SUGIYAMA, M. TANAKA & T. OZAWA. 1991. Age-associated accumulation of 8-hydroxydeoxyguanosine in mitochondrial DNA of human diaphragm. Biochem. Biophys. Res. Commun. **179:** 1023–1029.

42. HAYAKAWA, M., K. HATTORI, S. SUGIYAMA & T. OZAWA. 1992. Age-associated oxygen damage and mutations in mitochondrial DNA in human heart. Biochem. Biophys. Res. Commun. **189:** 979–985.

43. SAGAI, M. & T. ISHINOSE. 1980. Age-related changes in lipid peroxidation as measured by ethane, ethylene, butane and pentane in respired gases of rats. Life Sci. **27:** 731–738.

44. HRUSZKEWYCZ, A.M. 1992. Lipid peroxidation and mtDNA degeneration. A hypothesis. Mutat. Res. **275:** 243–248.

45. BANDY, B. & A.J. DAVISON. 1990. Mitochondrial mutations may increase oxidative stress: Implications for carcinogenesis and aging. Free Radic. Biol. Med. **8:** 5213–5239.

46. AGARWAL, S. & R.S. SOHAL. 1995. Differential oxidative damage to mitochondrial proteins during aging. Mech. Ageing Dev. **85:** 55–63.

47. DE LA ASUNCION, J.G., A. MILLAN, R. PLA, L. BRUSEGHINI, A. ESTERAS, F.V. PALLARDO, J. SASTRE & J. VINA. 1996. Mitochondrial glutathione oxidation correlates with age-associated oxidative damage to mitochondrial DNA. FASEB J. **10:** 333–338.

48. BOGENHAGEN, D. & D.A. CLAYTON. 1974. The number of mitochondrial deoxyribonucleic acid genomes in mouse L and human HeLa cells. J. Biol. Chem. **249:** 7991–7995.

49. ANDERSON, S., A.T. BANKIER, B.G. BARREL, M.H.L. DE BRUIJN, A.R. COULSON, J. DROUIN, I.C. EPERON, D.P. NIERLICH, B.A. ROE, F. SANGER, P.H. SCHREIER, A.J.H. SMITH, R. STADEN & I.G. YOUNG. 1981. Sequence and organization of the human mitochondrial genome. Nature **290:** 457–465.

50. ATTARDI, G. & G. SCHATZ. 1988. Biogenesis of mitochondria. Annu. Rev. Cell Biol. **4:** 289–333.

51. CLAYTON, D.A., J.N. DODA & E.C. FRIEDBERG. 1974. The absence of a pyrimidine dimer repair mechanism in mammalian mitochondria. Proc. Natl. Acad. Sci. USA **71:** 2777–2781.

52. Tomkinson, A.E., R.T. Bonk, J. Kim, N. Bartfel & S. Linn. 1990. Mammalian mitochondrial endonuclease activities specific for ultraviolet-irradiated DNA. Nucleic Acids Res. **18:** 929–935.

53. Driggers, W.J., S.P. Ledoux & G.L. Wilson. 1993. Repair of oxidative damage within the mitochondrial DNA of RINr 38 cells. J. Biol. Chem. **268:** 22042–22045.

54. Driggers, W.J., V.I. Grishko, S.P. Ledoux & G.L. Wilson. 1996. Defective repair of oxidative damage in the mitochondrial DNA of a Xeroderma pigmentosum group A cell line. Cancer Res. **56:** 1262–1266.

55. Hruszkewycz, A.M. 1988. Evidence for mitochondrial DNA damage by lipid peroxidation. Biochem. Biophys. Res. Commun. **153:** 191–197.

56. Ikebe, S., M. Tanaka, K. Ohno, W. Sato, K. Hattori, T. Kondo, Y. Mizuno & T. Ozawa. 1990. Increase of deleted mitochondrial DNA in the stratum in Parkinson's disease and senescence. Biochem. Biophys. Res. Commun. **170:** 1044–1048.

57. Cortopassi, G.A. & N. Arnheim. 1990. Detection of a specific mitochondrial DNA deletion in tissues of older humans. Nucleic Acids Res. **18:** 6927–6933.

58. Yen, T.C., J.H. Su, K.L. King & Y.H. Wei. 1991. Ageing-associated 5 kb deletion in human liver mitochondrial DNA. Biochem. Biophys. Res. Commun. **178:** 124–131.

59. Corral-Debrinski, M., G. Stepien, M. Shoffner, M.T. Lott, K. Kanter & D.C. Wallace. 1991. Hypoxemia is associated with mitochondrial DNA damage and gene induction. J. Am. Med. Assoc. **266:** 1812–1816.

60. Wei, Y.H. 1992. Mitochondrial DNA alterations as ageing-associated molecular events. Mutat. Res. **275:** 145–155.

61. Zhang, C., A.W. Linnane & P. Nagley. 1993. Occurrence of a particular base substitution (3243 A to G) in mitochondrial DNA of tissues of ageing humans. Biochem. Biophys. Res. Commun. **195:** 1104–1110.

62. Mänscher, C., T. Rieger, J. Müller-Höcker & B. Kadenbach. 1993. The point mutation of mitochondrial DNA characteristic for MERRF disease is found also in healthy people of different ages. FEBS Lett. **317:** 27–30.

63. Lee, H.C., C.Y. Pang, H.S. Hsu & Y.H. Wei. 1994. Ageing-associated tandem duplications in the D-loop of mitochondrial DNA of human muscle. FEBS Lett. **354:** 79–83.

64. Lee, H.C., C.Y. Pang, H.S. Hsu & Y.H. Wei. 1994. Differential accumulations of 4,977 bp deletion in mitochondrial DNA of various tissues in human ageing. Biochim. Biophys. Acta **1226:** 37–43.

65. Wei, Y-H., C-Y. Pang, B-J. You & H-C. Lee. 1996. Tandem duplications and large-scale deletions of mitochondrial DNA are early molecular events of human aging process. Ann. N. Y. Acad. Sci. **786:** 82–101.

66. Zhang, C., A. Baumer, R.J. Maxwell, A.W. Linnane & P. Nagley. 1992. Multiple mitochondrial DNA deletions in an elderly human individual. FEBS Lett. **297:** 34–38.

67. Yang, J.H., H.C. Lee & Y.H. Wei. 1995. Photoageing-associated mitochondrial DNA length mutations in human skin. Arch. Dermatol. Res. **287:** 641–648.

68. Fahn, H.J., L.S. Wang, R.H. Hsieh, S.C. Chang, S.H. Kao, M.H. Huang & Y.H. Wei. 1996. Age-related 4,977 bp deletion in human lung mitochondrial DNA. Am. J. Respir. Crit. Care Med. **154:** 1141–1145.

69. Shoffner, J.M., M.T. Lott, A.S. Voljavec, S.A. Soueidan, D.A. Costigan & D.C. Wallace. 1989. Spontaneous Kearns-Sayre/chronic external ophthalmoplegia plus syndrome associated with a mitochondrial DNA deletion: A slip-replication model and metabolic therapy. Proc. Natl. Acad. Sci. USA **86:** 7952–7956.

70. Holt, I.J., A.E. Harding & J.A. Morgan-Hughes. 1988. Deletions of mitochondrial DNA in patients with mitochondrial myopathies. Nature **331:** 717–719.

71. Zeviani, M., C.T. Moraes, S. DiMauro, H. Nakase, E. Bonilla, E.A. Schon & L.P. Rowland. 1988. Deletion of mitochondrial DNA in Kearns-Sayre syndrome. Neurology **38:** 1339–1346.

72. Wallace, D.C. 1992. Diseases of the mitochondrial DNA. Annu. Rev. Biochem. **61:** 1175–1212.

73. Hayakawa, M., K. Katsumata, M. Yoneda, M. Tanaka, S. Sugiyama & T. Ozawa. 1996. Age-related extensive fragmentation of mitochondrial DNA into minicircles. Biochem. Biophys. Res. Commun. **226:** 369–377.

74. YONEDA, M., K. KATSUMATA, M. HAYAKAWA, M. TANAKA & T. OZAWA. 1995. Oxygen stress induces an apoptotic cell death associated with fragmentation of mitochondrial genome. Biochem. Biophys. Res. Commun. **209:** 723–729.

75. MELOV, S., J.M. SHOFFNER, A. KAUFMAN & D.C. WALLACE. 1995. Marked increase in the number and variety of mitochondrial DNA rearrangements in aging human skeletal muscle. Nucleic Acids Res. **23:** 4122–4126.

76. REYNIER, P. & Y. MALTHIERY. 1995. Accumulation of deletions in mtDNA during tissue aging: Analysis by long PCR. Biochem. Biophys. Res. Commun. **217:** 59–67.

77. KOVALENKO, S.A., G. KOPSIDAS, J.M. KELSO & A.W. LINNANE. 1997. Deltoid human muscle mtDNA is extensively rearranged in old age subjects. Biochem. Biophys. Res. Commun. **232:** 147–152.

78. ADACHI, K., Y. FUJIURA, F. MAYUMI, A. NOZUHARA, Y. SUGIU, T. SAKANASHI, T. HIDAKA & H. TOSHIMA. 1993. A deletion of mitochondrial DNA in murine doxorubicin-induced cariotoxicity. Biochem. Biophys. Res. Commun. **195:** 945–951.

79. ZEVIANI, M. 1992. Nucleus-driven mutations of human mitochondrial DNA. J. Inherited Metab. Dis. **15:** 456–471.

80. BLOK, R.B., D.R. THORBURN, G.N. THOMPSON & H.H.M. DAHL. 1995. A topoisomerase II cleavage site is associated with a novel mitochondrial DNA deletion. Hum. Genet. **95:** 75–81.

81. SCHON, E.A., M. HIRANO & S. DIMAURO. 1994. Mitochondrial encephalomyopathies: Clinical and molecular analysis. J. Bioenerg. Biomembr. **26:** 291–299.

82. MECOCCI, P., U. MACGARVEY, A.E. KAUFMAN, D. KOONTZ, J.M. SHOFFNER, D.C. WALLACE & M.F. BEAL. 1993. Oxidative damage to mitochondrial DNA shows marked age-dependent increases in human brain. Ann. Neurol. **34:** 609–616.

83. HOU, J.H. & Y.H. WEI. 1996. The unusual structures of the hot-regions flanking large-scale deletions in human mitochondrial DNA. Biochem. J. **318:** 1065–1070.

84. HOU, J.H. & Y.H. WEI. 1998. AT-rich sequences flanking the 5′-end breakpoint of the 4977-bp deletion of human mitochondrial DNA are located between two bent-inducing DNA sequences that assume distorted structure in organello. Mutat. Res. **403:** 75–84.

85. FEIG, D.I. & L.A. LOEB. 1994. Oxygen radical induced mutagenesis is DNA polymerase specific. J. Mol. Biol. **235:** 33–41.

86. PANG, C.Y., H.C. LEE, J.H. YANG & Y.H. WEI. 1994. Human skin mitochondrial DNA deletions associated with light exposure. Arch. Biochem. Biophys. **312:** 534–538.

87. YANG, J.H., H.C. LEE, K.J. LIN & Y.H. WEI. 1994. A specific 4977-bp deletion of mitochondrial DNA in human ageing skin. Arch. Dermatol. Res. **286:** 386–390.

88. FAHN, H.J., L.S. WANG, S.H. KAO, S.C. CHANG, M.H. HUANG & Y.H. WEI. 1998. Smoking-associated DNA mutations and lipid peroxidation in human lung tissue. Am. J. Respir. Cell Mol. Biol. In press.

89. YEN, T.C., Y.S. CHEN, K.L. KING, S.H. YEH & Y.H. WEI. 1989. Liver mitochondrial respiratory functions decline with age. Biochem. Biophys. Res. Commun. **65:** 994–1003.

90. TROUNCE, I., E. BYRNE & S. MARZUKI. 1989. Decline in skeletal muscle mitochondrial respiratory chain function: Possible factor in ageing. Lancet **i:** 637–639.

91. COOPER, J.M., V.M. MANN & A.H. SCHAPIRA. 1992. Analyses of mitochondrial respiratory chain function and mitochondrial DNA deletion in human skeletal muscle: Effect of ageing. J. Neurol. Sci. **113:** 91–98.

92. HSIEH, R.H., J.H. HOU, H.S. HSU & Y.H. WEI. 1994. Age-dependent respiratory function decline and DNA deletions in human muscle mitochondria. Biochem. Mol. Biol. Int. **32:** 1009–1022.

93. LEZZA, A.M.S., D. BOFFOLI, S. SCACCO, P. CANTATORE & M.N. GADALETA. 1994. Correlation between mitochondrial DNA 4977-bp deletion and respiratory chain enzyme activities in aging human skeletal muscles. Biochem. Biophys. Res. Commun. **205:** 112–119.

94. PAPA, S. 1996. Mitochondrial oxidative phosphorylation changes in the life span. Molecular aspects and physiopathological implications. Biochim. Biophys. Acta **1276:** 87–105.

95. MÜLLER-HÖCKER, J. 1989. Cytochrome *c* oxidase deficient cardiomyocytes in the human heart—an age-related phenomenon. A histochemical ultracytochemical study. Am. J. Pathol. **134:** 1167–1173.

96. MÜLLER-HÖCKER, J. 1990. Cytochrome *c* oxidase deficient fibers in the limb muscle and diaphragm of man without muscular disease: An age-related alteration. J. Neurol. Sci. **100:** 14–21.

97. CLAYTON, D.A. 1982. Replication of animal mitochondrial DNA. Cell **28:** 693–705.
98. AMES, B.N., M.K. SHIGENAGA & T.M. HAGEN. 1993. Oxidants, antioxidants, and the degenerative diseases of aging. Proc. Natl. Acad. Sci. USA **90:** 7915–7922.
99. CALLEJA, M., P. PENA, C. UGALDE, C. FERREIRO, R. MARCO & R. GARESSE. 1993. Mitochondrial DNA remains intact during *Drosophila* aging, but the levels of mitochondrial transcripts are significantly reduced. J. Biol. Chem. **268:** 18891–18897.
100. HOLT, I.J., A.E. HARDING, R.K.H. PETTY & J.A. MORGAN-HUGHES. 1990. A new mitochondrial disease associated with mitochondrial DNA heteroplasmy. Am. J. Hum. Genet. **46:** 428–433.
101. CIAFALONI, E., E. RICCI, S. SHANSKE, C.T. MORAES, G. SILVESTRI, M. HIRANO, S. SIMONETTI, C. ANGELLINI, M.A. DONATI, C. GARCIA, A. MARTINUZZI, R. MOSEWICH, S. SERVIDEI, E. ZAMMARCHI, E. BONILLA, D.C. DE VIVO, L.P. ROWLAND, E.A. SCHON & S. DIMAURO. 1992. MELAS: clinical features, biochemistry, and molecular genetics. Ann. Neurol. **31:** 391–398.
102. CHOMYN, A., A. MARTINUZZI, M. YONEDA, A. DAGA, O. HURKO, D. JOHNS, S. T. LAI, I. NONAKA, C. ANGELINI & G. ATTARDI. 1992. MELAS mutation in mtDNA binding site for transcription termination factor causes defects in protein synthesis and in respiration but no changes in levels of upstream and downstream mature transcripts. Proc. Natl. Acad. Sci. USA **89:** 4221–4225.
103. NAKASE, H., C.T. MORAES, R. RIZZUTO, A. LOMBES, S. DIMAURO & E.A. SCHON. 1990. Transcription and translation of deleted mitochondrial genomes in Kearns-Sayre syndrome: Implications for pathogenesis. Am. J. Hum. Genet. **46:** 418–427.
104. WEI, Y.H. 1998. Oxidative stress and mitochondrial DNA mutations in human aging. Proc. Soc. Exp. Biol. Med. **217:** 53–63.

Tissue-specific Distribution of Multiple Mitochondrial DNA Rearrangements during Human Aging[a]

SERGEY A. KOVALENKO,[b] GEORGE KOPSIDAS,[b] JOANNE KELSO,[b] FRANKLIN ROSENFELDT,[c] AND ANTHONY W. LINNANE[b,d]

[b]Centre for Molecular Biology and Medicine, Epworth Hospital, 185–187 Hoddle Street, Richmond 3121, Melbourne, Victoria, Australia

[c]Cardiac Research Unit, Baker Institute, Melbourne, Victoria, Australia

ABSTRACT: Mitochondria, according to the free radical theory of aging, are the major source of reactive oxygen species (ROS). The results, presented in this paper, question the role of reactive oxygen species in contributing significantly to the extent of mitochondrial bioenergy degradation of the tissues, which can be correlated with mtDNA rearrangements. We report here that mtDNA rearrangements, including deletions and duplications, in tissues from human aged subjects, occur in levels ranging from very low in liver, to considerable in cardiac muscle, to almost total in skeletal muscle. The extent of mtDNA rearrangements is correlated at both the individual tissue and cell level with cytochrome oxidase (COX) activity as the exemplifier of cellular bioenergy capacity. Thus, the ROS proposal in its simplest form as it affects mtDNA and mitochondrial electron transport system is not supported by the available data.

A major proposal in the field of aging is that the reactive oxygen species (ROS) are important contributors to the total process.[1] Indeed, an age-related increase in the concentration of DNA-oxidized bases,[2] lipids,[3,4] and protein carbonyl content[5,6] has been reported in a number of model systems,[7–9] leading to the hypothesis that such damage, especially to mtDNA,[10] may play a critical role in the aging process.[11] However, the involvement of free radicals in aging as a consequence or cause of senescence has been critically examined.[12–14]

From the present study of three highly aerobic tissues, our results do not support the simple concept that ROS plays a major role in terms of mtDNA deletions, as shown by quantitative extra-long polymerase chain reaction (XL-PCR) and cytochrome-c oxidase (COX) activity. In the present communication we have shown that the mtDNA deletion/mutation profile differs not just between cells of the same tissue, demonstrated earlier,[15,16] but also that there are extreme differences observed among a number of tissues, as exemplified here by human skeletal muscle (deltoid), cardiac muscle (atrial appendage), and liver. Thus, multiple mtDNA mutations progressively accumulated with age in skeletal muscle to the extent that little or no full-length mtDNA was present in late-age subjects. By contrast, multiple mtDNA mutations progressively accumulated with age in the atrial

[a]This work was supported by ATTORI Ltd., Australia.

[d]Tel: 61-3-94264200; fax: 61-3-94264201; e-mail: tlinnane@cmbm.com.au

appendage, but to a lesser extent than in deltoid muscle, whereas in liver, an insignificant level of deletions/mutations was detected. Cytochrome oxidase activity as the indicator of cellular bioenergy capacity mirrored the extent of the mtDNA rearrangements.

MULTIPLE MITOCHONDRIAL DNA MUTATIONS IN AGING

Until recently, the analysis of mtDNA mutations was limited to the small DNA fragments that could be amplified by conventional PCR. Only the 5 kb (4977 bp) "common deletion" and point mutations linked to mitochondrial diseases have been intensively studied.[10,17–22] However, investigating any one single deletion or point mutation may result in a distorted perspective as to the extent of somatic mtDNA mutations accumulation in the aging process. Indeed, the low abundance of the 4977 bp common mtDNA deletion in aged skeletal muscle and brain striatum (less than 0.1% of the wild-type genome[23,24]) would be alone unlikely to be physiologically significant. The level of any one mtDNA mutation, be it a point mutation or deletion, is insufficient to result in the gross physiological and pathological changes associated with the decline of bioenergy capacity observed in aged tissues. We regard these mutations as illustrative of numerous and more extensive mutation.

Although the majority of the deletion studies have focused on a single 4977 bp human mtDNA deletion, several other age-associated mtDNA deletions have also been identified.[25] In the further development of our own studies, we have identified at least 10 different mtDNA deletions in human tissues of one aged individual, a 69-year-old female.[26] We later showed that such multiple deletions are also age associated.[25] The majority of these multiple mtDNA deletions involve short, direct-repeat sequences of mtDNA at which the break point (site of deletion) occurs. There are many thousands of short direct-repeat sequences in mtDNA, ranging from 4 bp to 13 bp,[27] thus illustrating the potential for multiple and varying deletions in human mtDNA.[28] Furthermore, some repeat pairs generate a family of closely related deletions,[26,29] further emphasizing the multiplicity of deletion events that can occur during aging. However, following sequence analysis of multiple rearrangements in mtDNA from senescent heart tissue of mice, we identified direct repeats of two or more nucleotides at the 5′ and 3′ break points in seven out 12 break points sequenced, suggesting that direct repeats are not necessary requirements for somatic mtDNA rearrangements to occur.[30]

The data presented here was generated through quantitative XL-PCR that amplifies the entire 16,569 bp human mtDNA sequence and putatively allows the collective assessment of all mtDNA insertion/deletion mutations.[31] We demonstrate here that aged human deltoid muscle accumulates a wide variety of mtDNA rearrangements (FIG. 1). The amount of full-length mtDNA progressively decreases of deltoid muscle mtDNA for a 5-year-old subject through 37-, 69-, 75-, and 84-year-old individuals. A marked decline in full-length mtDNA was even observed with a 37-year-old subject, whereas for the 84-year-old subject only a small amount of full-length mtDNA was still present. Earlier we reported for a 90-year-old subject that full-length mtDNA was either not detected or occurred in trace amounts.[31] In parallel with the decline in full-length mtDNA, there is an accompanying increase with age in the total amount and copy number of multiple mtDNA rearrangements. The presence of deleted mitochondrial genomes in postmitotic tissue from aged tissues has been confirmed by minicircle PCR (data not shown; see also ref. 32).

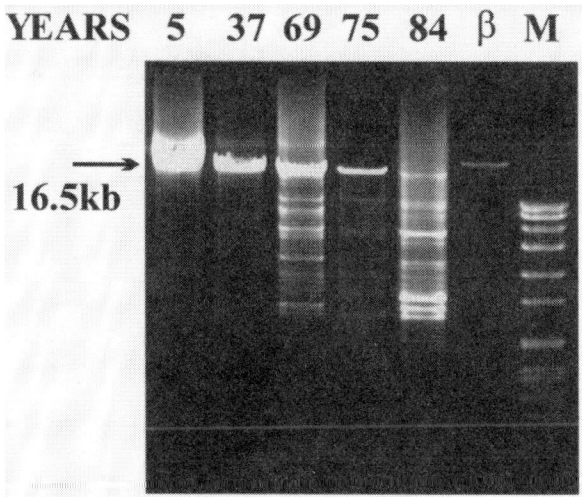

FIGURE 1. Extra-long PCR with mtDNA from human deltoid muscle, demonstrating a decrease in full-length mtDNA with a corresponding increase in extensively damaged mtDNA in old individuals.

A TISSUE-SPECIFIC DISTRIBUTION OF MULTIPLE mtDNA MUTATIONS

The nature and extent of the mtDNA rearrangements have been examined by quantitative XL-PCR in different tissues from the same subjects aged 5 to 72 years old (FIGURES 2 and 3). Comparison of the occurrence of full-length mtDNA in liver and deltoid muscle specimens from a 5- and 72-year-old subject showed that although there is the expected marked decrease in the muscle specimens, there is no detectable changes with age between the two liver samples (FIG. 2). Common 5 kb deletion analyses have earlier repeatedly shown insignificant levels of the mutation in old-age liver specimens.[19,33,34]

The results reported here show the outcome of XL-PCR analysis of mtDNA extracted from human skeletal (deltoid) muscle, cardiac (atrial appendage) muscle, and liver taken from individuals of various ages. These individuals did not demonstrate any clinical manifestations of mitochondrial diseases. The mtDNA from the 5 year old essentially gave a full-length 16.5 kb product from both liver and deltoid muscle (FIG. 2). As the age of the sampled individual increased, the concentration of full-length mtDNA decreased in skeletal muscle accompanied by a corresponding increase in DNA rearrangements. The trend seen with skeletal muscle was not observed in liver. Liver did not demonstrate a substantial decrease in full-length mtDNA and did not accumulate significant mtDNA rearrangements with respect to increasing age.

Right atrial appendage and skeletal muscle surrounding the mammary artery were obtained from a number of individuals undergoing heart bypass surgery. XL-PCR analysis of these tissues revealed that cardiac tissue did not demonstrate any marked age-associated decrease in the concentration of full-length mtDNA seen in skeletal muscle. Cardiac tissue appeared to maintain full-length mtDNA. Cardiac tissue also demonstrated age-associated mtDNA rearrangements, however, and showed substantially less mtDNA rearrangements

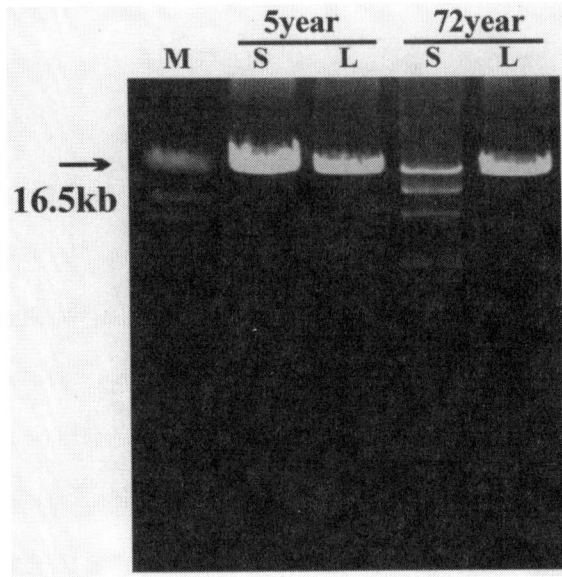

FIGURE 2. Tissue-specific age-associated accumulation of multiple mitochondrial DNA mutations by means of XL-PCR. Skeletal muscle (S) vs. liver (L) tissue. No significant structural age-associated mtDNA changes were observed in mtDNA extracted from liver with age in contrast to skeletal muscle from same subject. The semiquantitative analysis of copy number of mtDNA was performed by using nuclear encoded β-globin genes as an internal control. The amount of intact mtDNA from skeletal muscle of a young healthy individual was greater than in liver.

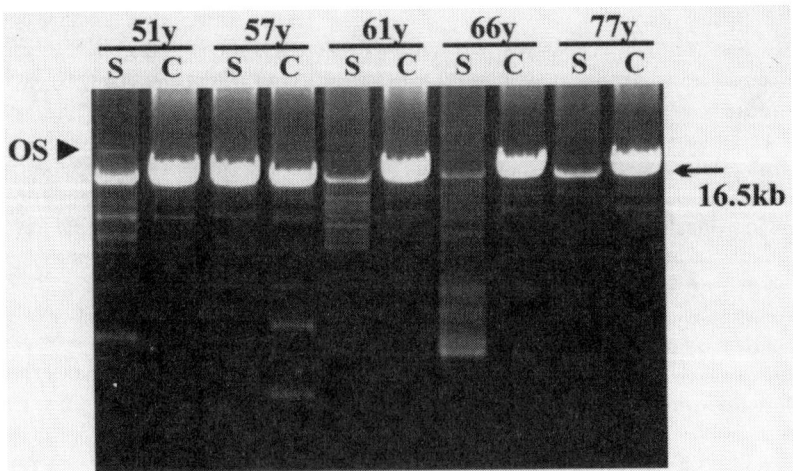

FIGURE 3. Tissue-specific age-associated accumulation of multiple mitochondrial DNA mutations by means of XL-PCR. Cardiac tissue (C) vs. skeletal muscle (S). Mitochondrial DNA from skeletal muscle is more mutated with age than mtDNA from cardiac tissue. The copy number of full-length mtDNA (16.5 kb) is more greatly reduced with age in skeletal muscle than cardiac muscle. There is an age-dependent effect of reduction in amount of full-length mtDNA from skeletal muscle, but not in mtDNA from heart. The occurrence of oversized mtDNA species is not a unique feature of skeletal muscle but also can be detected in cardiac tissue. OS = oversized DNA.

compared to skeletal muscle. Thus we have demonstrated that the multiple mtDNA mutations were distributed not evenly among tissue/organs of the same individual.

Another feature of interest is the appearance of oversized (more than 16.5 kb) mtDNA, which is shown to occur in the skeletal muscle of a 51-year-old subject as well as in the cardiac tissue of a 57-year-old subject. We documented earlier that the occurrence of oversized mtDNA depends on the tissue fragment selected for DNA isolation and the XL-PCR conditions,[31] whereas, oversized mtDNA species are often, but not always, amplified.

In order to perform quantitative analysis of mtDNA rearrangements, in addition to amplifying the mitochondrial genome, we routinely amplified a 23 kb portion of the β-globin gene from the same DNA sample. Amplifying a nuclear housekeeping gene such as the β-globin gene provided a semiquantitative measure of the XL-PCR reaction and allowed valid conclusions to be made about the relative concentrations of mtDNA templates used in each PCR analysis. Template concentrations were adjusted so that β-globin gene PCR products were comparable among different tissue extractions. This ensured that any differences seen in the amount of final mtDNA PCR product was due to an actual difference in the number of full-length mitochondrial templates between samples and not as a consequence of differing amounts of template added to the PCR reaction.

LOSS OF BIOENERGY CAPACITY DURING AGING

Significant declines with age in respiratory activity have been reported for many human tissues, including skeletal muscle[35] and heart.[36,37] An important observation was the experimental verification of the tissue bioenergy mosaic, in which discrete zones in particular tissues from elderly individuals showed loss of mitochondrial respiratory function as judged by histochemical staining.[36,38] Specifically, it was found that, although cardiac tissue from young subjects showed even and intense staining for the COX cardiac tissue from aged subjects, the cardiac tissue from aged subjects showed a mosaic pattern of staining, ranging from cells intensely stained to others showing very weak staining. In addition, the number of cells showing decreased COX activity increased with age.

In the present study we have demonstrated the strong correlation between the bioenergetic capacity of a given tissue by the assessment of number of cells with decreased COX activity and the total extent of mitochondrial DNA rearrangements. The skeletal muscle fibers and heart trabeculae enzyme-histochemically scored as strongly staining positive (++), positive (+), and negative (−) for cytochrome oxidase activity (TABLE 1). A very pronounced COX energy mosaic is observed in all aging subjects (40 years onwards), with cells with ++, +, or − activity being commonly observed. The mosaic was even observed with the skeletal muscle tissue of a 5-year-old subject; although ++ was clearly predominant some + and a few − cells were observed. Comparison of the skeletal muscle and cardiac muscle profiles shows that the decrease in COX is much greater and more extensive in the skeletal muscle with many more negative cells observed, whereas few negative cells are observed in the cardiac tissues of the older subjects. This would be expected in the subjects, as they still must possess a heart with adequate function to sustain life. Two subjects showed negative cardiac muscle cells, accounting for about 12–17% of total cells, whereas for skeletal muscle, all but one of the aged subjects showed more than 33% negative cells.

TABLE 1. Analysis of Cytochrome-*c* Oxidase Activity in Different Tissues from Human Subjects of Various Ages[a]

	Age (Years)	No. fibers or trabeculae counted	Scoring		
			++	+	−
Skeletal muscle	5	174	80	16	4
	40	156	42	21	37
	61	153	41	42	17
	69	189	29	38	33
	72	174	47	19	34
	90	425	19	43	38
Cardiac muscle	15	16	94	6	0
	36	6	100	0	0
	51	16	63	25	12
	51	9	100	0	0
	51	18	94	6	0
	57	6	67	33	0
	59	6	100	0	0
	69	6	50	50	0
	72	15	80	20	0
	74	12	8	75	17

[a]The cytochrome-*c* oxidase activity was used as a general indicator of the overall activity of mitochondrial enzymes of the respiratory chain. COX activity was examined on 10 μm skeletal muscle tissue sections according to the method outlined by Müller-Höcker.[36]

A CORRELATION BETWEEN COX ACTIVITY AND EXTENT OF mtDNA REARRANGEMENTS

A correlation between COX activity and extent of mtDNA rearrangements can be shown. The extent of the mtDNA rearrangements and COX activity in skeletal muscle taken from several of the subjects included in TABLE 1 are compared in FIGURE 4. The data are presented as the percentage of cells with measured COX (++) activity as compared with the changes in mtDNA. Simple inspection of XL-PCR products on agarose gels shows that the extent of mtDNA rearrangements correlates well with the number of strongly positive COX cells; the 69-year-old subject has only 29% strongly positive cells (whereby 71% are with decreased activity) and shows the most degraded mtDNA pattern. As would be expected, the 5-year-old subject has 80% strongly COX positive cells and shows only small changes in its mtDNA.

The same type of comparison is shown in FIGURE 5 for heart atrial appendage specimens taken from patients undergoing heart bypass surgery. The atrial appendage material was adjudged normal by conventional histochemical examinations of hematoxylin-eosin stained specimens prior to COX and mtDNA analyses. The results are of interest in two regards; one is that there is a general decrease with age of normal-length mtDNA and COX activity, but second, there can be considerable differences observed between individuals of the same age. The three 51-year-old subjects illustrate this situation. The COX activities range through 63, 94, and 100% and little change is observed in the amount of full-length mtDNA and extent of multiple rearrangements in the latter two specimens, whereas the mtDNA of the third subject has undergone substantial change. Comparison of the 72- (80% COX activity) and 74- (8% COX activity) year-old subject material further emphasizes this phenomenon. These results illustrate the paradigm that individuals age at different rates, and tissues are differentially affected from one individual to another. Thus, the data presented in FIGURES 4 and 5 confirm our earlier findings, clearly demonstrating that individuals with relatively low COX activity show a heterogenous population of rearranged mtDNA species with little full-length mtDNA.

In TABLE 2 the results presented in this paper and several others[33,39] are summarized. Multiple mtDNA deletions do occur but not to any significant extent in such young (under 5 years) subject tissues as liver, cardiac, and skeletal muscles. In aged subjects (60–90 years), for liver, there is no difference from the young, for cardiac muscle substantial changes are observed, whereas for skeletal muscle the changes are gross. The COX activities of the specimens parallel the mtDNA changes. The results of several other studies on the occurrence of the 5 kb deletion are tabulated and emphasize the different extents of occurrence of this mtDNA deletion, ranging from very low in kidney and cerebellum,[33,39] to substantial in brain cortex,[39] to large in the putamen of the brain.[39,40] These results emphasize the absolute importance and attention that must be paid to the particular organ/

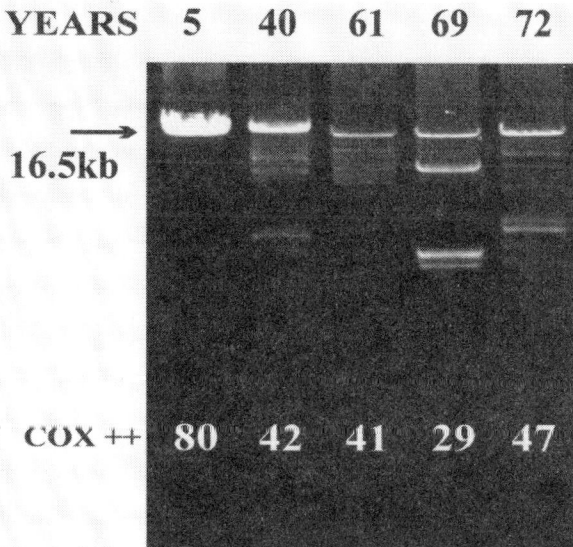

FIGURE 4. A correlation between COX activity and total extent of mtDNA rearrangements in skeletal muscle of human subjects of different ages.

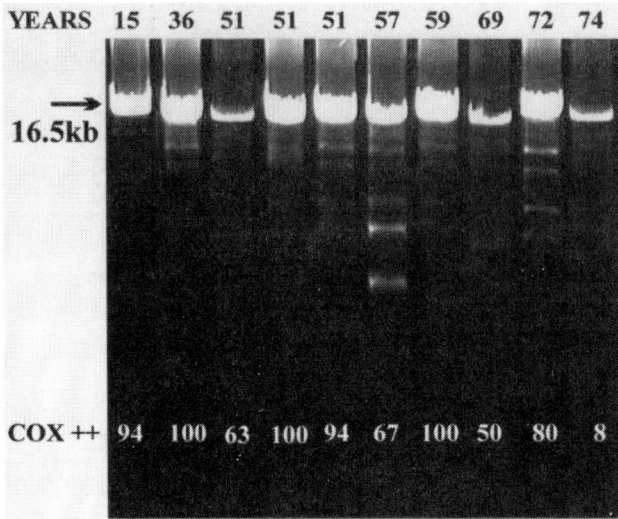

FIGURE 5. A correlation between COX activity and total extent of mtDNA rearrangements in heart (right atrial appendage) of different age human subjects undergoing heart bypass surgery.

tissue/cells when discussing and generalizing mtDNA changes as being consequential to the function and age of human (and other animal) subjects. The literature is replete with ill-founded generalizations based on limited single-tissue studies involving almost any species, for example, flies, birds, rats, and humans.

TABLE 2. Extent of Human Mitochondrial DNA Mutations and Cytochrome-c Oxidase Activity Varies with Age and the Specific Tissue

	Young		Aged	
Multiple Deletions	Mutations	COX	Mutations	COX
Liver	±	♦♦♦[a]	±	♦♦♦
Heart	±	♦♦♦	++	♦♦
Deltoid muscle	±	♦♦♦	+++	♦
4977 bp Deletion				
Kidney[33]	±		±	
Cortex (brain)[39]	±		+	
Cerebellum (brain)[39]	±		±	
Putaman (brain)[39]	±		+++	

[a] ♦, COX activity.

REACTIVE OXYGEN SPECIES AND mtDNA REARRANGEMENTS

It has been suggested that mitochondria, which consume about 90% of the cell's oxygen, metabolize 1–2% of the oxygen to superoxide and that, consequently, the mitochondria will suffer severe ROS damage.[41] However, more recently the extent of superoxide formation and the consequential *in vivo* formation of 8-OHdG has been questioned,[13] and it has been reported that the mitochondria of normal tissue do not produce significant amounts of superoxide.[14] In addition, many reports on the measurement of oxidative DNA damage, and especially mtDNA damage, show a considerable disagreement from one author to another as to the extent of damage.[42] Herein, we question the role of ROS on the extent of occurrence of mtDNA deletions. The mtDNA rearrangements, which we report, range from insignificant in liver cells (mitochondria) and intermediate in cardiac tissue (mitochondria), to very extensive in skeletal muscle (mitochondria); yet all are highly aerobic tissues. The extent of the mtDNA deletions correlate well with the level of tissue and cellular COX activity. It therefore follows that *in vivo* the mitochondrial electron transport system does not simply generate ROS, which are causative of the mtDNA rearrangements and loss of COX activity. Indeed, the evidence is in favor of the mtDNA replication process being largely responsible for the mtDNA deletions. The deletions arise as a consequence of mtDNA containing a large number of base-pair direct repeats, which together with the asynchronous replication of the heavy and light strands of mtDNA can lead to base sequence mismatches resulting in a wide range of deletions.

Although our results cast doubt on the significance of mitochondrially generated ROS as being responsible for the observed gross mtDNA rearrangements and bioenergy degradation of the cells, it is not suggested that ROS damage does not occur in tissues. Clearly, proteins,[5,6] lipids,[3,4] and DNA[2] are oxidized to some extent, but there is not an age-progressive accumulation of these products; as such they are repaired or degraded thereby establishing an equilibrium status quo between ongoing damage and rescue during the aging process. The outcome of this damage, it is suggested may not be of significance in an aging individual in the absence of other complicating factors. The normal repair and scavenger systems are presumably crucial to cellular well-being, but in normal aging individuals the defense systems are adequate. It is these systems that may warrant our attention as to whether they become inadequate to the task. Conceivably liver (mitochondria) cells, cardiac (mitochondria) muscle cells, and deltoid skeletal muscle (mitochondria) cells have different antioxidant defense capacities, or can repair mtDNA changes, or, in the case of liver, can replace the cells with extensive mtDNA rearrangements. Thus, mechanisms responsible for the accumulation of mitochondrial DNA rearrangements with age in tissue, leading to a decline in cellular bioenergy capacity, may be due to replicative error or some other process. In any case, this situation remains an open question.

REFERENCES

1. HARMAN, D. 1956. Ageing: A theory based on free radical and radiation chemistry. J. Gerontol. **11:** 298–300.
2. RICHTER, C., J.-W. PARK & B.N. AMES. 1988. Normal oxidative damage to mitochondrial and nuclear DNA is extensive. Proc. Natl. Acad. Sci. USA **85:** 6465–6467.

3. TAPPEL, A.L. 1973. Lipid peroxidation damage to cell components. Fed. Proc. **32**(8): 1870–1874.
4. FRAGA, C.G. & A.L. TAPPEL. 1988. Damage to DNA concurrent with lipid peroxidation in rat liver slices. Biochem. J. **252**(3): 893–896.
5. OLIVER, C.N., B.W. AHN E.J. MOERMAN, S. GOLDSTEIN & E.R. STADTMAN. 1987. Age-related changes in oxidized proteins. J. Biol. Chem. **262**(12): 5488–5491.
6. STADTMAN, E.R., P.E. STARKE-REED, C.N. OLIVER, J.M. CARNEY & R.A. FLOYD. 1992. Protein modification in aging. Exs **62**: 64–72.
7. SOHAL, R.S., I. SVENSSON, B.H. SOHAL & U.T. BRUNK. 1989. Superoxide anion radical production in different animal species. Mech. Ageing Dev. **49**(2): 129–135.
8. KU, H.H., U.T. BRUNK & R.S. SOHAL. 1993. Relationship between mitochondrial superoxide and hydrogen peroxide production and longevity of mammalian species. Free Radical Biol. Med. **15**(6): 621–627.
9. CHEN, Q., A. FISCHER, J.D. REAGAN, L.J. YAN & B.N. AMES. 1995. Oxidative DNA damage and senescence of human diploid fibroblast cells. Proc. Natl. Acad. Sci. USA **92**(10): 4337–4341.
10. HAYAKAWA, M., K. HATTORI, S. SUGIYAMA & T. OZAWA. 1992. Age-associated oxygen damage and mutations in mitochondrial DNA in human hearts. Biochem. Biophys. Res. Commun. **189**(2): 979–985.
11. SHIGENAGA, M.K., T.M. HAGEN & B.N. AMES. 1994. Oxidative damage and mitochondrial decay in aging. Proc. Natl. Acad. Sci. USA **91**(23): 10771–10778.
12. NOHL, H. & D. HEGNER. 1978. Do mitochondria produce oxygen radicals *in vivo*? Eur. J. Biochem. **82**(2): 563–567.
13. LINDAHL T. 1993. Instability and decay of primary structure of DNA. Nature **362**: 709–715.
14. NOHL H. 1993. Involvement of free radicals in aging: A consequence or cause of senescence. Br. Med. Bull. **49**(3): 653–667.
15. LINNANE, A.W., G. KOPSIDAS, J. KELSO & S.A. KOVALENKO. 1997. Occurrence of age-associated multiple mtDNA rearrangements in human skeletal muscle. FASEB J. **11**(9): A1450.
16. KOPSIDAS, G., S.A. KOVALENKO, J. KELSO & A.W. LINNANE. 1998. An age-associated correlation between cellular bioenergy decline and mtDNA rearrangements in human skeletal muscle. Mutat. Res. In press.
17. LINNANE, A.W., A. BAUMER, R.J. MAXWELL, H. PRESTON, C.F. ZHANG & S. MARZUKI. 1990. Mitochondrial gene mutation: The ageing process and degenerative diseases. Biochem. Int. **22**(6): 1067–1076.
18. TANAKA M., H. INO, K. OHNO, T. OHBAYASHI, S. IKEBE, T. SANO, T. ICHIKI, M. KOBAYASHI, Y. WADA & T. OZAWA. 1991. Mitochondrial DNA mutations in mitochondrial myopathy, encephalopathy, lactic acidosis, and stroke-like episodes (MELAS). Biochem. Biophys. Res. Commun. **174**(2): 861–868.
19. YEN, T.C., J.H. SU, K.L. KING & Y.H. WEI. 1991. Ageing-associated 5 kb deletion in human liver mitochondrial DNA. Biochem. Biophys. Res. Commun. **178**(1): 124–131.
20. OZAWA, T. 1995. Mechanism of somatic mitochondrial DNA mutations associated with age and diseases. Biochim. Bioph. Acta **1271**(1): 177–189.
21. KADENBACH, B., C. MUNSCHER, V. FRANK, J. MULLER-HOCKER & J. NAPIWOTZKI. 1995. Human aging is associated with stochastic somatic mutations of mitochondrial DNA. Mutat. Res. **338** (1–6): 161–172.
22. LIU, V.W.S., C. ZHANG, A.W. LINNANE & P. NAGLEY. 1997. Quantitative allele-specific PCR: Demonstration of age-associated accumulation in human tissue of the A to G mutation at nucleotide 3243 in mitochondrial DNA. Hum. Mutation **9**(3): 265–271.
23. SIMONETTA, S., X. CHEN, S. DI MAURO & E.A. SCHON. 1992. Accumulation of deletions in human mitochondrial DNA during normal aging: Analysis by quantitative PCR. Biochem. Biophis. Acta **1180**: 113–122.
24. OZAWA, T., M. TANAKA, S. IKEBE, K. OHNO, T. KONDO & Y. MIZUNO. 1990. Quantitative determination of deleted mitochondrial DNA relative to normal DNA in parkinsonian striatum by a kinetic PCR analysis. Biochem. Biophys. Res. Commun. **172**(2): 483–489.
25. LINNANE, A.W., C. ZHANG, A. BAUMER & P. NAGLEY. 1992. Mitochondrial DNA mutation and the ageing process: Bioenergy and pharmacological intervention. Mutat. Res. **275**(3–6): 195–208.
26. ZHANG C., A. BAUMER, R.J. MAXWELL, A.W. LINNANE & P. NAGLEY. 1992. Multiple mitochondrial DNA deletions in an elderly human individual. FEBS Lett. **297**: 4–8.

27. NAGLEY, P., C. ZHANG, R.D. MARTINUS, F. VAILLANT & A.W. LINNANE. 1993. Mitochondrial DNA mutation and human aging: molecular biology, bioenergetics and redox therapy. *In* Mitochondrial DNA in Human Pathology. S. DiMauro, D. Wallace, Eds.: 137–157. Raven Press. New York.

28. HAYAKAWA, M., K. KATSUMATA, M. YONEDA, M. TANAKA, S. SUGIYAMA & T. OZAWA. 1995. Mitochondrial DNA minicircles, lacking replication origins, exist in the cardiac muscle of a young normal subject. Biochem. Biophys. Res. Commun. **215**(3): 952–960.

29. BAUMER, A., C. ZHANG, A.W. LINNANE & P. NAGLEY. 1994. Age-related human mtDNA deletions: A heterogeneous set of deletions arising at a single pair of directly repeated sequences. Am. J. Hum. Genet. **54**(4): 618–630.

30. LEE, C.M., S.S. CHUNG, J.M. KACZKOWSKI, R. WEINDRUCH & J.M. AIKEN. 1993. Multiple mitochondrial DNA deletions associated with age in skeletal muscle of rhesus monkeys. J. Gerontol. **48**: B201–B205.

31. KOVALENKO, S.A., G. KOPSIDAS, J. KELSO & A.W. LINNANE. 1997. Deltoid human muscle mtDNA is extensively rearranged in old age subjects. Biochem. Biophys. Res. Commun. **232**: 147–152.

32. MELOV, S., D. HINERFELD, L. ESPOSITO & D.C. WALLACE. 1997. Multi-organ characterisation of mitochondrial genomic rearrangements in ad libidum and caloric restricted mice show striking somatic mitochondrial DNA rearrangements with age. Nucleic Acids Res. **25**(5): 974–982.

33. ZHANG, C., M. BILLS, A. QUIGLEY, R.J. MAXWELL, A.W. LINNANE & P. NAGLEY. 1997. Varied prevalence of age-associated mitochondrial DNA deletions in different species and tissues: A comparison between human and rat. Biochem. Biophys. Res. Commun. **230**(3): 630–635.

34. HAGEN, T.M., D.L. YOWE, J.C. BARTHOLOMEW, C.M. WEHR, K.L. DO, J.Y. PARK & B.N. AMES. 1997. Mitochondrial decay in hepatocytes from old rats: Membrane potential declines, heterogeneity and oxidants increase. Proc. Natl. Acad. Sci. USA **94**(7): 3064–3069.

35. TROUNCE, I., E. BYRNE & S. MARZUKI. 1989. Decline in skeletal muscle mitochondrial respiratory chain function: Possible factor in ageing. Lancet **ii**: 637–639.

36. MÜLLER-HÖCKER, J. 1989. Cytochrome-*c*-oxidase deficient cardimyocytes in the human heart. An age related phenomenon. Am. J. Pathol. **134**: 1167–1173.

37. CORRAL D.M., J.M. SHOFFNER, , M.T. LOTT & D.C. WALLACE. 1992. Association of mitochondrial DNA damage with aging and coronary atherosclerotic heart disease. Mutat. Res. **275**(3–6): 169–180.

38. LINNANE, A.W., S. MARZUKI, T. OZAWA & M. TANAKA. 1989. Mitochondrial DNA mutations as an important contributor to ageing and degenerative diseases. Lancet **1**: 642–645.

39. CORRAL-DEBRINSKI, M., T. HORTON, M.T. LOTT, J.M. SHOFFNER, M.F. BEAL & D.C. WALLACE. 1992. Mitochondrial DNA deletions in human brain: Regional variability and increase with advanced age. Nat. Genet. **2**: 324–329.

40. SOONG, N.W., D.R. HINTON, G. CORTOPASSI & N. ARNHEIM. 1992. Mosaicism for a specific somatic mitochondrial DNA mutation in adult human brain. Nat. Genet. **2**(4): 318–323.

41. RICHTER, C. 1992. Reactive oxygen and DNA damage in mitochondria. Mutat. Res. **275**(3–6): 249–255.

42. BECKMAN K.B. & B.N. AMES. 1996. Detection and quantification of oxidative adducts of mitochondrial DNA. Methods Enzymol. **264**: 442–453.

Influences of Caloric Restriction on Age-associated Skeletal Muscle Fiber Characteristics and Mitochondrial Changes in Rats and Mice[a]

CONNIE M. LEE,[b] LAUREN E. ASPNES,[c] SUSAN S. CHUNG,[b] RICHARD WEINDRUCH,[d,e,f] AND JUDD M. AIKEN[b,g]

Departments of [b]Animal Health and Biomedical Sciences, [c]Nutritional Sciences, and [d]Medicine, and the [e]Wisconsin Regional Primate Research Center, University of Wisconsin, Madison, Wisconsin 53706
[f]Geriatric Research, Education and Clinical Center, Wm. S. Middleton VA Medical Center, Madison, Wisconsin 53705

ABSTRACT: The effect of caloric restriction (CR) initiated in adult rats (17 months of age) on the abundance of deleted mitochondrial genomes, mitochondrial enzymatic abnormalities, and fiber number was examined in rat skeletal muscle. Vastus lateralis muscle from young (3–4 months) *ad libitum*-fed, old (30–32 months) restricted (35% and 50% CR, designated CR35 and CR50, respectively), and old *ad libitum*-fed rats (29 months) was studied. CR preserved fiber number and fiber-type composition in the CR50 rats. In the old rats from all groups, individual fibers were found with either no detectable cytochrome-*c* oxidase activity (COX⁻), hyperactive for succinate dehydrogenase activity (SDH⁺⁺), or both COX⁻ and SDH⁺⁺. Muscle from the CR50 rats contained significantly fewer COX⁻ and SDH⁺⁺ fibers than did the muscle from the CR35 rats. CR50 rats also had significantly lower numbers of mtDNA deletion products in two (adductor longus and soleus) of the four muscles examined compared to CR35 rats. These data indicate that CR begun in late middle age can retard age-associated fiber loss and fiber-type changes as well as lower the number of skeletal muscle fibers exhibiting mitochondrial enzyme abnormalities. CR can also decrease the accumulation of deleted mitochondrial genomes.

Caloric restriction (CR), a consistently proven method of extending both average and maximum life span and attenuating the development of age-related diseases, has been studied extensively in rodents.[1,2] The nutritional principle behind CR is a reduction in the total caloric intake while maintaining sufficient intake of all other dietary essentials. The mechanisms by which CR extends life span remain unknown; however, one hypothesis is that CR reduces the level of oxidative stress within an aerobic organism.[3,4] Mitochondria, as sites of free radical production from the electron transport system (ETS),[5] may foster a state of oxidative stress and contribute to an age-associated decline in tissues that require high levels of energy, such as muscle and nerve tissue. These tissues also share the feature

[a]This work was supported by Grants RO1 AG11604 (J.M. Aiken) and RO1 AG10536 (R. Weindruch) from the National Institutes of Health. This is publication number 97-02 from the Madison VA Geriatric Research, Education, and Clinical Center.
[g]To whom correspondence should be addressed: Judd M. Aiken, Department of Animal Health and Biomedical Sciences, 1655 Linden Drive, Madison, WI 53706, USA. Tel: 608/262-7362; fax: 608/262-7420; e-mail: JMA@ahabs.wisc.edu

of having a limited capacity to replace damaged cells. In this study, we examined the influence of CR on skeletal muscle from mice and rats to determine whether CR could affect the age-associated decline in skeletal muscle mass, or "sarcopenia." We also sought to explore the possible contribution of mitochondria to sarcopenia by evaluating the ability of CR to attenuate age-associated mitochondrial dysfunction in skeletal muscle.

MUSCLE FIBER CHARACTERISTICS OF RODENTS SUBJECTED TO CR

As with many age-related pathologies, the etiology of sarcopenia may prove to be multifactorial, and this is reflected in the proposed hypotheses that include contraction-induced injury,[6] deficient satellite cell recruitment,[7] motor unit decline,[8] and increased oxidative stress.[9] Studies aimed at determining the role of these factors in human sarcopenia have been accomplished using a variety of techniques, including measurements of body composition, computed tomography and magnetic resonance imaging, muscle biopsy, and whole sectioning of muscles.[10]

Sarcopenia has been studied in a number of mammalian animal models, of which the rat is most commonly used. The characteristics of sarcopenia in old rats are dependent on the strain, gender, muscle, and age group examined. Clearly, the extent of sarcopenia is highly variable among the muscles examined. Yu *et al.*[11] found that the mass of the gastrocnemius muscle peaked at 12 months and began declining at 18 months in male Fischer 344 (F344) rats. Studies in quadriceps mass describe an ~30% reduction between 10- and 28-month-old animals.[12,13] In female F344 rats, the soleus declined in mass by 12.5% from 18 to 24 months with an additional loss of 2.5% by 27 months.[14] Muscles have also been observed in the rat, such as the flexor digitorum longus (FDL), which do not appear to atrophy.[15] In this study, we determined the influence of CR on age-related changes in muscle fibers, including fiber number, fiber type composition, and the overall histological appearance of the tissue.

Fiber Number

We first determined fiber number in the vastus lateralis of two young (3–4 months) and two groups of 4 old (30–32 months) Lobund-Wistar rats subjected to 35% and 50% caloric restriction (CR_{35} and CR_{50}, respectively) since 17 months of age. These animals were described previously.[16] The rationale for these two diets was to avoid the use of "obese" controls in our study, and, therefore, two restriction levels were implemented to attain the approximate body weights for control (~450 g) and restricted (~300 g) as previously reported for male Lobund-Wistar rats.[17] Entire cross-sections from the vastus lateralis midbelly were counted from photographs of fibers stained for cytochrome c oxidase (COX).[18] Rats subjected to the 50% CR regimen exhibited less fiber loss than those on the milder 35% reduction in calories. A 30% decrease in fiber number was observed in the CR_{35} as compared to the CR_{50} rats ($p < 0.05$), and no significant difference was observed between the CR_{50} and the young rats.[16]

We also evaluated the effect of CR on fiber number in two different muscles from B6D2F1 hybrid mice. Five- and 30-month-old male mice fed *ad libitum* (young (Y), n = 3; AL, n = 3, respectively) and 30-month-old mice subjected to a 40% caloric restriction

TABLE 1. Effect of Age and CR on Fiber Number in B6D2F1 Hybrid Mice

Mouse Group	Age (mo.)	No. of fibers through midbelly of epitrochlearis				No. of fibers through midbelly of soleus			
		No. of mice	Mean	SEM	Range	No. of mice	Mean	SEM	Range
Young	5	3	515	21	473–538	3	816	34	764–879
Ad libitum	30	3	401	13	389–434	3	711	8	696–721
CR	30	3	468	23	437–512	3	762	29	728–820

p = 0.0390	p = 0.0509
Ad libitum < CR and young	Ad libitum < CR and young

(CR, n = 3) were purchased from the NIA colony at the National Center for Toxicological Research (Jefferson, AR). Mice were maintained as previously described.[19] The young and AL mice were given free access to the NIH-31 open formula diet (Purina Mills Inc., Richmond, IN). The CR mice started restriction at four months and received 60% of the AL cohort's NIH-31 open formula diet consumption but were fed a vitamin-supplemented diet.

The mean number of fibers was determined through the midbelly of epitrochlearis and soleus muscles of three Y, AL, and CR mice (TABLE 1). Significant differences were observed for the epitrochlearis fiber number in these mice (AL < CR < young; p = 0.039). Although a similar trend was observed in the soleus from the three groups of mice, the differences failed to show statistical significance (p = 0.0509).

Fiber Type

Fiber-type composition has been determined on various muscles in rats;[13,20–22] however, whether changes in fiber-type composition occur with age is controversial and appears to be dependent on the specific muscle examined. We evaluated the fiber-type composition in the vastus lateralis of the three diet groups of Lobund-Wistar rats. The fiber-type composition was determined using an antibody specific for the heavy chain of fast myosin (Sigma; St. Louis, MO). The vastus lateralis from two young, 15 CR_{35}, and 15 CR_{50} rats were analyzed. The number of type I fibers was observed to decline with age (mean of 320 and 33 type I fibers per vastus lateralis in young and old CR_{35} samples, respectively). This decline, however, was attenuated in the CR_{50} animals (mean of 222 type I fibers).[16] The percentages of type I fibers were still considerably different when the fiber number differences we observed were taken into account (4.8%, 0.7%, and 3.2% for young, old CR_{35}, and old CR_{50}, respectively). Because fiber number declined by ~1800 fibers between 4 and 30–32 months and only ~290 of those fibers were type I, type II fibers represented the majority of fibers that were lost with age. Inasmuch as the rat vastus lateralis is primarily composed of type II fibers (~90%),[23] however, one would expect the majority of fibers lost to be type II fibers in this muscle.

Histological Appearance

Other more qualitative, histological changes were also identified in the rat vastus lateralis among the three diet groups (FIG. 1). Sections stained for COX activities were evaluated for the degree of atrophy based on the scoring system of Fujisawa,[24] ranging from 0 to +++ as based on the following criteria: 0, no atrophy; +, appearance of small, angular fibers scattered singly or in very small groups; ++, amount of affected fibers increased with entire fasciculi atrophied or hypertrophied; +++, at least 75% of section atrophied, but normal and/or hypertrophic fibers remained in small, isolated groups. The young rats, as expected, displayed no identifiable atrophy. In the CR_{35} rats, however, only 20% of the animals (3 of 15) showed no evidence of atrophy; 33% were +, 33% were ++, and 13% were scored as +++. The CR_{50} rats displayed less overt indications of atrophy. Forty percent (6 of 15) of the sections showed no identifiable atrophy; 20% were +, 27% were ++, and 13% were +++ with severe atrophy. In both the CR_{35} and CR_{50} rats, the majority of atrophic fibers were detected in type II fibers. These differences in the degree of atrophy between the two CR groups were not large enough to attain statistical significance (Mantel-Haenszel chi-square, p = 0.23).[25]

CR AND MITOCHONDRIA

The mitochondrion is unique to mammalian cellular organelles in that it contains its own genome, an ~16.5 kilobase double-stranded circular DNA molecule with the two DNA strands referred to as heavy (H) and light (L). The genome encodes 22 tRNAs, 2 rRNAs, and 13 polypeptides of the ETS. These proteins include seven subunits of NADH dehydrogenase (complex I), cytochrome *b* of ubiquinol:cytochrome *c* oxidoreductase (complex III), three subunits of COX (complex IV), and ATPase 6 and ATPase 8 of ATP synthase (complex V). The genome replicates in an asynchronous, bidirectional manner,

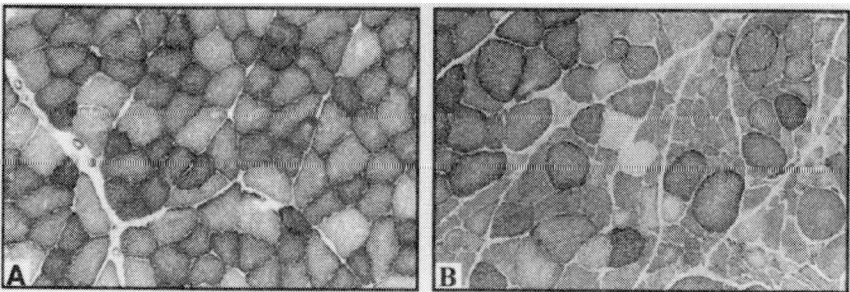

FIGURE 1. COX stains of vastus lateralis samples from a 4-month-old (**A**) and 32-month-old CR_{35} (**B**) Lobund-Wistar rat. Clusters of atrophic and hypertrophic fibers are observable in the CR_{35} vastus lateralis but not in the 4-month-old.

with DNA replication initiating at the heavy strand origin (O_H) in a clockwise direction until the light strand origin (O_L) is reached, at which point the light strand replication begins in a counterclockwise direction.[26] The smaller region between the O_H and the O_L is referred to as the "minor" arc (nucleotides 1 to 5752 in human mtDNA), whereas the larger area from O_L to O_H is the "major" arc (nucleotides 5742 to 16,569 in human mtDNA).[27]

mtDNA Deletions

Alterations of the mitochondrial genome, such as deletions or nucleotide point mutations and modifications, occur with normal aging. The most studied mitochondrial DNA (mtDNA) alteration, however, has been mtDNA deletions. Although initial gerontological studies of mtDNA deletions focused on a single specific deletion (mtDNA^{4977}) in humans, it soon became apparent that mtDNA deletions were not confined to a single specific deletion or regions of the genome, and that multiple mtDNA deletions occurred in many animal species.[28] The accumulation of mtDNA deletions with age has been proposed to contribute causally to aging processes;[29–31] however, their low abundance (typically < 0.1% of mitochondrial genomes) in analyses of tissue homogenates does not support the idea that deletions cause the physiological attrition of aging. On the other hand, analyses of defined numbers of muscle fibers,[32] as well as in situ hybridization studies,[33,34] have shown that mtDNA deletions are cellularly distributed in a mosaic fashion, accumulating to high levels in a subset of cells.

We have employed two different approaches to evaluate the levels of mtDNA deletions in rat skeletal muscle: polymerase chain reaction (PCR) and in situ hybridizations. The advantages of PCR analysis is that the numbers of different deletion products can be easily determined and that individual deletions can be characterized more thoroughly through subsequent DNA sequence analysis. Because mtDNA deletions appear to accumulate focally, however, PCR analysis does not address the levels of mtDNA deletions within individual cells. The advantage of histological techniques is that ETS enzymatic analyses can be performed, and on adjacent serial sections, in situ hybridizations can localize mtDNA deletions within individual cells. A more detailed analysis of the particular mtDNA deletion break point, however, cannot be obtained with standard in situ hybridization techniques.

We have analyzed the presence of mtDNA deletion products using PCR in four muscles from the young, CR_{35}, and CR_{50} Lobund-Wistar rats. The muscles examined were the adductor longus (ADL), soleus, extensor digitorum longus (EDL), and epitrochlearis. Total DNA was isolated from the muscles using a modified guanidinium method.[35] Southern blot analysis using probes specific to rat mtDNA was performed on all samples for normalization of mtDNA amounts. A 9.1 kilobase region of the mitochondrial genome was analyzed in a nested PCR reaction using primers positioned internal to the major arc of the mitochondrial genome.[16] PCR products were size fractionated on 4.0% NuSieve gels (FMC; Rockland, Maine) and visualized by ethidium bromide staining (FIG. 2). The number of different mtDNA deletion products was determined for each sample by counting visible bands on the ethidium bromide–stained gels. Statistically significant differences in numbers of deletion products were observed between the young and both the CR_{35} and CR_{50} rats in all muscle groups examined. Furthermore, the CR_{50} rats had signif-

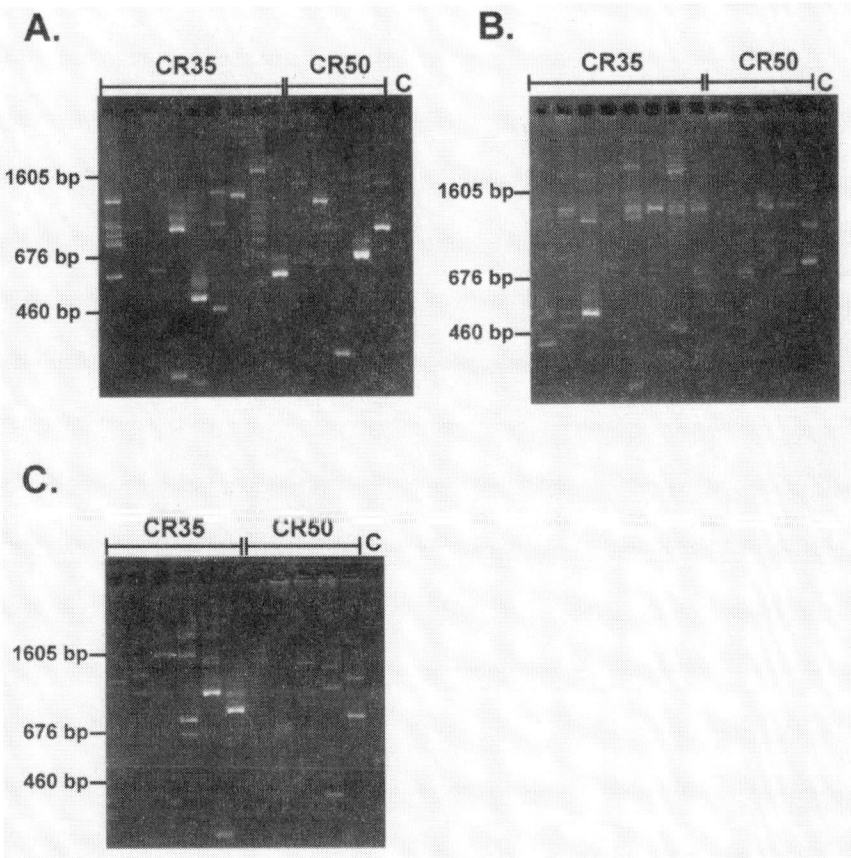

FIGURE 2. CR amplification of mtDNA deletion products from (**A**) CR_{35} (n = 9) and CR_{50} (n = 5) rat EDL; (**B**) CR_{35} (n = 8) and CR_{50} (n = 5) rat soleus; and (**C**) CR_{35} (n = 6) and CR_{50} (n = 5) rat epitrochlearis. Size markers (bps) are to the left of the gels. **C**: No template control.

icantly lower numbers of mtDNA deletion products in the ADL and soleus muscles as compared to the CR_{35} rats.[16] Primer shift experiments were performed on all samples to ensure that the PCR products seen were not the result of primer misannealing (data not shown).

The presence of deleted mitochondrial genomes has been demonstrated histologically by *in situ* hybridization experiments on various skeletal muscles in humans.[33,34] These studies have shown, by using mtDNA probes located either within or outside of deleted regions, that high levels of deleted mitochondrial genomes can be localized to individual cells. Serial sections adjacent to those used for *in situ* hybridizations were analyzed for mitochondrial enzymatic activities, and ETS abnormalities often, but not always, colocalized with the deleted mtDNA genomes.[33,34]

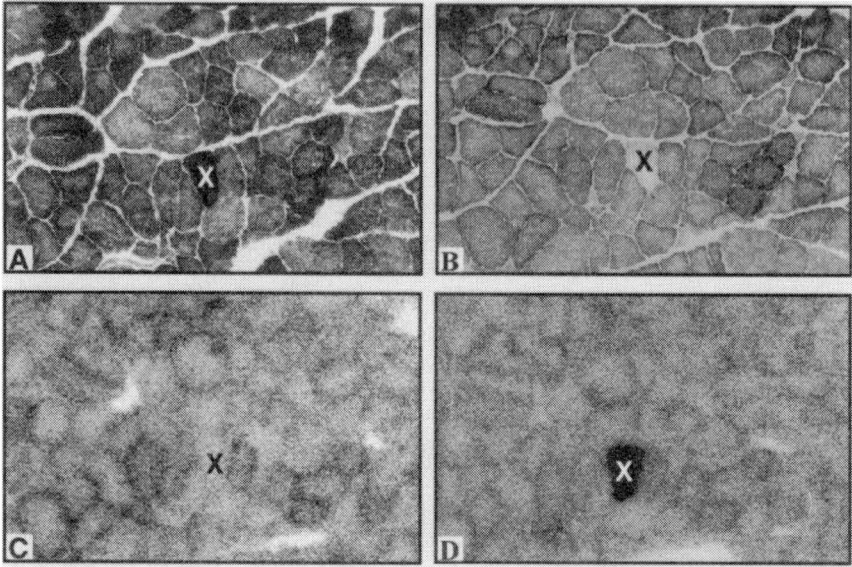

FIGURE 3. Histochemical staining (**A** and **B**) and *in situ* hybridizations (**C** and **D**) on frozen sections from a 29-month-old Lobund-Wistar rat vastus lateralis. A SDH^{++} fiber (indicated by an "X" in **A**) is also found to be deficient for COX activity (**B**). Serial sections were hybridized with ^{35}S-labeled antisense riboprobes specific for mitochondrial-encoded COX I (**C**) and cyt *b* (**D**). The cyt *b* probe hybridizes strongly with the specific fiber, whereas the COX I probe does not, indicating that a deletion event has occurred in that fiber.

We have performed *in situ* hybridizations on the vastus lateralis from the CR_{35} and CR_{50} rats. COX and succinate dehydrogenase (SDH) activities were first determined to identify sections that contained ETS abnormal fibers. Serial sections immediately adjacent were hybridized with antisense RNA probes to determine the levels of transcripts from the 16s rRNA, COX I, and cytochrome-*b* mitochondrial genes. Three sets of experiments were performed on the vastus lateralis from the CR_{35} animals. Four fibers were identified that had ETS abnormalities, and deletions in the major arc of the mitochondrial genome were detected in all four of these fibers (FIG. 3). None of the surrounding ETS normal fibers contained detectable levels of deleted mitochondrial genomes. These results support our findings from rhesus monkey quadriceps in which ~90% of ETS abnormal fibers (23 of 26) associated with mtDNA deleted molecules.[36]

ETS Enzymatic Activities

Evaluation of mitochondrial enzymatic activities histologically in human skeletal muscle has identified abnormalities in a broad class of neuromuscular disorders collectively known as mitochondrial myopathies.[27,37] "Ragged red" fibers (RRFs)[38] occur at high levels in patients with certain mitochondrial myopathies. RRFs are the result of an abnormal accumulation of mitochondria and are identifiable by staining for SDH (complex II of

ETS) activity. For example, the abundances of RRFs range from 0.5 to 18% of total fibers in patients with a disorder known as mitochondrial encephalomyopathy, lactic acidosis, and stroke-like episodes (MELAS)[39] and 1 to 32% in patients with chronic progressive external ophthalmoplegia (CPEO).[40] Deficiencies in the staining intensity complex IV of ETS give rise to so-called "COX⁻" fibers, which are also often detected in patients with mitochondrial myopathies and frequently colocalize with RRFs.

Although these mitochondrial defects reach relatively high levels in myopathy patients, they also are reported to occur at lower levels in normal aged humans without muscular disease.[41-43] RRFs were found to increase with age from 0.02% of the fibers in quadriceps from young persons (mean age = 25 years) to 0.33% of fibers in quadriceps biopsies of older individuals (mean age = 67 years).[43] Likewise, the number of COX⁻ fibers increased from 5 to 54 defects/cm^2 in limb muscle from younger individuals (< 60 yrs) compared with 80- to 90-year-olds.[41]

Although mitochondrial enzymatic activities have been studied biochemically as a function of age in laboratory animals,[44-48] they have been evaluated histologically far less extensively. We determined localized COX and SDH activities in the vastus lateralis of the young, CR_{35}, and CR_{50} rats and, in addition, analyzed the vastus lateralis from 5 *ad libitum* fed 29-month old Lobund Wistar rats. In the old CR_{35}, CR_{50}, and *ad libitum* rats, individual fibers were detected that were COX⁻, stained hyperreactive for SDH activity (SDH^{++}, indicative of a RRF), or were combined COX⁻ and SDH^{++} (COX^-/SDH^{++}). The CR_{35} vastus lateralis samples contained an average of 1.5 COX⁻, 2.0 SDH^{++}, and 1.1 COX^-/SDH^{++} fibers per section of vastus lateralis. The CR_{50} rats, however, contained 5- to 10-fold fewer of these abnormalities ($p \leq 0.012$ versus CR_{35} rats).[16] The 5 AL-fed 29-month-old rats examined had somewhat higher levels of COX and SDH abnormalities than did the CR_{35} rats, with an average of 2.0 COX⁻, 2.4 SDH^{++}, and 1.8 COX^-/SDH^{++} fibers per section of vastus lateralis. None of these abnormalities was detected in the vastus lateralis of the young rats.[16] It should be noted that the levels of COX⁻ and/or SDH^{++} fibers we detected may be considerable underestimates of the total number of ETS abnormalities, inasmuch as only two 8-μm cross-sections were characterized per muscle. This 16 μm distance represents only ~0.1% of the entire vastus lateralis.

CONCLUSIONS

Based on its capacity to greatly extend maximum life span and to keep animals physiologically "younger longer," most gerontologists share the view that CR is the most successful intervention tested, to date, in mammals.[1,2] Our analyses in mice subjected to CR early in life as well as in rats subjected to CR in late middle age demonstrated that CR diminishes age-associated fiber loss in certain muscles. In addition, CR initiated in the 17-month-old rats could also attenuate histological ETS enzymatic abnormalities and changes in fiber-type composition, delay the onset of histologic atrophy, and reduce the accumulation of mtDNA deletions. Collectively, these data indicate that CR can oppose several age-associated declines commonly observed in mammalian skeletal muscles.

REFERENCES

1. MASORO, E.J. 1988. Food restriction in rodents: An evaluation of its role in the study of aging. J. Gerontol. **43**: B59–B64.

2. WEINDRUCH, R. & R.L. WALFORD. 1988. The Retardation of Aging and Disease by Dietary Restriction. Charles C. Thomas. Springfield, IL.
3. WEINDRUCH, R. 1996. Caloric restriction and aging. Sci. Am. **274:** 46–52.
4. SOHAL, R.S. & R. WEINDRUCH. 1996. Oxidative stress, caloric restriction, and aging. Science **273:** 59–63.
5. CHANCE, B., H. SIES & A. BOVERIS. 1979. Hydroperoxide metabolism in mammalian organs. Phys. Rev. **59:** 527–605.
6. FAULKNER, J.A., S.V. BROOKS & E. ZERBA. 1995. Muscle atrophy and weakness with aging: Contraction-induced injury as an underlying mechanism. J. Gerontol. **50A:** 124–129.
7. CARLSON, B.M. 1995. Factors influencing the repair and adaptation of muscles in aged individuals: Satellite cells and innervation. J. Gerontol. **50A:** 96–100.
8. LARSSON, L. 1995. Motor units: Remodeling in aged animals. J. Gerontol. **50A:** 91–95.
9. WEINDRUCH, R. 1995. Interventions based on the possibility that oxidative stress contributes to sarcopenia. J. Gerontol. **50A:** 157–161.
10. HEYMSFIELD, S.B., D. GALLAGHER, M. VISSER, C. NUNEZ & Z.M. WANG. 1995. Measurement of skeletal muscle: Laboratory and epidemiological methods. J. Gerontol. **50:** 23–29.
11. YU, B.P., E.J. MASORO, I. MURATA, H.A. BERTRAND & F.T. LYND. 1982. Life span study of SPF Fischer 344 male rats fed ad libitum or restricted diets: Longevity, growth, lean body mass and disease. J. Gerontol. **37:** 130–141.
12. GARTHWAITE, S.M., H. CHENG, J.E. BRYAN, B.W. CRAIG & J.O. HOLLOSZY. 1986. Ageing, exercise and food restriction: Effects on body composition. Mech. Ageing Dev. **36:** 187-196.
13. HOLLOSZY, J.O., M. CHEN, G.D. CARTEE & J.C. YOUNG. 1991. Skeletal muscle atrophy in old rats: Differential changes in the three fiber types. Mech. Ageing Dev. **60:** 199-213.
14. ANSVED, T. & L. LARSSON. 1989. Effects of ageing on enzyme-histochemical morphometrical and contractile properties of the soleus muscle in the rat. J. Neuro. Sci. **93:** 105–124.
15. WALTER, T.J., H.L. SWEENEY & R.P. FARRAR. 1990. Aging does not affect contractile properties of type IIb FDL muscle in Fischer 344 rats. Am. J. Phys. **258:** C1031–C1035.
16. ASPNES, L.E., C.M. LEE, R. WEINDRUCH, S.S. CHUNG, E.B. ROECKER & J.M. AIKEN. 1997. Caloric restriction reduces fiber loss and mitochondrial abnormalities in aged rat muscle. FASEB J. **11:** 573–581.
17. SNYDER, D.L., M. POLLARD, B.S. WOSTMANN & P. LUCKERT. 1990. Life span, morphology, and pathology of diet-restricted germ-free and conventional Lobund-Wistar rats. J. Gerontol. **45:** B52–B58.
18. SELIGMAN, A.M., M.J. KARNOVSKY, H.L. WASSERKRUG & J.S. HANKER. 1968. Nondroplet ultrastructural demonstration of cytochrome oxidase activity with a polymerizing osmiophilic reagent, diaminobenzidine (DAB). J. Cell Biol. **38:** 1–14.
19. BLACKWELL, B.N., T.J. BUCCI, R.W. HART & A. TURTURRO. 1995. Longevity, body weight, and neoplasia in *ad libitum*-fed and diet-restricted C57BL/6 mice fed NIH-31 open formula diet. Toxicol. Pathol. **23:** 570–582.
20. TAUCHI, H., T. YOSHIOKA & H. KOBAYASHI. 1971. Age change of skeletal muscles of rats. Gerontologia **17:** 219–227.
21. CACCIA, M.R., J.B. HARRIS & M.A. JOHNSON. 1979. Morphology and physiology of skeletal muscle in aging rodents. Muscle & Nerve **2:** 202–212.
22. LARSSON, L. & L. EDSTRÖM. 1986. Effects of age on enzyme-histochemical fibre spectra and contractile properties of fast- and slow-twitch skeletal muscles in the rat. J. Neurol. Sci. **76:** 69–89.
23. ARMSTRONG, R.B. & R.O. PHELPS. 1984. Muscle fiber type composition of the rat hindlimb. Am. J. Anat. **171:** 259–272.
24. FUJISAWA, K. 1974. Some observations on the skeletal musculature of aged rats. J. Neurol. Sci. **22:** 353–366.
25. MANTEL, N. & W. HAENSZEL. 1959. Statistical aspects of the analysis of data from retrospective studies of disease. J. Natl. Cancer Inst. **22:** 719–748.
26. CLAYTON, D.A. 1982. Replication of animal mitochondrial DNA. Cell **28:** 693–705.
27. WALLACE, D.C. 1992. Diseases of the mitochondrial DNA. Annu. Rev. Biochem. **61:** 1175–1212.
28. LEE, C.M., R. WEINDRUCH & J.M. AIKEN. 1997. Age-associated alterations of the mitochondrial genome. Free Radical Biol. Med. **22:** 1259–1269.

29. LINNANE, A.W., S. MARZUKI, T. OZAWA & M. TANAKA. 1989. Mitochondrial DNA mutations as an important contributor to ageing and degenerative diseases. Lancet **1:** 642–645.

30. CORTOPASSI, G.A. & N. ARNHEIM. 1990. Detection of a specific mitochondrial DNA deletion in tissues of older humans. Nucleic Acids Res. **18:** 6927–6933.

31. WALLACE, D.C. 1992. Mitochondrial genetics: A paradigm for aging and degenerative diseases? Science **256:** 628–632.

32. SCHWARZE, S.R., C.M. LEE, S.S. CHUNG, E.B. ROECKER, R. WEINDRUCH & J.M. AIKEN. 1995. High levels of mitochondrial DNA deletions in skeletal muscle of old rhesus monkeys. Mech. Ageing Dev. **83:** 91–101.

33. MÜLLER-HÖCKER, J., P. SEIBEL, K. SCHNEIDERBANGER & B. KADENBACH. 1993. Different *in situ* hybridization patterns of mitochondrial DNA in cytochrome *c* oxidase-deficient extraocular muscle fibres in the elderly. Virchows Arch. A Pathol. Anat. Histol. **422:** 7–15.

34. JOHNSTON, W., G. KARPATI, S. CARPENTER, D. ARNOLD & E.A. SHOUBRIDGE. 1995. Late-onset mitochondrial myopathy. Ann. Neurol. **37:** 16–23.

35. LEE, C.M., S.S. CHUNG, J.M. KACZKOWSKI, R. WEINDRUCH & J.M. AIKEN. 1993. Multiple mitochondrial DNA deletions associated with age in skeletal muscle of rhesus monkeys. J. Gerontol. **48:** B201–B205.

36. LEE, C.M., M. LOPEZ, R. WEINDRUCH & J.M. AIKEN. Association of age-related mitochondrial abnormalities with skeletal muscle fiber atrophy. Manuscript submitted.

37. DIMAURO, S. 1993. Mitochondrial encephalomyopathies. *In* Molecular and Genetic Basis of Neurological Disease. R.N. Rosenberg, S.B. Prusiner, S. DiMauro, R.L. Barchi & L.M. Kunkel, Eds.: 665–694. Butterworth-Heinemann. Boston, MA.

38. ENGEL, W.K. & G.G. CUNNINGHAM. 1963. Rapid examination of muscle tissue. An improved trichrome method for fresh-frozen biopsy sections. Neurology **13:** 919–923.

39. MORAES, C.T., E. RICCI, E. BONILLA, S. DIMAURO & E.A. SCHON. 1992. The mitochondrial tRNA$^{Leu(UUR)}$ mutation in mitochondrial encephalomyopathy, lactic acidosis, and strokelike episodes (MELAS): Genetic, biochemical, and morphological correlations in skeletal muscle. Am. J. Hum. Genet. **50:** 934–949.

40. PRELLE, A., G. FAGIOLARI, N. CHECCARELLI, M. MOGGIO, A. BATTISTEL, G.P. COMI, P. BAZZI, A. BORDONI, M. ZEVIANI & G. SCARLATO. 1994. Mitochondrial myopathy: Correlation between oxidative defect and mitochondrial DNA deletions at single fiber level. Acta Neuropathol. **87:** 371–376.

41. MÜLLER-HÖCKER, J. 1990. Cytochrome *c* oxidase deficient fibres in the limb muscle and diaphragm of man without muscular disease: An age-related alteration. J. Neurol. Sci. **100:** 14–21.

42. MÜLLER-HÖCKER, J., K. SCHNEIDERBANGER, F.H. STEFANI & B. KADENBACH. 1992. Progressive loss of cytochrome *c* oxidase in the human extraocular muscles in ageing—A cytochemical-immunohistochemical study. Mutat. Res. **275:** 115–124.

43. RIFAI, Z., S. WELLE, C. KAMP & C.A. THORNTON. 1995. Ragged red fibers in normal aging and inflammatory myopathy. Ann. Neurol. **37:** 24–29.

44. TORII, K., S. SUGIYAMA, K. TAKAGI, T. SATAKE & T. OZAWA. 1992. Age-related decrease in respiratory muscle mitochondrial function in rats. Am. J. Respir. Cell. Mol. Biol. **6:** 88–92.

45. BOWLING, A.C., E.M. MUTISYA, L.C. WALKER, D.L. PRICE, L.C. CORK & M.F. BEAL. 1993. Age-dependent impairment of mitochondrial function in primate brain. J. Neurochem. **60:** 1964–1967.

46. SUGIYAMA, S., M. TAKASAWA, M. HAYAKAWA & T. OZAWA. 1993. Changes in skeletal muscle, heart and liver mitochondrial electron transport activities in rats and dogs of various ages. Biochem. Mol. Biol. Int. **30:** 937–944.

47. TAKASAWA, M., M. HAYAKAWA, S. SUGIYAMA, K. HATTORI, T. ITO & T. OZAWA. 1993. Age-associated damage in mitochondrial function in rat hearts. Exp. Gerontol. **28:** 269–280.

48. DESAI, V.G., R. WEINDRUCH, R.W. HART & R.J. FEUERS. 1996. Influences of age and dietary restriction on gastrocnemius electron transport system activities in mice. Arch. Biochem. Biophys. **333:** 145–151.

The Effects of Dietary Restriction on Mitochondrial Dysfunction in Aging[a]

RITCHIE J. FEUERS[b]

Department of Genetic Toxicology, National Center for Toxicological Research, 3900 NCTR Road, Jefferson, Arkansas 72079

ABSTRACT: Age-associated alterations in the mitochondrial electron transport system (ETS) may lead to free radical generation and contribute to aging. The complexes of the ETS were screened spectrophotometrically in gastrocnemius of young (10 month) as well as older (20 and 26 month) B6C3F1 female mice fed an *ad libitum* (AL) diet or a restricted (DR) in total calories diet (40% less food than AL mice). The activities of complexes I, III, and IV decreased significantly by 62%, 54%, and 74%, respectively, in old AL mice (AL_{20}) compared to young AL mice (AL_{10}). Complexes I, III, and IV from DR_{10} mice had activities that were significantly lower than those seen in AL_{10} mice (suggesting a lower total respiratory rate or improved efficiency). By contrast, complex II activity did not decrease with age (actually increased, but not significantly) in AL_{20} mice. Complex II was decreased across age in DR mice. K_m for ubiquinol-2 of complex III was significantly increased in AL_{10} animals (0.33 mM vs. 0.26 mM in DR_{10} mice) and was further increased with aging (0.44 mM in AL_{20} vs. 0.17 mM in DR_{20} mice). This suggests obstruction of binding, inhibition of electron flow in aging, which could yield premature product release as a free radical. Total complex IV by V_{max} was highest in AL_{10} mice, but the proportion of complex as high-affinity sites was lower (69%) than in either DR_{10} (80%) or DR_{20} (80%). The percentage of high-affinity sites decreased to only 45% in AL_{20} mice, and V_{max} was reduced by 75 percent. In AL_{26} mice high-affinity sites decreased to 33 percent. At physiologic concentration of reduced cytochrome *c*, significant dysfunction of complex IV in AL_{20} or AL_{26} mice would be expected with obstruction of overall electron transport. The age-associated loss of activity and function of complexes I, III, and IV may contribute to increased free radical production. Lack of sufficient DNA repair in mitochondria and juxtaposition to the ETS adds to susceptibility and accumulation of mtDNA and other mitochondrial macromolecular damage. DR seems to retard this deterioration of mitochondrial respiratory function by preserving enzymatic activities and function.

It was initially suggested by Harman[1] that free radicals may induce molecular damage and contribute to aging, and a growing body of evidence suggests that oxidative metabolism plays a central role in any mechanism of aging. The generation of ATP takes place in mitochondria where four electrons are added to O_2 with subsequent production of H_2O. However, it has been estimated that there may be as much as a 1–2% error for 1 electron additions.[2] Thus, electron transport (ETS) and oxidative phosphorylation have evolved as efficient systems, but a significant number of reactive oxygen species (ROS) may be produced over time within mitochondria.[3–7] In an escalating process, mitochondria may act as an initiating source of ROS, promoting their own disarray and destruction.[8]

[a]This work was supported in part by the National Center for Toxicological Research and the National Institute on Aging.
[b]Tel: 870/543-7437 or -7330; fax: 870/543-7136; e-mail: rfeuers@nctr.fda.gov

It can be argued that the rate of aging is dependent upon the balance between ROS and quantitative and qualitative aspects of cellular antioxidant systems. These opposing forces produce an acceptable steady state level of ROS due to antioxidant advantages in young animals.[9] However, this steady state level of ROS may increase with age as antioxidant systems are compromised.[10] For example, it has been shown that quantitative and qualitative activities of glucose 6-phosphate dehydrogenase, glutathione peroxidase, and catalase are compromised with age,[11,12] and in such oxidatively active tissues as skeletal muscle, antioxidant enzyme systems are induced within the mitochondrial fraction.[13]

Ames[14] has estimated that the number of oxidative lesions per cell per day to the DNA may be as high as 100,000 in the rat. This damage has been shown to be rather ubiquitous because it involves not only nuclear DNA damage, but damage to mtDNA, proteins,[15] membranes, and other macromolecules. However, much of the damage is repaired.[14] Unfortunately, as with the antioxidant systems, these repair processes are not perfectly efficient and limited, as is the case with mtDNA repair. It has been shown that as many as 2 million DNA lesions/cell may accumulate in aging (2 years) rat tissue,[16] and protein lesions also accumulate with age.[15] Inasmuch as oxidative damage increases and mito-chondrial function seem to decay with age,[14] a case may be made that mitochondrial dys-function plays a central role in any mechanism of aging. A growing literature suggests that alterations in mitochondrial ETS may be involved. Decreases with age in the activities of the four ETS complexes have been demonstrated for human skeletal muscle,[17–19] whereas declines in the activities of the ETS complexes with age in rat muscle have also been noted.[8,20,21]

Dietary restriction (DR) extends maximum life span and retards both the rate of biolog-ical aging and the development of age-associated degenerative diseases.[22–24] The mecha-nism of DR's action is not clear, but it has been speculated that, in part, DR may act by reducing mitochondrial free radical generation.[7,8,23–25]

MATERIAL AND METHODS

Animals and Housing

Female B6C3F1 mice were bred and raised in a specific pathogen-free animal facility at the National Center for Toxicological Research (NCTR), using procedures described previously.[26] This is a long-lived genotype. Females at the NCTR colony fed the NIH-31 diet *ad libitum* have an average life span of 30 months and a maximum life span (*i.e.*, the mean for the longest-lived decile) of 36 months. Animals were maintained individually in plastic cages with wire metal tops and hardwood chip bedding at 23°C. Cages were changed weekly, and fresh water was always available.

Experimental Design and Dietary Regimen

All mice were weaned at 3 weeks of age and fed the standard NIH-31 diet (containing 0.315% vitamin mixture and 0.185% mineral mixture) until they reached 14 weeks of age. The mice were then randomly assigned to either a control group, which was fed the stan-dard NIH-31 diet *ad libitum* (AL) or a group subjected to DR, which received 60% of the

AL intake of the NIH-31 diet supplemented with vitamins and a mineral mixture at 1.67 times that in the standard diet. In our colony, this DR regimen increases the average and maximum life spans to 41 and 47 months.

Tissue Collection and Preparation

A total of 40 mice were studied: 18 (n = 9 AL, n = 9 DR) were 10-month-old mice, 18 (n = 9 AL, n = 9 DR) were 20-month-old mice, and 4 were 26-month-old mice fed the AL diet. For two weeks prior to experimentation, DR animals were fed at the beginning of the dark span (1600–0400 h). All mice were killed by cervical dislocation between 1800 and 1900 hours. The gastrocnemius muscle was removed, frozen in liquid nitrogen, and then stored at –70°C. Later, the tissues were thawed on ice, weighed, minced, and homogenized using a Tekmar homogenizer. All tissues were homogenized at 4°C in 50 mM phosphate buffer (pH 7.4) in volumes 10 times the weight of tissue. Crude homogenates were centrifuged at 1000 g for 10 min at 4°C in a Beckman Model LS-75 centrifuge using a 50 Ti rotor. The supernatants were collected and centrifuged again at 12,000 g for 10 min at 4°C. The resultant pellets were suspended in 1 mL of buffer (250 mM mannitol, 70 mM sucrose, 1 mM EDTA, pH 7.4) and recentrifuged at 12,000 g for 10 min at 4°C. Finally, the 12,000 g pellets were resuspended in 0.5 mL of second buffer at –70°C until time of assay.

Enzyme Analyses

All four complexes of the mitochondrial ETS were measured spectrophotometrically on a COBAS FARA II autoanalyzer (Roche, New York) using modifications of the method of Ragan et al.[27] All the chemicals except ubiquinol were from Sigma (St. Louis, MO). Ubiquinol was graciously supplied by Dr. T. Ichien (Eisai Co., Tokyo, Japan). Complex I (NADH-ubiquinone oxidoreductase; EC 1.6.99.3) activity was measured by monitoring the oxidation of NADH to NAD by ubiquinone-1 at 30°C. The reaction mixture contained potassium phosphate buffer (KPB, 10 mM, pH 8.0), NADH (5 mM), lecithin (15 mg/mL), and the mitochondrial fraction (9 µL). The reaction was started by addition of ubiquinone-1 (10 mM) to the reaction mixture, and the decrease in absorbance was measured at 340 nm. Complex II (succinate-ubiquinone oxidoreductase; EC 1.3.5.1) activity was measured as a rate of reduction of ubiquinone-2 by succinate at 30°C followed by the secondary reduction of 2,6-dichlorophenolindophenol (DPIP) by the ubiquinol formed. The reaction mixture contained KPB (1 mM, pH 7.4) and distilled water. The decrease in absorbance was measured at 600 nm after the addition of a mixture of DPIP (1 mM), ubiquinone (2.5 mM), and the mitochondrial fraction (4 µL) to the above reaction mixture. Complex III (ubiquinol-cytochrome c oxidoreductase; EC 1.10.2.2) activity was assayed by following the rate of reduction of cytochrome c by ubiquinol-2 at 30°C. The reaction was started by addition of ubiquinol-2 to KPB (50 mM), EDTA (100 mM), cytochrome c (1 mM), and the mitochondrial fraction (5 µL) diluted 1:10 with sucrose (250 mM): Tris-HCl (10 mM) buffer (pH 7.8) containing potassium cyanide (50 mM). The reduced cytochrome c was measured as an increase in absorbance at 550 nm. Complex IV (cytochrome c oxidase; EC 1.9.3.1) activity was measured as a rate of oxidation of reduced cytochrome c (10 µM) by the enzyme at 30°C. The decrease in reduced cytochrome c was monitored at 550 nm. Protein concentrations were determined spectrophotometrically by the biuret method using a kit provided by COBAS FARA and bovine serum albumin as a standard.

Michaelis-Menten Kinetic Analysis

Complexes III and IV were measured at various ubiquinol-2 concentrations (0.03–3.0 mM) and reduced cytochrome c concentrations (0.5–50 µM), respectively, under the assay conditions described above. Mathematical analysis was performed using the methods described by Segel,[28] using a nonlinear curve-fitting program in Sigma Plot 5.0 (Jandel Scientific, Corte Madera, CA). K_m and V_{max} were calculated using the Hill equation for complex III and modifications of the Hill equation for a situation where two enzymes act on a single substrate for complex IV.

Statistical Analysis

All results are expressed as means ± standard error of the mean (SEM). Data were analyzed statistically by using the Student's independent or paired t-test, and p values of 0.05 or less were considered statistically significant. K_m and V_{max} values were calculated for each sample and then summed for each group prior to statistical analysis. Additionally, data from each sample were pooled prior to calculation of K_m and V_{max}. Statistical significance was similar regardless of approach.

RESULTS

The specific activity of complex I from gastrocnemius was determined in 10- and 20-month-old AL and DR mice.[8] The 20-month-old animals fed *ad libitum* (AL_{20}) displayed a significant decline (65%, $p < 0.05$) in enzyme activity compared to 10-month-old mice fed *ad libitum* (AL_{10}). Ten-month-old DR mice (DR_{10}) showed much lower enzyme activities compared to AL_{10} (71% decrease, $p < 0.05$). There was no age-associated change in activity for complex I for the 20-month-old DR mice (DR_{20}). For the AL-fed mice there was no significant age-associated change in specific activity of complex II. At 10 months of age there was no influence of diet on complex II activity. However, at 20 months of age mice on DR showed a significantly (69%, $p < 0.05$) lower enzyme activity compared to AL-fed mice. Similar to complex I, AL_{20} mice showed an overt age-associated decrease (55%, $p < 0.05$) in complex III activity compared to AL_{10} mice. At 10 months of age the activity of complex III was lower (40%, $p < 0.05$) in mice subjected to DR than in controls. Among the four complexes studied, complex IV showed the most severe reduction in activity between 10 and 20 months of age in the AL-fed mice (75%, $p < 0.001$). The activity observed in 26-month-old AL mice was also very low compared to AL_{10} mice. The influence of DR was age dependent. At 10 months of age, the activity of preparations from DR mice was less (36%, $p < 0.05$) than that from AL_{10} mice. However, at 20 months of age, the average activity of samples from DR mice exceeded (121%, $p < 0.05$) that of AL_{20} mice. The activities of DR_{10} and DR_{20} mice did not differ.[8]

TABLE 1. K_m and V_{max} Determinations for Complex III[8,a]

$Diet_{Age}$	V_{max}	K_m
AL_{10}	214.1 ± 7.7^a	$0.34 \pm .03^b$
DR_{10}	160.6 ± 12.2^b	$0.26 \pm .05^c$
AL_{20}	149.9 ± 7.6^b	$0.44 \pm .05^a$
DR_{20}	243.8 ± 8.3^a	$0.18 \pm .02^c$

[a]K_m and V_{max} were calculated directly from velocity versus ubiquinol-2 concentration data using the Michaelis-Menten equation ($v = V_{max} [S]/K_m + [S]$). V_{max} units = U/mg protein. K_m units = nM ubiquinol-2. Different superscript letters indicate statistically significant differences ($p < 0.05$).

Kinetic analysis of complex III indicated an age-associated reduction in the binding affinity of the enzyme to its substrate in AL mice (TABLE 1). A higher K_m value (0.44 ± 0.05 mM) was found for AL_{20} than for AL_{10} (0.34 ± 0.03 mM). In mice subjected to DR, the K_m values were significantly lower than in AL mice in both 10 months (0.26 ± 0.05 mM) and 20 months (0.18 ± 0.02 mM) of age. Total complex III present as indicated by V_{max} value suggests that old mice had higher amounts of complex III than all other groups ($p < 0.05$). The number of binding sites was calculated to be one for all age and diet groups.

Biphasic kinetic behavior, which is characteristic for complex IV, was observed. This results from the existence of both high- and low-affinity binding sites for complex IV. The K_m of the high-affinity binding site ranged from 5.8 to 8.9 μM, whereas the K_m of the low-affinity binding site was 21 μM in all 10- and 20-month groups (TABLE 2). At 26 months, the K_m for the high-affinity site increased to 14.0 μM. The K_m for the low-affin-

TABLE 2. K_m and V_{max} Determinations for Complex IV[8,a]

$Diet_{Age}$	High affinity		Low affinity		Percent High-affinity Sites
	V_{max}	K_m	V_{max}	K_m	
AL_{10}	41.2 ± 4.0^a	8.9 ± 1.2^a	19.8 ± 2.4^a	21.3 ± 0.2^a	68
DR_{10}	31.7 ± 5.1^b	7.6 ± 2.4^a	8.1 ± 2.8^b	21.5 ± 0.3^a	80
AL_{20}	7.2 ± 0.6^c	5.8 ± 1.8^a	8.5 ± 0.4^b	21.4 ± 0.2^a	46
DR_{20}	33.4 ± 3.2^b	7.9 ± 1.6^a	8.4 ± 1.3^b	21.4 ± 0.3^a	80
AL_{26}	3.9 ± 0.7^d	14.0 ± 0.4^b	8.0 ± 3.8^b	28.4 ± 32.2^a	33

[a]K_m and V_{max} values were calculated from velocity data versus reduced cytochrome c concentrations using the Hill equation modeled for two enzymes acting on a single substrate ($v = V_{max1} [S]^n/K_{m1} + [S]^n + V_{max2} [S]^n/K_{m2} + [S]^n$). The number of binding sites per complex (n in the Hill equation) is not reported but ranged between 1.4 and 1.8 in all cases except for low-affinity sites of AL mice where n = 12. V_{max} units = nmol/min/mg protein. K_m units = μM reduced cytochrome c. Different superscript letters indicate statistically significant differences ($p < 0.05$).

ity site for samples from AL_{26} mice could not be confidently determined using the Hill equation. For AL mice, the V_{max} attributed to high-affinity sites declined from 83% from 10 to 20 months of age and was 91% less than the 10-month value for AL_{26} mice. No age-associated changes in V_{max} for high-affinity sites was observed in the DR mice; however, these values were 19–23% less than values for AL_{10} mice. The V_{max} attributable to low-affinity sites was highest in the AL_{10} mice, with values for AL_{20}, AL_{26}, DR_{10}, and DR_{20} being ~50% lower. The fraction of complex IV binding sites of high affinity was significantly reduced to 46% in AL_{20} compared to values of 69 and 80% in AL_{10} and DR_{10} mice, respectively. The DR_{20} mice maintained this same level (80%) of high-affinity binding sites. The fraction of high-affinity sites for AL_{26} mice was 33 percent.

DISCUSSION

Major declines ranging from 54 to 74% in the specific activities of complexes I, III, and IV in mitochondria prepared from the gastrocnemius muscle of 20-month-old mice (late middle age in this long-lived strain) compared to 10-month old animals were observed. By contrast, complex II did not show statistically significant changes with age. Decreased complex I and IV activities agrees with the data of Torri et al.[20] who studied mitochondria from a limb muscle (psoas major) of young Wistar rats of three ages (~2, 8, and 13 months). The activities found for complexes I and IV in the 13-month-old rats were only 49 and 76%, respectively, that of the 2-month-old rats. Significant age-related changes in complexes II and III were not observed. Therefore, it is clear that large decreases in mitochondrial ETS activities occur many months before these animals are old. This group extended their findings to include older rats (21–23 months) and did not observe further changes in activities from those recorded at 13 months of age.[21] Human skeletal muscle has also been studied [29] for the influences of aging on ETS activities, with the most consistent declines reported for complexes I and IV.[17–19]

The present data[8] are the first to describe the influence of life span-prolonging DR on the ETS activities and kinetic properties in muscle mitochondria. Several strong changes were induced by DR. At 10 months of age, the activities of complexes I, III, and IV were 33–64% lower in DR mice than in AL mice. By contrast, at 20 months of age, the activities for DR mice for complexes I and III were not different from those of controls, and complex IV activity was 53% higher than the control value. This outcome at 20 months was largely a consequence of 54–75% declines in complex I, III, and IV activities between 10 and 20 months in AL mice, whereas no significant changes were occurring in the activities of DR mice. The kinetic analysis of complex III revealed a 29% increase in the K_m for ubiquinol-2 from 10 to 20 months in AL mice with no change in K_m in DR mice. A 90% decline was observed in the V_{max} calculated for the high-affinity site of complex IV from 10 to 26 months of age in AL mice, and through 20 months of age, this change was attenuated by DR. The K_m for high-affinity sites was uninfluenced by age or diet through 20 months but was approximately twofold higher for the 26-month AL mice. Therefore, this increase in K_m for complex IV represents an age-associated change for AL mice not occurring until after 20 months of age in this model. Another kinetic parameter, the percentage of total binding sites, which were of high affinity, fell progressively in AL mice from 68%

at 10 months, to 46% at 20 months and 33% at 26 months. This value was 80% for DR mice at both ages (10 and 20 months) studied.

To date, gerontologic studies of the ETS have emphasized activities of the complexes. Although valuable, activity data are less able than are kinetic data to elucidate mechanistic properties of these enzyme complexes. Complex I consists of 41 subunits, 7 of which are encoded by mtDNA, whereas complex II is unique in its being the smallest complex (4 subunits) all of nuclear origin.[30] With limited remaining sample, the kinetic properties of complexes III and IV were studied. Complex III consists of 11 subunits (10 of nuclear origin), with the one of mitochondrial origin (cytochrome b apoprotein) being a redox center and constituting 18% of the mass of complex III.[31,32] Complex IV consists of 13 subunits, 3 of which (subunits 1, 2, and 3) are encoded by mtDNA and constitute 55% of the mass of complex IV. Subunits 1 and 2 contain redox centers whereas subunit 3 acts in proton transport.[18,33] The binding site is formed by subunit 2 from one monomer and subunit 3 from the adjacent monomer.[31]

The age-associated reduction in the V_{max} for complex III from AL muscle suggests a decrease in total enzyme content. By contrast, V_{max} actually increased over 10 to 20 months in DR mice. This is especially impressive when coupled to the maintenance of a high affinity for substrate with age in the DR mice, suggesting not only more of the complex, but higher catalytic efficiency. Conversely, in the AL situation, K_m increased while V_{max} fell, indicating less of the complex with lower catalytic efficiency. A reduced potential for substrate binding, or the poorly controlled binding just described, may contribute to age-associated increases in superoxide and hydrogen peroxide production (which were ameliorated by DR), as Sohal's group recently reported.[10] A large part of mitochondrial free radical production is thought to arise between NADH dehydrogenase and ubiquinone/cytochrome b^2. If ubiquinol-2 binding is poor, two predictions might follow: (1) an increase in ubiquinol-2 levels would occur in order to achieve "youthful" activities and, (2) free radical production would increase due to premature release of a free radical product. Additionally, it is known that the cytochrome b apoprotein is associated with the binding site for ubiquinol-2. Changes in the ability of this molecule to bind substrate may imply structural changes within this mtDNA-coded subunit.

The active site of complex IV is composed by subunits of mtDNA origin. Therefore, it may be argued that the lack of age or dietary changes in K_m for either high- or low-affinity sites through 20 months of age indicates a lack of structural change in these subunits. This could suggest that the mtDNA regions encoding these proteins had accumulated insufficient damage to yield reduced substrate affinity. However, at 26 months, K_m did increase substantially in AL mice, which supports the possibility of mtDNA damage of consequence. The V_{max} data indicate that the amount of total enzyme complex fell with age in AL mice while the proportion of high-affinity sites decreased. In as much as the number of high-affinity sites decrease at a faster rate than the low-affinity sites, it can be argued that some catastrophic alteration (or array of alterations, including modifications of membrane structures affecting these membrane-bound complexes) results in eventual loss of activity at the high-affinity site. Interestingly, the proportion of total V_{max} derived from high-affinity sites was marginally higher at 10 months of age in DR mice (80% vs. 68%), and this same level was maintained in DR_{20} mice. In short, all of the kinetic changes observed with age in AL mice were opposed by DR. These differences that occur with DR would improve the ability of the complex to catalyze conversions of substrate to product at lower concentrations of reduced cytochrome c.

Age-associated ETS dysfunction, especially decreases in total complex coupled with decreased binding affinity, and the accumulation of a higher proportion of low-affinity sites provide a potential for a biochemical mechanism for age-associated increases in free radical generation. Additionally, alterations in catalysis would likely obstruct normal electron flow (a kind of electron "traffic jam"), further increasing the potential for free radical generation. Support for this idea comes from Sohal,[34] who found that partial inhibition of complex IV in *Drosophila* stimulated mitochondrial hydrogen peroxide production. Any of the age-associated alterations in ETS activities or kinetic parameters could cause reduced ATP synthesis and compromise cellular function. All of the age-associated alterations were opposed by DR.

The data presented herein illustrates a situation where free radicals are generated at low rates in the younger animal with abundant and efficient ATP generation. The data suggest that there is an obstruction of electron flow though complex I as the amount of the complex diminishes, and through complex III due to loss of total complex and increases in K_m with age. This builds a kind of a dam against electron flow (most notably at complex III). Thus, fewer electrons can pass through the ETS and ultimately to complex IV. During aging, high-affinity binding sites of complex IV are lost, further impeding electron flow. It is suggested that due to these problems more free radicals would be generated at these sites along the ETS. One can envision that as oxidative lesions accumulate, a mechanism exists where a single electron is passed, and due to poor binding, a premature free radical product would be released prior to the second electron passage. Caloric restriction resolves these qualitative and quantitative aging problems associated with complexes I, III, and IV, and this may be one mechanism through which caloric restriction limits free radical generation, leading to extension of maximum achievable life span.

It is suggested that future efforts should center on identification of new nutritional and endocrine interventions that act to improve enzyme efficiency. These types of interventions would obviously be compatible and perhaps act synergistically with emerging antioxidant interventions.

REFERENCES

1. HARMAN, D. 1956. Aging: A theory based on free radical and radiation chemistry. J. Gerontol. **11:** 298–300.
2. CHANCE, B., H. SIES & A. BOVERIS. 1979. Hydroperoxide metabolism in mammalian organs. Physiol. Rev. **59:** 527–603.
3. FLEMING, J.E., J. MIQUEL, S.F. COTTRELL, L.S. YENGOYAN & A.C. ECONOMOS. 1982. Is cell aging caused by respiration-dependent injury to the mitochondrial genome? Gerontology **28:** 44–53.
4. LINNANE, A.W., S. MARZUKI, T. OZAWA & M. TANAKA. 1989. Mitochondrial DNA mutations as an important contributor to ageing and degenerative diseases. Lancet **i:** 642–645.
5. BANDY, B. & A.J. DAVISON. 1990. Mitochondrial mutations may increase oxidative stress: Implications for carcinogenesis and aging. Free Radical Biol. Med. **8:** 523–539.
6. SOHAL, R.S. & U.T. BRUNK. 1992. Mitochondrial production of pro-oxidants and cellular senescence. Mutat. Res. **275:** 295–304.
7. FEUERS, R.J., R. WEINDRUCH & R.W. HART. 1993. Caloric restriction, aging and antioxidant enzymes. Mutat. Res. **295:** 191–200.
8. DESAI, V.G., R. WEINDRUCH, R.W. HART & R.J. FEUERS. 1996. Influence of age and dietary restriction on gastrocnemius electron transport system activities in mice. Arch. Biochem. Biophys. **333:** 145–151.

9. BOVERIS, A. 1997. Mitochondrial Production of Oxyradicals, mtDNA Damage and Aging [abstract]. The 16th Congress of the International Association of Gerontology, Association of Gerontology, Adelaide, Aus. **423**.

10. SOHAL, R.S., H.-H. KU, S. AGARWAL, M.J. FORSTER & H. LAL. 1994. Effect of age and caloric restriction on DNA oxidative damage in different tissues of C57BL/6 mice. Mech. Ageing Dev. **74:** 121–133.

11. ORIAKY, E.T., F. CHEN, V.G. DESAI, J.L. PIPKIN, J.G. SHADDOCK, R. WEINDRUCH, R.W. HART & R.J. FEUERS. 1997. A circadian study of liver antioxidant enzyme systems of female fischer-344 rats subjected to dietary restriction for six weeks. Age **20:** 221–227.

12. FEUERS, R.J., R. WEINDRUCH, J.E.A. LEAKEY, P.H. DUFFY & R.W. HART. 1997. Increased effective activity of rat liver catalase by dietary restriction. Age **20:** 228–232.

13. LUHTALA, T.A., E.B. ROECKER, R.J. FEUERS & R. WEINDRUCH. 1994. Dietary restriction attenuates age-related increases in rat skeletal muscle antioxidant enzyme activities. J. Gerontol. **49:** B231–B238.

14. AMES, B.N. & M.K. SHIGENAGA. 1992. Oxidants are a major contributor to aging. Ann. N.Y. Acad. Sci. **663:** 65–96.

15. BERLETT, B.D. & E.R. STADTMAN. 1997. Protein oxidation in aging, disease, and oxidative stress. J. Biol. Chem. **272**(33): 20313–20316.

16. AMES, B.N., M.K. SHIGENAGA & T.M. HAGEN. 1993. Oxidants, antioxidants, and the degenerative diseases of aging. Proc. Natl. Acad. Sci. **90:** 7915–7922.

17. TROUNCE, I., E. BARNEY & S. MARZUKI. 1989. Decline in skeletal muscle mitochondrial respiratory chain function: Possible factor in aging. Lancet **i:** 637–639.

18. COOPER, J.M., V.M. MANN & A.H.V. SCHAPIRA. 1992. Analysis of mitochondrial respiratory chain function and mitochondrial DNA deletions. In human skeletal muscle: Effect of aging. J. Neurol. Sci. **113:** 91–98.

19. BOFFOLI, D., S.C. SCACCO, R. VERGARI, G. SOLARINO, G. SANTACROCE & S. PAPA. 1994. Decline with age of the respiratory chain activity in human skeletal muscle. Biochem. Biophys. Acta **1226:** 73–82.

20. TORRI, K., S. SUGIYAMA, K. TAKAGI, T. SATAKE & T. OZAWA. 1992. Aging-associated deletions of human diaphragmatic mitochondrial DNA. Am. J. Respir. Cell. Mol. Biol. **6:** 88–92.

21. SUGIYAMA, S., M. TAKASAWA, M. HAYAKAWA & T. OZAWA. 1993. Changes in skeletal muscle, heart, and liver mitochondrial electron transport activities in rats and dogs of various age. Biochem. Mol. Biol. Int. **30:** 937–944.

22. MASORO, E.J. 1988. Food restriction in rodents: An evaluation of its role in the study of aging. J. Gerontol. Biol. Sci. **43:** B59–B64.

23. WEINDRUCH, R. & R.L. WALFORD. 1988. The Retardation of Aging and Disease by Dietary Restriction. Charles C.Thomas. Springfield, IL.

24. WEINDRUCH, R. 1996. Caloric restriction and aging. Sci. Am. **274**(i): 46–52.

25. YU, B.P., S. LAGANIERE & J.W. KIM. 1989. Influence of life-prolonging food restriction on membrane lipoperoxidation and antioxidant status. In Oxygen Radicals in Biology and Medicine. M.G. Simic, K.A. Taylor, J.F. Ward & T. Von Sonntag, Eds.: 1067–1073. Plenum. New York.

26. DUFFY, P.H., R.J. FEUERS, J.A. LEAKEY, K.D. NAKAMURA, A. TURTURRO & R.W. HART. 1989. Effect of chronic caloric restriction on the physiological variables related to energy metabolism in the male Fischer 344 rat. Mech. Ageing Dev. **48:** 117–133.

27. RAGAN, C.I., M.T. WILSON, V.M. DARLEY-USMAR & P.N. LOWE. 1987. Sub-fractionation of mitochondria and isolation of the proteins of oxidative phosphorylation. In Mitochondria: A Practical Approach. V.M. Darley-Usmar, D. Rickwood & M.T. WILSON, Eds.: 79–112. IRL Press. Oxford.

28. SEGEL, I.H. 1975. Enzyme Kinetics: Behavior and Analysis of Rapid Equilibrium and Steady-State Enzyme Systems. Wiley. New York.

29. HOLLOSZY, J.O., Ed. 1995. Workshop on sarcopenia: Muscle atrophy in old age. J. Gerontol. Biol. Sci. **50A**. Special issue.

30. GILLHAM, N.W. 1994. Organelle Genes & Genomes. Oxford University Press. New York.

31. CAPALDI, R.A., V. DARLEY-USMAR, S. FULLER & F. MILLETT. 1982. Structural and functional features of the interaction of cytochrome c with complex III and cytochrome c oxidase. FEBS Lett. **138:** 1–7.

32. GONZALEZ-HALPHEN, D., M.A. LINDORFER & R.A. CAPALDI. 1988. Subunit arrangement in beef heart complex III. Biochemistry **27**(18): 7021–7031.
33. KADENBACH, B., B. SCHNEYDER, O. MELL, S. STROH & A. REIMANN. 1991. Respiratory chain proteins. Rev. Neurol. **147:** 436–442.
34. SOHAL, R.S. 1993. Aging, cytochrome oxidase activity, and hydrogen peroxide release by mitochondria. Free Radical Biol. Med. **14:** 583–588.

The Universality of Bioenergetic Disease

Age-associated Cellular Bioenergetic Degradation and Amelioration Therapy

ANTHONY W. LINNANE,[a] S. KOVALENKO, AND E.B. GINGOLD

Centre for Molecular Biology and Medicine, Epworth Hospital, 185–187 Hoddle Street, Richmond, Melbourne, Victoria 3121, Australia

ABSTRACT: During the present century there has been a dramatic change in life expectancy in advanced societies, now exceeding 80 years. As distinct from life expectancy, life potential is said to be at least 120 years, so that the continuing increase in knowledge has the potential for further major changes in the survival of humans conceivably in the near future. This presentation will be concerned with one aspect of the development of biomedical advances related in part to a concept of an "age-related universality of bioenergetic disease," and its potential amelioration and proposed impact on age-related disease and lifestyle. Aging is a complex biological process associated with a progressive decline in the physiological and biochemical performance of individual tissues and organs, leading to age-associated disease and senescence. Consideration of the progressive accumulation of mitochondrial DNA mutation with age and the tissue/cellular bioenergy decline associated with the aging process has led us to the proposal of a "universality of bioenergetic disease" and the potential for a redox therapy for the condition. This concept envisages that a tissue-bioenergetic decline will be intrinsic to various diseases of the aged and thereby contribute to their pathology, in particular, heart failure, degenerative brain disease, muscle and vascular diseases, as well as other syndromes. The information and concepts embodied in this proposal will be reviewed under the following headings: (1) mitochondrial DNA deletion mutation in some tissue is very extensive and shows mosaicism; (2) age-associated tissue/cellular bioenergy mosaic closely corresponds to the mtDNA profile; (3) cellular bioenergy as a function of mitochondrial bioenergy, glycolysis, and plasma membrane oxidoreductase; (4) redox therapy for the reenergization of cells, tissues, and whole organs. A redox therapy based on coenzyme Q_{10} has demonstrated profound alteration in heart function of old rats; no significant effect was observed with young rats.

Aging is a highly complex biological process associated with a progressive decline in the performance of individual tissues and organs, leading to age-associated disease and senescence. We have proposed that a critical factor in this decline is the loss of bioenergetic capacity of the cells of aging tissue.[1] A key element of this proposal is that the loss of bioenergetic capacity is associated with mutations in the mitochondrial DNA (mtDNA) of the cells. As tissues age, mtDNA mutations accumulate in individual cells; eventually some cells will reach the point at which the ability to make the mtDNA-encoded components of the mitochondrial energy generation system is seriously impaired. If this occurs in a significant number of cells in a tissue, the function of that tissue will be compromised and consequentially will contribute to such age-associated pathologies as skeletal muscular and neurological degeneration, heart failure, strokes, and other diseases.

[a]Tel: 61-3-9426 4200; fax: 61-3-9426 4201; e-mail: tlinnane@cmbm.com.au

The human mitochondrial genome, a closed circular molecule of only 16,569 base pairs, is entirely concerned with producing a small number of proteins that are critical components of the respiratory chain and ATPase synthetase.[2] The mtDNA encoded genes are of two types, those that code directly for the proteins (13 genes), and those that code for the ribosomal (2 genes) and transfer RNA (22 genes) molecules needed to form the mitochondrial protein synthesis system. There is very little spacer DNA in the genome; hence any mutation is likely to be in a coding region and to have deleterious effects.

As each cell has over 1000 copies of the mtDNA molecule,[3,4] a mutation in any single molecule will not seriously affect the bioenergy capacity of that cell. However, as mutations accumulate when the cells age, a point will be reached where that cell cannot synthesize adequate amounts of mtDNA-encoded proteins and hence is unable to synthesize adequate levels of the respiratory complexes. The energy status of a tissue will depend on what proportion of its cells are bioenergetically depleted in this way.

An understanding of the processes involved in the depletion of the bioenergetic capacity of aging tissues has led us to propose a therapeutic intervention.[5] We have suggested that the bioenergetic decline with age will be ameliorated by dietary supplementation with coenzyme Q_{10}, the only readily exchangeable component of the respiratory chain.

The role of coenzyme Q_{10} as an essential redox compound of the mitochondrial electron transport chain functioning in the production of biochemical energy is well established, having been known for over 30 years.[6] Nonetheless, other roles have been postulated for this factor, including that of a redox component of extramitochondrial[7,8] and plasma membranes,[9,10] electron transfer chains, as an antioxidant,[11–13] and as a membrane stabilizer.[14] These observations have led earlier to the common impression that nonbioenergetic roles may be mainly responsible for reported therapeutic effects of coenzyme Q_{10}.[15] However, observations made in our own laboratory[5,16–18] have led us to conclude that its major therapeutic role arises out of its action as an electron carrier and proton conductor (redox compound) in membrane systems.

The concept of redox therapy is based on a number of significant observations made in our laboratory. (1) Cells from aging tissues show deficiencies in their respiratory function, and these deficiencies arise from mutations in mtDNA. (2) Cells in culture can grow anaerobically, but only if a redox sink, such as pyruvate, for the reoxidation of NADH is added to the growth media. (3) Cells lacking in mtDNA and thus respiratory function (*rho*[0] cells) also require a redox sink for growth. In some *rho*[0] cell lines either ferricyanide or coenzyme Q_{10} can substitute for pyruvate to provide the necessary redox support for growth. This support appears to be mediated through the plasma membrane oxidoreductase (PMOR). (4) Rat tissues showing deficient respiratory functions, resulting either from AZT treatment or aging, can have the effects of this deficiency ameliorated by the administration of coenzyme Q_{10}.

This paper will review the evidence that we have accumulated for these propositions and demonstrate how they support our comprehensive hypothesis concerning the contribution of mtDNA mutations to the human aging process and the beneficial role of redox therapy by coenzyme Q_{10}.

mtDNA MUTATION

The first clear demonstration of the association of mitochondrial disease with mtDNA mutation was reported by Morgan Hughes and colleagues.[19] A range of diseases have since been demonstrated as arising from either point mutations or deletions of the mtDNA, and these can vary greatly in their severity (for review, see refs. 20, 21). Although a large number of different symptoms can result from such mutations, commonly they affect either the neurologial system (leading, among other consequences, to visual impairment or deafness), muscle tissues (leading to various cardiomyopathies and to extreme exercise fatigue), or combinations of these areas. It is not clear why various deficiencies in the mitochondrial respiratory process can lead to such a divergent range of clinical outcomes. However, relevant factors presumably include the nature of the mutation (and whether it affects a single component or the whole respiratory system), the degree of heterozygosity existing in the cells, the tissue distribution of the mutant mtDNA molecules, and the specific metabolism of the individually affected tissues. A number of the syndromes are maternally inherited, whereas others arise as a result of early somatic mutation.

There is now abundant evidence both from our laboratory[1,22-25] and others[26,27] that mtDNA mutations also occur in normal subjects and that such mutations progressively accumulate with age. However, unlike the situation with mitochondrial disease in which a particular mutant mtDNA type can dominate the population, in aging tissues, individual mutant mtDNA types have rarely been reported as representing more than 0.1% of the total mtDNA.[28-30] Each individual mutation thus does not occur at a high enough frequency during the aging process to itself cause a decline in bioenergetic capacity of tissues and lead to the observed biochemical and physiological changes. Instead, we have proposed that it is the cumulative effect of a large number of different mtDNA mutations that leads to the observed decline.[5]

Until recently, evidence supporting this key proposal has been somewhat indirect, and there was some doubt if the overall level of mtDNA mutations would be sufficient to lead to the observed bioenergy decline. However, the recent procedure, extra-long PCR (XL-PCR) has allowed us to assess the overall integrity of the mitochondrial genome. Using this procedure, the PCR reaction is capable of amplifying full-length mtDNA molecules so that the total cellular level of full-length mtDNA can be determined. It will also reveal a wide range of different deletion mutations, or indeed other rearrangements, in a single reaction. Results from such analysis on skeletal (deltoid) muscle cells from old age human subjects dramatically demonstrated that extensive mtDNA mutation had taken place.[31] It can be estimated that less than 5% of the total mtDNA from the muscle tissue of a 90-year-old subject was still in the form of full-length mtDNA, the main population of mtDNA molecules being made up of both deletion products and oversized rearrangements. It would be expected that cells with this extent of mtDNA damage would be greatly compromised in their ability to synthesize mitochondrial proteins.

It is considered that mtDNA mutations arise as random events. Thus some cells in a tissue would, by chance, have large numbers of mutant mtDNA molecules and others a greater proportion of normal molecules. An important prediction arises from this: the tissue would be expected to be a "bioenergy mosaic," with some cells deficient in respiratory activity and other cells, in the same tissue, having normal levels of respiratory activity. This prediction has been confirmed by direct staining of the tissues for the level of the respiratory chain component cytochrome oxidase (COX). A tissue bioenergy mosaic is revealed, with some of the cells showing little staining for COX activity and other cells in the same tissue showing high levels of activity.[1,23,32] Furthermore, as would be predicted, the percentage of COX staining cells decreases in aging tissue.

We have now been able to extend this work to the level of individual skeletal muscle fibers and have shown a good correlation between the occurrence of mtDNA mutation and loss of COX activity.[33] In these experiments human deltoid skeletal muscle tissue samples, obtained from a 5-year-old and a 90-year-old subject, were sectioned for microscopic examination and then stained for COX activity. The COX activities shown for individual fibers varied greatly and ranged from fibers with normal activity levels, through those with intermediate levels of activity, to those with little COX activity. Congruous fibers from adjacent sections were then identified, dissected out, and subjected to XL-PCR analysis. Whereas the fibers showing high COX activities revealed a pattern of full-length mtDNA, with few, if any, deletions, COX-deficient fibers from the same tissues demonstrated a heterogeneous population of rearranged mtDNA molecules and very low levels of full-length mtDNA. It was observed that even the fibers with intermediate COX levels showed little full-length mtDNA, a result suggesting the possibility that these cells achieved their COX activity by intracellular mtDNA complementation. The same fiber-specific patterns were obtained from the tissues of both the 5-year-old and 90-year-old subjects, but the proportion of the fiber types varied greatly between these samples. Thus, although COX-deficient fibers were observed in the 5-year-old subject, they were rare. This result establishes the relationship between age-associated accumulation of mtDNA mutation and COX activity and provides compelling support for the hypothesis of mtDNA mutation-driven bioenergy degradation as a key feature of the aging process, at least in some tissues.

REDOX THERAPY

The identification of respiratory system deficiencies as underlying the bioenergy depletion of cells from aging tissue suggests possibilities for pharmacological intervention.[5] It is presently not possible to replace the deficient protein components of the respiratory complexes, but a knowledge of the nature of the deficiency may allow the circumvention of the blockage or the kinetic enhancement of the performance of the remaining active components. An intervention that boosts the ability of cells to produce ATP will be beneficial in restoring the functions of that cell.

The capacity of a tissue to generate energy is dependent both on its rate of mitochondrial respiration and glycolysis. A cell using glycolysis as its major source of ATP needs to reoxidize the NADH formed in the glycolytic pathway. If oxidative phosphorylation is functional, this system will reoxidize the NADH leading to reduction of oxygen to water and a concurrent synthesis of further ATP. However, if the mitochondrial system is sufficiently impaired, the reoxidation of NADH becomes a major difficulty for the cell. The aim of pharmacological intervention is to support the reoxidation of the NADH and hence enable the cells to continue to function despite their lower levels of respiratory chain components. We refer to such treatment as redox therapy.

The first successful example of the use of redox therapy was with a patient with a mitochondrial disease manifesting as a deficiency in complex III of the respiratory chain.[34] Two redox compounds, menadione (vitamin K_3, a naphthoquinone) and ascorbic acid, were administered, theoretically to bypass the defect putatively located in complex III. Although the actual mechanism of action of these compounds remains unclear (for discussion, see ref. 5), the patient's symptoms were substantially alleviated. Most importantly, ATP synthesis was increased, as evidenced by NMR studies that demonstrated an enhanced creatine phosphate-to-phosphate ratio in the skeletal muscle of the treated patients.

Since this pioneering work, there have been many studies using coenzyme Q_{10} as the therapeutic compound, and this work has reinforced the effectiveness of redox therapy for many patients with mitochondrial disease (bioenergy capacity defects)[35–38] as well as other disease states.[39–44] Coenzyme Q plays a central role in the respiratory chain, acting as electron acceptor for both complexes I and II and the electron donor to complex III. It is the only element of the chain whose external addition to the cell is possible. The actual mechanism of the coenzyme Q action is yet to be fully established. It has been reported that the normal level of coenzyme Q in the mitochondrial membrane is below that required for kinetic saturation of complex I.[45,46] It might be the case that coenzyme Q is a rate-limiting element in the respiratory chain of mitochondrially damaged tissues and thus that an excess of this component drives the deficient respiratory chain at its maximum possible rate. Indeed, it has been shown that coenzyme Q_{10} levels in the tissues of human subjects decrease with age.[47] Alternatively, or in addition, coenzyme Q_{10} might be generating superoxide radicals, and thus acting as a redox sink. As will be seen in the next section, however, the mitochondrial respiratory chain is not the only site at which coenzyme Q_{10} can be postulated to act.

When we first proposed that loss of cellular bioenergy capacity was a significant contributor to the process of aging, we speculated that redox therapies could be developed to attenuate the pathological consequences of the aging process.[1] In ensuing years, coenzyme Q_{10} dietary supplementation has gained increasing popularity in this role. Evidence that we have gathered to support this supposition is presented in the subsequent sections of this paper.

SURVIVAL OF CELLS WITHOUT RESPIRATORY FUNCTION

Deficiencies in the bioenergetic capacity of mitochondria have been established as occurring both in cases of mitochondrial disease and during the process of normal aging. We have been interested in the question of how cells in culture can dispense with mitochondrial function and still gain sufficient energy for survival from glycolysis. Our work has revealed that this is indeed possible, provided that the cells are given some means of reoxidizing the NADH that the glycolytic process generates.

In early studies, we demonstrated that cultured Namalwa cells can be grown anaerobically but that this growth was dependent on the addition of excess pyruvate to the glucose-based medium.[48] The pyruvate serves as a redox sink, being reduced to lactate and thus allowing the reoxidation of the NADH (effectively replacing oxygen). Such anaerobic cultures grew at a similar rate to cells growing under fully aerobic conditions. Furthermore, glucose utilization increased by a factor of only about twofold. Thus cells are able to make efficient use of glycolysis to support their energy requirements, provided that pyruvate is available to act as a redox sink for NADH oxidation.

Our laboratory first reported many years ago that mutants of the yeast *Saccharomyces cerevisiae* are able to dispense completely with mtDNA.[49] Such *rho*[0] mutants, which can arise as a result of ethidium bromide treatment, obtain their energy from glycolysis, with ethanol acting as the redox sink. More recently, it has been shown that analogous rho^0 human cell lines can be generated,[50,51] but in our hands human cells require very prolonged treatments with ethidium bromide, which is an inhibitor of mitochondrial DNA replication but also is a nuclear mutagen. Indeed, we speculate that human cells may be "petite negative" in the classical sense and require nuclear mutation for conversion to a petite positive state, as has been shown for petite negative yeasts.[52,53] Thus for the establishment of a stable rho^0 human cell line, nuclear mutation as well as inhibition of mtDNA replication is required. In any event, such mutants contain no mtDNA and hence are completely deficient in the mitochondrially coded components of the respiratory system. For these mammalian mutants pyruvate acted as a redox sink and allowed support of cellular survival and growth by glycolytically produced energy.

We have studied the requirements for the generation and survival of human rho^0 cell lines and have demonstrated that in some rho^0 lines pyruvate can be replaced in the growth media by either potassium ferricyanide or coenzyme Q_{10}.[16] Potassium ferricyanide is unable to penetrate the cell membrane and enter the cell, and thus the activity of this compound is clearly at the level of the membrane. This result strongly suggests the involvement of the PMOR complex in supporting glycolytic growth. The PMOR system enables the reoxidation of cytosolic NADH to NAD^+, with molecular oxygen acting as the electron acceptor.[9] Clearly, however, this activity is inadequate to support sufficient reoxidation to allow the growth of rho^0 cells in the absence of additional redox supplementation. Both potassium ferricyanide and coenzyme Q_{10} are known to stimulate this enzyme system and apparently allow it to reach activity levels at which such glycolytically supported growth can be sustained without the need for added pyruvate, at least in some rho^0 strains.

The PMOR system is clearly of critical importance to rho^0 cells. We have demonstrated that the activity of this system increases during the process of generating rho^0 cells; furthermore the increase in PMOR activity parallels the decline in mitochondrial respiratory activity.[54] This report also demonstrated that the addition of *p*-chloromercuriphenyl-sulfonic acid, an inhibitor of the PMOR system, was lethal to rho^0 cells, but not to rho^+ cells. Clearly, the PMOR system plays a major role in cells deprived of mitochondrial function.

The discovery of the role of the PMOR system provides another important parameter for understanding the role of coenzyme Q_{10} in redox therapy. Coenzyme Q_{10} is an integral functional component of the PMOR system. Evidence from Fred Crane and colleagues has demonstrated that added coenzyme Q_{10} can stimulate the growth of many cell lines in serum-free media, apparently via the PMOR.[9,55] Hence this system provides a site, additional to the mitochondrial respiratory chain, at which coenzyme Q_{10} can act to ameliorate the consequences of deficiencies in the mitochondrial respiratory chain.

The interplay between the NADH reoxidizing systems of the mitochondria and the plasma membrane is illustrated by our work on induction of apoptosis by inhibitors of these systems.[56] In particular, inhibitors of the mitochondrial system, rotenone, antimycin A, and oligomycin, induced an apoptotic response in a number of rho^+ mammalian cell lines, including some producing the protein Bcl-2, which blocks apoptosis induced by other stimuli. On the other hand, rho^0 cultures were unaffected by these drugs. Additionally, the key role of the PMOR is illustrated by the induction of apoptosis in cells treated with capsaicin, an inhibitor of the PMOR.[57] Interestingly, Bcl-2 positive cells do not experience apoptosis when treated with capsaicin.

MITOCHONDRIAL DEFICIENCIES IN RATS CAN BE AMELIORATED BY COENZYME Q_{10} AND ANALOGUES

AZT (zidovudine), a key drug used in the treatment of AIDS due to its property of inhibiting reverse transcriptase, also inhibits mitochondrial DNA polymerase. As a result, prolonged treatment of patients or animals with AZT causes a degeneration of the integrity of mitochondrial functions in a way that is reminiscent of those changes seen in mitochondrial disease and aging. In particular, AZT treatment can lead to severe skeletal and cardiac myopathies.[58,59]

We have used AZT-treated rats as a model system to study the effect of coenzyme Q_{10} in reversing the bioenergy decline associated with the degeneration of the mitochondrial system.[17] Rats were intraperitoneally injected with AZT for 85–90 days. Separate groups of rats were injected, in addition to the AZT, with either coenzyme Q_{10} or coenzyme Q_{10C} (decyl Q, a Q_{10} analogue with a 10 carbon saturated side chain) at doses comparable to those used therapeutically for human mitochondrial disease patients. Muscle performance was directly measured by exposing the soleus muscle of anesthetized rats, subjecting the muscle to electrical stimulation and measuring the contraction force with a strain gauge. The fatigue profile was determined in a cycle of stimulation for one third of a second and rest for two thirds.

Our experiments[17] revealed that the force exerted by the soleus muscle of young control rats (untreated with AZT) declined during the first 200 seconds of the experiment but thereafter reached a steady level of about 70% of initial force. This steady state has been shown to be maintained for over 8000 seconds. When the same measurement was made on AZT-treated rats, however, a dramatic decrease in the steady state force was observed, this being as low as 30–40% of the initial force. Hence AZT treatment clearly had a detrimental effect on the function of the soleus muscle.

When the young rats were treated with coenzyme Q_{10} or its analogue coenzyme Q_{10C}, it was shown that these compounds countered the effect of the AZT and resulted in muscle performances that approached those of untreated rats. Both coenzyme Q_{10} and coenzyme Q_{10C} restored muscle function and produced fatigue profiles similar to those of untreated rats, with the analogue coenzyme Q_{10C} being particularly effective in this role. By contrast, treatment with the short chain, and hence more hydrophilic analogue coenzyme Q_{3C}, did not significantly improve the AZT-affected muscle performance.

The decline in muscle function induced by AZT treatment and its restoration by simultaneous coenzyme Q_{10C} treatment were shown to be paralleled by similar changes in mitochondrial bioenergy functions.[17] ETP_H (electron transfer phosphorylating submitochondrial particles) were prepared from heart mitochondria isolated from the AZT-treated rats and the AZT plus coenzyme Q_{10C}–treated rats used for the study of muscle function. Membrane potential of these particles, generated in response to the electron transport chain substrates NADH and succinate, was monitored using the dye oxonol VI. It was found with NADH as the substrate that the potential generated by ETP_H from AZT-treated rats was lower than those for untreated rats and that cotreatment of the rats with coenzyme Q_{10C} served to restore the membrane potential generated to untreated levels.

Hence, although the AZT treatment was causing a clear decline in energy-producing capacity of the submitochondrial particles, this decline could be reversed by the coadministration of coenzyme Q_{10C}. Interestingly, AZT treatment had no effect on the membrane potential generated in response to the substrate succinate, a result suggesting that the major decline in mitochondrial energy-producing capacity in AZT-treated rats is in complex I.

We also compared membrane potentials of heart ETP_H from aged (34–36 months) and young (5–6 months) rats[17] and observed a dramatic fall in the membrane potential with the aged rats. In contrast to the results with AZT-treated rats, the aged rats showed a decline in membrane potential both with NADH and with succinate as substrate. Significantly, the concentration of particles prepared from the young rats and aged rats was adjusted to have the same rate of complex II activities. Hence it can be deduced that aged rats have deficiencies in complexes III and/or IV as well as in complex I.

COENZYME Q_{10} IMPROVES TOLERANCE TO PACING STRESS IN THE RAT HEART

In elderly patients the results of interventions that stress the myocardium, such as coronary bypass surgery and angioplasty, are inferior to those in the young. A possible contributing factor is an age-related reduction in cellular energy production. We have conducted a study using rats as a model system to ascertain whether coenzyme Q_{10} is able to ameliorate this age-associated decline in ability to withstand stress.[60]

Young and aged rats were randomly assigned to receive daily injections of either coenzyme Q_{10} or of the vehicle for six weeks. Hearts were then isolated, perfused, and subjected to conditions of severe aerobic pacing stress using the Lagendorff procedure. Measurements were then made to ascertain the ability of the isolated hearts to maintain work capacity following aerobic pacing stress.

Even before the applied stressing, there was a clear difference between the young and senescent hearts. Thus in senescent hearts cardiac work was 74% and oxygen consumption (MVO_2) 66% of that observed for young hearts. Coenzyme Q_{10} treatment abolished these differences, improving the levels of these parameters in senescent hearts to levels found in young hearts. Following pacing stress, both the work levels and the oxygen consumption of hearts from all groups dropped significantly, but the decline was much greater in senescent hearts. Expressing the results as a percentage of prepacing levels, poststress cardiac work levels were 17% in senescent hearts as compared to 45% in young hearts. Similarly, MVO_2 levels declined to 61% for senescent hearts and 74% for young hearts. Senescent rats treated with coenzyme Q_{10}, however, gave results similar to those from young rats. Hence, cardiac work recovered to 48% prepacing values and MVO_2 to 82 percent. It is notable that the hearts of young animals injected with coenzyme Q_{10} were unchanged by the treatment.

It is thus apparent that senescent heart tissue is less tolerant to aerobic stress than hearts of young rats and that treatment of the old animals with coenzyme Q_{10} improves their aerobic stress tolerance. These results support the concept that coenzyme Q_{10} treatment may be beneficial for senescent patients before cardiac intervention.[42,43]

CONCLUSIONS

During the aging process there is a progressive accumulation of mutations in mitochondrial DNA that results in a decrease in bioenergetic capacity of the tissues in which they occur. XL-PCR studies have revealed that these mutations, particularly in human deltoid skeletal muscle tissue, can be very extensive.

As mtDNA mutation is a random process, it would be expected to occur to different extents in individual cells making up a tissue, and thus to lead to a tissue mutation mosaic. Using cytochrome oxidase levels as an exemplifier of respiratory activity, we have demonstrated a bioenergetic mosaic in skeletal muscle. Furthermore, we have shown a close correlation between loss of cytochrome oxidase activity and mtDNA mutation in the individual muscle fibers. This observation provides powerful evidence for the proposal that mtDNA mutation plays a significant role in the cellular bioenergy status of aging individuals.

Cellular bioenergy can be seen as a function of mitochondrial bioenergy, glycolysis, and the PMOR. Studies on anerobic cell growth and rho^0 cultures have revealed that cells can survive in the absence of a functional mitochondrial respiratory system, provided that they have a means to reoxidize the NADH generated by glycolysis. The PMOR can play a central role in this process; indeed inhibitors of this system are lethal to rho^0 cells. Coenzyme Q_{10} can support the growth of rho^0 cells and seems likely to act by enhancing the action of the PMOR system.

There are a number of levels at which externally added coenzyme Q_{10} might act to provide support for the mitochondrially damaged cells making up senescent tissue. The raised concentrations of this factor might act directly on the deficient mitochondrial respiratory chain. On the other hand, or in addition, it may act, as it apparently does in rho^0 cell cultures, on the PMOR system, and thus enable the cells to survive on glycolytically produced energy. As coenzyme Q_{10} is found in most membranes within the cell, it may also act at such sites as the Golgi membrane oxidoreductase.[7] Which of these potential mechanisms operate *in vivo* has yet to be ascertained.

We have suggested the use of dietary supplements of coenzyme Q_{10} to ameliorate the degenerative consequences of the aging process. Evidence for this proposition that has been discussed in this review demonstrates beneficial effects of coenzyme Q_{10} administration on whole organs, tissues, and subcellular particles taken from senescent animals. Dietary supplementation with coenzyme Q_{10} is thus indicated as a treatment to improve the quality of life of aged individuals and to provide protection against such age-related conditions as heart failure and neurodegenerate disease.

REFERENCES

1. LINNANE, A.W., S. MARZUKI, T. OZAWA & M. TANAKA. 1989. Mitochondrial DNA mutations as an important contributor to ageing and degenerative diseases. Lancet **1:** 642–645.
2. ANDERSON, S., A.T. BANKIER, B.G. BARRELL, M.H. DE BRUIJN, A.R. COULSON, J. DROUIN, I.C. EPERON, D.P. NIERLICH, B.A. ROE, F. SANGER, P.H. SCHREIER, A.J. SMITH, R. STADEN & I.G. YOUNG. 1981. Sequence and organization of the human mitochondrial genome. Nature **290:** 457–465.
3. CLAYTON, D.A. 1982. Replication of animal mitochondrial DNA. Cell **28:** 693–705.
4. BOGENHAGEN, D. & D.A. CLAYTON. 1974. The number of mitochondrial deoxyribonucleic acid genomes in mouse L and human HeLa cells. Quantitative isolation of mitochondrial deoxyribonucleic acid. J. Biol. Chem. **249:** 7991–7995.

5. LINNANE, A.W., C. ZHANG, A. BAUMER & P. NAGLEY. 1992. Mitochondrial DNA mutation and the ageing process: bioenergy and pharmacological intervention. Mutat. Res. **275:** 195–208.

6. CRANE, F.L. 1989. Comments on the discovery of coenzyme Q: A commentary on 'Isolation of a Quinone from Beef Heart Mitochondria.' Biochim. Biophys. Acta **1000:** 358–361.

7. CRANE, F.L. & D.J. MORRE. 1977. Evidence for coenzyme Q function in Golgi membranes. *In* Biomedical and Clinical Aspects of Coenzyme Q, K. Folkers & Y. Yamamura, Eds.: Vol. 1: 3–14. Elsevier. Amsterdam.

8. KALEN, A., B. NORLING, E.L. APPELKVIST & G. DALLNER. 1987. Ubiquinone biosynthesis by the microsomal fraction from rat liver. Biochim. Biophys. Acta **926:** 70–78.

9. SUN, I.L., E.E. SUN, F.L. CRANE, D.J. MORRE, A. LINDGREN & H. LOW. 1992. Requirement for coenzyme Q in plasma membrane electron transport. Proc. Natl. Acad. Sci. USA **89:** 11126–11130.

10. LENAZ, G., R. FATO, C. CASTELLUCCIO, M.L. GENOVA, C. BOVINA, E. ESTORNELL, V. VALLS, F. PALLOTTI & G. PARENTI CASTELLI. 1993. The function of coenzyme Q in mitochondria. Clin. Invest. **71:** S66–70.

11. BEYER, R.E., K. NORDENBRAND & L. ERNSTER. 1987. The function of coenzyme Q in free radical production and as an antioxidant: A review. Chem. Scr. **27:** 145–153.

12. ERNSTER, L. & R.E. BEYER. 1991. Antioxidant functions of coenzyme Q: Some functional and pathophysiological implications. *In* Biomedical and Clinical Aspects of Coenzyme Q. K. Folkers, G.P. Littarru & Y. Yamagami, Eds.: Vol. 6: 45–58. Elsevier. Amsterdam.

13. KAGAN, V.E., H. NOHL & P.J. QUINN. 1996. Coenzyme Q: Its role in scavenging and generation of radicals in membranes. *In* Textbook of Antioxidants. E. Cadenas & L. Packer, Eds.: 157–201. M. Dekken Inc. New York.

14. OZAWA, T. 1985. Formation of oxygen radicals in the electron transport chain and antioxidant properties of coenzyme Q. *In* Coenzyme Q: Biochemistry, Bioenergetics and Clinical Applications of Ubiquinone. G. Lenaz, Ed.: 441–456. John Wiley. Chichester.

15. FOLKERS, K., G.P. LITTARRU & T. YAMAGAMI. 1991. Biomedical and clinical aspects of coenzyme Q, Vol. 6. Elsevier. Amsterdam.

16. MARTINUS, R.D., A.W. LINNANE & P. NAGLEY. 1993. Growth of rho^0 human Namalwa cells lacking oxidative phosphorylation can be sustained by redox compounds potassium ferricyanide or coenzyme Q_{10} putatively acting through the plasma membrane oxidase. Biochem. Mol. Biol. Int. **31:** 997–1005.

17. LINNANE, A.W., M. DEGLI ESPOSTI, M. GENEROWICZ, A.R. LUFF & P. NAGLEY. 1995. The universality of bioenergetic disease and amelioration with redox therapy. Biochim. Biophys. Acta **1271:** 191–194.

18. ESPOSTI, M.D., A. NGO, A. GHELLI, B. BENELLI, V. CARELLI, H. MCLENNAN & A.W. LINNANE. 1996. The interaction of Q analogs, particularly hydroxydecyl benzoquinone (idebenone), with the respiratory complexes of heart mitochondria. Arch. Biochem. Biophys. **330:** 395–400.

19. HOLT, I.J., A.E. HARDING & J.A. MORGAN-HUGHES. 1988. Deletions of muscle mitochondrial DNA in patients with mitochondrial myopathies. Nature **331:** 717–719.

20. WALLACE, D.C. 1994. Mitochondrial DNA mutations in diseases of energy metabolism. J. Bioenerg. Biomembr. **26:** 241–250.

21. SCHAPIRA, A.H.V. 1994. Respiratory chain adnomalities in human disease. *In* Mitochondria: DNA, Protein and Disease. V. Darley-Usmar & A.H.V. Schapira, Eds.: 241–278. Portland Press. London.

22. LINNANE, A.W., A. BAUMER, R.J. MAXWELL, H. PRESTON, C.F. ZHANG & S. MARZUKI. 1990. Mitochondrial gene mutation: The ageing process and degenerative diseases. Biochem. Int. **22:** 1067–1076.

23. NAGLEY, P., I.R. MACKAY, A. BAUMER, R.J. MAXWELL, F. VAILLANT, Z-X. WANG, C. ZHANG & A.W. LINNANE. 1992. Mitochondrial DNA mutation associated with aging and degenerative disease. Ann. N. Y. Acad. Sci. **673:** 92–102.

24. ZHANG, C., A. BAUMER, R.J. MAXWELL, A.W. LINNANE & P. NAGLEY. 1992. Multiple mitochondrial DNA deletions in an elderly human individual. FEBS Lett. **297:** 34–38.

25. ZHANG, C., A.W. LINNANE & P. NAGLEY. 1993. Occurrence of a particular base substitution (3243 A to G) in mitochondrial DNA of tissues of aging humans. Biochem. Biophys. Res. Commun. **195:** 1104–1110.

26. WALLACE, D.C. 1992. Mitochondrial genetics: A paradigm for aging and degenerative diseases? Science **256:** 628–632.
27. WEI, Y.H. 1992. Mitochondrial DNA alterations as ageing-associated molecular events. Mutat. Res. **275:** 145–155.
28. CORTOPASSI, G.A. & N. ARNHEIM. 1990. Detection of a specific mitochondrial DNA deletion in tissues of older humans. Nucleic Acids Res. **18:** 6927–6933.
29. SIMONETTI, S., X. CHEN, S. DIMAURO & E.A. SCHON. 1992. Accumulation of deletions in human mitochondrial DNA during normal aging: Analysis by quantitative PCR. Biochim. Biophys. Acta **1180:** 113–122.
30. ZHANG, C., L.E. PETERS, A.W. LINNANE & P. NAGLEY. 1996. Comparison of different quantitative PCR procedures in the analysis of the 4977-bp deletion in human mitochondrial DNA. Biochem. Biophys. Res. Commun. **223:** 450–455.
31. KOVALENKO, S.A., G. KOPSIDAS, J.M. KELSO & A.W. LINNANE. 1997. Deltoid human muscle mtDNA is extensively rearranged in old age subjects. Biochem. Biophys. Res. Commun. **232:** 147–152.
32. MULLER-HOCKER, J., K. SCHNEIDERBANGER, F.H. STEFANI & B. KADENBACH. 1992. Progressive loss of cytochrome *c* oxidase in the human extraocular muscles in ageing—A cytochemical-immunohistochemical study. Mutat. Res. **275:** 115–124.
33. KOPSIDAS, G., S.A. KOVALENKO, J.M. KELSO & A.W. LINNANE. 1998. Age associated bioenergy degradation: A definitive correlation between cellular bioenergy decline and extensive mtDNA rearrangements in human skeletal muscle. Mutat. Res. In press.
34. ELEFF, S., N.G. KENNAWAY, N.R. BUIST, V.M. DARLEY-USMAR, R.A. CAPALDI, W.J. BANK & B. CHANCE. 1984. ^{31}P NMR study of improvement in oxidative phosphorylation by vitamins K_3 and C in a patient with a defect in electron transport at complex III in skeletal muscle. Proc. Natl. Acad. Sci. USA **81:** 3529–3533.
35. BARBIROLI, B., C. FRASSINETI, P. MARTINELLI, S. IOTTI, R. LODI, P. CORTELLI & P. MONTAGNA. 1997. Coenzyme Q_{10} improves mitochondrial respiration in patients with mitochondrial cytopathies. An *in vivo* study on brain and skeletal muscle by phosphorous magnetic resonance spectroscopy. Cell. Mol. Biol. **43:** 741–749.
36. CHEN, R.S., C.C. HUANG & N.S. CHU. 1997. Coenzyme Q_{10} treatment in mitochondrial encephalomyopathies. Short-term double-blind, crossover study. Eur. Neurology **37:** 212–218.
37. GVOZDJAKOVA, A., J. KUCHARSKA & J. GVOZDJAK. 1994. Redox therapy in mitochondrial diseases using coenzyme Q_{10}. Bratisl. Lek. Listy **95:** 443–451.
38. SOBREIRA, C., M. HIRANO, S. SHANSKE, R.K. KELLER, R.G. HALLER, E. DAVIDSON, F.M. SANTORELLI, A.F. MIRANDA, E. BONILLA, D.S. MOJON, A.A. BARREIRA, M.P. KING & S. DIMAURO. 1997. Mitochondrial encephalomyopathy with coenzyme Q_{10} deficiency. Neurology **48:** 1238–1243.
39. DIGIESI, V., F. CANTINI, A. ORADEI, G. BISI, G.C. GUARINO, A. BROCCHI, F. BELLANDI, M. MANCINI & G.P. LITTARRU. 1994. Coenzyme Q_{10} in essential hypertension. Mol. Aspects Med. **15** Suppl: s257–263.
40. DUNLOP, I.S. & P. DUNLOP. 1995. Reversible ophthalmoplegia in CPEO. Aust. N.Z. J. Ophthalmology **23:** 231–234.
41. FOLKERS, K., R. BROWN, W.V. JUDY & M. MORITA. 1993. Survival of cancer patients on therapy with coenzyme Q_{10}. Biochem. Biophys. Res. Commun. **192:** 241–245.
42. JUDY, W.V., W.W. STOGSDILL & K. FOLKERS. 1993. Myocardial preservation by therapy with coenzyme Q_{10} during heart surgery. Clin. Invest. **71:** S155–161.
43. LANGSJOEN, P., P. LANGSJOEN, R. WILLIS & K. FOLKERS. 1994. Treatment of essential hypertension with coenzyme Q_{10}. Mol. Aspects Med. **15** Suppl: S265–272.
44. MORTENSEN, S.A. 1993. Perspectives on therapy of cardiovascular diseases with coenzyme Q_{10} (ubiquinone). Clin. Invest. **71:** S116–123.
45. LENAZ, G., R. FATO, C. CASTELLUCCIO, M. BATTINO, M. CAVAZZONI, H. RAUCHOVA & G. PARENTI CASTELLI. 1991. Coenzyme Q saturation kinetics of mitochondrial enzymes: Theory, experimental aspects and biomedical implications. *In* Biomedical and Clinical Aspects of Coenzyme Q. K. Folkers, G.P. Littarru & Y. Yamagami, Eds.: Vol. 6: 11–18. Elsevier. Amsterdam.
46. ESTORNELL, E., R. FATO, C. CASTELLUCCIO, M. CAVAZZONI, G. PARENTI CASTELLI & G. LENAZ. 1992. Saturation kinetics of coenzyme Q in NADH and succinate oxidation in beef heart mitochondria. FEBS Lett. **311:** 107–109.

47. KALEN, A., E.L. APPELKVIST & G. DALLNER. 1989. Age-related changes in the lipid compositions of rat and human tissues. Lipids **24:** 579–584.

48. VAILLANT, F., B.E. LOVELAND, P. NAGLEY & A.W. LINNANE. 1991. Some biochemical properties of human lymphoblastoid Namalwa cells grown anaerobically. Biochem. Int. **23:** 571–580.

49. LINNANE, A.W. & P. NAGLEY. 1978. Mitochondrial genetics in perspective: The derivation of a genetic and physical map of the yeast mitochondrial genome. Plasmid **1:** 324–345.

50. DESJARDINS, P., E. FROST & R. MORAIS. 1985. Ethidium bromide–induced loss of mitochondrial DNA from primary chicken embryo fibroblasts. Mol. Cell. Biol. **5:** 1163–1169.

51. KING, M.P. & G. ATTARDI. 1989. Human cells lacking mtDNA: Repopulation with exogenous mitochondria by complementation. Science **246:** 500–503.

52. CHEN, X.J. & G.D. CLARK-WALKER. 1993. Mutations in MGI genes convert Kluyveromyces lactis into a petite-positive yeast. Genetics **133:** 517–525.

53. CHEN, X.J. & G.D. CLARK-WALKER. 1995. Specific mutations in alpha- and gamma-subunits of F1-ATPase affect mitochondrial genome integrity in the petite-negative yeast Kluyveromyces lactis. EMBO J. **14:** 3277–3286.

54. LARM, J.A., F. VAILLANT, A.W. LINNANE & A. LAWEN. 1994. Up-regulation of the plasma membrane oxidoreductase as a prerequisite for the viability of human Namalwa rho[0] cells. J. Biol. Chem. **269:** 30097–30100.

55. SUN, I.L., E.E. SUN, F.L. CRANE & D.J. MORRE. 1990. Evidence for coenzyme Q function in transplasma membrane electron transport. Biochem. Biophys. Res. Commun. **172:** 979–984.

56. WOLVETANG, E.J., K.L. JOHNSON, K. KRAUER, S.J. RALPH & A.W. LINNANE. 1994. Mitochondrial respiratory chain inhibitors induce apoptosis. FEBS Lett **339:** 40–44.

57. LAWEN, A., R.D. MARTINUS, G.L. MCMULLEN, P. NAGLEY, F. VAILLANT, E.J. WOLVETANG & A.W. LINNANE. 1994. The universality of bioenergetic disease: The role of mitochondrial mutation and the putative inter-relationship between mitochondria and plasma membrane NADH oxidoreductase. Mol. Aspects Med. **15** Suppl: s13–27.

58. LAMPERTH, L., M.C. DALAKAS, F. DAGANI, J. ANDERSON & R. FERRARI. 1991. Abnormal skeletal and cardiac muscle mitochondria induced by zidovudine (AZT) in human muscle *in vitro* and in an animal model. Lab. Invest. **65:** 742–751.

59. DALAKAS, M.C., I. ILLA, G.H. PEZESHKPOUR, J.P. LAUKAITIS, B. COHEN & J.L. GRIFFIN. 1990. Mitochondrial myopathy caused by long-term zidovudine therapy [see comments]. N. Engl. J. Med. **322:** 1098–1105.

60. ROWLAND, M.A., P. NAGLEY, A.W. LINNANE & F.L. ROSENFELDT. 1998. Coenzyme Q_{10} treatment improves the tolerance of the senescent myocardium to pacing stress in the rat. Cardiovasc. Res. In press.

Mitochondrial Decay in Aging

Reversal through Supplementation of Acetyl-L-Carnitine and N-*tert*-Butyl-α-phenyl-nitrone[a]

TORY M. HAGEN[c], CAROL M. WEHR[d], AND BRUCE N. AMES[b]

Department of Molecular and Cell Biology, University of California at Berkeley, Berkeley, California 94720, USA

ABSTRACT: We show that mitochondrial function in the majority of hepatocytes isolated from old rats (24 mo) is significantly impaired. Mitochondrial membrane potential, cardiolipin levels, respiratory control ratio, and overall cellular O_2 consumption decline, and the level of oxidants increases. To examine whether dietary supplementation of micronutrients that may have become essential with age could reverse the decline in mitochondrial function, we supplemented the diet of old rats with 1% (w/v) acetyl-L-carnitine (ALCAR) in drinking water. ALCAR supplementation (1 month) resulted in significant increases in cellular respiration, mitochondrial membrane potential, and cardiolipin values. However, supplementation also increased the rate of oxidant production, indicating that the efficiency of mitochondrial electron transport had not improved. To counteract the potential increase in oxidative stress, animals were administered N-*tert*-butyl-α-phenyl-nitrone (30 mg/kg) (PBN) with or without ALCAR. Results showed that PBN significantly lowered oxidant production as measured by 2,7′-dichlorofluorescin diacetate (DCFH), even when ALCAR was coadministered to the animals. Thus, dietary supplementation with ALCAR, particularly in combination with PBN, improves mitochondrial function without a significant increase in oxidative stress.

Mitochondrial decay may be one of the principal underlying causes that lead to cellular decline in the aging process.[1,2] The factors involved in mitochondrial dysfunction remain to be elucidated; however, there is growing evidence that the mitochondria are ultimately the agents of their own decay. Mitochondrial electron transport is not completely efficient and a small yet detectable level of reactive oxygen species (ROS) are constantly being produced. Despite an impressive array of antioxidant defenses, the level of oxidative damage to key mitochondrial constituents is extensive. In particular, oxidative mtDNA damage is enormous compared to the levels found in nuclear DNA.[3] This high level of damage, if not repaired properly, could be converted into mutations and may be the cause of the 17-fold higher accumulation of mtDNA mutations compared to nuclear mutation levels with age.[4] The increased level of mutations would adversely affect mitochondrial electron transport and further increase the rate of ROS production. ROS may increasingly affect the inner mitochondrial membrane with age. The fatty acid components constituting

[a]This work was supported by a Grant from the Sandoz Gerontological Foundation (T.M.Hagen); by NCI Outstanding Investigator Grant CA, 39910; and NIEHS Center Grant, ES01896.

[b]To whom correspondence should be addressed. Tel: 510/642-5165; fax: 510/643-7935; e-mail: bnames@uclink4.berkeley.edu

[c]Current address: Linus Pauling Institute at Oregon State University, 571 Weniger Hall, Corvallis, OR 97331. Tel: 541/737-5083; fax: 541/737-5077; e-mail: tory.hagen@orst.edu

[d]wehr@uclink4.berkeley.edu

the phospholipids of the inner mitochondrial membrane become more unsaturated, making them even more prone to oxidative damage.[5] These changes could also alter membrane fluidity, which in turn, may change the conformation of transmembrane proteins embedded in the lipid bilayer. Alterations to membrane fluidity as well as oxidation of critical thiol groups on proteins may decrease substrate transport, reducing the ability of mitochondria to meet cellular energy demands or regulate calcium homeostasis.

Mitochondria may also be adversely affected by loss of key micronutrients supplied by the cell. Cardiolipin, an important phospholipid that serves as a cofactor for a number of critical mitochondrial transport proteins, declines significantly with age.[6,7] This decline may reflect enhanced oxidative damage and removal of cardiolipin from the membrane but may also be due to decreased *de novo* synthesis. Ubiquinone levels also decline with age, which may result in decreased mitochondrial electron transport.[8] Cellular and mitochondrial antioxidant status also becomes limited.[9] This loss may reflect increased use because of the heightened cellular oxidative stress in aged tissue or the decline in gastrointestinal absorption of ascorbate or precursors for glutathione synthesis. Loss of antioxidants, coupled with increased reactive oxygen species production, may lead to heightened susceptibility of mitochondria for oxidative damage. Finally, carnitine levels also become limited, which may deprive mitochondria of fatty acids for β-oxidation.[10]

Thus, there appear to be a number of interconnected mechanisms that eventually cause age-associated decline in mitochondrial dysfunction. Many of these pathways contribute to or result from constant oxidative damage to mitochondria, leading to accumulation of mtDNA mutations, damage to constituents of the inner mitochondrial membrane, and loss of critical micronutrients supplied by the cell. Although reversing the effects of mitochondrial mutations directly through genetic engineering is currently unfeasible, it may be possible to supply mitochondria with micronutrients through dietary supplementation and therefore maintain mitochondrial function. In the current study, we supplemented the diet of young (3–5 mo) and old (20–28 mo) rats with acetyl-L-carnitine (ALCAR) to determine whether this compound could positively affect mitochondrial function in hepatocytes isolated from old rats. We also sought to augment the antioxidant capacity of cells isolated from old rats by giving animals PBN, a potent hydrophilic antioxidant.

MATERIAL AND METHODS

The following chemicals were used: trypan blue, heparin (sodium salt), and rhodamine 123 (R123) (Sigma, St. Louis); 2′,7′-dichlorofluorescin diacetate (DCFH) (Molecular Probes, Eugene OR); collagenase (type D) (Boehringer Mannheim, Indianapolis); and *N*-tert-butyl-α-phenyl-nitrone (PBN) (Aldrich, Milwaukee). Acetyl-L-carnitine (chloride salt) (ALCAR) was provided by Sigma Tau (Pomezia, Italy). All other reagents were reagent grade or better. Double-distilled water was used throughout.

Rats (Fischer 344, virgin male, outbred albino) (3 to 5 mo) were obtained from Simonsen (Gilroy, CA). Old rats (identical strain; 20 to 28 mo) were from the National Institute on Aging animal colonies. All animals were fed Purina rodent chow and water *ad libitum*. Animals were acclimatized at the Berkeley animal facilities for at least one week prior to experimentation.

ALCAR Supplementation

Old and young rats were given a 1.5% (w/v) solution of ALCAR in their drinking water and allowed to drink *ad libitum* for one month prior to euthanasia and hepatocyte isolation. Both young and old rats typically drank ~25 mL/rat per day (data not shown), which would provide a daily ALCAR dose of approximately 0.5 g/kg body weight per day for old rats and 0.7 g/kg body weight per day for young rats.

PBN Administration

Some rats were administered 30 mg PBN/kg body weight (dissolved in DMSO, 15% (v/v), in physiological saline), twice daily by intraperitoneal injection for two weeks prior to sacrifice.

Cell Isolation and Separation

Liver tissue was dispersed into single cells by collagenase perfusion.[11] Cell number was assessed by using a hemocytometer, and viability was determined by trypan blue exclusion. Viability was usually greater than 90% in both age groups. Cells from old rats were separated into unique subpopulations, as described.[12]

Flow Cytometry

Hepatocytes (2.0×10^6 cells) were incubated with R123 (0.01 mg/mL) for 30 min at 37° C and then subjected to flow cytometry, as described.[12] Because different numbers of cells were analyzed in separate experiments, all data were normalized as to cell number.

DCFH Measurement

Formation of oxidants in cells were determined using DCFH.[13] Cellular oxygen consumption was measured using a Yellow Springs Instruments 5300 oxygen electrode and monitor. Cells (4.0×10^6 cells) were added to 3 mL of Krebs-Henseleit balanced salts medium supplemented with 1 mM glucose and 7 mM glutamate, pH 7.4, that had been previously equilibrated to 20° C, and oxygen consumption was monitored for at least 15 minutes.

Cardiolipin Measurement

Cardiolipin was separated from cellular lipid extracts using a Waters RCM 100 Radial Module and Radial Pak Resolve Silica cartridge (5 μm particle size; 0.8×10 cm) using a cyclohexane: 2-propanol: H_2O solvent system (45:50:5) at a flow rate of 1 mL/minute. [14] Cardiolipin was detected at 203 nm using a Kratos Spectraflow 773 UV detector and was quantified relative to standards.

Statistical Analysis

Statistical significance was determined using the paired Student's *t*-test or single factor ANOVA. Results are expressed as the mean ± SEM.

RESULTS

To date, most studies have used isolated mitochondria to determine age-related changes to the organelle. However, mitochondria become fragile and susceptible to lysis with age; thus purified mitochondria from old tissue do not truly represent that found *in vivo*.[15] Moreover, use of isolated mitochondria eliminates the possibility of distinguishing the consequences that age-related changes in their function may have on the cell as a whole. We have avoided these problems by assaying for mitochondrial changes in intact hepatocytes freshly isolated from young and old rats using noninvasive probes for mitochondrial function.

Functional Mitochondrial Decay

Isolated hepatocytes were incubated with R123, a fluorescent dye that accumulates specifically in the mitochondria (based on the inner membrane potential), as a probe to examine this key parameter of mitochondrial function. Results showed that at least two and often three distinct populations of hepatocytes develop with age (TABLE 1). The largest population of cells (accounting for 67% of the cells) from old rats had mitochondria with a significantly lower membrane potential than cells from young animals. The smaller sub-populations of cells from old rats had mitochondria that were moderately impaired or retained the same functional characteristics as seen in cells from young rats.[12] This type of age-related heterogeneity has not been reported before and thus provides a different per-

TABLE 1.

Experiment	Young	Old (total)	Fraction A	Fraction B	Fraction C
$\Delta\psi^{a}$	-154.3 ± 20.4	-101.1 ± 18.4	-70.4 ± 8.9	-92.6 ± 7.0	-154.4 ± 18.3
	$(3)^{e}$	(4)	(3)	(3)	(3)
O_2 consumption^{b}	480.1 ± 60.3	328.3 ± 42.3	305.8 ± 12.1	445.2 ± 22.8	500.7 ± 85.6
	(5)	(3)	(4)	(3)	(3)
RCR^{c}	8.9 ± 0.9	5.6 ± 0.9	5.4 ± 0.8	7.0 ± 0.7	4.5 ± 0.4
	(4)	(3)	(5)	(3)	(4)
DCFH^{d}	8.1 ± 1.2	14.5 ± 2.1	17.4 ± 1.8	10.2 ± 2.3	13.8 ± 2.9
	(4)	(4)	(4)	(4)	(4)

^{a}Average membrane potential in mV. Measured as described in ref.12.
^{b}In μM; O_2 per min per 10^7 cells.
^{c}Mitochondrial respiratory control ratio, as described in ref.12.
^{d}In fluorescence units/μM O_2/minute.
^{e}Numbers in parentheses denote n values.

spective as to mitochondrial dysfunction and the consequences of those changes to the cell during aging. We have successfully separated these cell subpopulations by centrifugal cell elutriation/cell sorting and have characterized some of the underlying events that may have caused the appearance of mitochondrial heterogeneity.

Cells containing mitochondria with the lowest functional ability also were the least metabolically active and had mitochondria that were more uncoupled and that released significantly higher levels of reactive oxygen species (per unit of oxygen consumed) than cells from young rats (TABLE 1). There was also an increase in cellular DNA-protein cross-links when compared to cells from young rats (data not shown). The other cell subpopulations from old rats also showed varying degrees of the same age-related alterations (TABLE 1). The major subpopulation of cells also had substantially lower levels of cardiolipin, a key phospholipid necessary for normal substrate transport in the mitochondrion (FIG. 2).

Rejuvenation through Dietary Supplementation

Because carnitine levels decline significantly with age,[10] we supplemented the diet of old rats with ALCAR, a derivative of L-carnitine, as a means of replenishing this key factor within the cell. Short-term ALCAR supplementation (1.5% (w/v) for 1 mo) resulted in substantial improvement in the average mitochondrial membrane potential (FIG. 1). This treatment also abolished the age-related appearance of the distinct cell subpopulations. Most importantly, ALCAR caused mitochondria in cells isolated from old rats to generate the same average mitochondrial membrane potential as cells isolated from untreated young animals (FIG. 1). ALCAR supplementation did not affect the average membrane potential in cells taken from young rats, indicating that ALCAR was exerting its effect by replenishing needed micronutrients necessary for proper mitochondrial function.

To determine whether the ALCAR-induced rejuvenation of mitochondrial membrane potential was due to reversal in the decline of mitochondrial biomolecules other than carnitine, we measured mitochondrial cardiolipin. Cardiolipin was chosen as a biomarker for changes in the mitochondrial membrane because of its critical role as an essential cofactor in mitochondrial transport proteins.[6] As shown in FIGURE 2, hepatocytes from old rats exhibit a marked decline in this phospholipid. Cardiolipin levels in hepatocytes dropped from 21.2 ± 2.5 (n = 5) to 10.5 ± 3.6 μg per 10^6 cells (n = 4) (p < 0.05) in cells from young rats versus cells from old rats. ALCAR supplementation reversed this loss in cardiolipin concentrations (FIG. 2). Thus, ALCAR supplementation not only facilitates increased fatty acid transport into mitochondria but also effectively reverses the age-associated decline in substrate transport due to loss of cardiolipin.

ALCAR also increased overall cellular respiration that had declined with age. Hepatocytes from old rats had an overall oxygen consumption rate of 328 ± 42 μM O_2/min per 10^7 cells (TABLE 1). By contrast, the oxygen consumption characteristics for hepatocytes from ALCAR-supplemented rats was 462 ± 18 μM O_2/min per 10^7 cells (n = 6) which was not significantly different from cells from young rats (TABLE 1).

Although ALCAR supplementation reversed many of the altered characteristics evident in mitochondrial metabolism with age, analysis of the rate of cellular oxidant flux (as measured by the rate of DCFH fluorescence) also showed that supplementation increased oxidant production. As shown in FIGURE 3, the rate of oxidants in hepatocytes from ALCAR-supplemented old rats was 18.2 ± 1.2 Fl. units/μM O_2 per 10^7 cells, approxi-

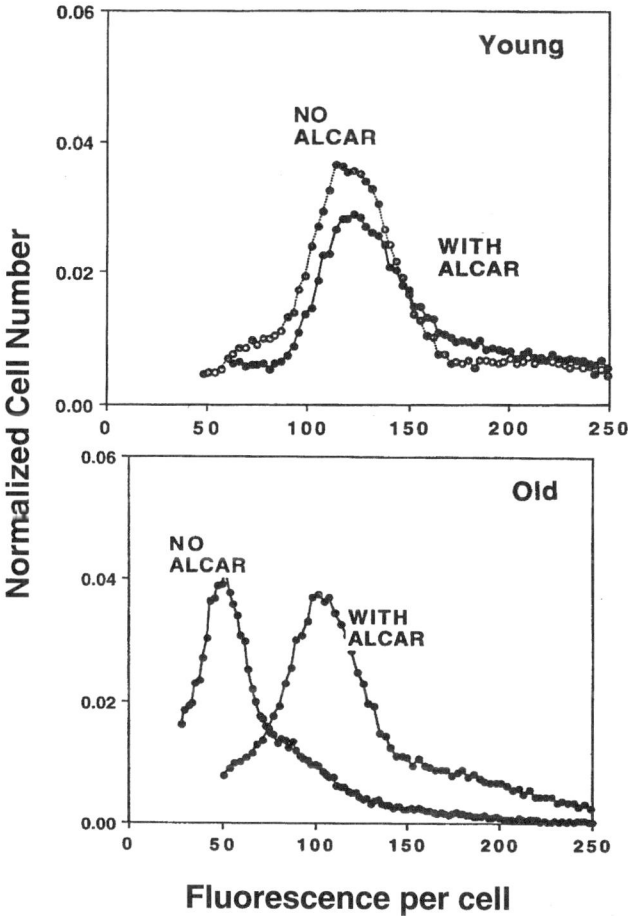

FIGURE 1. ALCAR reverses the age-associated decline in R123 fluorescence. Hepatocytes isolated from rats either supplemented with or without 1.5% (w/v) ALCAR for one month were incubated with R123 30 min before analysis by flow cytometry. Results show that ALCAR supplementation reverses the age-associated decline in mitochondrial membrane potential and abolishes the appearance of mitochondrial heterogeneity with age. Shown is a fluorogram typical of that seen in at least six experiments. (Hagen *et al.*[16] With permission from the National Academy of Sciences USA.)

mately a 30% increase over oxidant production in cells from untreated old rats. By contrast, ALCAR supplementation did not alter the rate of oxidant production in cells from young animals (data not shown). Although these results show that ALCAR markedly improves overall mitochondrial metabolism, it cannot mask the attendant accumulation of mitochondrial mutations that decrease the efficiency of electron transport and cause increased oxidant production in a more rapidly respiring mitochondrion.

Because ALCAR supplementation could result in increased oxidative stress to the cell, we also sought to establish whether administration of antioxidants could reduce the risk of

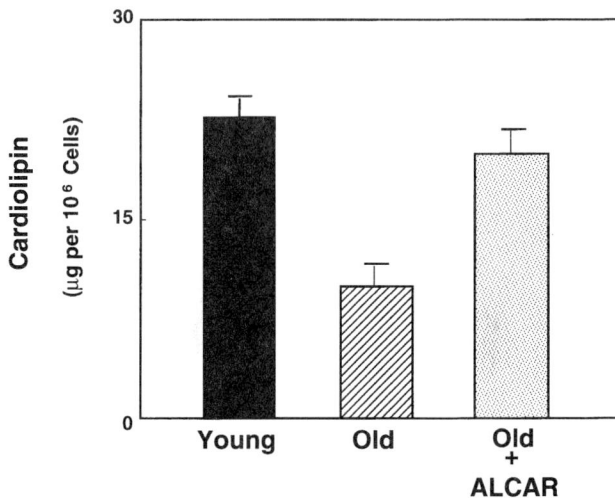

FIGURE 2. Cardiolipin, a key phospholipid necessary for mitochondrial substrate transport, was extracted from hepatocytes; the levels were assessed by HPLC. Results show that mitochondria in cells from old rats have significantly lower cardiolipin when compared to cells isolated from young rats. ALCAR supplementation reverses this decline. (Hagen et al.[16] With permission from the National Academy of Sciences.)

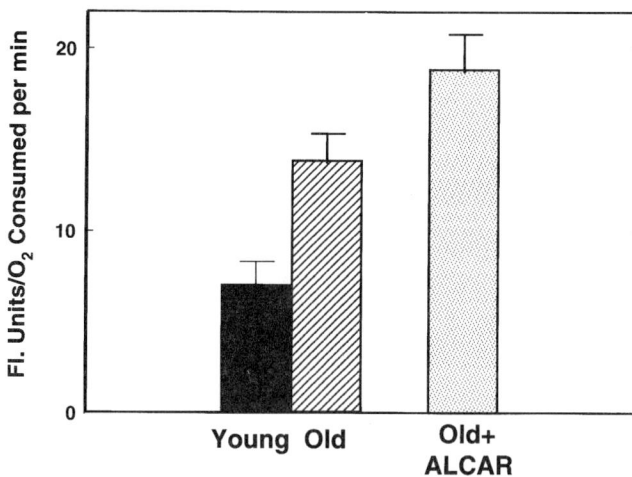

FIGURE 3. 2,7-Dichlorofluorescin diacetate (DCFH) was used as a probe to determine the rate of oxidant production in cells isolated from animals (with or without ALCAR). ALCAR supplementation causes an enhanced rate of oxidant production. These results indicate that ALCAR may improve mitochondrial function but may also increase the production of oxidants produced as by-products, due to increased inefficiencies in mitochondrial electron transport. (Hagen et al.[16] With permission from the National Academy of Sciences.)

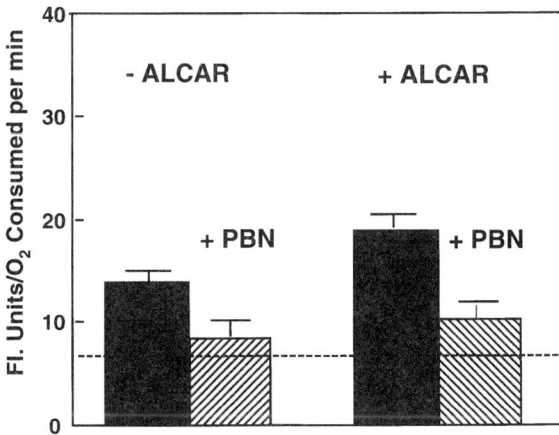

FIGURE 4. Hepatocytes isolated from 24-month-old rats that had been given PBN for two weeks with or without ALCAR have a significantly lower rate of oxidant production. Dashed line indicates the rate of oxidant production in cells from untreated young rats. Results show that PBN decreases the apparent rate of oxidant production, even in the presence of ALCAR.

oxidative injury. For this study, we administered PBN, a potent hydrophilic antioxidant, to rats with or without ALCAR. A twice daily dose of PBN (30 mg/kg) for two weeks significantly reduced the appearance of oxidants in cells taken from old rats without ALCAR supplementation (FIG. 4). The rate of oxidant flux in cells isolated from PBN-treated rats had an apparent oxidant rate of 9.2 ± 2.4 Fl. units/O_2 consumed/10^7 cells, which was no longer significantly different than cells from young animals ($n = 5$; $p < 0.04$). Moreover, the steady state levels of oxidative DNA damage markedly declined even when ALCAR was coadministered to the animals. In ALCAR- and PBN-treated rats, the rate of DCFH fluorescence was 11.6 ± 1.9 Fl. units/O_2 consumed/10^7 cells (FIG. 4). Again, this rate of oxidant production was not significantly different from controls ($n = 6$; $p < 0.05$). Thus, dietary supplementation with ALCAR, particularly in combination with PBN, improves mitochondrial function without a significant increase in oxidative stress.

DISCUSSION

As shown in the results of this paper, and in results previously published,[16] mitochondria within cells decay with age and could severely affect normal cellular metabolism. The degree of cellular dysfunction due to mitochondrial damage or how such changes could affect organ function is not yet known. This is partly due to difficulties in interpretation of results using isolated mitochondria because of the high degree of lysis that occurs when they are extracted from old animals. Furthermore, mitochondrial isolation from a whole organ cannot determine alterations that take place in specific cells within an organ. Therefore, isolating mitochondria, even if the fraction obtained accurately represented that found *in vivo*, would not reflect the extent of age-associated alterations at the cellular level.[15] We have avoided these problems by assaying for mitochondrial changes in cells freshly isolated from young and old rats.

The causes for the apparent heterogeneity in average mitochondrial membrane potential, respiratory control ratios, oxidant production, and cellular oxygen consumption (TABLE 1) are probably multifactorial. Both genetic and epigenetic changes to the mitochondria and to the cell might be involved. It can be assumed that the high level of mtDNA mutations with age may exert a physiological effect by creating mitochondria with altered components of the electron transport chain. These mutations may result in varying degrees of inefficient electron transport and a coincident increase in superoxide production. This concept is consistent with our findings of enhanced oxidant production in cells from aged rats, in general, and the varying degrees of oxidant production in the isolated cell subpopulations, in particular. Moreover, age-related inefficiency of electron transport would not be expected to improve with ALCAR supplementation, even though ALCAR apparently increases the flux of electrons through the electron transport chain. This may be the reason why ALCAR increases the rate of oxidant production, even though it reverses the decline in many parameters of mitochondrial function.

Such genetic changes that result in enhanced efflux of superoxide could cause epigenetic changes as well. Oxidants may cause increased damage and utilization of such critical micronutrients as ubiquinone or small molecular weight antioxidants. The significant loss of cardiolipin with age may, in part, be due to greater oxidative damage and turnover. Loss of cardiolipin, coupled with oxidation of critical thiol groups in key proteins, may adversely affect transport of substrates necessary for mitochondrial function. These changes could directly impact the ability of mitochondria to maintain their membrane potential and lead to the age-related changes observed.

The argument that epigenetic factors significantly affect mitochondrial function with age is supported by our ALCAR supplementation studies. ALCAR effectively abolished the heterogeneity in average mitochondrial membrane potential seen in cells from old rats and increased this key functional parameter back to the levels evident in cells from young animals. Moreover, ALCAR reversed the age-associated decline in cardiolipin levels, indicating that ALCAR not only can improve fatty acid transport but has a remarkable ability to maintain the mitochondrial inner membrane. It is presently not known how ALCAR exerts this effect. ALCAR may increase ATP production, which would allow increased synthesis and repair of damaged biomolecules. Measurement of ATP production and metabolite turnover studies in cells following ALCAR supplementation will be necessary to discern whether ALCAR works through these parameters. ALCAR supplementation also appears to benefit old animals in a variety of other ways. ALCAR prevents or slows age-related memory impairment through increased neurotransmitter production[17] and maintains synaptic contacts[18] and levels of certain hormonal receptors.[19] Thus, ALCAR can reverse many aspects of cellular dysfunction with age, principally through maintenance of mitochondrial function.

This report also shows that ALCAR supplementation increases the rate of oxidant production. However, because ALCAR also increases DNA repair, it is not clear whether this small but significant increase in oxidant flux translates into increased oxidative damage or DNA mutations. To ameliorate any possible increase in oxidative damage, we also administered PBN to rats. PBN is a well-characterized free radical spin-trapping agent that has also been shown to be a potent hydrophilic antioxidant. It effectively reduces oxidative injury in a variety of pathophysiological conditions.[20] PBN significantly lowered the age-associated increase in oxidant production, even when it was coadministered with ALCAR. Thus, ALCAR supplementation, particularly in combination with PBN, provides an effective means of reversing many parameters of mitochondrial dysfunction with age. Long-term administration of these compounds to animals is warranted to discern whether they can ameliorate mitochondrial decay with age and enhance the overall quality of life for a longer period of time.

ACKNOWLEDGMENT

We thank M.-H. Song, for her excellent technical assistance.

REFERENCES

1. HARMON, D. 1983. Age **6:** 86–94.
2. SHIGENAGA M.K. *et al.* 1994. Oxidative damage and mitochondrial decay in aging. Proc. Natl. Acad. Sci. USA **91:** 10771–10778.
3. RICHTER, C. 1995. Oxidative damage to mitochondrial DNA and its relationship to ageing. Int. J. Biochem. Cell Biol. **27:** 647–653.
4. WALLACE D.C. *et al.* 1987. Sequence analysis of cDNAs for the human and bovine ATP synthase beta subunit: Mitochondrial DNA genes sustain seventeen times more mutations. Curr. Genet. **12:** 81–90.
5. LAGANIERE S. & B.P. YU. 1993. Modulation of membrane phospholipid fatty acid composition by age and food restriction. Gerontology **39:** 7–18.
6. PARADIES, G. & F.M. RUGGIERO. 1990. Age-related changes in the activity of the pyruvate carrier and in the lipid composition in rat-heart mitochondria. Biochim. Biophys. Acta **1016:** 207–212.
7. HOCH, F.L. 1992. Cardiolipins and biomembrane function. Biochim. Biophys. Acta **1113:** 71–133.
8. SUGIYAMA, S. *et al.* 1995. Preservation of mitochondrial respiratory function by coenzyme Q_{10} in aged rat skeletal muscle. Biochem. Mol. Biol. Int. **37:** 1111–1120.
9. MEZZETTI, A. *et al.* 1996. Systemic oxidative stress and its relationship with age and illness. J. Am. Geriatr. Soc. **44:** 823–827.
10. HANSFORD, R.G. 1983. Bioenergetics in aging. Biochim. Biophys. Acta **726:** 41–80.
11. MOLDÉUS, P. *et al.* 1978. Isolation and use of liver cells. Methods Enzymol. **52:** 60–71.
12. HAGEN, T.M. *et al.* 1997. Mitochondrial decay in hepatocytes from old rats: Membrane potential declines, heterogeneity and oxidants increase. Proc. Natl. Acad. Sci. USA **94:** 3064–3069.
13. LEBEL, C.P. *et al.* 1992. Evaluation of the probe 2′,7′-dichlorofluorescin as an indicator of reactive oxygen species formation and oxidative stress. Chem. Res. Toxicol. **5:** 227–231.
14. ROBINSON, N.C. 1990. Silicic acid HPLC of cardiolipin, mono-, and dilysocardiolipin, and several of their chemical derivatives. J. Lipid Res. **31:** 1513–1516.
15. WILSON, P.D. & L.M. FRANKS. 1975. The effect of age on mitochondrial ultrastructure and enzymes. Adv. Exp. Med. Biol. **53:** 171–183.
16. HAGEN, T.M. *et al.* 1998. Acetyl-L-carnitine fed to old rats partially restores mitochondrial function and ambulatory activity. Proc. Natl. Acad. Sci. USA **95:** 9562–9566.
17. CARTA, A., M. CALVANI, D. BRAVI & S.N. BHUACHALLA. 1993. Acetyl-L-carnitine and Alzheimer's disease: Pharmacological considerations beyond the cholinergic sphere. Ann. N.Y. Acad. Sci. **695:** 324–326.
18. BERTONI-FREDDARI, C. *et al.* 1994. Dynamic morphology of the synaptic junctional areas during aging: The effect of chronic acetyl-L-carnitine administration. Brain Res. **656:** 359–366.
19. CASTORNICI, M. *et al.* 1994. Age-dependent loss of NMDA receptors in hippocampus, striatum, and frontal cortex of the rat: Prevention by acetyl-L-carnitine. Neurochem. Res. **19:** 795–798.
20. KURODA, S. *et al.* 1996. Delayed treatment with alpha-*N-tert*-butyl-nitrone (PBN) attenuates secondary mitochondrial dysfunction after transient focal ischemia in the rat. Neurobiol. Dis. **3:** 149–157.

Mitochondrial Free Radical Production and Aging in Mammals and Birds[a]

GUSTAVO BARJA[b]

Department of Animal Biology-II (Animal Physiology), Faculty of Biology, Complutense University, Madrid 28040, Spain

ABSTRACT: The mitochondrial rate of oxygen radical (ROS) production is negatively correlated with maximum life span potential (MLSP) in mammals following the rate of living theory. In order to know if this relationship is more than circumstantial, homeothermic vertebrates with MLSP different from that predicted by the body size and metabolic rate of the majority of mammals (like birds and primates) must be studied. Birds are unique because they combine a high rate of basal oxygen consumption with a high MLSP. Heart, brain, and lung mitochondrial ROS production and free radical leak (percent of total electron flow directed to ROS production) are lower in three species of birds of different orders than in mammals of similar body size and metabolic rate. This suggests that the capacity to show a low rate of ROS production is a general characteristic of birds. Using substrates and inhibitors specific for different segments of the respiratory chain, the main ROS generator site (responsible for those bird–mammalian differences) in state 4 has been localized at complexes I and III in heart mitochondria and only at complex I in nonsynaptic brain mitochondria. In state 3, complex I is the only generator in both tissues. The results also suggest that the iron–sulphur centers are the ROS generators of complex I. A general mechanism that allows pigeon mitochondria to show a low rate of ROS production can be the capacity to maintain a low degree of reduction of the ROS generator site. In heart mitochondria, this is supplemented with a low rate of oxygen consumption physiologically compensated with a comparatively higher heart size. A low rate of free radical production near DNA, together with a high rate of DNA repair, can be responsible for the slow rate of accumulation of DNA damage and thus the slow aging rate of longevous animals.

Many theories have been proposed during the last decades trying to explain aging. Among them, the free radical theory of aging[1] and its special relationship to mitochondria first signaled by D. Harman in 1972,[2] is gaining increasing support from available data.

The majority of previous work regarding that theory has concentrated on the possible role of antioxidants. Experiments in large samples of animals have studied the effect of increasing the tissue antioxidant levels on long-term survival. These manipulations have been performed in different ways, including dietary supplementation,[3–5] antioxidant induction,[6,7] or gene transfection.[8–10] The outcome in the majority of these studies is the same: the increases in tissue antioxidants can increase survival and mean life span, especially when the experimental conditions are not fully optimum for survival, but the maximum life span potential (MLSP) is not increased. The observed improvement in mean life span is in agreement with the increasingly held notion that antioxidants can unspecifically

[a]This work was supported in part by a Grant from the National Health Research Foundation (No. 96/1253).

[b]Mailing address: Departamento de Biología Animal-II (Fisiología Animal), Facultad de Biología, Universidad Complutense, Madrid 28040, Spain. Tel: 34-91-3944919; fax: 34-91-3944935.

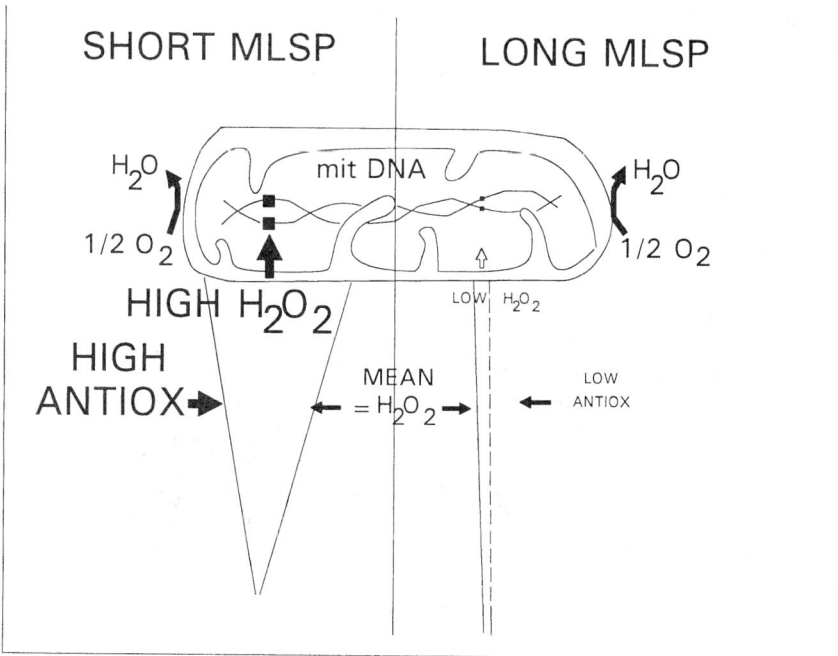

FIGURE 1. Animals with long MLSP show both low levels of tissue antioxidants and of mito-chondrial free radical production, whereas the contrary occurs in short-lived species. Similar levels of *mean* cellular H_2O_2 in short- and long-lived species are expected, but the local concentration of O_2 radicals near the places of free radical production would be much lower in long- than in short-lived animals. This will decrease oxidative damage at critical targets situated near the sites of free radical production, like the mitochondrial DNA. Modified from Barja *et al.*[23]

protect against the development of degenerative diseases, including cardiovascular diseases[11] and cancers.[12]

On the other hand, comparative studies in mammalian[13–20] or vertebrate[21–23] species with widely different MLSPs have shown that endogenous constitutive levels of tissue antioxidants are negatively correlated with MLSP. This led us to propose[21,23] that the reason for this is that the rate of free radical production near DNA (*e.g.*, at the inner mitochondrial membrane) is negatively correlated with MLSP (Fig.1). Longevous species would produce radicals slowly at specific sites located near DNA, and this would be partially responsible for their slow rate of accumulation of DNA damage and slow aging rate. Available data supporting this notion in mammals and birds, accumulated in recent years, are the subject of this review.

THE RATE OF LIVING THEORY

Classic studies have shown that the basal metabolic rate is generally related to aging and MLSP in the majority of animal species studied. This relationship is known as the "rate of living" theory of aging.[24] Even though the first global approach to this problem can be attributed to Pearl, the constancy of the total metabolic output (LEP = life energy potential; the number of total calories transformed per kg during the whole life span) in some mammalian species had already been described much earlier by Max Rubner.[25]

The basal specific rate of oxygen consumption (per gram weight) negatively correlates with body weight, with an exponent of –0.25 in mammals, whereas MLSP positively correlates with body weight, with an exponent of around +0.20, also in mammals. Thus, the LEP,[25,26] which is equal to the specific metabolic rate multiplied by the MLSP, is essentially a size-independent parameter in the majority of mammals inasmuch as both exponents (–0.25 and about +0.20) tend to cancel each other.

The constancy of LEP in the majority of mammals is not restricted to this particular animal group because MLSP also increases as a function of body weight in birds, with an exponent of +0.19 (FIG. 2). In other animals there are not enough reliable data, but present information suggests that the rate of living theory will also hold essentially true inasmuch as large animals in each phylogenetic group also tend to live longer, and the negative relationship between specific metabolic rate and body weight with an exponent of –0.25 is a universal characteristic in all vertebrate and invertebrate animal groups so far studied.[27,28] Furthermore, in poikilothermic animals the MLSP increases in proportion to the decrease in metabolic rate brought about by a decrease in the temperature of maintenance. The specific metabolic rate corresponds to the number of calories transformed per unit time and weight. Nevertheless, because it is closely related to the rate of oxygen consumption in aerobic animals, it is tempting to suggest that the rate of living theory arises from the possibility that if a given animal species consumes a large amount of O_2 at the mitochondrial level (to sustain a high basal metabolic rate), it would also produce a large number of oxygen radicals per unit time at these organelles. The basic process underlying the rate of living theory could then be the rate of production of O_2 radicals. However, we must critically consider that if the basal metabolic rate is faster and if the tissues consume more O_2 per

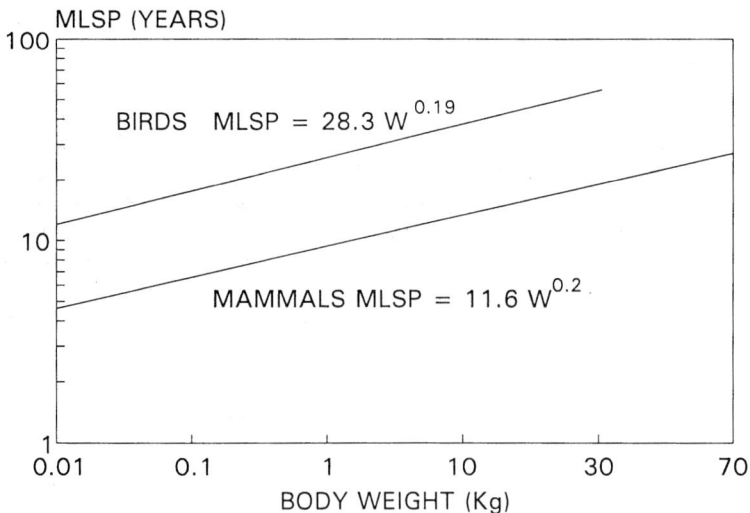

FIGURE 2. Relationship between body weight (W) and maximum life span potential (MLSP) in mammals and birds. Based on Calder[29] and Prinzinger.[28]

time unit, they will also synthesize and degrade many kinds of molecules faster, and a myriad of biochemical processes will also run at a quicker pace. Thus, the basic phenomenon underlying the rate of living theory could be the rate of O_2 radical production, but, in principle, it could also correspond to many other factors related to the metabolic rate.

In clarifying this problem, an interesting approach is to study animal species with different LEPs. When this occurs inside a phylogenetic group, the species implicated are considered as exceptions to the rate of living theory. Well-known examples are primates and humans, who have an LEP 2–4 times greater than that of the majority of mammals.[26] Humans show the longest MLSP and LEP among primates, and the human MLSP is four times higher than expected from the rate of living theory. The only two groups of animals in which the relationship between metabolic rate, body weight, and MLSP has been extensively studied, mammals and birds, also show different LEPs. As it is apparent in FIGURE 2, the slope of the line relating body weight and MLSP is similar in birds (exponent +0.19) and mammals (exponent +0.20), but the line is shifted upwards in birds in relation to mammals (the "a" coefficient is higher in birds). At a given size (and oxygen consumption), birds live around 3–4 times longer[29,30] than mammals (the LEP is higher in birds than in mammals). The causes of this extraordinary longevity are not known. The rate of living theory holds true inside both animal groups, but something occurred during evolutionary divergence from reptiles that allowed birds to show simultaneously high oxygen consumption values and high MLSPs, whereas from reptiles to mammals the "price" paid for an increased oxygen consumption was a decrease in MLSP (mammals of the same size as poikilothermic vertebrates usually show a much higher O_2 oxygen consumption and a shorter MLSP). Birds would represent a problem for the free radical theory of aging if their high basal rate of oxygen consumption would lead (as it is commonly believed) to a high rate of free radical generation in their mitochondria. If birds were to generate large numbers of free radicals in their tissues, the free radical theory would lack universal applicability, even for homeothermic vertebrates. On the contrary, if the free radical theory is correct, bird mitochondria must produce a small number of radicals per time unit, in spite of the high oxygen consumption and metabolic rate of these animals.

MITOCHONDRIAL FREE RADICAL PRODUCTION AND MLSP

Recent data show that mitochondrial free radical production is lower in the white-footed mouse (MLSP = 8 years) than in the house mouse (MLSP = 4 years) and that it inversely correlates with MLSP in five species of flies.[31] Mitochondrial oxygen radical production has also been compared among mammalian species following the rate of living theory and having different maximum longevities.[32,33] The results obtained showed a strong negative exponential correlation between liver mitochondrial O_2^- or H_2O_2 production ($r = -0.91$, FIG. 3) and MLSP. A similar negative relationship in mitochondria from kidney and heart of the same species was found afterwards by the same authors.[34] Nevertheless, because the species included followed the rate of living theory (decrease in MLSP as body size decreases and basal metabolic rate increases), the results obtained could also be interpreted as a correlate of that theory: the species with short MLSP could show high mitochondrial H_2O_2 production simply because their rates of mitochondrial O_2 consumption were also higher. A positive correlation between mitochondrial oxygen consumption and oxygen radical production and between mitochondrial oxygen radical production and

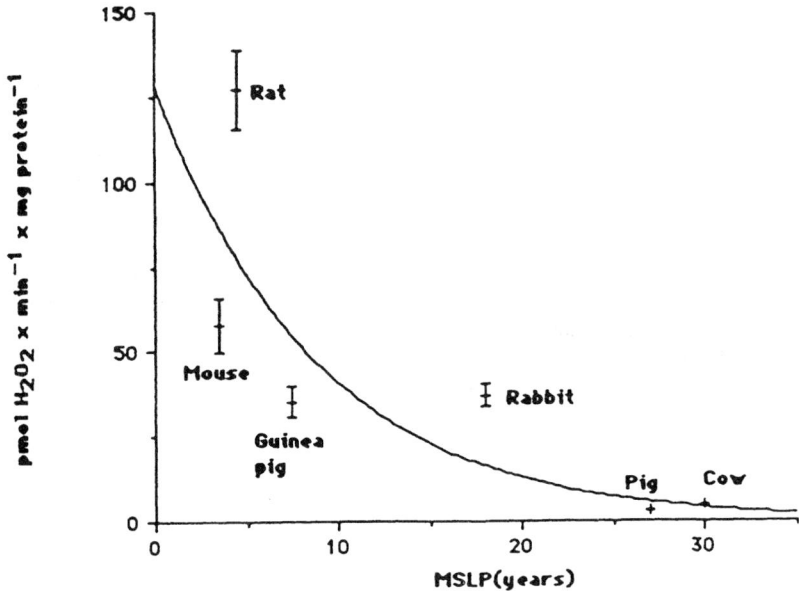

FIGURE 3. Negative correlation between liver mitochondrial H_2O_2 production and maximum life span (MLSP) in mammalian species following the rate of living theory. A similar relationship has been obtained in the same animal species for heart and kidney mitochondrial O_2^- and H_2O_2 production[34] and for liver mitochondrial O_2^- production.[32] $r = -0.91$. (Sohal et al.[33] With permission from *Mechanisms of Ageing and Development.*)

basal metabolic rate were indeed found in that work.[34] Thus, as explained above, these studies cannot discard the possibility that the correlation observed between mitochondrial oxygen radical production and MLSP were due to the correlation of mitochondrial oxygen radical production with the basal metabolic rate, this last parameter hypothetically correlating with other unknown factors causing aging. This is why the mitochondrial H_2O_2 production in birds, animals with both high oxygen consumption and high MLSP, was studied. If a low rate of free radical production contributes to slow aging rate in birds, their mitochondria should show a low rate of H_2O_2 production, in spite of their high respiratory activity. In an initial study, we have indeed shown that crude brain mitochondria and lung mitochondria show a rate of oxygen radical production substantially lower in pigeons (MLSP = 35 years) than in rats (MLSP = 4 years).[35] These two vertebrate homeotherms have a similar body size and basal metabolic rate (oxygen consumption). A lower mitochondrial H_2O_2 production in the pigeon than in the rat has been also independently reported for brain, heart, and kidney.[36] We also found that the percent of electrons out of sequence, which reduce oxygen to oxygen radicals at the respiratory chain (the percent free radical leak), instead of reducing oxygen to water at the terminal cytochrome oxidase, was lower in pigeon than in rat in crude brain mitochondria and lung mitochondria.[35] The simple idea that a high oxygen consumption necessarily leads to a high rate of oxygen radical production at mitochondria was not true, at least in the pigeon.

The above results obtained in crude pigeon mitochondria encouraged us to perform more detailed additional studies of free radical production in bird mitochondria of the same or different species, using a specific and sensitive fluorometric kinetic H_2O_2 detection method that does not alter the respiratory control index. This method instantaneously reacts to variations in H_2O_2 levels and does not show interference from mitochondrial antioxidants.[37] FIGURE 4 shows some of the results obtained, using substrates and inhibitors specific for different segments of the respiratory chain, in rat and pigeon mitochondria from a postmitotic tissue highly relevant for aging and having the richest mitochondrial density among vital organs, the heart. In both species, free radical production with pyruvate/malate was strongly increased to similar levels by addition of either rotenone or antimycin A, and further addition of myxothiazol decreased free radical production to the levels observed with pyruvate/malate alone (FIG. 4).[38] All the rates of free radical production shown in FIGURE 4 were higher in rat than in pigeon mitochondria. Other observations of that study were (1) free radical production was higher with pyruvate/malate than with succinate; (2) thenoyltrifluoroacetone did not increase succinate-supported free radical production; (3) both ethoxyformic anhydride and chloromercuribenzoate strongly depressed the rotenone-stimulated pyruvate/malate-supplemented rates of free radical production; and (4) the basal (substrate alone) and maximum (with pyruvate/malate plus rotenone or antimycin A) free radical leaks were higher in rat than in pigeon heart mitochondria. These observations localized the main free radical generator of both rat and pigeon heart mitochondria at complexes I and III of the respiratory chain, in agreement with previous studies, mainly performed in rat or cow heart mitochondria.[39–45] The results also suggest that the complex I free radical generator is situated between the sites of ferri-

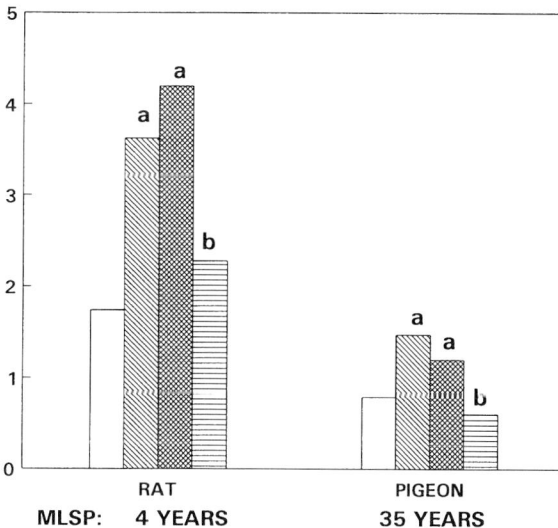

FIGURE 4. Heart mitochondrial H_2O_2 production in rats and pigeons with pyruvate/malate (PYR) as substrates (nmol H_2O_2/min/mg protein). □, PYR; ▨, PYR+AA; ▩, PYR+ROT; ▤, PYR+AA+MYX. Inhibitors: AA (antimycin A); ROT (rotenone); MYX (myxothiazol). MLSP = maximum life span potential. a: significantly different from pyruvate/malate alone; b: significantly different from PYR+AA. All analogous values were significantly higher in rats than in pigeons. Data from Herrero and Barja.[38]

cyanide and ubiquinone reduction and possibly corresponds to the iron-sulfur centers of this complex. In the case of heart mitochondria, both complex I and complex III would be responsible for the lower rate of free radical generation observed in the pigeon in relation to the rat.

Our initial rat–pigeon studies in crude brain mitochondria were also followed by more detailed studies in nonsynaptic brain mitochondria of these two animal species isolated with ficoll gradients.[46] Similar to what was obtained in heart mitochondria, rotenone and antimycin A strongly stimulated free radical production with pyruvate/malate, ethoxyformic anhydride and chloromercuribenzoate strongly depressed the rotenone-stimulated pyruvate/malate-supplemented rates of free radical production, thenoyltrifluoroacetone did not increase succinate-supported free radical production, and free radical production and leaks were higher again in the rat than in the pigeon. Nevertheless, this time myxothiazol did not decrease the antimycin-stimulated rate of free radical production with pyruvate/malate, and free radical production with succinate was not only lower than with pyruvate/malate, but it was almost undetectable. In addition, using ferrocytochrome c as substrate in hypotonically treated mitochondria, direct evidence of the lack of involvement of complex IV in free radical generation was obtained. All those results showed that the only free radical generator of nonsynaptic brain mitochondria is located at complex I in both rats and pigeons and probably corresponds again to the complex I iron–sulfur centers. This generator is thus exclusively responsible for the lower rate of free radical generation of pigeon versus rat nonsynaptic brain mitochondria.

The studies described above offered also some clues to the mechanism that allows pigeon mitochondria to show lower rates of free radical production than rat mitochondria. State 4 oxygen consumption was similar in rat and pigeon nonsynaptic brain mitochondria, whereas the free radical leak was lower in the pigeon (see above), and the addition of rotenone to pyruvate/malate-supplemented nonsynaptic brain mitochondria eliminated the differences in free radical production between both species.[46] In the case of heart mitochondria, however, state 4 oxygen consumption and free radical leak were lower in the pigeon than in the rat, and the difference in free radical generation between both species persisted after addition of rotenone to pyruvate/malate-supplemented mitochondria.[38] Addition of ADP caused a larger decrease in free radical leak in rat than in pigeon pyruvate/malate-supplemented mitochondria obtained from both tissues.[47] Thus, the relatively lower free radical production of nonsynaptic pigeon brain mitochondria seems to be due to a capacity of these mitochondria to maintain a lower degree of reduction of the complex I generator in the steady state. In the case of heart mitochondria, this mechanism is supplemented with a relatively low oxygen consumption and electron flow in the pigeon. This lower mitochondrial oxygen consumption (and thus probably lower rate of ATP generation) is compensated for with the much larger heart size (and stroke volume) of birds, in

---

FIGURE 5. Tracings showing the effect of ADP on free radical production of rat heart mitochondria with complex II- (A) or complex I-linked (B) substrates during the stimulation of oxygen consumption from state 4 to state 3. H_2O_2 increases the fluorescence at 312 nm excitation and 420 nm emission. PYR/MAL = pyruvate/malate; ADP (500 μM). Addition of ADP during the kinetic run stopped free radical production with succinate (A) but not with pyruvate/malate (B). Transient perturbation of the tracings at the moment of ADP addition are due to a slight opening of the sample compartment to add ADP. (Herrero and Barja.[47] With permission from the *Journal of Bioenergetics and Biomembranes*.)

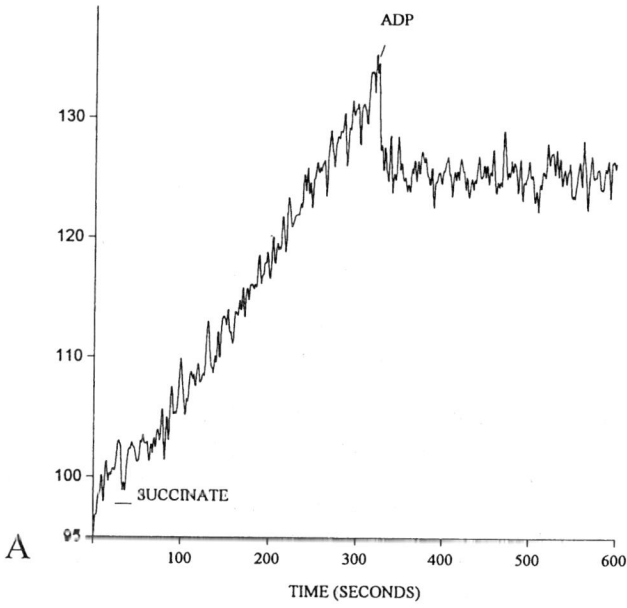

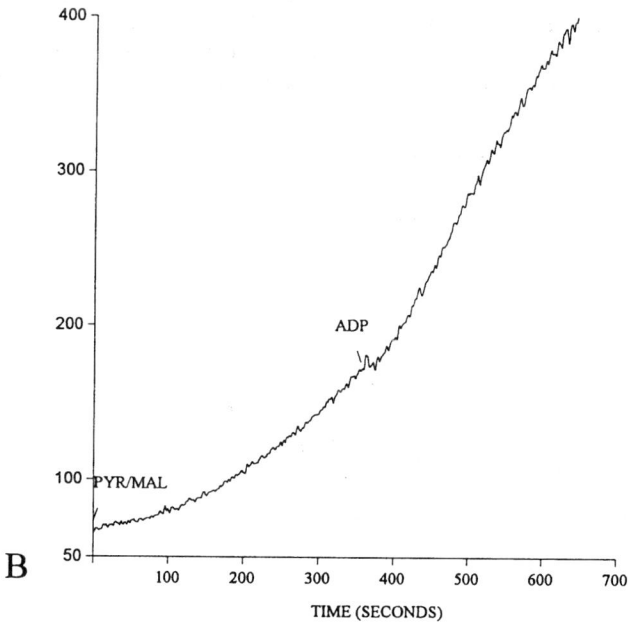

general (including pigeons), in relation to mammals, allowing a similar cardiac output per unit body mass in both groups.

The effect of ADP on the rate of oxygen radical generation of heart and nonsynaptic brain rat and pigeon mitochondria has also been recently studied.[47] In agreement with classic studies,[48,49] succinate-supplemented rat and pigeon heart mitochondrial free radical production was stopped by ADP additions causing the stimulation of respiration from state 4 to state 3 (FIG. 5A). Nevertheless, with complex I–linked substrates, mitochondria produce free radicals in state 3 at rates similar or somewhat higher than during resting respiration (FIG. 5B). The absence of sharp increases in free radical production in spite of a strong increase in oxygen consumption during oxidative phosphorylation (state 3) was possible due to strong decreases of free radical leak in that state. This shows again (like in the rat–pigeon comparison) that oxygen radical generation does not necessarily increase in proportion to increases in mitochondrial oxygen consumption. In fact, the decrease in free radical leakage during the state 4 to state 3 transition can help to explain two apparent paradoxes: (1) the lack of massive muscle oxidative damage and shortening of life span due to exercise, in spite of up to 23-fold increases of oxygen consumption, together with the very low levels of antioxidants present in muscles (and heart and brain); (2) the presence of some degree of oxidative stress during exercise, due to continuation or slight stimulation of complex I free radical production in state 3, in spite of the stop of mitochondrial free radical production by ADP with succinate as substrate. These results also showed that, whereas the complex III generator stops producing free radicals in state 3, this does not happen to the complex I generator. Taken together, the data point to a main role of complex I in free radical generation because, at variance with complex III, it is active in both nonsynaptic brain mitochondria and in heart mitochondria, and it is also active in the two main physiologically relevant states, state 3 and state 4.

The lower rates of free radical generation of pigeon versus rat mitochondria,[35,36,38,46,47] observed in at least four tissues (heart, brain, lung, and kidney) and occurring both in state 4 and state 3,[47] are in agreement with the free radical theory of aging, taking into account the MLSPs of the two species. Nevertheless, the lower free radical production of the

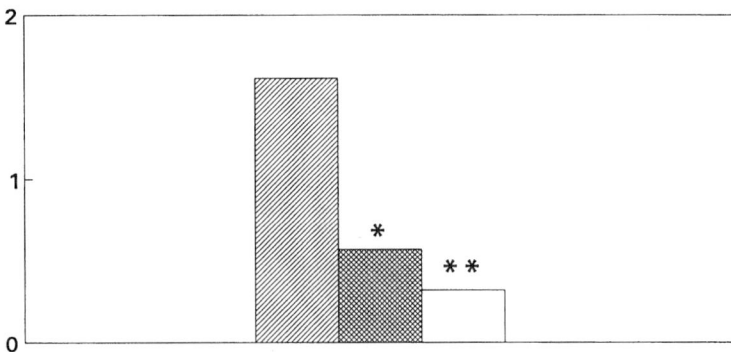

FIGURE 6. Heart mitochondrial H_2O_2 production in mice, parakeets (*Melopsittacus undulatus*), and canaries (*Serinus canarius*) with pyruvate/malate as substrates (nmol H_2O_2/min/mg protein). MLSP = maximum life span potential. * ($p < 0.05$) and ** ($p < 0.01$): significantly different from mouse H_2O_2 production. ▨, mouse (MLSP, 3.5 yr); ▩, parakeet (MLSP, 21 yr); ▭, canary (MLSP, 24 yr).

pigeon could also be due in principle to other characteristics of this particular animal species not related to its high MLSP. In order to extrapolate the low free radical production of the pigeon to birds, in general, we considered it important to study the same problem in other birds species appertaining to different orders than pigeons (Columbiformes). With this purpose, heart mitochondria were isolated from mice (MLSP = 3.5 years), canaries (*Serinus canarius*; MLSP = 24 years; order Passeriformes) and parakeets (*Melopsittacus undulatus*; MLSP = 21 years; order Psittaciformes). The three homeotherm species also show, like in the rat–pigeon comparison, similar values of basal-specific metabolic rate and body size. As is shown in FIGURE 6, oxygen radical production of heart mitochondria is also significantly lower in canaries and parakeets than in mice. This suggests that the capacity to show a low rate of ROS production is a general characteristic of bird mitochondria.

FREE RADICAL PRODUCTION, DNA, AND AGING

The lower free radical production of bird versus mammalian mitochondria strongly supports the free radical theory of aging. It also supports the notion that the inverse correlation observed between mitochondrial free radical production and MLSP in mammals following the rate of living theory is mechanistically related to their MLSPs, instead of being a simple correlate of their metabolic rates. The rate of living theory of aging has many exceptions, but the negative relationship between mitochondrial free radical production and MLSP occurs in all the animal species studied up to date, either mammals or birds. Longevous animals always show relatively low rates of mitochondrial free radical production, no matter if their metabolic rate is low (mammals of large body size) or high (birds). Our results also show that the rate of oxygen radical production of heart mitochondria is similar in rats and mice (see FIGURES 4 and 6), in agreement with their similar longevities (3.5 and 4 years), even though the basal-specific metabolic rate is two-fold higher in the mouse than in the rat. Here again the rate of free radical production, not the rate of oxygen consumption, correlates with MLSP and with the rate of aging. Available data indicate that a parameter relating oxidative stress to MLSP is the (mitochondrial) rate of free radical production.

If mitochondrial free radical production is important for aging, this will have a reflection in the mitochondria themselves. The possible relevance of these organelles for aging was first signaled by Denham Harman in 1972.[2] Recently, much evidence of the implication of mitochondria in the aging process has accumulated. More than 90% of the oxygen is consumed by mitochondria in healthy tissues. Thus, it is expected that the main free radical source in healthy tissues, not subjected to a higher than normal oxidative stress, will be the mitochondria. The mitochondrial DNA has a very high information density, lacks protective histones and polyamines, and shows a much lower repair than nuclear DNA. Furthermore, the mitochondrial DNA is situated in close vicinity to the main free radical generator of the healthy organism, the inner mitochondrial membrane, and occasionally even contacting it.[50,51] In agreement with the above, oxidative damage in mitochondrial DNA (oxo^{8}dG: 8-oxo-2′-deoxyguanosine, referred to dG) is around 15-fold higher than in nuclear DNA in rats.[51] The same is true for mitochondrial versus nuclear DNA deletions. Oxidative damage and deletions in mitochondrial DNA exponentially increase with age in the human heart.[52] Caloric restriction, the only experimental manipulation capable of slowing the rate of aging and of increasing the MLSP, decreases mitochondrial free radical production in various mice tissues without consistently changing antioxidant levels.[53] We have also found recently that liver mitochondria from longevous species constitutively have a lower degree of membrane fatty acid unsaturation than those of short-lived species. This is not due to a low polyunsaturated fatty acid (PUFA) content; instead, it is mainly due to a redistribution between types of PUFAs, from the highly unsaturated docosa-hexaenoic (22:6n-3) and arachidonic (20:4n-6) acids to the less unsaturated linoleic acid (18:2n-6) in longevous animals.[54] This leads to a much higher resistance to lipid peroxida-tion in mitochondria from longevous animals.[54]

Short-lived species, like laboratory rodents, have a large rate of generation of oxygen radicals and high levels of cellular antioxidants. These two characteristics would be com-pensatory in many parts of the cell. Longevous species have small rates of free radical generation, evolutionarily compensated for with low constitutive antioxidant levels. The mean cellular H_2O_2 concentration will thus be similar in both kinds of species. At the places of free radical generation (*e.g.*, at the inner mitochondrial membrane), however, the local concentration of free radicals will be much higher in short-lived than in long-lived animals (FIG. 1), possibly leading to a much higher oxidative damage to mitochondrial DNA at a given age in animals with short MLSPs.[21,23,55]

The majority of cellular genes are situated in the nucleus, not at the mitochondria. Even though all reports are not in agreement,[56] it has been described that oxo[8]dG increases with age in rat tissues.[57] Most importantly, liver and urine oxo[8]dG levels are inversely related to MLSP in mammals, humans showing the minimum levels and the largest longevity among studied species.[14,58] Recent data from our laboratory show that heart oxo[8]dG is also three-fold lower in pigeons than in rats.[59] This is consistent with the lower rates of mitochon-drial free radical generation of longevous species. How the mitochondrial-derived oxida-tive damage is transmitted from mitochondria to the nucleus is not known at present. Various possible explanations have been proposed,[60,61] but up to now, none of them has reached experimental demonstration. A complementary possibility is that differences among animal species in the rate of free radical production at critical sites (near DNA) occur not only at mitochondria but also at the nuclear membrane or at sites of binding of metals to nuclear DNA. Although much weaker in activity than the mitochondrial respira-tory chain, the nuclear membrane is known to contain an electron transport chain capable of free radical generation. Clarification of these possibilities must wait for the overcoming of methodological difficulties concerning measurements of free radical production at those nuclear sites in species with different longevities.

It is generally believed that DNA damage is of paramount importance for aging. Thus, the rate of production of free radicals and other damaging chemicals would be an impor-tant determinant of aging and MLSP but not the only one. The rate of DNA repair must also be important. Various works have indeed shown the existence of strong positive cor-relations between DNA repair systems and mammalian MLSP across species, including humans.[62–65] The combination of high levels of DNA repair with low rates of free radical production near DNA in longevous animals can contribute to explain their very small lev-els of steady-state oxidative DNA damage[14,58] and slow aging rate (FIG. 7), whereas the quantitative differences in these two factors between animals are not enough, when inde-pendently considered, to explain their differences in maximum longevity.

Much previous work has been centered on testing the possibility that antioxidants decrease with age and, more scarcely, on possible increases of free radical production with

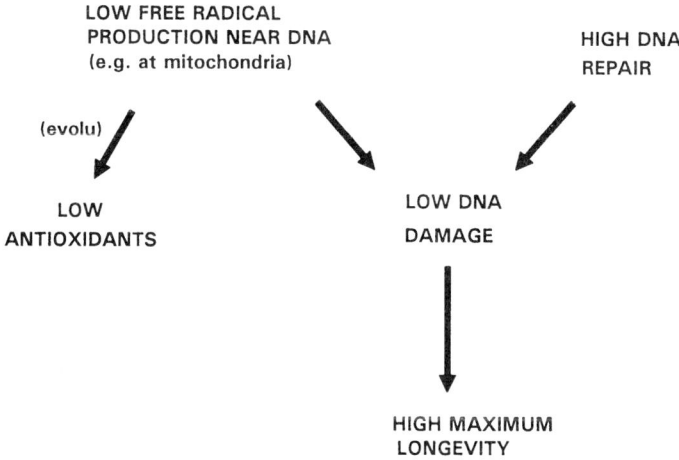

FIGURE 7. Model linking characteristics found in longevous animals (evolu: evolutionary relationship). The rest of the relationships linked by arrows are mechanistic.

age. Available data do not consistently support either the first[66-69] or the second[45,70-72] possibility, but neither of them are needed for the free radical theory of aging to be true. A constant rate of free radical production would cause an accumulation of oxidative DNA damage, because DNA repair cannot be 100% effective. The accumulation of oxidative DNA damage would be quicker in animals with a high rate of free radical production and/ or lower DNA repair, even if the intensity of these last two processes did not change during aging. It is the difference between animal species in the rate of these two processes in the adult animal (both young and old) that matters in explaining their differences in the rate of accumulation of DNA damage and rate of aging.

REFERENCES

1. HARMAN, D. 1956. Aging: A theory based on free radical and radiation chemistry. J. Gerontol. **11:** 298–300.
2. HARMAN, D. 1972. The biological clock: The mitochondria? J. Am. Geriatr. Soc. **20:** 145–147.
3. HARMAN, D. 1968. Free radical theory of aging: Effect of free radical inhibitors on the mortality rate of male LAF1 mice. J. Gerontol. **23:** 476–482.
4. COMFORT, A. 1971. Effects of ethoxyquin on the longevity of C3H mice. Nature **229:** 254–255.
5. MIQUEL, J. 1983. Determination of biological age in antioxidant-treated *Drosophila* and mice. *In* Intervention in the aging process. Part B: Basic Research and Preclinical Screening. A.R. Liss, Ed.: 317–358. Alan R. Liss. New York.
6. LÓPEZ-TORRES, M., R. PÉREZ CAMPO, C. ROJAS, S. CADENAS & G. BARJA DE QUIROGA. 1993. Simultaneous induction of superoxide dismutase, glutathione reductase, GSH and ascorbate in liver and kidney correlates with survival throughout the life span. Free Radical Biol. Med. **15:** 133–142.
7. LÓPEZ-TORRES M., R. PÉREZ-CAMPO, A. FERNÁNDEZ, C. BARBA & G. BARJA DE QUIROGA. 1993. Brain glutathione reductase induction increases early survival and decreases lipofuscin accumulation in aging frogs. J. Neurosci. Res **34:** 233–242.
8. SETO, N.O.L., S. HAYASHI & G.M. TENER. 1990. Overexpression of Cu-Zn superoxide dismutase in *Drosophila* does not affect life-span. Proc. Natl. Acad. Sci. USA **87:** 4270–4274.

9. STAVELEY, B.E., J.P. PHILLIPS & A. HILLIKER. 1990. Phenotypic consequences of copper-zinc superoxide dismutase overexpression in *Drosophila melanogaster.* Genome **33:** 867–872.
10. ORR, W.C. & R.S. SOHAL. 1992. The effects of catalase gene overexpression on life span and resistance to oxidative stress in transgenic *Drosophila melanogaster.* Arch. Biochem. Biophys. **297:** 35–41.
11. GEY, K.F., P. PUSKA, P. JORDAN & U.K. MOSER. 1991. Inverse correlation between plasma vitamin E and mortality from ischemic heart disease in cross-cultural epidemiology. Am. J. Clin. Nutr. **53:** 326–334.
12. BLOT, W.J., J.Y. LI, P.R. TAYLOR, W. GUO, S. DAWSEY, G.Q. WANG, C.S. YANG, S.F. ZHENG, M. GAIL, G.Y. LI, Y. YU, B. LIU, J. TANGREA, Y. SUN, F. LIU, J.F. FRAUMENI, Y.H. ZHANG & B. LI. 1993. Nutrition intervention trials in Linxian, China: Supplementation with specific vitamin/mineral combinations, cancer incidence, and disease-specific mortality in the general population. J. Natl. Cancer Inst. **85:** 1483–1491.
13. TOLMASOFF, J.M., T. ONO & R.G. CUTLER. 1980. Superoxide dismutase: Correlation with life-span and specific metabolic rate in primate species. Proc. Natl. Acad. Sci. USA **77:** 2777–2781.
14. CUTLER, R.G. 1991. Antioxidants and aging. Am. J. Clin. Nutr. **53:** 373S–379S.
15. ALTMAN, P. & D. DITTMER. 1974. Biology Data Book. 1804–1814. Fed. Am. Soc. Exp. Biol. Bethesda, MD.
16. DE MARCHENA, O., M. GUARNERI & G. MCKHANN. 1974. Glutathione peroxidase levels in brain. J. Neurochem. **22:** 773–776.
17. LAWRENCE A. & R.F. BURK. 1978. Species, tissue and subcellular distribution of non-Se-dependent glutathione peroxidase activity. J. Nutr. **108:** 211–215.
18. SUMMER, K.H. & H. GREIM. 1981. Hepatic glutathione S-transferases: Activities and cellular localization in rat, rhesus monkey, chimpanzee and man. Biochem. Pharmacol. **30:** 1719–1720.
19. ALTMAN, P. & D. DITTMER. 1961. Blood and Other Body Fluids. 89–95. Fed. Am. Soc. Exp. Biol. Washington, DC.
20. PASHKO, L.L. & A.G. SCHWARTZ. 1982. Inverse correlation between species life span and specific cytochrome P-448 content of cultured fibroblasts. J. Gerontol. **37:** 38–41.
21. LÓPEZ-TORRES, M., R. PÉREZ-CAMPO, C. ROJAS, S. CADENAS & G. BARJA. 1993. Maximum life span in vertebrates: Correlation with liver antioxidant enzymes, glutathione system, ascorbate, urate, sensitivity to peroxidation, true malondialdehyde, *in vivo* H_2O_2, and basal and maximum aerobic capacity. Mech. Ageing Dev. **70:** 177–199.
22. PÉREZ-CAMPO, R., M. LÓPEZ-TORRES, C. ROJAS, S. CADENAS & G. BARJA. 1994. Longevity and antioxidant enzymes, non-enzymatic antioxidants, oxidative stress, malondialdehyde, and *in vivo* H_2O_2 levels in the vertebrate lung: A comparative study. J. Comp. Physiol. **163:** 682–689.
23. BARJA, G., S. CADENAS, C. ROJAS, M. LÓPEZ-TORRES & R. PÉREZ-CAMPO. 1994. A decrease of free radical production near critical targets as a cause of maximum longevity. Comp. Biochem. Physiol. **108B:** 501–512.
24. PEARL, R. 1928. The rate of living. University of London Press. London.
25. RUBNER, M. 1908. Das Problem der Lebensdauer und seine Beziehungen zu Wachstum und Ernährung. R. Oldenburg, Ed., München.
26. CUTLER, R.G. 1984. Antioxidants, aging and longevity. *In* Free Radicals in Biology, Vol.VI. W.A. Pryor, Ed.: 371–428. Academic Press. New York.
27. SCHMIDT-NIELSEN, K. 1984. Scaling. Why is animal size so important? 56–74. Cambridge University Press. New York.
28. PRINZINGER, R. 1993. Life span in birds and the ageing theory of absolute metabolic scope. Comp. Biochem. Physiol. **105A:** 609–615.
29. CALDER, W.A. 1985. The comparative biology of longevity and lifetime energetics. Exp. Gerontol. **20:** 161–170.
30. HOLMES, D.J. & S. N. AUSTAD. 1995. The evolution of avian senescence patterns: Implications for understanding primary aging processes. Am. Zool. **35:** 307–317.
31. SOHAL, R.S. & R. WEINDRUCH. 1996. Oxidative stress, caloric restriction, and aging. Science **273:** 59–63.

32. SOHAL, R.S., I. SVENSSON, B.H. SOHAL & U.T. BRUNK. 1989. Superoxide anion radical production in different species. Mech. Ageing Dev. **49:** 129–135.
33. SOHAL, R.S., I. SVENSSON & U.T. BRUNK. 1990. Hydrogen peroxide production by liver mitochondria in different species. Mech. Ageing Dev. **53:** 209–215.
34. KU, H.H., U.T. BRUNK & R.S. SOHAL. 1993. Relationship between mitochondrial superoxide and hydrogen peroxide production and longevity of mammalian species. Free Radical Biol. Med. **15:** 621–627.
35. BARJA, G., S. CADENAS, C. ROJAS, R. PÉREZ-CAMPO & M. LÓPEZ-TORRES. 1994. Low mitochondrial free radical production per unit O_2 consumption can explain the simultaneous presence of high longevity and high aerobic metabolic rate in birds. Free Radical Res. **21:** 317–328.
36. KU, H.H. & R.S. SOHAL. 1993. Comparison of mitochondrial pro-oxidant generation and antioxidant defenses between rat and pigeon: Possible basis of variation in longevity and metabolic potential. Mech. Ageing Dev. **72:** 67–76.
37. BARJA, G. 1998. Kinetic measurement of free radical production. *In* Methods of Aging Research. B.P. Yu, Ed.: Chapt 23: 533–548. CRC Press. Boca Ratón, FL.
38. HERRERO, A. & G. BARJA. 1997. Sites and mechanisms responsible for the low rate of free radical production of heart mitochondria in the long-lived pigeon. Mech. Ageing Dev. **98:** 95–111.
39. BOVERIS, A., E. CADENAS & A.O.M. STOPPANI. 1976. Role of ubiquinone in the mitochondrial generation of hydrogen peroxide. Biochem. J. **156:** 435–444.
40. TAKESHIGE, K. & S. MINAKAMI. 1979. NADH- and NADPH-dependent formation of superoxide anions by bovine heart submitochondrial particles and NAD-ubiquinone-reductase preparation. Biochem. J. **180:** 129–135.
41. TURRENS, J.F. & A. BOVERIS. 1980. Generation of superoxide anion by the NADH dehydrogenase of bovine heart mitochondria. Biochem. J. **191:** 421–427.
42. TURRENS, J.F., A. ALEXANDRE & A.L. LEHNINGER. 1985. Ubisemiquinone is the electron donor for superoxide formation by complex III of heart mitochondria. Arch. Biochem. Biophys. **237:** 408–414.
43. NOHL, H. & W. JORDAN. 1986. The mitochondrial site of superoxide formation. Biochem. Biophys. Res. Commun. **138:** 533–539.
44. BEYER, R.E. 1992. An analysis of the role of coenzyme Q in free radical generation and as an antioxidant. Biochem. Cell Biol. **70:** 390–403.
45. HANSFORD, R.G., B. HOGUE & V. MILDAZIENE. 1997. Dependence of H_2O_2 formation by rat heart mitochondria on substrate availability and donor age. J. Bioenerg. Biomembr. **29:** 89–95.
46. BARJA, G. & A. HERRERO. 1998. Localization at complex I and mechanism of the higher free radical production of brain nonsynaptic mitochondria in the short-lived rat than in the longevous pigeon. J. Bioenerg. Biomembr. In press.
47. HERRERO, A & G BARJA 1997. ADP-regulation of mitochondrial free radical production is different with Complex I- or Complex II-linked substrates: Implications for the exercise paradox and brain hypermetabolism. J. Bioenerg. Biomembr. **29:** 243–251.
48. LOSCHEN, G., L. FLOHÉ & B. CHANCE. 1971. Respiratory chain-linked H_2O_2 production in pigeon heart mitochondria. FEBS Lett. **18:** 261–264.
49. BOVERIS, A., N. OSHINO & B. CHANCE. 1972. The cellular production of hydrogen peroxide. Biochem. J. **128:** 617–630.
50. SHEARMAN, C.W. & G.F. KALF. 1977. DNA replication by a membrane-DNA complex from rat liver mitochondria. Arch. Biochem. Biophys. **182:** 573–586.
51. RICHTER, CH., J.W. PARK & B.N. AMES. 1988. Normal oxidative damage to mitochondrial and nuclear DNA is extensive. Proc. Natl. Acad. Sci. USA **85:** 6465–6467.
52. OZAWA, T. 1995. Mitochondrial DNA mutations associated with aging and degenerative diseases. Exp. Gerontol. **30:** 269–290.
53. SOHAL, R.S., H.H. KU, S. AGARWAL, M.J. FORSTER & H. LAL. 1994. Oxidative damage, mitochondrial oxidant generation and antioxidant defenses during aging and in response to food restriction in the mouse. Mech. Ageing Dev. **74:** 121–133.
54. PAMPLONA, R., J. PRAT, S. CADENAS, C. ROJAS, R. PÉREZ-CAMPO, M. LÓPEZ-TORRES & G. BARJA. 1996. Low fatty acid unsaturation protects against lipid peroxidation in liver mitochondria from long-lived species: The pigeon and human case. Mech. Ageing Dev. **86:** 53–66.
55. BARJA, G. 1996. Los radicales libres mitocondriales como factores principales determinantes de la velocidad del envejecimiento. Rev. Esp. Gerontol. Geriatr. **31:** 153–161.

56. HIRANO, T., Y. HOMMA & H. KASAI. 1995. Formation of 8-hydroxyguanine in DNA by aging and oxidative stress. *In* Oxidative Stress and Aging. R.G. Cutler, L. Packer, J. Bertram & A. Mori, Eds.: 69–76. Birkhäuser. Basel.

57. FRAGA, C.G., M.K. SHIGENAGA, J.W. PARK, P. DEGAN & B.N. AMES. 1990. Oxidative damage to DNA during aging: 8-Hydroxy-2'-deoxyguanosine in rat organ DNA and urine. Proc. Natl. Acad. Sci. USA **87:** 4533–4537.

58. ADELMAN, R., R.L. SAUL & B.N. AMES. 1988. Oxidative damage to DNA: Relation to species metabolic rate and life span. Proc. Natl. Acad. Sci. USA. **85:** 2706–2708.

59. HERRERO, A. & G. BARJA. 1998. Heart and brain oxo^{8}dg in the genomic DNA of rats and pigeons and maximum longevity. Ann. N.Y. Acad. Sci. This volume.

60. MIQUEL, J. 1988. An integrated theory of aging as a result of mitochondrial-DNA mutation in differentiated cells. Arch. Gerontol. Geriatr. **12:** 99–117.

61. RICHTER, CH. 1988. Do mitochondrial DNA fragments promote cancer and aging? FEBS Lett. **241:** 1–5.

62. HART, R.W. & R.B. SETLOW. 1974. Correlation between deoxyribonucleic acid excision-repair and life-span in a number of mammalian species. Proc. Natl. Acad. Sci. USA **71:** 2169–2173.

63. HALL, K.Y., R.W. HART, A.K. BENIRSCHKE & R.L. WALFORD. 1984. Correlation between ultraviolet-induced DNA repair in primate lymphocytes and fibroblasts and species maximum achievable longevity. Mech. Ageing Dev. **24:** 164–173.

64. BÜRKLE, A., K. GRUBE & J.H. KÜPPER. 1992. Poly ADP-ribosylation: Its role in inducible DNA amplification, and its correlation with the longevity of mammalian species. Exp. Clin. Immunogenet. **9:** 230–240.

65. FRANCIS, A.A., W.H. LEE & J.D. REGAN. 1981. The relationship of DNA excision repair of ultraviolet-induced lesions to the maximum life span of mammals. Mech. Ageing Dev. **16:** 181–189.

66. BARJA, G., R. PÉREZ & M. LÓPEZ. 1990. Antioxidant defenses and peroxidation in liver and brain of aged rats. Biochem. J. **272:** 247–250.

67. BARJA DE QUIROGA, G., M. LÓPEZ-TORRES & R. PÉREZ-CAMPO. 1992. Relationship between antioxidants, lipid peroxidation and aging. *In* Free Radicals and Aging. I. Emerit & B. Chance, Eds.: 109–123. Birkhäuser. Basel.

68. PÉREZ, R., M. LÓPEZ & G. BARJA DE QUIROGA. 1991. Aging and lung antioxidant enzymes, glutathione, and lipid peroxidation in the rat. Free Radical Biol. Med. **10:** 35–39.

69. BENZI, G. & A. MORETTI. 1995. Age- and peroxidative stress-related modifications of the cerebral enzymatic activities linked to mitochondria and the glutathione system. Free Radical Biol. Med. **19:** 77–101.

70. YU, B.P. 1995. Modulation of oxidative stress as a means of life-prolonging action of dietary restriction. *In* Oxidative Stress and Aging. R.G. Cutler, L. Packer, J. Bertram & A. Mori, Eds.: 331–342. Birkhäuser. Basel.

71. ZHAN, H., C.P. SUN, C.G. LIU & J.H. ZHOU. 1992. Age-related change of free radical generation in liver and sex glands of rats. Mech. Ageing Dev. **62:** 111–116.

72. GUARNIERI, C., C. MUSCARI & C.M. CALDARERA. 1992. Mitochondrial production of oxygen free radicals in the heart muscle during the life span of the rat: Peak at middle age. *In* Free Radicals and Aging. I. Emerit & B. Chance, Eds.: 73–77. Birkhäuser. Basel.

Oxidative Damage in the Senescence-accelerated Mouse

AKITANE MORI,[a,d] KOZO UTSUMI,[b] JIANKANG LIU,[a,e] AND MASANORI HOSOKAWA[c]

[a]Okayama University Medical School, Okayama 700-0914, Japan
[b]Institute of Medical Science, Center for Adult Diseases,
Kurashiki, Kurashiki 710-0824, Japan
[c]Chest Disease Research Institute, Kyoto University, Kyoto 606-8397, Japan

ABSTRACT: The senescence-accelerated mouse (SAM) exhibited a shortened life span (about 18 months) and early manifestation of various signs of senescence, including changes in physical activity, skin, and spinal curvature. The mechanism of senescence acceleration in SAM is thought to be related to free radical damage. Oxidative phosphorylation was estimated in liver mitochondria from SAMPS and the senescence-resistant subtrain, SAMR1. The respiratory control ratio decreased during aging, and the ATP/O, an index of ATP synthesis, was depressed at 18 months of age in SAMPS. DNP-dependent uncoupled respiration in liver mitochondria was markedly decreased, and active uptake of calcium was markedly dysfunctional with aging. These findings suggest that the functional disorders in mitochondria may be closely related to the shorter life span of SAMPS. White-footed (WF) mice can live at least to 5.5 years, when some animals are still capable of reproducing and their external body condition remains healthy. The mitochondrial functions were examined in the same way as in the SAM experiments. However, no particular finding responsible for their longevity was observed in WF mice at 3 and 12 months old. More comprehensive examinations on more aged WF mice are needed for explanation of their greater longevity.

The senescence-accelerated mouse (SAM) was established by Takeda *et al.* as a murine model of accelerated aging.[1–3] SAM-prone (SAMP) mice exhibit a shortened life span (about 18 months) and early manifestation of various signs of senescence, including a moderate to severe decline of activity, hair loss, lack of hair glossiness, skin coarseness, periophthalmic lesions and cataracts, and increased lordokyphosis of the spine. SAMP strains include SAMP1, SAMP2, SAMP3, SAMP6, SAMP7, SAMP8, SAMP9, SAMP10, and SAMP11. Symptoms of accelerated senescence are a common characteristic in SAM strains. The SAMP8 strain was separated from SAMP2 and is characterized by age-related learning and memory deficits and, therefore, is thought to be a good model for aging.[4–6] Otherwise, SAMR strains (*i.e.*, SAMR1, SAMR4, and SAMR5) are senescence-resistant inbred strains.[6]

Recently, we observed age-associated biochemical changes in the brain and liver in SAMP8, which strongly suggest that the mechanism of senescence acceleration in SAM mice is related to free radical damage.[7] Free radicals are thought to be an important factor

[d]Corresponding author: Dr. Akitane Mori, Department of Neuroscience, Institute of Molecular and Cellular Medicine, Okayama University Medical School, Okayama 700-0914, Japan. Tel: 81-86-276-6006; fax: 81-86-276-6061.
    [e]Present address: Division of Biochemistry & Molecular Biology, University of California at Berkeley, 94720-3200, USA.

in aging.[8] Especially oxidative damage by free radicals to DNA, proteins, and lipids in mitochondria may play a key role in aging.[9] Age-associated dysfunction in energy coupling and oxidative phosphorylation was found in liver mitochondria of SAMP8. It was also observed that the spin-trap N-tert-alpha-phenyl-butylnitrone (PBN) prolongs the life span of SAMP8 mice.[10] Based on these findings, a hypothesis suggesting involvement of free radicals in the mechanism of accelerated senescence in SAM mice was proposed.

OXIDATIVE DAMAGE AND A DEFENSE SYSTEM IN SAM MICE

Oxidative damage to DNA, protein, and lipids is implicated in a variety of degenerative diseases and aging. Nakamoto et al. reported that SAMP1 showed higher carbonyl content, an index of oxidatively modified proteins, in the liver and brain compared with that of SAMR1.[11] Komura et al. observed markedly increased serum peroxide levels until 8 months of age in SAMP1 as compared with SAMR1.[12] A similar temporary increased level of lipid peroxide was recognized in the skin of SAMP1.[13] Tobita et al. demonstrated increased DNA single-strand breaks in the brain of 9- and 12-month-old SAMP8 mice and suggested that one of causal factors responsible for age-related DNA damage was that induced by free radicals.[14]

The age-associated changes in superoxide dismuatse (SOD) activity, thiobarbituric acid–reactive substances (TBARS) as an index of lipid peroxidation, and glutathione (GSH) levels in the brain and liver in SAMP8 mice were investigated and compared with results obtained in ddY mice (FIG. 1).[7]

Both at 3 and 11 months of age, SAM mice had higher SOD activity in the brain than the corresponding ddY mice. SAM mice at 11 months of age showed higher SOD activity in the liver than at 3 months of age and also the ddY mice at 11 months of age. No difference in the SOD activity was seen between 3- and 11-month-old ddY mice. On the other hand, both SAM and ddY mice at 11 months showed higher TBARS levels in the brain and liver than the younger animals. A significant difference in liver TBARS levels was observed between SAM and ddY mice at the age of 3 months, but not at 11 months.

The relationship between SOD activity and aging is controversial. Vanella et al.,[15] Mavelli et al.,[16] and Hiramatsu et al.[17] reported that SOD activity increased significantly with aging both in the cerebrospinal fluid of patients and in the rat brain. SOD activity increases may be the expression of a self-protective mechanism from superoxide-generating radicals operating in aging, because the production of superoxide radicals in the brain has been reported to increase with age.[18] Therefore, these findings indicate that SAM mice may have more superoxide radical production, at least at 3 and 11 months of age. By contrast, Yagi et al. reported that SAMP8 liver homogenates showed less SOD activity than those of SAMR1.[19] Further decreased SOD activity in neuronal and glial cells of the aged rat cortex[20] and unchanged SOD actvity in rat and mice brain have been reported.[21,22] The effect of 100% oxygen on TBARS production and SOD activity in the rat brain was stud-

FIGURE 1. Age-related changes in SOD, TBARS, and GSH in the brain. Superoxide dismutase (SOD) activity (**A**), thiobarbituric acid–reactive substances (TBARS) levels (**B**), and reduced glutathione (GSH) levels (**C**) in the brain of SAM and ddY mice. The values represent the mean ± SD with 6–11 animals. $*p < 0.05$ and $**p < 0.01$ vs. 3-month-old mice; a, $p < 0.01$, SAM vs. ddY mice (one-way ANOVA). ▨, SAM; □, ddY.

A

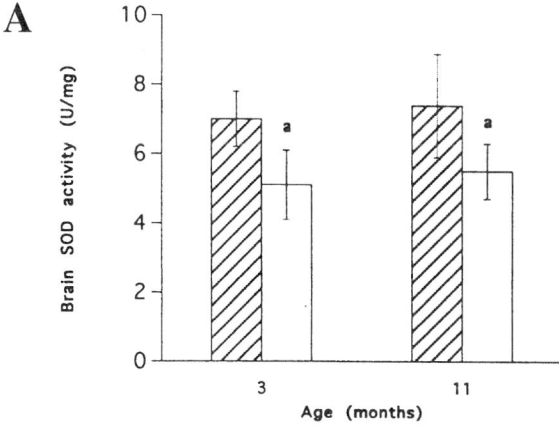

B

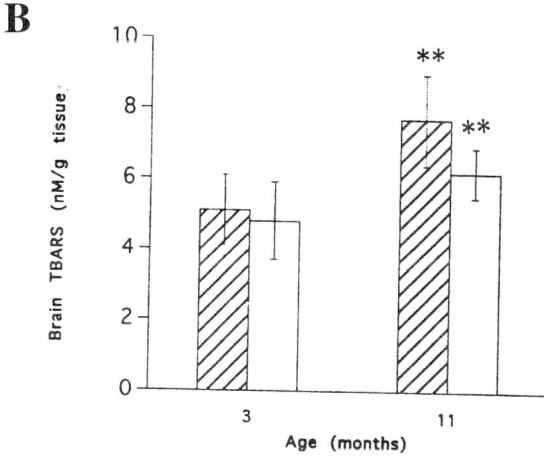

C

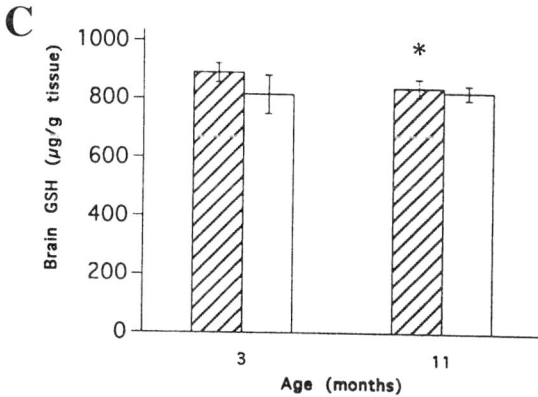

ied one, two, and three weeks after injection with ferric chloride.[23] It was found that oxygen inhalation accelerated mitochondrial SOD activity one week after and inhibited it three weeks after the injection, whereas inhibiting cytosolic SOD activity one week after and accelerating it three weeks after injection. Oxygen inhalation accelerated TBARS production 1 week after and decreased it 3 weeks after the injection. These biphasic changes may be brought about by time-dependent and complicated biological reactions. For example, increased mitochondrial SOD activity one week after and cytosolic SOD activity three weeks after may be related to a compensatory mechanism against the increased oxygen free radicals, and decreased mitochondrial SOD activity three weeks after may reflect damaged mitochondrial structure at this stage. Alternatively, increased TBARS one week after may be related to higher peroxidation of lipids and other biomolecules, and decreased TBARS three weeks after may be explained by diminished lipid peroxidation reaction in severely damaged cells. Changes of TBARS, observed in SAM mice, may reflect such a situation.

SAM mice at 11 months of age showed lower GSH levels both in the brain and liver, as compared with values at three months of age. Significantly lower GSH levels in the liver were observed in SAM mice as compared to ddY mice. GSH is known to play an important role in antioxidant defense by acting in combination with glutathione peroxidase to break down hydrogen peroxide and lipid peroxides. It also scavenges a wide variety of free radicals, including carbon-centered peroxyl, phenoxy-, and semiquinone radicals.[24,25] The lower GSH levels may be due to a higher consumption of GSH for scavenging the higher production of free radicals in SAM mice. The reduced levels of GSH with aging in the brain and liver in SAM mice suggest an increased susceptibility to oxidative damage with accelerated aging.

MITOCHONDRIAL DYSFUNCTION IN SAM MICE

Accelerated mutations in mitochondrial DNA caused by free radicals in such terminally differentiated cells as nerve, cardiac muscle, and skeletal muscle are thought to be important in the mechanism of aging.[26] Decreased activity of mitochondrial electron transport complexes has been demonstrated in Parkinson's disease and in aging.[27] Moreover, mitochondria are the principal superoxide generating organelles in cells. Mitochondrial SOD, which maintains a low level of superoxide, should minimize injury by oxygen free radicals and H_2O_2 production by mitochondria. Nevertheless, oxidative stress may induce functional damage to mitochondria and may lead to apoptosis and/or accelerated aging. Recently, studies have shown that mitochondria are affected by a cell death program that is related to active oxygen generation and that dysfunction of mitochondria may be involved in apoptotic cell death.[28–31] For this reason it was of interest to investigate changes in mitochondrial function in the liver and heart of SAMP8 mice during the course of aging (FIG. 2).[32]

Oxidative phosphorylation was measured in isolated liver mitochondria[33] of SAMR1 and SAMP8 at 6, 12, and 18 months of age. Mitochondrial respiration (calculated as μmoles O_2/min/mg protein) in the presence of succinate and inorganic phosphate (state 4 respiration) was increased by adding ADP (state 3 respiration) and then decreased ADP (state 4 respiration), corresponding to completion of a temporary period of ATP synthesis.

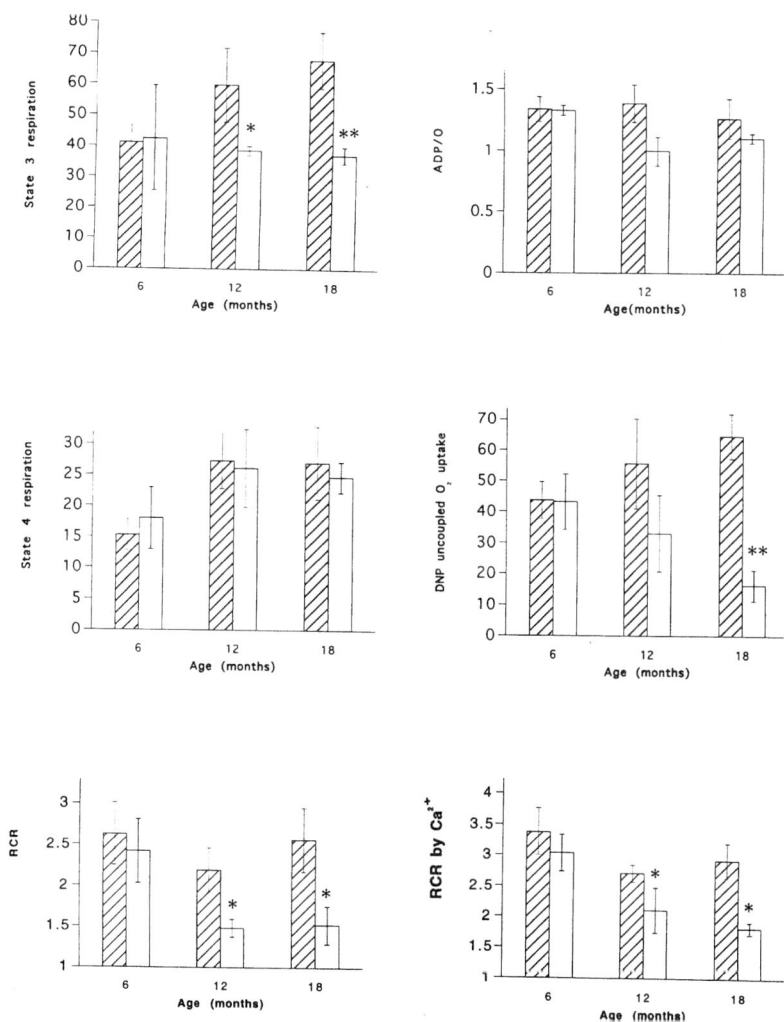

FIGURE 2. Functional characteristics of liver mitochondria from SAMP8 and SAMR1 mice. $n = 3$, 7, and 6 for 6, 12, and 18 months of age for SAMR1 mice, respectively; and $n = 3$, 6, and 3 for 6, 12, and 18 months of age for SAMP8 mice, respectively. $*p < 0.05$, $**p < 0.01$: significant difference from SAMR1. ▨, SAMR1; ▢, SAMP8.

The respiratory control ratio (RCR: state 3 respiration/state 4 respiration) and the ADP/O ratio after adding 200µM ADP were determined as described by Estabrook.[34] No significant difference in state 3 and state 4 respiration was observed between SAMP8 and SAMR1 mice at six months. However, state 4 respiration was markedly increased, and RCR tended to decline for SAMP8 after 12 months and remained at this value at 18 months of age. This decrease was due to an increase in state 4 respiration, whereas state 3 respiration remained constant. The ADP/O ratio declined at 12 months of age in SAMP8 mice and remained depressed at 18 months of age in SAMP8 mice.

Mitochondrial respiration is uncoupled by agents such as dinitrophenol (DNP), causing the rate of oxygen consumption to increase, reflecting an accelerated electron transport activity. Uncoupled respiration is paralleled with state 3 respiration, which is the rate of phosphorylating respiration induced by addition of ADP. DNP-induced uncoupling respiration declined at 12 months of age in SAMP8 mice mitochondria and dramatically declined at 18 months to less than a half of state 4 respiration at this stage.

Ca^{2+} is actively taken up into mitochondria through the inner membrane, using energy transfer reactions, which induce a transitory accelerated respiration when small amounts of calcium are added to mitochondria. The RCR induced by Ca^{2+} in liver mitochondria had a high value at six months of age, decreased markedly at 12 months of age, and persisted at a low rate up to 18 months of age. However, uncoupled respiration induced by Ca^{2+} in SAMP8 mice at 18 months of age was still very high. This suggests that the enzymes themselves of the electron transfer system were not damaged at this stage and rather that the activity of the Ca^{2+} transport was altered, perhaps because of membrane potential changes. By 12 months of age, SAMP8 mice mitochondria displayed lower oxidative phosphorylation. Age-associated changes in SOD activity, TBARS, and GSH in the brain and liver in SAMP8, together with other reports on SAM mice,[19] suggest that oxidative stress, possibly due to free radicals, may be a cause of the observed mitochondrial dysfunction. However, at present, direct demonstration of increased generation of oxygen free radicals in mitochondria is necessary to confirm an involvement of free radicals in mitochondrial dysfunction, as changes in SOD and/or TBARS in tissue do not always parallel free radical generation.[23]

Decrease in the ADP/O ratio in the SAMP8 mice at 12 and 18 months can be explained by an age-related uncoupling. This may explain the functional decline in this mouse strain, inasmuch as less ATP is made per unit of oxygen consumed.

DNP induces uncoupled respiration in mitochondria. However, the increased rate of oxygen consumption induced by DNP in liver mitochondria SAMP8 mice at 18 months of age declined to less than half that seen in liver motochondria from SAMP8 mice at six months of age. This suggests that the coupling mechanism for energy transport reactions of the electron transport system may be altered in SAMP8 at 18 months of age. This may be due to damage to the components of the electron and energy transfer system.

Bcl-X is an oncogenic product of the Bcl-family protein,[31,35] distributed on the outer membrane of liver mitochondria, which inhibits apoptotic cell death. To investigate the involvement of the protective action of Bcl-X against the mitochondrial dysfunction in SAMP8, expression of the Bcl-Xs/L in liver mitochondria of SAMP8 was studied by an immunoblotting method using anti-Bcl-Xs/L antibody. A decrease in Bcl-Xs/L of liver mitochondria of SAMP8 was observed.[32] Bcl-2 is known to protect cells from hydrogen peroxide- and menadione-induced oxidative death.[31] Bcl-X belongs to the Bcl-2 family of proteins and is distributed on the outer membrane of mitochondria.[35] Because SAMP8 mice were found to have a lower expression of Bcl-X, this suggests that the decreased expression of Bcl-X may result in the dysfunction of mitochondria in SAMP8 mice.

A SPIN TRAP PROLONGS THE LIFE SPAN OF SAM MICE

PBN is a widely used spin-trapping agent.[36] PBN has been successfully applied to the investigation of the free radical reaction, such as in ischemia-reperfusion injury *in vitro*[37,38] and *in vitro* [39–43] and is known to reduce the mortality of rats exposed to trauma or endotoxin, situations in which oxygen radicals are produced.[44,45] Furthermore, PBN has been reported to normalize the age-related biochemical parameters and physiological functions, such as memory.[46,47]

PBN (30 mg/kg, ip) was administered daily to SAMP8 mice after they reached maturity at three months of age.[10] PBN markedly prolonged the life span of the mice. Overall, the 50% survival rate was 42 weeks in the control and 56 weeks in the PBN administered group. This finding supports Harman's free radical theory that "the aging process may be simply the sum of the deleterious radical reactions going on continuously throughout life in cells and tissues."[8] Recently, Kumari *et al.* observed that β-CATECHIN, an antioxidant drink, containing ascorbate, green tea extract, dunaliella carotene, vitamin E, and sunflower seed extract, also prolonged the life span of SAMP8 mice.[48] Antioxidant nutrients, especially vitamin C, vitamin E, and carotenoids, are thought to play an important role in health maintenance and disease prevention, including many age-related disorders.[49] Synthetic antioxidants, such as spin traps, may also in the future be therapeutically helpful for prevention and/or treatment of age-related disorders induced by free radicals.

COMPARISON BETWEEN SAM MICE AND WHITE-FOOTED MICE

Sacher and Hart pointed out the value of the white-footed mice as a model for human aging.[50] White-footed mice, *Peromyscus leucopus*, can live up to 5.5 years, when some animals are still capable of reproducing and their external body condition remains healthy. This species is thought to be a useful model to examine aging in comparison with other mice having lower longevity.[51] Sohal *et al.* estimated rates of prooxidant generation, antioxidant defense, and protein oxidative damage in the heart and brain from 3.5-month-old male mice by comparing the house mice (*Mus musculus*) with white-footed mice.[52] White footed mice had a greater than 2-fold life span and metabolic potential than house mice. This comparative study showed that a long life span and the higher metabolic potential of white footed mice correlated with a low rate of mitochondrial $O_{2\cdot-}$ and H_2O_2 generation, high activities of catalase and glutathione peroxidase, and low levels of protein oxidative damage, as well as low susceptibility to oxidative damage. On the other hand, Guo *et al.* reported that in white-footed mice, except for a decrease of glutathione peroxidase in male mice and SOD in female mice, the activities of other antioxidant enzymes did not change significantly even in very old mice, that is, at 24–46 months old.[53]

Oxidative phosphorylation in liver mitochondria isolated from white-footed mice was examined at 12 months of age and compared with SAM mice of the same age. Markedly

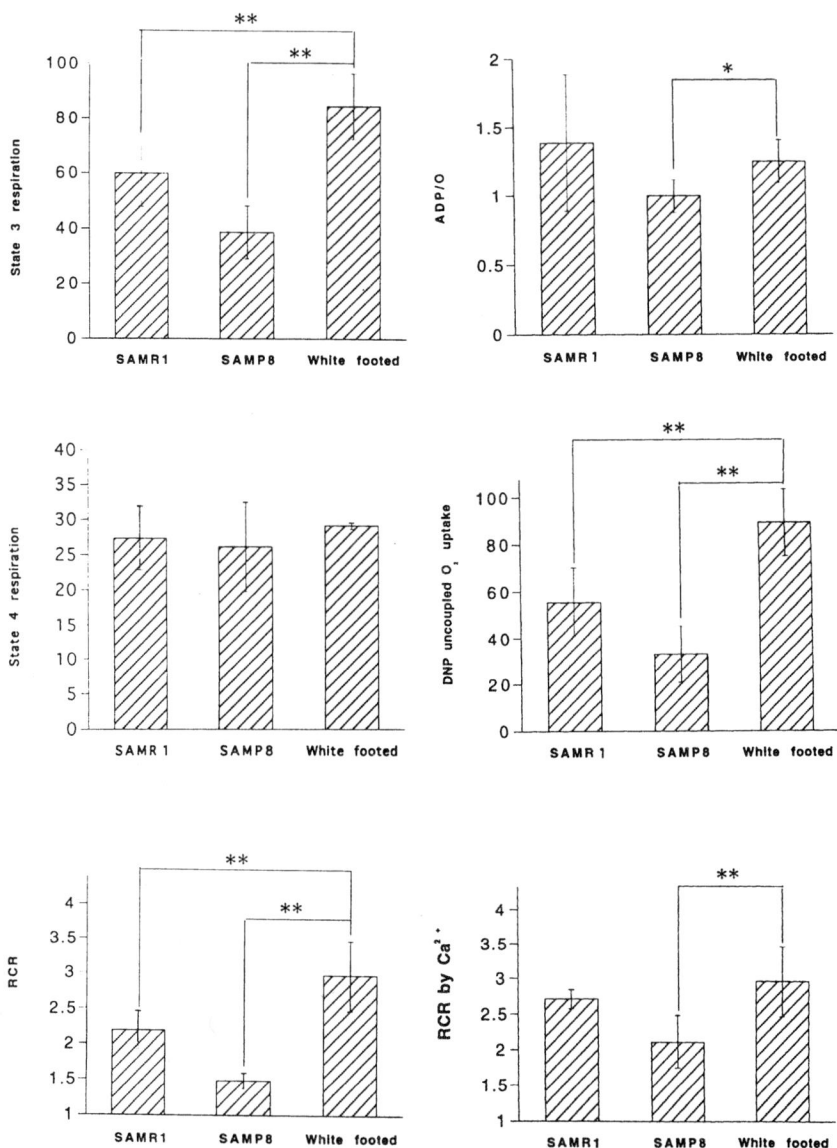

FIGURE 3. Functional characteristics of liver mitochondria from SAMR1 and SAMP8, and white-footed mice at 12 months of age. $n = 7$ for SAMR1 mice, $n = 6$ for SAMP8 mice, and $n = 13$ for white-footed mice. $*p < 0.05$; $**p < 0.01$.

higher state 3 respiration, RCR, and DNP-uncoupled O_2 uptake were observed in liver mitochondria of white-footed mice as compared to SAMP8 or SAMR1 mice. The ADP/O ratio and RCR by Ca^{2+} in liver mitochondria of white-footed mice were also significantly higher than those of SAMP8 mice. These findings suggest that the respiration rate and level of ATP sysnthesis are higher in mitochondria of white-footed mice as compared with both SAMP8 and SAMR1 mice (Fig. 3). No significant difference was observed in SOD activity or TBARS in liver mitochondria of white-footed mice or SAMP8 mice at 12 months of age. More comprehensive studies are needed to clarify the age-related changes in functional characteristics of mitochondria.

SUMMARY

The senescence-accelerated mouse (SAM) exhibits a shorter life span of about 18 months and early manifestation of various signs of senescence, including decreased activity, hair loss, lack of hair glossiness, skin coarseness, periophthalmic lesions and cataracts, and increased lordokyphosis of the spine. During aging, when compared with ddY mice, SAMP8 mice showed a higher superoxide dismutase activity in the brain (at both 3 and 11 months old), lower levels of reduced glutathione level in the liver (at 3 months old), and higher thiobarbituric acid reactive substances (TBARS) in the liver (at 3 months old). These findings suggest that the mechanism of senescence acceleration in SAM mice may be related to free radical damage.

Oxidative phosphorylation was estimated in liver mitochondria from SAMP8 mice and the senescence-resistant substrain. In SAMP8 mice, the ATP/O, an index of ATP synthesis, was depressed at 18 months of age in SAMP8 mice. Dinitrophenol-induced uncoupled respiration in mitochondria and active uptake of calcium was markedly decreased with aging.

The spin-trap *N-tert*-alpha-phenyl-butylnitrone (PBN, 30mg/kg, ip) was administered daily to SAMP8 mice after they reached maturity at three months of age. PBN markedly prolonged life span.

White-footed mice can live up to 5.5 years. At this age some animals are still capable of reproducing, and their external body condition remains healthy. When the liver mitochondrial functions were examined in the same way as studied in the SAM experiments, these preparations were observed to exhibit higher state 3 respiration, respiratory control ratio, and much higher dinitrophenol-induced uncoupled O_2 uptake in liver mitochondria of white-footed mice at 12 months old. No significant difference in superoxide dismutase activity or TBARS in liver mitochondria was found between white-footed mice and SAMP8 mice. More comprehensive studies are needed for a clearer explanation of the short longevity of SAM mice and greater longevity of white-footed mice. These preliminary studies suggest that oxidative damage by free radicals and/or their defense system may be factors involved in the different life spans of these species.

ACKNOWLEDGEMENT

We thank Professor Lester Packer, University of California at Berkeley, for stimulating discussion and critical reading of the manuscript.

REFERENCES

1. TAKEDA, T., M. HOSOKAWA & K. HIGUCHI. 1991. Sensescence-accelerated mouse (SAM): A novel murine model of accelerated senescence. J. Am. Geriatr. Soc. **39:** 911–919.
2. HOSAKAWA, M., R. KASAI, K. HIGUCHI, S. TAKESHITA, K. SHIMIZU, H. HAMAMOTO, A. HONMA, M. IRINO & K. TODA. 1984. Grading score system: A method for evaluation of the degree of senescence in senescence accelerated mouse (SAM). Mech. Ageing Dev. **26:** 91–102.
3. HOSOKAWA, M., S. TAKESHITA, K. HIGUCHI, K. SHIMIZU, M. IRINO, K. TODA, A. HONMA, A. MATSUMURA, K. YASUHIRA & T. TAKEDA. 1984. Cataract and other ophthalmic lesions in senescence accelerated mouse (SAM): morphology and incidence of senescence associated ophthalmic changes in mice. Exp. Eye Res. **38:** 105–114.
4. MIYAMOTO, M., Y. KIYOTA, N. YAMAZAKI, A. NAGAOKA, T. MATSUO, Y. NAGAWA & T. TAKEDA. 1986. Age-related changes in learning and memory in the senescence-accelerated mouse (SAM). Physiol. Behav. **38:** 399–406.
5. YAGI, H., S. KATHO, I. AKIGUCHI & T. TAKEDA. 1988. Age-related deterioration of ability of acquisition in memory and learning in senescence accelerated mouse: SAM-P/8 as an animal model of disturbances in recent memory. Brain Res. **474:** 86–93.
6. TAKEDA, T., M. HOSOKAWA & K. HIGUCHI. 1994. Senescence-accelerated mouse (SAM): A novel murine model of aging. *In* The SAM Model of Senescence. T. Takeda, Ed.: 15–22. Elsevier Science B. V. Amsterdam.
7. LIU, J. & A. MORI. 1993. Age-associated changes in superoxide dismutase activity, thiobarbituric acid reactivity and reduced glutathione level in the brain and liver in senescence accelerated mice (SAM): Acomparison with ddY mice. Mech. Aging Dev. **71:** 23–30.
8. HARMAN, D. 1981. The aging process. Proc. Natl. Acad. Sci. USA **78:** 7124–7128.
9. SHIGENAGA, M.K., T. HAGEN & B.N. AMES. 1994. Oxidative damage and mitochondrial decay in aging. Proc. Natl. Acad. Sci. USA **91:** 10771–10778.
10. EDAMATSU, R., A. MORI & L. PACKER. 1995. The spin-trap N-*tert*-α-phenyl-butylnitrone prolongs the life span of the senescence accelerated mouse. Biochem. Biophys. Res. Commun. **211:** 847–849.
11. NAKAMOTO, H., A. NAKAMURA, S. GOTO, M. HOSOKAWA, H. FUJISAWA & T. TAKEDA. 1994. Acceleration of oxidatively modified proteins in senescence-accelerated mouse SAMP11 and SAMR1. *In* The SAM Model of Senescence. T. Takeda, Ed.: 137–140, Elsevier Science. B. V. Amsterdam.
12. KOMURA, S., K. YOSHINO, N. OHISHI & K. YAGI. 1994. Serum lipid peroxide level of the senescence-accelerated mouse. *In* The SAM Model of Senescence. T. Takeda Ed.: 141–144. Elsevier Science. B. V. Amsterdam.
13. YOSHINO, K., S. KOMURA, T. OKADA, N. OHISHI & K. YAGI. 1994. Skin lipid peroxide level of the senescence-accelerated mouse. *In* The SAM Model of Senescence. T. Takeda, Ed.: 145–148. Elsevier Science. B. V. Amsterdam.
14. TOBITA, M., S. NAKAMURA, I. NAGANO, A. ITOH & Y. ITOYAMA. 1994. DNA single-strand breaks in hippocampal regions of senescence accelerated mice (SAMP8/Ta) detected by modified *in situ* nick translation procedure. *In* The SAM Model of Senescence. T. Takeda, Ed.: 125–128. Elsevier Science. B. V. Amsterdam
15. VANELLA, A., E. GOREMIA, G. D'URSO, P. TIRIOLO, I.D. SILVESTRO, R. GRIMALDI & R. PINTURO. 1982. Superoxide dismutase activity in aging rat brain. Gerontology **228:** 108–113.
16. MAVELLI, L., B. MONDOV, R. FEDERICO & G. ROTILLO. 1978. Superoxide dismutase activity in developing rat brain. J. Neurochem. **31:** 363–364.
17. HIRAMATSU, M., M. KOHNO, R. EDAMATSU, K. MITSUTA & A. MORI. 1992. Increased superoxide dismutase activity in aged human cerebrospinal fluid and rat brain determined by electron spin resonance spectrometry using spin trap method. J. Neurochem. **58:** 116–1164.
18. SAWADA, M. & J.C. CARLSON. 1987. Changes in superoxide radical and lipid peroxide formation in the brain, heart and liver during the life time of the rat. Mech. Aging Dev. **41:** 125–137.
19. YAGI, K., K. YOSHINO, S. KOMURA, K. KONDO & N. OHISHI. 1988. Superoxide level in the senescence-accelerated mouse. J. Clin. Biochem. Nutr. **5:** 21–27.
20. GEREMIA, E., S.D. BARATTA, S. ZAFARANA, R. GIODANA, M.R. PINIZZOTTO, M.G. LA ROSA & A. GAROZZO. 1990. Antioxidant enzymatic systems in neuronal and glial cell-enriched fractions of rat brain during aging. Neurochem. Res. **15:** 719–723.

21. KELLOGG, E.W. & I. FRIDOVICH. 1976. Superoxide dismutase in the rat and mouse as a function of age and longevity. J. Gerontol. **31:** 405–408.

22. DE QUIROGA, B.C., R. PEREZ-CAMPO & M. LOPEZ TORRES. 1990. Antioxidant defenses and peroxidation in liver and brain of aged rats. Biochem. J. **272:** 247–250.

23. MORI, A., M. HIRAMATSU & I. YOKOI. 1992. Post-traumatic epilepsy, free radicals and antioxidant therapy. *In* Free Radicals in the Brain. L. Packer, L. Prilipko & Y. Christen, Eds.: 109–122. Springer-Verlag. Berlin & Heiderberg.

24. HALLIWELL, B. & J. M.C. GUTTERIDGE, Eds. 1989. Free Radicals in Biology and Medicine. 2nd ed. Clarendon Press. Oxford.

25. MOLDEUS, P. & B. JERNSTROM. 1983. Interaction of glutathione with reactive intermediates. *In* Functions of Glutathione: Biochemical, Physiological, Toxicological and Clinical Aspects. A. Larsson, S. Orrenius, H. Holmgren & B. Mannervik, Eds.: 99–108. Raven Press. New York.

26. SUGIYAMA, S., K. HATTORI, M. HAYAKAWA & T. OZAWA. 1991. Quantitative analysis of age-associated accumulation of mitochondrial DNA with deletion in human hearts. Biochem. Biophys. Res. Commun. **180:** 894–899.

27. ROOT, R.K., J. METCALF, N. OSHINO & B. CHANCE. 1975. H_2O_2 release from human granulocyte during phagocytosis. I. Documentation, quantitation, and some regulating factors. J. Clin. Invest. **55:** 945–955.

28. SCHULZE-OSTHOFF, K., A.C. BAKKER, B. VANHAESEBROECK, R. BEYAERT & W.A. JACOB. 1992. Cytotoxic activity of tumor necrosis factor is mediated by early damage of mitchondrial function. J. Biol. Chem. **267:** 5317–5323.

29. LIU, X., C.N. KIM, J. YANG, R. JEMMERSON & X. WANG. 1996. Induction of apoptotic program in cell-free extracts: Requirement for dATP and cytochrome *c*. Cell **86:** 147–157.

30. KRIPPNER, A., A. MATSUMOTO-YAGI, A. GOTTLIEB & M. RABIOR. 1996. Loss of function of cytochrome *c* in Jurkat cells undergoing fas-mediated apoptosis. J. Biol. Chem. **271:** 21629–21636.

31. HOCKENBERY, D.M., Z.N. OLTAVAI, X.-M. YIN, C.L. MILLIMAN & S.J. KORSMEYER. 1993. Bcl-2 functions in an antioxidant pathway to prevent apoptosis. Cell **75:** 241–251.

32. NAKAHARA, H., T. KANO, Y. INAI, K. UTSUMI, M. HIRAMATSU, A. MORI & L. PACKER. 1998. Mitochondrial dysfunction in the senescence accelerated mouse (SAM). Free Radical Biol. Med. **24:** 85–92.

33. HOGEBOOM, G.H. 1955. Fractionation of cell components of animal tissues. *In* Methods in Enzymology. S.P. Colowick & N.O. Kaplan, Eds.: **1:** 16–19. Academic Press. New York and London.

34. ESTABROOK, R.W. 1967. Mitochondrial respiratory control and the polarographic measurement of ADP:O ratios. *In* Methods in Enzymology, R.W. Estabrook & M.E. Pullman, Eds.: **10:** 41–47. Academic Press. New York and London.

35. KRAJEWSKI, S., M. KRAJEWSKA, A. SHABALK, H.-G. WANG, S. IRIE, L. FONG & J.C. REED. 1994. Immunohistochemical analysis of *in vivo* patterns of Bcl-X expression. Cancer Res. **54:** 5501–5507.

36. DAVIES, M.J. 1989. Direct detection of radical production in the ischaemic and reperfused myocardium: Current status. Free Radical Res. Comm. **7:** 275–284.

37. GARLICK, P.B., M.J. DAVIES, D.J. HEARSE & T.F. SLATER. 1987. Direct detection of free radicals in the reperfused rat heart using ESR spectroscopy. Circ. Res. **61:** 757–763.

38. SAKAMOTO, A., S.T. OHNISHI, T. OHNISHI & R. OGAWA. 1991. Protective effect of a new anti-oxidant on the rat brain exposed to Ischemia-reperfusion injury: Inhibition of free radical formation and lipid peroxidation. Free Radical Biol. Med. **11:** 385–391.

39. BOLLI, R. & P.B. MCCAY. 1990. Use of spin traps in intact animals undergoing myocardial ischemia/reperfusion: A new approach to assessing the role of oxygen radicals in myocardial "stunning." Free Radical Res. Comm. **9:** 169–180.

40. PINEMAIL, J., J.O. DEFRAIGNE, C. FRANSSEN, T. DEFECHEREUX, J.L. CANIVET, C. PHILIPPART & M. MEURISSE. 1990. Evidence of *in vivo* free radical generation by spin trapping with α-phenyl *N-tert*-butyl nitrone during ischemia/reperfusion in rabbit kidneys. Free Radical Res. Comm. **9:** 181–186.

41. BOLLI, R., B.S. PATEL, M.O. JEROUDI, X.Y. LI, J.F. TRIANA, E.K. LAI & P.B. MCCAY. 1990. Iron-mediated radical reactions upon reperfusion contribute to mycardial "stunning." Am. J. Physiol. **259:** H1901–H1911.

42. JOTTI, A., L. PARACCHINI, G. PERLETTI & F. PICCONINI. 1992. Cardiotoxicity induced by doxorubicin *in vivo*: Protective activity of the spin trap alpha-phenyl-tert-butyl nitrone. Pharmacol. Res. **26:** 143–150.

43. BRADAMANTE, S., E. MOUTI, L. PARACCHINI, E. LAZZARNI & F. PICCNINI. 1992. Protective activity of the spin trap tert-butyl-alpha-phenyl nitrone (PBN) in reperfused rat heart. J. Mol. Cell. Cardiol. **24:** 375–386.

44. HAMBURGER, S.A. & P.B. McCAY. 1989. Endotoxin-induced mortality in rats is reduced by nitrones. Circ. Shock **29:** 329–334.

45. NOVELLI, G.P., P. ANGLLOLINL, R. TANI, G. CONSALE & L. BORDL. 1985. Phenyl-*t*-butyl-α-phenylnitrone is active against traumatic shock in rat. Free Radical Res. Commun. **1:** 321–327.

46. CARNEY, J.M., P.E. STARKE-REED, C.N. OLLIVER, R.W. LANDUM, M.S. CHENG, J.F. WU & R.A. FLOYD. 1991. Reversal of age-related increase in brain protein oxidation, decrease in enzyme activity, and loss in temporal and spatial memory by chronic administration of the spin-trapping compound *N-tert*-butyl-*a*-phenylnitrone. Proc. Natl. Acad. Sci. USA **88:** 3633–3636.

47. SOCCI, D.J., B.M. CANDALL & G.W. ARENDASH. 1995. Chronic antioxidant treatment improves the cognitive performance of aged rats. Brain Res. **693:** 85–94.

48. KUMARI, M.V.R., T. YONEDA & M. HIRAMATSU. 1997. Effect of "β-CATECHIN" on the life of senescence accelerated mice (SAM-P8 Strain). Biochem. Mol. Biol. Int. **41:**1005–1011.

49. PACKER, L. 1996. Antioxidant defenses in biological systems: An overview. *In* Proceedings of the International Symposium on Natural Antioxidants, Molecular Mechanisms and Health Effects. L. Pakcer, M.G. Traber & W. Xin, Eds.: 9–23. AOCS Press. Champaign, Illinois.

50. SACHER, G.A. & R.W. HART. 1978. Longevity, aging and comparative cellular and molecular biology of the house mouse, *Mus musculus*, and the white-footed mouse, *Peromyscus leucopus*. Birth Defect **14:** 71–96.

51. BURGER, J. & M. GOCHFELD. 1992. Survival and reproduction in *Peromyscus leucopus* in the laboratory: Viable model for aging studies. Growth Dev. Aging **56:** 17–22.

52. SOHAL, R.A., H.-H. KU & S. AGARWAL. 1993. Biochemical correlates of longevity in two closely related rodent species. Biochem. Biophys. Res. Commun. **196:** 7–11.

53. GUO, Z., M. WANG, G. TIAN, J. BURGER, M. GOCHFELD & C.S. YANG. 1993. Age- and gender-related variations in the activities of drug-metabolizing and antioxidant enzymes in the white-footed mouse (*Peromyscus leucopus*). Growth Dev. Aging **57:** 85–100.

Roundtable Discussion

How Best to Ameliorate the Normal Increase in Mitochondrial Superoxide Formation with Advancing Age

CHAIR: ANTHONY W. LINNANE[a]

PARTICIPANTS: SYMPOSIUM SPEAKERS PLUS LARS ERNSTER, HANS NOHL, AND STEN ORRENIUS

[a]Centre for Molecular Biology and Medicine, Epworth Hospital, 185–187 Hoddle Street, Richmond, Melbourne, Victoria 3121, Australia

Aging is related to the rate of oxygen consumption. Over 90% of oxygen is used by the mitochondria; of this, 1 to 3% is diverted to superoxide radical formation. Although other sources of superoxide radicals (and H_2O_2) are present in the cell, their contribution should be small compared to that of the mitochondria.

Mitochondrial-derived superoxide radicals can produce mitochondrial DNA changes. These changes are reflected in the synthesis of defected respiratory chain components that further increase the rate of superoxide radical formation and decrease the rate of ATP production. The decline in mitochondrial function with time is believed to contribute to the progressive increase in the chance of disease and death with increasing age. Thus, decreasing the rate of mitochondrial superoxide radical formation should be a significant focus of efforts to increase the span of healthy, useful life.

Today, mitochondrial superoxide radical production can be decreased by caloric restriction, and probably by nitrone and nitroso compounds—if so, then also by nitroxides and hydroxylamines. Birds divert a smaller fraction of the O_2 they use to the formation of superoxide radicals than do mammals; the same is true of the long-lived white-footed mouse compared to normal mice. Can means be found to do the same in humans? Another approach to the problem would be to search for compounds that can associate with electron-rich areas of the respiratory chain—possibly buckminsterfullerene and its derivatives (free radical sponges)—so as to block access by O_2 to these areas.

Denham Harman

ANTHONY W. LINNANE CHAIR (*Epworth Hospital, Melbourne, Victoria, Australia*): I would like to make a very brief introductory comment. When information is being given, it is important to identify the source of the information, that is, what animal, what tissue, and the appropriateness in the commentator's view of the methodologies.

Insufficient attention has been paid in the reporting of results regarding individual tissues and organs under study. Indeed, do these tissues age? Thus, with regard to systemic disease, the liver is essentially an immortal organ, and studies on the liver probably have nothing whatsoever to do with the aging process. Yet a very large number of experiments have been carried out on the liver, and the results have been treated as an exemplifier of many other tissues in aging studies.

Human subjects do not die of an aging liver; on the contrary, we do die as a consequence of the changes that occur in aging hearts and brains. Only a few moments ago I learned from Sten Orrenius that a number of the observations he has made in heart effectively do not occur in tissues like liver, in his view, because of the high SOD and catalase concentrates in this organ. Presumably different organs have different defense mechanisms or extents thereby for dealing with putative age changes, and, of course, different organs have different basic metabolisms.

I am now going to ask each member of the roundtable to take a few minutes to summarize their position and experience on the question of mitochondrial superoxide formation with advancing age.

STEN ORRENIUS (*Karolinska Institutet, Stockholm*): I will start with some remarks. My first remark is to agree with our chair, that one has to be very careful with regard to which tissues are relevant and in which tissues these phenomena have been shown. Another aspect that has not been discussed in great detail is that the escape of reactive oxygen species under these conditions will be the result of both formation and the metabolism of these reactive oxygen species. Particularly in the mitochondria, the glutathione system is very important, and we have to consider differences between various cell types and tissues with regard to both generation and capacity to inactivate and trap these reactive species.

For my second remark, I would like to pick up on Dr. Ozawa's presentation, in which he showed that the result of the mutations and the increase in active oxygen production in certain instances is the initiation of apoptosis with fragmentation of cells. I am going to discuss the role of mitochondria in cell death in greater detail tomorrow, but I think it is important to emphasize that a possible result of this impaired respiration with increased generation of reactive oxygen species is that the apoptotic process can be activated. This was clear from Dr. Ozawa's presentation, and our chair, Dr. Linnane, reported several years ago, that acute treatment with inhibitors of the respiratory chain will produce apoptosis.

LINNANE: Thank you, Sten. Takayuki, would you like to make a comment in response to those remarks?

TAKAYUKI OZAWA (*University of Nagoya, Nagoya, Japan*): Genetics and bioenergetics studies demonstrate unexpected fragility of mitochondrial (mt) DNA in human tissues. Its guanine residue forms an adduct with the hydroxyl radical (8-OH-dG), leading to random large deletions. Hence, in such stable tissues as brain and muscle, the great majority of wild-type mtDNA that are fragmented into hundreds of types of deleted mtDNA correlated with oxidative damage, resulting in pleioplasmic defects in the mt energy-transducing system. The mtDNA fragmentation to this level is demonstrated in cardiac myocytes of normal subjects over age 80 and in mtDNA of a cultured cell line under hyperbaric oxygen stress for three days, leading a majority of cells to apoptotic death. The extreme fragility of mtDNA could be ameliorated by protecting oxygen stress and by externally administered redox substances such as coenzyme Q (CoQ). In our animal experiments, complex I and IV of the rat skeletal muscle respiratory chain declined with age down to *ca.* 50% at 55 weeks old compared with 7 weeks old. In rats fed with CoQ-enriched laboratory chow (0.2% w/w), the activities were significantly mitigated compared with the normal control. Similar protective effects of ubiquinone were observed in rat mt respiratory chain activities diminished with administration of fried beef–derived mutagenic factor or that of an antitumor drug, doxorubicin, having an adverse effect due to lipid peroxidation. In clinical studies (reviewed by Mortensen in *Clinical Investigator*, 1993), double-blind placebo-controlled trials definitely confirmed that CoQ has a place as adjunctive treatment in heart failure with beneficial effects on the clinical outcome, the patients' physical activity, and their quality of life.

LINNANE: Thank you, Takayuki. Are there any comments from the audience before we continue?

LI LI JI (*University of Wisconsin, Madison*): Our understanding of the aging effect is that superoxide production in the mitochondria is hinged greatly on the accurate measurement of superoxide anion production in mitochondria. From this session, we have heard about such different methods as dichlorofluorescein, SOD coupled assays, and homovanillic acid. I would like the speakers to address the advantages and limitations of the various assays available.

LINNANE: Thank you for that question. I will ask each of the presenters to make a comment.

HANS NOHL (*Veterinary University of Vienna, Vienna, Austria*): The major question we have to address concerns the biological promoter of aging that brings about the transformation from adolescence to senescence. What evidence do we have that oxygen radicals play a role in the underlying mechanism that triggers the biological process of aging? In line with the lectures this morning there are a great number of papers that show an imbalance between antioxidants and prooxidants, in favor of the latter for the progression of life. The question to be answered is whether the gradual establishment of oxidative stress is the cause or the consequence of aging! We have to keep in mind that any disturbance of metabolism of compartmentation and of structural integrity ends up in the disturbance of oxygen salvage as well as necessarily giving rise to the formation of oxygen radicals. The effect of the feeding of antioxidants on life span was often taken as an indirect indicator for the estimation of the potential role oxygen radicals play in the phenomenon of aging. Positive effects resulting from elevation of the antioxidant status were considered as evidence of the key role of oxygen radicals in the aging process. However, the validity of this type of study is limited due to distribution patterns of the antioxidants that do not always follow the sites of radical formation, due to rate constants that are often below those of radicals with natural biomolecules, and due to undesired side effects. Also overexpression of SOD and catalase in flies, which was reported to increase life span, does not answer the question about the role of oxygen radicals. The activity of these enzymes is diffusion limited, so that normal enzyme contents can also sufficiently control steady state levels of superoxide radicals and H_2O_2 in the cell. In addition, these antioxidant enzymes are not homogeneously distributed in the cell but exist in particular compartments. This may indicate the need of controlled steady state levels of oxygen radicals and H_2O_2 to maintain biological homeostasis.

A more promising approach for evaluating the possible role of O_2 radicals in the biological process of aging as initiator, pacemaker, or promotor requires the identification of the site as well as the knowledge of the regulating mechanism of O_2 radical formation. Due to its particular role as the most active oxygen reductant of the cell, mitochondria were generally considered as the site of O_2-radical generation. However, there is increasing agreement that mitochondria normally do not contribute to cellular prooxidant formation. Conditions used so far to establish a one-e^- leak from the respiratory chain to oxygen out of sequence were nonphysiological, namely the inhibition of mitochondrial respiration with antimycin.

However, analysis of mitochondria reveals that lipids, proteins, and DNA are increasingly faced with strong oxidant attacks with the progression of life. In particular, the oxidative alterations of mitochondrial DNA, which is a constituent of the matrix located close to the electron transfer chain, elucidate that something happens to the respiratory chain during aging that transforms these coupled redox systems to oxygen-radical generators. A theoretical onset to explain this seems to be possible, taking into consideration that life is established as an open metabolic flow system that will readily incorporate and accumulate environmental minihits. We have recently developed model systems in which mitochondria were exposed to various environmental pollutants (such as food additives and insecticides) and also to transient hypoxia as a model for variations of organ supply with oxygenated blood. Exposure to the various minihits was found to cause unequilibrated minidefects at the level of the respiratory chain, making electron leakage to oxygen out of sequence possible. Our analytical studies on the mechanism of this deviation of electrons from the regular energy-linked pathway to dioxygen revealed that O_2 radical release was always linked to a fall in the efficiency of energy gain from these mitochondria. This was, however, not a result of the misdirection of electrons from the energy-linked pathway, but was due to the decrease of the proton-motive force that drives ATP synthesis. This means that in addition to the initiation of oxidative stress, the cell is faced with a decrease of energy supply from mitochondria that is both linked to and in turn stimulated by the establishment of oxidative stress. This situation may explain the transition from a homeostatic system (adolescence) to a stochastic system that ultimately makes organization and maintenance of life impossible.

It appears, therefore, that a more promising strategy to slow down mitochondrial contribution to increased O_2 radical formation is the prevention of accumulating minihits that are responsible for deranged electron transfer from the respiratory chain to oxygen. However, it cannot be forgotten that also natural factors of our living space necessarily in contact with biological systems may develop the potency of minihits, with the consequence of the disturbance of homeostasis linked to O_2 formation.

LINNANE: Thank you very much Dr. Nohl. I knew that you might have a different point of view.

LARS ERNSTER (*University of Stockholm*): My questions relate to what Dr. Nohl has just elaborated on: What is the primary event in causing mitochondrial damage, if we accept the concept that mitochondria are involved in the process of aging and that oxidative processes and especially oxidation of mitochondrial DNA are involved, plus the notion that this process increases with age in an exponential manner? Suppose that we have a xenobiotic or a nuclear gene mutation that causes a deficiency in the synthesis of a mitochondrial enzyme that is encoded in a nuclear gene, for example, pyruvate dehydrogenase. What will then happen is that we get a limitation in mitochondrial activity, for example, a diminished activity of the respiratory chain, leading to what Dr. Beckman was referring to this morning as a "feedback effect" or, as we may also refer to it, a "vicious circle." The inhibition of the respiration may lead to an increased formation of reactive oxygen species, which will damage mitochondrial DNA and lead to further injury to respiratory chain enzymes and increased generation of reactive oxygen species. Such a vicious cycle may explain the exponential nature of the aging process. However, it does not answer the question as to where the process starts: at the level of the nuclear DNA or the mitochondrial DNA? It is now almost dogmatically believed that the mitochondrial respiratory chain, and in particular the autoxidation of the semiquinone of ubiquinone, is the main source of the formation of reactive oxygen species in mammalian cells. I think there is a great deal of evidence against this concept. First, ubisemiquinone in mitochondria is known to be stabilized against autoxidation by being bound to specific ubiquinone-binding proteins that are components of complexes I, II, and III of the respiratory chain. Any formation of reactive oxygen species by the respiratory chain most probably occurs through other components of these complexes; for example, we saw good evidence this afternoon that complex I can generate reactive oxygen species without the involvement of ubiquinone. Second, ubiquinone is known to occur, in addition to mitochondria, also in other cellular membranes and in blood plasma and has been shown to act in its reduced form as an efficient antioxidant in all of these locations. Third, it is well known that mammalian cells contain a number of enzymes that generate reactive oxygen species under physiological conditions; the monoamine oxidases in the outer membrane of mitochondria, xanthine oxidase in the cytosol and the plasma membrane, and various peroxisomal flavin oxidases generate hydrogen peroxide (and xanthine oxidase, also superoxide radical). These are probably the main sources of reactive oxygen species formed in animal cells under physiological conditions. In addition, various xenobiotics can give rise to such species, for example, through oxidation by way of the cytochrome P-450 system. A final point that I would like to raise in this context, not directly related to the above questions, concerns the mechanism behind the beneficial effect of caloric restriction, a topic dealt with in several lectures at this meeting. It is usually assumed that this effect is due to a restriction of substrate supply and thereby of the generation of reactive oxygen species as a by-product of the respiratory chain. An alternative explanation, in my opinion, could be that a limitation in the supply of nutrients, especially sugar and fat, results in a restriction of substrates, glucose and fatty acids, for the peroxisomal flavin oxidases and thereby of the generation of hydrogen peroxide through these enzymes.

LINNANE: Thank you, Lars. Dr. Beckman will now make a comment.

KENNETH BECKMAN (*University of California, Berkeley*): It seems that the title of this session has changed from "How can we ameliorate mitochondrial superoxide generation that occurs normally with age?" to "Does mitochondrial superoxide generation increase with age?" I'm quite happy with that change of focus, inasmuch as I think it's premature to talk about ameliorating a problem that we don't know for sure exists. I agree with Dr. Ernster that the assumption that mitochondrial oxidants are the most important oxidants in the cell's aging was hasty. For instance, I think there is evidence in the case of some age-related pathologies that oxidants formed by other processes may be more important. I think that interspecies comparisons of mitochondrial oxidants may end up answering both the question of whether mitochondrial oxidants are the most important source of reactive oxygen species, and the question of how a catastrophic vicious cycle may be initiated. Regarding Dr. Nohl's comment, I think that differences in species-specific life span are unlikely to be due to exogenous sources of oxidants. If you have large differences in life span between, for instance, birds and rodents, then to talk about minihits that are exogenous as opposed to minihits that are endogenous doesn't make a whole lot of sense to me. Of course, I like the idea of minihits, small incidents of oxidative stress that can damage mitochondrial DNA or proteins and lead to a vicious cycle. However, it seems to me that if you have organisms that have a 10-fold difference in maximum life span potential, as do rats and humans, then this difference is not going to be due to irradiation, for instance.

As far as questions that I think are still important, they are as follows. (1) Does this vicious cycle that's been proposed actually exist? There is no evidence to say that damage to mitochondrial DNA necessarily is going to lead to altered proteins that increase radical production. I found interesting today the evidence that there is a protein product resulting from genetic fusion following mtDNA deletion that is in fact synthesized *in vivo*. That is interesting and novel because, so far, people haven't looked at the functional significance of mtDNA deletions in terms of oxygen generation. Certainly, people have shown that you can get a decrease in oxidative phosphorylation in mitochondria that possess deleted molecules, but what I think would be interesting would be to show that there is also an increase in oxygen generation. Incidentally, my experience with ρ^o cells does not suggest that that is the case. On the other hand, ρ^o cells are extreme instances that are unlikely to occur *in vivo*. (2) As I said before, I think that species comparisons of mitochondrial oxidative metabolism will be very important. (3) Last, there have been some papers in the last year showing that signal transduction, for instance, the induction of mitogenic stimuli in diploid fibroblasts by ras, increases superoxide generation by a plasma membrane-bound oxidase. So now there is a whole new series of endogenous minihits to consider, which raises the possibility that mitogenic stimulation may be more rapid in short-lived species. If this were the case, it might be that oxidants are important, but that they are not only mitochondrial in origin, but may also result from a higher rate of mitogenesis in short-lived animals. In conclusion, it seems to me that before we talk about ameliorating mitochondrial superoxide generation, we have to find out where this superoxide generation is from, and if it's mitochondrial, we need to ask whether or not there is a vicious cycle occurring.

LINNANE: Would you like to comment, Lars?

ERNSTER: I don't really mean that it cannot come from outside the cell. But it may also come from outside the mitochondria within the cell. There are a number of extramitochondrial enzymes that generate ROS, for example, xanthine oxidase.

NOHL: I agree that apart from mitochondria we also have to consider a variety of other potential superoxide-radical sources in the body. In contrast to mitochondria these generators are not present in all cells; nor do they fundamentally contribute to the function and body mass of the cell as mitochondria do. The turnover rate of oxygen through mitochondria concerns more than 90% of total cellular oxygen. All components involved in mitochondrial electron transfer are capable of transferring single electrons, which is a prerequisite for superoxide formation. The accumulating traces of oxidative damage in mitochondria from aging animals also support the idea that mitochondria are the main generators of superoxide radicals in the cell, inasmuch as radical generation distant from these organelles is unlikely to exert this oxidative alteration due to the instability of the oxygen radicals.

LINNANE: There is, it appears to me, an assumption in much of the discussion that radical damage and mitochondrial DNA mutation associated with age are synonymous. I suggest such conclusions are not warranted. Rather, the extensive mtDNA mutation demonstrated to be age-associated is deletion mutation; ROS-induced oxidative mutation has not been shown to be extensive or formally correlated with deletion mutations.

Indeed, examination of a number of tissues with a high oxidative activity, such as human liver and skeletal muscles, shows no correlation between the extensive deletion mutations observed in skeletal muscle versus the virtual complete absence of deletion mutations from liver mtDNA, for which age-associated oxidative damage (albeit small) has been reported.

OZAWA: I think that there is a question as to where the active oxygen is generated under physiological conditions. In 1946, Professor Michaelis theoretically predicted that one- and two-electron reductants of oxygen, namely active oxygen, are essential intermediates to generate its four-electron reductant, H_2O. His prediction was verified by Dr. B. Chance in 1976. One-, two-, and four-electron reduction of oxygen occurs sequentially in the active site of cytochrome oxidase at physiological 50 µM oxygen in a tissue. Higher oxygen concentration from not only the terminal part of the respiratory chain, but the middle and the initial part of the chain, and other organelles than mitochondria, could generate the active oxygen, depending on their K_m to oxygen. Hence, the amount of the active oxygen generated is dependent on the tissue examined and on the condition, such as ischemia-reperfusion transition, where a vast amount of the active oxygen is generated transiently.

JUDD M. AIKEN (*University of Wisconsin, Madison*): I really have nothing astute to add to this discussion other than to say I think that the questions raised emphasize that there is a real lack of knowledge about basic mitochondrial functioning. We keep stumbling on this every time we run a series of experiments trying to determine the mitochondrial replication rate. Responding to Dr. Linnane's point, we've certainly seen mtDNA deletions in the liver. Their abundance in liver I cannot comment on, as we have not studied it sufficiently. Certainly, one thing to keep in mind with these deletions is that they're affected by the number of cells that are being analyzed, and the more cells you put in the mix, the greater the number of amplification cycles that you need to identify.

SALVATORE PEPE (*Baker Medical Research Institute, Melbourne, Victoria, Australia*): I too do not have much more to add except with regard to the original question of amelioration. I agree with Dr. Beckman and Dr. Aiken, too, in that it is quite premature and difficult to prescribe anything when we are struggling with our understanding of mitochondrial function and all of the various factors that come together to modulate the end point that we are measuring. So at this stage it is important to make sure that we really are always comparing "apples with apples" with regard to comparison of data between the various research groups and experimental models that we are discussing here today.

RITCHIE J. FEUERS (*National Center for Toxicological Research, Jefferson, Arkansas*): I think it may be premature to hang everything on oxidative damage, as indicated by the introductory slides shown by Dr. Harman that presented a fairly exhaustive list of the many different theories of aging. It is likely that oxidative damage does play an important role in what we call aging, but oxidative damage may not be responsible for all of aging. If each of us were to write our definition of what aging might be, I suspect we would have as many different definitions as we have people in the audience. So I think it is important to keep this in mind. Certainly, I don't want to de-emphasize the importance of what the role of free radicals might be in a mechanism of aging.

This morning I showed a slide that I called the "obligatory survival curve slide." One of the fundamental pieces of information provided on the slide was to show that a mechanism of aging does exist, inasmuch as maximum achievable life span can be manipulated (in a positive manner), in this case by caloric restriction. I think this is compelling because from this we know that there is at least one way to ameliorate the aging process. With this knowledge there is the suggestion that there are additional interventions, such as various antioxidant therapies and nutritional modifications, through which a mechanism of aging may be manipulated.

A question was raised concerning caloric restriction and how it might produce changes that affect substrate levels, and then how that might ultimately influence basic metabolism and other biochemical events. For the last four or five years we have known that when you calorically restrict animals biochemical reactions become more efficient. This is achieved when enzyme-substrate affinity is increased, thus producing catalysis at much lower substrate levels. As an example, lower cellular glucose levels can still be used to fuel the glycolytic pathway, and an enzyme such as pyruvate kinase will bind lower levels of phospho*enol*pyruvate and ADP to catalyze their conversion to pyruvate and ATP as the product.

In general, during aging the efficiency of enzyme systems decreases. The improvement of efficiency associated with caloric restriction is often the result of improved (or maintenance of) enzyme regulation. It is in fact often a case of producing a reduced K_m that suggests that the enzyme can do more with less, and do it more efficiently. The cell can use less fuel to produce as much or more necessary metabolites and ATP.

These responses to aging and the resultant effects of caloric restriction seem to extend to mitochondrial function. Our data suggest that there is an obstruction of electron flow through complex I as the amount of the complex diminishes, and through complex III due to loss of total complex and increases in K_m with age. This builds a kind of a damn against electron flow (most notably at complex III). Thus, fewer electrons can pass through electron transport and ultimately to complex IV. During aging, high-affinity binding sites of complex IV are lost, further impeding electron flow. I suggest that due to these problems more free radicals would be generated at these sites along electron transport. One can envision that as lesions accumulate a mechanism might exist where a single electron is passed, and due to poor binding, a premature free radical product would be released prior to the second electron passage. Caloric restriction resolves these qualitative and quantitative aging problems associated with complexes I, III, and IV, and this may be one mechanism through which caloric restriction limits free radical generation leading to extension of maximum achievable life span. These aging effects were not seen for complex II, which is totally of nuclear origin. The fact of the matter is that changes do occur in I, III, and IV, which are in part encoded from mtDNA.

It seems to me that future efforts should center on identification of new nutritional and endocrine interventions that act to improve enzyme efficiency. These types of interventions would obviously be compatible and perhaps act synergistically with emerging antioxidant interventions.

Few free radicals are generated in the efficient electron transport of the young; the mtDNA are in a healthy, lesion-free state; membranes and proteins are all in good condition; and there is sufficient ATP generation. With time (or aging) there is a reversal of this situation. In the aging mitochondria, a minimum amount of ATP production is achieved, whereas huge increases in production of free radicals occurs. I suggest perhaps some of these events occur through mechanisms I have discussed, that is, decreased efficiency of electron transport. The result of this situation would be production of oxidative lesions to the mtDNA and other mitochondrial components, yielding an ever decreasing efficiency of electron transport. What I am suggesting is that a downward spiral or vicious circle is set into motion with an imbalance between antioxidant defenses and free radical production, which may be a key participant in a mechanism of aging.

Yau-Huei Wei (*National Yang-Ming University, Taipei, Taiwan*): I have one comment to add. The oxygen-free radical itself may not be that detrimental, but lipid peroxides and some organic free radicals are the most damaging species in aging tissues. They often bring about chain reactions of free radicals to amplify oxidative damage to biomolecules. I have shown you that the production of lipid peroxides in the submitochondrial particles is increased dramatically with age. The ever-increasing generation of lipid peroxides in mitochondrial membranes are very damaging to mtDNA molecules, which are attached, at least transiently, to the inner membrane of mitochondria. Thus, mtDNA and some membrane proteins get hit very frequently by lipid peroxides. Another important point is that the ability and capacity of the tissue cell to dispose of lipid peroxides and organic free radicals are decreased during aging. These two compounding factors lead to an age-dependent increase of oxidative damage to mitochondria in various human tissues.

Akitane Mori (*Okayama University Medical School, Okayama, Japan*): Recently, antioxidant protection from oxygen free radicals by polyphenols has been of current interest; that is, polyphenols are powerful antioxidants found in fruits, vegetables, and herbs. Packer's laboratory has documented that the flavonoids in the herb *Gingko biloba* can neutralize superoxide and hydroxyl radicals. Sastre *et al.* have reported that the oral administration of *Gingko biloba* extract prevented the age related impairment in mitochondrial morphology and also prevented the age-related impairment in mitochondrial function, inasmuch as it prevented the age-associated impairment in mitochondrial function in state 4 respiration that was linked to increased mitochondrial peroxide release (*In* Proceedings of the International Symposium on Natural Antioxidants, Molecular Mechanisms and Health Effect. L. Packer *et al.,* Eds., 1996). Meanwhile synthetic antioxidants may be useful for preventing the age-associated free radical damage in cells. EPC-K1 is composed of ascorbate and vitamin E joined by a phosphodiester linkage and has potent hydroxyl radical scavenging activity (Mori *et al.* 1989. Neurosciences **15:** 371–376). EPC-K1 is soluble in both water and organic solvent. A commercial cosmetic (Shiseido Co., Ltd.), containing 0.2% EPC-K1, is reported to normalize the turnover time of damaged skin, and it is widely used for skin care. Now, much evidence related to beneficial effects of EPC-0K1 on free radical–induced damage, for example, in ischemia/reperfusion and organ transplantation is accumulating. We demonstrated that a spin trap, *N-tert-α*-phenyl-butylnitrone (PBN), prolongs the life span of senescence-accelerated mice (SAM mice). PBN has been reported to normalize the age-related biochemical parameter and physiological functions, such as memory. These synthetic antioxidants may be possible therapeutics for ameliorating the normal increase in mitochondrial superoxide formation with advancing age.

Gustavo Barja (*Complutense University, Madrid, Spain*): I agree with our Dr. Linnane that it is not clear if mitochondrial free radical production increases with age. I remember now around six publications on the subject, the majority from heart mitochondria, and 50% of them support the idea but not the other 50 percent. Furthermore, the most frequently cited paper supporting the concept of an increase with age compares rats of 3 months of age to old ones. Thus, this increase can be maturation, not aging. In any case, I think that the increase with age is not the most important question. A very important question is if different rates of mitochondrial reactive oxygen generation (ROS), although constant with age, can explain, in part, the different rates of aging of different animals. Aging is a physiological phenomenon, and many physiological capacities (especially maximum ones) progressively decrease with age. A constant rate of free radical production would be consistent with these progressive losses of function. My message is that there is no need for an increase in ROS production with age for the free radical theory to be true. If ROS production increases with aging, then one could expect the rate of aging to be quicker in the old, which does not seem to be the case. What should accumulatively increase with aging is the damage (*e.g.*, oxidative DNA damage), not its cause. I would also like to comment on the scheme that Dr. Nohl showed us. It is, in principle, possible that hypoxia–ischemia or radiation can alter the mitochondria so that they produce more radicals, but another typical characteristic of aging is its intrinsic character. External factors like radiation cannot explain why different animals age at different rates. Some live only one year, whereas others live 100 years in similar environments. Concerning the comment of Dr. Ernster about our localization of the free radical production site in complex I, we consider it a main free radical source, and not only because of our own results. Looking back at the literature, complex III has been much emphasized. I think that this was at least, in part, due to the debate about which of two possible complex III generators were responsible. This led to the frequent use of succinate, which bypasses complex I and thus makes it very difficult to observe its contribution. We see ROS generation in complex I and III in state 4 in heart mitochondria, but in nonsynaptic brain mitochondria we see it only at complex I. And in state 3, complex I seems to be the only generator in these two kinds of mitochondria. Classic reports by Takeshigue and Minakami or Boveris in the 1980s or even by the group of Britton Chance as early as in 1967 (Hinckle *et al.*) showed ROS generation at complex I, and a recent (1997) paper from Dr. Hansford also reached the same conclusion. So, we should not forget about complex I ROS production. Last, I agree with the idea that free radicals will not be the only cause of aging. For instance, whenever we compare the rates of ROS production between species, the differences are always less than their differences in longevity. It seems obvious to me that the rate of repair must also be important. This has been perhaps little emphasized up to now in this meeting. There is consistent evidence showing that animals with a slow rate of aging have both a low rate of generation of damaging agents (like ROS) and a high rate of repair (*e.g.*, DNA repair).

ORRENIUS: This may not be so relevant any longer, but my remark is related to the cellular source, or site of generation, of reactive oxygen species. Although several other systems can generate reactive oxygen species, they would still, I think, have difficulties in competing with the mitochondria. For example, some of the cytochrome P-450 isozymes contribute to cellular superoxide generation but at rates that are probably much slower than that of the mitochondria. Similarly, the peroxisomal system in some tissues, particularly under induced conditions, can also be a generator of reactive oxygen species of quantitative importance. However, in the cell types that we discuss here, like muscle or heart, I think that both the P-450 system and the peroxisomal system may play relatively minor roles. Additional support for the role of mitochondrial superoxide generation, is the importance of the manganese superoxide dismutase and the detrimental effects of interference with this activity on cell survival, which has been observed. These observations underline both the importance of mitochondrial superoxide production and that this production can be locally controlled by mitochondrial superoxide dismutase activity.

LINNANE: I would like to make a statement concerning that. The mitochondria are very important in terms of the percentage oxygen they consume and, therefore, potentially radical generation, but information is growing on the recognition of the occurrence of other ROS-generating systems. Thus the P-450 system is now recognized in many membranes of the cell. Work is needed on the quantitation of ROS generated in the cell other than that by the mitochondria.

I would like to put a question to the panel. It's meant to be provocative and you'll excuse me, but I think it's a significant question. How many proteins, if any, have been demonstrated to lead to some catastrophic outcome because of free radical damage? As I understand it, most oxidized proteins are degraded and removed from the metabolic pool. For me, a lot of the present discussion goes round and round in circles searching for an outcome of ROS damage. The observation that a phenomenon occurs, without consideration of downstream subsequent events, does not make the phenomenon clinically important. What I have said for proteins I suggest may also be said for lipid peroxidation: with increasing age there appears to be no or little progressive increase in ROS cellular oxidized lipids. I invite the panel to comment; do I overstate the situation? Dr. Beckman, would you make a comment.

BECKMAN: I think I'll sidestep it slightly. I think that you're right: there haven't been many examples of the effect of damaging a given protein.

LINNANE: Or any other system.

BECKMAN: True, I was only answering the question you asked about proteins. However, I'll suggest that a potential place to look is in the realm of signal transduction. I have come up with two terms to help myself think about the roles of different oxidants: "legitimate" versus "illegitimate" oxidants. Legitimate oxidants are those you are supposed to generate, and illegitimate oxidants are those generated by mistake. Legitimate oxidants, for instance, would be those you've generated as a result of ras stimulation. (This is assuming that the recent results on ras are correct.) Illegitimate oxidants, however, may result from mitochondrial leakage. So what I'm saying is that generation of oxidants may be damaging *not* because they damage a certain protein or system but because they're mimicking legitimate oxidants. It may be that if an oxidative system exists that is going awry, mitochondria, for instance, it may send faulty signals to cells; of course, one of the hallmarks of aging is dysregulation of various systems. Also, I think it's important that people try to look at the effects of oxidant damage on individual proteins. For instance, if you damage a polymerase, does it induce mutations, for instance?

LINNANE: Signal transduction studies warrant more detailed study, and it appears that ROS may possibly play a regulatory role with regard to mitochondrial activities. Clearly there is mitochondrial and nuclear intercommunication; thus, for example, in a number of mitochondrial DNA diseased patients, decreases in cytochrome oxidase (COX) activity are accompanied by increased complex II activity that is nuclear encoded. Lars, would you like to make a comment?

ERNSTER: You wanted to have an example of oxidatively damaged proteins. Earl Stadtman has published a number of papers about oxidatively damaged proteins that increase with age. Glutamine synthase is such a protein.

LINNANE: The question becomes, To what extent is glutamine synthase damaged, and Has it been shown there is a significant metabolic effect as a consequence? Let me tell you one thing that does have an outcome, and that is mitochondrial DNA mutation and the consequential loss of the enzymes associated with the mutations, with the resultant impairment of the bioenergy system. That is a real, measurable physiological meaningful outcome.

My earlier remarks referred to ROS possibly playing a role in signal transduction. Thus calcium signaling may well be disturbed by oxidative stress affecting mitochondria. Changes in calcium flux would be a downstream consequence of changes in mitochondrial proton gradients, perhaps ROS induced.

OZAWA: The most serious outcome of mitochondrial DNA mutations associated with age is the collapse of the bioenergy, electron-chemical gradient in the form of transmembrane potential that sustains a vast amount of calcium inside of the mitochondria. Its collapse results in the efflux of calcium into cytosol, causing tremendous effects, including the activation of calcium-dependent nucleases, leading the nucleus into the small pieces of nucleosomes. Physiologically, our body needs to eliminate unwanted cells or damaged ones. Hence, there must be a signaling system to initiate cellular apoptosis; however, the details remain unclear.

LINNANE: So calcium is our signal, to some extent, surely.

OZAWA: Right.

LINNANE: Anyone else, please?

A.R. HIPKISS (*King's College, London*): I would like to point out that advanced glycosylation end products (AGEs) accumulate with age and can provoke a hyperoxic response in many cells, especially if they have receptors for AGEs (RAGEs). The beta-amyloid peptide fragment also binds to RAGEs and provokes a similar hyperoxic condition. It is therefore debatable as to which is the primary initiating event, the accumulation of AGEs, decreased antioxidant defenses, failure to degrade the aberrant polypeptide, or increased mitochondrial free-radical generation?

THOMAS VON ZGLINICKI (*Humboldt University, Berlin*): I would like to add yet another possible result of oxidative stress, namely telomere shortening and telomere damage. We have quite good evidence that telomere shortening is one specific result of oxidative stress, and, at the same time, it signals senescence in proliferating cells. So there is a clear connection between oxidative stress and senescence, at least in this one model of aging.

CHRISOPHER DRIVER (*National Institute of Aging, Parkville, Victoria, Australia*): I would like to make some comments about *Drosophila* in mitochondria and aging. There seems to be some data coming from work on *Drosophila* indicating that the production of radicals has the characteristics more of being part of a dying phase, rather than something that is there at the beginning, producing damage. At the end of the life span of *Drosophila*, there is a fairly substantial rise in radical production. At the same time, there is actually a fall in radical protection mechanisms, and this is a coordinated event. If we can look at the work of Bob Arking, who has produced strains selected for longevity, the long-lived strains are not that much more radical resistant, nor do they seem to make fewer radicals. What happens is that the radical crisis is simply delayed. I think that it is fair to say that mitochondria are involved in this process but in a late stage, rather than producing early-stage degeneration.

UNIDENTIFIED SPEAKER: I am not a scientist or a free radical person. I have heard that hormones are very important in the aging process, and I have been thinking today about mitochondria in endocrine tissues. What is known is that they age because they are very high metabolic tissues. We know that, for example, with type II diabetes, we have a decrease in insulin production by pancreatic beta cells, so I've been wondering whether the mitochondria and endocrine tissues could be very important. What is known about defective estrogen, testosterone, and DHEA on the health of mitochondria? What is known about mitochondria in endocrine tissues, and what happens to them with aging? I would like to know, because, clinically, hormones have been shown to delay some of the effects of aging.

LINNANE: Few, if any, studies have been made on endocrine tissues basically because of the limited availability of tissues, especially human tissues and the small number of studies yet undertaken on mtDNA in aged subjects.

UNIDENTIFIED SPEAKER: Concerning the question of what is known about free fatty acids and mitochondrial function and hormones, it is known that thyroxine has an uncoupling effect, and free fatty acids also, small free fatty acids up to 10 to 11 carbon atoms long, they have uncoupling effects on mitochondria. This means there is a decrease in the energy you gain; that's what is known.

LINNANE: Thank you.

S. GOTO (*Funabashi, Japan*): Regarding the possible involvement of mitochondria in oxidative stress of cells, I would like to make some comments based on our recent preliminary findings on protein oxidation. This may have something to do with Dr. Linnane's comment in terms of protein oxidation of the organelle. We have been involved in the study of oxidative stress from the point of carbonyl formation in tissue proteins, on which Professor Stadtman and his collaborators have made a lot of excellent contributions. What we have done is the immunoblot detection of protein carbonyls, in which we used antibodies against 2,4-dinitrophenylhydrazine to detect the protein modifications by which proteins of tissue or cell extracts were treated prior to polyacrylamide gel electrophoresis. With this method we found no detectable difference of carbonylation of proteins, in general, between mutant cells, called rho zero cells, which lack in functional mitochondria, and those of wild type cells. This finding suggests that as far as carbonyls can be regarded as a marker of protein oxidation, there is no major contribution of mitochondria to the oxidation of cellular proteins.

JEAN-PAUL CURTAY (*Paris*): So brainstorming about mitochondria and aging: I wonder if we couldn't be inspired by two acute situations, the one that we know, I think, better now is infarct. We know that there is an intoxication by calcium of the mitochondria and production of energy breakdown; so why not think of mild deficiency of magnesium in a chronic setting, which is a real fact when you look at the nutritional surveys. For instance, in France, 60% of the population doesn't get the required magnesium, and stress produces an overutilization of magnesium. So could a chronic magnesium deficiency help an excess calcium influx to mildly intoxicate other mitochondria? Maybe someone would be interested in seeing different regimens of magnesium, in view of what's happening. Another situation is chemotherapy or radiotherapy in the resistant cancers. You give BSO, and, in that case, as you know, you have lots of breakages of DNA, which doesn't kill the cell, but then the cell dies out of exhaustion from ATP and NAD. So obviously, GSH is very important. What about trying to booster GSH in the mitochondria from the precursor, as well as cysteine, or the reductants B_2 or vitamin C, as Alton Meister says. What would it give in the long run to sustain mitochondrial function?

JI: Not that I disagree with the idea that free radicals may inhibit the signal transduction pathway, but I have some evidence that I think might be to the contrary. One example is exercise training. I presented some data concerning skeletal muscle that antioxidant enzymes could be induced with training. But more interestingly, we did a study where rats swam every day for six hours for three months (published in *Acta Physiologica Scandinavica*). We found that such mitochondrial enzymes as citrate synthase and malate dehydrogenase in the heart were induced. They are not normally in the heart but only in the skeletal muscle with training. It seems that free radicals could induce antioxidant enzymes as well as mitochondrial oxidative capacity. Other evidence would be provided by hyperthyroidism. If you administer high levels of thyroid hormones, they could induce antioxidant enzymes and cause oxidative damage; but they also induce mitogenesis. So I am puzzled as to whether the idea is uniformly true.

BECKMAN: Let me rephrase my comment. I do *not* mean to say that oxidants would decrease signal transduction. What I am proposing is that a cell is constantly dealing with an onslaught of both legitimate and illegitimate oxidants. The point is, if the production of illegitimate oxidants is higher in an older tissue or older cell, and if the response to that increase in illegitimate oxidants requires the induction of antioxidants, it may be that the cell is no longer sensitive to its own legitimate oxidant-mediated signal-transduction pathways. That's purely hypothetical, of course. In any case, I wasn't proposing that oxidants would *decrease* signal transduction, but rather that they might *mask* it. Sure, illegitimate oxidants are a signal in their own right—they induce antioxidant defenses. But they may be a source of "noise." If the cell is essentially being deafened by its own illegitimate oxidant generation, then how is a legitimate signal transduction event that has to be mediated by a small amount of hydrogen peroxide going to be heard? I know that I'm telling a fairy tale, but such are the kind of mechanisms that I can imagine taking place. In which direction oxidants would affect signal transduction, I don't care to guess. I'm just proposing that oxidants may interfere with a cell indirectly. Rather than being directly damaging to a protein, they may have an amplified effect because they're interfering with a mechanism that has pleiotropic effects, such as signal transduction.

Comment

Hans Nohl
Institute of Pharmacology and Toxicology, Veterinary University Vienna, Veterinärplatz 1, A-1210 Vienna, Austria

There is a great deal of indirect evidence supporting the concept of the free radical theory of aging. It is unequivocally accepted that the imbalance between oxygen radicals and the control of undesired reactions is due to an increased formation rather than decreased radical scavenging activities. If the development of oxidative stress is significant for the transition from adolescence to senescence, the control of steady state levels of oxygen radicals should extend life span. Support for this concept came from the observation that an overexpression of SOD plus catalase in *Drosophila melanogaster* significantly increased the life span of these flies.

Although this observation allows the conclusion that O_2 radicals play a role in aging, it does not answer the key question about the underlying process causing increased levels of oxygen radicals with the progression of life. The knowledge of this mechanism is important for an interference in the biological transition to senescence if oxygen radicals are involved as suggested by the free radical theory of aging.

Mitochondria are considered to be mainly responsible for the gradual age-related increase of cellular prooxidant levels finally exceeding the control through natural antioxidants. Although isolated mitochondria from senescent rats exhibit alterations indicative of an imbalance between prooxidant formation and antioxidant activities, it is not clear how deviation of single e⁻ from the respiratory chain is regulated. Mitochondria, when carefully isolated from adult rats, do not produce superoxide radicals as by-products of regular respiration. External interference into mitochondrial superoxide release, which appears to increase with the progression of age, requires the understanding of the underlying mechanisms which transform the respiratory chain to a permanent superoxide generator. From our extensive studies on this question, we have gained insight into thermodynamic and kinetic conditions causing a loss of e⁻ from the respiratory chain to O_2 out of sequence. The understanding of these interrelationships led to the establishment of a general concept of alterations that is likely to occur in biological organisms in contact with their natural environment. This compulsory connection of biological organisms, being open metabolic flow systems, necessarily leads to the incorporation and preservation of minidefects that accumulate with the progression of age. Many of these natural minihits were found to affect the regular mitochondrial e⁻ flow, such that e⁻ leak to oxygen out of sequence and superoxide radical formation occurs.

Application of antioxidants to downregulate oxygen radicals in the cell caused limited effects only. Extension of mean life expectancy reported by several authors may be due to the control of free radical mechanisms involved in the etiology of typical age-related diseases normally limiting life span. There are two possibilities to explain the failure to efficiently interfere with the process of aging: (1) free radicals are not exclusively involved in the onset and propagation of senescence, and (2) external application of antioxidants does not sufficiently influence the free radical status of the cell because the overall antioxidant capacity remains unchanged, and antioxidants are not homogeneously distributed (compartmentalized).

Imbalance of radical levels established during senescence is a consequence of increased radical formation rather than the impairment of antioxidant control. It therefore appears to be more efficient to decrease the activity of the biological generator than to increase the various radical scavengers.

Comment

Yau-Huei Wei
Department of Biochemistry and Center for Cellular and Molecular Biology, National Yang-Ming University, Taipei 112, Taiwan

There are three ways to ameliorate the normal age-dependent increase of superoxide anions and other oxygen-free radicals in mitochondria of human tissue cells. First, we may decrease the caloric intake by diet restriction. It has been clearly shown that the rates of production of superoxide anions and other reactive oxygen species (ROS) in animal tissues are positively correlated with the cellular metabolic activity, which is determined by the rate of oxygen consumption of mitochondria.[1,2] Sohal and co-workers[3] have demonstrated that by decreasing the cellular levels of both superoxide anions and hydrogen peroxide one can increase the life span of the animal. Because more than 90% of oxygen used by the tissue cells for metabolism of energy-rich fuel molecules is consumed by the mitochondria, the best way to decrease the ROS production of mitochondria is to decrease the metabolic activity of tissue cells by diet restriction.[4] In fact, ample evidence now indicates that diet restriction decreases the oxidative damage to vital biomolecules in tissue cells and extends the life span of the animal.[4,5] Second, we may increase the capability and capacity of tissue cells to cope with the oxidative stress imposed by the continuous generation of ROS from mitochondria during aerobic metabolism. Ingestion of sufficient and balanced amounts of small-molecular-weight antioxidants, including vitamins C and E, beta-carotene, and such trace elements as selenium and zinc from the diet or vitamin/mineral supplements would provide us with competent free radical scavenging systems to cope with oxidative stress.[6] Particularly important is to avoid excess intake of one or some of the vitamins, as they may cause malabsorption of the other essential dietary elements and lead to an imbalance between prooxidants and antioxidants. Mild exercise has been demonstrated to be able to decrease the age-related changes of the structure and function of mitochondria in the muscle.[7] Exercise was also found to be able to upregulate gene expression of such free radical scavenging enzymes as manganese-superoxide dismutase, glutathione peroxidase, and catalase.[8,9] Last, we should avoid environmental insults (e.g., light irradiation, air pollutants, heavy metal exposure) and addiction to drugs, alcohol, and cigarette smoking. We have previously shown that long-term exposure to sunshine significantly increases the frequency of occurrence and abundance of the mutation of mtDNA of human skin.[10]

Alcohol drinking has long been known to enhance oxidative stress and damage (e.g., lipid peroxidation) of the liver.[11,12] It was also recently reported that oxidative damage and mutation of mtDNA in the liver tissues are much more extensive in alcoholic and ethanol-fed rats, compared with the normal controls.[13,14] Moreover, we found in a recent study that the incidence and abundance of mtDNA deletions in both the lung tissues and hair follicles of smokers are significantly higher than those of nonsmokers.[15,16] Moreover, it has been well documented that smoking deprives the blood circulation of antioxidants.[17] These observations suggest that the oxidative stress and damage elicited by the exogenous or environmental insults could be extensive and detrimental as well. Thus, we may ameliorate the production of superoxide anions and other free radicals so as to decrease the oxidative damage in our tissue cells by the aforementioned suggestions.

REFERENCES

1. SOHAL, R.S., H.H. KU, S. AGARWAL, M.J. FORSTER & H. LAL. 1994. Oxidative damage, mitochondrial oxidant generation and antioxidant defenses during aging and in response to food restriction in the mouse. Mech. Ageing Dev. 74: 121–133.
2. YU, B.P. & R. YANG. 1996. Critical evaluation of the free radical theory of aging: A proposal for the oxidative stress hypothesis. Ann. N.Y. Acad. Sci. 786: 1–11.
3. ORR, W.C. & R.S. SOHAL. 1994. Extension of life-span by overexpression of superoxide dismutase and catalase in Drosophila melanogaster. Science 263: 1128–1130.
4. YU, B.P. 1987. Update on food restriction and aging. In Review of Biological Research in Aging. M. Rothstein, Ed. Vol. IV: 495–505. Raven Press. New York.
5. SOHAL, R.S., S. AGARWAL, M. CANDAS, M. FORSTER & H. LAL. 1994. Effect of age and caloric restriction on DNA oxidative damage in different tissues of C57BL/6 mice. Mech. Ageing Dev. 76: 215–224.
6. JI, L.L. 1995. Oxidative stress during exercise: Implication of antioxidant nutrients. Free Radical Biol. Med. 18: 1079–1086.
7. BRIERLEY, E.J., M.A. JOHNSON, O.F. JAMES & D.M. TURNBULL. 1996. Effects of physical activity and age on mitochondrial function. Q.J. Med. 89: 251–258.

8. LEEUWENBURGH, C., R. FIEBIG, R. CHANDWANEY & L.L. JI. 1994. Aging and exercise training in skeletal muscle: Responses of glutathione and antioxidant enzyme systems. Am. J. Physiol. **267:** R439–445.

9. LAWLER, J.M., S.K. POWERS, H. VAN DIJK, T. VISSER, M.J. KORDUS & L.L. JI. 1994. Metabolic and antioxidant enzyme activities in the diaphragm: Effect of acute exercise. Respir. Physiol. **96:** 139–149.

10. PANG, C.Y., H.C. LEE, J.H. YANG & Y.H. WEI. 1994. Human skin mitochondrial DNA deletions associated with light exposure. Arch. Biochem. Biophys. **312:** 534–538.

11. KUKIELKA, E., E. DICKER & A.I. CEDERBAUM. 1994. Increased production of reactive oxygen species by rat liver mitochondria after chronic ethanol treatment. Arch. Biochem. Biophys. **309:** 377–386.

12. ROUACH, H., V. FATACCIOLI, M. GERTIL, S.W. FRENCH, M. MORIMOTO & R. NORDMANN. 1997. Effect of chronic ethanol feeding on lipid peroxidation and protein oxidation in relation to liver pathology. Hepatology **25:** 351–355.

13. CAHILL, A., X. WANG & J.B. HOEK. 1997. Increased oxidative damage to mitochondrial DNA following chronic ethanol consumption. Biochem. Biophys. Res. Commun. **235:** 286–290.

14. MANSOURI, A., B. FROMENTY, A. BERSON, M.A. ROBIN, S. GRIMBERT, M. BEAUGRAND, S. ERLINGER & D. PESSAYRE. 1997. Multiple hepatic mitochondrial DNA deletions suggest premature oxidative aging in alcoholic patients. J. Hepatol. **27:** 96–102.

15. LIU, C.S., S.H. KAO & Y.H. WEI. 1997. Smoking-associated mitochondrial DNA mutations in human hair follicles. Environ. Mol. Mutagen. **30:** 47–55.

16. FAHN, H.J., L.S. WANG, S.H. KAO, S.C. CHANG, M.H. HUANG & Y.H. WEI. 1998. Smoking-associated mitochondrial DNA mutations and lipid peroxidation in human lung tissues. Am. J. Respir. Cell. Mol. Biol. in press.

17. FRAGA, C.G., P.A. MOTCHNIK, A.J. WYROBEK, D.M. REMPEL & B.N. AMES. 1996. Smoking and low antioxidant levels increase oxidative damage to sperm DNA. Mutat. Res. **351:** 199–203.

Membrane and Receptor Modifications of Oxidative Stress Vulnerability in Aging

Nutritional Considerations

J. A. JOSEPH,[a,c] N. DENISOVA,[a] D. FISHER,[a] B. SHUKITT-HALE,[a] P. BICKFORD,[b] R. PRIOR,[a] AND G. CAO[a]

[a]USDA-ARS Human Nutrition Research Center on Aging at Tufts University,
Room 919, 711 Washington Street,
Boston, Massachusetts 02111, USA
[b]VA Medical Center,
Denver, Colorado 80262, USA

ABSTRACT: Evidence suggests that oxidative stress (OS) may contribute to the pathogenesis of age-related decrements in neuronal function and that OS vulnerability increases as a function of age. In addition to decreased endogenous protection, increases in OS vulnerability may result from changes in membrane lipids and distribution of receptor subtype. Using a PC-12 cell model system, we have shown that H_2O_2 or dopamine (DA) exposure induced deficits in the cell's ability to clear (extrude/sequester, E/S) Ca^{2+} that are similar to those seen in aging. When plasma membrane concentrations of sphingomyelin (SPM) were used, the SPM metabolite, sphingosine-1-phosphate was increased to the same levels as those seen in aging, and enhancement of OS-induced decreases in calcium E/S following KCL depolarization was observed. Differential decreases in CA^{2+} E/S were also seen following DA-induced OS in COS-7 cells transfected with one of five muscarinic receptor subtypes. Cells transfected with either M1, M2, or M4 receptors showed significantly greater vulnerability to OS (as expressed by greater decrements in calcium E/S and cell death) than those transfected with M3 or M5 receptors. The vitamin E analogue, Trolox, and the nitrone-trapping agent, PBN, were not effective in altering E/S decrements but were effective in preventing cell death 24 h after OS exposure. These findings suggest that putative regional (e.g., striatum and hippocampus) increases in OS vulnerability and loss of neuronal function in aging may be dependent upon membrane SPM concentration and receptor subtype. In related studies, attempts were made to determine whether increased OS protection via nutritional increases in antioxidant levels in rats [using diets supplemented with vitamin E (500IU/kg), strawberry extracts (9.4 g/kg dried aqueous extract, DAE), spinach (6.7 g/kg DAE), or blueberry extracts (10 g/kg DEA for six weeks)] would protect against exposure to 100% O_2 (a model of accelerated neuronal aging). Results indicated that these diets were effective in preventing OS-induced decrements in several parameters (e.g., nerve growth factor decreases), suggesting that although there may be increases in OS vulnerability in aging, phytochemicals present in antioxidant-rich foods may be beneficial in reducing or retarding the functional central nervous system deficits seen in aging or oxidative insult.

It is well known that in aging there are numerous declines in central neuronal functioning that can occur even in the absence of neurodegenerative disease. These decrements include such factors as decreases in neurotransmitter receptor concentrations, alterations in neuronal signaling, and decrements in Ca^{2+} homeostasis. All of these variables can lead

[c]Tel: 617/556-3178; fax: 617/556-3222; e-mail: joseph_ne@hnrc.tufts.edu

to the observed loss of sensitivity in numerous receptor systems, including (1) muscarinic (mAChR) (*e.g.*, memory processing[1]); (2) beta adrenergic (receptor-mediated relaxation of vasculature;[2,3] cardiovascular function[4]); (3) noradrenergic (inhibitory efficacy of norepinephrine on electrophysiological responses of cerebellar Purkinje's cells[5,6]); (4) GABAergic (inferior colliculus[7]); (5) hypothalamic peptidergic;[8] and (6) dopaminergic (*e.g.*, reduced motor behavioral responses to amphetamine in senescent rats[9]).

Importantly, these decrements can be expressed, ultimately, as alterations in motor and cognitive function. Age-related changes in motor behavior are generally observed as deficits in balance, strength, and coordination in both animals (*e.g.*, see ref. 10) and humans.[11] Cognitive dysfunctions in aging in both humans (*e.g.*, see refs.12,13) and animals (*e.g.*, see refs.14–16) include deficits in retrieval and short-term memory processing. Indeed, as is well known, these deficits become further exacerbated in neurodegenerative diseases such as Alzheimer's disease (AD)[11,12] and Parkinson's disease.[17] Although there has been a great deal of research that has attempted to delineate the mechanisms involved in these age-related declines in memory and motor functions, determination of the final "common pathway," if such exists, has been elusive, and attempts to reverse or retard these declines have been, with very few exceptions, singularly unsuccessful (especially with regard to motor function).

However, evidence derived from other systems showing changes with age (*e.g.*, cardiovascular[18]) has suggested that the major factor that may be involved in these neuronal deficits may be age-related decrements in the ability to mitigate the deleterious effects of life-long exposure to oxidative stress (OS).

It is only recently that oxidative stress is beginning to be specified with respect to the nervous system, and findings suggest that, indeed, the central nervous system may show greater vulnerability to the effects of OS than other sites, inasmuch as it is relatively deficient in free radical protection and uses high amounts of oxygen.[19] Findings concerned with neurodegenerative disease indicate that OS may be a primary factor in their etiology (*e.g.*, Alzheimer's disease[20,21] and Parkinson's disease[21]). In aging, within the past several years, there is an abundance of evidence accumulating that indicates that OS stress may be an important mediator of age-related declines in neuronal function. For example, it appears that several indices of antioxidant protection appear to be reduced in the aging brain. The following are examples. (1) There are increases in the ratio of oxidized to total glutathione.[22,23] (2) Significant lipofuscin accumulation with bcl-2 increases, and membrane lipid peroxidation has been observed as a function of age[24] in lipofuscin-containing vacuoles of neurons and glia and vascular cells.[25] (3) Reduced glutamine synthetase was observed in aging (which could be reflected in lowered glutathione levels[26]). Although controversial, it also appears that there may be increases in free radical oxidation of brain proteins as a function of age (*e.g.*, see ref. 27).

Besides the decreases in efficacy of the endogenous protection mechanisms, there also seem to be indications of decreases in repair (*e.g.*, see ref. 28). These factors may act in concert with accumulated advanced glycation end (AGE) products[29] to produce a cascade of effects leading to increased vulnerability to insult and, ultimately, to neuronal functional decline.

Although a great deal of research has been directed toward determining the relationship of OS to neuronal decline in aging and neurodegenerative disease, there have been few studies directed toward demonstrating age-related increases in sensitivity to OS on parameters that are known to change with age. In this regard, two tests have been used in our laboratory to carry out such evaluations (oxotremorine-enhancement of K^+-evoked dopamine release from striatal slices and carbachol-stimulated GTPase activity), which are age and OS sensitive,[30] in that they can assess the loss in sensitivity expressed by muscarinic receptors as a function of age in the striatum. The striatum shows significant functional and morphological declines as a function of age. The findings have indicated that oxotremorine (oxo) enhancement of dopamine (DA) release (K^+-ERDA) (in superfused striata) shows significant declines with aging[31] and that K^+-ERDA deficits are exacerbated by OS in an age-dependent manner.[30] Additional studies have shown that these deficits are primarily the result of alterations in signal transduction early in the transduction process[32,33] because carbachol stimulation of GTPase activity (an indicator of receptor-G protein coupling/uncoupling) was also found to be reduced with aging[33] and following OS.[30,34] In fact, the OS-induced reductions in oxo-enhanced K^+-ERDA and GTPase activity were greater in the tissue from the old animals than the young, indicating significant increases in vulnerability.

Thus, it appears that the changes we have observed are OS and age sensitive. If this is the case, then it is critical to determine if there are additional mechanisms in addition to decreased antioxidant protection that would contribute to the increased OS sensitivity in aging. Recent ongoing work from our laboratory has suggested that vulnerability increases also may be the result of increases in membrane lipids, particularly sphingomyelin (SPM), and may be dependent to a large extent on the composition of the receptor population in a particular brain area. In this effort we employed cell model systems to examine the effects of these variables on Ca^{2+} homeostasis and cell death. Since the mid-1970s Ca^{2+} dysfunction has been suggested to be involved in neuronal deficits seen in normal aging and age-related neurodegenerative diseases (*e.g.*, see discussions in refs. 35,36). Although the mechanisms involved in Ca^{2+} tissue injury are unknown, it has been postulated that they may also involve free radical activation through xanthine oxidase activation[37] and the generation of nitric oxide synthase and phospholipases.[38] Therefore, it may be that inasmuch as it is known that because OS appears to induce a deficit in Ca^{2+} regulation (*e.g.*, see ref. 39), an OS-induced insult that alters the extrusion/sequestration (E/S) of intracellular calcium ($[Ca^{2+}]_I$) may lead to the generation of additional prooxidants and compromise cell function, especially in an aged system.

In aging there are significant age-related increases in membrane cholesterol (CHL)[40] and SPM.[41] We hypothesized that the effects of OS upon membranes may be enhanced in aging, inasmuch as previous findings indicate that, among its other effects, OS increases membrane rigidity.

MEMBRANE AND RECEPTOR EFFECTS ON OXIDATIVE STRESS

In order to examine in relative isolation the interaction of membrane lipids and OS on specific indices of Ca^{2+} flux, we employed PC-12 cells as a model and examined these parameters in control, CHL-, SPM-, or SPM/CHL-treated cells that were then exposed to oxidative stress. Following CHL and SPM treatment, the levels of these lipids in the cell membranes were equivalent to those seen in aging.

These cells were incubated in maintenance media (RPMI-1640 with 2 mM glutamine, 10% horse serum, 5% fetal bovine serum (FBS), and 120 U/mL penicillin/streptomycin). Using fluorescence imaging, pixel-by-pixel comparisons of the captured images were made, and a ratio of Ca^{2+}-bound fura (340 nm excitation) to unbound fura (380 nm excitation) was generated for each pair of images. Three parameters of Ca^{2+} flux were examined in control or OS (300 µM H_2O_2)-treated control, SPM (500 µM), CHL (660 µM), or CHL/SPM (1 h at 37°C)-treated PC-12 cells: baseline, (pre-KCl Ca^{2+} levels), depolarization (expressed as percent of Ca^{2+} increase to 30 mM KCl), and recovery ([Ca^{2+}]$_i$ extrusion). Results showed that baseline Ca^{2+} levels were significantly increased by H_2O_2 treatment (*e.g.*, 300 µM, 200%), whereas the rise in free intracellular Ca^{2+} following KCl stimulation (*i.e.*, peak) was decreased (*e.g.*, 300 µM, 50%) and Ca^{2+} recovery time following depolarization was significantly increased and led eventually to cell death 24 h after H_2O_2 (Live/Dead Eukolight Kit, Molecular Probes).[42]

In cells in which membrane lipids (SPH, CHL) were increased, the results indicated that SPH significantly increased recovery in the presence or absence of H_2O_2 by 50%, and the level of conjugated dienes was significantly increased by 750 percent. These effects were not seen with CHL pretreatment. These results suggested that membrane SPM could be a critical factor in determining OS vulnerability and Ca^{2+} translocation in membranes.[43] This may be especially important in aging where there is increased membrane SPH and significant loss of calcium homeostasis (see above). In additional experiments, attempts were made to determine the particular metabolite of SPM that might be contributing to the increased effectiveness of H_2O_2 on recovery. In these studies, membrane SPM was depleted by incubating the PC-12 cells in either 100 mU/mL *Staphylococcus aureus* sphingomyelinase (Sase) or 2 mM L-cycloserine (L-CS), followed by incubation in H_2O_2 or incubation medium alone. Ca^{2+} recovery was then examined. Results showed that endogenously induced depletion of SPH by L-CS significantly decreased the recovery time of the cells. However, when SPM was depleted via Sase, similar effects were seen with respect to recovery that were observed with increases in the vulnerability to H_2O_2. These findings indicated that a metabolite of SPM was involved in these increases in OS vulnerability, because L-CS primarily antagonized the synthesis of SPM and, thus, decreased membrane SPM metabolite levels. Examinations of the metabolites that could have contributed to the increased OS vulnerability indicated that when cells were treated with sphingosine-1-phosphate (S-1-P, 1 µM) and exposed to H_2O_2 the results were similar to those seen with SPM pretreatment with S-P-1P, significantly increasing the vulnerability of cells to H_2O_2-induced decreases in recovery. No other SPM metabolite tested produced this effect (*e.g.*, C-2 ceramide, 100 µM). The nature of this effect is not clear. It seems that there are at least two different SPM pools (plasma membrane and newly synthesized in *cis-medial* Golgi stacks). Both pools affect Ca^{2+} homeostasis, but only the newly synthesized SPH was able to significantly decrease cell vulnerability to OS. Thus, these findings indicate that one factor that is important in determining OS vulnerability is membrane lipid content, especially those of SPM and its metabolite SP-1-P.

A second factor that is important in the increased vulnerability to oxidative stress in aging is the type and receptor number that are present in various brain regions. For example, it has been known for many years that there are region-specific variations in brain aging and that some areas show more deleterious effects than others (*e.g.*, striatum). Therefore, if OS is involved in inducing these changes, selective regional vulnerability to OS may reflect the qualitative makeup of the receptor populations. We attempted to set up a model to examine this by transfecting COS-7 cells with one of five muscarinic receptor subtypes (M_1–M_5) and to examine the alterations in vulnerability to OS, as described above for the PC-12 cells, except that oxotremorine (500 µM), a muscarinic agonist, was used to depolarize the cells instead of KCL. COS-7 cells do not normally express muscarinic receptors. We examined intracellular Ca^{2+} levels prior to and following depolarization (oxo), following OS exposure, by either H_2O_2 or DA incubation. Cell death was also examined at 4 h and 24 h post-OS treatment. Following H_2O_2 exposure the number of cells showing oxo-induced depolarization and Ca^{2+} recovery varied as a function of transfected mAChR subtype. The percent of cells showing the greatest decreases in responding (depolarizing) to oxo were those transfected with the M_1 (300 or 500 μM H_2O_2, 30%) and M_2 (45%) subtypes, whereas M_3, M_4, and M_5 cells showed no significant decreases in responding with H_2O_2. For recovery (the ability to E/S Ca^{2+}, see above), M_1, M_2, or M_4 transfected cells showed the greatest decreases following H_2O_2 (50–100%) or DA (25–50%), whereas recovery in M_3- and M_5-transfected DA- or H_2O_2-exposed cells did not show any significant decreases in E/S. Similar patterns to those seen for recovery were observed in subsequent examinations of the degree of alterations in cell viability (Live/Dead Eukolight kit) in cells transfected with M_1 at 4 h and 24 h post-DA treatment. Approximately 10% of 40% of the M_1-transected cells that died were apoptotic following DA exposure (Apo-Tag Kit). Conversely, no cell death was observed in M_3-transfected, DA-exposed cells. Thus, it may be that regional differences in OS vulnerability may be determined in part by differential receptor susceptibility to OS insult. For example, in the striatum the receptor populations most susceptible to OS are the most abundant, and the striatum shows profound changes with age. In this region, M_1 receptor protein is expressed in 78% of the neurons, whereas M_4 receptors are localized to 44% of striatal cells,[44] and M_2 receptors may be the most predominant. In addition, there are high concentrations of DA (which form a variety of OS-based toxic products (e.g., see ref. 45). Thus, the profound age-related changes in the striatum may be reflective of the interactions between DA and specific populations of vulnerable receptors experiencing OS.

In an effort to determine if this might be the case in specific receptor populations in neuronal tissue, we examined the differential vulnerability of D_1 and D_2 striatal DA receptor populations known to decrease with age and to be involved in the regulation of motor behavior to DA-induced oxidative stress in tissue obtained from young and old animals. The results indicated that even though DA uptake was reduced in the striatal tissue from the old animals, the D_2 receptors showed decrements in responding (quinelorane-stimulated GTPase activity) that were the same as those seen in the tissue from the young, whereas D_1 receptors showed even greater decrements in responding (SKF 38393-stimulated GTPase activity) than that seen in the young.[46] An important component contributing to the increased vulnerability of DA receptors may be glutamate activation, which results in the destabilization of Ca^{2+} regulation. It is known that NMDA receptors are located on dopaminergic axon terminals.[47] In addition, a great deal of evidence suggests the involvement of glutamate receptor activation with subsequent neurotoxicity (e.g., see ref. 48). More importantly, stimulation of AMPA receptors can also produce neurotoxicity (e.g., see ref. 49) through free radical generation[50] and destabilization of Ca^{2+} homeostasis[47] leading to death in dopaminergic neurons.[47]

As has been pointed out above, in aging there are already decrements in Ca^{2+} homeostasis,[35] and both the Na^+/Ca^{2+} exchanger[51] and the Ca^{2+}-ATPase pump[52] have been found to be compromised in aging. Thus, in aging, further disregulation in Ca^{2+} homeostasis by indirect AMPA activation in DA neurons may increase their vulnerability to oxidative stress, especially in old animals. It also may be that differential vulnerability to oxidative stress among various receptor systems may involve variations in the locations of such critical amino acids as cysteine or proline in the transmembrane segments of the receptors. It is known, for example, that cysteine, which can increase autoxidation and the formation of 5-S-cysteinyldopamine (*e.g.*, see ref. 53), is present in the binding crevice of the D_2 receptor (*e.g.*, see ref. 54). Recent evidence also indicates that prolines are peroxidized with a similar efficiency to proteins[55] and can prime the stimulation of superoxide generation in human neutrophils.[56]

At present, we are attempting to specify the interaction of these amino acids with DA-induced oxidative stress in relation to their locus in DA receptors. However, it may be that given the increased vulnerability that has been postulated to occur in aging it will be necessary to increase the level of antioxidant protection. One method to accomplish this may be through nutritional intervention.

NUTRITIONAL INCREASE OF ANTIOXIDANT PROTECTION IN AGING

Some indirect evidence that it might be possible to dietarily increase the antioxidant protection in aging can be derived from studies in rats and humans, showing positive effects of consumption of the flavonoid glycosides of *Ginkgo biloba* on the following: memory impairment,[57] difficulties in concentration,[58,59] decreases in the projection field of infrapyramidal mossy fibers in the hilus region of the hippocampus,[60] and the Ca^{+2}-induced increases in the oxidative metabolism of brain neurons.[61,62]

Research from our laboratory has indicated that animals maintained on diets containing fruits or vegetables that are high in antioxidant activity are more resistant to the deleterious effects of 48 h of 100% oxygen exposure on several neuronal parameters that also show declines in aging. Thus, control diets or those supplemented with either vitamin E (500 IU/kg), strawberry extracts (SE) (9.4 g/kg dried aqueous extract, DAE), spinach (SPN) (6.7 g/kg DAE), or blueberry extracts (BL) (10 g/kg DAE) were fed to 6- to 8-month-old F344 rats for eight weeks prior to O_2 exposure. Results showed that vitamin E, SE, BL, and SPN diets were all effective in preventing O_2-induced decreases in nerve growth factor in basal forebrain, as well as effective in striatal oxotremorine enhancement of K^+-ERDA and β-adrenergic receptor function in the cerebellum. Subsequent experiments have indicated that there may also be some prevention of these alterations in neuronal function (*e.g.*, striatal oxo-enhanced K^+-ERDA, cerebellar β-adrenergic receptor function) in rats maintained on similar diets from the age of 6 months to 15 months. We are presently attempting to determine if these diets will also be effective in reversing the deleterious effects of aging on the above parameters as well as on motor and cognitive behaviors. However, these findings suggest that nutritional intervention with fruits and vegetables may play an important role in preventing or perhaps even reversing the effects of oxidative stress in aging and brain function. We are presently attempting to delineate the mechanisms involved in these interventions.

REFERENCES

1. ABDALLAH, E.A.M., W.S. POU & E.E. EL-FAKAHANY. 1990. Aging does not alter muscarinic receptor-mediated inhibition of cyclic AMP formation in the striatum and hippocampus. Brain Res. **534:** 234–236.

2. Deisher, T.A., S. Mankani & B.B. Hoffman. 1989. Role of cyclic AMP-dependent protein kinase in the diminished beta adrenergic responsiveness of vascular smooth muscle with increasing age. J. Pharmacol. Exp. Ther. **249:** 812–819.

3. Hiremath, A.N., R.A. Pershe, B.B. Hoffman & T.F. Blaschke. 1989. Comparison of age-related changes in prostaglandin E1 and beta-adrenergic responsiveness of vascular smooth muscle in adult males. J. Gerontol. **44:** M13–17.

4. Lakatta, E.G. 1986. Diminished beta-adrenergic modulation of cardiovascular function in advanced age. Cardiol. Clin. **4:** 185–200.

5. Hoffer, B.J., G. Rose, K. Parfitt, R. Freedman & P.C. Bickford-Wimer. 1988. Age-related changes in cerebellar noradrenergic function. *In* Central Determinants of Age-Related Declines in Motor Function. Annals of the New York Academy of Sciences. J.A. Joseph Ed.: **515:** 269–286. New York, NY.

6. Gould, T.J. & P.C. Bickford. 1994. The effects of chronic treatment with *N*-tert-butyl-alpha-phenylnitrone on cerebellar noradrenergic receptor function in aged F344 rats. Brain Res. **660:** 333–336.

7. Caspary, D.M., J.C. Milbrandt & R.H. Helfert. 1995. Central auditory aging: GABA changes in the inferior colliculus. Exp. Gerontol. **30:** 349–360.

8. Sadow, T.F. & R.T. Rubin. 1992. Effects of hypothothalamic peptides on the aging brain. Psychoneuroendocrinology **17:** 293–314.

9. Joseph, J.A., R.E. Berger, B.T. Engel & G.S. Roth. 1978. Age-related changes in the nigrostriatum: A behavioral and biochemical analysis. J. Gerontol. **33:** 643–649.

10. Kluger, A., J. G. Gianutsos, J. Golomb, S.H. Ferris, A.E. George, E. Frannssen & B. Reisberg. 1997. Patterns of motor impairment in normal aging, mild cognitive decline, and early Alzheimer's disease. J. Gerontol. **52:** 28–39.

11. Joseph, J.A. & G.S. Roth. 1993. Hormonal regulation of motor behavior in senescence. J. Gerontol. **48:** 51–55.

12. Muir, J.L. 1997. Acetylcholine, aging, and Alzheimer's disease. Pharmacol. Biochem. Behav. **56:** 687–696.

13. West, R.L. 1996. An application of pre-frontal cortex function theory to cognitive aging. Psychol. Bull. **120:** 272–292.

14. Bartus, R.T., R.L. Dean, B. Beer & A.S. Lippa. 1982. The cholinergic hypothesis of geriatric memory dysfunction. Science **214:** 408–417.

15. Riekkinen, M., S. Kemppainen & P. Riekkinen Jr. 1997. Effects of stimulation of alpha1-adrenergic and NMDA/glycine-B receptors on learning deficits in aged rats. Psychopharmacology **131:** 49–56.

16. Ingram, D.K., M. Jucker & E.L. Spangler. 1994. Behavioral manifestations of aging. Pathol. Aging Rat **2:** 149–170.

17. Adams, J.D. & I. Odunze. 1991. Oxygen free radicals and Parkinson's disease. Free Radical Biol. Med. **10:** 161–191.

18. Halliwell, B. 1994. Free radicals and antioxidants: A personal view. Nutr. Rev. **52:** 253–265.

19. Olanow, C.W. 1992. An introduction to the free radical hypothesis in Parkinson's disease. Ann. Neurol. **32:** S2–S9.

20. Finch, C.E. & D.M. Cohen. 1997. Aging, metabolism, and Alzheimer's disease: Review and hypotheses. Exp. Neurol. **143:** 82–102.

21. Jenner, P. 1996. Oxidative stress in Parkinson's disease and other neurodegenerative disorders. Pathol. Biol. **44:** 57–64.

22. Ohkuwa, T., Y. Sato & M. Naoi. 1997. Glutathione status and reactive oxygen generation in tissues of young and old exercised rats. Acta Physiol. Scand. **159:** 237–244.

23. Zhang, J.R., P.K. Andrus & E.D. Hall. 1993. Age-related regional changes in hydroxyl radical stress and antioxidants in gerbil brain. J. Neurochem. **61:** 1640–1647.

24. Yu, B.P. 1994. Cellular defenses against damage from reactive oxygen species. Physiol. Rev. **76:** 139–162.

25. Migheli, A., P. Cavalla, R. Piva, M.T. Giordana & D. Schiffer. 1994. Bcl-2 protein expression in aged brain and neurodegenerative diseases. Neuroreport **5:** 1906–1908.

26. Carney, J.M., C.D. Smith, A.M. Carney & D.A. Butterfield. 1994. Aging- and oxygen-induced modifications in brain biochemistry and behavior. Ann. N.Y. Acad. Sci. **738:** 44–53.

27. BUTTERFIELD, D.A., B.J. HOWARD, S. YATIN, K.L. ALLEN & J.M. CARNEY. 1997. Free radical oxidation of brain proteins in accelerated senescence and its modulation by *N*-tert-butyl-alpha-phenylnitrone. Proc. Natl. Acad. Sci. USA **94:** 674–678.

28. ADAMS, J.D., S.K. MUKHERJEE, L.K. KLAIDMAN, M.L. CHANG & R. YASHAREL. 1996. Apoptosis and oxidative stress in the aging brain. **786:** 135–151.

29. MUNCH, G., J. THOME, P. FOLEY, R. SCHINZEL & P. RIEDERER. 1997. Advanced glycation end products in aging and Alzheimer's disease. Brain Res. Rev. **23:** 134–143.

30. JOSEPH, J.A., R. VILLALOBOS-MOLINA, N. DENISOVA, S. ERAT, R. CUTLER & J.G. STRAIN. 1996. Age differences in sensitivity to H_2O_2- or NO-induced reductions in K^+-evoked dopamine release from superfused striatal slices: Reversals by PBN or Trolox. Free Radical Biol. Med. **20:** 821–830.

31. JOSEPH, J.A., T.K. DALTON & W.A. HUNT. 1988. Age-related decrements in the muscarinic enhancement of K^+-evoked release of endogenous striatal dopamine: An indicator of altered cholinergic-dopaminergic reciprocal inhibitory control in senescence. Brain Res. **454:** 140–148.

32. JOSEPH, J.A., T.K. DALTON, G.S. ROTH & W.A. HUNT. 1988. Alterations in muscarinic control of striatal dopamine autoreceptors in senescence: A deficit at the ligand-muscarinic receptor interface? Brain Res. **454:** 149–155.

33. YAMAGAMI, K., J.A. JOSEPH & G.S. ROTH. 1992. Decrement of muscarinic receptor-stimulated low K_m GTPase activity in striata and hippocampus from aged rat. Brain Res. **576:** 327–331.

34. JOSEPH, J.A., R. VILLALOBOS-MOLINA, B.M. RABIN, T.K. DALTON, A. HARRIS & S. KANDASAMY. 1994. Reductions of ^{56}Fe heavy particle irradiation-induced deficits in striatal muscarinic receptor sensitivity by selective cross activation/inhibition of second messenger systems. Radiat. Res. **139:** 60–66.

35. LANDFIELD, P.W. & J.C. ELDRIDGE. 1994. The glucocorticoid hypothesis of age-related hippocampal neurodegeneration: Role of dysregulated intraneuronal calcium. Ann. N. Y. Acad. Sci. **746:** 308–321.

36. PAGLIUSI, S.R., P. GERRARD, M. ABDALLAH, D. TALABOT & S. CATISICAS. 1994. Age-related changes in expression of AMPA-selective glutamate receptor subunits: Is calcium-permeability altered in hippocampal neurons? Neuroscience **61:** 429–433.

37. LEE, K.S., S. FRANK, P. VANDERKLISH, A. ARAI & G. LYNCH. 1991. Inhibition of proteolysis protects hippocampal neurons from ischemia. Proc. Natl. Acad. Sci. USA **88:** 7233–7237.

38. LYNCH, D.R. & T.M. DAWSON. 1994. Secondary mechanisms in neuronal trauma. Curr. Opin. Neurol. **7:** 507–509.

39. CHENG, Y., P. WIXOM, M.R. JAMES-KRACKE & A.Y. SUN. 1994. Effects of extracellular ATP on Fe^{2+}-induced cytotoxicity in PC-12 cells. J. Neurochem. **66:** 895–902.

40. CHOE, M., C. JACKSON & B.P. YU. 1995. Lipid peroxidation contributes to age-related membrane rigidity. Free Radical Biol. Med. **18:** 977–984.

41. GIUSTO, N., M. ROQUE, M.E. IIINCHETA & M.G. DE BOSCHERO. 1992. Effect of aging on the content, composition and synthesis of sphingomyelin in the central nervous system. Lipids **27:** 835–839.

42. JOSEPH, J.A., J. STRAIN & N.D. JIMENEZ. 1997. Oxidant injury in PC-12 cells: A possible model of calcium "dysregulation" in aging I: Selectivity of antioxidant protection. J. Neurochem. **69:** 1252–1258.

43. DENISSOVA, N., J. STRAIN & J.A. JOSEPH. 1997. Oxidant injury in PC-12 cells: A possible model of calcium "dysregulation" in aging. II. Interactions with membrane lipids. J. Neurochem. **69:** 1259–1266.

44. HERSCH, S.M., C.A. GUTEKNUST, H.D. REES, C.J. HEILMAN & A.I. LEVEY. 1994. Distribution of m1-m4 muscarinic receptor proteins in the rat striatum: Light and electron microscopic immunocytochemistry using subtype-specific antibodies. J. Neurosci. **14:** 3351–3363.

45. HASTINGS, T.G. & M.J. ZIGMOND. 1994. Identification of catechol-protein conjugates in neostriatal slices incubated with [^{3}H] dopamine: Impact of ascorbic acid and glutathione. J. Neurochem. **63:** 1126–1132.

46. JOSEPH, J.A., S. ERAT, N. DENISOVA & R. VILLALOBOS-MOLINA. 1998. Receptor- and age-selective effects of dopamine oxidation on receptor-G protein interactions in the striatum. Free Radical Biol. Med. **24:** 827–834.

47. DE ERAUSQUIN, G.A., E. COSTA & I. HANBAUER. 1994. Calcium homeostasis, free radical formation and trophic factor dependence mechanisms in Parkinson's disease. Pharmacol. Rev. **46:** 467–482.

48. FORLONI, G., E. LUCCA, N. ANGERETTI, R. CHIESA & A. VEZZANI. 1997. Neuroprotective effect of somatostatin on non-apoptotic NMDA-induced neuronal death: Role of cyclic GMP. J. Neurochem. **68:** 319–327.

49. CHEN, Q., C. HARRIS, C.S. BROWN, A. HOWE, D.J. SURMEIER & A. REINER. 1995. Glutamate-mediated excitotoxic death of cultured striatal neurons is mediated by non-NMDA receptors. Exp. Neurol. **136:** 212–224.

50. SIMPSON, J.R. & O. ISACSON. 1993. Mitochodrial impairment reduces the threshold for *in vivo* NMDA-mediated neuronal death in the striatum. Exp. Neurol. **121:** 57–64.

51. MARTINEZ-SERRANO, A., P. BLANCO & J. SATRUSTEGUI. 1992. Calcium binding to cytosol and calcium extrusion mechanisms in intact synaptosomes and their alterations with aging. J. Biol. Chem. **267:** 4672–4679.

52. BENZI, G.A., R. GORINI, B. ARNABOLDI, GHIGINI & R.F. VILLA. 1994. Age-related alterations by chronic intermittent hypoxia on cerebral synaptosomal ATPase activities. J. Neural Transm. Suppl. **44:** 159–171.

53. PALUMBO, A., M. D'ISCHIA, G. MISURACA, L. DE MARTINO & G. PROTA. 1995. Iron- and peroxide-dependent conjugation of dopamine with cysteine: Oxidative routes to the novel brain metabolite 5-S-cysteinyldopamine. Biochim. Biophys. Acta **1245:** 255–261.

54. JAVITCH, J.A., D. FU & J. CHEN. 1996. Differentiating dopamine D_2 ligands by their sensitivities to modification of the cysteine exposed in the binding-site crevice. Mol. Pharmacol. **49:** 692–698.

55. GEBICKI, S. & J.M. GEBICKI. 1993. Formation of peroxides in amino acids and proteins exposed to oxygen free radicals. Biochem. J. **289:** 743–749.

56. WATANABE, Y., Y. SAGARA, K. SUGAHARA & H. KODAMA. 1994. Iminodipeptides containing proline with C-terminal and N-terminal residues prime the stimulation of human neutrophil superoxide generation by fMLP. Biochem. Biophys. Res. Commun. **205:** 758–764.

57. RAI, S.S., C. SHOVLIN & K.A. WESNES. 1991. A double-blind placebo controlled study of *Ginkgo biloba* extract ('tanakan') in elderly outpatients with mild to moderate memory impairment. Curr. Med. Res. Opin. **12:** 350–355.

58. KLEIJNEN, J. & P. KNIPSCHILD. 1992. *Ginkgo biloba.* Lancet **340:** 1136–1139.

59. KLEIJNEN, J. & P. KNIPSCHILD. 1992. *Ginkgo biloba* for cerebral insufficiency. Br. J. Clin. Pharmacol. **34:** 352–358.

60. BARKATS, M., P. VENAULT, Y. CHRISTEN & C. COHEN-SALMON. 1994. Effect of long-term treatment with Egb 761 on age-dependent structural changes in the hippocampi of three inbred mouse strains. Life Sci. **56:** 213–222.

61. OYAMA, Y., A. HAYASHI & T. UEHA. 1993. Ca^{+2}-induced increase in oxidative metabolism of dissociated mammalian brain neurons: Effect of extract of *Ginkgo biloba* leaves. J. Pharmacol. Jpn. **61:** 367–370.

62. OYAMA, Y., P.A. FUCHS, N. KATAYAMA & K. NODA. 1994. Myricetin and quercetin, the flavonoid constituents of *Ginkgo biloba* extract, greatly reduce oxidative metabolism in both resting and Ca^{+2}-loaded brain neurons. Brain Res. **635:** 125–129.

Immunohistochemical and ELISA Assays for Biomarkers of Oxidative Stress in Aging and Disease[a]

JOELLE M. ONORATO,[b] SUZANNE R. THORPE,[b,c] AND JOHN W. BAYNES[b,c,d]

[b]Departments of Chemistry and Biochemistry, and [c]Ophthalmology, University of South Carolina, Columbia, South Carolina 29208

ABSTRACT: Oxidative stress is apparent in pathology associated with aging and many age-related, chronic diseases, including atherosclerosis, diabetes mellitus, rheumatoid arthritis, and neurodegenerative diseases. Although it cannot be measured directly in biological systems, several biomarkers have been identified that provide a measure of oxidative damage to biomolecules. These include amino acid oxidation products (methionine sulfoxide, *ortho*-tyrosine (*o*-tyr) and dityrosine, chlorotyrosine and nitrotyrosine), as well as chemical modifications of protein following carbohydrate or lipid oxidation, such as N^ε-(carboxymethyl)lysine and N^ε-(carboxyethyl)lysine, and malondialdehyde and 4-hydroxynonenal adducts to amino acids. Other biomarkers include the amino acid cross-link pentosidine, the imidazolone adducts formed by reaction of 3-deoxyglucosone or methylglyoxal with arginine, and the imidazolium cross-links formed by the reaction of glyoxal and methylglyoxal with lysine residues in protein. These compounds have been measured in short-lived intracellular proteins, plasma proteins, long-lived extracellular proteins, and in urine, making them valuable tools for monitoring tissue-specific and systemic chemical and oxidative damage to proteins in biological systems. They are normally measured by sensitive high-performance liquid chromatography or gas chromatography-mass spectrometry methods, requiring both complex analytical instrumentation and derivatization procedures. However, sensitive immunohistochemical and ELISA assays are now available for many of these biomarkers. Immunochemical assays should facilitate studies on the role of oxidative stress in aging and chronic disease and simplify the evaluation of therapeutic approaches for limiting oxidative damage in tissues and treating pathologies associated with aging and disease. In this article we summarize recent data and conclusions based on immunohistochemical and ELISA assays, emphasizing the strengths and limitations of the techniques.

Oxidative stress (OxS) is implicated in the pathogenesis of aging and many age-related, chronic diseases, including atherosclerosis, diabetes mellitus, rheumatoid arthritis, and neurodegenerative disease[1]. Two general mechanisms contribute to an increase in OxS, either excessive generation of reactive oxygen species (ROS) or inadequate antioxidant defenses. The resultant increase in levels of ROS cannot be measured directly, but products of OxS, that is, oxidative damage, can be detected in target biomolecules. These products include a range of oxo, hydroxy, nitro, and chloro derivatives of amino acids (FIG. 1); products of lipid peroxidation, such as endo- and hydroperoxides and conjugated dienes; and hydroxy and oxo derivatives of nucleic acids. Compounds formed on oxidation of car-

[a]Research in the authors' laboratory was supported by National Institutes of Health research Grants DK-19971 (J.W. Baynes) and AG-11472 (S.R. Thorpe).

[d]Address for correspondence: Department of Chemistry and Biochemistry, Physical Sciences Center, 730 S. Main Street, Columbia, SC 29208. Tel. and fax: 803/777-7272; e-mail: baynes@psc.sc.edu

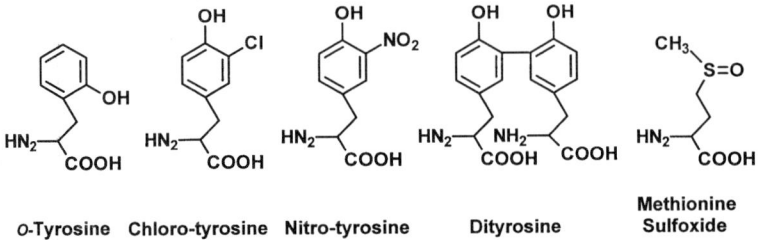

FIGURE 1. Protein oxidation products that have been detected in tissue proteins. MetSO is the most sensitive of these indicators of oxidative stress, but oxidation of Met may be reversible.

bohydrates and lipids also react with protein to form glycoxidation (FIG. 2) and lipoxidation (FIG. 3) products,[2–4] such as N^ε-(carboxymethyl)lysine (CML) and N^ε-(carboxyethyl)lysine (CEL), and malondialdehyde (MDA) and 4-hydroxynonenal (HNE) adducts to amino acids. Some of these compounds are indirect indicators of OxS, such as products of the reaction of protein with dicarbonyl compounds—the imidazolones formed by reaction of 3-deoxyglucosone (3DG) or methylglyoxal with arginine, and the imidazolium cross-links glyoxal-lysine dimer (GOLD) and methylglyoxal-lysine dimer (MOLD) (FIG. 2) formed by reaction of glyoxal (GO) or methylglyoxal (MGO) with lysine. Although the dicarbonyl precursors of these compounds may be formed nonoxidatively, for example, 3DG by rearrangement and hydrolysis of Amadori compounds and MGO by elimination of phosphate from triose phosphates, reactive aldehydes and dicarbonyls are detoxified by antioxidant pathways involving NADPH and glutathione (GSH), such as aldehyde and aldose reductases and glyoxalase pathway enzymes. Thus, steady state levels of dicarbonyl precursors reflect indirectly on the efficiency of antioxidant systems involved in carbonyl detoxification.[5,6] Some of these biomarkers of OxS are also products of mixed enzymatic and nonenzymatic chemistry, inasmuch as ROS, such as HOCl, NO, H_2O_2 and lipid peroxides, may be formed by enzymatic reactions.

Despite the range of oxidation products formed and increasing evidence for the involvement of OxS in aging and disease,[5–8] the assessment of the steady state level of OxS and evaluation of the effectiveness of antioxidant therapy is hampered by the complexity of assays for measuring specific oxidation products in tissues. In addition to the range of products formed, assays for these compounds typically employ acid hydrolysis of

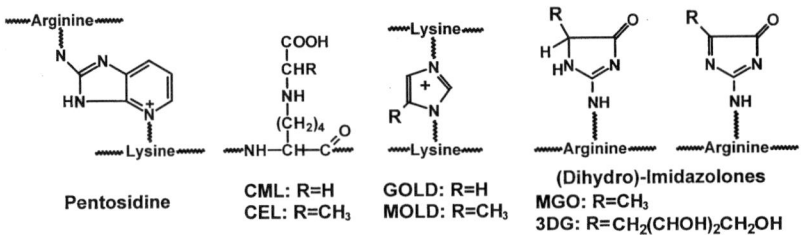

FIGURE 2. Glycoxidation products that have been detected in tissue proteins. As noted in the text, CML, CEL, GOLD, and MOLD may be derived from either glycoxidation or lipoxidation reactions. The dihydro-imidazolone adducts of MGO and 3DG to Arg undergo spontaneous oxidation to the imidazolones.

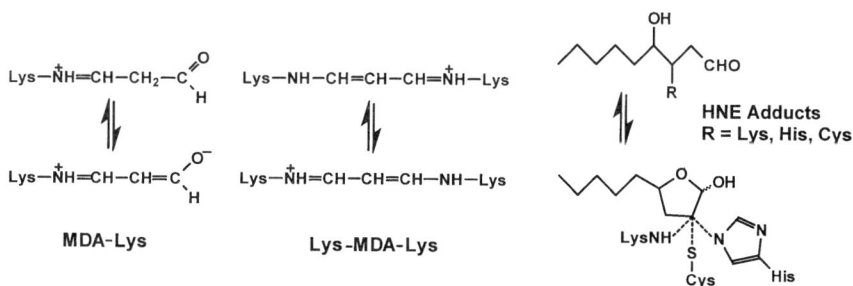

FIGURE 3. Lipoxidation products that have been detected in tissue proteins. Resonance-stabilized forms of MDA-Lys and MDA-Lys-MDA are shown. MDA-Lys exists primarily as the enolate anion at physiological pH. HNE forms adducts by Michael addition reactions with Lys, His, and Cys residues. These adducts exist primarily in the cyclic hemiacetal conformation.

the protein and use sophisticated analytical techniques, for example, multistep HPLC and gas chromatography-mass spectrometry (GC/MS) methodologies, which require sample derivatization and are generally limited by slow sample processing rates. Faster and more sensitive immunohistochemical and ELISA assays are now being introduced for detecting and measuring oxidation products in tissue proteins and should facilitate the assessment of OxS and the study of antioxidant supplements and therapies. In this article we will summarize recent advances in this direction, with emphasis on both the utility and limitations of immunological techniques. Review articles will be cited, when available, for summaries of published information.

CHEMICAL ANALYSIS OF OXIDATION PRODUCTS IN TISSUE PROTEINS

Amino Acid Oxidation Products

Amino acid oxidation products (FIG. 1) increase in proteins exposed to H_2O_2, metal-catalyzed oxidation, peroxynitrite, or HOCl *in vitro*.[9] In recent work we reported that methionine sulfoxide (MetSO) and *ortho*-tyrosine (*o*-Tyr) increase with age in human skin collagen. Age-adjusted levels of these biomarkers were not increased in diabetes, leading to the conclusion that OxS was not systemically increased in diabetes.[10] However, oxidation products in collagen may reflect only on the status of OxS in the extravascular compartment. Changes in OxS may be tissue specific, and studies on vascular and renal tissue are required to establish the role of OxS in diabetes. The association between oxidative stress and disease is clearer in the case of atherosclerosis. Levels of chlorotyrosine (Cl-Tyr), nitrotyrosine (NO_2-Tyr), and dityrosine are increased several fold in lipoproteins isolated from atherosclerotic plaque, compared to circulating lipoproteins,[11–13] supporting an association between OxS and atherosclerosis. The absence of increase in *o*- and *m*-Tyr, products of metal-catalyzed oxidation[14] in lipoproteins from plaque, led Heinecke and colleagues to conclude that myeloperoxidase, which was detectable in atherosclerotic plaque by immunohistochemistry, was responsible for oxidative modification of plaque lipoproteins.[13] Cytosolic protein carbonyls, measured by their reaction with dinitrophenylhydrazine (DNPH), increase with age in tissues, and their steady state levels appear to increase logarithmically with age and in response to OxS.[15] The increase with age is consistent with either an age-dependent increase in OxS or an age-dependent decrease in the rate of protein turnover. The structures of the major protein carbonyls in tissues are unknown.

Glycoxidation Products

The term "glycoxidation product" was introduced to describe advanced glycation end products (AGEs), which require both glycation and oxidation for their formation from glucose[2] (FIG. 2). The N-carboxyalkyl derivatives of lysine (CML and CEL), and GOLD and MOLD, and the fluorescent cross-link pentosidine are known to increase in lens protein and tissue collagen with age.[4,8,16] They may be formed following oxidative cleavage of Schiff base or Amadori adducts on protein or from ambient dicarbonyl compounds under nonoxidizing conditions. GO reacts with lysine to form CML and GOLD, and MGO yields CEL and MOLD; pentosidine may be formed by reaction of 3DG or dehydroascorbate with protein. 3DG may be formed nonenzymatically from Amadori compounds or enzymatically by decomposition of fructose-3-phosphate, a product of the fructose-3-kinase reaction.[17] The biochemical origin and pathways of metabolism of GO are unknown, but MGO is formed by both enzymatic pathways (from threonine and acetone) and by nonenzymatic pathways (decomposition of triose phosphates). The increase in age-adjusted levels of glycoxidation products in skin collagen in diabetes is consistent with the increases in substrate (glucose, 3DG, GO, and MGO) concentrations in plasma.[18,19]

Proteins react with GO, MGO, and 3DG *in vitro*, forming imidazolone adducts to arginine residues.[3] However, these adducts have not been detected in tissues by chemical assay,[20] possibly because of their equilibration into a mixture of epimers and tautomers. Further, although the initial product, a dihydroimidazolone, is stable to acid hydrolysis, it gradually oxidizes, at least *in vitro*, to form an acid-labile imidazolone. The extent to which these modifications are reversible, or evolve into other products *in vivo*, has not been rigorously studied, but Thornalley has shown that (dihydro)imidazolones are recognized by AGE-receptors on cells.[21]

Lipoxidation Products

Lipids are among the most readily oxidized substrates in both food products and biological systems. A vast array of products are produced on oxidation of polyunsaturated fatty acids (PUFA) and cholesterol in lipoproteins.[22] Although metal-catalyzed oxidation of low-density lipoprotein (LDL) leads to complete destruction of PUFAs and extensive modification of lysine residues, only a few protein modifications formed during lipoxidation reactions have been characterized.[3] Malondialdehyde (MDA) and 4-hydroxynonenal (HNE) are two of the characteristic, reactive aldehydes formed in these reactions. MDA and HNE adducts to Lys and the Lys-MDA-Lys cross-link have been measured by GC/MS and increase during metal-catalyzed oxidation of LDL *in vitro*, but they account for only about 1% of total lysine modification in oxidized LDL.[23] MDA-Lys, but not HNE-Lys, is detectable in plasma LDL and is 2- to 3-fold elevated in LDL isolated from atherosclerotic plaque (unpublished). We have recently shown that the glycoxidation products, CML (and CEL, unpublished), are also formed during oxidation of PUFAs in the presence of proteins.[24] CML was produced in amounts comparable to MDA-Lys and HNE-Lys during metal-catalyzed oxidation of LDL. Despite evidence that glycation enhances the susceptibility of proteins to oxidation *in vitro*,[25] the Amadori adduct on LDL was not a precursor of CML[24] during metal-catalyzed oxidation *in vitro*. CML and CEL are also increased 5- to 10-fold in plaque versus circulating lipoproteins (unpublished). Although both CML and CEL are readily detected in skin collagen and increase with age, MDA-Lys and HNE-Lys adducts are not detectable, possibly because of reversal of the imine linkage in MDA-Lys and the greater reactivity of HNE with histidine and cysteine residues in proteins. Characterization of other lipoxidative chemical modifications of protein is essential for progress in studies on the role of lipid peroxidation in aging and disease.

IMMUNOCHEMICAL DETECTION OF OXIDATION PRODUCTS IN TISSUES

Several biomarkers of protein oxidation, glycoxidation, and lipoxidation reactions have now been identified and can be measured in tissue proteins by chemical methods. However, a major limitation to progress in research on the role of these reactions in aging and disease is the difficulty in measuring trace concentration of individual oxidation products in tissue proteins by relatively sophisticated chemical assay procedures. ELISA assays provide a sensitive, automated, high throughput alternative for measuring a variety of biological analytes. For example, compared to HPLC and GC/MS analysis of hydrolyzed proteins, analysis of CML in intact proteins by an ELISA technique would greatly simplify its detection and provide a valuable tool for comparative analysis of the CML content of proteins. The same antibodies can also be used for immunohistochemical detection and comparison of levels of oxidation products in tissues. In this section we review recent progress in the development and application of ELISA and immunohistochemical assays for oxidation products, followed by discussion of their benefits and limitations.

Amino Acid Oxidation Products

Only three of the amino acid oxidation products shown in FIGURE 1 have been studied by immunochemical techniques: Cl-Tyr, NO_2-Tyr, and protein carbonyls. HOCl-modified proteins are detectable in atherosclerotic plaque by immunohistochemistry[26] and in renal inflammatory lesions.[27] A 30 fold increase in Cl-Tyr in plaque versus circulating LDL was confirmed by chemical analysis.[11] Based on immunohistochemical analysis, NO_2-Tyr is also increased primarily at sites of inflammation or OxS, for example, in the eye in experimental autoimmune uveitis,[28] in the lung following ischemic injury,[29] and in foam cells in atherosclerotic plaque.[30] Heinecke and colleagues also confirmed by chemical analysis a nearly 100-fold increase in NO_2-Tyr in plaque versus circulating LDL.[12] The chemical measurement of protein carbonyls (PCs) involves their reaction with DNPH, followed by spectrophotometric detection.[31] Antibodies to DNPH-hapten were also used for detecting PCs in protein by Western blot analysis,[31] and Winterbourn and colleagues[32] recently introduced an ELISA assay for PCs, using anti-DNPH antibody. The ELISA assay is sufficiently sensitive to detect PCs in plasma proteins and detects significant increases in PCs in plasma of critically ill patients. Unfortunately, the structures of the major PCs formed *in vivo* have not been described, so that little is known regarding the possible role of glycoxidation and lipoxidation reactions in their formation.

AGEs and Glycoxidation Products

The first immunochemical assays for AGEs were introduced in 1992. Araki *et al.*[33] used ELISA assays with both polyclononal and monoclonal antibodies to AGE-proteins to demonstrate an age-dependent increase in AGEs in lens proteins. CML was later identified as the epitope recognized by this antibody,[34] confirming previous chemical analyses of Dunn *et al.*[35] Makita *et al.*[36] used an ELISA assay with a polyclonal anti-AGE antibody of unknown specificity to demonstrate an increase in AGEs in serum proteins of diabetic patients and in nondiabetic patients with end-stage renal disease. AGEs on plasma proteins were decreased during hemodialysis and following renal transplantation, suggesting a role for the kidney in removal of AGE proteins or AGE precursors from plasma.[37] Using this antibody, Beisswenger *et al.*[38] were able to measure increases in AGEs in skin collagen associated with the early stages of renal and retinal disease in diabetic patients. Surprisingly, they observed no increase in collagen AGEs with age in nondiabetic subjects, in conflict with previous studies that have shown strong correlations between age and tissue levels of other AGEs, CML, and pentosidine.[19,39] Nakamura *et al.*[40] also used this antibody to demonstrate that the thiazolidine derivative, OPB-9195, inhibited AGE formation and AGE-derived cross-linking in skin collagen of diabetic rats, based on both immunohistochemical and ELISA assays for AGEs in collagen. Using immunohistochemistry, Rumble *et al.*[41] concluded that aminoguanidine also inhibited the increase in AGEs in the kidney of diabetic rats. In this case, CML was identified as at least one epitope recognized by the anti-AGE antibody. In 1994, Papanastasiou *et al.*[42] introduced a time-resolved fluorescence ELISA assay for AGE proteins in serum, significantly increasing the sensitivity of absorbance-based assays. They confirmed the results of earlier studies on changes in serum AGEs in diabetes and renal failure and also observed a correlation of serum AGEs with serum creatinine, but not with chronological age or blood glucose, in diabetic patients without clinically apparent renal disease. These results suggest that levels of AGEs in serum may serve as an early indicator of renal damage in diabetes. Schleicher *et al.*[43] also reported recently, using an anti-CML-protein antibody, that CML was increased in kidney (by immunochemistry) and in plasma (about 2-fold, by ELISA) of diabetic patients, independent of the presence of nephropathy.

Chemical characterization of the epitope specificity of antibodies is essential for understanding the role of oxidative stress and the mechanism(s) of protein modification in aging and disease. In 1995 Reddy *et al.*[44] reported that CML was the dominant epitope recognized by several polyclonal antibodies prepared in our laboratory and suggested that CML was a major immunogenic structure on the surface of AGE proteins. The CML specificity of anti-AGE antibodies was later confirmed by Ikeda *et al.*,[34] Rumble *et al.*,[41] and Nakayama *et al.*[45] (personal communication). Degenhardt *et al.*[46] also reported recently that the antibodies of Papanastasiou *et al.*[42] were monospecific for CML and that there was a strong correlation between results of assays of serum AGEs by ELISA and serum CML by GC/MS. Berg *et al.*[47] have also reported that the antibody of Makita *et al.*[37] recognizes CML. The fact that CML is formed during both glycoxidation and lipoxidation reactions raises some questions about the specificity of many "anti-AGE" antibodies. Using the CML-specific 6D12 antibody,[33,34] Kume *et al.*[48] detected an increase in AGEs in atherosclerotic plaque of nondiabetic subjects. It is possible that these "AGEs" in plaque are derived from peroxidation of PUFAs in plaque lipids, rather than from glycoxidation reactions. Anti-AGE antibodies with specificity for CML have also been used to detect increased levels of AGEs in amyloid plaque in dialysis-related amyloidosis,[49] and it is likely that CML is also a primary AGE epitope in Alzheimer's plaque.[50,51] Horiuchi and colleagues[52] also reported increased levels of CML in elastin of sun-exposed, compared to unexposed, skin of subjects with actinic elastosis. These observations about atherosclerotic and amyloid plaque and skin of normoglycemic individuals suggest that CML may be derived from PUFA and may be a general biomarker of oxidative stress. Even in diabetes, inflammation induced by metabolic stresses may cause oxidative damage in such tissues as the kidney and vascular wall, and the CML detected in these tissues may be derived from lipids, rather than glucose, especially in patients with hyperlipidemia.

Obviously, not all antibodies to AGE-proteins are specific for CML. Horiuchi and colleagues have absorbed the CML-specific fraction from a polyclonal anti-AGE antibody using an affinity column containing CML bovine serum albumin.[34] The residual antibody recognized AGE proteins, but the specificity of this antibody fraction is still unknown, nor is it certain that the non-CML-specific antibodies actually recognize AGEs in biological samples. Dean and colleagues[53] have also shown that amino acid hydroperoxides and dihydroxyphenylalanine are formed during the preparation of AGE proteins *in vitro*, so that numerous non-AGE epitopes may exist on model AGE proteins. However, some anti-AGE antibodies are clearly specific for AGEs, other than CML. Niwa *et al.*[54] described a monoclonal antibody specific for (dihydro)imidazolones formed in reaction of 3DG with arginine and reported that erythrocyte levels of these adducts were significantly elevated in diabetic patients. This antibody was also used in histochemical studies and detected increased levels of 3DG-Arg adducts in vascular and renal tissue from diabetic patients. In addition to preparation of anti-AGE antibodies using AGE proteins as immunogens, Monnier and colleagues have prepared antibodies against specific AGEs, pyrraline and pentosidine, linked as haptens to carrier macromolecules.[55,56] Hayase *et al.*[55] used the antipyrraline antibody in a radioimmunoassay procedure to demonstrate increased levels of pyrraline in plasma proteins from diabetic subjects. The antibody to pentosidine hapten was used in an ELISA assay,[56] showing that pentosidine was also increased in plasma proteins and skin collagen in diabetes. The results of this ELISA assay correlated strongly with results of chemical analyses for pentosidine.

Lipoxidation Products

In studies on the role of lipoprotein oxidation in atherogenesis, Haberland *et al.*[57] used antibodies to MDA-modified proteins and first reported the detection of MDA-protein epitopes in atherosclerotic plaque. Subsequent studies have shown that both MDA-Lys and HNE-Lys epitopes are present in atherosclerotic plaque[58] and that MDA-protein epitopes are enriched in Alzheimer's plaque,[59] HNE epitopes in amyloid deposits in patients with systemic amyloidosis,[60] and MDA and HNE epitopes in liver and plasma proteins of rats exposed to OxS by iron overload.[61] More recently, Kim *et al.* also[62] noted the presence of epitopes of lipid peroxide–modified proteins in atherosclerotic plaque and showed that their antibody could also detect lipid peroxide–modified proteins in normal human plasma by Western blot analysis. This antibody did not recognize MDA-modified protein, but its epitope specificity is unknown.

Bucala et al.[63] detected both lipid-linked and protein-bound AGEs on human LDL by ELISA using an anti-AGE antibody. They found increased levels of both protein and lipid AGEs in LDL from diabetic and uremic patients and a strong correlation among levels of AGE-protein, AGE-lipid, and lipid peroxidation products (thiobarbituric acid assay) in LDL. Reddy et al.[44] proposed that the AGE-lipids were probably products of carboxymethylation of phosphatidylethanolamine. Requena et al.[64] recently confirmed the presence of N-(carboxymethyl)-phosphatidylethanolamine and N-(carboxymethyl)-phosphatidylserine in oxidized LDL and in red cell membranes.[64] These results suggest that N-(carboxymethyl)phosphatidylethanolamine is a likely candidate for the AGE-lipid detected in LDL. Because MDA-Lys, CML, and CM lipids are produced together during lipoprotein oxidation in vitro, and because levels of MDA-Lys and CML are increased in uremic plasma and in LDL isolated from atherosclerotic plaque, it is not surprising that levels of these compounds and CM lipids would increase in concert with one another in circulating LDL in diabetes and uremia.

LIMITATIONS OF IMMUNOCHEMICAL ASSAYS
FOR OXIDATION PRODUCTS

Immunohistochemistry

Immunohistochemistry has proven to be an invaluable tool for identifying oxidation products in tissues and has demonstrated that a range of products of oxidation, glycoxidation, and lipoxidation reactions are clustered together at sites of inflammation. There are limitations, however, to the identification of specific products because of the cross-reactivity of antibodies. We have shown, for example, that N-(carboxymethyl)ethanolamine, derived from either glycoxidation or lipoxidation of phosphatidylethanolamine, cross-reacts with CML-specific antibody prepared against AGE protein.[64] Depending on the method of preparation of the tissue section, the AGEs detected in tissues may be lipid bound rather than protein-bound epitopes. The distinctions may be important, especially if there are significant differences in the turnover of lipid and protein in plaque—if protein bound, the AGEs may accumulate on inert protein and reflect on the history of the deposits, whereas if lipid bound, the AGEs may turn over and present a useful approximation of ongoing or steady state levels of OxS. It should be noted that MDA adducts to phosphatidylethanolamine have been detected in red cell membranes,[65] so that even "MDA-protein" epitopes in plaque may also be lipid derivatives. It is not possible to resolve this issue without chemical analysis of tissue components.

A second concern about cross-reactivity in immunohistochemical studies applies especially to polyclonal antibodies that contain a heterogeneous population of antibodies against a wide range of epitopes, even if a homogeneous hapten is used for immunization and the range of specificity is thought to be limited. Histochemical staining with antibodies does not necessarily mean that the epitope detected is the same as the primary specificity of the antibody. A minor cross-reactive fraction of antibody could "light up" the tissue section. It is helpful to isolate an epitope-specific fraction of the antibody by immunoaffinity chromatography using immobilized hapten and essential to demonstrate competition by protein modified with known haptens before concluding that a specific epitope is present in the tissue. Even in this case, as well as with monoclonal antibodies, the immunizing epitope might be bound more tightly by the antibody than a related epitope in tissues, and a determinant with low cross-reactivity may still lead to a positive reaction. The similarities in the core structures of furanyl-furoyl-imidazole (FFI)[66] and of the imidazolium cross-links, GOLD and MOLD (refs. 4 and 67, and FIG. 2), suggest that the detection of FFI in tissues may result from cross-reaction of the antibody with GOLD and MOLD, which have been chemically identified in tissues. This might explain the conflicting evidence from chemical and immunohistochemical studies regarding the presence of FFI in tissues.[66,68–70] Similarities in the pyridinium ring structures of pentosidine,[39] cross-lines,[71] and AGE-X$_1$[72] may also yield cross-reacting antibodies, compromising immunohistochemical evidence for their presence in tissue proteins. Proof of formation of these and other AGEs *in vivo*, including (dihydro)imidazolones derived from 3DG and MGO,[5,20,21,53,73] awaits confirmation by rigorous chemical methods. In other cases, a highly specific monoclonal antibody may be of limited use for detecting AGE epitopes on a variety of proteins. Giardino *et al.*[74] recently detected an increase in intracellular AGEs in endothelial cells exposed to a high glucose medium. Western blot analysis showed only a limited number of AGE-modified proteins in the cell, suggesting either that AGEs were formed on a limited set of intracellular proteins or, more likely, that only a limited range of AGE proteins was recognized by the monoclonal anti-AGE antibody.

Immunohistochemistry has been used in a semiquantitative fashion to detect disease-related changes in oxidation products in tissue proteins. The results of histochemical studies showing increased levels of Cl-Tyr, NO$_2$-Tyr, and CML in atherosclerotic plaque have been confirmed by chemical analyses (see above). However, a possible limitation to this comparative immunohistochemistry is the assumption that epitopes are equally exposed on the surface of proteins. This may not always be true, especially if tissue damage or turnover and repair mechanisms are activated by the disease process or if there are disease-related changes in the composition of tissue proteins or such associated molecules as proteoglycans. In some cases, tissue sections are treated by enzymatic digestion to induce exposure of epitopes and enhance sensitivity for immunostaining. This procedure may also differentially affect control and diseased tissues, yielding misleading results. The manifold limitations to rigorous qualitative and quantitative interpretations of immunohistochemical experiments highlight the importance of confirming immunological observations by chemical analysis.

ELISA Assays

ELISA assays using antibodies to oxidized proteins are more convenient than chromatographic procedures. They are especially useful when samples of similar composition are being compared and when absolute quantitation is not required. Although variations in antibody specificity and avidity lead to variability between laboratories, comparative data are often sufficient, for example, to determine whether levels of AGEs or lipoxidation products are higher or lower in patients, compared to control subjects, or in response to therapy. A major limitation of ELISA assays is that solubilization of protein is required for the analysis, especially for the detection of cross-link structures. Araki *et al.*[33] used alkaline treatment to denature and solubilize lens proteins, whereas others have used enzymatic digestion for assay of the AGE content of skin collagen.[38,41,76] Taneda and Monnier[56] used a pentosidine-specific antibody in an ELISA assay to measure the age-dependent increase in pentosidine in skin collagen in nondiabetic patients and an increase in pentosidine in plasma proteins of diabetic and uremic patients. Digestion of AGE proteins by either enzymatic or acid hydrolysis enhanced the sensitivity of the assay by 10- to 100-fold, consistent with the expectation that most of the pentosidine is buried in inaccessible intermolecular cross-links between proteins. The enhancement of sensitivity by hydrolysis was observed with both insoluble collagen and soluble plasma proteins, suggesting that hydrolysis of proteins, in general, may enhance the sensitivity of ELISA assays for some oxidation products. By contrast, CML is recognized with over 1,000-fold higher avidity in the peptide-bound form, compared to the free amino acid.[44] This variability might result from differences in the size of the epitope or the antigenic determinant on protein, so that each hapten must be independently evaluated.

CONCLUSION

Both chemical and immunochemical assays have shown that oxidation products accumulate in tissue proteins during aging and at sites of inflammation and deposition of plaque in age-related diseases. The accumulation of these oxidation products provides growing support for both the free radical theory of aging and the role of oxidative stress in chronic disease. Immunochemical techniques are becoming an increasingly important part of the methodology for detection and measurement of oxidation products in tissues, often providing the preliminary data and rationale for the design of more definitive, chemical studies. The combined application of immunochemical and chemical methods will undoubtedly lead to a better understanding of the role of oxidative stress in aging and disease and facilitate the evaluation of therapies designed to limit oxidative damage in tissues and to control pathologies associated with aging and disease.

REFERENCES

1. HALLIWELL, B. & J.M.C. GUTTERIDGE. 1989. Free Radicals in Biology and Medicine, 2nd ed. Clarendon Press. Oxford.
2. BAYNES, J.W. 1991. Role of oxidative stress in development of complications in diabetes. Diabetes **40:** 405–412.
3. REQUENA, J.R., M.X. FU, M.U. AHMED, A.J. JENKINS, T.J. LYONS & S.R. THORPE. 1996. Lipoxidation products as biomarkers of oxidative damage to proteins during lipid peroxidation reactions. Nephrol. Dial. Transplant. **11** (Suppl 1): 48–53.
4. WELLS-KNECHT, K.J., E. BRINKMANN, M.C. WELLS-KNECHT, J.E. LITCHFIELD, M.U. AHMED, S. REDDY, D.V. ZYZAK, S.R. THORPE & J.W. BAYNES. 1996. New biomarkers of Maillard reaction damage to proteins. Nephrol. Dial. Transplant. **11** (Suppl 1): 41–47.
5. THORNALLEY, P.J. 1996. Advanced glycation and the development of diabetic complications. Unifying the involvement of glucose, methylglyoxal and oxidative stress. Endocrinol. Metab. **3:** 149–166.
6. DEGENHARDT, T.P, E. BRINKMANN-FRYE, S.R. THORPE & J.W. BAYNES. 1998. Role of carbonyl stress in aging and age-related diseases. *In* The Maillard Reaction in Foods and Medicine. J.

O'Brien, H.E. Nursten, M.J.C. Crabbe & J.M. Ames, Eds.:3–10. The Royal Society of Chemistry. London.
7. HORIUCHI, S. & N. ARAKI. 1994. Advanced glycation end products of the Maillard reaction and their relation to aging. Gerontology **40**:10–15.
8. THORPE, S.R. & J.W. BAYNES. 1996. Role of the Maillard reaction in diabetes mellitus and diseases of aging. Drugs & Aging **9**: 69–77.
9. DEAN, R.T., S. FU, R. STOCKER & M.J. DAVIES. 1997. Biochemistry and pathology of radical-mediated protein oxidation. Biochem. J. **324**: 1–18.
10. WELLS-KNECHT, M.C., T.J. LYONS, D.R. MCCANCE, S.R. THORPE & J.W. BAYNES. 1997. Age-dependent increase in *ortho*-tyrosine and methionine sulfoxide in human skin collagen is not accelerated in diabetes: Evidence against a generalized increase in oxidative stress in diabetes. J. Clin. Invest. **100**: 839–846.
11. HAZEN, S.L. & J.W. HEINECKE. 1997. 3-Chlorotyrosine, a specific marker of myeloperoxidase-catalyzed oxidation, is markedly elevated in low density lipoprotein isolated from human atherosclerotic intima. J. Clin. Invest. **99**: 2075–2081.
12. LEEUWENBURGH, C., M.M. HARDY, S.L. HAZEN, P. WAGNER, S. OH-ISHI, U.P. STEINBRECHER & J.W. HEINECKE. 1997. Reactive nitrogen intermediates promote low density lipoprotein oxidation in human atherosclerotic intima. J. Biol. Chem. **272**: 1433–1436.
13. LEEUWENBURGH, C., J.E. RASMUSSEN, F.F. HSU, D.M. MUELLER, S. PENNATHUR & J.W. HEINECKE. 1997. Mass spectrophotometric quantitation of markers of protein oxidation by tyrosyl radical, copper and hydroxyl radical in low density lipoprotein isolate from human atherosclerotic plaques. J. Biol. Chem. **272**: 3520–3526.
14. HUGGINS, T.G., M.C. WELLS KNECHT, N. DETORIE, S.R. THORPE & J.W. BAYNES. 1993. Formation of *o*-tyrosine and dityrosine in proteins during radiolytic and metal-catalyzed oxidation of proteins. J. Biol. Chem. **268**:12341–12347.
15. STADTMAN, E.R. 1995. The status of oxidatively modified proteins as a marker of aging. *In* Molecular Aspects of Aging. K. Esser & G.W. Martin, Eds.: 129–143. John Wiley. New York.
16. AHMED, M.U., E. BRINKMANN-FRYE, T.P. DEGENHARDT, S.R. THORPE & J.W. BAYNES. 1997. N$^\varepsilon$-(carboxyethyl)lysine, a product of chemical modification of proteins by methylglyoxal, increases with age in human lens proteins. Biochem. J. **324**: 565–570.
17. LAL, S., B.S. SZWERGOLD, A.H. TAYLOR, W.C. RANDALL, F. KAPPLER, K. WELLS-KNECHT, J.W. BAYNES & T.R. BROWN. 1995. Metabolism of fructose-3-phosphate in the diabetic rat lens. Arch. Biochem. Biophys. **318**: 191–199.
18. THORNALLEY, P.J. 1990. The glyoxalase system: New developments toward functional characterization of a metabolic pathway fundamental to life. Biochem. J. **269**: 1–11.
19. DYER, D., J. DUNN, S. THORPE, K. BAILIE, T. LYONS, D. MCCANCE & J. BAYNES. 1993. Accumulation of Maillard reaction products in skin collagen in diabetes and aging. J. Clin. Invest. **91**: 2463–2469.
20. ZYZAK, D.V. 1995. Studies on the Maillard reaction: Reaction of 3-deoxyglucosone with arginine residues in protein. Ph.D. thesis, University of South Carolina. Columbia, SC.
21. WESTWOOD, M.E., O.K. AGIROV, E.A. ABORDO & P.J. THORNALLEY. 1997. Methylglyoxal-modified arginine residues—a signal for receptor-mediated endocytosis and degradation of proteins by monocytic THP-1 cells. Biochim. Biophys. Acta **1356**: 84–94.
22. ESTERBAUER, H. & P. RAMOS. 1996. Chemistry and pathophysiology of oxidation of LDL. Rev. Physiol. Biochem. Pharmacol. **127**: 31–64.
23. REQUENA, J.R., M.X. FU, M.U. AHMED, A.J. JENKINS, T.J. LYONS, J.W. BAYNES & S.R. THORPE. 1997. Quantification of malondialdehyde and 4-hydroxynonenal adducts to lysine residues in native and oxidized human low-density lipoprotein. Biochem. J. **322**: 317–325.
24. FU, M.X., J.R. REQUENA, A.J. JENKINS, T.J. LYONS, J.W. BAYNES & S.R. THORPE. 1996. The advanced glycation end-product, N$^\varepsilon$-(carboxymethyl)lysine (CML), is a product of both lipid peroxidation and glycoxidation reactions. J. Biol. Chem. **271**: 9982–9986.
25. BOWIE, A., D. OWENS, P. COLLINS, A. JOHNSON & G.H. TOMKIN. 1993. Glycosylated low density lipoprotein is more sensitive to oxidation: Implications for the diabetic patient. Atherosclerosis **102**: 60–67.
26. HAZELL, L.J., L. ARNOLD, D. FLOWERS, G. WAEG, E. MALLE & R. STOCKER. 1996. Presence of hypochlorite-modified proteins in human atherosclerotic lesions. J. Clin. Invest. **97**: 1535–1544.

27. MALLE, E., C. WOENCKHAUS, G. WAEG, H. ESTERBAUER, E.F. GRONE & H.J. GROENE. 1997. Immunological evidence of hypochlorite-modified proteins in human kidney. Am. J. Pathol. **150:** 603–615.

28. WU, G.S., J. ZHANG & N.A. RAO. 1997. Peroxynitrite and oxidative damage in experimental autoimmune uveitis. Invest. Ophthalmol. Vis. Sci. **38:** 1333–1339.

29. ISCHIROPOULOS, H., A.G. AL-MEHDI & A.B. FISHER. 1995. Reactive species in ischemic rat lung injury: Contribution of peroxynitrite. Am. J. Physiol. **269** (Part 1): L158–L164.

30. BECKMAN, J.S., Y.Z. YE, P.G. ANDERSON, J. CHEN, M.A. ACCAVITTI, M.M. TARPEY & C.R. WHITE. 1994. Extensive nitration of protein tyrosines in human atherosclerosis detected by immunochemistry. Biol. Chem. Hoppe-Seyler **375:** 81–88.

31. LEVINE, R.L., J.A. WILLIAMS, E.R. STADTMAN & E. SHACTER. 1994. Carbonyl assays for determination of oxidatively modified proteins. Methods Enzymol. **233:** 346–357.

32. BUSS, H., T.P. CHAN, K.B. SLUIS, N.M. DOMIGAN & C.C. WINTERBOURN. 1997. Protein carbonyl measurement by a sensitive ELISA method. Free Radical Biol. Med. **23:** 361–366.

33. ARAKI, N., N. UENO, B. CHAKRABARTI, Y. MORINO & S. HORIUCHI. 1992. Immunochemical evidence for the presence of advanced glycation end products in human lens proteins and its positive correlation with age. J. Biol. Chem. **267:** 10211–10214.

34. IKEDA, K., T. HIGASHI, H. SANO, Y. JINNOUCHI, M. YOSHIDA, T. ARAKI, S. UEDA & S. HORIUCHI. 1996. N^ε-(carboxymethyl)lysine protein adduct is a major immunological epitope in proteins modified with advanced glycation end products of the Maillard reaction. Biochemistry **35:** 8075–8081.

35. DUNN, J.A., J.S. PATRICK, S.R. THORPE & J.W. BAYNES. 1989. Oxidation of glycated proteins: Age-dependent accumulation of carboxymethyllysine in lens proteins. Biochemistry **28:** 9464–9468.

36. MAKITA, Z., H. VLASSARA, A. CERAMI & R. BUCALA. 1992. Immunochemical detection of advanced glycosylation end products *in vivo*. J. Biol. Chem. **267:** 5133–5138.

37. MAKITA, Z., S. RADOFF, E.J. RAYFIELD, Z. YANG, E. SKOLNIK, V. DELANEY, E.A. FRIEDMAN, A. CERAMI & H. VLASSARA. 1991. Advanced glycosylation end products in patients with diabetic nephropathy. N. Engl. J. Med. **325:** 836–842.

38. BEISSWENGER, P., Z. MAKITA, T. CURPHEY, L. MOORE, S. JEAN, T. BRINCK-JOHNSEN, R. BUCALA & H. VLASSARA. 1995. Formation of immunochemical advanced glycosylation end products precedes and correlates with early manifestations of renal and retinal disease in diabetes. Diabetes **44:** 824–829.

39. SELL, D.R. & V.M. MONNIER. 1989. Structure elucidation of a senescence cross-link from human extracellular matrix: Implication of pentoses in the aging process. J. Biol. Chem. **264:** 21597–21602.

40. NAKAMURA, S., Z. MAKITA, S. ISHIKAWA, K. YASUMURA, W. FUJII, K. YANAGISAWA, T. KAWATA & T. KOIKE. 1997. Progression of nephropathy in spontaneous diabetic rats is prevented by OPB-9195, a novel inhibitor of advanced glycation. Diabetes **46:** 895–899.

41. RUMBLE, J.R., M.E. COOPER, T. SOULIS, A. COX, L. WU, S. YOUSSEF, M. JASIK, G. JERUMS & R.E. GILBERT. 1997. Vascular hypertrophy in experimental diabetes: Role of advanced glycation end products. J. Clin. Invest. **99:** 1016–1027.

42. PAPANASTASIOU, P., L. GRASS, H. RODELA, A. PATRIKAREA, D. OREOPOULOS & E. DIAMANDIS. 1994. Immunological quantification of advanced glycosylation end-products in the serum of patients on hemodialysis or CAPD. Kidney Int. **46:** 216–222.

43. SCHLEICHER, E.D., E. WAGNER & A.G. NERLICH. 1997. Increased accumulation of the glycoxidation product N^ε-(carboxymethyl)lysine in human tissues in diabetes and aging. J. Clin. Invest. **99:** 457–468.

44. REDDY, S.R., J. BICHLER, K.J. WELLS-KNECHT, S.R. THORPE & J.W. BAYNES. 1995. N^ε-(Carboxymethyl)lysine is a dominant advanced glycation end product (AGE) antigen in tissue proteins. Biochemistry **34:** 10872–10878.

45. NAKAYAMA, H., S. TANEDA, T. MITSUHASHI, S. KUWAJIMA, S. AOKI, Y. KURODA, K. MISAWA, K. YANAGISAWA & S. NAKAGAWA. 1991. Characterization of antibodies to advanced glycosylation end products in protein. J. Immunol. Methods **140:** 119–125.

46. DEGENHARDT, T.P., L. GRASS, S. REDDY, S. THORPE, E. DIAMANDIS & J.W. BAYNES. 1997. The serum concentration of the advanced glycation end-product N^ε-(carboxymethyl)lysine is increased in uremia. Kidney Int. **52:** 1064–1067.

47. BERG, T.J., P.A. TORJESEN, J.T. CLAUSEN, K. DAHL-JORGENSEN, H.-J. BANGSTAD, H. VOGT & K.F. HANSSEN. 1997. Carboxymethyllysine (CML) is increased in serum of children and adolescents with IDDM. Diabetologia 40 (Suppl. 1): A586.

48. KUME, S., M. TAKEYA, T. MORI, N. ARAKI, H. SUZUKI, S. HORIUCHI, T. KODAMA, Y. MIYAUCHI & K. TAKAHASHI. 1995. Immunohistochemical and ultrastructural detection of advanced glycation end products in atherosclerotic lesions of human aorta with a novel specific monoclonal antibody. Am. J. Pathol. 147: 654 667.

49. MIYATA, T., R. INAGI, Y. IIDA, M. SATO, N. YAMADA, O. ODA, K. MAEDA & H. SEO. 1994. Involvement of β_2-microglobulin modified with advanced glycation end products in the pathogenesis of hemodialysis-associated amyloidosis. J. Clin. Invest. 93: 521–528.

50. VITEK, M.P., K. BHATTACHARYA, J.M. GLENDENING, E. STOPA, H. VLASSARA, R. BUCALA, K. MANOGUE & A. CERAMI. 1994. Advanced glycation end products contribute to amyloidosis in Alzheimer disease. Proc. Natl. Acad. Sci. USA 91: 4766–4770.

51. SMITH, M.A., A. TANEDA, P.L. RICHEY, S. MIYATA, S.-D. YAN, D. STERN, L. SAYRE, V.M. MONNIER & G. PERRY. 1994. Advanced Maillard reaction end products are associated with Alzheimer Disease. Proc. Natl. Acad. Sci. USA 91: 5170–5174.

52. MIZUTARI, K., T. ONO, K. IKEDA, K. KAYASHIMA & S. HORIUCHI. 1997. Photo-enhanced modification of human skin elastin in actinic elastosis by N^ε-(carboxymethyl)lysine, one of the glycoxidation products of the Maillard reaction. J. Invest. Dermatol. 108: 797–802.

53. NIWA, T., T. KATSUZAKI, S. MIYAZAKI, T. MIYAZAKI, Y. ISHIZAKI, F. HAYASE, N. TATEMICHI & Y. TAKEI. 1997. Immunohistochemical detection of imidazolone, a novel advanced glycation end product, in kidneys and aortas of diabetic patients. J. Clin. Invest. 99: 1272–1280

54. FU, S., M-X. FU, J.W. BAYNES, S.R. THORPE & R.T. DEAN. 1998. Presence of DOPA and amino acid hydroperoxides in protein modified with advanced glycation end products (AGEs): Amino acid oxidation products as possible source of oxidative stress induced by AGE-proteins. Biochem. J. 330: 233–239.

55. HAYASE, F., R.H. NAGARAJ, S. MIYATA, F.G. NJOROGE & V.M. MONNIER. 1989. Aging of proteins: Immunological detection of a glucose-derived pyrrole formed during Maillard reaction *in vivo*. J. Biol. Chem. 263: 3758–3764.

56. TANEDA, S. & V.M. MONNIER. 1994. ELISA of pentosidine, an advanced glycation end product, in biological specimens. Clin. Chem. 40: 1766–1773.

57. HABERLAND, M.E., D. FONG & L. CHENG. 1988. Malondialdehyde-altered protein occurs in atheroma of Watanabe heritable hyperlipidemic rabbits. Science 241: 215–218.

58. YLÄ-HERTTUALA, S., W. PALINSKI, M.E. ROSENFELD, S. PARTHASARATHY, T.E. CAREW, S. BUTLER, J.L. WITZTUM & D. STEINBERG. 1989. Evidence for the presence of oxidatively modified low density lipoprotein in atherosclerotic lesions of rabbit and man. J. Clin. Invest. 84: 1086–1095

59. YAN, S.-D., X. CHEN, A.M. SCHMIDT, J. BRETT, G. GODMAN, Y.-S. ZOU, C.W. SCOTT, C. CAPUTO, T. FRAPPIER, M.A. SMITH, G. PERRY, S.-H. YEN & D. STERN. 1994. Glycated tau protein in Alzheimer disease: A mechanism for induction of oxidative stress. Proc. Natl. Acad. Sci. USA 91: 7787–7791.

60. ANDO, Y., N. NYHLIN, O. SUHR, G. HOLMGREN, K. UCHIDA, M. EL SAHLY, T. YAMASHITA, H. TERESAKI, M. NAKAMURA, M. UCHINO & M. ANDO. 1997. Oxidative stress is found in amyloid deposits in systemic amyloidosis. Biochem. Biophys. Res. Commun. 232: 497–502.

61. HOUGLUM, K., M. FILIP, J.L. WITZTUM & M. CHOJKIER. 1990. Malondialdehyde and 4-hydroxynonenal adducts in plasma and liver of rats with iron overload. J. Clin. Invest. 86: 1991–1998.

62. KIM, J.G., G. SABBAGH, N. SANTAMAN, J.N. WILCOX, R.M. MEDFORD & S. PARTHASARATHY. 1997. Generation of a polyclonal antibody against lipid peroxide modified proteins. Free Radical Biol. Med. 23: 251–259.

63. BUCALA, R., Z. MAKITA, T. KOSCHINSKY, A. CERAMI & H. VLASSARA. 1993. Lipid advanced glycosylation: Pathway for lipid oxidation *in vivo*. Proc. Natl. Acad. Sci. USA 90: 6434–6438.

64. REQUENA, J.R., M. AHMED, C.W. FOUNTAIN, T.P. DEGENHARDT, S. REDDY, C. PEREZ, T.J. LYONS, A.J. JENKINS, J.W. BAYNES & S.R. THORPE. 1997. Carboxymethylethanolamine, a biomarker of phosphoplipid modification during the Maillard reaction *in vivo*. J. Biol. Chem. 272: 17473–17479.

65. JAIN, S.K., R. McVIE, J. DUETT & J.J. HERBST. 1989. Erythrocyte membrane lipid peroxidation and glycosylated hemoglobin in diabetes. Diabetes **38:** 1539–1543.

66. PONGOR, S., P.C. ULRICH, F.A. BENCSATH & A. CERAMI. 1984. Aging of proteins: Isolation and identification of a fluorescent chromophore from the reaction of polypeptides with glucose. Proc. Natl. Acad. Sci. USA **81:** 2684–2688.

67. NAGARAJ, R.H., I.N. SHIPANOVA & F.M. FAUST. 1996. Protein cross-linking by the Maillard reaction: Isolation, characterization, and *in vivo* detection of a lysine-lysine cross-link derived from methylglyoxal. J. Biol. Chem. **271:** 19338–19345.

68. NJOROGE, F.G., A.A. FERNANDES & V.M. MONNIER. 1988. Mechanism of formation of the putative advanced glycosylation end product and protein crosslink 2-(2-furoyl)-4(5)-(2-furanyl)-1*H*-imidazole. J. Biol. Chem. **263:** 10646–10652.

69. HORIUCHI, S., M. SHIGA, N. ARAKI, K. TAKATA, M. SAITOH & Y. MORINO. 1988. Evidence against *in vivo* presence of 2-(2-furoyl)-4(5)-(2-furanyl)-1*H*-imidazole, a major fluorescent advanced end product generated by nonenzymatic glycosylation. J. Biol. Chem. **263:** 18821–18826.

70. PALINSKI, W., T. KOSCHINSKY, S.W. BUTLER, E. MILLER, H. VLASSARA, A. CERAMI & J.L. WITZTUM. 1995. Arterioscler. Thromb. Vasc. Biol. **15:** 571–582.

71. IENAGA, K., H. KAKITA, T. HOCHI, K. NAKAMURA, Y. NAKAZAWA, Y. FUKUNAGA, S. AOKI, G. HASEGAWA, Y. TSUTSUMI, Y. KITAGAWA & K. NAKANO. 1996. Crossline-like structure accumulates a fluorescent advanced glycation end product in renal tissues of rats with diabetic nephropathy. Proc. Jpn. Acad. **72** (Ser B): 78–84.

72. HORIUCHI, S., N. ARAKI, K. NAKAMURA, K. IKEDA, Y. FUKUNAGA & K. IENAGA. 1996. A new fluorescent crosslinker (AGE-X1) isolated from AGE-lysine derivatives. Diabetes **45:** 186A.

73. KONISHI, Y., F. HAYASE & H. KATO. 1994. Novel imidazolone compound formed by the advanced Maillard reaction of 3-deoxyglucosone and arginine residues in protein. Biosci. Biotech. Biochem. **58:** 1953–1955.

74. GIARDINO, I., D. EDELSTEIN & M. BROWNLEE. 1994. Nonenzymatic glycosylation *in vitro* and in bovine endothelial cells alters basic fibroblast growth factor activity. A model for intracellular glycosylation in diabetes. J. Clin. Invest. **94:** 110–117.

75. MENG, J., N. SAKATA, S. TAKEBAYASHI, T. ASANO, T. FUTATA, N. ARAKI & S. HORIUCHI. 1996. Advanced glycation end products of the Maillard reaction in aortic pepsin-insoluble and pepsin-soluble collagen from diabetic rats. Diabetes **45:** 1037–1043.

Assessing the Effects of Deprenyl on Longevity and Antioxidant Defenses in Different Animal Models

KENICHI KITANI,[a,e] SETSUKO KANAI,[b] GWEN O. IVY,[c] AND MARIA CRISTINA CARRILLO[d]

[a]National Institute for Longevity Sciences, 36-3, Gengo, Moriokacho, Obu-shi, Aichi, Japan 474

[b]Department of Clinical Physiology, Tokyo Metropolitan Institute of Gerontology, 35-2, Sakae-cho, Itabashi-ku, Tokyo, Japan 173

[c]Life Sciences Division, University of Toronto at Scarborough, 1265 Military Trail, Scarborough, Ontario, Canada M1C 1A4

[d]Institute de Fisiologia Experimental, Suipacha 570, 2000 Rosario, Universidad Nacional de Rosario, Argentina

ABSTRACT: Among many pharmaceuticals that have been tested for their effects on longevities of different animal rodents, deprenyl is unique in that its effects on longevity has been tested in at least four different animal species by independent research groups and that the effect has been postulated to be due to its effect of raising such antioxidant enzyme activities as superoxide dismutase (SOD) and catalase (CAT) in selective brain regions. Thus far, in all four species of animals examined (rats, mice, hamsters, and dogs), a positive effect was demonstrated, although the extent of its effect is quite variable. Our group has examined the effect on longevities in rats and mice and on antioxidant enzymes in rats, mice, and dogs. Although in rats of both sexes, we have obtained positive effects on longevity, two studies with different doses in mice did not reveal a significantly positive effect. We have observed, however, significantly positive effects on SOD (in Cu, Zn-, and Mn-) as well as CAT (but not glutathione peroxidase) activities in the brain dopaminergic system such as in the S. nigra and striatum (but not in hippocampus) in all rats, mice, and dogs, although the effects were quite variable, depending on the doses used. In mice, however, a long-term administration (3x/w, 3 months) caused a remarkable decrease in the magnitude of activity as well as a narrowing of the effective dose range, which may explain a relatively weak effect of the drug on mouse longevity. Further, a recent study on aging beagle dogs by Ruehl et al. showed a remarkable effect on longevity, which agrees with our SOD study in dogs. Although deprenyl has been claimed to have several other effects, such as a radical scavenging effect and a neuroprotective effect, past reports on its effects on longevities and antioxidant defenses are compatible with the notion that the drug prolongs the life span of animals by reducing the oxidative damage to the brain dopaminergic system during aging. Further, our studies on F-344 rats as well as a dog study by Ruehl et al. suggest that the drug may at least partially prolong the life span of animals by enhancing immune system function and preventing tumor development in animals.

Attempts at prolonging the human life span with pharmaceuticals, nutrients, and many other means have been made throughout the long history of humans since ancient days. However, all of these attempts have been in vain, at least in modern scientific terms.

[e]Tel: +81-562-45-0183; fax: 81-562-45-0184; e-mail: kitani@nils.go.jp

Although the general consensus in experimental gerontology is that the only reliable means for prolonging the life span of rodents is by dietary restriction, recent attempts at pharmacological intervention in aging and age-associated pathologies in experimental animals are becoming more and more realistic. Nonetheless, these observations still lack solid scientific explanations.

Most of the efforts in the past as well as at present involve administration of so-called antioxidants; some of the past results are summarized elsewhere.[1] Although it has been suggested that (-)deprenyl works as a free radical scavenger,[2] as will be discussed in detail later, the drug is unique in that it also modifies endogenous antioxidant enzyme activities.[3–9] Since the initial report by Knoll,[3] that chronic administration of this drug caused a drastic increase in the life span of aging male rats, at least eight studies in four different animal species have been published. Except for mice, on which negative results were observed in several earlier studies, in all other studies in rats,[10,11] hamsters,[12] and dogs,[13] significantly positive results were reported for prolongation of life spans. It is not clear, however, how deprenyl worked to prolong the life span of these animals. Although other possibilities are not excluded, our own observations on the ability of deprenyl to increase antioxidant enzyme activities in selective brain regions, such as the substantia nigra (S. nigra) and striatum (but not the hippocampus or cerebellum), appear to include one likely possibility with regard to a mechanism underlying its effect on life span.

In this review, we summarize some of the past reports on the effects of deprenyl on the life span of animals as well as on antioxidant enzymes and will discuss the possible causal relationship between these effects. Furthermore in relation to these effects, we will also discuss several pharmacological effects of this interesting drug and its future potential.

EFFECTS OF DEPRENYL ON LIFE SPANS OF DIFFERENT ANIMAL SPECIES

Rats

The first study of deprenyl's effect on life span was reported by Knoll[3] in rats. In his study, he began administering deprenyl, sc 3 times a week, at a dose of 0.25 mg/kg to male Logan-Wistar rats at the age of 24 months.[3] In his initial study, he demonstrated that the remaining life expectancy of treated male rats after 24 months of age was more than two times longer than that of saline-treated control rats. The second study[10] on male Fischer rats given the same dose of deprenyl beginning at 24 months of age, as in Knoll's study, however, resulted in only a 16% increase in the remaining life span of treated animals.[10] In the third study, the drug administration was started in male Fischer rats at the age of 18 months at a dose 0.5 mg/kg (3 times a week, sc) and revealed an increase of 34% in the remaining life expectancy after 24 months of age[11] (FIG. 1). These studies are discussed in detail in our previous reviews.[14,15] To the knowledge of the authors, no further work has been thus far published on rats.

Mice

At least two studies have been previously published.[16,17] Interestingly, neither of these earlier studies obtained a significantly positive effect of deprenyl on longevity of the animals. The results of our own unpublished work on mice using two different doses of deprenyl (0.25 mg/kg, 0.5 mg/kg, 3 times a week) also failed to obtain significantly posi-

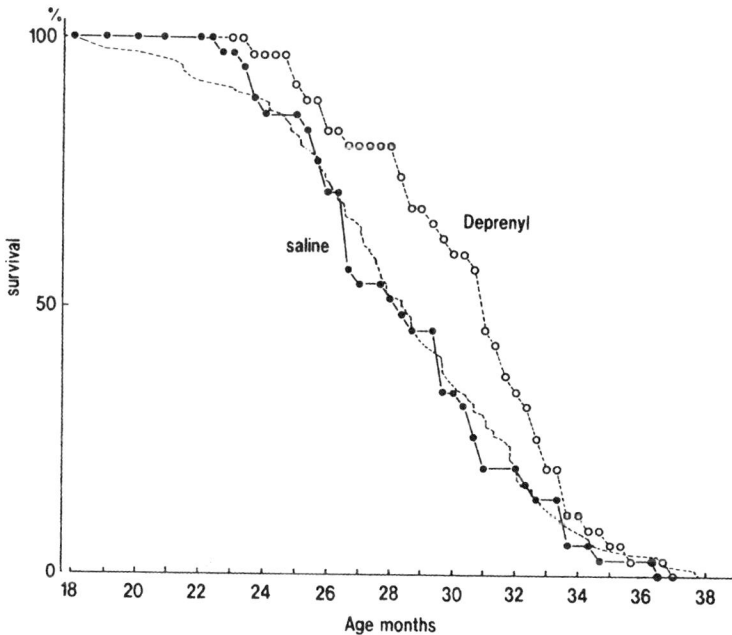

FIGURE 1. Survival curves of control (closed circles) and deprenyl-treated (open circles) rats as expressed from pooled data of three cohorts. Broken line without symbols indicates data from 100 animals raised in the specific pathogen-free farm of the institute as reported previously. (Kitani *et al.*[11] With permission from *Life Sciences*.)

tive results, although the 0.5 mg/kg dose could extend life span to some extent. Only one recent study has shown that deprenyl is also effective in significantly increasing the life span of this animal species.[18] However, mice appear to be the animal species in which a positive result is not easily obtained. For this reason our subsequent discussion will focus mainly on mice.

Hamsters

Stoll *et al.*[12] recently reported the results of their study on hamsters of both sexes (FIG. 2). Interestingly, they could demonstrate a significant effect of deprenyl on life span in female but not in male animals, although in males too, there were some positive modifications of the survival curve by the drug.

Dogs

Perhaps the most impressive result so far reported, except for the initial study by Knoll[3] on rats, is a recent study by Ruehl *et al.*[13] involving 82 beagle dogs. Although they did not see any significant difference between the life spans of young treated and control

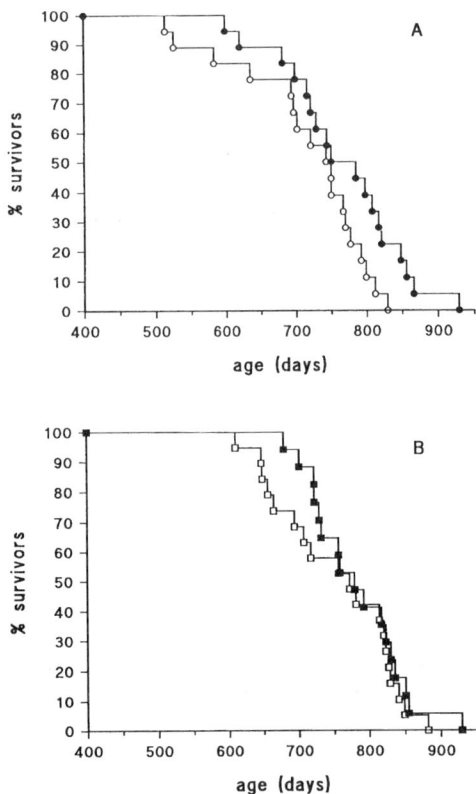

FIGURE 2. Survival curves of female (A, O, controls; ●, selegiline) and male (B, □, controls; ■, selegiline) Syrian hamsters (female control: n = 18, female deprenyl: n = 18, male control: n = 19, male deprenyl: n = 17). Selegiline was provided at a dose of 0.05 mg selegiline per kg body weight per day via the diet beginning at the age of 13 months until natural death. The slopes of the curves of the female hamsters were significantly different (p < 0.05; chi-square = 3.85; 1, DF), analyzed by the log rank test. (Stoll et al.[12] With permission from the *Neurobiology of Aging*.)

dogs, when a subset of dogs older than 10 years of age was observed, there was a significantly positive effect of deprenyl on the survivorship of animals (FIG. 3). Within 800 days of observation, dogs treated with deprenyl at a dose of 1 mg/kg (oral administration, every day) could enjoy remarkably longer survivals, as shown in FIGURE 3, in comparison with placebo-treated control dogs. Twelve of 15 deprenyl-treated dogs (80%) survived to the conclusion of the study, whereas only 7 of 18 control animals (39%) survived. It is noteworthy that a remarkable difference in pathologies between the two groups was a decrease in the incidence of mammary tumors in treated animals, which is known to be prevalent in this beagle dog strain. Indeed, although no dog treated with deprenyl died of neoplasia during the observation period, 11 elderly dogs in the control group died of different kinds of neoplasia. Among seven female dogs that died of malignancies, four died of malig-

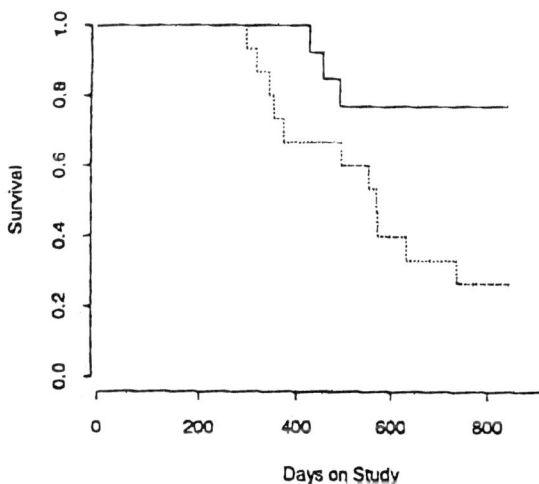

FIGURE 3. Survival of dogs between 10 and 15 years old at the start of the study and treated with deprenyl for at least 6 months. $p < 0.05$. ——, deprenyl; - - -, placebo. (Ruehl *et al.*[13] With permission from *Life Sciences*.)

nancy of mammary gland origin. We will come back to this point later, that is, the prevention of tumor occurrence by deprenyl.

EFFECTS OF DEPRENYL ON ANTIOXIDANT ENZYME ACTIVITIES

Again, Knoll was the first to point out that superoxide dismutase (SOD) activity was significantly elevated in the striatum of rats by administration of the drug for three weeks.[3] He reported, however, that catalase activities were not significantly raised. Further he showed that glutathione peroxidase (GSHPx) activities were also increased to some extent.[3] He later, however, found no significant effect of the drug on SOD activities in another rat strain and suggested that this effect of deprenyl may not always occur.[19]

SUMMARY OF OUR PAST OBSERVATIONS

Our group has worked quite extensively on the effect of deprenyl on antioxidant enzyme activities in various brain regions. Since these results have been reported elsewhere in detail,[4-9] our results will be only briefly summarized below. (1) Activities of any species of SOD (Cu, Zn-, and Mn-) can be increased by an appropriate dose of the drug, but the effect is generally greater for Mn-SOD.[4-9] (2) The effect is selective to such brain regions as S. nigra, striatum, and to some extent cerebral cortex; however, it is not selective in hippocampus, cerebellum, or the liver.[5,6] (3) To achieve an optimal increase in SOD activities, three weeks of treatment are needed, and the CAT increase is somewhat slower than that of SOD.[7] (4) Unlike the study of Knoll,[3] we found that CAT activities can be increased significantly in brain regions in which SOD activities are increased; however,

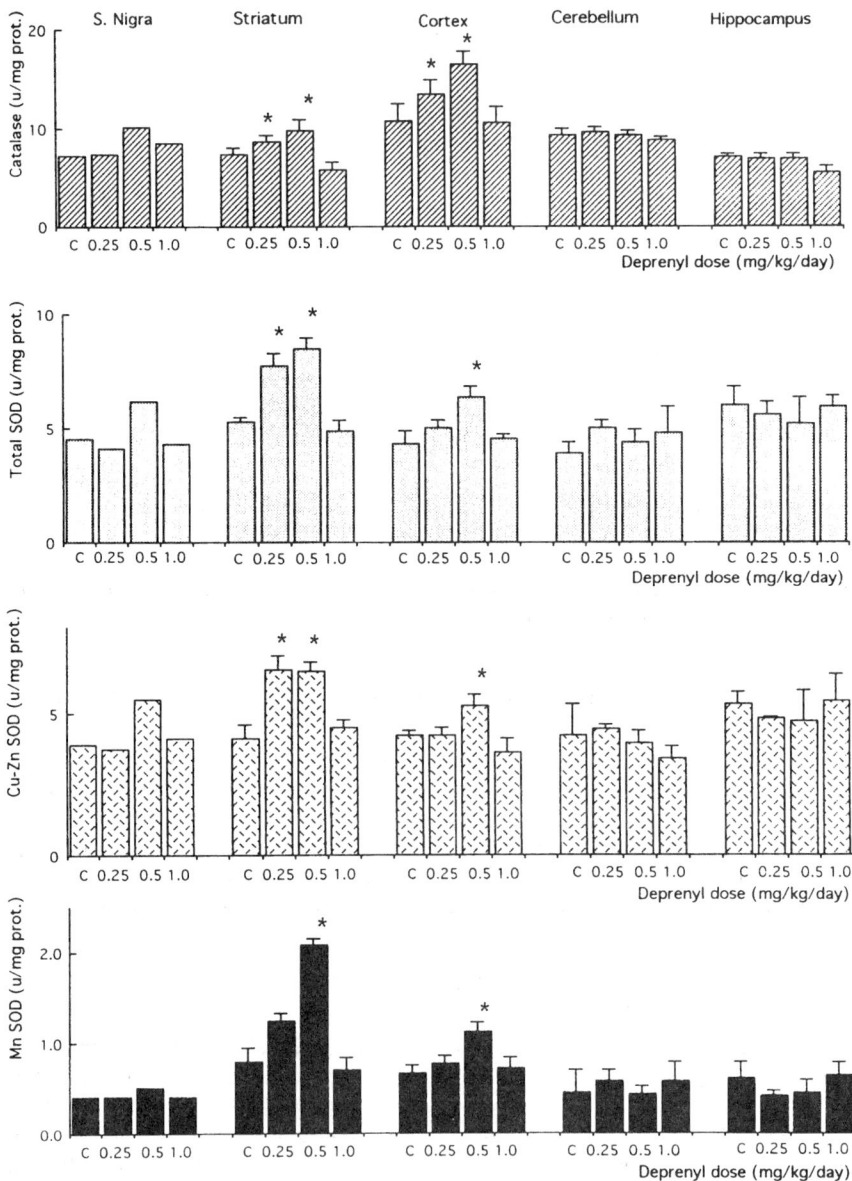

FIGURE 4. Catalase and superoxide dismutase activities in five different brain regions in old male mice treated with different doses of deprenyl for three months, starting at the age of 26 months. Doses indicated are those per injection, and sc injections were done 3 times a week for 3 months. *Significantly different from respective control values (p < 0.05, ANOVA + Scheffe's test). Number of animals in each group is 3 or 4.[22] C = control.

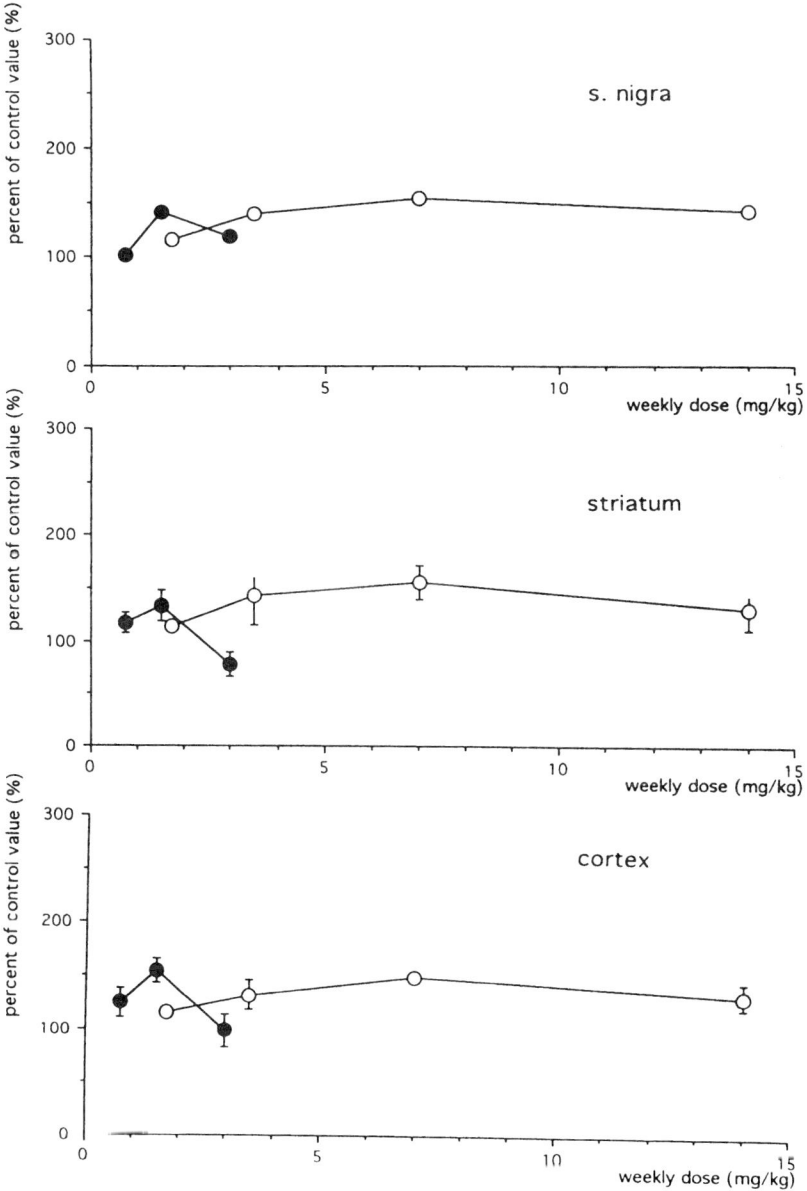

FIGURE 5. Relative catalase activities in three different brain regions in old mice treated with deprenyl for 3 weeks (open circles) or 3 months (closed circles). Enzyme activities in treated animal groups were expressed as percentages of respective values in control groups. In the short-term study,[20] deprenyl was continuously infused sc for 3 weeks (via osmotic minipump), whereas in the long-term study,[22] deprenyl was injected sc three times a week for 3 months. For the purpose of comparison, doses were recalculated as weekly doses for the two studies. (Carrillo *et al.*[22] With permission from *Life Sciences.*)

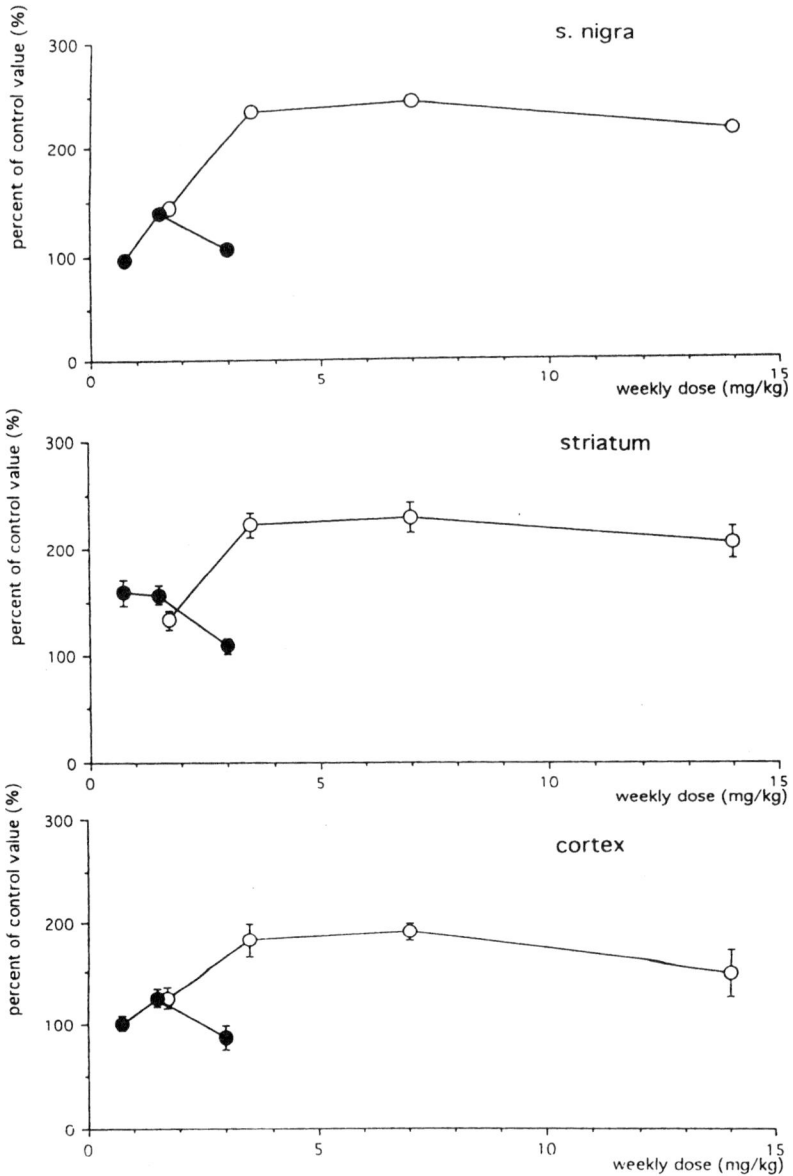

FIGURE 6. Relative Cu, Zn-SOD activities in three different brain regions in old male mice treated with deprenyl for 3 weeks (open circles) or 3 months (closed circles). Enzyme activities in treated animal groups are expressed as percentages of respective values in control groups. In the short-term study, deprenyl was continuously infused sc for 3 weeks,[20] whereas in the long-term study,[22] deprenyl was injected sc three times a week for 3 months. For the purpose of comparison, doses were recalculated as weekly doses for the two studies. (Carrillo et al.[22] With permission from Life Sciences.)

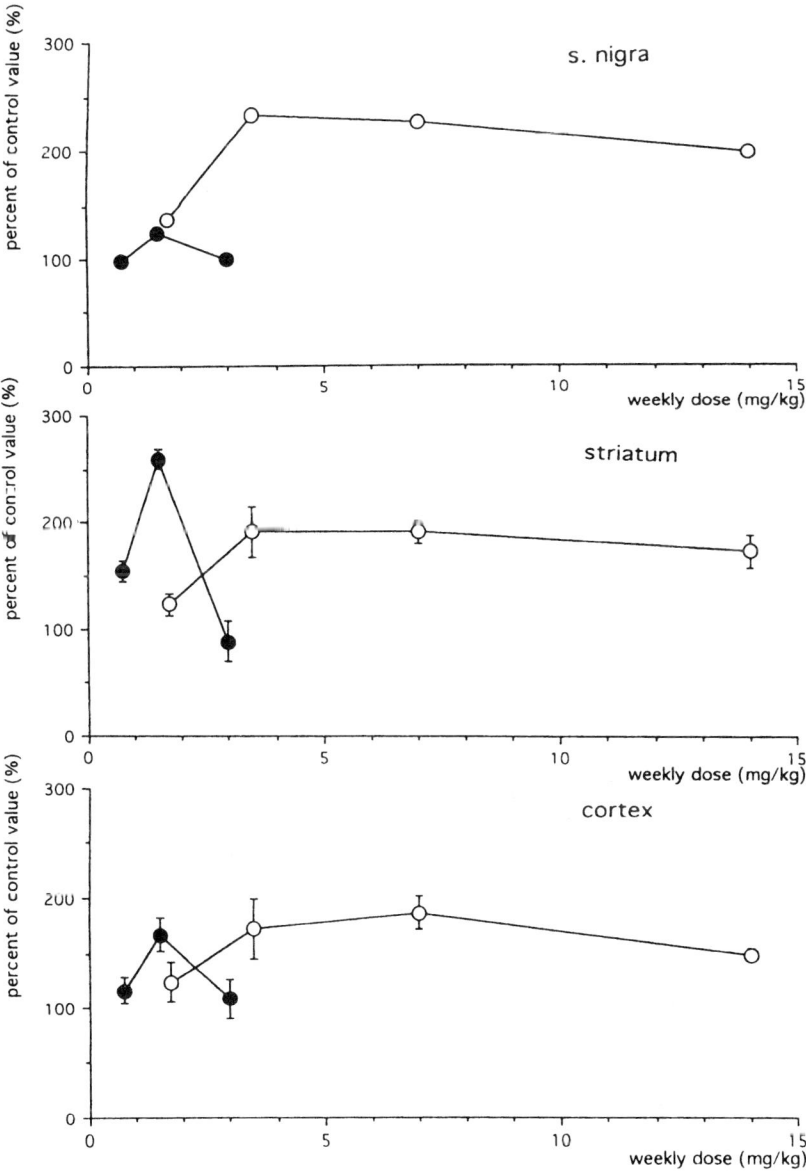

FIGURE 7. Relative Mn-SOD activities in three different brain regions in old mice treated with deprenyl for 3 weeks (open circles) or 3 months (closed circles). Enzyme activities in treated animal groups are expressed as percentages of respective values in control groups. In the short-term study,[20] deprenyl was continuously infused sc for 3 weeks, whereas in the long-term study,[22] deprenyl was injected sc three times a week for 3 months. For the purpose of comparison, doses were recalculated as weekly doses for the two studies. (Carrillo *et al.*[22] With permission from *Life Sciences*.)

there is no effect on GSHPx activities.[5,6] (5) These effects of deprenyl were demonstrated not only in rats but in mice[20] and dogs[21] as well. (6) The dose effect is remarkable. If the dose is too small, it does not cause a significant effect, and, as the dose is increased, it shows a dose-dependent increase in its effect; however, if a dose becomes excessive, the effect becomes less efficient and if it is further increased, it causes a decrease, rather than an increase in the activities, thus demonstrating a typical inverted U-shaped effect.[6,8] (7) To further complicate matters, the optimal dose is different depending on the animal species, strain, and especially the sex of rats, such that young male rats require a 10-fold higher dose (2.0 mg/kg) in comparison to young female rats (0.2 mg/kg) in order to demonstrate an optimal increase in antioxidant enzymes.[6,8] This effect may not hold for other species due to the specifics of the rat liver P450 system. (8) It should be emphasized that the age of animals is also a key factor in determining an optimal dose. In male rats, old rats require a lower dose to have an optimal effect than do young rats; however, in females, age has an opposite effect, resulting in an increase in the optimal dose with aging. Several possible causes for the variability of an optimal dose by these factors have been discussed previously.[6,8,14,15] (9) Our recent study on mice[22] has found that the duration of treatment is also an important factor, at least in this animal species. Thus, we will discuss this point in more detail.

Although a significant increase in SOD and CAT activities was previously demonstrated by us with a short-term treatment of three weeks in mice,[20] we wanted to know the effects of long-term treatment, as this would be crucial for performing a life span study.

FIGURE 4 summarizes the effects of a long-term (3 months) treatment with deprenyl in aging C57BL male mice, which began receiving injections of the drug at the age of 26 months (sc, 3 times a week). We have found that deprenyl significantly increases SOD and CAT activities; however, the effective dose range and the magnitude of increase in enzyme activities were much smaller than in a short-term treatment for three weeks.

FIGURES 5–7 summarize the comparison of the effects for a short-term (3 weeks) and a long-term (3 months) treatment with the drug. In the short-term treatment, we administered the drug continuously for 21 days, whereas in the long-term study we injected the drug three times a week. The comparison of the two studies was made on a weekly dose basis. It is apparent that long-term treatment caused (1) a narrowing of the effective dose range for both SOD and CAT activities; (2) a decrease in the magnitude of increase of various enzyme activities of SOD, with the exception of the Mn-SOD in striatum; and (3) a decrease in the optimal dose of deprenyl by a factor of 5 to 10. Although a similar tendency can be seen in rats too, especially for a decrease in the optimal dose, the narrowing of the effective dose range was not as remarkable in rats as it was in mice.[9]

A POSSIBLE CAUSAL RELATIONSHIP BETWEEN THE TWO EFFECTS OF DEPRENYL

Deprenyl is primarily a monoamine oxidase B (MAO B) inhibitor.[23] Further, numerous pharmacological effects have been reported for this drug. Some of these are summarized in TABLE 1. It should be emphasized that some of the effects itemized in this TABLE may be interrelated and may therefore not be independent. However, the prolongation of the average life span (in at least two studies, the maximum survivals were also extended[3,10]), which up to now has been demonstrated in at least four different animal species (rats,

TABLE 1. Pharmacology of Deprenyl

1. MAO B inhibition[19]
2. Antidepressant effect[23]
3. Increase in antioxidants (SOD, CAT, GSH)[3–9]
4. Catecholaminergic activity enhancer[19]
Increase of noradrenaline release
Inhibition of reuptake of dopamine
Recovery of sexual capability of old male rats
5. Life span extension[3,10–13,18]
6. Neuroprotective effect[34]
7. Radical scavenging effect[2]
8. Immunomodulation[25]

mice, hamsters, and dogs) may be the result of complex pharmacological effects of the drug and is not easily understood.

The physiological significance of increase in SOD and CAT enzyme activities in selective brain regions has remained totally unclear until recently. Some have suggested that it may have been the mere result of an oxygen crisis, possibly caused by the administration of the drug, and thus may not have had any beneficial effect on the organism, although we have insisted that such a possibility is very unlikely.[7,14,15] A recent study reported by Knollema *et al.*[24] has provided an answer to this question. After deprenyl administration for three weeks in male F-344 rats, animals were exposed to an acute hypoxia–ischemia episode followed by reperfusion. Twenty-four hours after the hypoxia treatment, animals were sacrificed, and a histological examination was done in several brain regions. They found that the previous deprenyl pretreatment had a definitive effect in preventing the neuronal damage caused by hypoxia and reperfusion in selective brain regions, such as the S. nigra and striatum, but not in the hippocampus. It is interesting that the deprenyl treatment also prevented neuronal damage in the thalamus. Although they did not examine activities of SOD and CAT, the selectivity of its effect on the dopaminergic system suggests that it is most likely that these preventive effects were due to the increase in SOD and CAT enzyme activities by deprenyl pretreatment. Although such an effect of deprenyl may have limited value in the treatment of stroke (because it requires an advance treatment with the drug prior to a stroke episode), regarding the longevity of animals, this work may have a significant impact, inasmuch as the brain dopaminergic system, especially S. nigra and striatum, is believed to be exposed to perpetual oxidative stress during aging. Thus deprenyl treatment may also be effective in preventing tissue damage due to chronic oxidative stress during aging.

HOW CAN THE PROTECTION OF THE DOPAMINERGIC SYSTEM AGAINST CHRONIC OXIDATIVE DAMAGE PROLONG THE LIFE SPAN OF ANIMALS?

The discussion herein is very speculative, with no direct proof. However, we want to raise the possibility of one mechanism of the action of deprenyl. The brain dopaminergic system is known to maintain such fundamental functions of the organism as locomotion

and reproduction. It has also been suggested that the dopaminergic system regulates many
humoral factors, such as TNF and growth factor. Also, there are some suggestions that
deprenyl is involved with the regulation of several interleukins.[25] Thus, once the dopaminergic system is better preserved during aging by deprenyl treatment, it may have a significant impact on the functions of the organism. For example, the dopaminergic system may
work to prevent the development of tumors, possibly by releasing TNF and other antitum-
origenic factors.

In our previous study on the life span of F-344 male rats, we observed one interesting
phenomenon, that is, the smaller variability of body weight of deprenyl-treated rats during
aging than in control rats, despite quite comparable average body weights throughout the
observation period (FIG. 8).[11] Interestingly, a similar tendency can be seen in a previous
Canadian study using the same strain of rat.[10] This strain of rat (Fischer 344) is notorious
for developing many kinds of subcutaneous tumors late in life. Although they are mostly
benign in nature, they grow so big that sometimes their weight becomes greater than the
rest of the rat body, resulting in a tremendous increase in the apparent body weight. On the
other hand, this strain is known to lose body weight after 24 months of age due to emaciation. These two opposing factors cause a continuous increase in the variation of body
weight of animals as they become older (FIG. 8). Thus the smaller variation in body weight
in deprenyl-treated rats with a comparable average body weight to control rats may mean a
slower decrease of body weight with age due to emaciation and the slower development of

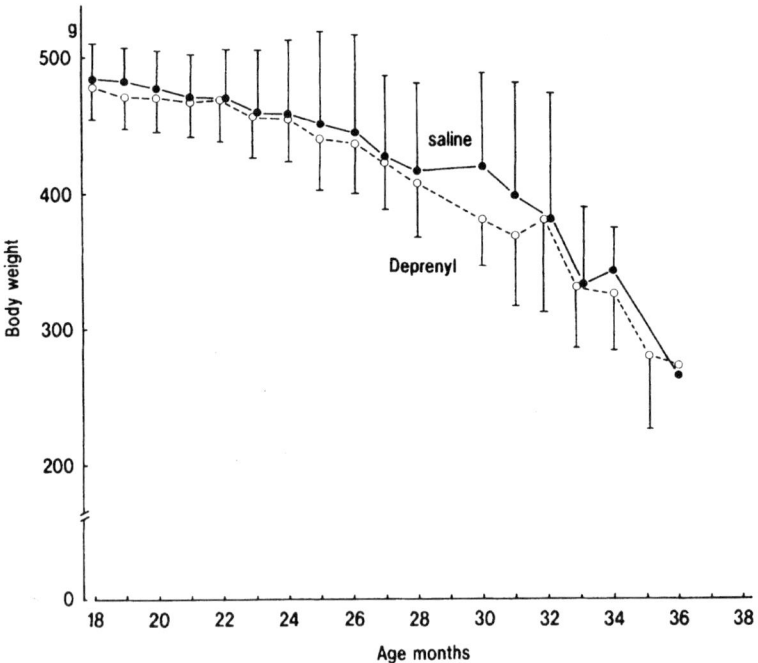

FIGURE 8. Sequential changes in body weights of control (closed circles) and deprenyl-treated
(open circles) rats. Despite comparable average body weights for both groups, the variation in body
weight in the deprenyl-treated group is constantly smaller than that in the control group. Vertical
bars indicate 1 SD. (Kitani et al.[11] With permission from *Life Sciences*.)

tumors during aging in deprenyl-treated animals. We have previously suggested that one factor that caused a longer life span in deprenyl-treated F-344 rats would be the prevention or slowing of the occurrence (and development) of subcutaneous tumors, which can be an indirect cause of the death of animals.[7,11,15]

The data reported by Ruehl *et al.*[13] strongly support our previous contention that in some way deprenyl treatment can prevent tumor growth in the organism. It remains totally unknown whether this effect of deprenyl is the sole mechanism for the prolongation of the life span of animals. In the study of Stoll *et al.* on hamsters,[12] they demonstrated a significant prolongation of the averge life span despite comparable body weights. However, assuming that the increase in SOD and CAT activities is at least a partial cause for the prolongation of the life span of animals, we have a very reasonable explanation for the fact that in mice we have seen thus far the greatest difficulty in significantly prolonging their life span by the drug, inasmuch as we have shown that it is extremely difficult to maintain the higher SOD and CAT activities for a long period with deprenyl in this animal species. Although a 0.5 mg/kg (3 times a week) dose was shown to be effective in maintaining the SOD and CAT activities at levels higher than control mice,[22] it remains totally unknown how long deprenyl can maintain its effect when administration is continued for a longer time. Further, dosage could also explain discrepant results reported by Stoll *et al.*[17] between male and female hamsters, because it is quite possible that the dose used in their study was not equally effective regarding the elevation of SOD and CAT activities between the two sexes.

Other indirect evidence that the increase in SOD and CAT activities may be a causal factor for the life span prolongation of animals by deprenyl is that when the dose of 1.0 mg/kg (3 times a week) was administered in aging F-344 rats, after 13 months of treatment, the treated groups had shorter life spans than did control rats, and there was no effect of deprenyl on SOD and CAT activities in animals that survived, despite the fact that this dose was quite effective in increasing these activities, at least for one month of treatment.[26]

FUTURE PERSPECTIVES

The attempts to prolong the life span of animals by means of administration of pharmaceuticals and nutrients will very likely continue. For so-called antioxidant chemicals, which have a direct radical scavenging effect, the pharmacokinetics should be the key factor for the drug's success.[14] Deprenyl has also been claimed to be effective as a radical scavenger.[2] At present, there is no way to exclude the possibility that deprenyl's effect on the life span of animals is due to its direct effect as a radical scavenger. However, we believe that another possibility, as discussed here, is that deprenyl may be modifying the life span of animals by indirectly modulating endogenous antioxidant defense mechanisms. This, then, provides another approach to pharmacological intervention in aging and age-associated pathologies, that is, the modification of endogenous antioxidant defenses. In this context, deprenyl may be a prototype of drugs that possess such properties.

Finally, the clinical significance of deprenyl (and/or its analogues) may need some discussion. In terms of the life-prolonging effect of the drug in humans, we do not as yet have any solid optimistic evidence. Although the initial retrospecive study by Birkmayer *et al.*[27]

revealed a significantly prolonged remaining life expectancy in Parkinson's disease patients treated with deprenyl and levodopa in comparison with patients treated with levodopa only, to the knowledge of the authors, no well-conrolled study in patients has ever confirmed the significant effect of the drug regarding the life expectancy of subjects. Rather, one study from the United Kingdom[28] has reported a shorter remaining life span as well as a lesser effect of the drug in patients with Parkinson's disease. However, as far as the effect of the drug is concerned, this is the only study that reported a negative effect of the drug on the progression of Parkinson's disease, which is in contrast with a number of other well-controlled studies on Parkinson's disease[29–31] that all confirmed a beneficial effect of the drug on the course of this disease. The discrepancy between the study from the UK[28] and others[29–31] needs to be clarified in the future. However, it is the belief of the authors that the result of a single study from Britain should not be taken as evidence, for various reasons, against the further clinical trials of this most interesting drug.

The mechanisms(s) of how the drug counteracts the progression of Parkinson's disease must be carefully reconsidered. The primary idea that the drug inhibits MAO B oxidation, resulting in the decrease in the formation of such toxic dopamine metabolite(s) as 6-hydroxydopamine, so that the oxidative damage of striatal neurons is saved by deprenyl in Parkinson's disease, was originally a plausible one, inasmuch as MPTP induced parkinsonian-like disorders in experimental animals and this effect was efficiently prevented by deprenyl pretreatment.[32] However, the discovery that the drug can prevent disorders caused by MPP^+ (an oxidative metabolite of MPTP)[2,33] has blown up the initial idea. Several other suggestions are now being considered regarding the effect of the drug on Parkinson's disease. (1) There appears to be a radical scavenging effect. Wu et al.[2] have shown that the drug can effectively reduce the formation of salicylate adducts caused by infusion of MPP^+, suggesting that the drug may have a radical scavenging effect. This mechanism may result in the prevention of the loss of striatal neurons in Parkinson's disease. (2) Salo and Tatton have repeatedly shown that the drug has a peculiar neuroprotective effect,[34] which also may work to protect the deterioration of the striatal system in Parkinson's disease. (3) Modifications (upregulation) of SOD and CAT in the dopaminergic system, including the striatum, as discussed in this article may also have to do with the significant effect of deprenyl for Parkinson's disease.

Up to now, there is convinving evidence that the drug is not effective for amyotrophic lateral sclerosis (ALS).[35] By contrast, many arguments are now ongoing regarding its efficacy in Alzheimer's disease (AD). There have been many clinical trials on its efficacy in AD. Although some studies have reported a beneficial effect of the drug, at least on the symptomatology of AD, some other reports have argued against its effect. Only recently, one carefully controlled double-blind prospective study has reported a significant effect of the drug on the progression of AD, especially on its functional loss.[36] This study is also unique in its affirmative results on a significant effect of α-tocopherol on the progression of AD. It remains to be clarified whether these two drugs worked by means of the same mechanism(s), possibly as a direct radical scavenger, or by some other means.

Regardless of the underlying mechanism(s) of these drugs, it does not seem too optimistic to suggest that we appear to be having a clue for real pharmacological intervention in aging and age-associated disorders, which thus encourages our future endeavors in this kind of approach.

ACKNOWLEDGMENT

The skillful secretarial work of Mrs. T. Ohara is gratefully acknowledged.

REFERENCES

1. HARMAN, D. 1994. Free-radical theory of aging. Increasing the functional life span. *In* Pharmacology of Aging Processes. Methods of Assessment and Potential Interventions. I. Zs.-Nagy, D. Harman & K. Kitani, Eds.: **717:** 1–15. Annals of the New York Academy of Sciences. New York.
2. WU, R.M., C.C. CHIUEH, A. PERT & D.L. MURPHY. 1993. Apparent antioxidant effect of L-deprenyl on hydroxyl radical formation and nigral injury elicited by MPP^+ *in vivo*. Eur. J. Pharmacol. **243:** 241–247.
3. KNOLL, J. 1988. The striatal dopamine dependency of life span in male rats: Longevity study with (-)deprenyl. Mech. Ageing Dev. **46:** 237–262.
4. CARRILLO, M.C., S. KANAI, M. NOKUBO & K. KITANI. 1991. (-)Deprenyl induces activities of both superoxide dismutase and catalase but not of glutathione peroxidase in the striatum of young male rats. Life Sci. **48:** 517–521.
5. CARRILLO, M.C., K. KITANI, S. KANAI, Y. SATO & G.O. IVY. 1992. The ability of (–)deprenyl to increase superoxide dismutase activities in the rat is tissue and brain region selective. Life Sci. **50:** 1985–1992.
6. CARRILLO, M.C., S. KANAI, M. NOKUBO, G.O. IVY, Y. SATO & K. KITANI. 1992. (–)Deprenyl increases activities of superoxide dismutase and catalase in striatum but not in hippocampus: The sex and age-related differences in the optimal dose in the rat. Exp. Neurol. **116:** 286–294.
7. CARRILLO, M.C., S. KANAI, Y. SATO, G.O. IVY & K. KITANI. 1992. Sequential changes in activities of superoxide dismutase and catalase in brain regions and liver during (–)deprenyl infusion in male rats. Biochem. Pharmacol. **44:** 2185–2189.
8. CARRILLO, M.C., S. KANAI, Y. SATO, M. NOKUYBO, G.O. IVY & K. KITANI. 1993. The optimal dosage of (–)deprenyl for increasing superoxide dismutase activities in several brain regions decreases with age in male Fischer 344 rats. Life Sci. **52:** 1925–1934.
9. CARRILLO, M.C., K. KITANI, S. KANAI, Y. SATO, K. MIYASAKA & G.O. IVY. 1994. The effect of a long term (6 months) treatment of (-)deprenyl on antioxidant enzyme activities in selective brain regions in old female Fischer 344 rats. Biochem. Pharmacol. **47:** 1333–1338.
10. MILGRAM, N.W., R.J. RACINE, P. NELLIS, A. MENDONCA & G.O. IVY. 1990. Maintenance of L-deprenyl prolongs life in aged male rats. Life Sci. **47:** 415–420.
11. KITANI, K., S. KANAI, Y. SATO, M. OHTA, G.O. IVY & M.C. CARRILLO. 1993. Chronic treatment of (–)deprenyl prolongs the life span of male Fischer 344 rats: Further evidence. Life Sci. **52:** 281–288.
12. STOLL, S., U. HAFNER, B. KRÄNZLIN & W.E. MÜLLER. 1997. Chronic treatment of Syrian hamsters with low-dose selegiline increases life span in females but not males. Neurobiol. Aging **18:** 205–211.
13. RUEHL, W.W., T.L. ENTRIKEN, B.A. MUGGENBURG, D.S. BRUYETTE, W.G. GRIFFITH & F.F. HAHN. 1997. Treatment with L-deprenyl prolongs life in elderly dogs. Life Sci. **61:** 1037–1044
14. KITANI, K., S. KANAI, M.C. CARRILLO & G.O. IVY. 1994. (–)Deprenyl increases the life span as well as activities of superoxide dismutase and catalase but not of glutathione peroxidase in selective brain regions in Fischer rats. *In* Pharmacology of Aging Processes. Methods of Assessment and Potential Interventions. I. Zs.-Nagy, D. Harman & K. Kitani, Eds.: **717:** 60–71. Annals of the New York Academy of Sciences. New York.
15. KITANI, K., K. MIYASAKA, S. KANAI, M.C. CARRILLO & G.O. IVY. 1996. Upregulation of antioxidant enzyme activities by deprenyl: Implications for life span extension. *In* Pharmacological Intervention in Aging and Age-Associated Disorders. K. Kitani, A. Aoba & S. Goto, Eds.: **726:** 391–409. Annals of the New York Academy of Sciences. New York.
16. INGRAM, D.K., H.L. WIENER, M.E. CHACHICH, J.M. LONG, J. HENGEMIHLE & M. GUPTA. 1993. Chronic treatment of aged mice with L-deprenyl produces marked MAO-B inhibition but no

beneficial effects on survival, motor performance, or nigral lipofuscin accumulation. Neurobiol. Aging **14:** 431–440.

17. PIANTANELLI, L., A. ZAIA, G. ROSSOLINI, C. VITICECHI, R. TESTA, A. BASSO & A. ANTOGNINI. 1994. Influence of L-deprenyl treatment on mouse survival kinetics. *In* Pharmacology of Aging Processes. Methods of Assessment and Potential Interventions. I. Zs.-Nagy, D. Harman & K. Kitani, Eds.: **717:** 72–78. Annals of the New York Academy of Sciences. New York.

18. ARCHER, J.R. & D.E. HARRISON. 1996. L-Deprenyl treatment in aged mice slightly increases life spans, and greatly reduces fecundity by aged males. J. Gerontol. Biol. Sci. **31A:** B448–B453.

19. KNOLL, J. 1990. Advances in Neurology. Vol. 53. Parkinson's Disease: Anatomy, Pathology and Therapy. 425–429. Raven Press. New York.

20. CARRILLO, M.C., K. KITANI, S. KANAI, Y. SATO, K. MIYASAKA & G.O. IVY. 1994. (–)Deprenyl increases activities of superoxide dismutase and catalase in certain brain regions in old male mice. Life Sci. **54:** 975–981.

21. CARRILLO, M.C., N.W. MILGRAM, P. WU, G.O. IVY & K. KITANI. 1994. (–)Deprenyl increases activities of superoxide dismutase (SOD) in striatum of dog brain. Life Sci. **54:** 1483–1489.

22. CARRILLO, M.C., K. KITANI, S. KANAI, Y. SATO, G.O. IVY & K. MIYASAKA. 1996. Long term treatment with (–)deprenyl reduces the optimal dose as well as the effective dose range for increasing antioxidant enzyme activities in old mouse brain. Life Sci. **59:** 1047–1057.

23. KNOLL, J., Z. ECSERI, K. KELEMEN, J. NIEVEL & B. KNOLL. 1965. Phenylisopropylmethylpropinylamine (E-250), a new spectrum psychic energizer. Arch. Int. Phamacodyn. Ther. **155:** 154–164.

24. KNOLLEMA, S., W. AUKEMA, H. HOM, J. KORF & G.J.T. HORST. 1995. L-Deprenyl reduces brain damage in rats exposed to transient hypoxia-ischemia. Stroke **26:** 1883–1887.

25. RUEHL, W.W., D. BICE, B. MUGGENBURG, D.D. BRUYETTE & D.R. STEVENS. 1994. L-Deprenyl and canine longevity: Evidence for an immune mechanism, and implications for human aging. 2nd Conference on Anti-Aging Medicine, Las Vegas. Abstract p. 25–26.

26. KITANI, K., M.C. CARRILLO, S. KANAI, Y. SATO & G.O. IVY. 1993. Chronic treatment of (–)deprenyl abolishes the dose effect on the superoxide dismutase (SOD) and catalase (CAT) in striatum in the rat. The XVth Congress of the International Association of Gerontology. Abstract p. 164.

27. BIRKMAYER, W., J. KNOLL, M.B.H. RIEDER, V.H. YOUDIM & J. MARTON. 1985. Improvement of life expectancy due to L-deprenyl addition to Madopar treatment in Parkinson's disease: A long-term study. J. Neural. Transm. **64:** 113–127.

28. LEES, A.J. 1995. Comparison of therapeutic effects and mortality data of levodopa and levodopa combined with selegiline in patients with early, mild Parkinson's disease. Br. Med. J. **311:** 1602–1607.

29. THE PARKINSON STUDY GROUP. 1989. Effect of deprenyl on the progression of disability in early Parkinson's disease. N. Engl. J. Med. **321:** 1364–1371.

30. TETRUD, J.W. & J.W. LANGSTON. 1989. The effect of deprenyl (selegiline) on the natural history of Parkinson's disease. Science **245:** 519–522.

31. CALZOTTI, S., A. NEGRATTI & A. CASSIO. 1995. L-Deprenyl as an adjunct to low-dose bromocriptine in early Parkinson's disease: A short-term, double blind, and prospective study. Clin. Neuropharmacol. **18:** 250–257.

32. COHEN, G., P. PASIK, B. COHEN, A. LEIST, C.Y. MYTILINEOU & M.D. YAHR. 1984. Pargyline and (–)deprenyl prevent the neurotoxicity of 1-methyl-4-phenyl-1,2,3,6-tetrahydropyridine (MPTP) in monkeys. Eur. J. Pharmacol. **106:** 209–210.

33. VIZUETC, M.L., V. STEFFEN, A. AYALA, J. CANO & A. MACHADO. 1993. Protective effect of deprenyl against 1-methyl-4-phenylpyridinium neurotoxicity in rat striatum. Neurosci. Lett. **152(1–2):** 113–116.

34. SALO, P.T. & W.G. TATTON. 1992. Deprenyl reduces the death of motoneurons caused by axotomy. J. Neurosci. Res. **31:** 394–400.

35. MAZZINI, L., D. TESTA, C. BALZARINI & G. MORA. 1994. An open-randomized clinical trial of selegiline in amyotrophic lateral sclerosis. J. Neurol. **241:** 223–227.

36. SANO, M., C. ERNERSTO, R.G. THOMAS, M.R. KLAUBER, K. SCHAFFER, M. GRUNDMAN, H.P. WOODBURY, J. GROUDON, C.W. COTMAN, E. PFEIFFER, L.S. SCHNEIDER & L.J. THAL. 1997. A controlled trial of selegiline, alpha-tocopherol, or both as treatment for Alzheimer's disease. N. Engl. J. Med. **336:** 1216–1222.

Age-associated Memory Impairment

Assessing the Role of Nitric Oxide

ROBERT C. MEYER,[a] EDWARD L. SPANGLER,[a] HIDEKI KAMETANI,[b] AND DONALD K. INGRAM[a,c]

[a]*Molecular Physiology and Genetics Section, Nathan W. Shock Laboratories, Gerontology Research Center,[d] National Institute on Aging, National Institutes of Health, Johns Hopkins Bayview Campus, 5600 Nathan Shock Lane, Baltimore, Maryland 21224, USA*

[b]*Department of Psychology, Fukuoka Prefectural, University, Tagawa, Fukuoka, Japan*

ABSTRACT: Several neurotransmitter systems have been investigated to assess hypothesized mechanisms underlying the decline in recent memory abilities in normal aging and in Alzheimer's disease. Examining the performance ot F344 rats in a 14-unit T-maze (Stone maze), we have focused on the muscarinic cholinergic (mACh) and the *N*-methyl-D-aspartate (NMDA) glutamate (Glu) systems and their interactions. Maze learning is impaired by antagonists to mACh or NMDA receptors. We have also shown that stimulation of mACh receptors can overcome a maze learning deficit induced by NMDA blockade, and stimulation of the NMDA receptor can overcome a similar blockade of mACh receptors. No consistent evidence in rats has been produced from our laboratory to reveal significant age-related declines in mACh or NMDA receptor binding in the hippocampus (HC), a brain region that is greatly involved in processing of recent memory. Thus, we have directed attention to the possibility of a common signal transduction pathway, the nitric oxide (NO) system. Activated by calcium influx through the NMDA receptor, NO is hypothesized to be a retrograde messenger that enhances presynaptic Glu release. Maze learning can be impaired by inhibiting the synthetic enzyme for NO, nitric oxide synthase (NOS), or enhanced by stimulating NO release. However, we have found no age-related loss of NOS-containing HC neurons or fibers in rats. Additionally, other laboratories have reported no evidence of an age-related loss of HC NOS activity. In a microdialysis study we have found preliminary evidence of reduced NO production following NMDA stimulation. We are currently working to identify the parameters of this phenomenon as well as testing various strategies for safely stimulating the NO system to improve memory function in aged rats.

Impaired ability to form new memories is a cardinal symptom of early Alzheimer's disease[1] and also occurs in normal aging, but to a much lesser degree.[2] An underlying assumption of this observation is that the memory impairment observed in Alzheimer's disease and advanced aging involves similar mechanisms.

The search for the neurochemical mechanisms of geriatric memory impairment has focused on several neurotransmitter systems, including norepinephrine,[3] serotonin,[4] opioid,[5] gamma-amino butyric acid,[6] acetylcholine,[7] and glutamate[8] systems. By far, the cholinergic hypothesis of geriatric memory impairment[7] has garnered the most research interest. Following several years of failed attempts to produce effective cholinergic ago-

[c]To whom all correspondence should be addressed. Tel: 410/558-8180; fax: 410/558-8323; e-mail: doni@vax.grc.nia.nih.gov

[d]Accredited by the American Association for the Accreditation of Laboratory Animal Care.

nists for clinical use,[9,10] this effort has produced two drugs, tacrine[11] (Cognex) and E2020[12] (Aricept), for treating memory dysfunction in patients with Alzheimer's disease.

The glutamate hypothesis of geriatric memory impairment[8] has generated research to a lesser degree than the cholinergic hypothesis, but it remains a key hypothesis.[13] A major reason for the interest in the glutamate system goes beyond its involvement in memory formation, due largely to its role in excitotoxicity.[14] Specifically, excess calcium entry through the glutamate receptor, identified as the N-methyl-D-aspartate (NMDA) gated cationic channel has been linked to neurodegeneration associated with ischemic stroke and other trauma.[15] Thus, an important issue regarding the development of therapies aimed at this neurotransmitter system is to avoid overstimulation resulting in excitotoxicity and neuronal death.

MAZE LEARNING AS A MODEL

Our laboratory has conducted studies using rodent models to identify specific neurotransmitter systems involved in age-related memory impairment and to test possible treatment strategies.[16–18] To this end, a 14-unit T-maze (FIG. 1) has been used to evaluate various hypotheses. This behavioral paradigm, also known as the Stone maze, has provided robust evidence of age-related decline in memory acquisition in a wide range of rodent species.

FIGURE 1. Configuration of 14-unit T-maze. Arrows denote correct pathway. Errors are scored as any deviation from this pathway. G, goal area; → S, start area; ——, guillotine door; _ _, false guillotine door.

Training is conducted in two phases. The first phase consists of pretraining in a straight runway. The response contingency for the animal is to learn to move to a goal box to avoid the onset of a mild foot shock. After reaching a criterion of successful avoidances of foot shock, the animal is then provided 15–20 training trials in the 14-unit T-maze. The response contingency in the maze remains the same as in pretraining, that is, the animal must successfully negotiate five segments of the maze (each within 10 s) to avoid the onset of foot shock enroute to the goal box. As the maze is equipped with a series of infrared photosensors, errors can be scored automatically as deviations from the correct pathway.

CHOLINERGIC HYPOTHESIS

Involvement of cholinergic systems for accurate performance in the 14-unit T-maze has been demonstrated.[18–21] Moreover, this effort has lead to the development of a potent, long-acting, and highly specific acetylcholinesterase inhibitor, phenserine.[22,23] Chronic treatment with this indirect cholinergic agonist has significantly improved the performance of aged rats in the 14 unit T maze.[24]

GLUTAMATE HYPOTHESIS

Despite the recent emerging success of other studies supporting the cholinergic hypothesis, it remains important to consider the involvement of other neurotransmitter systems. Such an approach recognizes that normal memory processing involves diverse neurotransmitter systems and that Alzheimer's disease is also characterized by deficits in multiple neurotransmitter systems.[25] Thus, we have also directed attention to the NMDA glutamate system and to a particular signal transduction event that operates through this system, specifically, the action of nitric oxide (NO). As depicted in FIGURE 2, calcium entry through the NMDA receptor-gated cationic channel stimulates calmodulin, which in turn modulates production of NO from arginine via the synthetic enzyme, nitric oxide synthase (NOS). As a rapidly diffusing gaseous radical, NO serves as a retrograde neurotransmitter reacting in the presynaptic terminal with heme moieties of guanyl cyclase to stimulate presynaptic glutamate release via the action of cGMP.[26] Although several studies have questioned the involvement of NO in acquisition of new memories,[27,28] many studies have supported its role in memory processing[29–32] as well as in the generation of long-term potentiation (LTP) as evidence of modified synaptic transmission.[33,34]

Involvement of the NMDA receptor in learning has been shown in several previous studies using the 14 unit T-maze. A number of sites on the receptor can be targeted to block calcium entry through the cationic channel or to stimulate calcium entry. A learning impairment can be induced by intraperitoneal (ip) injection of (±)-3-(2-carboxypiperazine-4-yl)-propyl-1-phosphonic acid (CPP, 9 mg/kg), which is a competitive antagonist of the glutamate binding site on the NMDA receptor.[35] As depicted in FIGURE 3, male Fischer-344 (F344) rats (3–4 mo old) given CPP show a higher number of errors compared to saline-injected controls. The CPP-induced impairment can be attenuated by stimulating either of two modulatory sites, the glycine site or the polyamine site. The glycine receptor agonist, D-cycloserine (30–40 mg/kg ip), can clearly improve learning in CPP-treated rats. In rodent models and in patients with Alzheimer's disease, this drug and other glycine

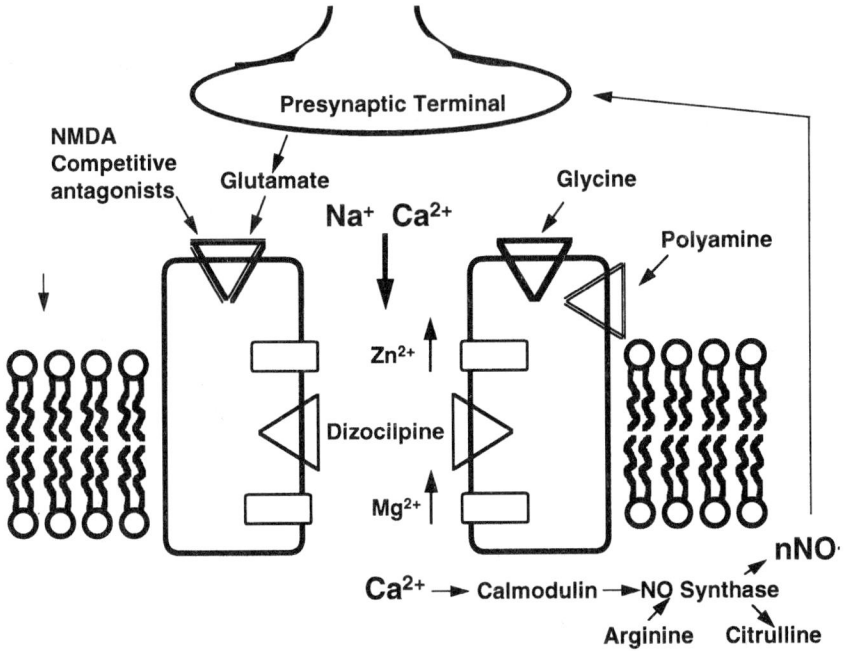

FIGURE 2. Schematic model of the NMDA receptor–ion channel complex, indicating cation influx through the channel, various binding domains, and stimulation of nitric oxide synthesis.

agonists have demonstrated efficacy for improving performance in memory tasks.[36–39] Also evident in FIGURE 3 is that stimulation of the polyamine site by spermine (2.5–5 mg/kg ip) can attenuate the CPP-induced learning impairment. In a preliminary study of aged rats in the 14-unit T-maze, we demonstrated that systemic treatment (ip) with another polyamine agonist, spermidine, could improve performance.[40] Thus, the polyamine site clearly is capable of modulating NMDA receptor activity.

In previous studies, a compound that acts as an antagonist within the NMDA calcium channel, dizocilpine, was also shown to impair learning of young F344 rats in the 14-unit T-maze.[41] Moreover, as measured by an increased number of maze errors, aged F344 rats (24 mo) were more sensitive to the antagonist effects of dizocilpine compared to young rats (3 mo).[42] This observation suggested that aged rats might have reduced concentrations of NMDA receptors, which would make them more sensitive to the antagonistic effects of dizocilpine. Examination of hippocampal [^{3}H]glutamate binding supported this hypothesis.[42] Compared to young (3 mo) counterparts, aged rats (24 mo) showed reduced binding presumably to NMDA receptors. The results of several follow-up studies using more specific ligands for the NMDA receptor, including dizocilpine, did not support this conclusion.[43,44] Furthermore, a recent review of the literature indicated a lack of consistent support for the hypothesis of a loss of hippocampal binding sites during normal aging in both rodents and humans (including the brains of Alzheimer's disease patients).[45]

If a decline in receptor binding sites does not underlie reduced maze learning in aged F344 rats, then other mechanisms would have to be considered. Impaired glycine or

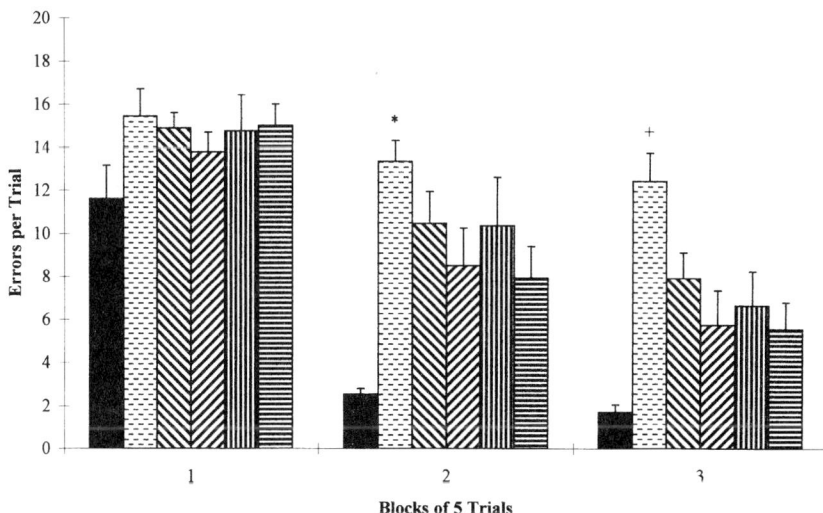

FIGURE 3. Mean errors (SEM) made by young F344 rats in the 14-unit T-maze following treatment with saline, (±)-3-(2-carboxypiperazine-4-yl)-propyl-1-phosphonic acid (CPP), or CPP plus D-cycloserine (DCS) or CPP plus spermine. *, significantly different from saline control group, $p < 0.05$. +, significantly different from all other groups, $p < 0.05$. ■, saline control; ▨, CPP, 9 mg/kg; ▨, CPP+DCS, 30 mg/kg; ▨, CPP+DCS, 40 mg/kg; ▥, CPP+SPM, 2.5 mg/kg; ▤, CPP+SPM, 5 mg/kg.

polyamine modulation of the NMDA receptor might be involved in reducing signal transduction through this receptor;[46,47] however, we have directed attention to other signal transduction events that possibly underlie the deficient learning observed in aged rats. Specifically, we have focused on NO involvement in maze learning.

NITRIC OXIDE HYPOTHESIS

NOS inhibitors can be used to block NO production.[48] Results of previous studies in the 14-unit T-maze have shown that the NOS inhibitor, N^ω-nitro-L-arginine (NARG), delivered either ip[49] or intracerebroventricularly (icv),[50] impaired maze learning in young (3–5 mo) rats. Moreover, treatment with the NO donor, sodium nitroprusside (1 mg/kg ip), attenuated the NARG-induced impairment.[49,50] Whereas these results provide evidence of NO involvement in maze learning, other problems regarding their interpretation can be identified. NARG is not specific to the neuronal form of NOS and also produces marked hypertension when delivered ip or icv.[51,52] The hypertension was produced because of NARG's action on the endothelial form of NOS.[53,54] NOS also exists in an inducible form in macrophages.[55] Thus, the use of the neuronal-specific NOS inhibitor, 7-nitroindazole (7-NI), can overcome these possible confounds.[56]

Results presented in FIGURE 4 show a maze-learning impairment induced by 7-NI.[57] When young F344 rats (4–6 mo) were treated with 7-NI (70 mg/kg ip), they made significantly more errors compared to saline-treated controls. In addition, when groups of 7-NI-

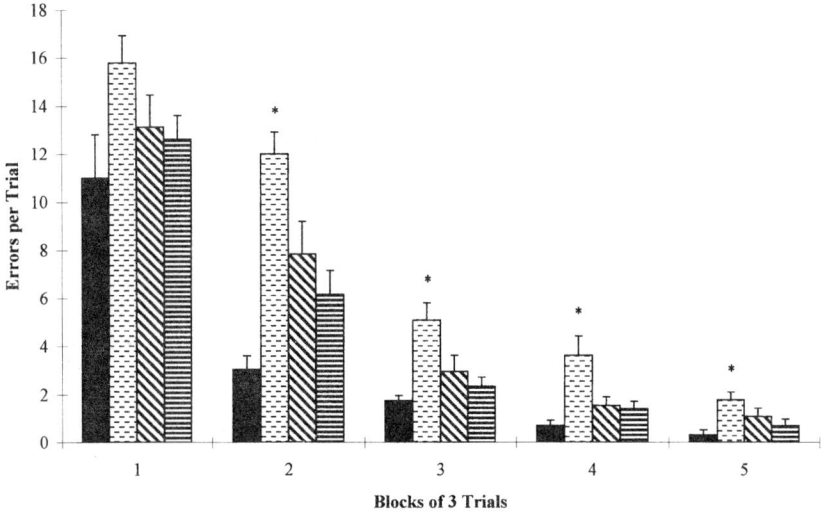

FIGURE 4. Mean errors (SEM) made by young F344 rats in the 14-unit T-maze following treatment with saline, 7-nitroindazole (7-NI), or 7-NI plus molsidomine (MOL). *, significantly different from all other groups, p < 0.05. ■, saline control; ▨, 7-NI, 70 mg/kg; ▨, 7-NI+MOL, 2 mg/kg; ▤, 7-NI+MOL, 4 mg/kg.

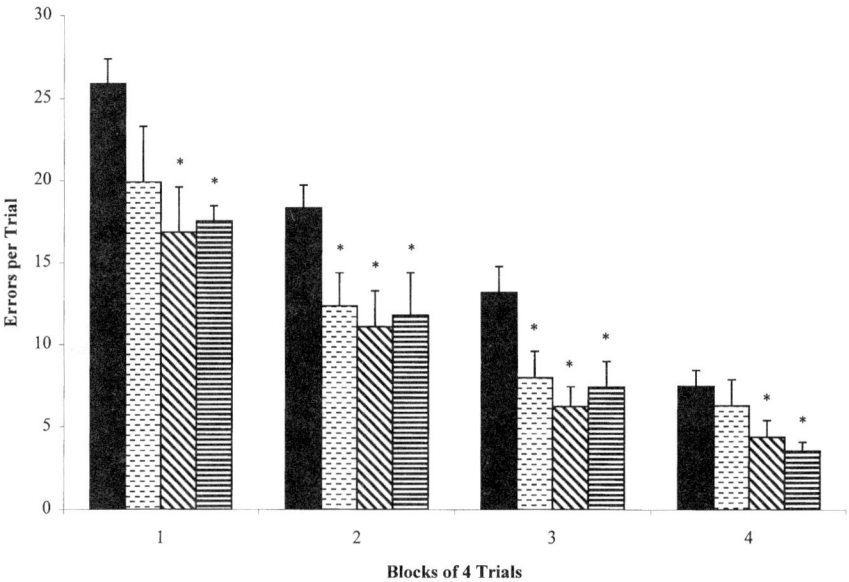

FIGURE 5. Mean errors (SEM) made by aged F344 rats in the 14-unit T-maze following treatment with saline or molsidomine. *, significantly different from saline control group, p < 0.05. ■, saline control; ▨, MOL, 2 mg/kg; ▨, MOL, 4 mg/kg; ▤, MOL, 8 mg/kg.

treated rats were also injected with the NO donor, molsidomine (2–4 mg/kg ip), their learning impairment was significantly attenuated.[57] These results provide strong evidence of NO involvement in learning in this behavioral paradigm.

To test the hypothesis that the age-related impairment in memory formation was linked to NO activity, we have begun testing the effects of molsidomine in aged rats. FIGURE 5 presents results of a preliminary study in which a marked improvement in maze learning was observed in aged rats following treatment with molsidomine (4 mg/kg ip). Thus, a strategy for examining the therapeutic effects of this and other NO donors appears to be a viable approach.

Despite the interest generated by the behavioral results outlined above, identification of any age-related defect in NMDA-mediated NO signal transduction has remained elusive. Using NADPH histochemistry, Kuo *et al.*[58] and Yamada *et al.*[59] have examined age-related loss of NOS-containing neurons and neuropil in F344 and Wistar rats, respectively. Both studies reported a significant age-related decline in density of striatal NADPH-positive neurons, and Kuo *et al.* noted a reduction in NADPH staining of the striatal neuropil, which indicates a loss of NOS-containing fibers. Yamada *et al.* also reported a significant age-related decline in NADPH-positive neurons in the cerebral cortex; however, Kuo *et al.* found no significant loss of NADPH-positive hippocampal neurons nor any decrease in NADPH staining in the molecular layer of the dentate gyrus. The latter finding did not support those of Rebeck *et al.*, who observed a marked loss of NADPH staining in this hippocampal region from brains of individuals with Alzheimer's disease.[60] These fibers represent input from the entorhinal cortex, another brain region severely affected in Alzheimer's disease.

A few laboratories have also examined age-related changes in NOS activity in specific brain regions using enzymatic assays. Yamada *et al.* observed significant age-related changes in NOS activity in the cerebellum but not in the hippocampus, striatum, or cere-

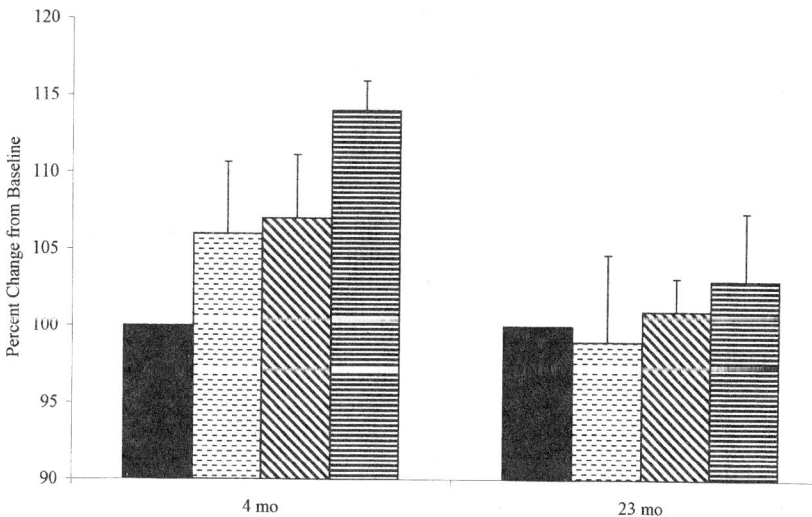

FIGURE 6. Nitrate concentration in microdialysis samples obtained from the striatum of young and aged F344 rats following NMDA stimulation. ■, baseline; ▤, two hours; ▨, four hours; ▤, six hours.

bral cortex.[59] Thus, current data would suggest that NOS activity is intact in the aged rat brain. So the question remains as to the loci in NO signal transduction that might underlie the age-related impairment in maze learning.

To provide more precise control over neurochemical events, we have recently begun studies aimed at measuring NO production using *in vivo* microdialysis. FIGURE 6 presents preliminary results in which we have examined NMDA-stimulated production of nitrate, which is an oxidative by-product of NO and that is highly correlated with NO production. Comparing young (4mo) and aged (23mo) F344 rats, these findings indicate a decline in NMDA-stimulated nitrate production in samples obtained from probes placed in the striatum. We are currently extending these studies to examine hippocampal sites. This approach should permit us to study various aspects of NO signal transduction and to identify specific age-related changes. Results from previous microdialysis studies in rats indicate an age-related decline in cGMP following NMDA-receptor stimulation.[61,62]

CONCLUSIONS AND FUTURE DIRECTIONS

Learning in the 14-unit T-maze appears to involve NMDA receptor-mediated signal transduction, with considerable evidence supporting the involvement of NO. Preliminary results indicate an age-related decline in NMDA-stimulated NO production and also indicate that maze learning can be improved in aged rats treated with a NO donor. Additional *in vivo* microdialysis studies are planned to confirm the loci of any age-related defects in NO signal transduction. We are also examining the possible therapeutic effects of several novel NO donors. Hopefully, some of these compounds will prove to be consistently effective in our model system; however, we will then have to turn our attention to the long-term safety of such treatments, as elevating levels of NO in brain may produce long-term neurodegenerative effects secondary to glutamate excitotoxicity.

REFERENCES

1. BELLEVILLE, S., I. PERETZ & D. MALENFANT. 1996. Examination of the working memory components in normal aging and in dementia of the Alzheimer's type. Neuropsychologia **34:** 195–207.
2. CROOK, T., R.T. BARTUS, S. FERRIS, P. WHITEHOUSE, G. COHEN & S. GERSHON. 1986. Age-associated memory impairment: Proposed diagnostic criteria and measures of clinical change—Report of a National Institute of Mental Health Work Group. Dev. Neuropsychiatry **2:** 261–276.
3. ARNSTEN, A.F. & P.S. GOLDMAN-RAKIC. 1987. Noradrenergic mechanisms in age-related cognitive decline. J. Neural. Transm. Suppl. **24:** 317–324.
4. ALTMAN, H.J. & H.J. NORMILE. 1988. What is the nature of the role of the serotonergic nervous system in learning and memory: Prospects for developent of an effective treatment strategy for senile dementia. Neurobiol. Aging **9:** 627–638.
5. NAGAHARA, A.H., T.M. GILL, M. NICOLLE & M. GALLAGHER. 1996. Alterations in opiate receptor binding in the hippocampus of aged Long-Evans rats. Brain Res. **707:** 22–30.
6. MOORE, H., M. SARTER & J.P. BRUNO. 1992. Age-dependent modulation of *in vivo* cortical acetylcholine release by benzodiazepine receptor ligands. Brain Res. **596:** 17–29.
7. BARTUS, R.T., R.L. DEAN, B. BEER & A.S. LIPPA. 1982. The cholinergic hypothesis of geriatric memory dysfunction. Science **217:** 408–412.
8. GREENAMYRE, J.T. & A.B. YOUNG. 1989. Excitatory amino acids and Alzheimer's disease. Neurobiol. Aging **10:** 593–602.

9. POMPONI, M., E. GIACONINI & M. BRUFANI. 1990. Present state and future development of the therapy of Alzheimer's disease. Aging **2:** 125–153.

10. HAGAN, J.J. 1994. The status of the cholinergic hypothesis of dementia. *In* Anti-Dementia Agents. D. Nicholson, Ed.: 86–138. Academic Press. San Diego, CA.

11. EAGGER, S.A., R. LEVY & B.J. SAHAKIAN. 1991. Tacrine in Alzheimer's disease. Lancet **337:** 989–992.

12. KAWAKAMI, Y., A. INOUE, T. KAWAI, M. WAKITA, H. SUGIMOTO & A.J. HOPFINGER. 1996. The rationale for E2020 as a potent acetylcholinesterase inhibitor. Bioorg. & Med. Chem. **4:** 1429–1446.

13. STONE, T.W. 1994. Excitatory amino acids and dementia. *In* Anti-Dementia Agents. D. Nicholson, Ed.: 229–249. Academic Press. San Diego, CA.

14. CHOI, D.W. & S.M. ROTHMAN. 1990. The role of glutamate neurotoxicity in hypoxia-ischemic neuronal death. Annu. Rev. Neurosci. **13:** 171–182.

15. LYNCH, D.R. & T.M. DAWSON. 1994. Secondary mechanisms in neuronal trauma. Curr. Opin. Neurol. **7:** 510–516.

16. INGRAM, D.K. 1988. Complex maze learning in rodents as a model of age-related memory impairment. Neurobiol. Aging **9:** 475–485.

17. INGRAM, D.K., E.L. SPANGLER, S. IIJIMA, H. IKARI, H. KUO, N.H. GREIG & E.D. LONDON. 1994. Rodent models of memory dysfunction in Alzheimer's disease and normal aging: Moving beyond the cholinergic hypothesis. Life Sci. **55:** 2037–2049.

18. INGRAM, D.K., A. SHIMADA, E.L. SPANGLER, H. IKARI, J. HENGEMIHLE, H. KUO & N. GREIG. 1996. Cognitive enhancement: New strategies for stimulating cholinergic, glutamatergic, and nitric oxide systems. Ann. N.Y. Acad. Sci. **786:** 348–361.

19. SPANGLER, E.L., P. RIGBY & D.K. INGRAM. 1986. Scopolamine impairs learning performance of rats in a 14-unit T-maze. Pharmacol. Biochem. Behav. **25:** 673–679.

20. SPANGLER, E.L., M.E. CHACHICH, N.J. CURTIS & D.K. INGRAM. 1989. Age-related impairment in complex maze learning in rats: Relationship to neophobia and cholinergic antagonism. Neurobiol. Aging **10:** 133–141.

21. KAMETANI, H., E.L. SPANGLER, E.L. BRESNAHAN, S. KOBAYOSHI, J.M. LONG & D.K. INGRAM. 1993. Impaired acquisition in a 14-unit T-maze following septal lesions in rats is correlated with lesion size and hippocampal acetylcholinesterase staining. Physiol. & Behav. **53:** 221–228.

22. IIJIMA, S., N.H. GREIG, P. GAROFALO, E.L. SPANGLER, B. HELLER, A. BROSSI & D.K. INGRAM. 1993. Phenserine: A physostigmine derivative that is a long-acting inhibitor of cholinesterase and demonstrates a wide dose range for attenuating a scopolamine-induced learning impairment of rats in a 14-unit T-maze. Psychopharmacology **112:** 415–420.

23. GREIG, N.H., X-F. PEI, T.T. SUNCKANI, D.K. INGRAM & A. BROSSI. 1995. Phenserine and ring C hetero-analogues: Drug candidates for the treatment of Alzheimer's disease. Med. Res. Rev. **15:** 3–31.

24. IKARI, H., E.L. SPANGLER, N.H. GREIG, X-F. PEI, A. BROSSI, D. SPEER, N. PATEL & D.K. INGRAM. 1995. Maze learning in aged rats is enhanced by phenserine, a novel anticholinesterase. Neuroreport. **6:** 481–484.

25. PALMER, A.M. & S. GERSON. 1990. Is the neuronal basis of Alzheimer's disease cholinergic or glutamatergic? FASEB J. **4:** 2745–2752.

26. GARTHWAITE, J. 1991. Glutamate, nitric oxide, and cell–cell signaling in the nervous system. Trends Neurosci. **14:** 60–67.

27. BANNERMAN, D.M., P.F. CHAPMAN, P.A.T. KELLY, S.P. BUTCHER & R.G.M. MORRIS. 1994. Inhibition of nitric oxide synthase does not impair spatial learning. J. Neurosci. **14:** 7404–7414.

28. BARNES, C.A., B.L. MCNAUGHTON, D.S. BREDT, C.D. FERRIS & S. SNYDER. 1994. Nitric oxide synthase inhibition *in vivo*: Lack of effect on hippocampal synaptic enhancement or spatial memory. *In* Long-term Potentiation: A Debate of Current Issues. M. Baudry & J. Davis, Eds.: 37–43. MIT Press. Cambridge, MA.

29. CHAPMAN, P.F., C.M. ATKINS, M.T. ALLEN, J.E. HALEY & J.E. STEINMETZ. 1992. Inhibition of nitric oxide synthase impairs two different forms of learning. Neuroreport **3:** 567–570.

30. ESTALL, L.B., S.J. GRANT & G.A. CICALA. 1993. Inhibition of nitric oxide (NO) production selectively impairs learning and memory in the rat. Pharmacol. Biochem. Behav. **46:** 959–962.

31. Mogensen, J., G. Wortwein, A. Hasman, P. Nielsen & Q. Wang. 1995. Functional and neuro-chemical profile of place learning after L-nitro-arginine in the rat. Neurobiol. Learn. Mem. **63:** 54–65.

32. Mogensen, J., G. Wortwein, B. Gustafson & P. Ermens. 1995. L-Nitro-arginine reduces hippocampal mediation of place learning in the rat. Neurobiol. Learn. Mem. **64:** 17–24.

33. Bohme, G.A., C. Bon, J-M. Stutzmann, A. Doble & J-C. Blanchard. 1991. Possible involvement of nitric oxide in long-term potentiation. Eur. J. Pharmacol. **199:** 379–381.

34. Madison, D.V. & E.M. Schuman. 1991. A requirement for the intercellular messenger nitric oxide in long-term potentiation. Science **254:** 1503–1506.

35. Meyer, R.C., J.K. Knox, D. Purwin, E.L. Spangler & D.K. Ingram. 1997. Combined modulation of the glycine and polyamine sites of the NMDA receptor attenuates CPP-induced learning deficits of rats in a 14-unit T-maze. Psychopharmacology **135:** 196–304.

36. Baxter, M.G., T.H. Lanthorn, K.M. Frick, S. Goldski et al. 1994. D-Cycloserine, a novel cognitive enhancer, improves spatial memory in aged rats. Neurobiol. Aging **15:** 207–213.

37. Flood, J.F., J.E. Morley & T.H. Lanthorn. 1992. Effects on memory processing by D-cycloserine, an agonist of the NMDA/glycine receptor. Eur. J. Pharmacol. **221:** 249–254.

38. Herting, R.L. 1991. Milacemide and other drugs active at glutamate NMDA receptors as potential treatment for dementia. Ann. N.Y. Acad. Sci. **640:** 237–240.

39. Schwartz, B.L., R.L. Hashtroudi, H. Herting, H. Anderson & S.I. Deutcsh. 1991. Glycine prodrug facilitates memory retrieval in humans. Neurology **41:** 1341–1343.

40. Shimada, A., E.D. London, A. Mukhin, E.L. Spangler & D.K. Ingram. 1994. Polyamine modulation of NMDA receptors as a strategy for cognitive enhancement in aged rats. Soc. Neurosci. Abst. **21:** 198.

41. Spangler, E.L., E.L. Bresnahan, P. Garofalo, N.J. Muth, B. Heller & D.K. Ingram. 1991. NMDA receptor channel antagonist by dizocilpine (MK801) impairs performance of rats in aversively-motivated complex maze tasks. Pharmacol. Biochem. Behav. **40:** 949–958.

42. Ingram, D.K., P. Garofalo, E.L. Spangler, C.R. Manitone, I. Odano & E.D. London. 1992. Reduced density of NMDA receptors and increased sensitivity to dizocilpine-induced learning impairment in aged rats. Brain Res. **580:** 273–280.

43. Kusztos, R.D., D.K. Ingram & E.D. London. 1995. Hippocampal binding of [^{3}H]CGS19755: Effects of aging and chronic nimodipine. Neurobiol. Aging **17:** 453–457.

44. Shimada, A., A. Mukhin, D.K. Ingram & E.D. London. 1997. N-Methyl-D-aspartate receptor binding in brains of rats of different ages. Neurobiol. Aging **18:** 329–333.

45. Ingram, D.K., A. Shimada, A. Mukhin & E.D. London. 1996. Current assessment of the glutamate hypothesis of geriatric memory impairment: Focus on hippocampal NMDA receptors. Perspect. Brain Aging Res. **1:** 30–35.

46. Palmer, A.M. & M.A. Burns. 1994. Preservation of redox, polyamine, and glycine modulatory domains of the N-methyl-D-aspartate receptor in Alzheimer's disease. J. Neurochem. **62:** 187–196.

47. Steele, J.E., A.M. Palmer, G.C. Stratemann & D.M. Bowen. 1989. The N-methyl-D-aspartate receptor complex in Alzheimer's disease: Reduced regulation by glycine but not zinc. Brain Res. **500:** 369–373.

48. Dwyer, M.A., D.S. Bredt & S.H. Snyder. 1991. Nitric oxide synthase: Irreversible inhibition by N^G-nitro-L-arginine in brain in vitro and in vivo. Biochem. Biophys. Res. Commun. **176:** 1136–1141.

49. Ingram, D.K., E.L. Spangler, H. Kametani, R.C. Meyer & E.D. London. 1997. Intracerebroventricular injection of the nitric oxide synthase inhibitor, N^G-nitro-L-arginine, in rats impairs learning in a 14-unit T-maze. Eur. J. Pharmacol., **341:** 11–16.

50. Ingram, D.K., E.L. Spangler, R.C. Meyer & E.D. London. 1997. Learning in a 14-unit T-maze is impaired in rats following systemic treatment with the nitric oxide synthase inhibitor, N^G-nitro-L-arginine. Eur. J. Pharmacol. **341:** 1–9.

51. Cabrera, C. & D. Bohr. 1995. The role of nitric oxide in the central control of blood pressure. Biochem. Biophys. Res. Commun. **206:** 77–81.

52. Hajj-Ali, A.F., T.M. Reilly & P.C. Wong. 1994. Modulation of renal vasoconstrictor effect of N^G-nitro-L-arginine in rabbit by angiotensin II and alpha-I adrenergic receptor blockade. J. Pharmacol. Exp. Ther. **270:** 1152–1157.

53. NANAEV, A., K. CHWALISZ, H.G. FRANK, G. KOHNEN, C. HEGELE-HARTUNG & P. KAUFMANN. 1995. Physiological dilation of uteroplacental arteries in the guinea pig depends on nitric oxide synthase activity of extravillous trophoblast. Cell Tissue Res. **282:** 407–421.

54. NEMADE, R.V., A.I. LEWIS, M. ZUCCARELLO & J.T. KELLER. 1995. Immunohistochemical localization of endothelial nitric oxide synthase in vessels of the dura mater of the Sprague-Dawley rat. Neurosci. Lett. **197:** 78–80.

55. TORRES, D., G. FERRARIO, G. BONETTA, L. PERVERSI & F. SPERANZA. 1996. *In vitro* and *in vivo* induction of nitric oxide by murine macrophages stimulated with *Bordetella pertussis.* FEMS Immunol. Med. Microbiol. **13:** 95–99.

56. KELLY, P.A., I.M. RITCHIE & G.W. ARBUTHNOTT. 1995. Inhibition of neuronal nitric oxide synthase by 7-nitroindazole: Effects upon local cerebral blood flow and glucose use in the rat. J. Cereb. Blood Flow Metab. **15:** 766–773.

57. MEYER, R.C., E.L. SPANGLER, N. PATEL, E.D. LONDON & D.K. INGRAM. 1997. Impaired learning in rats on a 14-unit T-maze by 7-nitroindazole, a neuronal nitric oxide synthase inhibitor, is attenuated by the nitric oxide donor, molsidomine. Eur. J. Pharmacol. **341:** 17–22.

58. KUO, H., J. HENGEMIHLE & D.K. INGRAM. 1997. Nitric oxide synthase in rat brain: Age comparisons quantitated with NADPH-diaphorase histochemistry. J. Gerontol: Biol. Sci. **52A:** B146–151.

59. YAMADA, K., Y. NODA, Y. KOMORI, H. SUGIHARA, T. HASEGAWA & T. NABESHIMA. 1996. Reduction in the number of NADPH-diaphorase-positive cells in the cerebral cortex and striatum in aged rats. Neurosci. Res. **24:** 393–402.

60. REBECK, G.W., W.K. MARZLOGG & B.T. HYMAN. 1993. The pattern of NADPH-diaphorase staining, a marker of nitric oxide synthase activity, is altered in the perforant pathway terminal zone in Alzheimer's disease. Neurosci. Lett. **152:** 165–168.

61. VALLEBUONA, F. & M. RAITERI. 1995. Age-related changes in the NMDA receptor/nitric oxide/cGMP pathway in the hippocampus and cerebellum of freely moving rats subjected to transcerebral microdialysis. Eur. J. Neurosci. **7:** 694–701.

62. VALLEBUONA, F. & M. RAITERI. 1994. Extracellular cGMP in the hippocampus of freely moving rats as an index of nitric oxide (NO) synthase activity. J. Neurosci. **14:** 134–139.

Telomeres: Influencing the Rate of Aging

THOMAS VON ZGLINICKI[a]

Institute of Pathology, Charité, Humboldt University, Schumannstr. 20/21, 10098 Berlin, Germany

ABSTRACT: Evidence is reviewed that suggests a central role for telomeres in one major model of biological aging, namely, proliferative senescence. Telomeric shortening with each cell division does not only act as a biological clock, but appears to trigger the ultimate loss of proliferative ability via activation of the p53-dependent check point system.

Oxidative stress induces single-stranded damage in telomeric DNA. It is not clear yet whether this damage occurs in the form of single-stranded gaps or overhangs or as arbitrarily distributed single-stranded breaks. However, in contradiction to the rest of the genome, this damage is not repaired in telomeres. It is, therefore, the major cause of telomere shortening even under standard *in vitro* cell culture conditions. Therefore, controlling the oxidative load onto DNA, in general, and, especially, onto telomeres might become a major factor to influence the rate of aging.

Further experiments demonstrate that G-rich single-stranded telomeric DNA fragments do activate the p53 check point control, leading to an inhibition of proliferation in wild-type p53 cells. Not only the shortening of telomeres down to a "signal value," but accumulation of telomeric single-stranded DNA fragments, as well, could be relevant triggers for proliferative senescence.

BIOLOGICAL CLOCKS, THE HAYFLICK LIMIT, AND THE TELOMERES

The question of the biological clock is a fascinating one to biogerontologists and to the general public alike. Is there one single mechanism that governs the rate of aging, that determines the biological age of an organism or, at least, a cell or a cell clone? If such a mechanism exists, can it be used to influence the rate of aging? In other words, can the pace of the clock be altered, or can the clock be reset altogether?

Denham Harman was among the first who suggested a plausible molecular basis for a biological clock.[1] He proposed that continuous generation of deleterious oxygen free radicals as a by-product of mitochondrial oxidative metabolism would eventually lead to intolerable damage and thereby determine cellular life span. This idea has stimulated extremely fruitful research conceptions. It has grown out into numerous branches, including, for instance, the rate of living theory or the caloric restriction theory,[2] and has become probably the most important molecular theory of aging today.

At about the same time, Leonhard Hayflick set an important foundation for empirical aging research by detecting the limited proliferative life span of human fibroblasts.[3] Fibroblasts, as well as other human somatic cells *in vitro*, loose their proliferative capacity after a certain number of population doublings (the Hayflick limit), which appears to be fixed within rather narrow limits for any cell strain tested so far. Although cells after passing the Hayflick limit still live for a long time as postmitotic cells, they cannot be stimulated to proliferate again except by transformation.

[a]Tel. +4930 2802 3178; fax: +4930 2802 3371; e-mail: zglinick@rz.charite.hu-berlin.de

A strong relationship to aging *in vivo* was established by showing correlations of the Hayflick limit with donor age, species maximum life span, and such premature aging syndromes as Hutchinson-Gilford and Werner's.[4] Finally, senescent cells were detected in human tissues *in vivo*.[5]

Therefore, proliferative senescence is one of the very few experimentally well-defined cellular models of aging. As such, it has been studied extensively. However, the events that counted cell division and that would eventually signal the final exit from the cell cycle were not found for a long time.[6]

In 1990, however, it was demonstrated that telomeres in human fibroblasts shorten by a constant amount with each cell division.[7] Telomeres were originally functionally defined as structures capping the ends of all chromosomes that are indispensable for successful chromosome segregation and for protection of chromosome ends against degradation.[8] Later, their molecular structure was shown to consist of long stretches of simple repeats that allow their identification and sizing by hybridization in Southern blots of enzymatically restricted genomic DNA. The repeat sequence in humans is $5'-(TTAGGG)_n-3'$, and the number of uninterrupted repeats in telomeres of human chromosomes from embryonic or germline cells is about 2.3×10^3, with a high degree of length polymorphism. Between 30 and 200 bp on average, or between 5 and 30 repeats, are lost with each division of human cells.

Telomeres shorten both with replicative age *in vitro*[7] and with donor age *in vivo*.[9,10] Telomere length can be used to predict the replicative capacity of a cell strain.[11] It was clear from the very definition of a telomere that its loss must lead to gross impairment of cell growth and survival.[8] Accordingly, telomeric shortening has been regarded as a very likely candidate for a biological clock.[12]

TELOMERIC MAINTENANCE, SENESCENCE, AND IMMORTALITY

Shortening of chromosome ends with each cell division was not unexpected: As soon as the semiconservative nature of DNA replication was discovered, it became clear that the distal 3′ end of the lagging strand of a linear DNA molecule could not completely be copied (the so-called end replication problem).[13] The resulting chromosome erosion is buffered sufficiently well in mortal cells by the presence of repetitive telomeric sequences at the ends of all chromosomes. However, nature had to find another solution in immortal unicellular organisms or in the cells of the germ line.

The common solution is activated telomerase, a ribonucleoprotein reverse transcriptase that uses an internal RNA as template to elongate the G-rich 3′-end of telomeres.[14–16] Telomerase activity has been demonstrated in all unicellular organisms tested so far, in the vast majority of immortal human cell lines, in human germline cells, and in about 85% of all tested human cancers.[17,18] Weak telomerase activity has also been found in stem cells of actively proliferating human tissues, but not in "normal" somatic tissue.[18] Other, probably recombination-based mechanisms to maintain the telomeric length, in spite of continuous shortening in dividing cells, are standard in *Drosophila* and are possible in some immortal human cell lines as well.[19]

Taken together, all available data suggest that active telomeric length maintenance is a necessary, albeit probably not sufficient, condition for immortality. On the other hand, telomeric loss is a sufficient, but formally not necessary, condition for cell senescence.

Studies using human fibroblasts, epithelial cells, and keratinocytes[20,21] have shown that transformation of human cells to immortality requires at least two distinct steps. The first is to overcome the Hayflick limit or the mortality stage I[22] by the action of tumor virus antigens like SV40T or HPV E6 and/or E7. Although other factors might be involved, it is clear that blocking of the tumor suppressors p53 and pRb is a major mode of action of these viral antigens. p53 and pRb are known to be control components of a cellular check point system that blocks the cell cycle in response to DNA damage in the form of strand breaks[23] as well as to other, partially yet ill-defined stimuli. After viral transfection, cells pass the Hayflick limit by about 10.20 population doublings. During that time, telomeres continue to shorten, and telomerase activity is typically absent. Eventually, the culture runs into crisis (mortality stage II), where most of the cells die. However, immortal clones may arise from here by mutational event(s). Telomeric length is stabilized in these clones, quite often at a level below that measured at the Hayflick limit.

From these results, a double role for telomeres in cell growth control was suggested.[12] First, shortening down of telomeres to some intermediate level might be recognized as DNA damage by a cellular check point control and might signal permanent cell cycle exit (*i.e.*, proliferative senescence). Second, if the check point system is overridden by muta- tion or viral infection, as is the case in most tumor cells, some further telomeric shortening is tolerated. However, more or less complete loss of telomeres would lead to insustainable genomic instability and cell death. At this point, a strong selective pressure for activation of telomerase develops.

OXIDATIVE STRESS, SINGLE-STRANDED DAMAGE, AND THE MECHANISM OF TELOMERIC SHORTENING

When the loss of telomeric repeats with proliferative age in diploid human cells was experimentally established, it seemed clear without doubt that only the end replication problem (see above) could be its cause. In fact, this result was a strong argument for ascribing proliferative senescence solely to genetic cause(s). This view, however, was in contradiction with both classic and recent experiments showing a dependency of the Hay- flick limit on the amount of oxidative stress applied to the cell cultures.[24–27] It was also in contradiction with experiments that clearly established a dependency of aging in postmi- totic cells on oxidative stress.[2,28] On the other hand, most oxidative stress experiments, and especially those applying chronic oxidative stress, suffer from the difficulty of defin- ing clearly the amount of oxidative stress applied and the prime target molecules involved. Accordingly, a role of oxidative damage for the proliferative senescence was repeatedly questioned[29,30] and could not be established unambiguously for a long time.

In 1995, we could demonstrate that both the rate of aging and the rate of telomeric shortening of fibroblasts *in vitro* could be accelerated in parallel by chronic mild oxidative stress.[31] Treatment with 40% normobaric oxygen, which enhances mitochondrial genera- tion of reactive oxygen species but is itself not cytotoxic, induced a senescent phenotype in WI-38 fibroblasts after about three population doublings (PD) at PD 28, whereas con- trol cells under normoxia proliferated at least to PD 45. The senescent phenotype was irre- versible and was indistinguishable from proliferative senescence regarding morphology, at both the light and the electron microscope level, and regarding nuclear DNA content, mitochondrial respiration and water content,[32] and cellular amount of lipofuscin. Most importantly, telomeres in prematurely senescent cells under hyperoxia were as short as those in cells reaching the Hayflick limit. A comparison of the data revealed about a five-fold increase of the telomeric shortening rate per population doubling under hyperoxia as compared to normoxic controls. Even under continued hyperoxia, however, telomeres did not shorten any further in cells that did no longer proliferate. Instead, we found an accumulation of sites sensitive to S1 nuclease, a single-stranded specific nuclease, specifically in telomeres of nonproliferating cells.[31]

To establish whether the S1-sensitive sites resemble single-stranded sites in telomeres rather than, for instance, loops or mismatches, we treated confluent human fibroblasts with hydrogen peroxide or the alkylating agent, N-methyl-N'-nitro-N-nitrosoguanidine (MNNG).[33] We measured both the length of double-stranded telomeric restriction fragments following treatment with S1 nuclease (which converts all single-stranded sites into double-strand breaks) and the length of the single-stranded fragments in the alkaline gel. Although alkaline gels from telomeres could not be quantified due to interstrand interactions of G-rich stretches, data from both techniques were in good qualitative agreement, demonstrating that S1 nuclease detects truly single-stranded regions in telomeres.

The frequency and/or the size of single-stranded regions in telomeres increased following treatment with hydrogen peroxide, with hyperoxia, and with MNNG. The time course of induction and repair was followed in fibroblasts after a single bolus dose of hydrogen peroxide:[33] The treatment induced single-strand breaks all over the genome. However, in the bulk of the genome, as well as in nontranscribed interstitial minisatellite sequences, proficient repair diminished the frequency of single-stranded breaks down to the background frequency of about one per mega–base pair within less than 24 hours. On the other hand, a significant fraction of telomeric single-stranded sites remained unrepaired, even if followed for up to 20 days.

Oxidative damage to telomeres is not restricted to situations of enhanced oxidative stress but occurs under standard normoxic cell culture conditions as well.[34] If fibroblasts are held at confluence under normoxic conditions for extended periods of time, they accumulate S1-sensitive sites in their telomeres as well. Moreover, after release from the confluent proliferation block, the telomeric shortening rate increases transiently but goes back to the control level after one to two population doublings. This result suggests that telomeric shortening under all conditions might be governed by telomere-specific single-stranded damage rather than by the end replication problem.

It is not clear yet what the telomeric single-stranded sites really are. There are at least two possibilities. (1) They could be "classic" single-stranded breaks similar to the rest of the genome, distributed at random along the length of the telomere. This would be the most obvious idea from the point of view of oxidative DNA damage. The open question, then, would be why there is a telomere-specific deficiency in the late steps of base excision repair? In this model, telomeric shortening in one of the daughter chromosomes would occur due to the loss of the strand distal to a single stranded break during DNA replication. If F would be the frequency of single-stranded breaks within all 92 telomeres at the end of G1 in a single cell, R = FT/(4 × 92) (T, telomeric length) would be the resulting telomeric shortening rate per population doubling.[34]

(2) It was shown recently that telomeres contain rather long (between 45 and more than 200 bases) stretches of single-stranded DNA in the G-rich strand, most probably as terminal single-stranded overhangs.[35,36] It is not known yet whether or how oxidative stress could contribute to the induction or elongation of those stretches. However, the fact that

these overhangs were found at all telomeres is in clear contradiction to the end replication model of telomeric shortening and confirms that other mechanisms must contribute significantly to telomeric shortening. In this model, the telomeric shortening rate would simply be equal to half of the overhang length, that is, the difference between the telomeric fragment length as measured without and with S1 nuclease treatment.

In fact, both models can predict telomeric shortening rates as measured under different conditions reasonably well (TABLE 1). At present, any combination of single-stranded breaks, gaps, or overhangs might be possible to become induced or enhanced in telomeres following oxidative stress, therefore. Whatever the specific form of single-stranded damage might be, it is not repaired and will contribute to the shortening of telomeres during the next round of DNA replication. Moreover, single-stranded telomeric DNA fragments will occur in the nuclei, either as overhangs or gaps or due to the liberation of single-stranded stretches of DNA distal to a single-stranded break during replication.

Although it is quite clear that both telomeric shortening and proliferative senescence are accelerated by increased oxidative stress, it is not known yet whether the correlation between telomeric shortening rate and rate of aging holds under conditions of decreased oxidative stress as well. There is good experimental evidence that decreasing the oxidative load slows down the aging process of fibroblast populations. Both chronic hypoxia and treatment with the spin trap α-phenyl-t-butyl-nitrone (PBN) were used to this end.[24–27] However, telomeric length has not yet been measured under those conditions. It is not known whether telomeric shortening slows down as well and whether the final inhibition of proliferation occurs at the same average telomere length in strains aging under conditions of low oxidative stress. Measurements are in progress to clarify this point.

Taken together, it appears that telomeres are hypersensitive to oxidative stress. They accumulate a "damage record" even under conditions that do not leave unrepaired damage

TABLE 1. Telomeric Shortening Rate as Predicted by Different Models of Telomeric Single-stranded Damage[a]

Treatment	ΔT (bp)	F_{random}	R_{random} (bp/PD)	$R_{overhang}$ (bp/PD)	$R_{measured}$ (bp/PD)
Control[34]	305 ± 95	4.3 ± 1.1	82 ± 21	153 ± 48	124 ± 13
Confluent[34]	1099 ± 136	17.9 ± 2.0	340 ± 38	550 ± 68	345 ± 92
Control[33]	328 ± 44	4.9 ± 0.7	80 ± 11	164 ± 22	90 ± 10
Hyperoxia[33]	813 ± 243	14.0 ± 4.0	228 ± 65	407 ± 122	560 ± 55

[a]The difference (ΔT) between the (double-stranded) telomere length and the telomeric length after degradation of single-stranded sites by S1 nuclease was measured in proliferating control fibroblasts (Control), in confluent fibroblasts (Confluent),[34] and in fibroblasts under chronic hyperoxia (Hyperoxia).[33] The random damage model assumes that single-stranded sites are generated randomly along the telomere. With this assumption, the frequency of single-stranded sites in the telomeric compartment per cell (F_{random}) and the telomeric shortening rate (R_{random}) can be calculated.[33] On the other hand, the overhang model assumes that the length of a single-stranded overhang is dependent on external factors. The telomeric shortening rate ($R_{overhang}$) is estimated as half the overhang length. $R_{measured}$ is the measured telomeric shortening rate.

in the bulk of the cellular DNA. This property opens up the possibility that telomeres might act as stress sensors, triggering cell cycle arrest both under conditions of acute but minor damage (transient cell cycle arrest for repair) and under conditions of chronic damage reaching a certain threshold (proliferative senescence).

TELOMERIC SHORTENING AND THE CELL CYCLE BLOCK

It has been speculated repeatedly that telomeric shortening not only counts cell division but, after reaching some critical level, might trigger the senescent cell-cycle inhibition.[12,37,38] Different mechanisms have been suggested, including deletion of critical gene(s) located near telomeres[13] or silencing of subtelomeric genes by heterochromatin shifting.[22] However, the most promising hypothesis at present seems to be the one that attempts to explain senescence by telomeric shortening–driven activation of the tumor suppressor p53; p53 is activated by DNA damage in the form of strand breaks.[23,39,40] Activated p53 blocks cell proliferation primarily in G1, and it is believed that transactivation of p21 expression, an inhibitor of cyclin-dependent kinases, plays a major role in this process. In senescent cells, p53 is activated and p21 levels are increased,[41–43] The question, then, is how telomere shortening might activate p53? This one is hard to answer because it is completely unclear how DNA damage by, for instance, γ-irradiation or hydrogen peroxide treatment triggers the activation of p53. The product of the Ataxia Telangiectasia gene ATM has recently been implicated in the signal transfer from damaged DNA to p53; however, the possible mechanisms are as unclear as before.[44]

Because telomere length is highly polymorphic, telomeric shortening might lead to a complete loss of uninterrupted telomeric repeats on at least one chromosome, and this loss might be recognized as a double-stranded break by the p53 system. This suggestion by Calvin Harley[12] gains credit as we learn more about telomere-specific protein binding and DNA structure.[45,46] Another suggestion attempts to explain the generation of DNA-stranded breaks by fusion-breakage-fusion cycles starting from telomere–telomere dicentrics or ring chromosomes.[37]

It appears possible that p53 can be activated by only one double-stranded break per cell.[23] On the other hand, many types of DNA lesions are processed via single-stranded DNA intermediates, and single-stranded DNA seems to be a possible direct signal for cell-cycle arrest in *Xenopus*,[47] *Saccharomyces cerevisiae*,[48,49] and human cells.[23,50] Different types of DNA damage might trigger a check point response, for instance, at different phases of the cell cycle.[51] In yeast, elimination of a telomere causes cell-cycle arrest mediated by the RAD9 gene, a functional homologue of the p53 gene.[52] In cdc13 mutants this cell-cycle arrest is triggered by the occurrence of single-stranded telomeric DNA.[48,49]

In light of the above-cited recent results on the role of single-stranded DNA and single-stranded damage for telomeric shortening,[33–36] it seemed reasonable to test whether telomeric single-stranded DNA might be able to trigger the p53 response. In a study using both fibroblasts and wild-type p53 (wtp53) tumor cell cultures, we confirmed that mild oxidative treatment induces single-stranded damage specifically in telomeres. It increased p53 and p21 levels and blocked cell proliferation in G1. More importantly, we found that p53 and p21 levels became elevated, and a G1 block was executed by treating the cells with short G-rich telomeric single-stranded DNA fragments as well.[53] The cell-cycle block was dependent on the presence of wtp53. It was transient in wtp53 tumor cells that harbor active telomerase, and release from the block was accompanied by stimulation of telomerase activity and elongation of telomeres.[53] These data suggest that telomeric G-rich single-stranded fragments, which are obviously produced in the course of telomeric shortening (see above), might be involved in the activation of the p53-dependent cell-cycle arrest. Again, the intermediates in the signaling pathway are not known. However, it is known that binding of single-stranded fragments to the C-terminal domain of p53 enhances its transactivation ability.[50] The possibility exists that signaling might be a very direct one, therefore.

CONCLUSIONS AND FUTURE DIRECTIONS

It is suggested that the telomere hypothesis of proliferative senescence should be modified with respect to the causes of telomeric shortening. Telomere-specific single-stranded damage rather than the end-replication problem should be regarded as the prime cause of telomeric shortening with each cell division. This modification allows the unification of genetic and environmental views of aging. In addition, it offers a possible explanation of how p53 might be activated by telomeric shortening.

A number of open questions still remain. This concerns, for instance, the mechanisms leading to single-stranded damage and telomeric shortening. Work is under progress to examine the strand specificity of oxidative damage and to clarify whether oxidative stress elongates single-stranded overhangs or whether it acts arbitrarily along the telomere. Also, it has still to be demonstrated whether telomeric shortening can be slowed down by anti-oxidative treatments.

Another important aspect is the correlation of the telomeric hypothesis with the results of cell fusion studies.[4] At present, a correlation between telomerase activity and any of the four complementation groups for cellular immortality could not be demonstrated.[54] Deeper insight into the molecular architecture of telomerase[16] and into further mechanisms for telomere maintenance[19] might, we hope, be helpful in this respect.

A central question is the generality of the telomere model of senescence. Telomeric shortening does not only correlate to aging in different human cells, but telomere length maintenance via a functional telomerase activity has been demonstrated to be necessary for cell viability and the immortal phenotype in yeast and in *Tetrahymena*.[55] On the other hand, some rodent strains have very long telomeres and active telomerase in somatic cells. It has been suggested that these properties might contribute to the high probability of neoplastic transformation in rodent fibroblasts.[56] It is clear that the "classical" model of telomeric shortening down to a certain limit as a trigger of senescence does evidently not hold here. If, however, the accumulation of single-stranded telomeric fragments triggers senescence as a permanent p53-dependent cell-cycle block, one could imagine that such a mechanism functions in cells with very long telomeres as well. The antioxidant activity in rodent cells is much lower than in humans, and this property could be important for a higher rate of telomeric damage.

Finally, the telomeric model of proliferative senescence has still to be tested with respect to its importance for the aging of organisms as a whole. Telomeres shorten during aging of replicating tissues *in vivo*, and the possibility exists that exhaustion of the proliferative potential of cells of the immune system, the skin, or the endothelium might contribute to age-related diseases.[5,55,57] In the future, those tissues might become possible targets for interventions aimed at slowing down the rate of telomere loss.

[NOTE ADDED IN PROOF: While this article was in press, the successful transfection of the catalytic subunit of human telomerase in human somatic cells was reported.[58] Clones that express high telomerase activity were able to counteract telomere shortening and proliferate presently far beyond their normal Hayflick limit. These data provide formal proof for the suggested casual role of telomeres in replicative senescence.]

REFERENCES

1. HARMAN, D. 1972. The biological clock: The mitochondria? J. Am. Geriatr. Soc. **20:** 145–147.
2. SOHAL, R.S. & R. WEINDRUCH. 1996. Oxidative stress, caloric restriction, and aging. Science **273:** 59–63.
3. HAYFLICK, L. 1965. The limited *in vitro* lifetime of human diploid cell strains. Exp. Cell Res. **37:** 614–636.
4. HENSLER, P.J. & O.M. PEREIRA-SMITH. 1995. Human replicative senescence. A molecular study. Am. J. Pathol. **147:** 1–9.
5. DIMRI, G.P., X. LEE, G. BASILE, M. ACOSTA, G. SCOTT, C. ROSKELLEY, E.E. MEDRANO, M.H. LINSKENS, I. RUBELJ, O.M. PEREIRA SMITH et al. 1995. A biomarker that identifies senescent human cells in culture and in aging skin *in vivo*. Proc. Natl. Acad. Sci. USA **92:** 9363–9367.
6. GOLDSTEIN, S. 1990. Replicative senescence: The human fibroblast comes of age. Science **249:** 1129–1133.
7. HARLEY, C.B., A.B. FUTCHER & C.W. GREIDER. 1990. Telomeres shorten during ageing of human fibroblasts. Nature **345:** 458–460.
8. GALL, J.G. 1995. Beginning of the end: Origins of the telomere concept. E.H. Blackburn & C.W. Greider, Eds.: 1–10. Cold Spring Harbor Laboratory Press. New York.
9. HASTIE, N.D., M. DEMPSTER, M.G. DUNLOP, A.M. THOMPSON, D.K. GREEN & R.C. ALLSHIRE. 1990. Telomere reduction in human colorectal carcinoma and with ageing. Nature **346:** 866–868.
10. SLAGBOOM, P.E., S. DROOG & D.I. BOOMSMA. 1994. Genetic determination of telomere size in humans: A twin study of three age groups. Am. J. Hum. Genet. **55:** 876–882.
11. ALLSOPP, R.C., H. VAZIRI, C. PATTERSON, S. GOLDSTEIN, E.V. YOUNGLAI, A.B. FUTCHER, C.W. GREIDER & C.B. HARLEY. 1992. Telomere length predicts replicative capacity of human fibroblasts. Proc. Natl. Acad. Sci. USA **89:** 10114–10118.
12. HARLEY, C.B. 1991. Telomere loss: Mitotic clock or genetic time bomb? Mutat. Res. **256:** 271–282.
13. OLOVNIKOV, A.M. 1973. A theory of marginotomy: The incomplete copying of template margin in enzymic synthesis of polynucleotides and biological significance of the phenomenon. J. Theor. Biol. **41:** 181–190.
14. GREIDER, C.W. & E.H. BLACKBURN. 1985. Identification of a specific telomere terminal transferase activity in *Tetrahymena* extracts. Cell **43:** 405–413.
15. FENG, J., W.D. FUNK, S.S. WANG, S.L. WEINRICH, A.A. AVILION, C.P. CHIU, R.R. ADAMS, E. CHANG, R.C. ALLSOPP, J. YU et al. 1995. The RNA component of human telomerase. Science **269:** 1236–1241.
16. LINGNER, J., T.R. HUGHES, A. SHEVCHENKO, M. MANN, V. LUNDBLAD & T.R. CECH. 1997. Reverse transcriptase motifs in the catalytic subunit of telomerase. Science **276:** 561–567.
17. KIM, N.W., M.A. PIATYSZEK, K.R. PROWSE, C.B. HARLEY, M.D. WEST, P.L. HO, G.M. COVIELLO, W.E. WRIGHT, S.L. WEINRICH & J.W. SHAY. 1994. Specific association of human telomerase activity with immortal cells and cancer. Science **266:** 2011–2015.
18. KIM, N.W. 1997. Clinical implications of telomerase in cancer. Eur. J. Cancer **33:** 781–786.
19. BRYAN, T.M. & R.R. REDDEL. 1997. Telomere dynamics and telomerase activity in *in vitro* immortalised human cells. Eur. J. Cancer **33:** 767–773.

20. Shay, J.W., B.A. Van Der Haegen, Y. Ying & W.E. Wright. 1993. The frequency of immortalization of human fibroblasts and mammary epithelial cells transfected with SV40 large T-antigen. Exp. Cell Res. **209:** 45–52.
21. Klingelhutz, A.J., S.A. Foster & J.K. McDougall. 1996. Telomerase activation by the E6 gene product of human papillomavirus type 16. Nature **380:** 79–81.
22. Wright, W.E. & J.W. Shay. 1992. The two-stage mechanism controlling cellular senescence and immortalization. Exp. Gerontol. **27:** 383–389.
23. Huang, L.C., K.C. Clarkin & G.M. Wahl. 1996. Sensitivity and selectivity of the DNA damage sensor responsible for activating p53-dependent G1 arrest. Proc. Natl. Acad. Sci. USA **93:** 4827–4832.
24. Packer, L. & K. Fuehr. 1977. Low oxygen concentration extends the lifespan of cultured human diploid cells. Nature **267:** 423–425.
25. Balin, A.K., D.P.B. Goodman, H. Rasmussen & V.J. Cristofalo. 1977. The effect of oxygen and vitamin E on the lifespan of human diploid cells *in vitro*. J. Cell Biol. **74:** 58–67.
26. Chen, Q., A. Fischer, J.D. Reagan, L.J. Yan & B.N. Ames. 1995. Oxidative DNA damage and senescence of human diploid fibroblast cells. Proc. Natl. Acad. Sci. USA **92:** 4337–4341.
27. Saito, H., A.T. Hammond & R.E. Moses. 1995. The effect of low oxygen tension on the *in vitro* replicative life span of human diploid fibroblast cells and their transformed derivatives. Exp. Cell Res. **217:** 272–279.
28. Shigenaga, M.K., T.M. Hagen & B.N. Ames. 1994. Oxidative damage and mitochondrial decay in aging. Proc. Natl. Acad. Sci. USA **91:** 10771–10778.
29. Balin, A.K., D.P.B. Goodman, H. Rasmussen & V.J. Cristofalo. 1978. Oxygen-sensitive stages of the cell cycle of human diploid cells. J. Cell Biol. **78:** 390–400.
30. Poot, M. 1991. Oxidants and antioxidants in proliferative senescence. Mutat. Res. **256:** 177–189.
31. von Zglinicki, T., G. Saretzki, W. Döcke & C. Lotze. 1995. Mild hyperoxia shortens telomeres and inhibits proliferation of fibroblasts: A model for senescence? Exp. Cell Res. **220:** 186–193.
32. von Zglinicki, T. & C. Schewe. 1995. Mitochondrial water loss and aging of cells. Cell Biochem. Funct. **13:** 181–187.
33. Petersen, S., G. Saretzki & T. von Zglinicki. 1998. Preferential accumulation of single-stranded regions in telomeres of human fibroblasts. Exp. Cell Res. **239:** 152–160.
34. Sitte, N., G. Saretzki & T. von Zglinicki. 1998. Accelerated telomere shortening in fibroblasts after extended periods of confluency. Free Radical Biol. Med. **24:** 885–893.
35. Makarov, V.L., Y. Hirose & J.P. Langmore. 1997. Long G tails at both ends of human chromosomes suggest a C strand degradation mechanism for telomere shortening. Cell **88:** 657–666.
36. McElligott, R. & R.J. Wellinger. 1997. The terminal DNA structure of mammalian chromosomes. EMBO J. **16:** 3705–3714.
37. Vaziri, H. & S. Benchimol. 1996. From telomere loss to p53 induction and activation of a DNA-damage pathway at senescence: The telomere loss/DNA damage model of cell aging. Exp. Gerontol. **31:** 295–301.
38. von Zglinicki, T. & G. Saretzki. 1997. Die molekularen Mechanismen der Seneszenz in Zellkultur. Z. Gerontol. Geriatr. **30:** 24–28.
39. Nelson, W.G. & M.B. Kastan. 1994. DNA strand breaks: The DNA template alterations that trigger p53-dependent DNA damage response pathways. Mol. Cell. Biol. **14:** 1815–1823.
40. DiLeonardo, A., S.P. Linke, K. Clarkin & G.M. Wahl. 1994. DNA damage triggers a prolonged p53-dependent G1 arrest and long-term induction of Cip1 in normal human fibroblasts. Genes & Dev. **8:** 2540–2551.
41. Noda, A., Y. Ning, S.F. Venable, O.M. Pereira Smith & J.R. Smith. 1994. Cloning of senescent cell-derived inhibitors of DNA synthesis using an expression screen. Exp. Cell Res. **211:** 90–98.
42. Atadja, P., H. Wong, I. Garkavtsev, C. Veillette & K.D. Riabowol. 1995. Increased activity of p53 in senescing fibroblasts. Proc. Natl. Acad. Sci. USA **92:** 8348–8352.
43. Bond, J., M. Haughton, J. Blaydes, V. Gire, D. Wynford-Thomas & F. Wyllie. 1996. Evidence that transcriptional activation by p53 plays a direct role in the induction of cellular senescence. Oncogene **13:** 2097–2104.

44. BROWN, K.D., Y. ZIV, S.N. SADANANDAN, L. CHESSA, F.S. COLLINS, Y. SHILOH & D.A. TAGLE. 1997. The ataxia-telangiectasia gene product, a constitutively expressed nuclear protein that is not up-regulated following genome damage. Proc. Natl. Acad. Sci. USA **94:** 1840–1845.

45. MARCAND, S., E. GILSON & D. SHORE. 1997. A protein-counting mechanism for telomere length regulation in yeast. Science **275:** 986–990.

46. WELLINGER, R.J. & D. SEN. 1997. The DNA structures at the ends of eukariotic chromosomes. Eur. J. Cancer **33:** 735–749.

47. KORNBLUTH, S., C. SMYTHE & J.W. NEWPORT. 1992. *In vitro* cell cycle arrest induced by using artificial DNA templates. Mol. Cell. Biol. **12:** 3216–3223.

48. GARVIK, B., M. CARSON & L.H. HARTWELL. 1995. Single-stranded DNA arising at telomeres in cdc13 mutants may constitute a specific signal for the RAD9 checkpoint. Mol. Cell. Biol. **15:** 6128–6138.

49. LIN, J.J. & V.A. ZAKIAN. 1996. The *Saccharomyces* CDC13 protein is a single-strand TG1-3 telomeric DNA-binding protein *in vitro* that affects telomere behavior *in vivo*. Proc. Natl. Acad. Sci. USA **93:** 13760–13765.

50. JAYARAMAN, L. & C. PRIVES. 1995. Activation of p53 sequence-specific DNA binding by short single strands of DNA requires the p53 C-terminus. Cell **81:** 1021–1029.

51. PAULOVITCH, A.G., D.P. TOCZYSKI & L.H. HARTWELL. 1997. When checkpoints fail. Cell **88:** 315–321.

52. SANDELL, L.L. & V.A. ZAKIAN. 1993. Loss of a yeast telomere: Arrest, recovery, and chromosome loss. Cell **75:** 729–739.

53. SARETZKI, G., N, SITTE, U. MERKEL, T.D. NGUYEN, R. WURM & T. VON ZGLINICKI. 1998. Telomere shortening triggers a p53-dependent cell cycle arrest via accumulation of g-rich single stranded DNA fragments. Oncogene. Submitted.

54. WHITAKER, N.J., T.M. BRYAN, P. BONNEFIN, A.C.M. CHANG, E.A. MUSGROVE, A.W. BRAITHWAITE & R.R. REDDEL. 1995. Involvement of RB-1, p53, p16INK4 and telomerase in immortalisation of human cells. Oncogene **11:** 971–976.

55. CHIU, C.P. & C.B. HARLEY. 1997. Replicative senescence and cell immortality: The role of telomeres and telomerase. Proc. Soc. Exp. Biol. Med. **214:** 99–106.

56. PROWSE, K.R. & C.W. GREIDER. 1995. Developmental and tissue-specific regulation of mouse telomerase and telomere length. Proc. Natl. Acad. Sci. USA **92:** 4818–4822.

57. CHANG, E. & C.B. HARLEY. 1996. Telomere length and replicative aging in human vascular tissues. Proc. Natl. Acad. Sci. USA **92:** 11190–11194.

58. BODNAR, A.G., M. OUELLETTE, M. FROLKIS, S.H. HOLT, C.P. CHIU, G.B. MORIN, C.B. HARLEY, J.W. SHAY, S. LICHTSTEINER & W.E. WRIGHT. 1998. Extension of life-span by introduction of telomerase into normal human cells. Science **279:** 349–352.

Redox Regulation of the Caspases during Apoptosis[a]

MARK B. HAMPTON,[c] BENGT FADEEL, AND STEN ORRENIUS[b]

Institute of Environmental Medicine, Division of Toxicology,
Karolinska Institutet, Box 210, S-171 77 Stockholm, Sweden

ABSTRACT: Apoptosis is now widely recognized as being a distinct process of importance both in normal physiology and pathology. In the current paradigm for apoptotic cell death, the activity of a family of proteases, caspases, related to interleukin-1β-converting enzyme (ICE) orchestrates the multiple downstream events, such as cell shrinkage, membrane blebbing, glutathione (GSH) efflux, and chromatin degradation that constitute apoptosis. Recent studies suggest that mitochondria could be the principle sensor and that the release of mitochondrial factors, such as cytochrome *c*, is the critical event governing the fate of the cell. One of the most reproducible inducers of apoptosis is mild oxidative stress, although it is unclear how an oxidative stimulus can activate the caspase cascade. Oxidative modification of proteins and lipids has also been observed in cells undergoing apoptosis in response to nonoxidative stimuli, suggesting that intracellular oxidation may be a general feature of the effector phase of apoptosis. The caspases themselves are cysteine-dependent enzymes and, as such, appear to be redox sensitive. Indeed, our recent work on hydrogen peroxide–mediated apoptosis suggests that prolonged or excessive oxidative stress can actually prevent caspase activation. A physiological example of this is the NADPH oxidase-derived oxidants generated by stimulated neutrophils that prevent caspase activation in these cells. Pursuant to these findings, stimulated neutrophils appear to use a specialized caspase-independent pathway to initiate phosphatidylserine (PS) exposure and subsequent phagocytic clearance. The possible implications of these dual roles for reactive oxygen species in apoptosis, that is, induction and inhibition of caspases, are discussed in the present review.

Reactive oxygen species (ROS), also called oxygen free radicals or oxidants, are generated in all aerobic cells. They are produced during normal mitochondrial respiration, they are used by specialized phagocytic cells to destroy invading pathogens, and they are by-products of the intracellular metabolism of toxic drugs and environmental agents. ROS can react rapidly with a range of biological molecules, making them highly destructive. Fortunately, aerobic cells are endowed with a complex network of antioxidants to scavenge ROS or repair the damage; however, the overwhelming of these defenses, termed oxidative stress, is implicated in numerous disease states.[1]

The gradual accumulation of oxidative damage has also been hypothesized to contribute to aging.[2] For example, it was recently reported that in liver cells from old rats the mitochondrial membrane potential declines and mitochondrial heterogeneity and production of oxidants increase.[3] Similar findings have been made in mouse lymphocytes, suggesting that a population of lymphocytes from old mice are more susceptible to activation

[a]M. Hampton was supported by the New Zealand Foundation for Research, Science and Technology.
[b]Corresponding author: Tel: (46 8) 728 7590; fax: (46 8) 32 90 41; e-mail: sten.orrenius@imm.ki.se
[c]Current address: Department of Pathology, Christchurch School of Medicine, P.O. Box 4345, Christchurch, New Zealand.

of the mitochondrial permeability transition, which would be compatible with the suggestion that mitochondrial dysfunction contributes to immunosenescence.[4]

Although it is known that acute oxidative stress kills cells, the mode of cell death has been largely ignored. Today, the study of cell death itself is one of the most rapidly growing areas in biomedical research. It is recognized that many compounds, rather than being directly cytotoxic, actually cause sublethal damage that triggers an innate suicide program in the cell. This process, termed apoptosis, involves cell shrinkage, nuclear condensation and fragmentation, and the bundling up of the cellular material into apoptotic bodies. Changes on the outer surface of the plasma membrane, such as the exposure of phosphatidylserine (PS), label these bodies for removal by neighboring cells or phagocytes, without the leakage of cellular material. Apoptosis is essential during development in maintaining tissue homeostasis in the adult organism and also for the removal of damaged, infected, or potentially neoplastic cells.[5]

The possible role of increased apoptotic cell death in aging is less well established, although a number of reports suggest that this may well be the case. For example, aging in the circulation is associated with an increased susceptibility to apoptosis in neutrophils,[6,7] and age-associated immune dysfunction has been suggested to correlate with defects in T-cell apoptosis.[8] Aging has also been reported to be associated with a progressive increase in the proportion of terminal deoxynucleotidyltransferase-mediated dUTP nick end labeling (TUNEL)-positive hepatocytes,[9] and sensory hair cells in the inner ear exhibit age-related cell death by apoptosis.[10] Further, myocyte cell death, apoptotic and necrotic in nature, constitutes an important determinant of the aging process.[11] However, in contrast to these reports, a recent study of apoptosis in mice following gamma irradiation has found a striking age-dependent decrease in radiation-induced apoptosis in splenic lymphocytes but not in colorectal epithelial cells,[12] and Zhou *et al.*[13] showed a decrease in Fas-mediated apoptosis in thymocytes and peripheral T cells of aged mice. Thus, the overall relationship between apoptosis and aging is still unclear.

Apoptosis can be triggered by numerous factors, including receptor-mediated signals, removal of growth factors, and cell damage. Each of these stimuli has its own specific pathway that leads to activation of the apoptotic process; however, all appear to converge at a highly conserved sequence of events. The caspases are a central component of this apoptotic program. This family of cysteine-dependent enzymes cleaves its target substrates on the carboxyl side of aspartate residues. Caspases exist in the cytoplasm as inactive zymogens (procaspases). Conversion to the active form involves zymogen proteolysis and assembly and joining of two identical heterodimers. Caspases cleave a defined number of target proteins, and these changes lead to the characteristic features of apoptosis and organized disassembly of the cell. Caspase activity has been detected at an early stage in numerous models of apoptosis, and specific peptide inhibitors of the caspases can block almost all of the subsequent apoptotic changes.[14,15]

The ability of caspases to cleave and activate other cytoplasmic procaspases sets off a destructive cascade of events.[14,15] Although this proteolytic cascade is likely to be common among different models of apoptosis, the sequence of events leading to activation of the first caspases will vary depending on the apoptotic stimulus. The best characterized systems are the cytotoxic T-lymphocyte component granzyme B, which can directly cleave specific procaspases, and the Fas (CD95/APO-1) surface receptor, which sequesters and activates procaspases via interactions between intracellular "death domains."[16]

The majority of other models seem more complex. The challenge of contemporary apoptosis research will be to elucidate the sequence of events leading to caspase activation in these systems.

ACTIVATION OF APOPTOSIS BY ROS

Incubating cells with exogenous oxidants, or adding redox-active compounds, can trigger apoptosis.[17,18] How then does a broad-spectrum biological oxidant initiate apoptosis? One possibility is that extensive oxidation of cellular components could trigger general protection systems, and these sensors detect the damage and activate the apoptotic machinery. A classic example of this is p53 detection of DNA damage. An alternate mechanism is that oxidants could activate specific signaling pathways that lead to apoptosis. Low-level oxidative stress is already known to activate different protein kinase pathways and transcription factors,[19] and the ability of nitric oxide to rapidly react with the heme group of guanylate cyclase has been used as a message in several cellular pathways.[20]

We have recently focused on the ability of hydrogen peroxide to trigger apoptosis in Jurkat T lymphocytes. Hydrogen peroxide is a relatively ineffective oxidant, and we suspected that selective pathways were involved in hydrogen peroxide–mediated apoptosis. Initial experiments showed that 50 μM hydrogen peroxide (10^6 cells/mL) optimally triggered caspase activation and subsequent apoptosis.[21] This occurred several hours after the initial bolus of hydrogen peroxide had been consumed by the cells,[21] and addition of hydrogen peroxide to cytoplasmic extracts does not activate the caspases directly (Hampton, unpublished data), indicating that other factors were involved in the pathway to activation.

Mitochondria are receiving considerable attention in apoptosis research for their ability to release proapoptotic factors.[22] In particular, the export of cytochrome c to the cytoplasm has been detected in various systems, and a groundbreaking study showed that cytochrome c, in combination with other cytosolic factors, is capable of triggering the caspase cascade.[23] Kluck et al.[24] reported cytochrome c release from peroxide-treated T lymphoblastoid CEM cells, and in our Jurkat cell model cytochrome c was detected in the cytoplasm within one to two hours after treatment with hydrogen peroxide, which was at least one hour before significant caspase activation could be detected.[25] Specific inhibition of the as yet unknown export mechanism will be important to determine the necessity of cytochrome c release for hydrogen peroxide–mediated caspase activation.

Another important issue is to determine the initial target of the hydrogen peroxide and how this initiates cytochrome c export. Mitochondria are well known targets of oxidative stress, with alterations in cellular calcium metabolism as a prominent result of oxidant challenge.[26] Opening of the membrane permeability transition pore is a potential mechanism for releasing mitochondrial factors into the cytoplasm,[27] and nitric oxide has recently been reported to trigger apoptosis by pore transition.[28] However, this event, which causes loss of the mitochondrial membrane potential, is apparently not necessary for cytochrome c release in several systems.[24,29] It is therefore possible that cells may have a specific mechanism for exporting mitochondrial cytochrome c to the cytoplasm.

The importance of dissecting the pathway from oxidative stress to apoptosis is not restricted to models where cells are exposed to exogenous oxidants. Intracellular oxidant production has been detected in cells incubated with a wide range of seemingly independent apoptotic agents, and some of these changes are suggested to occur suffi-ciently early to be intricately involved in the activation of apoptosis.[30] A recent example is p53-mediated apoptosis, which is hypothesized to occur by the increased transcription of prooxidant factors, leading to caspase activation and apoptosis by the mechanisms discussed above.[31] If aging is shown to be associated with increased apoptosis, then the increased oxidant production from mitochondria may be a prominent mechanism. Fur-ther support for the involvement of ROS in apoptosis comes from the widespread inhib-itory effect of antioxidants.[32] However, more comprehensive research is required to understand the role and mechanisms of endogenous oxidant production in apoptosis.

INHIBITION OF APOPTOSIS BY ROS

The caspases are cysteine-dependent enzymes and, as such, may be expected to be sensitive to the redox status of the cell. Other proposed effector molecules, such as calpain and tissue transglutaminase, also have active site thiol groups. Therefore, one must consider the possibility that a switch to an oxidizing environment may affect the activity of the apoptosis-effector enzymes that require a reduced thiol group to function

The lag phase in hydrogen peroxide–mediated apoptosis led us to investigate the effect of costimulating Jurkat T cells with hydrogen peroxide and anti-Fas antibody.[21] Concentrations of hydrogen peroxide that could eventually trigger apoptosis were able to block Fas-mediated caspase activity and apoptosis. The block was only temporary, how-ever, and the cells appeared to be able to repair or replace the caspases, and normal activ-ity and apoptosis resumed within one hour.[21] However, the caspases remained nonfunctional at higher concentrations of hydrogen peroxide, and necrosis ensued. A dose-dependent switch from apoptosis to necrosis is common with oxidants and redox-active compounds.[33] We believe that extensive oxidative inactivation of the caspases pre-vents the apoptotic pathway from being used, and necrosis occurs by default.

Nitric oxide has also been reported to inhibit apoptosis, via caspase inactivation, and direct S-nitrosylation of the active site thiol group was implicated.[34–37] We have studied extensively the effect of dithiocarbamates on apoptosis.[38] These have been used widely as antioxidants, and their ability to block apoptosis in a variety of models has served as evi-dence for the involvement of ROS in apoptosis. However, we have shown them to be potent thiol oxidizing compounds that inhibit procaspase processing as well as mature caspase activity in the cell.[39] As was the case for hydrogen peroxide, the inhibitory effect of dithiocarbamates is only temporary, and over extended periods of time they go on to induce apoptosis themselves.

NEUTROPHIL OXIDANTS AND APOPTOSIS

Neutrophils are phagocytic blood cells whose primary function is to ingest and kill invading pathogens. One prominent component of their antimicrobial arsenal is the NADPH oxidase, which is assembled in activated neutrophils and directs large amounts of oxidants at the ingested pathogen.[40] A range of antimicrobial and digestive compounds, stored in cytoplasmic granules, are also used to destroy the invading microorganisms.

Therefore, more than any other cell, it is crucial that neutrophils are removed from the tissue without releasing their toxic contents.

Neutrophils in circulation are short-lived, with an average life span of less than 24 hours.[41] In culture, it has been shown that they die by spontaneous apoptosis,[42] and we have recently found that caspase activation occurs under these conditions.[43] However, it was the activated neutrophil, with its concurrent oxidative burst, that was of particular interest to us. Neutrophils are essentially *kamikaze* cells. They are rendered ineffective shortly after performing their ultimate function, and it is likely that the oxidants used to destroy the ingested prey are also responsible for the death of the neutrophil. We found that phorbol ester–stimulated neutrophils died rapidly, although in the absence of the classical hallmarks of apoptosis, such as DNA fragmentation and characteristic morphological changes. Furthermore, no signs of caspase activity were detected. However, we could clearly measure PS exposure on the surface of these cells. This event will be particularly important for neutrophil clearance by macrophages arriving at an inflammatory site. Inhibition of the NADPH oxidase blocked PS exposure, indicating that the oxidants themselves are a crucial component of this specialized pathway. We hypothesize that caspases are not able to function under the conditions of intense oxidant generation in the stimulated neutrophil; therefore, the stimulated cells have their own specialized pathway for cell death and subsequent clearance, independent of caspase activity. The mechanisms of oxidant-induced PS exposure in neutrophils remain to be determined.

CONCLUDING REMARKS

It is becoming apparent that the redox status of a cell can have complex and multilayered effects on apoptosis (FIG. 1). Addition of exogenous oxidants is sufficient to trigger the apoptotic caspases; however, the initial targets and mechanism of caspase activation remain unclear. Recent studies suggest mitochondria could be the principle sensor and that the release of such mitochondrial factors as cytochrome c is the critical event leading to caspase activation.

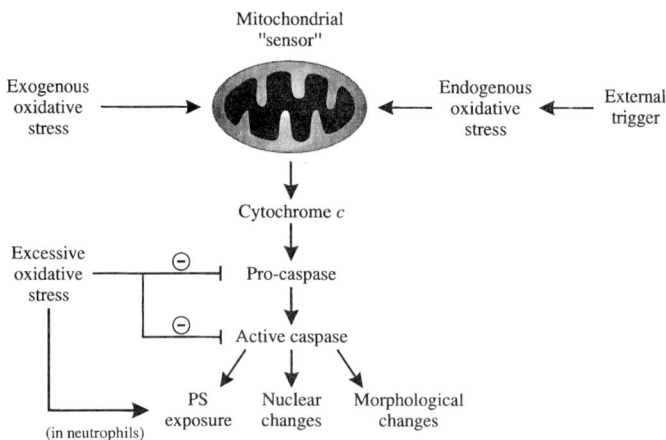

FIGURE 1. Induction and inhibition of caspase activity by ROS. See text for explanation.

The effectors of apoptosis, in particular the caspases, are redox sensitive, and the cell needs to maintain a reducing environment for these to function. Therefore, apoptosis cannot occur in cells subjected to excessive oxidative stress. Indeed, the stimulated neutrophil, whose death is triggered by its own oxidants, appears to have a specialized pathway for its clearance that is independent of caspase activity. However, the ability of oxidants to inhibit caspase function need not be incompatible with oxidant-dependent caspase activation. First, low levels of oxidants appear sufficient for optimal caspase activation, suggesting specific signaling pathways rather than widespread oxidation. Second, the observed ability of cells to repair or replace oxidized caspases indicates that the full complement of apoptotic effector molecules can be returned at some time after the initial oxidative stress.

It remains to be determined whether aberrant redox regulation of apoptosis contributes to different human disease states. Elucidation of the pathways of oxidant-induced apoptosis may provide alternate therapies to scavenging of the initial oxidant in those systems where excessive oxidative stress leads to cell death. Notably, the prooxidant status of some tumor cell lines suggests that their insensitivity to apoptosis may result from oxidative impairment of caspase function.[44]

REFERENCES

1. HALLIWELL, B. & J.M.C. GUTTERIDGE. 1989. Free Radicals in Biology and Medicine. Clarendon Press. Oxford.
2. OZAWA, T. 1997. Genetic and functional changes in mitochondria associated with aging. Physiol. Rev. **77:** 425–464.
3. HAGEN, T.M., D.L. YOWE, J.C. BARTHOLOMEW, C.M. WEHR, K.L. DO, J.Y. PARK & B.N. AMES. 1997. Mitochondrial decay in hepatocytes from old rats—membrane potential declines, heterogeneity and oxidants increase. Proc. Natl. Acad. Sci. USA **94:** 3064–3069.
4. ROTTENBERG, H. & S. WU. 1997. Mitochondrial dysfunction in lymphocytes from old mice: Enhanced activation of the permeability transition. Biochem. Biophys. Res. Commun. **240:** 68–74.
5. WYLLIE, A.H., J.F.R. KERR & A.R. CURRIE. 1980. Cell death: The significance of apoptosis. Int. Rev. Cytol. **68:** 251–306.
6. MATSUBA, K.T., S.F. VANEEDEN, S.G. BICKNELL, B.A.M. WALKER, S. HAYASHI & J.C. HOGG. 1997. Apoptosis in circulating PMN—increased susceptibility in L-selectin deficient PMN. Am. J. Physiol. **41:** H2852–H2858.
7. FULOP, T., C. FOUQUET, P. ALLAIRE, N. PERRIN, G. LACOMBE, J. STANKOVA, M. ROLAPLESZCZYNSKI, D. GAGNE, J.R. WAGNER, A. KHALIL & G. DUPUIS. 1997. Changes in apoptosis of human polymorphonuclear granulocytes with aging. Mech. Ageing Dev. **96:** 15–34.
8. HERNDON, F.J., H.C. HSU & J.D. MOUNTZ. 1997. Increased apoptosis of CD450RO(-) T cells with aging. Mech. Ageing Dev. **94:** 123–134.
9. HIGAMI, Y., I. SHIMOKAWA, T. OKIMOTO, M. TOMITA, T. YUO & T. IKEDA. 1997. Effect of aging and dietary restriction on hepatocyte proliferation and death in male F344 rats. Cell Tissue Res. **288:** 69–77.
10. USAMI, S., Y. TAKUMI, S. FUJITA, H. SHINKAWA & M. HOSOKAWA. 1997. Cell death in the inner ear associated with aging is apoptosis. Brain Res. **747:** 147–150.
11. KAJSTURA, J., W. CHENG, R. SARANGARAJAN, P. LI, B.S. LI, J.A. NITAHARA, S. CHAPNICK, K. REISS, G. OLIVETTI & P. ANVERSA. 1996. Necrotic and apoptotic myocyte cell death in the aging heart of Fischer 344 rats. Am. J. Physiol. **40:** H1215–H1228.
12. POLYAK, K., T.T. WU, S.R. HAMILTON, K.W. KINZLER & B. VOGELSTEIN. 1997. Less death in the dying. Cell Death Differ. **4:** 242–246.
13. ZHOU, T., C.K. EDWARDS & J.D. MOUNTZ. 1995. Prevention of age-related T cell apoptosis defect in CD2-fas transgenic mice. J. Exp. Med. **182:** 129–137.
14. ZHIVOTOVSKY, B., D.H. BURGESS, D.M. VANAGS & S. ORRENIUS. 1997. Involvement of cellular proteolytic machinery in apoptosis. Biochem. Biophys. Res. Commun. **230:** 481–488.

15. COHEN, G.M. 1997. Caspases: The executioners of apoptosis. Biochem. J. **326:** 1–16.
16. WALLACH, D. 1997. Placing death under control. Nature **388:** 123–126.
17. SLATER, A.F.G., C.S.I. NOBEL, D.J. VAN DEN DOBBELSTEEN & S. ORRENIUS. 1996. Intracellular redox changes during apoptosis. Cell Death Differ. **3:** 57–62.
18. JACOBSON, M.D. 1996. Reactive oxygen species and programmed cell death. Trends Biochem. Sci. **21:** 83–86.
19. SUZUKI, Y.J., H.J. FORMAN & A. SEVANIAN. 1997. Oxidants as stimulators of signal transduction. Free Radical Biol. Med. **22:** 269–285.
20. IGNARRO, L.J. 1991. Signal transduction mechanisms involving nitric oxide. Biochem. Pharmacol. **41:** 485–490.
21. HAMPTON, M.B. & S. ORRENIUS. 1997. Dual regulation of caspase activity by hydrogen peroxide: Implications for apoptosis. FEBS Lett. **414:** 552–556.
22. KROEMER, G., N. ZAMZAMI & S.A. SUSIN. 1997. Mitochondrial control of apoptosis. Immunol. Today **18:** 44–51.
23. LIU, X., C.N. KIM, J. YANG, R. JEMMERSON & X. WANG. 1996. Induction of apoptotic program in cell-free extracts: Requirement for dATP and cytochrome c. Cell **86:** 147–157.
24. KLUCK, R.M., E. BOSSY-WETZEL, D.R. GREEN & D.D. NEWMEYER. 1997. The release of cytochrome c from mitochondria: A primary site for Bcl-2 regulation of apoptosis. Science **275:** 1132–1136.
25. STRIDH, H., M. KIMLAND, D.P. JONES, S. ORRENIUS & M.B. HAMPTON. 1998. Cytochrome c release and caspase activation in hydrogen peroxide- and tributyltin-induced apoptosis. FEBS Lett. **429:** 351–355
26. ORRENIUS, S. 1994. Mechanisms of oxidative cell damage: An overview. *In* Oxidative processes and antioxidants. Pp. 53–71. Raven Press. New York.
27. ZORATTI, M. & I. SZABO. 1995. The mitochondrial permeability transition. Biochim. Biophys. Acta **1241:** 139–176.
28. HORTELANO, S., B. DALLAPORTA, N. ZAMZAMI, T. HIRSCH, S.A. SUSIN, I. MARZO, L. BOSCA & G. KROEMER. 1997. Nitric oxide induces apoptosis via triggering mitochondrial permeability transition. FEBS Lett. **410:** 373–377.
29. YANG, J., X. LIU, K. BHALLA, C.N. KIM, A.M. IBRADO, J. CAI, T. PENG, D.P. JONES & X. WANG. 1997. Prevention of apoptosis by Bcl-2: Release of cytochrome c from mitochondria blocked. Science **275:** 1129–1132.
30. BUTTKE, T.M. & P.A. SANDSTROM. 1995. Redox regulation of programmed cell death in lymphocytes. Free Radical Res. **22:** 389–397.
31. POLYAK, K., Y. XIA, J.L. ZWEIER, K.W. KINZLER & B. VOGELSTEIN. 1997. A model for p53-induced apoptosis. Nature **389:** 300–305.
32. SLATER, A.F.G., C.S.I. NOBEL & S. ORRENIUS. 1995. The role of intracellular oxidants in apoptosis. Biochim. Biophys. Acta **1271:** 59–62.
33. DYPBUKT, J.M., M. ANKARCRONA, M. BURKITT, Å. SJÖHOLM, K. STRÖM, S. ORRENIUS & P. NICOTERA. 1994. Different prooxidant levels stimulate growth, trigger apoptosis, or produce necrosis of insulin-secreting RINm5F cells. J. Biol. Chem. **269:** 30553–30560.
34. DIMMELER, S., J. HAENDELER, M. NEHLS & A.M. ZEIHER. 1997. Suppression of apoptosis by nitric oxide via inhibition of interleukin-1β-converting enzyme (ICE)-like and cysteine protease protein (CPP)-32-like proteases. J. Exp. Med. **185:** 601–607.
35. OGURA, T., M. TATEMICHI & H. ESUMI. 1997. Nitric oxide inhibits CPP32-like activity under redox regulation. Biochem. Biophys. Res. Commun. **236:** 365–369.
36. MOHR, S., B. ZECH, E.G. LAPETINA & B. BRUNE. 1997. Inhibition of caspase-3 by S-nitrosation and oxidation caused by nitric oxide. Biochem. Biophys. Res. Commun. **238:** 387–391.
37. MELINO, G., F. BERNASSOLA, R.A. KNIGHT, M.T. CORASANITI & G.N.A. FINAZZI-AGRO. 1997. S-nitrosylation regulates apoptosis. Nature **388:** 432–433.
38. ORRENIUS, S., C.S.I. NOBEL, D.J. VAN DEN DOBBELSTEEN, M.J. BURKITT & A.F.G. SLATER. 1996. Dithiocarbamates and the redox regulation of cell death. Biochem. Soc. Trans. **24:** 1032–1038.
39. NOBEL, C.S.I., D.H. BURGESS, B. ZHIVOTOVSKY, M.J. BURKITT, S. ORRENIUS & A.F.G. SLATER. 1997. Mechanism of dithiocarbamate inhibition of apoptosis: Thiol oxidation by dithiocarbamate disulfides directly inhibits processing of the caspase-3 proenzyme. Chem. Res. Toxicol. **10:** 636–643.

40. Winterbourn, C.C. 1990. Neutrophil oxidants: Production and reactions. *In* Oxygen Radicals: System Events and Disease Processes. Pp. 31–70. Karger. Basel.

41. Boggs, D.R. 1967. The kinetics of neutrophilic leukocytes in health and in disease. Semin. Hematol. **4:** 359–386.

42. Savill, J.S., A.H. Wyllie, J.E. Henson, M.J. Walport, P.M. Henson & C. Haslett. 1989. Macrophage phagocytosis of aging neutrophils in inflammation. Programmed cell death in the neutrophil leads to its recognition by macrophages. J. Clin. Invest. **83:** 865–875.

43. Fadeel, B., A. Åhlin, J-I. Henter, S. Orrenius & M.B. Hampton. 1998. Involvement of caspases in neutrophil apoptosis: Regulation by reactive oxygen species. Blood. In press.

44. Clement, M. & I. Stamenkovic. 1996. Superoxide anion is a natural inhibitor of Fas-mediated cell death. EMBO J. **15:** 216–225.

Nutrition, Osteoporosis, and Aging

B. E. CHRISTOPHER NORDIN,[a,b,c,e] ALLAN G. NEED,[a] TRACY STEURER,[a]
HOWARD A. MORRIS,[a] BARRY E. CHATTERTON,[d] AND MICHAEL HOROWITZ[c]

[a]Division of Clinical Biochemistry, Institute of Medical and Veterinary Science,
P.O. Box 14 Rundle Mall, Adelaide 5000, South Australia, Australia
[b]Department of Pathology, The University of Adelaide
[c]Department of Medicine, Royal Adelaide Hospital
[d]Department of Nuclear Medicine, Royal Adelaide Hospital

ABSTRACT: Loss of bone is an almost universal accompaniment of aging that proceeds at an average rate of 0.5–1% per annum from midlife onwards. There are at
least four nutrients involved in this process: calcium, salt, protein, and vitamin D, at
least in women. The pathogenesis of osteoporosis in men is more obscure. Calcium is
a positive risk factor because calcium requirement rises at the menopause due to an
increase in obligatory calcium loss and a small reduction in calcium absorption that
persist to the end of life. A metaanalysis of 20 calcium trials shows that this process
can generally be arrested by calcium supplementation, although there is some doubt
about its effectiveness in the first few years after menopause. Salt is a negative risk
factor because it increases obligatory calcium loss; every 100 mmol of sodium takes 1
mmol of calcium out of the body. Restricting salt intake lowers the rate of bone
resorption in postmenopausal women. Protein is another negative risk factor;
increasing animal protein intake from 40 to 80 g daily increases urine calcium by
about 1 mmol/day. Low protein intakes in third world countries may partially protect against osteoporosis. Vitamin D (sometimes called a nutrient and sometimes a
hormone) is important because age-related vitamin D deficiency leads to malabsorption of calcium, accelerated bone loss, and increased risk of hip fracture. Vitamin D
supplementation has been shown to retard bone loss and reduce hip fracture incidence in elderly women.

Loss of bone is an almost universal accompaniment of aging in all mammals. In humans, it
starts at about the age of 50 in men and at the menopause in women and proceeds at an
average rate of 0.5–1% per annum to the end of life. It occurs rather faster in women than
in men and more rapidly in some bones than others (FIG. 1). At some point in this progression, the bone or bones become "osteoporotic," but there is doubt as to whether this
threshold should be set at 2 or 2.5 standard deviations below the young normal mean.[1,2]
Fracture risk is a continuous, inverse, exponential function of bone density (FIG. 2), but it
is common practice to assume a "fracture threshold" to coincide with the "osteoporosis
threshold."

The "bone density," which defines the degree of osteoporosis and the fracture risk, is
generally determined by dual-energy X-ray absorptiometry (DXA) but can also be measured by computerized axial tomography (CT), by radiological techniques (morphometry
or densitometry), or by bone biopsy. The DXA technique measures mineral only and is
two-dimensional rather than three-dimensional. It is expressed in g/cm^2 and extrapolated
to whole bone areal density on the assumption that the mineral content of the bony tissue
itself is normal, that is, that osteomalacia is not present. Areal density does not, of course,

[e]Tel: (61-8) 8222 3000; fax: (61-8) 8222 3538; e-mail: christopher.nordin@imvs.sa.gov.au

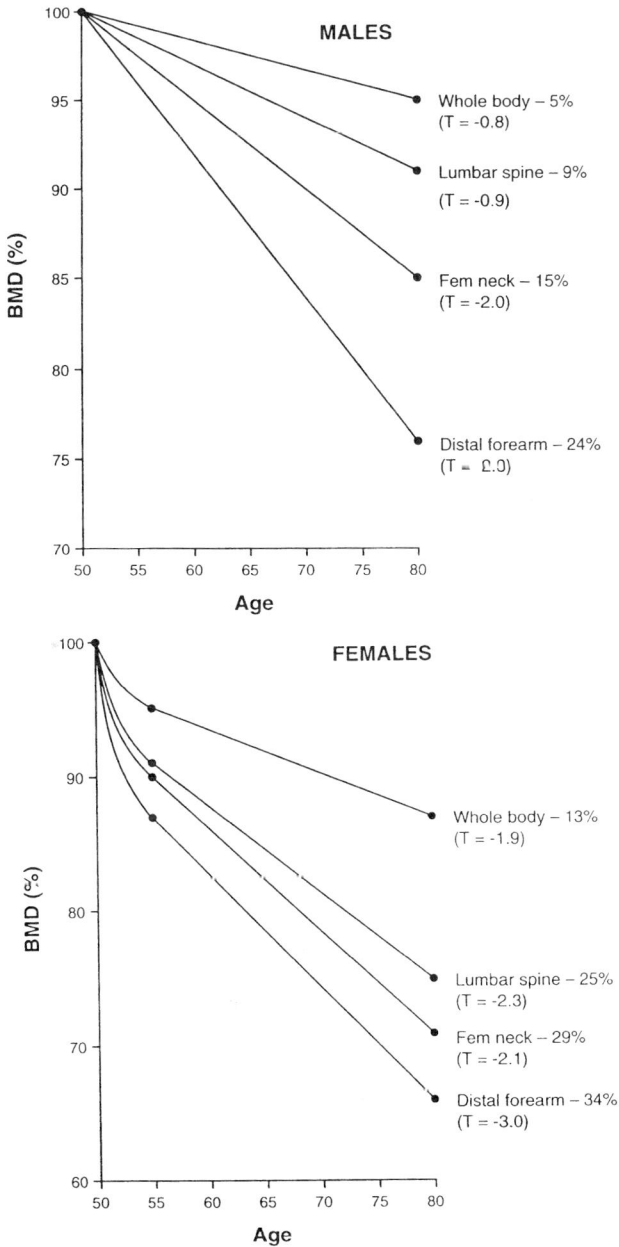

FIGURE 1. Semidiagrammatic representation of bone loss in men and women from the age of 50 to 80 (approximated from reference lines provided by Lunar, Hologic, and Norland). BMD, bone mineral density. fem, femoral.

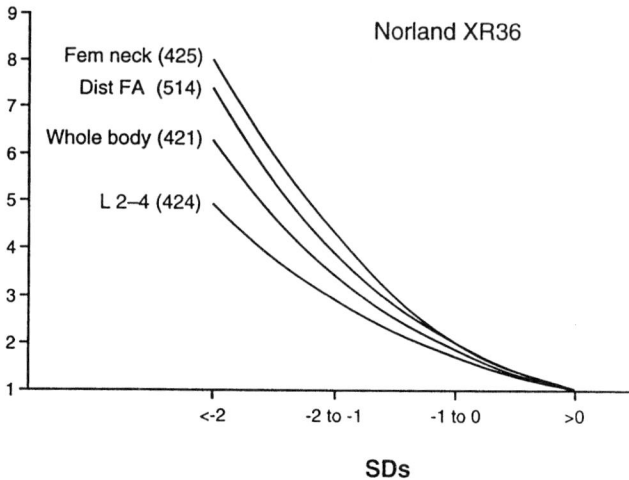

FIGURE 2. Odds ratios (any fracture vs. no fracture) in unselected women as a function of bone density at 4 different skeletal sites. All p < .001. Dist FA = distal forearm.

fully correct for bone size but is an adequate measure of bone density in most situations. However, it tends to underestimate volumetric bone density in small subjects and to overestimate it in large ones.

The bone density of any given individual is a function of their midlife bone mineral density (BMD) and their subsequent rate of bone loss. Because the range of bone densities in the young normal population is relatively wide (coefficient of variation about 12%), the peak bone density is the major determinant of BMD in the early stages of bone loss but becomes progressively less important with advancing age, because individuals lose bone at different rates. The crossover point of these two determinants of BMD occurs in women at about age 65,[3] after which the rate of bone loss becomes increasingly dominant. The main determinants of peak bone density are genetic,[4] but there is some evidence that calcium intake and exercise also make a contribution.[5]

The present paper will deal essentially with the bone-losing rather than the bone-gaining end of the life cycle. Two considerations need to be borne in mind. First, all bone loss must be due to an imbalance between the rates of bone formation and bone resorption—reduced bone formation, increased bone resorption, or a combination of the two. Second, the hormonal status of women is so different than that of men that the sexes need to be considered separately with respect to intrinsic hormonal and aging factors but can be considered together with respect to nutritional factors that are common to them. Before dealing with the subject in this way, it will be necessary to refer briefly to experimental osteoporosis in animals.

EXPERIMENTAL OSTEOPOROSIS

Any analysis of osteoporosis in humans needs to start from the well-established fact that it can be induced by calcium deficiency in experimental animals, as documented by observations going back to the turn of the century[6] and amply confirmed in recent years.[7]

There can be little doubt about the mechanism. When calcium intake falls below obligatory calcium loss (via the urine, feces, and skin), calcium is mobilized from the skeleton to maintain the ionized calcium concentration in the extracellular fluid, probably through the mediation of the parathyroid glands, and whole bone is destroyed *pari passu*. This contrasts with the effect of true vitamin D deficiency, which lowers the ionized calcium in the extracellular fluid (from loss of the calcemic action of vitamin D on bone)[8] and so causes unequivocal secondary hyperparathyroidism, hypophosphatemia, and failure to mineralize new bone as it is being formed. Simple calcium deficiency causes loss of whole bone, whereas true vitamin D deficiency reduces the mineral content of the bony tissue itself. However, these two nutritional deficiencies cannot be completely separated because of the malabsorption of calcium, which is the first manifestation of vitamin D insufficiency. It is therefore inherently probable—though it has not been established experimentally—that a mild degree of vitamin D insufficiency leads to osteoporosis from calcium malabsorption, whereas true deficiency of vitamin D, involving loss of its calcemic action, leads to rickets and osteomalacia.

As in humans, osteoporosis can also be induced experimentally by administration of corticosteroids (the main effect of which is to depress bone formation)[9] and by ovariectomy (which increases bone resorption).[10] However, the effect of ovariectomy on bone is itself a function of calcium intake[11,12] and can also be accelerated by sodium feeding, which increases urinary calcium loss.[13] This raises the possibility that estrogen deficiency exerts its effect on bone indirectly by increasing calcium requirement rather than by a direct action on bone cells, although there is evidence of the latter from *in vitro* studies.[14]

OSTEOPOROSIS IN WOMEN

The acute loss of bone that follows the menopause in women (FIG. 1) affects the trabecular component in particular but involves most of the skeleton and continues more or less linearly to the end of life. The associated biochemical changes include rises in the complexed fraction of the plasma calcium (secondary to an unexplained rise in plasma bicarbonate), rises in plasma alkaline phosphatase and urinary hydroxyproline (representing increased bone resorption followed by a compensatory increase in bone formation), increased obligatory calcium loss in the urine[15–18] and a small, but significant fall in calcium absorption.[19,20] Although these changes are most marked close to the menopause, they persist to the end of life and are all reversible with hormone treatment.[21,22] There is some uncertainty about whether serum parathyroid hormone (PTH) changes at the menopause[17,23,24] but there is general agreement that the serum calcitriol level does not alter.[25] It is widely assumed that the primary event is an increase in cytokine-mediated bone resorption secondary to estrogen deficiency[14] and that the other events are all secondary, though this does not explain the electrolyte changes. An alternative view is that the increased loss of calcium in the urine and fall in calcium absorption are the primary events and that the bone loss is due to negative calcium balance secondary to an increase in calcium requirement.[20] In favor of the former view is the undoubted efficacy of estrogen treatment in preventing bone loss and even in producing some gain of bone.[26,27] On the other hand, calcium trials in postmenopausal women have shown convincing inhibition of bone loss (though no bone gain),[20] and it could be argued that the combined actions of estrogen on calcium excretion[21,22] and calcium absorption[28] would be likely to be more effective than simple calcium supplementation. Whatever the truth may be, there is general agreement that the calcium allowance needs to be substantially increased for postmenopausal women, and two recent consensus conferences have both recommended an intake of 1500 mg.[29,30]

It is relevant to the current debate to note that bone resorption in postmenopausal women can be suppressed not only by hormone treatment but also by calcium supplementation,[31] by thiazide administration (which reduces calcium excretion[32]), and even by simple salt restriction, which acts by reducing obligatory calcium loss.[33] It should also be noted that calcium absorption is low in some 50% of women with spinal osteoporosis,[34] that the high bone resorption in these cases can be suppressed by treatment with calcitriol, which improves calcium absorption,[35] and that calcitriol treatment has been shown to reduce the vertebral fracture rate in women.[36] All this evidence points to a role for negative calcium balance in the genesis of postmenopausal osteoporosis while not discounting the possible contribution of an age-related decline in bone formation.[37]

OSTEOPOROSIS IN MEN

In men, the situation is rather different. The bone loss that starts at about the age of 50[38,39] is not associated with any rise in bone resorption markers[38] but appears rather to be linked to an age-related decline in gonadal function and to a rise from a decline in bone formation rather than an increase in bone resorption.[40]

Thus present evidence suggests that calcium deficiency in the widest sense (*i.e.*, negative calcium balance due to low calcium intake and/or low calcium absorption and/or high obligatory calcium loss) plays a much larger role in the genesis of age-related osteoporosis in women than it does in men. Nonetheless, there can be little doubt that, regardless of sex, any factors that tend to create or aggravate a state of negative calcium balance must aggravate a preexisting trend to bone loss and increase the risk of osteoporosis. There are at least three such nutritional factors, in addition to calcium itself, that are equally applicable to both sexes. They are protein and salt (for their effect on calcium excretion), and vitamin D status (for its effect on calcium absorption). These will be considered in turn.

NUTRITIONAL FACTORS

Protein Intake

There is a positive correlation between protein intake and urinary calcium.[41] This effect has been attributed to the acid load resulting from protein breakdown and/or to the complexing of calcium in the kidneys by the sulfate ion[42] but is more probably due to the calcium-complexing effect of the phosphate ions excreted after metabolism of proteins.[15] The magnitude of the effect is such that 40 g of animal protein (representing about 800 mg (26 mmol) of elemental phosphorus) takes out about 40 mg (1 mmol) of calcium.[43] The effect of this on calcium requirement depends on the prevailing associated calcium intake

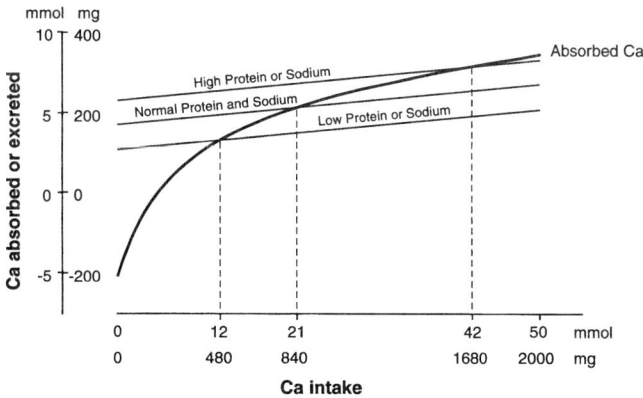

FIGURE 3. The relation between calcium intake and calcium absorbed (curved line) and calcium excreted via urine and skin (diagonal line). The middle line intersection represents the theoretical calcium requirement of normal subjects on an average western diet (see ref. 20). The upper line represents the effect of increasing protein by 40 gm or sodium by 2.3 gm, and the lower line represents the effect of reducing protein or sodium by the same amount.

because of the curvilinear relationship between calcium ingested and calcium absorbed (FIG. 3). At a calcium intake of 800 mg (20 mmol), increasing the animal protein intake from 60 to 100 g would increase the calcium requirement from 840 mg (21 mmol) to 1680 mg (42 mmol), whereas reducing the animal protein intake from 60 to 20 g would reduce the calcium requirement from 840 mg (21 mmol) to 480 mg (12 mmol) (FIG. 3). This may help to explain the apparently higher calcium requirement of western nations as compared to the third world.

Sodium Intake

There is a strongly positive correlation between sodium and calcium excretion, such that every 2.3 g (100 mmol) of sodium ingested takes out 40 mg (1 mmol) in the urine.[15,44,45] The effect of increasing sodium intake by 2.3 g (100 mmol) or reducing it by the same amount, therefore, has the same effect on calcium requirement as the 40 g change in animal protein intake described above. This variation is within the range reported between different nations.[46] If these effects are additive (which is likely but unproven), then the combination of only half the sodium and half the protein difference would be enough to change the requirement in the manner outlined above.

VITAMIN D

Calcium Absorption

It is widely assumed that calcium absorption falls with age, although this belief is not very well documented and relies rather heavily on one early publication.[47] However, our current data suggest that it is true and starts at about age 60, at least in women (FIG. 4), and

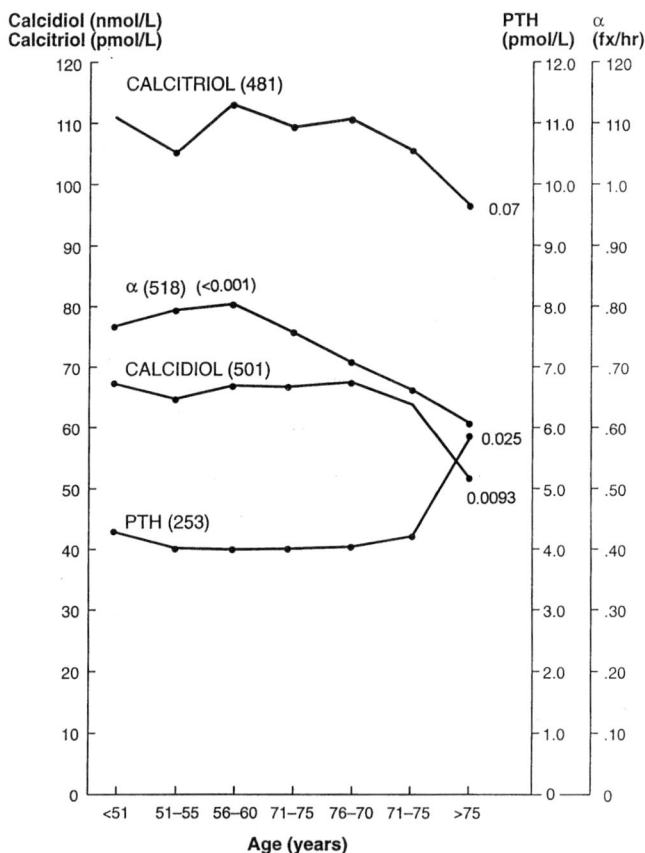

FIGURE 4. Serum vitamin D metabolites, radiocalcium absorption (α), and serum PTH as a function of age in "normal" postmenopausal women. The numbers in each set and the p values are shown. Only α falls significantly with age in the whole set, but PTH is significantly higher and calcidiol significantly lower in the oldest group than in the rest of the set. The corresponding difference in calcitriol is not quite significant. fx, fraction.

may be quite distinct from the small fall at the menopause mentioned above. At least three processes can be invoked to explain this phenomenon. The first would be an intrinsic decline in calcium absorptive capacity with age due to some change in the calcium transport mechanism itself, as may occur at the menopause. Such an abnormality has been identified in elderly women with spinal osteoporosis in whom calcium absorption tended to be lower than would have been expected from the serum calcitriol level.[34] A similar change has been reported in aging rats[48] and in elderly normal women.[49] FIGURE 4 also suggests that the fall in calcium absorption precedes the decline in vitamin D status. Moreover, age remains a significant determinant of calcium absorption after vitamin D has been allowed for (TABLE 1). The importance of calcium absorption in determining the calcium requirement is illustrated in FIGURE 5, which shows the effect on the calcium requirement

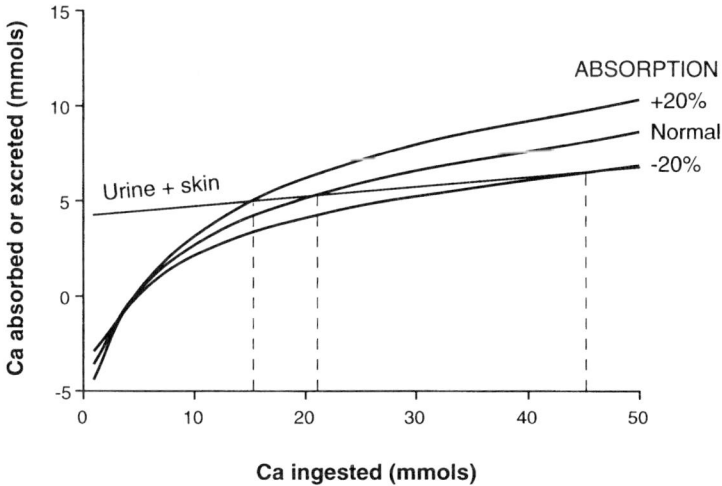

FIGURE 5. The effect on calcium requirement of increasing or reducing calcium absorption by 20% based on the same primary data as shown in FIG. 3. A 20% increase in calcium absorption reduces requirement from 21 to 15 mmol (840–600 mg), but a 20% decrease in absorption increases the requirement to 46 mmol (1840 mg).

of increasing or reducing calcium absorption by 20 percent. The other two possible explanations are age-related declines in serum calcidiol and serum calcitriol.

Serum Calcidiol

The calcidiol concentration in the serum tends to fall with age (FIG. 4) for at least two reasons: one is a decline in skin thickness[50,51] (FIG. 6) and the other is a decline in exposure to sunlight associated with decreasing outdoor activity due to physical infirmity. The well-known seasonal variation in serum calcidiol and the less well-documented decline in serum calcidiol with latitude[52] indicate how dependent this variable is on the hours of sunlight exposure. There is evidence from many countries of declining serum calcidiol levels

TABLE 1. Regression of Radiocalcium Absorption (α) on Serum Calcitriol (pmol/L) and Calcidiol (nmol/L) in 442 "Normal" Postmenopausal Women

	t	p
$\alpha = 0.00221 \times$ calcitriol	6.9	<0.001
$-0.00068 \times$ calcidiol	-1.3	0.18
$-0.0051 \times$ age	-3.3	0.001
$+0.86$/hour	7.9	<0.001
$R^2 = 11.9\%$		

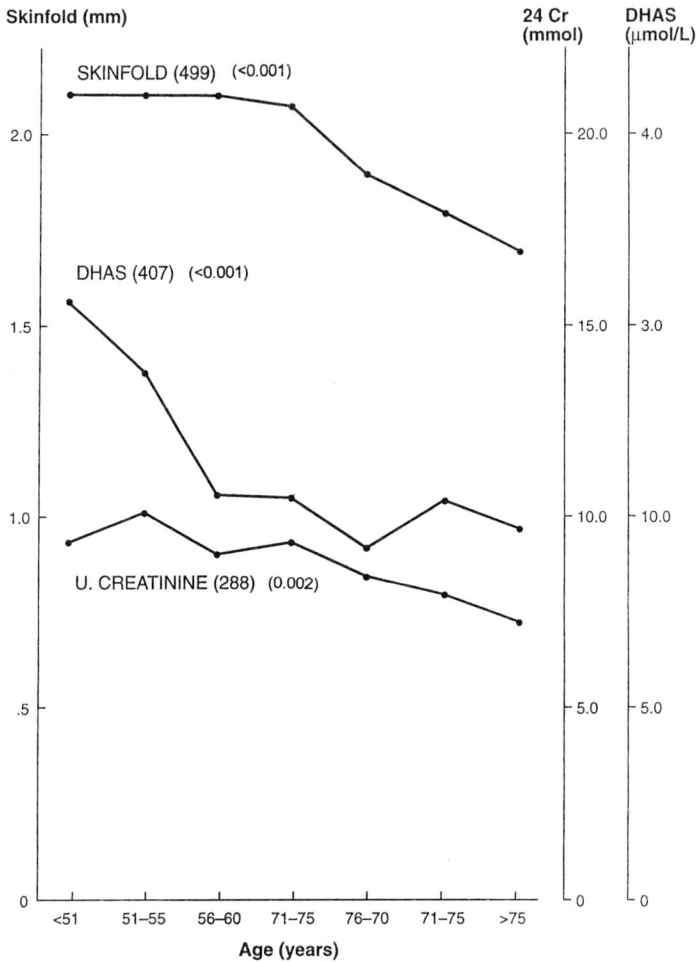

FIGURE 6. Skinfold thickness (mean value of three measurements on the dorsum of each hand), serum dehydroepiandrosterone sulfate (DHAS), and urinary creatinine as a function of age in normal postmenopausal women. The numbers in each set are shown and followed by the p value, representing the significance of the regression on age in each set.

with age[53–55] and strong evidence of vitamin D insufficiency (and thin skin) in hip fracture cases in western nations[56–59] (FIG. 7). Moreover, low-dose vitamin D treatment has been shown to reduce the rate of bone loss in elderly women in England,[60] and low-dose vitamin D with calcium has reduced hip fracture risk in nursing-home residents in France.[61] There is therefore a strong presumption that vitamin D insufficiency accelerates bone loss and contributes to hip fracture. The effect of vitamin D insufficiency on muscle function[62] may also contribute to the link between vitamin D status and hip fracture by predisposing

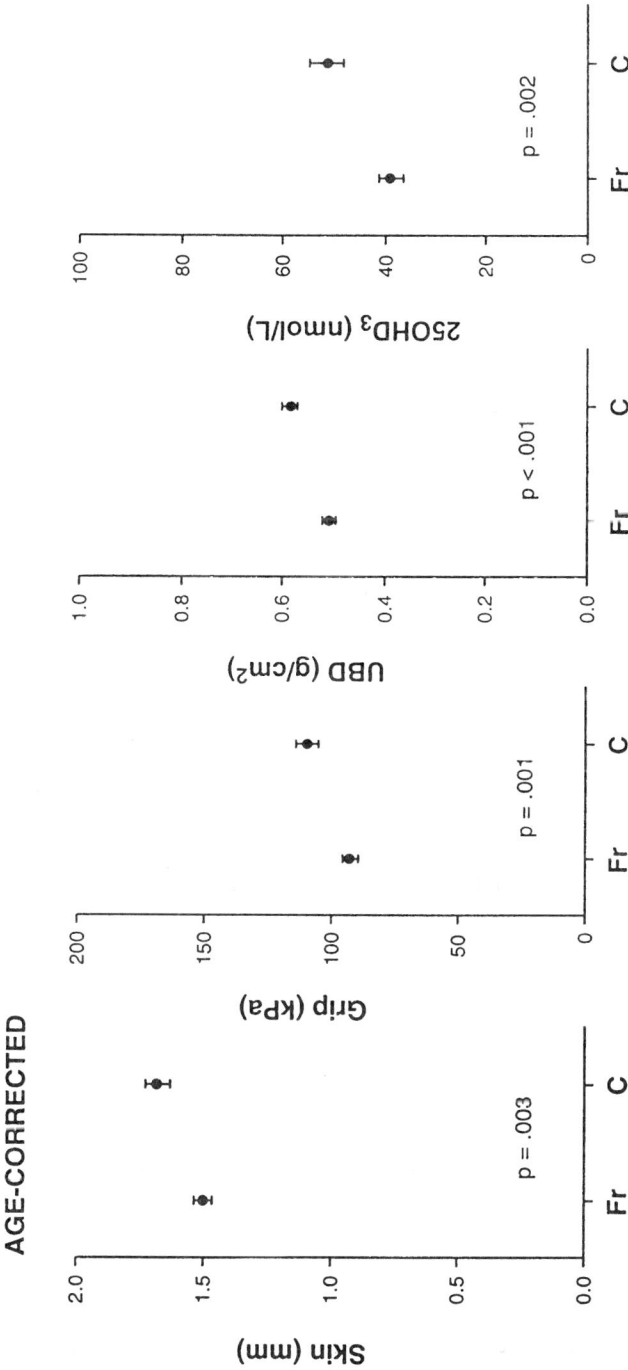

FIGURE 7. Mean age-corrected (± SE) values of skinfold thickness (dorsum of hands), grip strength, middistal ulna bone density, and serum calcidiol in 89 women hospitalized for hip fracture (mean age 81.9 years) and 47 free-living controls (mean age 76.4 years). Fr, fracture; C, control; kPa: kilopascals; 25-OHD₃: calcidiol.

TABLE 2. Regression of Ulnar Bone Density (UBD) on Age and Grip Strength (kPa) in 89 Women Hospitalized for Fracture of the Proximal Femur (mean age 81.9 years: range 61–99)

	t	p
UBD = 0.00070 × age	0.38	0.70
+0.00172 × grip	3.9	<0.001
+0.294 g/cm^2	1.8	0.084
R^2=16.0%		

to falls. Bone density and muscle strength are also reduced in female hip fracture cases[63] (FIG. 7) and are internally related (TABLE 2), but it has not been shown that these abnormalities are related to vitamin status or that they respond to treatment with vitamin D. There is, of course, a decline in muscle mass with age, reflected in the decline in urine creatinine, and a marked decline in adrenal androgen production (FIG. 6), which may be as important as the decline in serum calcidiol. Very little work, however, has been done on the hormonal status of hip fracture cases.

Serum Calcitriol

It is tempting to assume that the age-related fall in serum calcidiol leads directly to malabsorption of calcium and accelerated bone loss. However, the effect of vitamin D on calcium absorption is probably mediated by calcitriol rather than calcidiol (TABLE 1) and it is therefore this variable that must be looked at.

There is in fact only a small fall in serum calcitriol with age (probably after age 70) (FIG. 4). It is secondary to the decline in serum calcidiol but less significant because of the intervention of secondary hyperparathyroidism (TABLE 3 and FIG. 4). The result is that serum PTH in postmenopausal women is negatively related to serum calcidiol but positively related to serum calcitriol. Thus, calcidiol drives PTH, which in turn drives calcitriol (TABLE 4). Nonetheless, some fall in serum calcitriol does occur with age and is probably sufficient to account for the fall in calcium absorption. Although it is primarily due to the age-related decline in serum calcidiol, declining renal function may also contribute towards the end of life.

TABLE 3. Regression of Serum Calcitriol on Serum Calcidiol (nmol/L) and Serum PTH (pmol/L) in 460 Normal Postmenopausal Women

	t	p
Calcitriol = 0.373 × calcidiol	3.1	0.002
+5.74 × PTH	3.5	<0.001
+75.9 pmol/L	6.3	<0.001
R^2=7.3%		

TABLE 4. Regression of Serum PTH on Serum Calcitriol (pmol/L) and Calcidiol (nmol/L) in 207 Normal Postmenopausal Women

	t	p
PTH = 0.0080 × calcitriol	3.3	0.001
−0.0219 × calcidiol	−5.1	<0.001
+4.47 pmol/L	11.3	<0.001
R^2=13.8%		

CONCLUSIONS

The interplay between the primary effects of "aging" on bone and the secondary effects of other age-related changes is complex and is represented in FIGURE 8. The factors that increase bone resorption tend to dominate the scene, perhaps because bone resorption is easier to monitor with current biochemical markers than bone formation and nutritional factors that regulate bone resorption are easier to identify than those that may regulate

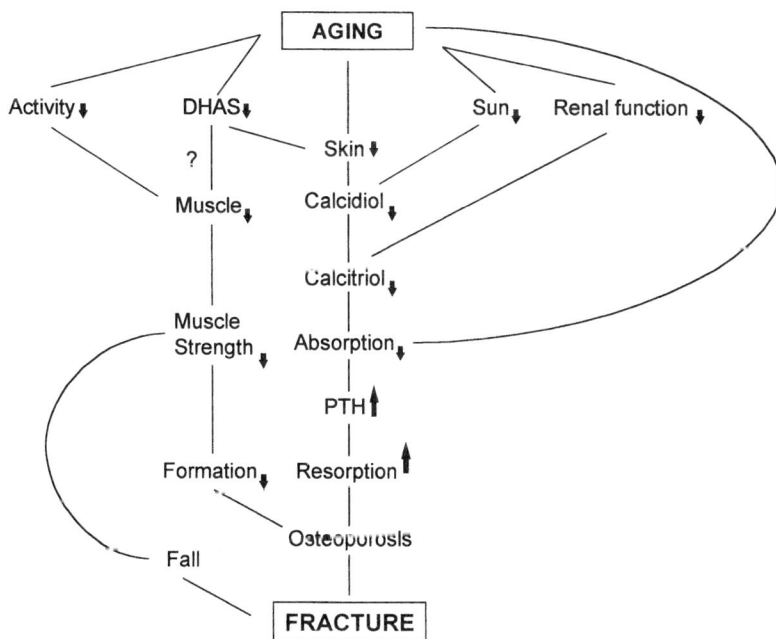

FIGURE 8. Diagrammatic representation of the principal pathways to hip fracture, as described in the text.

bone formation. Paramount among the former is probably the age-related decline in calcium absorption that itself is multifactorial in origin. It arises in part from intrinsic changes in the calcium transport mechanism and in part from a decline in vitamin D status. The latter is in turn due to declining exposure to sunlight, declining dermal capacity to use sunlight for synthesis of vitamin D and declining renal function. Irrespective of the immediate roles of sex hormones in determining bone status, these extrinsic factors are calculated to increase calcium requirement with age in both sexes and so predispose to negative calcium balance and osteoporosis.

On the other side of the bone turnover equation are those factors that tend to lower bone formation, notably declining muscular strength and activity possibly in turn related to declining androgen production and/or to vitamin D insufficiency and/or to such disabilities as arthritis and heart disease. These are the factors that not only affect bone adversely but also predispose to falls and are therefore doubly harmful. Vitamin D deficiency, however, may also feed into the pathway through its effect on muscle function and may for that reason be even more important than would appear at first sight. A schema such as this is, however, only an incomplete appreciation. There are many other risk factors for falls (such as failing eyesight) that may be statistically more important than those in the diagram. Moreover, this paper only deals with those nutritional aspects of the problem that now seem important, other, more important, ones may yet emerge.

REFERENCES

1. NORDIN, B.E.C. 1987. Editorial: The definition and diagnosis of osteoporosis. Calcif. Tissue Int. **40:** 57–58.
2. KANIS, J.A. 1994. WHO Study Document. Assessment of fracture risk and its application to screening for postmenopausal osteoporosis: Synopsis of a WHO Report. Osteoporosis Int. **4:** 368–381.
3. NORDIN, B.E.C. 1993. Keynote Address: Bone mass, bone loss, bone density and fractures. Osteoporosis Int. Suppl. **1:** S1–7.
4. SLEMENDA, C.W., J.C. CHRISTIAN, C.J. WILLIAMS, J.A. NORTON & C.C. JOHNSTON. 1991. Genetic determinants of bone mass in adult women: A reevaluation of the twin model and the potential importance of gene interaction on heritability estimates. J. Bone Miner. Res. **4:** 469–475.
5. HALIOUA, L. & J.J.B. ANDERSON. 1989. Lifetime calcium intake and physical activity habits: Independent and combined effects on the radial bone of healthy women. Am. J. Clin. Nutr. **49:** 534–541.
6. NORDIN, B.E.C. 1960. Osteomalacia, osteoporosis and calcium deficiency. Clin. Orthop. **17:** 235–258.
7. WU, D.D., R.D. BOYD, T.J. FIX & D.B BURR. 1990. Regional patterns of bone loss and altered bone remodeling in response to calcium deprivation in laboratory rabbits. Calcif. Tissue Int. **47:** 18–23.
8. DELUCA, H.F. 1980. Vitamin D. revisited. Clin. Endocrinol. Metab. **9:** 1–26.
9. ORTOFT, G. & H. OXLUND. 1988. Reduced strength of rat cortical bone after glucocorticoid treatment. Calcif. Tissue Int. **43:** 376–382.
10. WRONSKI, T.J., P.L. LOWRY, C.C. WALSH & L.A. IGNASZEWSKI. 1985. Skeletal alterations in ovariectomized rats. Calcif. Tissue Int. **37:** 324–328.
11. HODGKINSON, A., J.E. AARON, A. HORSMAN, M.S.F. McLACHLAN & B.E.C. NORDIN. 1978. Effect of oophorectomy and calcium deprivation on bone mass in the rat. Clin. Sci. Mol. Med. **54:** 439–446.
12. SHEN, V., R. BIRCHMAN, R. XU, R. LINDSAY & D.W. DEMPSTER. 1995. Short-term changes in histomorphometric and biochemical turnover markers and bone mineral density in estrogen- and/or dietary calcium-deficient rats. Bone **16:** 149–156.

13. GOULDING, A. & D. CAMPBELL. 1983. Dietary NaCl loads promote calciuria and bone loss in adult oophorectomized rats consuming a low calcium diet. J. Nutr. **113:** 1409–1414.

14. PACIFICI, R. 1996. Review: Estrogen, cytokines, and pathogenesis of postmenopausal osteoporosis. J. Bone Miner. Res. **11:** 1043–1051.

15. NORDIN, B.E.C. & K.J. POLLEY. 1987. Metabolic consequences of the menopause. A cross-sectional, longitudinal, and intervention study on 557 normal postmenopausal women. Calcif. Tissue Int. **41:** S1–S60.

16. STEPAN, J.J., J. POSPICHAL, J. PRESL & V. PACOVSKY. 1987. Bone loss and biochemical indices of bone remodeling in surgically induced postmenopausal women. Bone **8:** 279–284.

17. PRINCE, R.L. *et al.* 1995. The effects of menopause and age on calcitropic hormones: A cross-sectional study of 655 healthy women aged 35 to 90. J. Bone Miner. Res. **10:** 835–842.

18. KELLY, P.J., N.A. POCOCK, P.N. SAMBROOK & J.A. EISMAN. 1989. Age and menopause-related changes in indices of bone turnover. J. Clin. Endocrinol. Metab. **69:** 1160–1165.

19. HEANEY, R.P., R.R. RECKER, M.R. STEGMAN & A.J. MOY. 1989. Calcium absorption in women: Relationships to calcium intake, estrogen status, and age. J. Bone Miner. Res. **4:** 469–475.

20. NORDIN, B.E.C. 1997. Calcium and osteoporosis. Nutrition **13:** 664–686.

21. HOROWITZ, M., J.M WISHART, A.G. NEED, H.A. MORRIS & B.E.C. NORDIN. 1993. Effects of norethisterone on bone related biochemical variables and forearm bone mineral in postmenopausal osteoporosis. Clin. Endocrinol. **39:** 649–655.

22. PRINCE, R.L., I. SCHIFF & R.M. NEER. 1990. Effects of transdermal estrogen replacement on parathyroid hormone secretion. J. Clin. Endocrinol. Metab. **71:** 1284–1287.

23. PRINCE, R.L., I. DICK, P. GARCIA-WEBB & R.W. RETALLACK. 1990. The effects of the menopause on calcitriol and parathyroid hormone: Responses to a low dietary calcium stress test. J. Clin. Endocrinol. Metab. **70:** 1119–1123.

24. SOKOLL, L.J., F.D. MORROW, D.M. QUIRBACH & B. DAWSON-HUGHES. 1988. Intact parathyrin in postmenopausal women. Clin. Chem. **34:** 407–410.

25. FALCH, J.A., H. OFTEBRO & E. HAUG. 1987. Early postmenopausal bone loss is not associated with a decrease in circulating levels of 25-hydroxyvitamin D, 1,25-dihydroxyvitamin D, or vitamin D-binding protein. J. Clin. Endocrinol. Metab. **64:** 836–841.

26. PRINCE, R.L. *et al.* 1991. Prevention of postmenopausal osteoporosis: A comparative study of exercise, calcium supplementation and hormone-replacement therapy. N. Engl. J. Med. **325:** 1189–1195.

27. CHRISTIANSEN, C. & B.J. RIIS. 1990. 17β-Estradiol and continuous norethisterone: A unique treatment for established osteoporosis in elderly women. J. Clin. Endocrinol. Metab. **71:** 836–841.

28. CIVITELLI, R., D. AGNUSDEI, P. NARDI, F. Zacchei, L.V. Avioli & C. Gennari. 1988. Effects of one-year treatment with estrogens on bone mass, intestinal calcium absorption, and 25-hydroxyvitamin D-1α-hydroxylase reserve in postmenopausal osteoporosis. Calcif. Tissue Int. **42:** 77–86.

29. NIH Consensus Conference. 1994. Optimal Calcium Intake. J. Am. Med. Assoc. **272:** 1942–1948.

30. Consensus Development Conference. 1993. Diagnosis, prophylaxis, and treatment of osteoporosis. Am. J. Med. **94:** 646–650.

31. NEED, A.G., M. HOROWITZ, H.A. MORRIS & B.E.C. NORDIN. 1991. Effects of three different calcium preparations on urinary calcium and hydroxyproline excretion in postmenopausal osteoporotic women. Eur. J. Clin. Nutr. **45:** 357–361.

32. PEH, C.A., M. HOROWITZ, J.M. WISHART, A.G. NEED, H.A. MORRIS & B.E.C. NORDIN. 1993. The effect of chlorothiazide on bone-related biochemical variables in normal postmenopausal women. J. Am. Geriatr. Soc. **41:** 513–516.

33. NEED, A.G., H.A. MORRIS, D.B. CLEGHORN, D. DENICHILO, M. HOROWITZ & B.E.C. NORDIN. 1991. Effect of salt restriction on urine hydroxyproline excretion in postmenopausal women. Arch. Intern. Med. **151:** 757–759.

34. MORRIS, H.A., A.G. NEED, M. HOROWITZ, P.D. O'LOUGHLIN & B.E.C. NORDIN. 1991. Calcium absorption in normal and osteoporotic postmenopausal women. Calcif. Tissue Int. **49:** 240–243.

35. NEED, A.G., B.E.C. NORDIN, M. HOROWITZ & H.A. MORRIS. 1990. Calcium and calcitriol therapy in osteoporotic postmenopausal women with impaired calcium absorption. Metabolism **39** (Suppl 1): 53–54.

36. TILYARD, M., G.F.S SPEARS, J. THOMSON & S. DOVEY. 1992. Treatment of postmenopausal osteoporosis with calcitriol or calcium. N. Engl. J. Med. **326:** 357–362.

37. HAN, Z.-H., S. PALNITKAR, D. SUDHAKER RAO, D. NELSON & A.M. PARIFITT. 1997. Effects of ethnicity and age or menopause on the remodeling and turnover of iliac bone: Implications for mechanisms of bone loss. J. Bone Miner. Res. **12:** 498–508.

38. WISHART, J.M., A.G. NEED, M. HOROWITZ, H.A. MORRIS & B.E.C. NORDIN. 1995. Effect of age on bone density and bone turnover in men. Clin. Endocrinol. **42:** 141–146.

39. MEIER, D.E., E.S. ORWOLL, E.J. KEENAN & R.M. FAGERSTROM. 1987. Marked decline in trabecular bone mineral content in healthy men with age: Lack of correlation with sex steroid levels. J. Am. Geriatr. Soc. **35:** 189–197.

40. BROCKSTEDT, H., M. KASSEM, E.F. ERIKSEN, L. MOSEKILDE & F. MELSEN. 1993. Age- and sex-related changes in iliac cortical bone mass and remodeling. Bone **14:** 681–691.

41. LINKSWILER, H.M., M.B. ZEMEL, M. HEGSTED & S. SCHUETTE. 1981. Protein-induced hypercalciuria. Fed. Proc. Fed. Am. Soc. Exp. Biol. **40:** 2429–2433.

42. SCHUETTE, S.A., M. HEGSTED, M.B. ZEMEL & H.M. LINKSWILER. 1981. Renal acid, urinary cyclic AMP, and hydroxyproline excretion as affected by level of protein, sulfur amino acid, and phosphorus intake. J. Nutr. **111:** 2106–2116.

43. NORDIN, B.E.C., H.A. MORRIS, A.G. NEED & M. HOROWITZ,. 1996. Dietary calcium and osteoporosis. *In* Proceedings of the 2nd WHO Symposium on Health Issues for the 21st Century: Nutrition and Quality of Life, Kobe, Japan 1993. P. Pietinen, C. Nishda & N. Khaltaev, Eds: 181–198. World Health Organization. Geneva.

44. NORDIN, B.E.C., A.G. NEED, H.A. MORRIS & M. HOROWITZ. 1993. The nature and significance of the relation between urine sodium and urine calcium in women. J. Nutr. **123:** 1615–1622.

45. GOULDING, A. & P.E. LIM. 1983. Effects of varying dietary salt intake on the fasting excretion of sodium, calcium and hydroxyproline in young women. N.Z. Med. J. **96:** 853–854.

46. Intersalt Cooperative Research Group. 1988. Intersalt: An international study of electrolyte excretion and blood pressure. Results for 24 hour urinary sodium and potassium excretion. Br. Med. J. **297:** 319–328.

47. BULLAMORE, J.R., J.C. GALLAGHER, R. WILKINSON & B.E.C. NORDIN. 1970. Effect of age on calcium absorption. Lancet **2:** 535–537.

48. WOOD, R.J., C.L. THEALL, J.H. CONTOIS & I.H. ROSENBERG. 1988. Intestinal end-organ resistance to 1,25 dihydroxyvitamin D stimulation of calcium absorption in the senescent rat. *In* Vitamin D: Molecular, Cellular and Clinical Endocrinology. A.W. Norman, K. Schaefer, H-G. Grigoleit & D. vanHerrath, Eds.: 907–908. Walter de Gruyter. Berlin.

49. EBELING, P.R., M.E. SANDGREN, E.P. DIMAGNO, A.W. LANE, H.F. DELUCA & B.L. RIGGS. 1992. Evidence of an age-related decrease in intestinal responsiveness to vitamin D: Relationship between serum 1,25-dihydroxyvitamin D_3 and intestinal vitamin D receptor concentrations in normal women. J. Clin. Endocrinol. Metab. **75:** 176–182.

50. MACLAUGHLIN, J. & M.F. HOLICK. 1985. Aging decreases the capacity of human skin to produce vitamin D_3. J. Clin. Invest. **76:** 1536–1538.

51. NEED, A.G., H.A. MORRIS, M. HOROWITZ & B.E.C. NORDIN. 1993. Effects of skin thickness, age, body fat, and sunlight on serum 25-hydroxyvitamin D. Am. J. Clin. Nutr. **58:** 882–885.

52. DUBBELMAN, R., J.H.P. JONXIS, F.A.J. MUSKIET, & A.E.C. SALEH. 1993. Age-dependent vitamin D status and vertebral condition of white women living in Curaçao (The Netherlands Antilles) as compared with their counterparts in The Netherlands. Am. J. Clin. Nutr. **58:** 106–109.

53. van der WIELEN, R.P. *et al.* 1995. Serum vitamin D concentrations among elderly people in Europe. Lancet **346:** 207–210.

54. GLOTH, F.M., C.M. GUNDBERG, B.W. HOLLIS, J.G. HADDAD & J.D. TOBIN. 1995. Vitamin D deficiency in homebound elderly persons. J. Am. Med. Assoc. **275:** 838–839.

55. OOMS, M.E. *et al.* 1995. Vitamin D status and sex hormone binding globulin: Determinants of bone turnover and bone mineral density in elderly women. J. Bone Miner. Res. **10:** 1177–1184.

56. AARON, J.E., J.C. GALLAGHER, J. ANDERSON, L. STASIAK, E.B. LONGTON & B.E.C. NORDIN. 1974. Frequency of osteomalacia and osteoporosis in fractures of the proximal femur. Lancet **2:** 229–233.

57. BAKER, M.R., H. McDONNELL, M. PEACOCK & B.E.C. NORDIN. 1979. Plasma 25-hydroxyvitamin D concentrations in patients with fractures of the femoral neck. Br. Med. J. **1:** 589.

58. MORRIS, H.A. G.W. MORRISON, M. BURR, D.W. THOMAS & B.E.C. NORDIN. 1984. Vitamin D and femoral neck fractures in elderly South Australian women. Mcd. J. Aust. **140:** 519–521.

59. LIPS, P. *et al.* 1982. Histomorphometric profile and vitamin D status in patients with femoral neck fracture. Metab. Bone Dis. & Relat. Res. **4:** 85–93.

60. NORDIN, B.E.C., M.R. BAKER, A. HORSMAN, & M. PEACOCK. 1985. A prospective trial of the effect of vitamin D supplementation on metacarpal bone loss in elderly women. Am. J. Clin. Nutr. **42:** 470–474.

61. CHAPUY, M.C. *et al.* 1992. Vitamin D_3 and calcium to prevent hip fractures in elderly women. N. Engl. J. Med **327:** 1637–1642.

62. PEACOCK, M. 1993. Osteomalacia and rickets. *In* Metabolic Bone and Stone Disease. Third Edition. B.E.C. Nordin, A.G. Need & H.A. Morris, Eds.: 83–118. Churchill Livingstone. Edinburgh.

63. COOPER, C., D.J.P. BARKER & C. WICKHAM. 1988. Physical activity, muscle strength, and calcium intake in fracture of the proximal femur in Britain. Br. Med. J. **297:** 1443–1446.

The Effect of Long-term Dietary Supplementation with Antioxidants[a]

M. MEYDANI,[b] R. D. LIPMAN, S. N. HAN, D. WU, A. BEHARKA, K. R. MARTIN, R. BRONSON, G. CAO, D. SMITH, AND S. N. MEYDANI

JM USDA-Human Nutrition Research Center on Aging at Tufts University, 711 Washington Street, Boston, Massachusetts, USA

ABSTRACT: The impact of diet and specific food groups on aging and age-associated degenerative diseases has been widely recognized in recent years. The modern concept of the free radical theory of aging takes as its basis a shift in the antioxidant/prooxidant balance that leads to increased oxidative stress, dysregulation of cellular function, and aging. In the context of this theory, antioxidants can influence the primary "intrinsic" aging process as well as several secondary age-associated pathological processes. For the latter, several epidemiological and clinical studies have revealed potential roles for dietary antioxidants in the age-associated decline of immune function and the reduction of risk of morbidity and mortality from cancer and heart disease. We reported that long-term supplementation with vitamin E enhances immune function in aged animals and elderly subjects. We have also found that the beneficial effect of vitamin E in the reduction of risk of atherosclerosis is, in part, associated with molecular modulation of the interaction of immune and endothelial cells. Even though the effects of dietary antioxidants on aging have been mostly observed in relation to age-associated diseases, the effects cannot be totally separated from those related to the intrinsic aging process. For modulation of the aging process by antioxidants, earlier reports have indicated that antioxidant feeding increased the median life span of mice to some extent. To further delineate the effect of dietary antioxidants on aging and longevity, middle-aged (18 mo) C57BL\6NIA male mice were fed *ad libitum* semisynthetic AIN-76 diets supplemented with different antioxidants (vitamin E, glutathione, melatonin, and strawberry extract). We found that dietary antioxidants had no effect on the pathological outcome or on mean and maximum life span of the mice, which was observed despite the reduced level of lipid peroxidation products, 4-hydroxynonenol, in the liver of animals supplemented with vitamin E and strawberry extract (1.34 ± 0.4 and 1.6 ± 0.5 nmol/g, respectively) compared to animals fed the control diet (2.35 ± 1.4 nmol/g). However, vitamin E–supplemented mice had significantly lower lung viral levels following influenza infection, a viral challenge associated with oxidative stress. These and other observations indicate that, at present, the effects of dietary antioxidants are mainly demonstrated in connection with age-associated diseases in which oxidative stress appears to be intimately involved. Further studies are needed to determine the effect of antioxidant supplementation on longevity in the context of moderate caloric restriction.

[a]This project has been funded at least in part with Federal funds from the U.S. Department of Agriculture, Agricultural Research Service, under contract number 53-K06-01. The contents of this publication do not necessarily reflect the views or policies of the U.S. Department of Agriculture, nor does mention of trade names, commercial products, or organizations imply endorsement by the U.S. Government.
[b]Corresponding author: Tel: 617/556-3126; fax: 617/556-3224; e-mail: meydani_vbl@hnrc.tufts.edu

Aging is associated with changes in physical characteristics and a decline of many physiological functions in humans. Although the phenomenon of aging is well known, the basic nature of the aging process is not well understood. Among the several theories of aging, the free radical theory of aging, first proposed in 1956 by Harman,[1] has received increasing attention in the last two decades.[2] The theory postulates that the free radical reaction is a single common process that might be responsible for aging and intimately involved in many age-associated disorders. The modern concept of this theory is supported by accumulated data in recent years elucidating the time-dependent shift in the antioxidant/prooxidant balance in favor of oxidative stress that may lead to dysregulation of cellular function, aging, and functional deterioration in later life. Mitochondria, the subject of several reviews in this conference, have been identified as a major source of oxygen free radicals that may contribute to the aging process. Other sources of free radicals, such as plasma membranes, producing free radicals through the activity of such enzymes as lipoxygenase, cyclooxygenase, and NADPH oxidase have also been recognized as potential producers of free radicals. The microsomal electron transport chain, cytochromes P450 and b5, peroxisomal oxidases, flavoproteins, and xanthine oxidases are also other cellular sources of free radicals.[3]

In addition to direct measurement or modulation of free radicals in the aging process and cytotoxicity, indirect evidence regarding changes in the antioxidant defense system has supported the free radical theory of aging. Several mechanisms of free radical–induced oxidative damage, cellular aging, and cytotoxicity are reviewed by other speakers in this conference. Here, the evidence for the contributory role of antioxidant defense mechanisms to longevity and health status are briefly reviewed.

A complex intrinsic antioxidant defense mechanism present in most aerobic organisms scavenges free radicals and reduces oxidative stress, providing compelling evidence that, if unchecked, free radicals and oxidative stress will lead to cytotoxicity and contribute to the aging process. A strong positive correlation between maximum life span of species and their potential antioxidant capacity has been determined in several oxidative defense mechanisms.[4] Moreover, a positive correlation between tissue concentrations of specific antioxidants and the inherent life span in mammals has been elucidated. In testing such relationships, Cutler and associates[5] have determined strong positive correlations in the concentration of superoxide dismutas (SOD) per specific metabolic rate in liver of primates with maximal life span potential.[4]

Increased longevity in animal models by caloric restriction is attributed in part to the modulation of free radical production.[6] In this paradigm, changes in oxidative stress status and activity of antioxidant enzyme systems were suggested to be contributing factors for increasing longevity and thus support the free radical hypothesis of aging. Additional supporting evidence has recently emerged from modulation of enzymatic antioxidant defense mechanisms in transgenic flies, where the accumulation of oxidative damage to DNA and proteins retarded and increased the mean and median life span by 30%.[7]

DIETARY ANTIOXIDANTS AND AGING

The impact of diet and specific food groups on modulation of free radicals and thus aging and age-associated degenerative disease has been widely recognized in recent years. A strong positive correlation has been identified between maximum life span and the con-

centrations of such nonenzymatic antioxidants as tocopherol, carotenoids, and other molecules with antioxidant capacity, such as urate in plasma.[4] Inasmuch as evidence has
indicated that free radical reactions, in part, are the cause of aging and are involved in
many age-related degenerative diseases, nutritional supplementation, especially with antioxidants, are promoted for health benefits and increasing longevity. Thus, many researchers have focused on the role of antioxidants on the prevention and treatment of age-related
diseases or physiological deterioration associated with aging, providing new insights into
the mechanism of action in which free radicals and oxidative stress are known to be
involved (see below).

Earlier studies testing the free radical hypothesis of aging have examined the effect on
their longevity of increasing the antioxidant capacity of organisms by antioxidant supplementation. Addition of 1% by weight of 2-mercaptoethylamine to the diet of male LAF1
mice has shown increased average life span from 24.5 to 31.6 months.[8] Increase of average life span by 18–20% has been observed in C3H mice supplemented with ethoxyquin.[9]
Feeding mice with butylated hydroxytoluene (BHT), a strong synthetic antioxidant, have
shown no effect on average or maximum life span.[10] By contrast, BHT in BALB/c mice
increased the average life span.[11] Feeding autoimmune NZB mice a diet supplemented
with vitamin E at a dose of 2,500 ppm increased the life span in these animals by a small
fraction (7%), whereas ethoxyquin at a dose of 10,000 ppm substantially (32%) increased
the life span in this mouse.[12] These and other studies have indicated that increased intake
of antioxidants may contribute to the increase of median and maximum life span by affecting free radical production.[13] However, the increases in average life span observed with
synthetic antioxidants are related in part to decreased food intake and loss of body weight
in animals. Thus, increase in the average life span in these studies may be attributed to the
caloric restriction effect of these synthetic antioxidants.

To further delineate the effect of dietary antioxidants on aging and longevity, we
recently conducted a study where middle-aged (18-month-old) C57BL/6NIA male mice
were fed diets supplemented with several antioxidants. We used mice raised in barrier conditions at our animal facility at the Jean Mayer USDA Human Nutrition Research Center
on Aging at Tufts University and derived from stock originally obtained from the National
Institute on Aging (NIA) colony. Mice were fed *ad libitum* a modified semisynthetic
AIN76 diet supplemented with one of dietary antioxidants, including vitamin E (500
ppm), glutathione (0.5%), vitamin E plus glutathione (500 ppm + 0.5%, respectively),
melatonin (11 ppm), and strawberry extract (1%). One group of mice as a control group
was fed basal AIN76 diet without any supplement. The mice consumed these diets from
the age of 18 months until sacrificed at 24 months of age (cross-sectional study) or until
the time when 50% of the mice in each group had died (longitudinal study).

The selection of these antioxidants was based on positive health benefit effects that we
had observed in previous studies or those published in the literature. Previously, we had
shown that old C57/BL6NIA mice had an improved immune response when supplemented
with 500 ppm vitamin E for 3 months.[14] We have also demonstrated that feeding a diet
supplemented with 1% glutathione improved the immune response in old mice.[15] In as
much as we have observed positive effects on immune response of mice with these two
antioxidants, we sought to test the synergistic or additive effects of vitamin E- and glu-
tathione-supplemented diets in animals. Recently, melatonin, in addition to functioning as
a regulator of circadian rhythms, has been postulated to have strong biological antioxidant
activity that may improve the immune system of aged animals as well as retard aging.[16,17]

A melatonin-supplemented group was included in this study to determine whether melatonin levels *in vivo* can be modulated by dietary intake and if increased melatonin consumption would affect longevity and disease in old mice. Last, the beneficial effects of fruits and vegetables on the prevention of age-related degenerative diseases have also been posited as potentially enhancing health status. The antioxidant capacity of the strawberry has been determined using the oxygen radical absorbance capacity (ORAC) assay and shown to contain an appreciably higher ORAC value than other fruits and vegetables.[18] Therefore, a water-soluble fraction of strawberry extract was selected as a supplement added to the diet to compare its efficacy with the other antioxidants in this study.

We found no effect of experimental diets on the longevity of the middle-aged C57/BL male mice, as measured by the average age of death. This observation was in agreement with the recent study by Bexlepkin *et al.*,[19] showing that supplementing the diet of middle-aged and old male C57BL/6 with a mixture of β-carotene, alpha-tocopherol, ascorbic acid, rutin, selenium, and zinc had no effect on the longevity. It was of interest that all of the mice in this study gained additional weight when they started consuming the semisynthetic modified AIN76 diet at the age of 18 months, when the body weight in male C57/BL mice normally reaches its apex.[20,21] Therefore, although all the mice in this study were heavier than expected, no difference in weight was observed between dietary groups.

Improvement of immune function in the aged can substantially contribute to an individual's quality of life and perhaps to longevity. We sought to compare the efficacy of various dietary antioxidants on improved immune function in middle-aged mice by infecting them with influenza virus and monitoring their response. The increased susceptibility of the aged to influenza virus infection can be attributed to several factors, including an age-associated decline of cell-mediated and humoral immune response and increased oxidative stress with aging.[22–24] To test whether augmented antioxidant consumption would ameliorate these age-associated changes in immune function, mice were nasally infected with influenza virus after being fed the various diets described above for six months. Five days following viral exposure, the mice were sacrificed, and pulmonary viral titer was measured. Only those mice fed a diet supplemented with vitamin E had significantly lower pulmonary viral titer compared to those fed the control diet. Vitamin E–supplemented mice also maintained their body weight after viral infection, whereas the other group manifested significant loss in weight. The weight loss in the other groups correlated with decreased food intake. The vitamin E–supplemented group had a significantly higher food intake five days postinfection. These positive effects of vitamin E were in accordance with our previous studies showing that vitamin E enhances cell-mediated and humoral immune responses in aged mice and elderly humans.[14,25,26] Interestingly, in mice supplemented with vitamin E plus glutathione, the positive effect of vitamin E alone on the immune function was diminished with the supplemental glutathione, despite reduced liver levels of 4-hydroxynonenal (4-HNE) and malondialdehyde (MDA), indices of lipid peroxidation in liver.

All mice that were sacrificed after six months of being fed dietary antioxidants were thoroughly examined for lesions, abnormalities, and tumors. Antioxidant supplementation started during middle-age did not alter age-associated lesion patterns in these older mice. Fatty liver, focal kidney atrophy, and renal tubular proteinaceous casts were observed more frequently in this study, compared to the reference population of C57/BL mice.

In conclusion, dietary supplementation with antioxidants initiated during middle-age did not appear to affect longevity, lesion patterns, or burden. However, vitamin E was

effective in enhancing immune response when aged mice were challenged with a pathogen. It is possible that if the animals had been started on these antioxidant supplementation
regimens earlier in, our data would have shown an increased life span, as reported by Bexlepkin et al.[19]

DIETARY ANTIOXIDANTS AND AGE-ASSOCIATED DISEASE

Free radicals, in addition to being one of the contributing factors to the aging proces,
have a hypothesized role in the pathogenesis of many age-related degenerative diseases. [27]
The presence of free radical metabolites has been identified in diseased tissues; however,
it is not clear whether free radicals are the cause or the result of pathologic conditions. The
association of pathology with decreased antioxidant concentrations or, alternately, reduced
risk of disease with higher antioxidant capacity, provides indirect evidence for the role of
free radicals and oxidative stress in many age-related diseases.[28,29]

Several studies have shown an inverse association of fruit and vegetable consumption
with risk of morbidity and mortality from such degenerative diseases as cancer and cardiovascular disease.[30,31] It is interesting to note that the antioxidant components of fruits and
vegetables have been postulated to reduce risk of these diseases that are prevalent in middle to late age. A recent burst of research activity focusing on the role of antioxidants in
the treatment and prevention of age-related diseases in which free radicals and oxidative
stress are involved has produced new insights into their mechanism of action. The antioxidant vitamins E and C and β-carotene (provitamin A) have received considerable attention
for their potential role in the prevention of such degenerative diseases as cancer and cardiovascular disease. The totality of evidence from several epidemiological studies and
clinical trials indicates that intake of vitamins E and C above the recommended dietary
allowances may improve age-associated decline of the immune system and reduce risk of
certain types of cancer and cardiovascular disease.

Certain cancers occur more frequently with advancing age, and research suggests that
cancer pathogenesis can be modulated by diet. Vitamins E and C as well as carotenoids
have routinely been shown to have a chemopreventive role in carcinogenesis by inhibiting
mutagenesis and cell transformation in vitro through the quenching of free radicals. However, epidemiological studies of intakes of these vitamins have yielded inconsistent results
for the prevention of total and site-specific cancers. Recent studies have found a strong
association between decreased risk of cancer and dietary vitamin C, fruit, and vegetable
intake. Several epidemiological studies have found significant protection against cancer
with increased intake of vitamin C. Furthermore, epidemiological evidence suggests that a
high intake of vitamin C and/or vitamin C–rich foods may decrease the risk of oral, esophageal, gastric, pancreatic, and colorectal cancers.[32] A large body of research has over-
whelmingly demonstrated that increased intake of foods rich in β-carotene is associated
with reduced risk of cancer of the cervix, esophagus, and stomach.[33] An overview of
numerous epidemiological studies has shown that high intakes of fruits and vegetables
rich in carotenoids or elevated blood concentrations of β-carotene are associated with a
decreased risk of cancer at several cancer sites. Supplementation of individuals with 15
mg β-carotene, 30 mg vitamin E, and 50 μg selenium for 5.25 years substantially
decreased total and cancer mortality in a large, high-risk population in Linxian, China, by
13%, primarily by lowering the rates of stomach and esophageal cancer.[34] However, long-

term supplemental intake of β-carotene by smokers and nonsmokers in several large studies failed to demonstrate a reduction in lung cancer.[35–37]

Oxygen-derived free radicals and lipid peroxidation are now widely accepted as major contributors to the etiology of atherosclerosis and its chronic disorders, including coronary heart disease, stroke and ischemic dementia.[38] Epidemiology generally supports an inverse relationship between dietary antioxidants and cardiovascular disease (CVD). Cross-cultural studies of European populations have found a strong correlation between low plasma vitamin E and, to a lesser extent, β-carotene and vitamin C, and increased rates of ischemic heart disease mortality rates.[39] Several recent epidemiological studies have shown a strong association between dietary vitamin E and reduced risk of CVD.[40–43] Clinical findings also support the preventive effect of vitamin E on the progression of coronary artery atherosclerosis.[44,45]

Dietary antioxidants may contribute to the decrease of CVD by reduction of free radical formation as well as oxidative stress in general, by protecting against LDL oxidation and platelet aggregation, and by inhibiting synthesis of proinflammatory cytokines. Dietary antioxidants may also decrease the interaction and activation of immune, endothelial, and smooth muscle cells of vessel walls. In atherosclerosis, indiscriminate uptake of oxidized LDL by scavenger receptors of macrophages results in accumulation of cholesterol-laden foam cells and fatty streak formation. Recent data have elucidated an important role for antioxidants in the reduction of adhesive interaction of immune and endothelial cells in the early stage of atherogenesis.[46,47]

Aging is associated with a decline of immune response, resulting in increased infection and other chronic disorders. Evidence indicates that altered regulation of immune function in the elderly is associated with increased production of oxygen free radicals. Dietary antioxidants have been shown to augment several immunologic parameters, including increased neutrophil mobility, increased delayed-type hypersensitivity (DTH) response, and stimulated lymphocyte proliferation. An insufficient level of vitamin E causes immune cell membranes to become unstable and enhances production of immunosuppressors, such as prostaglandins, which can be reversed with vitamin E supplementation.[33,48] Meydani and colleagues[25, 26] have shown that both short- and long-term vitamin E supplementation increase several indices of immune function, including increases in T-cell proliferation and IL-2 production, DTH, and antibody response.

Numerous studies have shown increased free radical formation and lipid peroxidation as a causative factor in various neurological disorders. Recently, Sano *et al.*[49] reported that supplemental intake of vitamin E (2000 IU/day) for two years delayed progression of Alzheimer's disease. Earlier, Fahn[50] showed that high doses of vitamin E (3200 mg/day) and vitamin C (3000 mg/day) extended the time to medicinal intervention in patients with Parkinson's disease by 2.5 years. Thus, for these two major age-related neurological diseases in which oxidative damage is believed to play a role, large doses of antioxidant vitamins have been shown to be effective in delaying disease progression.

CONCLUSION

We conclude that there is strong experimental evidence to support the free radical theory of aging. As a result of increased understanding of the characteristics and production of free radicals, an increasing number of diseases and disorders, as well as the aging process itself, demonstrate a link either directly or indirectly to these reactive and potentially destructive molecules. We predict that reduction of free radicals or decreasing their rate of production may delay aging and the onset of degenerative conditions associated with aging. Evidence for the efficacy of antioxidant supplements, started at middle-age, on longevity and increasing median and maximum life span is not convincing. However, compelling evidence indicates that increased consumption of dietary antioxidants or fruits and vegetables containing nutritive and nonnutritive compounds with antioxidant properties may contribute to the improvement of the quality of life by delaying onset and reducing the risk of degenerative diseases associated with aging.

ACKNOWLEDGMENT

The authors would like to thank Timothy S. McElreavy, M.A., for preparation of this manuscript.

REFERENCES

1. HARMAN, D. 1956. Aging: A theory based on free radical and radiation chemistry. J. Gerontol. **11:** 198–300.
2. YU, B.P. 1993. Free Radicals in Aging. CRC Press. Boca Raton, FL.
3. YU, B.P. 1993. In Free Radicals in Aging. B.P. Yu, Ed.:57–88. CRC Press. Boca Raton, FL.
4. CUTLER, R.G. 1991. Antioxidants and aging. Am. J. Clin. Nutr. **53:** 373S–379S.
5. TOLMASOFF, J. M. *et al.* 1980. Superoxide dismutase: Correlation with life span and specific metabolic rate in primate species. Proc. Natl. Acad. Sci. USA **77:** 2777–2781.
6. SOHAL, R.S. & WEINDRUCH, R. 1996. Oxidative stress, caloric restriction, and aging. Science **273:** 59–63.
7. SOHAL, R.S., *et al.* 1995. Simultaneous overexpression of copper- and zinc-containing superoxide dismutase and catalase retard age-related oxidative damage and increases metabolic potential on *Drosophila melanogaster.* J. Biol. Chem. **270:** 15671–15674.
8. HARMAN, D. 1968. Free radical theory of aging: Effect of free radical inhibitors on the mortality rate of male LFA mice. J. Gerontol. **23:** 476–482.
9. COMFORT, A. 1971. Effect of ethoxyquin on the longevity of C3H mice. Nature **229:** 254–255.
10. KOHN, R.R. 1971. Effect of antioxidants on life span of C57/BL mice. J. Gerontol. **26:** 376–380.
11. CLAPP, N.K. *et al.* 1979. Effects of the antioxidant buthylated hydroxytoluene (BHT) on mortality in BALB/c mice. J. Gerontol. **34:** 497–501.
12. HARMAN, D. 1980. Free radical theory of aging: Beneficial effect of antioxidants on the life span of male NZB mice; role of free radical reactions in the deterioration of the immune system with age and in the pathogenesis of systemic lupus erythematosus. Age **3:** 64–73.
13. HARMAN, D. 1994. Aging: Prospect for further increases in the functional life span. Age **17:** 119–146.
14. MEYDANI, S.N. *et al.* 1986. Vitamin E supplementation suppresses prostaglandin E2 synthesis and enhances the immune response of aged mice. Mech. Aging Dev. **34:** 191–201.
15. FURUKAWA, T. *et al.* 1987. Reversal of age-associated decline in immune responsiveness by dietary glutathione supplementation in mice. Mech. Aging Dev. **38:** 107–117.
16. REITER, R.J., M.I. PABLOS, T.T. AGAPITO & J.M. GUERRERO. 1996. Melatonin in the context of the free radical theory of aging. Ann. N.Y. Acad. Sci. **786:** 362–378.
17. PIERPAOLI, W. & W. REGELSON. 1994. Pineal control of aging: Effect of melatonin and pineal grafting on aging mice. Proc. Natl. Acad. Sci. USA **91:** 787–791.
18. WANG, H. *et al.* 1996. Total antioxidant capacity of fruits. J. Agric. Food Chem. **44:** 701–705.
19. BEXLEPKIN, V.G. *et al.* 1996. The prolongation of survival in mice by dietary antioxidants depends on their age by the start of feeding this diet. Mech. Aging. Dev. **92:** 227–234.

20. GOODRICK, C.L. *et al.* 1990. Effects of intermittent feeding upon body weight and life span in inbred mice: Interaction of genotype and age. Mech. Aging. Dev. **55:** 69–87.

21. TURTURRO, A.K. BLANK, D. MURASKO & R. HART. 1994. Mechanisms of caloric restriction affecting aging and disease. Ann. N.Y. Acad. Sci. **719:** 159–170.

22. HENNET, T. *et al.* 1992. Alterations in antioxidant defenses in lung and liver of mice infected with influenza A virus. J. Gen. Virol. **73:** 39–46.

23. PETERHANS, E. 1994. *In* Natural Antioxidants in Human Health and Disease. B. Frei, Ed.: 489–514. Academic Press. San Diego.

24. BUFFINTON, G.D. *et al.* 1992. Oxidative stress in lung of mice infected with influenza A virus. Free Radical Res. Commun. **16:** 99–110.

25. MEYDANI, S.N. *et al.* 1990. Vitamin E supplementation enhances cell-mediated immunity in healthy elderly subjects. Am. J. Clin. Nutr. **52:** 557–563.

26. MEYDANI, S.N. *et al.* 1997. Vitamin E supplementation enhances *in vivo* immune response in healthy elderly: A dose-response study. J. Am. Med. Assoc. **277:** 1380–1386.

27. HALLIWELL, B. 1994. Free radicals and antioxidants: A personal view. Nutr. Rev. **52:** 253–265.

28. HALLIWELL, B. & J.M.C. GUTTERIDGE. 1995. Free Radicals in Biology and Medicine. Oxford University Press. Oxford.

29. FREI, B. 1994. Natural Antioxidants in Human Health and Disease. Academic Press. San Diego.

30. AMES, B.N. *et al.* 1993. Oxidants, antioxidants, and the degenerative diseases of aging. Proc. Natl. Acad. Sci. USA **90:** 7915–7922.

31. MEYDANI, M. 1995. Vitamin E. Lancet **345:** 170–175.

32. BLOCK, G. 1991. Vitamin C and cancer prevention: The epidemiologic evidence. Am. J. Clin. Nutr. **53:** 270S–282S.

33. MEYDANI, S.N. *et al.* 1995. Antioxidants and the immune response in aged persons: Overview of the present evidence. Am. J. Clin. Nutr. **62:** 1462S–1476S.

34. BLOT, W.J. *et al.* 1995. The Linxian Trials: Mortality rates by vitamin-mineral intervention group. Am. J. Clin. Nutr. **62**(suppl): 1425S–1426S.

35. The Alpha-Tocopherol Beta Carotene Cancer Prevention Study Group. 1994. The effect of vitamin E and beta carotene on the incidence of lung cancer and other cancers in male smokers. N. Engl. J. Med. **330:** 1029–1035.

36. GOODMAN, G.E. *et al.* 1996. The association between participant characteristics and serum concentrations of beta-carotene, retinol, retinyl palmitate, and alpha-tocopherol among participants in the Carotene and Retinol Efficacy Trial (CARET) for prevention of lung cancer. Cancer Epidemiol. Biomarks Prev. **5:** 815–821.

37. HENNEKENS, C.H. *et al.* 1996. Lack of effect of long-term supplementation with beta carotene on the incidence of malignant neoplasms and cardiovascular disease. N. Engl. J. Med. **334:** 1145–1149.

38. KNIGHT, J.A. 1995. Diseases related to oxygen-derived free radicals. Ann. Clin. Lab. Sci. **25:** 111–121.

39. GEY, K.F. *et al.* 1987. Plasma levels of antioxidant vitamins in relation to ischemic heart disease and cancer. Am. J. Clin. Nutr. **45:** 1368–1377.

40. RIMM, E.B. *et al.* 1993. Vitamin E consumption and the risk of coronary heart disease in men. N. Engl. J. Med. **328:** 1450–1456.

41. STAMPFER, M.J. *et al.* 1993. Vitamin E consumption and the risk of coronary disease in women. N. Engl. J. Med. **328:** 1444–1449.

42. LOSONCZY, K.G. *et al.* 1996. Vitamin E and vitamin C supplementation use and risk of all-cause and coronary heart disease mortality in older persons: The Established Populations for Epidemiologic Studies of the Elderly. Am. J. Clin. Nutr. **64:** 190–196.

43. KUSHI, L.H. *et al.* 1996. Dietary antioxidant vitamins and death from coronary heart disease in postmenopausal women. N. Engl. J. Med. **334:** 1156–1162.

44. HODIS, H.N. *et al.* 1995. Serial coronary angiographic evidence that antioxidant vitamin intake reduces progression of coronary artery atherosclerosis. J. Am. Med. Assoc. **273:** 1849–1854.

45. STEPHENS, N.G. *et al.* 1996. Randomized, controlled trial of vitamin E in patients with coronary disease: Cambridge Heart Antioxidant Study (CHAOS). Lancet **347:** 781–786.

46. FARUQI, R. *et al.* 1994. α-Tocopherol inhibits against induced monocytic cell adhesion to cultured human endothelial cells. J. Clin. Invest. **94:** 592–600.

47. MARTIN, A. *et al.* 1997. Vitamin E inhibits low density lipoprotein-induced adhesion of monocytes to human aortic endothelial cells *in vitro*. Arterioscler. Thromb. Vasc. Biol. **17:** 429–436.

48. MEYDANI, S.N. & R.P. TENGERDY. 1992. In Vitamin E in Health and Disease. L. Packer & J. Fuchs, Eds.: 549–562. Marcel Dekker. New York.

49. SANO, M. *et al.* 1997. A controlled trial of selegiline, alpha-tocopherol, or both as treatment for Alzheimer's disease. N. Engl. J. Med. **336:** 1216–1222.

50. FAHN, S. 1991. An open trial of high-dosage antioxidants in early Parkinson's disease. Am. J. Clin. Nutr. **53:** 380S–382S.

B Vitamins and Homocysteine in Cardiovascular Disease and Aging

DAVID E. L. WILCKEN[a,c] AND BRIDGET WILCKEN[b]

[a]*Department of Cardiovascular Medicine, University of New South Wales, The Prince Henry and Prince of Wales Hospitals, Sydney*
[b]*The New Children's Hospital, Westmead, Sydney*

ABSTRACT: The sulfur-containing amino acid, homocysteine, is formed from the essential amino acid methionine, and a number of B vitamins are involved in methionine metabolism. Pyridoxine, vitamin B_6, is a cofactor for cystathionine β synthase, which mediates the transformation of homocysteine to cystathionine, the initial step in the transsulfuration pathway and the urinary excretion of sulfur. In a normal diet there is conservation of the carbon skeleton, and about 50% of the homocysteine formed is remethylated to methionine via steps that require folic acid and vitamin B_{12}. A deficiency of any of these three vitamins leads to modest homocyst(e)ine elevation, as does diminished renal function, both of which are common in the elderly. It is also established that homocyst(e)ine elevation of this order is associated with increased cardiovascular risk but is also associated with most established risk factors, although it is thought to be an independent contributor.

In the inborn error of metabolism homocystinuria due to cystathionine β synthase deficiency there is greatly increased circulating homocyst(e)ine and a clear association with precocious vascular disease. In about 50% of these patients there is a vascular event before the age of 30 years. The homocysteine-induced adverse vascular changes appear to result from endothelial and smooth muscle cell effects and increased thrombogenesis. We have documented a highly significant reduction in the occurrence of vascular events during 539 patient years of treatment in 32 patients with cystathionine β synthase deficiency (mean age 30 years, range 9–66 years) by aggressive homocyst(e)ine lowering with pyridoxine, folic acid, and B_{12} (p = 0.0001). The 15 pyridoxine nonresponsive patients also received oral betaine.

Although a cause and effect relationship is postulated for the increased cardiovascular risk associated with mild homocysteine elevation, a common cause of this elevation is the methylenetetrahydrofolate reductase C677T mutation. Homozygotes occur in about 11% of Caucasian populations. However, the mutation is not associated with increased coronary risk. Since mild homocysteine elevation is easily normalized by B vitamin supplementation, usually with folic acid, it remains for controlled clinical trials of this inexpensive therapy to determine whether normalizing mild homocyst(e)ine elevation reduces cardiovascular risk.

It is now generally accepted that there is a definite association between mildly elevated circulating levels of the sulfur-containing plasma amino acid, homocysteine, and the occurrence of vascular disease. The results of a recent meta-analysis by Boushey and colleagues[1] of the many studies undertaken to explore this issue suggest that modest elevation of homocysteine accounts for 10% of the population risk for coronary artery disease (CAD). The findings indicated that for a 5 μmol/L increase in total homocysteine the odds

[c]Address for correspondence: Professor David Wilcken, Cardiovascular Genetics Laboratory, South Wing, Ground Floor, Edmund Blacket Building, Prince of Wales Hospital, Avoca Street, Randwick, NSW 2031 Australia. Tel: 61 2 9382 4832 (or 5); fax: 61 2 9382 4826; e-mail: d.wilcken@unsw.edu.au

ratio for an increase in CAD risk in men was 1.6 (95% confidence interval (C.I.), 1.4–1.7) and in women 1.8 (95% C.I., 1.3–1.9). For cerebrovascular disease it was 1.5 (95% C.I., 1.3–1.9); for the combined sexes and for peripheral vascular disease it was 6.8 (95% C.I., 2.9–15.8).[1] Very similar results were obtained in the more recent European Concerted Action Project, which involved 750 cases of atherosclerotic disease in both men and women aged less than 60 years and 800 age- and sex-matched controls.[2] There was also a dose-response effect between the increase in total homocysteine and risk in this study, which the investigators concluded was independent of other risk factors, although there were interactions with them, an observation that will be discussed further below.

The most recent contribution to the documentation of the homocysteine hypothesis is of a mortality study during a 4- to 6-year follow-up of coronary patients assessed angiographically in Norway during 1991 and 1992.[3] There were 587 patients of median age 62 years, among whom 64 had died, 50 of a cardiovascular, mainly coronary, cause. For the whole series there was a positive relationship between plasma homocysteine and age, as observed by others,[4] and levels were higher in men than in women, confirming our original observations.[5] After controlling for these, the strongest predictors of plasma homocysteine in descending order of significance were serum folate, serum creatinine, serum uric acid, serum B_{12}, and left ventricular ejection fraction ($p < 0.001$ for each). The key finding of the study was "a strong, graded dose-response relation between total homocysteine level and overall mortality;" this relationship was slightly stronger when the analysis was confined to the 50 patients who had died of cardiovascular disease.

The European Concerted Action Project involved only patients and controls younger than 60 years, as did many of those included in the meta-analysis of Boushey and colleagues.[1] However, the Norwegian prospective study was of patients over the age of 60 in the main. That there is a positive correlation between homocysteine levels and increasing age is due in part to relative shortages of B vitamins.[4] As will be discussed, B vitamins make important contributions to the metabolism of homocysteine. Shortages, most commonly of folic acid, and less commonly vitamin B_{12}, and pyridoxine (vitamin B_6) increase circulating homocysteine, as the recent Norwegian study confirms.[3] These shortages are much more common in the elderly.[4] The association in elderly subjects of increased prevalence of atherosclerotic carotid lesions, with both increased circulating total homocysteine and reduced B vitamin levels, mainly folic acid and B_{12}, is well documented in an extensive investigation by Selhub and colleagues.[6]

In view of these compelling findings, it is apposite to review the determinants of homocysteine metabolism and to consider further the evidence that supports an important role for increased circulating homocysteine in the pathogenesis of vascular disease. It is also important to assess the available therapeutic strategies to lower elevated levels and, most importantly, to explore the effects of lowering levels on vascular outcomes and the reduction of vascular disease.

THE EVOLUTION OF THE CONCEPT

The development of the concept that elevated levels of the circulating sulfur-containing amino acid, homocysteine, contributes to the pathogenesis of vascular disease began in 1962 with the identification in Ireland[7] and in the USA[8] of a new group of inborn errors of metabolism, now described collectively as homocystinuria. These studies identified high concentrations of homocystine, the oxidized form of homocysteine, in the urine in some children with mental retardation. It was shown that these children also had high circulating concentrations of homocysteine, measured as the oxidized compound homocystine, and the mixed disulfide, cysteine-homocysteine. Homocystine is not normally detected in human urine, but it is identified in measurable quantities when plasma levels are only slightly ele vated, as it is not avidly reabsorbed by the kidney.[9] During the three years following the dis-covery of homocystinuria there came the recognition that it was associated with severe vascular disease and that this was the usual cause of premature death.[10–12] It was not until 1969 however, that McCully[13] first postulated that vascular complications were the result of elevated homocysteine rather than of any of the other complex metabolic changes occurring in this group of disorders. The reasoning was as follows.

Homocysteine is formed from the essential sulfur-containing amino acid, methionine, via transmethylation reactions, as shown in FIGURE 1. The initial step in the metabolism of methionine is the formation of S-adenosyl-methionine, a major methyl donor, and then subsequently homocysteine. The homocysteine formed is either further metabolized by the transsulfuration pathway to cystathionine and cysteine with ultimately the excretion of sulfur in the urine or it is remethylated with the formation again of methionine. In classical homocystinuria there is reduced or absent activity of the enzyme cystathionine β-synthase, which mediates the step from homocysteine to cystathionine. Consequently a block in the pathway at that point leads to markedly elevated circulating homocysteine and its oxidized

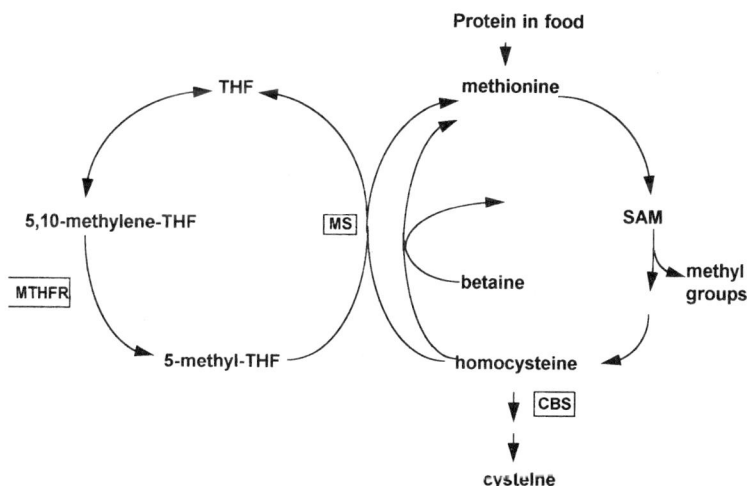

FIGURE 1. The methionine degradation pathway. Methionine is metabolized via S-adenosyl methionine (SAM) and S-adenosyl-homocysteine to homocysteine in the course of producing methyl groups for use in synthetic processes. In a normal diet about 50% of the homocysteine formed is metabolized via the transsulfuration pathway (see text). The first step involves the enzyme cystathionine β-synthase (CBS), for which pyridoxine (B_6) is the cofactor, and deficiences of this enzyme result in the usual form of homocystinuria. The remaining 50% of formed homocysteine is remethylated to methionine and requires 5-methyltetrahydrofolate as substrate, and methyl cobal-amin as a cofactor. Remethylation by trimethylglycine (betaine) may also occur via a separate path-way. MS, methionine synthase; MTHFR, 5,10 methylenetetrahydrofolate reductase.

products and also markedly elevated circulating levels of methionine. Pyridoxal phosphate (vitamin B_6) is the cofactor for cystathione β-synthase and the formation of cystathionine from homocysteine.

In normal subjects receiving a normal diet, there is conservation of the carbon/nitrogen/ sulfur skeleton so that about 50% of the homocysteine normally formed is remethylated to methionine by reactions that depend upon both folic acid and vitamin B_{12}.[9] The discovery of disorders of the remethylating pathway that also result in an accumulation of homocysteine and homocystinuria but with low methionine levels as opposed to the high concentrations in cystathionine β-synthase deficiency, provided the clue, as these remethylating disorders were also found to be associated with precocious vascular disease. Thus McCully suggested that elevated homocysteine was responsible for the vascular changes, as it was the common factor in both remethylating and transsulfuration disorders.[13]

MEASUREMENT OF PLASMA HOMOCYSTEINE

This is measured either as total "free" homocysteine in the supernatant after removal of the plasma protein, using an amino acid analyzer, or as total homocysteine, including the protein-bound moiety by HPLC. With the former approach, the homocysteine is oxidized to both homocystine and homocysteine-cysteine mixed disulfide during sample handling and preparation. Total unbound free homocysteine is then calculated as twice the molar concentration of homocystine plus the concentration of cysteine-homocysteine. Normal plasma levels after an overnight fast, as reported by different laboratories, are between 0.9 ± 0.3 and 3.5 ± 0.5 μmol/L (means $\pm$ SD). Alternatively "total" homocysteine, to include both free and protein-bound homocysteine, is measured most commonly using HPLC methodology after chemical reduction of homocystine and mixed disulfide. Because these latter measurements are easily automated this has become the preferred approach. Normal values recorded by different laboratories after an overnight fast lie between 5.7 ± 1.2 and 12.1 ± 4.0 μmol/L (means $\pm$ SD). Because with both methods the measured values represent the sum of circulating homocystine and cysteine-homocysteine, it is recommended that these total levels be reported as homocyst(e)ine concentrations although homocysteine is still commonly used. In healthy adults protein-bound homocysteine accounts for over 70% of the total measured homocysteine and for a smaller fraction in patients with the greatly elevated levels found in homocystinuria.[14] For either measurement the collected blood should be separated rapidly (within an hour). For measurement of free homocysteine, storage at –80°C is necessary until analysis. Storage at –20°C is quite adequate for measurement of total homocysteine.

THE NATURAL HISTORY OF VASCULAR DISEASE IN HOMOCYSTINURIA AND THE EFFECTS OF B VITAMIN TREATMENT

To establish that elevated circulating homocysteine is of itself functional in the pathogenesis of vascular disease, it is mandatory to show that reducing plasma levels by appropriate intervention also reduces cardiovascular risk. A landmark study in 1985 by Mudd and colleagues documented the natural history of vascular disease in 629 patients with homocystinuria due to cystathionine β-synthase (EC 4.2.1.22) deficiency by combining data provided by 113 physicians throughout the developed world.[15] Because the information was gathered in 1982 and early 1983, this study was able to define outcomes in untreated patients and confirm that the usual cause of premature death was vascular disease, with thromboembolism a major cause of morbidity. It is these findings that provide a basis for the assessment of the effects of current treatment regimens on vascular outcomes.

We have assessed the effects of treatment on vascular events in 40 patients with homocystinuria due to cystathionine β-synthase deficiency, these all being patients identified with this disorder in the state of NSW, Australia, and followed long term.[16] Five recently diagnosed patients are not included. The 40 patients are from 25 sibships, and they were diagnosed on the basis of characteristic clinical features, elevated levels of plasma methionine and free homocysteine, together with low free cysteine. We have monitored surviving patients with regular plasma amino acid analyses, usually twice yearly or more frequently, using Beckman or Jeol analyzers. Plasma was separated within 20 minutes of venesection, and either immediately deproteinized, using sulfosalicylic acid, or stored at −80°C until analysis, when deproteinization was achieved by ultrafiltration. The patients were characterized as being pyridoxine-responsive or pyridoxine-nonresponsive according to the level of total free plasma homocysteine after treatment with pyridoxine and folate only. Patients who maintained total free plasma homocysteine levels below 20 μmol/L after treatment (see below) were classified as pyridoxine responsive. Total plasma free homocysteine was calculated as the sum of twice the molar concentration of homocystine plus the concentration of cysteine-homocysteine mixed disulfide, expressed as μmol/liter.

TREATMENT REGIMENS

The standard treatment regimen for all patients was pyridoxine (100–200 mg/day), folic acid (5 mg/day), and intermittent hydroxocobalamin by injection, according to the serum B_{12} level measured usually twice yearly. Pyridoxine nonresponsive patients all received additional 6–9 g oral trimethylglycine (betaine), given in two divided doses.

Pyridoxine is a cofactor for the enzyme cystathionine β-synthase, deficiency of which is responsible for classical homocystinuria. In pyridoxine-responsive homocystinuric patients there is some residual enzyme activity that can be enhanced by pyridoxine treatment in the doses we have used. The effect is to increase the depressed activity of the transsulfuration pathway, thereby decreasing elevated circulating homocysteine and methionine. Folic acid enhances the remethylation of homocysteine to methionine, thereby further reducing the elevated homocysteine levels. The point of entry of folate coenzymes into the remethylation of homocysteine is via 5,10-methylenetetrahydrofolate reductase (MTHFR) (EC 1.1.1.68), which catalyzes the reduction of 5,10-methylenetetrahydrofolate to 5-methyl tetrahydrofolate. Cystathionine β synthase patients frequently become mildly B_{12} deficient and because vitamin B_{12} is also a cofactor for the remethylation of homocysteine to methionine, vitamin B_{12} deficiency reduces remethylation with a consequent elevation of homocysteine. This is avoided by periodic monitoring of vitamin B_{12} levels with repletion as indicated. The use of oral betaine has greatly improved the control of pyridoxine nonresponsive patients by enhancing further the remethylation of homocysteine to methionine by a different pathway.[17] It has proved to be a safe and effective therapy during a 16-year treatment period.[16]

RESULTS

We documented the occurrence of vascular events in the 40 patients we have studied with long-term follow-up in relation to whether or not they were receiving treatment at the time of the event. To assess the effect of treatment on the occurrence of vascular events, we compared the findings of our treated patients with those in the untreated patients reported by Mudd and colleagues.[15] In this way we were able to relate our findings in individual patients to the age-to-event curves for untreated patients documented by Mudd *et al.*[15] In their large series, maximum risk occurred after the age of 10 years; thereafter there was one event per 25 years. We compared our risk data with those findings. This provides a conservative estimate, as two of our patients had already had vascular events before treatment was started, and from the data of Mudd *et al.* this enhances the risk in untreated patients to about one event in 10 years.[15]

In the whole series of 40 patients there were 10 deaths. Eight of these were early cases, and six were never treated, whereas two had no effective treatment. Those never treated died at the ages of 2, 5, 6,10, 20, and 23 years, three of cerebral thrombosis and two suddenly of a presumed vascular cause; in one the cause of death was unknown. In the two patients for whom no biochemically effective treatment was given, both died of cerebral thrombosis at the age of 10 years. There were two deaths in patients who had received effective treatment. One death was an accidental death in a 15-year-old boy and unrelated to his disorder; the other death was in a 30-year-old woman who was found dead and at autopsy had had a massive pulmonary embolus. This woman had had psychiatric problems and lived remote from Sydney; it was uncertain whether or not she was actually taking her prescribed therapy at the time of her death.

Thirty-two of the 40 patients received effective treatment. In the 17 of these who were pyridoxine responsive, during a follow-up period of 281 patient years of treatment, plasma total free homocyst(e)ine levels were maintained consistently less than 20 μmol/liter. There were two vascular events. One patient described above died from a documented pulmonary embolus. The second patient was diagnosed at the age of 41 years and had had four vascular events before starting treatment at that time. He had a second (nonfatal) myocardial infarction at the age of 55 years. From the data of Mudd and colleagues,[15] in the 17 patients, if untreated, 11 events would have been expected, relative risk = .017 (95% C.I., 0.04–0.80), $\chi^2 = 4.94$, p = 0.026.

During 258 patient years of treatment in the 15 pyridoxine-nonresponsive patients there were no vascular events. The mean total free homocyst(e)ine levels in these patients (± SD) was 33 ± 17 μmol/liter. In 15 untreated pyridoxine nonresponsive patients treated for this time, at least 10 events would have been expected, $\chi^2 = 8.0$, p = 0.005. Thus during a total of 539 patient years of treatment in all 32 patients, there were two vascular events, whereas 21 events would have been expected in untreated patients. These results indicate that treatment markedly reduced the relative risk to 0.09 (95% C.I., 0.02–0.38), $\chi^2 = 14.22$, p = 0.0001.

These studies in an uncommon disorder, occurring in about 1 in 50,000 of the Australian population, establish that marked homocysteine elevation is associated with precocious vascular disease and that effective lowering of the elevated homocysteine, even to suboptimal levels, appreciably reduces cardiovascular risk. This then leads to the broader question of the relevance to cardiovascular risk of mild homocysteine elevation, which is not uncommon in the general population and, as mentioned, particularly in the older age groups.

VASCULAR DISEASE AND MILD HOMOCYSTEINE ELEVATION: SO-CALLED HYPERHOMOCYSTEINEMIA

In a significant proportion of patients with premature vascular disease, it is difficult to explain the occurrence and the extent of the disease in terms of standard risk factors, particularly in those with coronary artery disease. This was the stimulus for our initial study to explore the possibility that mild elevation of circulating homocysteine could contribute.[18] In a group of patients with early-onset coronary disease, we were able to show that in a significant proportion of these there was an increase in circulating homocysteine after the challenge of a methionine load.[18] The many studies that have confirmed these initial observations have demonstrated that either an increase in plasma homocysteine after an overnight fast or an abnormal increase after the challenge of a methionine load are common in patients with coronary, cerebral, and peripheral vascular disease, as referred to above.[1,2] However, the key question is whether these modestly elevated levels are contributing directly to the pathogenesis or whether they are a reflection of the effects of other risk factors with functional significance.

The recent very large Hordaland homocysteine study, involving 7591 men and 8585 women aged 40 to 67 years of age without a history of hypertension, diabetes, coronary artery, or cerebrovascular disease revealed positive associations between total homocysteine levels and increased cholesterol, increased blood pressure, increased resting heart rate, increase in age, male gender, and particularly to smoking.[19] There was also an inverse relationship with physical activity. It is difficult to control adequately for all these factors in studies such as the European Concerted Action Project[2] and that of mortality in the Norwegian coronary patients.[3] This is particularly so for smoking, for which quantitation as lifetime smoking dose (in pack years) appears to be the most relevant measurement.[20] In the Norwegian study, 75% of the patients were current or former smokers. There was also the positive relationship with serum creatinine in that study. Reduced renal function increases circulating homocysteine, and folic acid lowers these levels.[21] Thus the increase of homocysteine with age could reflect an age-related decline in renal function, amplified by reduced B vitamin intake because of poor nutritional status.

Further information that argues against small homocysteine increases contributing directly to an increase in cardiovascular risk comes from the identification by Frosst and colleagues of the common MTHFR mutation.[22] Those homozygous for the mutation occur with a frequency of about 11% in Caucasian populations.[23] They have reduced activity of the enzyme and a need for an above-average oral folate intake to ensure normal remethylation of homocysteine to methionine (FIG. 1); and when tissue folate stores, as assessed by red blood cell concentrations, fall below median population levels, there is a modest increase in circulating homocysteine.[24] Despite this, in the large number of coronary patients and control subjects in whom the mutation has been assessed in the United States and Australia, no association between the mutation and either the occurrence or the extent of coronary artery disease, or of a history of myocardial infarction, has been found, even although there was modest homocysteine elevation associated with the mutation.[25–27]

These data argue against mild hyperhomocysteinemia contributing directly to cardiovascular risk, unless the mutation is in some way cardioprotective.

For modest homocysteine elevation to be an independent contributor to increased cardiovascular risk, a biologically plausible mechanism to account for the effect should be evident. The mechanisms whereby high levels of homocysteine predispose to vascular disease are thought to include homocysteine-induced endothelial damage and smooth muscle cell proliferation and associated enhanced thrombogenesis.[9] Whether the modest elevations documented in vascular patients of about 5 μmol/L can have significant functional effects, however, must be considered in relation to the fact that only about 30% is free homocysteine. Thus one is left with the difficulty of providing a plausible mechanism for the deleterious effects of a 1 or 2 μmol/L increase in free circulating homocysteine on vascular function and endothelial integrity, when it is known that directly caused vascular pathology and relevant outcome data can be demonstrated only with increases that are larger, perhaps by 50- to 100-fold.

Notwithstanding these observations, when Tawakol and colleagues[28] compared 26 hyperhomocysteinemic individuals aged 60 to 80 years with total homocysteine levels ≥ 16 μmol/L with 15 age- and sex-matched controls with levels < 11 μmol/L in relation to noninvasively assessed endothelial function, flow-mediated endothelium-dependant vasodilatation was distinctly reduced in the former group who had a higher homocysteine level. Furthermore there is good evidence that in elderly patients, mild homocysteine elevation is related to carotid disease detected by ultrasound and that the elevation is associated with reduced circulating B vitamin levels, predominantly folic acid.[6] Thus these findings are consistent with those of Tawakol and colleagues,[28] that there is endothelial dysfunction in elderly patients with modest homocysteine increases of this order and that this is manifested by reduced endothelium-dependent vasodilatation. At the other end of the age spectrum, these same functional changes can be demonstrated in children with homocystinuria due to cystathionine β-synthase deficiency who have greatly elevated homocysteine levels.[29]

In summary the B vitamins, folic acid, vitamin B_{12}, and pyridoxine (B_6) have pivotal roles in methionine metabolism and in determining circulating homocysteine levels. In the elderly, for nutritional reasons, deficiencies of these vitamins are common, particularly folic acid deficiency. Consequently so too is mild homocysteine elevation. There is now clear evidence that in the elderly modest homocysteine elevation is associated with altered endothelial function, potentially at an early stage of atherogenesis, and that there are clear associations between modest homocysteine elevation and the occurrence of coronary, cerebral, peripheral, and carotid vascular disease; there is also an increased risk of further vascular events in coronary patients. An association of modest homocysteine elevation with increasing age, decline in renal function, and with other standard risk factors can also be demonstrated. There is now definite evidence that reducing the high homocysteine levels found in homocystinuria due to cystathionine β-synthase deficiency also reduces cardiovascular events in these high-risk patients. It remains to be established whether normalizing the mild homocysteine elevation commonly found in the general population and particularly in the elderly, and which is easily reversible with vitamin supplementation, also reduces vascular risk. Appropriate trials using folic acid are about to begin in relevant populations to determine this, but it will be some years yet before we have the answer to this key current question.

REFERENCES

1. BOUSHEY, C.J., S.A.A. BERESFORD, G.S. OMENN & A.G. MOTULSKY. 1995. A quantitative assessment of plasma homocysteine as a risk factor for vascular disease. J. Am. Med. Assoc. **274**: 1049–1057.

2. GRAHAM, I.M., L.E. DALY, H.M. REFSUM et al. 1997. Plasma homocysteine as a risk factor for vascular disease. J. Am. Med. Assoc. **277**: 1775–1781.

3. NYGARD, O., J.E. NORDREHAUG, H. REFSUM, P.M. UELAND, M. FARSTAD & S.E. VOLLSET. 1997. Plasma homocysteine levels and mortality in patients with coronary artery disease. N. Engl. J. Med. **337**: 230–236.

4. SELHUB, J., P.F. JACQUES, P.W.F. WILSON, D. RUSH & I.H. ROSENBERG. 1993. Vitamin status and intake as primary determinants of homocysteinemia in an elderly population. J. Am. Med. Assoc. **270**: 2693–2698.

5. WILCKEN, D.E.L. & V.J. GUPTA. 1979. Cysteine-homocysteine mixed disulphide: Differing plasma concentrations in normal men and women. Clin. Sci. **57**: 211–215.

6. SELHUB, J., P.F. JACQUES, A.G. BOSTOM et al. 1995. Association between plasma homocysteine concentrations and extra-cranial carotid-artery stenosis. N. Engl. J. Med. **332**: 286–291.

7. CARSON, N.A.J. & D.W. NEILL. 1962. Metabolic abnormalities detected in a survey of mentally backward individuals in Northern Ireland. Arch. Dis. Child. **37**: 505–513.

8. GERRITSEN, T., J.G. VAUGHN & H.A. WAISMAN. 1962. The identification of homocystine in the urine. Biochem. Biophys. Res. Commun. **9**: 493–496.

9. MUDD, S.H., H.L. LEVY & F. SKOVBY. 1995. Disorders of transsulfuration. In The metabolic and molecular basis of inherited disease, 7th edition. C.R. Scriver, A.L. Beaudet, W.S. Sly & D. Vale, Eds.: 1279–1327. McGraw-Hill. New York.

10. CARSON, N.A.J., C.E. DENT, C.M.B. FIELD & G.E. GAULL. 1965. Homocystinuria: Clinical and pathological review of ten cases. J. Pediatr. **66**: 565–583.

11. SCHMIKE, R.N., V.A. MCKUSICK, T. HUANG & A.D. POLLACK. 1965. Homocystinuria: Studies of 20 families with 38 affected members. J. Am. Med. Assoc. **193**: 711–719.

12. WHITE, H.H., L.P. ROWLAND, S. ARAKI, H.L. THOMPSON & D. COWEN. 1965. Homocystinuria. Arch. Neurol. **13**: 455–470.

13. MCCULLY, K.S. 1969. Vascular pathology of homocysteinemia: Implications for the pathogenesis of arteriosclerosis. Am. J. Pathol. **56**: 111–128.

14. WILEY, V.C., N.P.B. DUDMAN & D.E.L. WILCKEN. 1988. Interrelations between plasma free and protein-bound homocysteine and cysteine in homocystinuria. Metabolism **37**: 191–195.

15. MUDD, S.H., F. SKOVBY & H.L. LEVY et al. 1985. The natural history of homocystinuria due to cystathionine β-synthase deficiency. Am. J. Hum. Genet. **37**: 1–31.

16. WILCKEN, D.E.L. & B. WILCKEN. 1997. The natural history of vascular disease in homocystinuria and the effects of treatment. J. Inherited Metab. Dis. **20**: 295–300.

17. WILCKEN, D.E.L., B. WILCKEN, N.P.B. DUDMAN & P.A. TYRRELL. 1983. Homocystinuria: The effects of betaine in the treatment of patients not responsive to pyridoxine. N. Engl. J. Med. **309**: 448–453.

18. WILCKEN, D.E.L. & B. WILCKEN. 1976. The pathogenesis of coronary artery disease. A possible role for methionine metabolism. J. Clin. Invest. **57**: 1079–1082.

19. NYGARD, O., S.E. VOLLSET, H. REFSUM, I. STENSVOLD, A. TVERDAL, J.E. NORDREHAUG et al. 1995. Total plasma homocysteine and cardiovascular risk profile. J. Am. Med. Assoc. **274**: 1526–1533.

20. WANG, X.L., C. TAM, R.M. MCCREDIE & D.E.L. WILCKEN. 1994. Determinants of severity of coronary artery disease in Australian men and women. Circulation **89**: 1974–1981.

21. WILCKEN, D.E.L., N.P.B. DUDMAN, P.A. TYRRELL & M.R. ROBERTSON. 1988. Folic acid lowers elevated plasma homocysteine in chronic renal insufficiency: Possible implications for prevention of vascular disease. Metabolism **37**: 697–701.

22. FROSST, P., H.J. BLOM & R. MILOS et al. 1995. A candidate genetic risk factor for vascular disease: A common mutation in methylenetetrahydrofolate reductase. Nat. Genet. **10**: 111–113.

23. WILCKEN, D.E.L. 1997. MTHFR 677C→T mutation, folate intake, neural tube defect, and cardiovascular risk. Lancet **350:** 603–604.

24. JACQUES, P.F., A.G. BOSTOM & R.R. WILLIAMS *et al.* 1996. Relation between folate status, a common mutation in methylenetetrahydrofolate reductase, and plasma homocysteine concentrations. Circulation **93**: 7–9.
25. WILCKEN, D.E.L., X.L. WANG, A.S. SIM & R.M. MCCREDIE. 1996. Distribution in healthy and coronary populations of the methylenetetrahydrofolate reductase (MTHFR) C677T mutation. Arterioscler. Thromb. Vasc. Biol. **16**: 878–882.
26. VAN BOCKXMEER, F.M., C.D.S. MAMOTTE, S.D. VASIKARAN & R.R. TAYLOR. 1997. Methylenetetrahydrofolate reductase gene and coronary artery disease. Circulation **95**: 21–23.
27. MA, J., M.J. STAMPFER, C.H. HENNEKINS *et al.* 1996. Methylenetetrahydrofolate reductase polymorphism, plasma folate, homocysteine, and risk of myocardial infarction in US physicians. Circulation **94**: 2410–2416.
28. TAWAKOL, A., T. OMLAND, M. GERHARD, J.T. WU & M.A. CREAGER. 1997. Hyperhomocyst(e)inemia is associated with impaired endothelium-dependent vasodilation in humans. Circulation **95**: 1119–1121.
29. CELERMAJER, D.S., K. SORENSEN, M. RYALLS *et al.* 1993. Impaired endothelial function occurs in the systemic arteries of children with homozygous homocystinuria but not in their heterozygous parents. J. Am. Coll. Cardiol. **22**: 854–858.

Nutrition, Cancer, and Aging

IVOR E. DREOSTI[a]

CSIRO Division of Human Nutrition, P.O. Box 10041, Gouger Street, Adelaide 5000, South Australia

ABSTRACT: The parallel increase in cancer risk with advancing age is well recognized, and several pathophysiological mechanisms common to both conditions have been proposed to explain this interrelationship. The importance of nutrition, both in delaying the aging process and in protecting against cancer is also well recognized, and it is therefore of interest to compare the relative impact several of the more widely studied dietary manipulations may have on each of these conditions.

For example, caloric restriction, which putatively reduces oxidative stress and effectively increases life span in animals also seems to reduce the incidence of many cancers, possibly due to diminished mitogenesis. Likewise, oxidative damage to DNA appears to be common to both processes but may be more important in the mitochondria with respect to aging and in the nucleus in relation to cancer. Inadequate dietary folate and impaired DNA methylation status are closely associated with increased cancer risk, and recently defective somatic cell methylation and accumulated genetic instability have been proposed as key mechanisms contributing to senescence.

Several other well-established anticancer dietary strategies, which include increased fiber intake and the consumption of more fruits and vegetables, have not been studied extensively in relation to aging, although many of the phytochemicals considered important as chemopreventive agents for cancer may well contribute to delaying the aging process.

Although not directly related to nutrition, but nevertheless highly relevant, is the question of physical activity, which has been strongly linked to a reduction in risk of some cancers. Although less is known with respect to exercise and biological markers of aging, physical activity does appear to retard the age-related decline in the muscle strength and in the bone density.

CANCER AND AGING

Cancer is a disease strongly associated with age, as is evident from the very low occurrence of cancer in the first decade of life, which rises to an incidence of about 30% in the seventh decade.[1] In this regard it may be considered a disease of the elderly, although from another viewpoint the multistep nature of carcinogenesis suggests that the disease actually originates much earlier in life. In a similar vein, many functional deficits associated with senescence accumulate progressively throughout life, which raises the question as to whether there may be a common mechanism that underlies the etiology of both aging and cancer.[2]

Clearly, some cancers can be linked directly to an inherited genetic defect, whereas others can be attributed largely to exposure to environmental carcinogens. Many cancers, however, develop spontaneously as a result of accumulated random damage to genetic material and other critical cellular structures, and the resulting functional deficits not infrequently bear a resemblance to those associated with aging. In particular, oxidative

[a]Tel: +61 (08) 8303 8837; fax: +61 (08) 8303 8899; e-mail: ivor.dreosti@dhn.csiro.au

damage to key biomolecules suggests a possible common mechanism underlying the progressive cellular dysfunction characteristic of both conditions,[3] which has led researchers in oncology and gerontology to propose strategies aimed at reducing oxidative tissue damage in a bid to delay aging and to compress the time frame of chronic disease.

NUTRITION AND CANCER

A strong link between nutrition and cancer has been recognized for many years, and authorities nowadays consider that approximately 35–50% of all cancers have a dietary link in their etiology, arising either from the consumption of carcinogens and cancer-promoting compounds in food or more commonly from inadequate intake of anticarcinogens and protective factors in the diet.[4]

Cancer-causing Dietary Factors

Several factors potentially capable of aggravating cancerogenesis are nowadays recognized to be strongly diet related and include the consumption of excessive energy, alcohol, fats, salt, and cured foods, as well as the presence of carcinogens derived from charred (polycyclic hydrocarbons) and heavily browned (heterocyclic amines) animal products and mycotoxins from moldy foods (TABLE 1).[5]

Cancer-protective Dietary Factors

With respect to cancer protection, dietary fiber, folate, and the phytochemicals and phytoestrogens found in fruits and vegetables have received most attention. Collectively they are considered able to reduce the risk of many cancers by up to 50% (TABLE 1).[5,6]

Fiber

Insoluble or indigestible fiber as typically occurs in wheat bran has been found to offer significant protection against colon cancer, due mainly to the influence of the fiber as

TABLE 1. Dietary Factors and Cancer

Cancer-causing Factors	Cancer-protective Factors
Excessive energy intakes	Fiber, insoluble and soluble
Alcohol	Calcium
Excessive dietary fat	Folate
Excessive salt and cured foods	Phytochemicals
Moldy food	Phytoestrogens
Carcinogenic pyrolysis products	Antioxidants

roughage on gut transit time, and due to the capacity of this material to reduce exposure of the gut epithelial cells to carcinogens in the digesta of the large bowel.[5] More recently, an important role has been recognized for soluble or digestible fiber, including resistant starch, in protecting against bowel cancer, due in part to the capacity of this material to be fermented in the colon and to act as a prebiotic in promoting the growth of favorable microflora in the lumen of the large bowel, as well as releasing several short chain fatty acids that contribute favorably to colonic epithelial cell health.[7] In parallel, interest has grown concerning the potential benefit to bowel health that may arise from consumption of probiotic foods, in order to colonize the large bowel with the optimum microflora population.[8]

Calcium

Accumulating, but not yet definitive, evidence suggests that increased levels of dietary calcium may offer protection specifically against colon cancer, possibly by precipitating carcinogenic secondary bile acids and by suppressing proliferation of the colonic epithelium.[9]

Folate

The importance of folate in cancer protection was first demonstrated in animal studies when folate deficiency was linked to enhanced chemically induced carcinogenesis and subsequently in humans in relation to cervical dysplasia and colon cancer.[10,11] More recently, the importance of folate for the maintenance of genetic stability and for control of gene expression has further highlighted the potential importance of folate in cancer protection. Mechanistically, folate deficiency may enhance carcinogenesis through the role of folate in maintaining the level of S-adenosyl methionine and consequently the production of deoxythymidine monophosphate for DNA synthesis, and because it is needed for the methylation of cytosine in cytosine-guanine sequences, which, if hypomethylated, may lead to enhanced expression of specific oncogenes.[10,11]

Phytochemicals

The anticancer role of the many putative phytochemicals is an area of burgeoning research interest, as many of these minor dietary factors, although not regarded at present as essential nutrients, do nevertheless appear to offer protection against the development of several major cancers. The site of action of this protection may vary between phytochemicals and may occur anywhere in the multistep sequence of events associated with carcinogenesis that range from enhanced protection and repair of genetic material, through accelerated carcinogen detoxification and inhibition of oncogene expression, to differentiation of neoplastic cells and improved immunosurveillance and destruction of transformed cancer cells (TABLE 2).[12,13] Included among these protective mechanisms is the important capacity of antioxidants to reduce damage by active molecular species of oxygen to genomic DNA and to other key biological macromolecules.

TABLE 2. Possible Sites of Action of Phytochemicals in Cancer Protection

Stages of Carcinogenesis	Mechanism Involved
Initiation	Prevention of carcinogen formation
	Carcinogen binding and sequestration
	Blocking of carcinogen action:
	• Reduced carcinogen activation
	• Increased carcinogen detoxification
	• Antioxidant protection
	• Enhanced DNA repair
Promotion	Suppression of cancer expression
	• Oncogene/suppressor gene control
	• Modulation of cell signaling
	• Maintenance of cell–cell communication
Progression	Enhanced cell differentiation
	Reduced angiogenesis
	Enhanced immunosurveillance
	Reduced metastasis

Antioxidants

Although many dietary anticarcinogens with fundamentally different modes of action have been implicated as chemopreventive agents, oxidative free radical damage to DNA and protection by antioxidants remains a central theme in relation to nutrition and cancer. With regard to antioxidants it needs to be recognized that these compounds fall into three broad categories (TABLE 3): (1) endogenous antioxidants that are synthesized in the body and are not readily influenced by dietary manipulation, for example, urate, glutathione, and lipoic acid; (2) exogenous, nutrient antioxidants that constitute an important component of both water-soluble and lipid-soluble defense mechanisms and that are derived entirely from external sources and are easily influenced by dietary manipulation, for example, antioxidant vitamins and antioxidant trace elements; and (3) exogenous nonnutrient antioxidant phytochemicals derived from the diet, many of which are stronger antioxidants *in vitro* than the nutrient antioxidants but which are relatively poorly characterized in terms of tissue distribution and pharmacokinetics, for example, bioflavonoids, carotenoids, and polyphenols.

A wealth of literature exists with respect to the protection afforded by antioxidants against injury in both humans and animals.[14] However, attention has often focused on the effect of supplementation with individual nutrient antioxidants at levels very much higher than can be obtained even from a diet rich in fruits and vegetables, which currently reflects the protective dietary paradigm.[15] Positive outcomes in these intervention studies are therefore difficult to interpret in practical terms and tend to suggest that for optimum effectiveness antioxidants need to be consumed as a complex mixture of nutrient and nonnutrient antioxidants, such as are found in many plant foods.

TABLE 3. Components of the Antioxidant Defense System

Endogenous—Synthesized	Exogenous—Dietary	
	Nutrient	Nonnutrient
Uric acid	Vitamin A	Carotenoids
Sacrificial substrates—albumin	Vitamin E	Flavonoids
Glutathione	Vitamin C	Polyphenols
Ubiquinone	Cu, for superoxide dismutase	Tocotrienols
Lipoic acid	Zn, for superoxide dismutase	Terpenes
Metallothionein	Mn, for superoxide dismutase	Phytate
Metal-binding proteins	Se, for glutathione peroxidase	Melanin
Ferroxidases	β–carotene	Carnosine

At the biochemical level, oxidative tissue damage may occur anywhere within the cell and may involve any macromolecule, depending on where the active oxygen-centered radical is produced. In relation to cancer, oxidative damage to DNA probably represents a major carcinogenic event and may occur by random attack on DNA by active oxygen species produced within the nucleoplasm or in a site-specific manner by hydroxyl radicals released by Fenton chemistry in the near vicinity of such bound, redox-active divalent cations as iron and copper. DNA damage of this nature generally results in base hydroxylation and DNA strand breakage of the deoxyribose backbone, which in turn can lead to mutations and increase the risk of carcinogenesis.[16,17] Recently it has been shown that many of the carcinogenic mutations arising from oxidative injury to DNA involve activation of several oncogenes encoded in chromosomal DNA,[18] which suggests that the anticancer effect of antioxidants with respect to DNA mainly involves protection of nuclear DNA and reduced mutagenesis in the nuclear genome.

NUTRITION AND AGING

The potential importance of nutrition to the aging process has been a topic of interest for many decades and has encompassed not only the effect of diet on cell senescence and death but also the role nutrients may play in delaying the decline in immunocompetence, skeletal integrity, cognitive function, hormonal balance, and other bodily functions often associated with aging organisms. Several nutrients, including vitamins and essential minerals, have been linked beneficially with these disorders, but in terms of dietary control of the aging process most attention by far has focused on energy restriction.

Energy Restriction

The capacity of energy restriction (30–40%) to prolong the average life span in animals by about one-third to one-half is well established.[12,19,20] Mechanistically, the beneficial effect of this practice has been attributed largely to reduced oxidative stress on cellular structures, especially the mitochondrion, which results in less mitochondrial dysfunction arising from oxidatively induced mutations in mitochondrial DNA and consequent defects in the tricarboxylic acid cycle and the electron transport chain.[21] Also, although less definitive in outcome, high levels of antioxidant supplementation in several experimental animal studies have been reported to increase the average life span, which lends some further support to the notion of oxidative damage as a central mechanism in the aging process.[3,21]

CANCER, AGING, AND THE NUTRITIONAL INTERFACE

The interface between many noncommunicable degenerative diseases of the elderly, which include cancer and the aging process, is multifaceted. Sociologically, successful aging implies freedom from degenerative disease for as long as possible. In many cases, however, the strategy to ensure this outcome may not be diet related, or if it is, it may bear no relevance to the aging process and to the extension of the average life span. Indeed it has been theorized that even if mortality from cancer were to be totally eliminated, no meaningful increase would occur in the average life span, unless the underlying mechanisms for the remaining fatal degenerative diseases of aging and aging itself were also addressed.[2,3] Thus although reductions in the incidence of diseases like cancer, osteoporosis, cataract, and arthritis are critical for successful aging, life span itself will not be affected by reducing these conditions unless the central mechanism underlying their development is shared also by the aging process.

In this regard, and in relation to nutrition and cancer, it is unlikely that many of the dietary strategies known to affect cancer risk (*e.g.*, soluble and insoluble fiber, salt, and meat mutagens) will bear directly on the aging process. However, on present evidence, oxidative stress to biological macromolecules appears to be a prime candidate mechanism contributing to the process of aging and to the pathogenesis of many degenerative diseases. With respect to cancer, the macromolecule mainly affected is probably nuclear DNA, whereas aging may be related more to damage to mitochondrial DNA. Indeed it is probable that in both cases reduced oxidative stress associated with dietary restriction and increased antioxidant protection obtained from the diet may contribute to a reduction in the progression of both conditions. In a similar vein, defective methylation of DNA and the attendant genetic instability associated with inadequate dietary folate is another potential mechanism that warrants further study in relation to diet, cancer, and the aging nexus.

REFERENCES

1. AMES, B.N., M.K. SHIGENAGA & T.M. HAGAN. 1993. Oxidants, antioxidants and the degenerative diseases of aging. Proc. Natl. Acad. Sci. USA **90:** 7915–7922.
2. HARMAN, D. 1991. The aging process: Major risk factor for disease and death. Proc. Natl. Acad. Sci. USA **88:** 5360–5363.
3. CUTLER, R.G. 1991. Human longevity and aging: Possible role of reactive oxygen species. *In* Annals of the New York Academy of Sciences. W. Pierpaoli & N. Fabris, Eds.: **621:** 1–28. New York Academy of Sciences. New York.
4. DOLL, R. 1992. The lessons of life: Keynote address to the nutrition and cancer conference. Cancer Res. (Suppl) **52:** 2024S–2029S.
5. KIM, Y-I. & J.B. MASON. 1996. Nutrition chemoprevention of gastrointestinal cancers: A critical review. Nutr. Rev. **54:** 259–279.

6. JACOBS, M.M. 1993. Diet, nutrition and cancer research: An overview. Nutr. Today **27:** 19–23.

7. KRITCHEVSKY, D. 1995. Epidemiology of fibre, resistant starch and colorectal cancer. Eur. J. Cancer Prev. **4:** 345–352.

8. PLAYNE, M.J. 1995. Probiotic microorganisms. Recent Adv. Microbiol. **3:** 215–254.

9. PENCE, B.C. 1993. Role of calcium in colon cancer prevention: Experimental and clinical studies. Mutat. Res. **290:** 87–95.

10. GIOVANUCCI, E. *et al.* 1993. Folate, methionine and alcohol intake and risk of colorectal adenoma. J. Natl. Cancer Inst. **85:** 875–883.

11. MASON, J.B. 1994. Folate and colonic carcinogenesis: Searching for a mechanistic understanding. J. Nutr. Biochem. **5:** 170–175.

12. CARAGAY, A.B. 1992. Cancer-preventive foods and ingredients. Food Technol. **46**(4)**:** 65–68.

13. DREOSTI, I.E. 1996. Bioactive ingredients: Antioxidants and polyphenols in tea. Nutr. Rev. **54:** 551–558.

14. DORGAN, J.F. & A. SCHATZKIN. 1991. Antioxidant micronutrients in cancer prevention. Hematol./Oncol. Clin. N. Am. **5:** 43–68.

15. DRAGSTED, L.O., M. STRUBE & J.C. LARSEN. 1993. Cancer protective factors in fruits and vegetables: Biochemical and biological background. Pharmacol. & Toxicol. **72** (Suppl.1): 116–135.

16. FLOYD, R.A. 1990. The role of 8-hydroxyguanine in carcinogenesis. Carcinogenesis **11:** 1447–1450.

17. DREOSTI, I.E. 1991. Free radical pathology of the genome. *In* Trace Elements, Micronutrients and Free Radicals. I.E. Dreosti, Ed.: 149–169. Humana Press. New Jersey.

18. CLAYSON, D.B., R. MEHTA & F. IVERSON. 1994. International Commission for Protection Against Environmental Mutagens and Carcinogens. Oxidative DNA damage the effects of certain genotoxic and operationally non-genotoxic carcinogens. Mutat. Res. **317:** 25–42.

19. SOHAL, R.S. *et al.* 1994. Oxidative damage, mitochondrial oxidant generation and antioxidant defenses during aging and in response to food restriction in the mouse. Mech. Ageing Dev. **74:** 121–133.

20. LUFT, R. 1994. The development of mitochondrial medicine. Proc. Natl. Acad. Sci. USA **91:** 8731–8738.

21. HARMAN, D. 1983. Free radical theory of aging: Consequences of mitochondrial aging. Age **6:** 86–94.

Combined Exercise and Dietary Intervention to Optimize Body Composition in Aging

MARIA A. FIATARONE SINGH[a]

Jean Mayer USDA Human Nutrition Research Center, Tufts University, Exercise and Nutrition Laboratory, 711 Washington Street, Boston, Massachusetts 02111, USA

ABSTRACT: Concomitant losses of skeletal muscle and bone mass along with gradual accretion of adipose tissue typify usual human aging. Recent investigations have attempted to modify these processes with various combinations of dietary and exercise intervention in older adults. Complete nutritional supplements given with weight-lifting exercise have been shown to augment muscle and fat gains in healthy older men, but have merely suppressed habitual dietary intake when administered to frail sedentary elders, and have not altered body composition responses to strength training in this population. Protein supplementation at twice the RDA does not improve skeletal muscle function or increase muscle mass in healthy elderly weight lifters compared to those on a normal diet. Calcium supplementation during one year of aerobic training has an independent beneficial effect on cortical bone density at the femoral neck in postmenopausal women, whereas the exercise is associated with trabecular bone increases in the lumbar vertebrae. Hypocaloric dieting, with or without aerobic exercise, results in losses of weight, fat and lean mass in obese elderly men and women. By contrast, resistance training during hypocaloric dieting augments lean mass while further reducing fat mass. Low protein, isoenergetic diets result in muscle atrophy in older women. Current studies will determine the ability of resistance training to offset these catabolic effects on skeletal muscles of a low-protein (0.6 g/kg/day) diet prescribed for elderly with chronic renal failure. More long-term studies of efficacy and feasibility of diet and exercise combinations are needed in the aged to optimize the potential for healthful shifts in body composition.

Changes in body composition with aging have been increasingly recognized as a potentially modifiable factor in the quest for optimal health, function, and longevity. Aging is typically associated with a gradual increase in weight throughout midlife, predominantly due to the accretion of adipose tissue. This adipose tissue storage parallels the age-related decline in energy expenditure in physical activity[1] and occurs despite decreases in average energy consumption with increasing age. This increase in fat mass is accompanied by decreases in both muscle (sarcopenia) and bone (osteopenia) compartments of lean tissue, beginning in midlife and continuing into extreme old age.[2–7]

Loss of muscle mass poses significant health risks to the elderly, including impairment of maximal aerobic capacity,[8] glucose intolerance,[9] lower resting metabolic rate,[10] immune dysfunction,[11,12] slower gait velocity,[13] and functional dependency.[14] Increased and centralized adiposity, on the other hand, has been associated with excess morbidity and mortality from cardiovascular disease, diabetes, hypertension, stroke, arthritis, sleep apnea, and mobility impairment.[15–20] Loss of bone mass is a primary risk factor for osteoporotic fractures and their associated mortality, institutionalization, functional decline, chronic pain, and mobility impairment.[21,22]

---

[a]Current address: 59 Bundarra Ave. N., Wahroonga NSW 2076, Australia. e-mail: mafiat@ibm.net

Taken together, these adverse changes in body composition represent a substantial burden of disease and disability for the elderly. Although genetic factors are certainly involved in the changes observed, it is suggested by epidemiologic studies[23] that lifestyle modifications in the areas of nutritional intake and energy expenditure in physical activity are potentially of great importance in the achievement of optimal body composition with aging. It has therefore been a primary focus of investigations in our laboratory over the last decade to answer the following questions in relation to body composition and aging: (1) Can dietary modification augment the adaptation to exercise training? and (2) Can exercise training minimize the adverse consequences of dietary inadequacy? Three primary goals of this approach can be identified: prevention/treatment of sarcopenia, reduction/redistribution of adipose tissue, and prevention/treatment of osteopenia. The major studies that have been reported to date are reviewed in the sections that follow. Gaps in our current knowledge and suggestions for the direction of future research efforts are outlined at the conclusion of this paper.

SARCOPENIA

There are now numerous studies in normal healthy older adults that indicate that high-intensity resistance training is associated with increases in fat-free mass or muscle area and slight decreases in percent body fat, usually with minimal alteration of total body weight.[24–28] The observed adaptive response of skeletal muscle to hypertrophy in these studies is quite variable and is influenced by the intensity and duration of the intervention, subject characteristics, and the precision of the measurement.[29] Typically, whole body changes after programs of 12- to 52-week duration have been modest, with 1–2 kg gains in muscle and 1–2% decreases in body fat observed.[30] By contrast, aerobic training does not normally increase muscle mass and therefore is not the modality of choice when body composition shifts to counteract age-related sarcopenia are the primary goal of training.

In the first investigation to demonstrate significant change in muscle area with resistance training in the elderly, Frontera reported that 12 men in their 60s and 70s had a 28 34% increase in fiber areas in the vastus lateralis after 12 weeks of knee extensor and flexor progressive resistance training, despite no overall changes in body weight or composition.[31] However, this study was also designed to test the interaction of diet and exercise on body composition changes. Six of the men were administered a 560 kcal/day (17% protein) liquid supplement in addition to their habitual diet. Although strength changes were similar between groups, the supplemented men had two to three times the gain in muscle area by CT scan (see Fig. 1) compared to the unsupplemented exercisers, as well as an increase in regional adipose tissue[32] and creatinine excretion, implying whole body increases in muscle mass following the combined intervention. Because the supplement contained extra energy as well as protein, it was not clear from this study which was the active dietary component leading to more skeletal muscle hypertrophy.

We theorized that the subjects in this study may have been too healthy and well-nourished to maximally benefit from complete nutritional supplementation. We therefore applied a similar combined intervention to a group of frail, marginally nourished long-term care patients.[13] Our working hypothesis was that there would be a small effect of nutritional supplementation or resistance training alone on muscle mass accretion, but

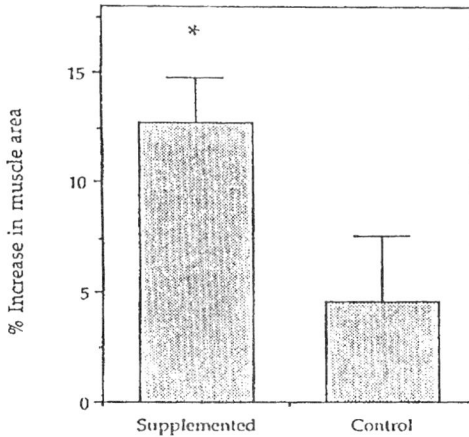

FIGURE 1. Relative increases in muscle cross-sectional area by computerized tomography scans of the midthigh after resistance training with or without nutritional supplementation in healthy older men.[32] There was a significantly greater gain in muscle area in the supplemented group (p < 0.05) compared to the unsupplemented exercisers. The nutritional supplement provided 560 kcal/ day in a liquid formula, which was 17% protein, 43% carbohydrate, and 40% fat.

that the combined intervention would stimulate the greatest adaptation, both functionally and in terms of body composition. We studied 100 frail nursing home residents of average age 86 years in a randomized controlled trial of resistance exercise, complete nutritional supplementation (360 kcal/day, 17% protein), both interventions, or a double placebo control condition. Body composition was assessed by anthropometric measurements, total body water, bioelectric impedance, whole body potassium, and computerized tomography of the midthigh. The supplement alone had no beneficial effect on weight, any measure of body composition, or muscle function, likely due to the fact that it was associated with a compensatory decrease in energy intake from the normal diet, so that no net increase in total energy intake occurred (see FIG. 2). Exercisers who were not supplemented had no change in total energy intake, and although they more than doubled their muscle strength, no significant gains in lean tissue were seen. By contrast, the exercisers who also had access to the nutritional supplement had a net gain in dietary energy intake of approximately 300 kcal/day (20% above baseline), presumably as an adaptation to their increased energy needs imposed by adoption of resistive exercise. However, there was no additional benefit of this supplementary energy and protein in terms of body composition or function beyond that attributable to exercise alone.

Several conclusions may be drawn from this study. First, without a reversal of the conditions associated with an extremely sedentary lifestyle (low muscle mass, resting metabolic rate, and energy expenditure in physical activity), attempts to augment lean mass with nutritional supplementation alone may fail, which is consistent with clinical experience in this area. Second, resistance training, because it can modify all of these factors,[33] and also has been shown to result in higher levels of activity outside of the exercise

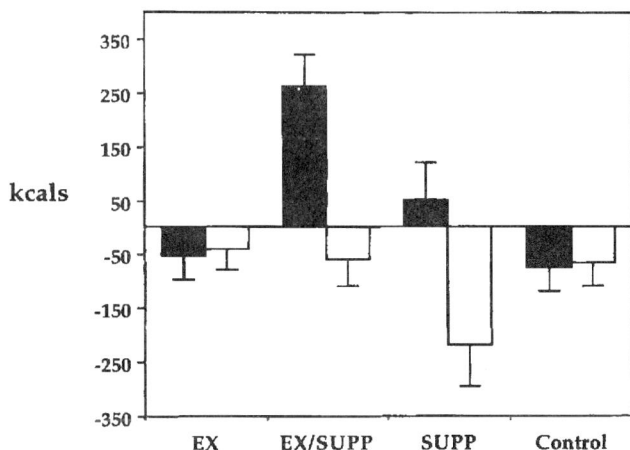

FIGURE 2. Changes in energy intake after resistance training and/or nutritional supplementation in frail elderly nursing home residents.[13] Dietary intake calculated from 3 days' food weighing, data expressed as mean ± SEM. Dietary energy intake included the energy content of meals, snacks, and supplements other than the supplement or placebo used in the study. Total energy intake included dietary energy intake plus the energy content of the study supplement. Exercise significantly blunted the decline in dietary energy intake during the trial ($p = 0.04$). The study supplement only augmented total energy intake in subjects who also exercised. ■, total intake; □, diet.

class,[13,30] may offer the best chance to alter energy balance and improve dietary intake in this population. Third, longer-term or possibly larger changes in energy balance may be needed to result in an improved body composition adaptation to exercise when frailty and sarcopenia are very advanced.

Although the study by Meredith did demonstrate an additive effect of diet and exercise on muscle mass accretion in elderly men,[32] the multicomponent nature of the nutritional supplement used did not allow identification of the active dietary constituent (energy vs. protein). Therefore, a double-blind placebo-controlled study was conducted[34] in which 12 healthy elderly men and women (aged 56–80 yrs) were randomly assigned to diets that provided either 0.8 or 1.6 g protein/kg/day within a weight-maintaining diet during 12 weeks of whole body progressive resistance training. There was an increase in fat-free mass of 1.4 ± 0.4 kg ($p < 0.01$) and a decrease in fat mass (1.8 ± 0.4 kg, $p < 0.001$) in the group as a whole but no effect of dietary protein status on any measure of body composition or functional adaptation to the exercise training. The increase in fat-free mass appeared to be due to an increase in total body water rather than protein or mineral mass. These results, taken together with the earlier study by Frontera and Meredith,[31,35] would suggest that provision of additional energy may augment body composition shifts during strength training in the elderly, whereas protein intake at twice the R.D.A. offers no benefit over a normal diet providing the R.D.A. of protein. It should be noted that the loss of fat mass in Campbell's subjects[34] suggests that they were in slight negative energy balance during the study, which may explain their lack of anticipated increases in body cell mass secondary to resistive exercise.

At the other end of the spectrum, resistance training may be looked at as a way of counteracting the negative effects of a low protein or hypocaloric diet. For example, Castaneda[11] has shown that low protein intake (0.45 g/kg/day) results in significant losses of lean tissue, immune response, and muscle function in eight weeks despite adequate energy intake to maintain body weight (see FIG. 3). There are certain clinical situations in which low protein intake is required (chronic renal or liver failure) or difficult to change (severe anorexia), and the adverse effects of such diets, which will exacerbate age-related sarcopenia, have not been adequately addressed in current practice. We are therefore currently conducting a trial of progressive resistance training during a low-protein diet, prescribed for older individuals with chronic renal failure, to determine whether the exercise intervention can counteract the deleterious effects of the low-protein diet on body composition.

Finally, hypocaloric diets for weight loss are known to result in losses of both fat and lean (muscle and bone) tissue.[36] Ballor[37,38] has shown in both young and older women that resistance training provided concurrently or subsequent to hypocaloric dieting in obese subjects results in loss of weight and fat mass but preservation or increase in lean mass. This is of considerable importance in older women, whose repetitive cycles of dieting may result in substantial reductions in muscle and bone tissue, thus exposing them to many health risks. Aerobic training during hypocaloric dieting does not alter the loss of lean tissue,[38] and so is not as useful as a combined modality for the treatment of obesity in the elderly when preservation of lean tissue is also a goal.

In summary, resistance training typically results in shifts in body composition (increased muscle, decreased fat) in the elderly if the interventions are conducted at relatively high intensity and for an adequate length of time. Excess dietary protein has not been shown to augment these body composition changes beyond those attained on a normal diet. Energy needs during resistance training increase by approximately 15%[34] in healthy elderly, and therefore weight stability requires some modulation of energy intake. A more positive energy balance will increase both muscle and fat mass during training in healthy older subjects but has not been shown to be effective in very elderly frail individ-

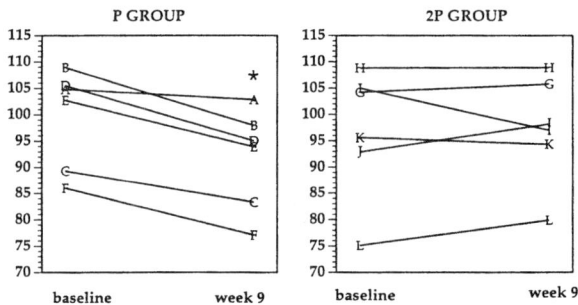

FIGURE 3. Changes in total body potassium in variable protein diets in elderly women. (Castaneda et al.[11] With permission from the American Journal of Clinical Nutrition.) The changes over time in the P group (0.45 g/kg/day protein) were significant (p < 0.002). There were no significant changes in the 2P group (.092 g/kg/day protein).

uals. In combination with an intentionally hypocaloric diet for weight loss, resistance training can alter the composition of the weight loss, such that adipose tissue is lost while lean body mass is maintained or increased. The potential for resistance training to offset the catabolic effects of a low-protein diet, such as that prescribed for chronic renal failure, is the focus of ongoing research. The combined exercise and dietary approach to both the prevention and treatment of sarcopenia in the elderly thus consists of moderate to high-intensity progressive resistance training and adequate energy intake. There is currently no data supporting the use of other dietary modifications or aerobic training for this purpose.

ADIPOSE TISSUE ACCRETION

Major reviews and meta-analyses indicate little evidence of the ability of exercise to significantly modify body weight and overall composition as an isolated intervention in normal elders.[39-41] In a meta-analysis of 53 such studies (up to 36 wk in duration), Ballor reported that, in men, aerobic exercise resulted in an average loss of 1.2 kg of body weight compared to a 1.2 kg gain in weight-lifting studies.[41] Body fat decreased regardless of exercise modality, by 1.5 kg (1.7%) compared to controls. Weight training was significantly better at increasing fat free mass (FFM) (2.2 kg) compared to 0.8 kg with cycling and no difference from controls in walking/jogging studies. In women, only walking/jogging resulted in a decrease in body weight (0.6 kg) compared to controls, as well as significant reductions in body fat (1.3 kg, 1.7%) and no change in FFM. Weight and fat loss was greatest in those with high body fat initially and the highest exercise-related energy expenditure, which may explain the more potent effect seen in men.

By comparison, hypocaloric dieting, if adhered to, can substantially reduce fat mass in obese elders. Katzel recently reported the results of a randomized trial of hypocaloric dieting versus aerobic exercise training in obese sedentary men 46–80 years of age who were selected for being 120–160% of ideal body weight.[36] Subjects (mean age 61 year, BMI 30 kg/m^2) were randomized to nine months of hypocaloric dieting (300–500 kcal/day deficit) or aerobic exercise (45 min of treadmill or cycling at 70–80% of heart rate reserve for 30–45 min, 3 days/wk). As can be seen in FIGURE 4, weight loss (average 9.5 kg) and decrease in total and abdominal obesity occurred only in the diet group. Seventy-five percent of the weight lost was fat tissue and was accompanied by improvements in fasting insulin levels and glucose tolerance, cholesterol, and blood pressure. Compared to younger subjects, the older men lost significantly less weight and percent body fat. Insulin sensitivity and aerobic capacity were the only outcomes that changed in the exercise group.

Thus, the body composition, fat distribution, and metabolic changes seen in some previous exercise studies of nonobese or younger individuals were not evident.[47] Presumably, this is because the exercisers increased their dietary intake to remain weight stable, and lean mass was not altered by this modality of exercise.

In summary, there is no evidence to date from randomized clinical trials in obese elderly that aerobic exercise without dietary restriction can significantly lower body weight, percent body fat, central adiposity, or lipid profiles. Other reasons to advocate such exercise in this group, however, include increases in aerobic fitness[43] and insulin sensitivity,[44] which may occur independently of weight loss in the elderly.

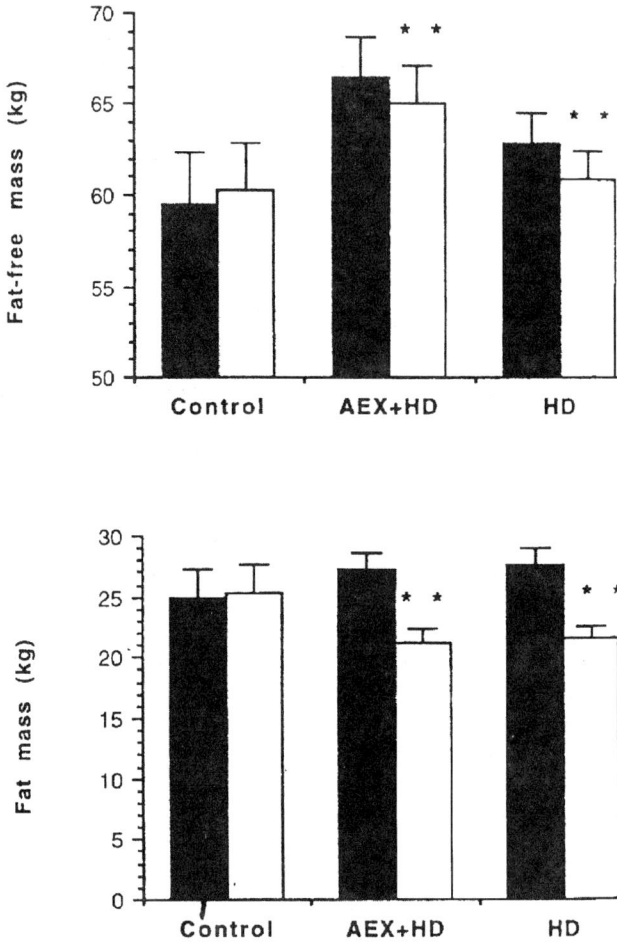

FIGURE 4. Body composition changes following weight loss by hypocaloric diet (HD) vs. aerobic exercise (AEX) in obese older men. (Dengel *et al.*[48] With permission from *Medicine & Science in Sports & Exercise*.) Measurements made by hydrostatic weighing. Solid bars are initial values, and open bars are final values after 10 months of intervention or control condition. * indicates significant (p < 0.0001) difference from initial values within group. Values are expressed as mean ± SEM.

Hypocaloric dieting is the most potent means to achieve short-term weight loss. For example, in a meta-analysis of 89 studies involving 1800 type II diabetics conducted over the past 30 years, diet alone had a significantly greater effect on weight loss (9 kg) and glycosolated hemoglobin levels, compared to mean losses of 3.8 kg in exercise, diet, and behavioral interventions.[45] However, hypocaloric dieting alone, although effective if adhered to in creating an energy deficit and therefore loss of body weight and fat, has undesirable consequences in the elderly, including exacerbation of age-related losses of lean tissue, decreased metabolic rate, and risk for micronutrient deficiencies. Both endurance and resistance training are associated with increased energy expenditure through the cost of activity, increased basal metabolic rate, increased thermic effect of a meal, and increased lean body mass (resistance) and can induce small losses of body weight and total fat as well mobilization of fat from abdominal sites in older men and women. The reduction in resting energy expenditure secondary to dieting may be prevented or attenuated by concurrent exercise,[38] although this is not uniformly seen.[46] Finally, the increase in aerobic fitness itself has many physiologic and psychological consequences that cannot be achieved by dieting alone and may enhance long-term behavioral adaptations to minimize weight cycling. It makes sense, therefore, that a combination of diet and exercise modalities in obese individuals may produce the largest losses of weight while attaining more desirable body composition ratios, functional gains, and metabolic profiles than either treatment alone.

Despite the theoretical argument advanced above, there are no randomized controlled trials of weight loss comparing hypocaloric dieting to diet plus exercise in obese elders. However, two nonrandomized studies have been reported in this population. Dengel[47,48] studied obese 60-year-old men who were nonrandomly assigned by preference to hypocaloric dieting and behavioral modification, diet plus aerobic exercise for 10 months (3 days per week at 50–85% of VO_{2max}), or a control group that received brief dietary instructions only. Results were presented only for those subjects who completed the study and lost at least 3 kg (33% of the diet group, 62% of the diet plus exercise, and 57% of controls). Weight loss averaged 8–9 kg, body fat decreased by approximately 5%, and FFM decreased by 1–2 kg by underwater weighing, with no difference between treatment groups. As expected, only the exercisers exhibited an increase in aerobic capacity, but there was no difference in lipid lowering or decrease in skinfold thicknesses (preferentially truncal) between groups.

Thus, no additional benefit to weight loss, body composition, lipid profile, or fat distribution was attributable to exercise. However, it should be noted that more than twice as many exercisers as dieters achieved the minimum study goal (loss of 3 kg), and this may be one of the most important findings of this study. Unfortunately, the quasiexperimental design of this study limits the generalizability of these findings to individuals who may be unwillingly prescribed such treatments by their physicians. Overall, the increase in aerobic fitness has independent predictive value for functional independence and mortality in the elderly,[49] and is therefore a worthy goal in obese elders who are at higher risk of death and disability than their leaner peers.

In the only other study combining diet and aerobic exercise in older subjects, Fox reported that 41 healthy obese women of average age 66 years were nonrandomly assigned to diet groups that reduced intake by 500 or 700 kcal/day, or a diet plus exercise group with a combined deficit of 700 kcal/day.[50] Exercise consisted of walking for 1 h 3 days/week and resistive exercises 2 days/week. After 24 weeks, weight loss (6.5 kg), body fat, lean body mass (by dual photon absorptiometry), and fasting glucose decreased significantly over time with all treatments, whereas fat distribution (by anthropometrics), lipids, and fasting insulin levels did not change in any group. Thus exercise, even combining aerobic and resistive components, did not offer any additional benefit to these women.

By contrast with these two studies in the elderly, Wood found that exercise increased the loss of body fat and abdominal girth[51] and resulted in higher HDL and lower LDL/HDL ratios after one year, compared to dieting alone in overweight middle-aged men, but had no additive effect in women aged 25–49. Thus, the apparent resistance of these middle-aged women and older men[47,48] and women[50] to the benefits of exercise in weight control compared to trials in younger men may be due to gender-specific effects or the less intensive nature of the energy deficit prescribed.

Another approach to adipose tissue loss is the combination of resistance training and hypocaloric dieting. Resistance training studies in the healthy elderly suggest favorable shifts in energy balance, even in the absence of weight loss or increases in fat-free mass. Campbell has reported that total energy requirements for weight maintenance are increased approximately 15% after 12 weeks of resistance training in older men and women, primarily due to increases in resting metabolic rate.[34] Treuth has also described increased resting energy expenditure and fat oxidation after resistance training in post-menopausal women.[52] Over the long term, such acute adaptations may significantly affect energy balance and contribute to the maintenance of a healthful body weight.

Although resistance training has been shown to reduce total and abdominal fat, dieting is a more potent way to achieve these goals. The primary purpose of combining weight lifting and diet in obese individuals is to stimulate accretion of lean mass (muscle and bone) in the face of an energy deficit. There is no randomized trial of hypocaloric dieting versus diet plus resistance training in the elderly. There is one study in young women in which Ballor reported on 41 obese young women randomly assigned to diet, resistive exercise, diet plus exercise, or control groups for 8 weeks.[37] Dieting significantly decreased body weight and fat mass. Resistance training significantly increased arm muscle area, strength, and lean body weight and decreased percent body fat. The combined therapy group, therefore, ended up with a better body composition profile (increased lean, decreased fat) than either isolated intervention could achieve. Importantly, hypocaloric dieting (1000 kcal/day) did not appear to limit the functional adaptation or hypertrophy of muscle secondary to resistance training.

There is only one randomized controlled trial comparing resistance training to endurance training after hypocaloric dieting in obese older subjects. After 11 weeks of dieting, Ballor randomly allocated 18 older subjects to moderate intensity resistance or aerobic exercise training for an additional 12 weeks.[38] As can be seen in FIGURE 5, weight remained stable in the weight lifters, as the decline in fat mass was masked by a gain in FFM. By contrast, the endurance training group lost fat and FFM, which combined to produce a significant drop in total body weight. Only the resistance-trained group had increases in resting energy expenditure and the thermic effect of a meal, attenuating declines in these parameters that had accompanied their hypocaloric dieting.

In the long-term control of body weight, these body composition and resting energy expenditure adaptations seen with resistance training may be extremely important in minimizing the tendency for weight and fat to be regained after dieting.[53] In addition, the preservation or accretion of lean tissue and muscle strength, which is only seen with resistive exercise, assumes ever greater importance with age.

In summary, the limited literature to date suggests that although the short-term weight loss associated with diet plus exercise is usually not different from diet treatments alone, there are several extremely important reasons to coadminister these therapies. First, hypocaloric dieting alone results in a decline in total energy expenditure due to decreases in resting metabolic rate, the thermic effect of feeding, and the energy cost of physical activities (as body weight is lost), as well as decreased spontaneous activity levels in ani-

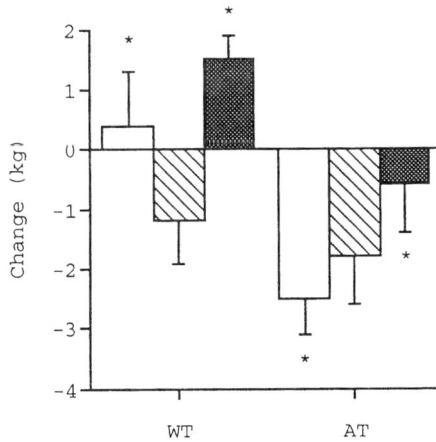

FIGURE 5. Changes in body weight (open bars) fat mass (hatched bars), and fat-free mass (solid bars) following 12 weeks of exercise training. (Ballor *et al.*[38] With permission from *Metabolism: Clinical & Experimental*) WT = resistance training; AT = aerobic training. Measurements made by hydrostatic weighing. Values are mean ± SEM. * indicates AT group change significantly different (p < 0.05) from WT group.

mal models. The simultaneous use of exercise may offset these trends by increasing resting metabolic rate, the thermic effect of a meal, and adding the energy cost of the specific activity prescribed. In addition, resistance training (but not aerobic training) has been shown to result in increased spontaneous or habitual activity levels in the elderly,[13,30] thus further increasing total energy expenditure. By replacing a portion of the desired energy deficit with increased energy expenditure, the allowable dietary energy intake can be liberalized slightly, which should help compliance over time.

Second, the composition of the weight loss that is incurred through hypocaloric dieting may range from 10–50% lean tissue, depending upon the severity of the energy restriction. In the elderly, this is particularly worrisome, because it would be occurring at a time when biological aging, chronic disease, and disuse atrophy are already contributing to sarcopenia. However, resistance training has been shown to change the composition of weight loss during dieting, such that lean mass is gained while body weight and fat are lost.[37,38]

Third, although hypocaloric dieting can result in substantial losses of weight, fat mass, and improvement of cardiovascular risk factor profiles in the short term, it does not result in improved aerobic capacity, muscle strength, balance, and other aspects of fitness that are seen in combined programs of diet and exercise.

Fourth, there appears to be a better adherence to such other positive lifestyle modifications as diet and smoking cessation if exercise behavior is encouraged simultaneously. This may be due to exercise-induced increases in self-efficacy that generalize to other behaviors, alleviation of coexistent depression or low self-esteem, or positive feedback from early success in reaching goals. Whatever the mechanism, the long-term efficacy of all weight loss interventions requires the adoption of lifelong behavioral changes, and exercise appears to aid in this process.

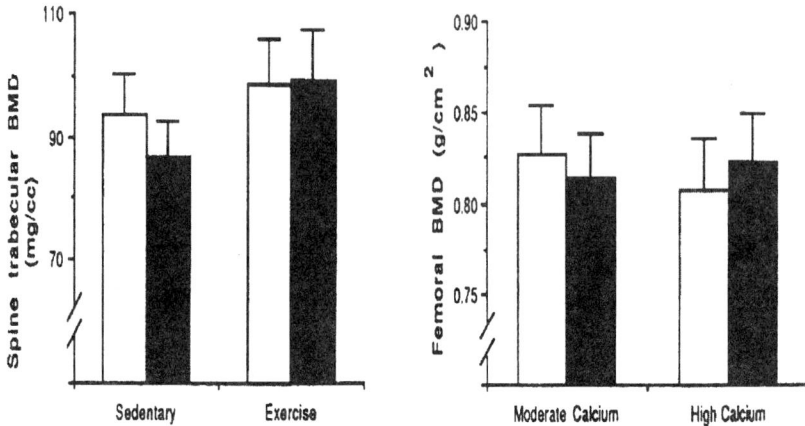

FIGURE 6. Changes in bone mineral density (BMD) at the lumbar spine and femoral neck after 12 months of walking or calcium supplementation in postmenopausal women. (Nelson *et al.*[67] With permission from the *American Journal of Clinical Nutrition.*) Baseline measurements are shown in open bars, 12-month measurements in solid bars; data are presented as the mean ± SEM. Exercise had a significant effect at the lumbar spine, as measured by quantitative computerized tomography (p = 0.028). Calcium supplementation was associated with a significant improvement in femoral neck bone mineral density, measured by dual photon absorptiometry (p = 0.014). There was no interaction between dietary and exercise interventions.

OSTEOPENIA

Many studies of both aerobic and resistance training have been shown to stabilize or increase bone density in postmenopausal women.[54–62] Similarly, calcium and vitamin D supplementation can augment bone density and may prevent fracture, particularly when dietary intake or vitamin status is marginal or deficient at baseline.[63–66] It is possible that these interventions may be synergistic when applied in combination, if the mechanisms of action involve complementary pathways.

Nelson[67] tested the hypothesis that a combination of aerobic exercise and calcium supplementation would prove superior to exercise alone in the treatment of postmenopausal osteopenia. Thirty-six women of average age 60 were assigned to a one-year walking program or usual activities and then randomly administered a drink containing either 831 or 41 mg/day calcium in addition to their diet. There was a significant improvement in lumbar spine bone mineral density secondary to exercise (0.5% increase vs. 7.0% decrease in sedentary women) and an improvement in femoral neck density in the high-calcium group (2.0% vs. 1.1% decrease in moderate calcium group). Thus the two modalities acted independently of each other at different bone sites in this study as shown in FIGURE 6.

In a subsequent study of weight-lifting exercise in postmenopausal women, we showed significant differences in both trabecular bone of the lumbar spine as well as cortical bone at the femoral neck compared to sedentary controls after one year.[30] This apparently more generalized effect of resistance training compared to our earlier studies of aerobic training raises the question of whether calcium supplementation during resistance training might further augment the gains seen at the femoral neck. This combination of diet and exercise therapy has not yet been tested, and other combinations of calcium, vitamin D, and exercise also require further study to delineate possible potentiating effects.

SUMMARY

Concomitant losses of skeletal muscle (sarcopenia) and bone mass (osteopenia) along with gradual accretion and centralization of adipose tissue typify usual human aging, and these body composition changes impact substantially upon disease expression and functional status in old age. Recent investigations have therefore attempted to modify these processes with various combinations of dietary and exercise intervention in older adults.

Sarcopenia may be attributable, in part, to disuse as well as to inadequate dietary intake of protein or energy. Low protein, isoenergetic diets result in muscle atrophy in older women in just eight weeks. Resistance training in both fit and frail elders can increase muscle mass within this same time period. Complete (energy and protein) supplements administered concurrently with resistance training exercise augmented muscle and fat gains in healthy older men. However, protein supplementation alone did not improve skeletal muscle function or increase muscle mass in healthy elderly weight lifters compared to those on a normal diet. Complete (energy and protein) supplements merely suppressed habitual dietary intake when administered to frail sedentary elders and did not alter body composition responses to resistance training in this population, in contrast to the healthy elderly subjects. It is unknown whether longer-term supplementation would enhance body composition or functional adaptations to resistance training in frail elders. Current studies will determine the ability of resistance training to offset the catabolic effects of a low-protein diet, such as that prescribed for chronic renal failure in the elderly.

The optimal combination of diet and exercise modality for control of body weight and reduction in adipose tissue has been controversial. Hypocaloric dieting results in losses of weight, fat, central adiposity, and lean mass in obese elderly men and women. Endurance exercise alone may reduce central adipose stores but has only small effects on overall body composition. Adding aerobic exercise to an energy-restricted diet does not prevent the loss of lean tissue or markedly augment the fat losses. By contrast, resistance training during hypocaloric dieting augments lean mass while further reducing fat mass.

Both aerobic and resistance training have been shown to maintain or enhance bone density, as have calcium and vitamin D supplements in some studies. Combinations of these modalities have been less well studied. Calcium supplementation in vitamin D replete postmenopausal women during one year of aerobic training had an independent beneficial effect on cortical bone density at the femoral neck, while the exercise was associated with increases in trabecular bone in the lumbar vertebrae. The effect of hypocaloric dieting on bone mass in the elderly requires study, as does the question of whether aerobic or resistance training would be able to offset this unwanted side effect. Similarly, the effect of concurrent resistance training and calcium or vitamin D supplementation on bone mass has not yet been reported, and further research is indicated in this area of combined exercise and dietary approaches to osteoporosis prevention and treatment.

More long-term studies of the efficacy and feasibility of diet and exercise combinations are needed in the aged to optimize the potential for healthful shifts in body composition through lifestyle modification.

REFERENCES

1. PATE, R.R., M. PRATT, S.N. BLAIR, W.L. HASKELL *et al.* 1995. Physical activity and public health: A recommendation from the Centers for Disease Control and Prevention and the American College of Sports Medicine. J. Am. Med. Assoc. **273:** 402–407.
2. MAZESS, R. 1982. On aging bone loss. Clin. Orthop. **165:** 239–252.
3. ORWOLL, E.S., D.C. BAUER, T.M. VOGT & K.M. FOX. 1996. Axial bone mass in older women. Study of Osteoporotic Fractures Research Group. Ann. Intern. Med. **124:** 187–196.
4. PRIOR, J.C., Y.M. VIGNA, S.I. BARR, S. KENNEDY, M. SCHULZER & D.K. LI. 1996. Ovulatory premenopausal women lose cancellous spinal bone: A five year prospective study. Bone **18:** 261–267.
5. RICHELSON, L. 1984. Relative contributions of aging and estrogen deficiency to postmenopausal bone loss. N. Engl. J. Med. **311:** 1273–1275.
6. RIGGS, B., H. WAHNER, W. DUNN, R. MAZESS, K. OFFORD & L. MELTON. 1981. Differential changes in bone mineral density of the appendicular and axial skeleton with aging: Relationship to spinal osteoporosis. J. Clin. Invest. **67:** 328–335.
7. RIGGS, B., H. WAHNER, L. MELTON, L. RICHELSON, H. JUDD & K. OFFORD. 1986. Rates of bone loss in appendicular and axial skeletons of women: Evidence of substantial vertebral bone loss before menopause. J. Clin. Invest. **77:** 1487–1491.
8. FLEG, J. & E. LAKATTA. 1988. Role of muscle loss in age-associated reduction in VO$_2$ max. J. Appl. Physiol. **65:** 1147–1151.
9. PRATLEY, R., J. HAGBERG, E. ROGUS & A. GOLDBERG. 1995. Enhanced insulin sensitivity and lower waist-to-hip ratio in master athletes. Am. J. Physiol. **268:** E484–E490.
10. TZANKOFF, S.P. & A.H. NORRIS. 1978. Longitudinal changes in basal metabolic rate in man. J. Appl. Physiol. **33:** 536–539.
11. CASTANEDA, C., J. CHARNLEY, W. EVANS & M. CRIM. 1995. Elderly women accommodate to a low-protein diet with losses of body cell mass, muscle function, and immune response. Am. J. Clin. Nutr. **62:** 30–39.
12. MORLEY, J. 1990. Geriatric Nutrition: A Comprehensive Review. Raven Press, Ltd. New York.
13. FIATARONE, M.A., E.F. O'NEILL, N.D. RYAN *et al.* 1994. Exercise training and nutritional supplementation for physical frailty in very elderly people. N. Engl. J. Med. **330:** 1769–1775.
14. WAGNAR, E.H., A.Z. LACROIX, D.M. BUCHNER & E.B. LARSON. 1992. Effects of physical activity on health status in older adults. Ann. Rev. Public Health **13:** 451–468.
15. COLDITZ, G., W. WILLETT, A. ROTNITZKY & J. MANSON. 1995. Weight gain as a risk factor for clinical diabetes mellitus in women. Ann. Intern Med. **122:** 481–486.
16. HARRIS, M.I., W.C. HADDEN, W.C. KNOWLER & P.H. BENNETT. 1987. Prevalence of diabetes and impaired glucose tolerance and plasma glucose levels in U.S. population aged 20–74 yr. Diabetes **36:** 523–534.
17. HARRIS, T., G.M. KOVAR, R. SUZMAN, J.C. KLEINMAN & J.J. FELDMAN. 1989. Longitudinal study of physical ability in the oldest-old. Am. J. Public Health **79:** 698–702.
18. MANSON, J., E. RIMM, M. STAMPFER *et al.* 1991. Physical activity and incidence of non-insulin-dependent diabetes mellitus in women. Lancet **338:** 774–778.
19. MANSON, J., D. NATHAN, A. KROLEWSKI, M. STAMPFER, W. WILLETT & C. HENNEKENS. 1992. A prospective study of exercise and incidence of diabetes among U.S. male physicians. J. Am. Med. Assoc. **268:** 63–67.
20. MONTGOMERY, I., J. TRINDER, S. PAXTON, D. HARRIS, G. FRASER & I. COLRAIN. 1988. Physical exercise and sleep: The effect of the age and sex of the subjects and type of exercise. Acta Physiol. Scand. **133:** 36–40.
21. CUMMINGS, S. 1985. Epidemiology of osteoporosis and osteoporotic fractures. Epidemiol. Rev. **7:** 178–208.

22. NEVITT, M.C., S.R. CUMMINGS, S. KIDD & D. BLACK. 1989. Risk factors for recurrent nonsyncopal falls: A prospective study. J. Am. Med. Assoc. **261:** 2663–2668.
23. DIPIETRO, L., D.F. WILLIAMSON, C.J. CASPERSEN & E. EAKER. 1993. The descriptive epidemiology of selected physical activities and body weight among adults trying to lose weight: The Behavioral Risk Factor Surveillance System survey, 1989. Int. J. Obes. & Relat. Metab. Disord. **17:** 69–76.
24. CARTEE, G.D. 1994. Aging skeletal muscle: Response to exercise. Exercise Sports Sci. Rev. **22:** 91–120.
25. CHARETTE, S., L. MCEVOY, G. PYKA, C. SNOW-HARTER, D. GUIDO, R. WISWELL & R. MARCUS. 1991. Muscle hypertrophy response to resistance training in older women. J. Appl. Physiol. **70:** 1912–1916.
26. FIATARONE, M.A., E.C. MARKS, N.D. RYAN, C.N. MEREDITH, L.A. LIPSITZ & W.J. EVANS. 1990. High-intensity strength training in nonagenarians. Effects on skeletal muscle. J. Am. Med. Assoc. **263:** 3029–3034.
27. LILLEGARD, W.A. & J.D. TERRIO. 1994. Appropriate strength training. Med. Clin. North Am. **78:** 457–477.
28. MCCARTNEY, N., A. HICKS, J. MARTIN & C. WEBBER. 1995. Long-term resistance training in the elderly: Effects on dynamic strength, exercise capacity, muscle, and bone. J. Gerontol. **50A:** B97–B104.
29. NELSON, M., M. FIATARONE, J. LAYNE et al. 1996. Analysis of body-composition techniques and models for detecting change in soft tissue with strength training. Am. J. Clin. Nutr. **63:** 678–686.
30. NELSON, M., M. FIATARONE, C. MORGANTI, I. TRICE, R. GREENBERG & W. EVANS. 1994. Effects of high-intensity strength training on multiple risk factors for osteoporotic fractures. J.Am. Med. Assoc. **272:** 1909–1914.
31. FRONTERA, W.R., C.N. MEREDITH, K.P. O'REILLY, H.G. KNUTTGEN & W.J. EVANS. 1988. Strength conditioning in older men: Skeletal muscle hypertrophy and improved function. J. Appl. Physiol. **64:** 1038–1044.
32. MEREDITH, C.N., W.R. FRONTERA & W.J. EVANS. 1992. Body composition in elderly men: Effect of dietary modification during strength training. J. Am. Geriatr. Soc. **40:** 155–162.
33. PRATLEY, R., B. NICKLAS, M. RUBIN, J. MILLER, A. SMITH, M. SMITH, B. HURLEY & A. GOLDBERG. 1994. Strength training increases resting metabolic rate and norepinephrine levels in healthy 50- to 65-yr-old men. J. Appl. Physiol. **76:** 133–137.
34. CAMPBELL, W.W., M.C. CRIM, V.R. YOUNG & W.J. EVANS. 1994. Increased energy requirements and changes in body composition with resistance training in older adults. Am. J. Clin. Nutr. **60:** 167–175.
35. MEREDITH, C.N., W.R. FRONTERA & W.J. EVANS. 1988. Effect of diet on body composition changes during strength training in elderly men. Am. J. Clin. Nutr. **47:** 767.
36. KATZEL, L.I., E.R. BLEECKER, E.G. COLMAN, E.M. ROGUS, J.D. SORKIN & A.P. GOLDBERG. 1995. Effects of weight loss vs. aerobic exercise training on risk factors for coronary disease in healthy, obese, middle-aged and older men. A randomized controlled trial [see comments]. J. Am. Med. Assoc. **274:** 1915–1921.
37. BALLOR, D.L., V.L. KATCH, M.D. BECQUE & C.R. MARKS. 1988. Resistance weight training during caloric restriction enhances lean body weight maintenance. Am. J. Clin. Nutr. **47:** 19–25.
38. BALLOR, D.L., J.R. HARVEY-BERINO, P.A. ADES, J. CRYAN & J. CALLES-ESCANDON. 1996. Contrasting effects of resistance and aerobic training on body composition and metabolism after diet-induced weight loss. Metab. Clin. Exp. **45:** 179–183.
39. EPSTEIN, L. & R. WING. 1980. Aerobic exercise and weight. Addictive Behav. **5:** 371–388.
40. THOMPSON, J., G. JARVIE & R. LAHEY. 1982. Exercise and obesity: Etiology, physiology and intervention. Psychol. Bull. **91:** 55–79.
41. BALLOR, D. & R. KEESEY. 1991. A meta-analysis of the factors affecting exercise-induced changes in body mass, fat mass, and fat-free mass in males and females. Int. J. Obesity **15:** 717–726.
42. SCHWARTZ, R.S., K.C. CAIN, W.P. SHUMAN et al. 1992. Effect of intensive endurance training on lipoprotein profiles in young and older men. Metab. Clin. Exp. **41:** 649–654.
43. RUOTI, R.G., J.T. TROUP & R.A. BERGER. 1994. The effects of nonswimming water exercises on older adults. J. Orthop. Sports Phys. Ther. **19:** 140–145.

44. HERSEY, W.C.r., J.E. GRAVES, M.L. POLLOCK *et al.* 1994. Endurance exercise training improves body composition and plasma insulin responses in 70- to 79-year-old men and women. Metab. Clin. Exp. **43:** 847–854.

45. BROWN, S., S. UPCHURCH, R. ANDING, M. WINTER & G. RAMIREZ. 1996. Promoting weight loss in type II diabetes. Diabetes Care **19:** 613–624.

46. BALLOR, D.L. & E.T. POEHLMAN. 1993. Exercise intensity does not affect depression of resting metabolic rate during severe diet restriction in male Sprague-Dawley rats. J. Nutr. **123:** 1270–1276.

47. DENGEL, D.R., J.M. HAGBERG, P.J. COON, D.T. DRINKWATER & A.P. GOLDBERG. 1994. Effects of weight loss by diet alone or combined with aerobic exercise on body composition in older obese men. Metab. Clin. Exp. **43:** 867–871.

48. DENGEL, D.R., J.M. HAGBERG, P.J. COON, D.T. DRINKWATER & A.P. GOLDBERG. 1994. Comparable effects of diet and exercise on body composition and lipoproteins in older men. Med. Sci. Sports Exercise **26:** 1307–1315.

49. BLAIR, S.N., H.W. KOHL, R.S. PAFFENBARGER *et al.* 1989. Physical fitness and all-cause mortality: A prospective study of healthy men and women. J. Am. Med. Assoc. **262:** 2395–2401.

50. FOX, A.A., J.L. THOMPSON, G.E. BUTTERFIELD, U. GYLFADOTTIR, S. MOYNIHAN & G. SPILLER. 1996. Effects of diet and exercise on common cardiovascular disease risk factors in moderately obese older women. Am. J. Clin. Nutr. **63:** 225–233.

51. WOOD, P.D., M.L. STEFANICK, P.T. WILLIAMS & W.L. HASKELL. 1991. The effects on plasma lipoproteins of a prudent weight-reducing diet, with or without exercise, in overweight men and women. N. Eng. J. Med. **325:** 461–466.

52. TREUTH, M., G. HUNTER, R. WEINSIER & S. Kell. 1995. Energy expenditure and substrate utilization in older women after strength training: 24-h calorimeter results. J. Appl. Physiol. **78:** 2140–2146.

53. VAN Dale, D. & W. Saris. 1989. Repetitive weight loss and weight regain: Effects on weight reduction, resting metabolic rate, and lipolytic activity before and after exercise and/or diet treatment. Am. J. Clin. Nutr. **49:** 409–416.

54. ALOIA, J., S. COHN, J. OSTUNI, R. CANE & K. ELLIS. 1978. Prevention of involutional bone loss by exercise. Ann. Intern. Med. **89:** 356–358.

55. CHOW, R., J.E. HARRISON & C. NOTARIUS. 1987. Effect of two randomised exercise programes on bone mass of healthy postmenopausal women. Br. Med. J. **295:** 1441–1444.

56. DALSKY, G., K. STOCKE, A. EHSANI, E. SLATOPOLSKY, W. LEE & S. BIRGE. 1988. Weight-bearing exercise training and lumbar bone mineral content in postmenopausal women. Ann. Intern. Med. **108:** 824–828.

57. KRALL, E.A. & B. DAWSON-HUGHES. 1994. Walking is related to bone density and rates of bone loss. Am. J. Med. **96:** 20–26.

58. KROLNER, B., B. TOFT, S.P. NIELSEN & E. TONDEVOLD. 1983. Physical exercise as prophylaxis against involutional vertebral bone loss: A controlled trial. Clin. Sci. **64:** 541–546.

59. MENKES, A., R. MAZEL, R. REDMOND *et al.* 1993. Strength training increases regional bone mineral density and bone remodeling in middle-aged and older men. J. Appl. Physiol. **74:** 2478–2484.

60. NIELSEN, H.K., K. BRIXEN, L.P. KRISTENSEN, H.P. PEDERSEN & E. SANDAGER. 1992. Effects of different kinds of exercise on bone mass and bone metabolism in elderly women. Eur. J. Exp. Musculoskel. Res. **1:** 41–46.

61. PRUITT, L.A., R.D. JACKSON, R.L. BARTELS & H.J. LEHNHARD. 1992. Weight-training effects on bone mineral density in early postmenopausal women. J. Bone Miner. Res. **7:** 179–185.

62. SMITH Jr., E.L., W. REDDAN & S.P.E. 1981. Physical activity and calcium modalities for bone mineral increase in aged women. Med. Sci. Sports Exercise **13:** 60–64.

63. DAWSON-HUGHES, B., G.E. DALLAL, E.A. KRALL, L. SADOWSKI, N. SAHYOUN & S. TANNENBAUM. 1990. A controlled trial of the effect of calcium supplementation on bone density in postmenopausal women. N. Engl. J. Med. **323:** 878–883.

64. HEANEY, R. 1982. Calcium nutrition and bone health in the elderly. Am. J. Clin. Nutr. **36:** 986–1013.

65. PRINCE, R.L., M. SMITH, I.M. DICK, R.I. PRICE, P.G. WEBB, N.K. HENDERSON & M.M. HARRIS. 1991. Prevention of postmenopausal osteoporosis: A comparative study of exercise, calcium supplementation, and hormone-replacement therapy. N. Eng. J. Med. **325:** 11189–11195.

66. RECKER, R. & R. SAVILLE. 1977. Effects of estrogen and calcium carbonate on bone loss in post-menopausal women. Ann. Int. Med. **87:** 649–655.

67. NELSON, M., E. FISHER, F. DILMANIAN, G. DALLAL & W. EVANS. 1991. A 1-y walking program and increased dietary calcium in postmenopausal women: Effects on bone. Am. I. Clin. Nutr. **53:** 1304–1311.

The Biochemical, Pathophysiological, and Medical Aspects of Ubiquinone Function

HANS NOHL,[a] LARS GILLE, AND KATRIN STANIEK

Institute of Pharmacology and Toxicology, Veterinary University of Vienna, Josef Baumann-Gasse 1, A-1210 Vienna, Austria

ABSTRACT: Ubiquinone (Q) shares its biological implication in membrane-associated redox reactions with a variety of other redox carriers, such as dehydrogenases, non-heme-iron proteins, and cytochromes. Peculiarities arise from the lack of transition metals, which in contrast to the other electron carriers do not participate in redox-shuttle activities of Q. Another peculiarity is the lipophilicity of Q, which allows free movement between reductants and oxidants of a membrane. The chemistry of Q reduction and ubiquinol oxidation requires the stepwise acceptance and transfer of two single electrons associated with the addition or release of two single H^+. These special qualities are widely used in biological membranes for linear electron transfer and transmembranous H^+ translocation. In mitochondria it was long reported that under certain conditions linear e^- transfer from the semireduced form (SQ·) to native oxidants of the respiratory chain may run out of control, thereby establishing a permanent source of oxygen radical release. It should be mentioned that in mitochondria e^- transfer to dioxygen out of sequence requires a particular treatment with inhibitors and uncouplers of the respiratory chain. Nevertheless, it is generally assumed that Q is mainly involved in mitochondrial $O_2^{·-}$ generation and that mitochondria represent the major source of $O_2^{·-}$ radicals under physiological and various pathophysiological conditions. The ever-increasing application of coenzyme Q as an antioxidant for the prophylaxis and treatment of a great variety of functional disorders, including senescence, has considerably stimulated our interest in the potential prooxidative potency of this natural electron carrier. Experimental evidence will be presented that under physiological conditions Q implicated in mitochondrial e^- transfer of the respiratory chain is not involved in cellular oxygen activation. It will also be shown that alterations of Q from an e^- carrier to an active radical promotor is possible under various conditions. In addition, reaction products emerging from the antioxidant activity of ubiquinol were found to stimulate the formation of inorganic as well as organic oxygen radicals.

COENZYME Q IN BIOCHEMISTRY

Coenzyme Q (ubiquinone) is a biological compound that is widely distributed in all plants, all animals, and in most microorganisms. It is present in all tissues associated with biomembranes. However, its biological function is not clear so far. In accordance with the chemistry of redox-cycling ubiquinone, one may assume that this compound acts both as an electron carrier and proton translocator.[1] In mitochondria, coenzyme Q is involved in energy-linked redox processes.[2]

Due to its lipophilicity ubiquinone (Q) interacts with dehydrogenases and shuttles a pair of two single electrons to cytochromes by diffusion (FIG. 1). This bioactivity requires two consecutive deprotonation steps, that result in the transfer of two protons into the

[a]Corresponding author: Tel.: +43 1 25077 4400; fax: +43 1 25077 4490; e-mail: hans.nohl@vuwien.ac.at

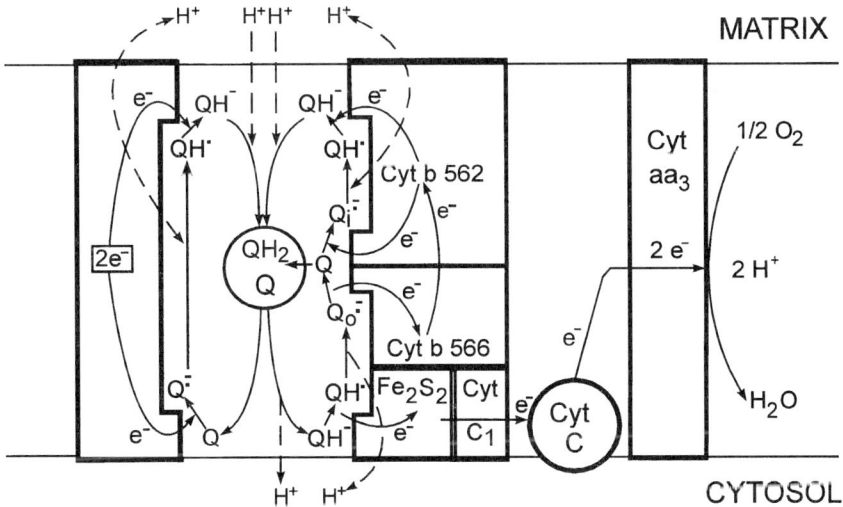

FIGURE 1. Scheme of coenzyme Q functions as electron carrier and proton translocator in the mitochondrial respiratory chain.

cytosol. In contrast to the protein compounds of complex I, the bc_1 complex, and cytochrome oxidase (cyt aa_3), which all contribute to the establishment of a transmembranous proton gradient by a well-controlled proton conductance, redox-cycling Q releases protons by deprotonation under the control of *pK* values. These peculiarities are involved in the regulation of thermodynamic and kinetic conditions, allowing a bifurcation of one electron cycling through b-type cytochromes while the other electron is linearly transferred to cytochrome oxidase.[3,4] The transmembranous establishment of the proton gradient is the energy source used for ATP synthesis. The backflow of protons through the proton channel of the ATP synthase results in ATP release through conformational changes of the enzyme. The complex regulation of the transfer of reducing equivalents through the Q cycle makes Q particularly susceptible to derangements of energy-linked respiration. In addition, the K_M values for the interaction of Q with dehydrogenases reveal that respiratory activity is under the control of Q available for shuttling reducing equivalents to cytochromes. Extensive studies of Lenaz *et al.* exclude diffusion rates as being a limiting factor.[5,6] The decrease of energy gain from energy-linked respiration of aged rats[7] may, therefore, indicate limitation of Q interaction with dehydrogenases due to pool levels below K_M values. Lenaz *et al.* reported that mitochondrial Q levels were slightly reduced in senescent rats.[8] However, supplementation to normal levels did not improve the efficiency of energy gain.

PATHOPHYSIOLOGICAL ASPECTS OF COENZYME Q

Apart from the involvement in mitochondrial energy conservation redox-cycling Q of the respiratory chain was long suggested to release single electrons to dioxygen out of sequence.[9,10] This unexpected electron pathway was considered to be of pathophysiologi-

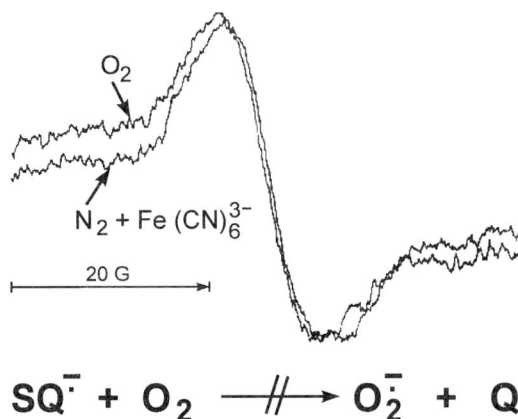

$$SQ^{\bar{\cdot}} + O_2 \quad \xlongequal{\quad//\quad}\!\!\!\!\longrightarrow \quad O_2^{\bar{\cdot}} + Q$$

FIGURE 2. The effect of oxygen on the ubisemiquinone-related ESR signal in native rat heart mitochondria. Ubisemiquinone pools of intact mitochondria are insensitive to oxygen. $Fe(CN)_6^{3-}$ was used as an artificial electron acceptor to allow mitochondrial electron transfer in the absence of oxygen. The equation demonstrates the insensitivity of redox-cycling SQ· to oxygen. For details, see Nohl et al.[13]

cal significance and contribute to the age-related establishment of oxidative stress.[7,11,12] We have tested the existence of this pathway in freshly isolated mitochondria.

Redox-cycling ubisemiquinones (SQ·) were followed by means of ESR spectroscopy in the absence and presence of oxygen (FIG. 2). Ubisemiquinones were found to be insensitive to O_2 in intact mitochondria, clearly excluding redox-cycling ubisemiquinones as a physiological generation site for $O_2^{\cdot-}$ radicals.[13] Our further studies on the existence of an electron-leakage pathway from mitochondrial ubisemiquinones to oxygen revealed that under certain conditions SQ· may readily undergo autoxidation.

The following conditions were found to be required for autoxidation: (1) adjustment of redox potentials of SQ· to values allowing one-electron transfer to O_2; (2) shift of redox-cycling SQ· pools to the deprotonated (anionic) form. Redox-cycling SQ· associated with the cytosolic part of complex III (SQ·$_{out}$) closely interacts with the redox center of cyt b$_{566}$ (see FIG. 1). It was assumed that the midpoint potential of this SQ· species is amenable to electrostatic interaction with the heme center of cyt b$_{566}$, thereby being changed to strong negative values sufficient for one electron transfer to oxygen.[14] However, autoxidation is prevented as long as electron transfer occurs in the intact phospholipid bilayer. This was clearly demonstrated in mitochondria in which the molecular order of the phospholipid bilayer was changed. The derangement of the structural order (induced by toluene accumulation) was assessed from spin-labeling experiments, allowing the calculation of order parameters (FIG. 3).[15]

Structural derangement of the mitochondrial inner membrane, where electron carriers of the respiratory chain are operating, results in the interaction of O_2 with SQ· (FIG. 4).

The susceptibility of redox-cycling SQ· to O_2 increases with increasing disorder of the phospholipid bilayer. The transfer of the odd electron from redox-cycling SQ· to O_2 is getting clear if one correlates the susceptibility of SQ· with the amount of $O_2^{\cdot-}$ released (FIG. 5).[16] The underlying chemistry of the autoxidation is demonstrated by the requirement of

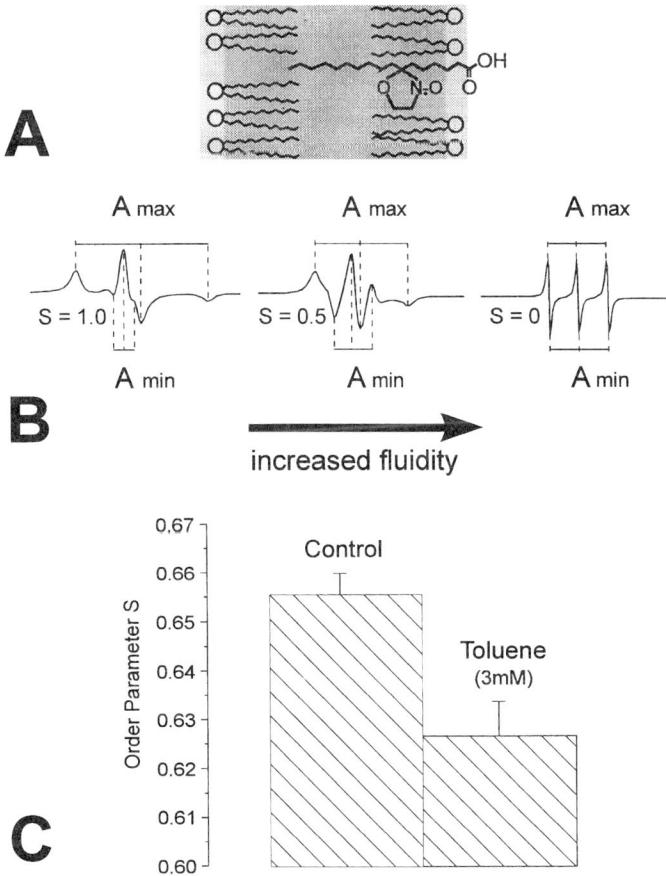

FIGURE 3. The effect of toluene insertion on order parameters of the mitochondrial membrane. **A**: The spin label used to determine the order parameters was a 5-doxyl-stearic acid. Due to the long hydrocarbon chain of this spin label, it is expected to take an orientation parallel to the fatty acid chains of phospholipids in the bilayer of the inner mitochondrial membrane. **B**: The more the membrane becomes fluid, the more isotropic appears the ESR spectrum of the spin label, and the derived order parameter (S) is decreased. **C**: The order parameter of the mitochondrial membrane after toluene insertion (3 mM) assessed by spin-label measurements is significantly lower as compared to untreated mitochondria (control). For details, see Nohl *et al.*[15]

H_2O for an interaction of SQ· with O_2 (FIG. 6). The effect of water on the SQ·-related ESR signal in the absence of oxygen indicates acceleration of disproportionation of the protonated species.

If one transfers this observation to redox-cycling SQ· operating in intact or deranged phospholipid membranes, FIGURE 7 may represent the situation under which SQ· of the respiratory chain shuttle electrons to O_2 out of sequence. $O_2^{\bullet-}$ radical release increases lin-

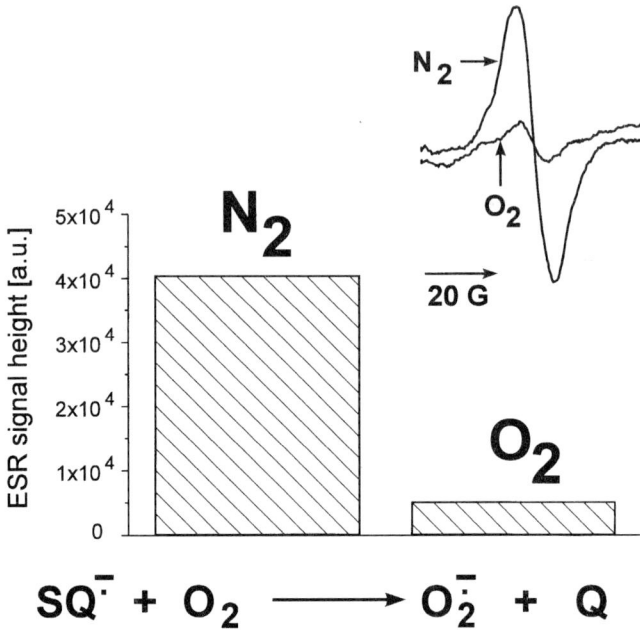

FIGURE 4. Mitochondrial ubisemiquinone pools become sensitive to dioxygen when incubated with toluene. After toluene insertion the ESR signal intensity of the ubisemiquinone pools is considerably diminished by oxygen (O_2), whereas in the absence of oxygen the ESR signal remains unchanged. $Fe(CN)_6^{3-}$ was used again to replace for oxygen as an artificial electron acceptor under anaerobic conditions (N_2). For details, see Nohl et al.[13]

earily with the accessibility of water from the cytosol to redox-cycling SQ· in the inner mitochondrial membrane. Derangement of this membrane was achieved by the insertion of toluene. Contact to the water phase of redox-cycling SQ· was also found to be established when mitochondria were deprived of oxygen for a period of time (ischemia). Transient hypoxia may periodically occur in senescent individuals, resulting in the accumulation of lactate, which is in equilibrium with cytosolic NADH. NADH is directly oxidized by the exogenous NADH dehydrogenase of rat heart mitochondria, which shuttles the reducing equivalents via complex I to cytochrome oxidase. This electron pathway is not associated with ATP generation and leads to an alkalization of the cytosol.[17] Furthermore, an SQ· species is involved in this unusual electron pathway that appears not to be active under normal respiratory conditions.

Relaxation of this SQ· species in the cavity of an ESR spectrometer is impeded, which can be seen from the low microwave power required for saturation conditions (FIG. 8). By contrast, SQ· species taking part in regular energy-linked respiration only saturate at microwave power values beyond 20 mW (TABLE 1).

This peculiarity indicates an impaired physical interaction with the redox partner of this SQ· species also suggesting a shift of its midpoint potential to more negative values.

TABLE 1. ESR Characteristics of Ubisemiquinone Species in Intact Mitochondria and in Isolated Respiratory Complexes Detected by ESR Spectroscopy and Their Sensitivity to Inhibitors of the Respiratory Chain[a,15,19,20,21]

ESR Parameters	Nohl/Gille $SQ_l^{\cdot\,-}$	Suzuki/King $SQ_n^{\cdot\,-}$	Ohnishi/Trumpower $SQ_s^{\cdot\,-}$	Slater/de Vries $SQ_{C,out}^{\cdot\,-}$	Slater/de Vries $SQ_{C,in}^{\cdot\,-}$
	Heart mitochondria + NADH	isolated complex 1	isolated complex 2	isolated complex 3	
Line width	6.7G	6.8G (<5 mW)	12G (<100 mW)	8.3G (<10 μW)	10.0G (<10 μW)
Pmax	2 mW	5/20 mW	>100 mW	~20 mW	~20 mW
T	200°K	232°K	50°K	50°K	50°K
Sensitive to	rotenone	rotenone	TTFA	myxothiazol/antimycin A	

[a]Line width, peak-to-peak line width of the ESR signal; Pmax, microwave power when the maximal ESR signal amplitude is achieved; T, measurement temperature; TTFA, thenoyltrifluoroacetone.

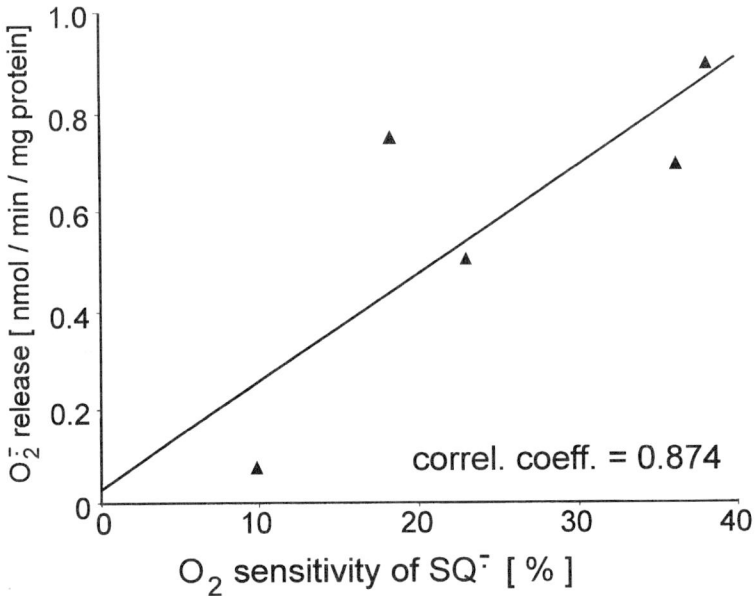

FIGURE 5. Correlation between oxygen sensitivity of redox-cycling ubisemiquinones and super-oxide radical release from respiring mitochondria. For details, see Nohl *et al.*[16]

The interaction of this SQ· species with the water phase (cytosol) was demonstrated by an interaction of the water-soluble paramagnetic Cr^{3+} salt, which causes the disappearance of the saturation effect on the respective ESR signal due to spin–spin exchange reactions (FIG. 8). We, therefore, conclude that the redox-cycling SQ· species involved in the non-energy-linked electron pathway operates in contact with the water phase, allowing autoxidation and the release of $O_2^{\bullet-}$ radicals. Mitochondria isolated from senescent rats release $O_2^{\bullet-}$ radicals as a byproduct of respiration. As shown in the case of membrane alteration and hypoxic NADH accumulation, redox-cycling SQ· are also involved in senescent animals in the leakage of electrons to oxygen out of sequence (FIG. 9).

This became evident from the susceptibility of mitochondrial SQ· to oxygen. Thermodynamic conditions required for this nonphysiological electron transfer are established during aging by a combination of physical membrane alterations and the activation of the NADH-related non-energy-linked electron transfer pathway. Uncontrolled electron leak from the regular respiratory pathway affects the efficiency of energy gain irrespective of the mechanism that triggers $O_2^{\bullet-}$ formation. A rationale that may link these effects on a molecular base is the efficiency of the establishment and preservation of the transmembranous proton gradient used for ATP synthesis. The slight decrease of electron flow rates passing the three phosphorylation sites during $O_2^{\bullet-}$ release is not critical in this respect, considering the relationship of O_2 consumption for $O_2^{\bullet-}$ and H_2O formation ($\cong 2/98$).

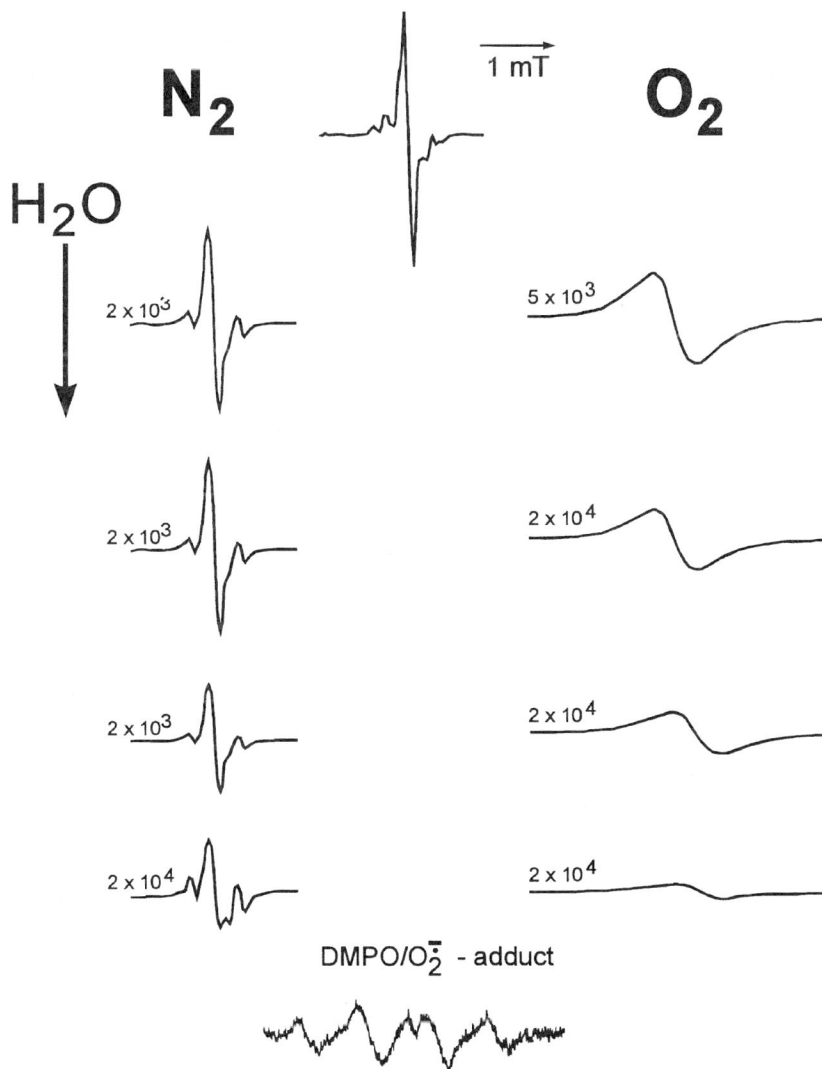

FIGURE 6. The ESR signal intensity of ubisemiquinone anion radicals in water-free acetonitrile after stepwise addition of water in the absence (N_2) and in the presence of oxygen (O_2). Autoxidation products of ubisemiquinones were detected by the ESR spin trapping with 5,5-dimethyl-1-pyrroline *N*-oxide (DMPO). Due to the characteristic splitting pattern of the ESR signal obtained, the DMPO/ $O_2^{\bullet-}$ adduct was identified. For details, see Nohl *et al.*[22]

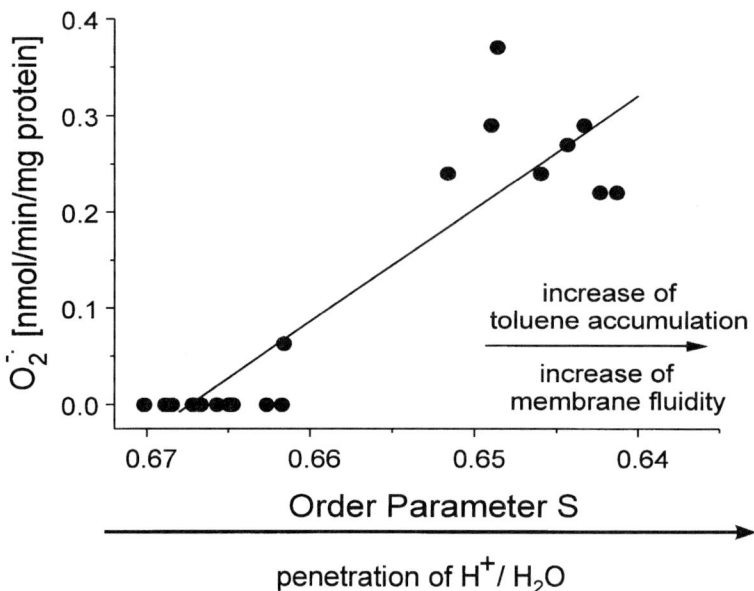

FIGURE 7. Correlation between the toluene-induced fluidity increase (decrease of the order parameter S) of the mitochondrial membrane and the release of superoxide radicals as a side product of respiration $r = -0.925$; $p < 0.001$. For details, see Nohl et al.[15]

COENZYME Q IN MEDICINE

In recent years coenzyme Q has increasingly been assumed to exert antioxidant functions in the various biological membranes. In homogenous solutions coenzyme Q in its fully reduced form (ubiquinol, QH_2) was found to be as active in scavenging peroxyl radicals as vitamin E (FIG. 10, left graph).[18] By contrast, vitamin E was a more efficient terminator of lipid peroxidation when associated with lipid membranes (of phosphatidylcholine liposomes) when compared to QH_2 (FIG. 10, right graph). The most likely explanation for this contradiction can be derived from the general equation describing the interaction of an antioxidant (AH) with a radical (X·).

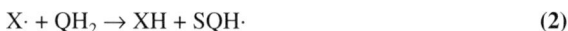

$$X\cdot + AH \rightarrow XH + A\cdot \tag{1}$$

$$X\cdot + QH_2 \rightarrow XH + SQH\cdot \tag{2}$$

Analogous to this reaction, SQ· formation is expected. Due to the instability of SQ· intermediates expected to be formed as the QH_2-derived reaction product, the existence of SQ· in peroxidizing membranes at 77°K was followed (FIG. 11, inset). While QH_2 is oxidized in peroxidizing liposomes, SQ· are formed as intermediate oxidation products (FIG. 11). The strong hydrophilic Gd^{3+} salt, which has a high spin density, was used to probe the localization of SQ· evolved from the antioxidant activity of QH_2 in the lipid bilayer. Spin-spin interactions revealed that parts of SQ· formed exist close to the water phase (FIG. 12). According to conditions earlier described for autoxidation, this SQ· fraction underwent

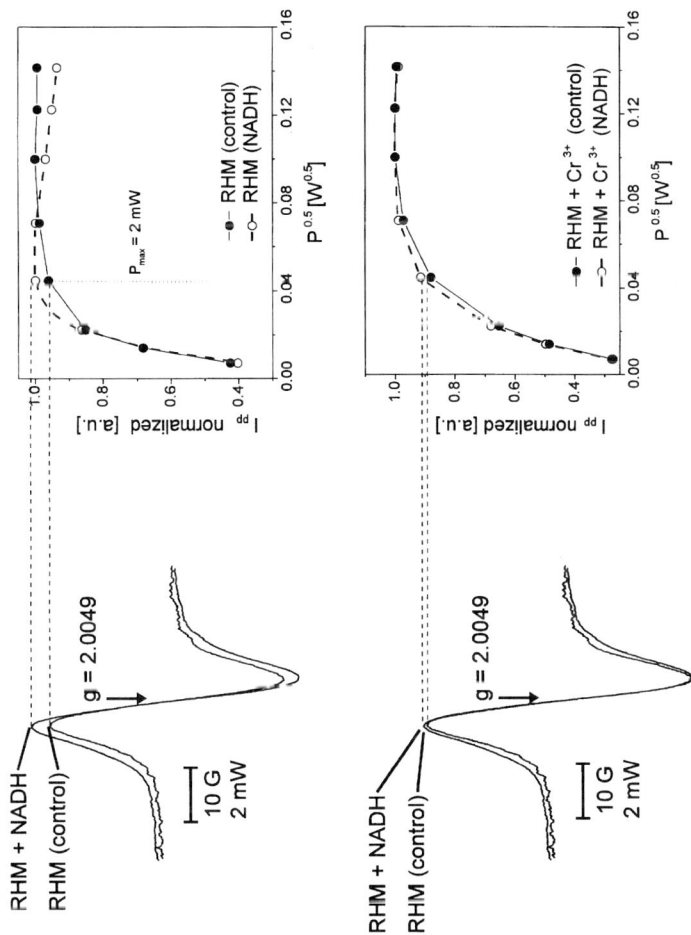

FIGURE 8. Power saturation behavior of the ubisemiquinone ESR signal in succinate/fumarate-respiring rat heart mitochondria (RHM) after preincubation with exogenous NADH (NADE preincubated). Control mitochondria were preincubated analogously without NADH (control) (upper graph). After the addition of Cr^{3+} the power saturation differences disappear (lower graph). The intensities of the individual curves were normalized based on the maximum intensity of each individual curve for comparison of the saturation behavior at the same level of intensity. I_{pp}, peak-to-peak intensities of the ESR signal; $P^{0.5}$, square root of microwave power; g, g-factor of the ESR signal. For details, see Nohl *et al.*[15]

Parameters changed	Incorp. of lipophilic xenobiotics		Activation of non-energy-linked respiration (Ischemia)		Senescence	
	P/O	RC	P/O	RC	P/O	RC
Efficiency of energy gain	⇩	⇩	⇩	⇩	⇩	⇩
SQ˙ contact with the water phase	membrane disorder ⇑		e⁻ transfer close to the aqueous phase		- membrane disorder - e⁻ transfer close to the aqueous phase	
O₂- sensitivity of redox-cycling ubisemiquinone	(waveform) Contr.		(waveform) Contr.		(waveform) Contr.	
Effective O₂⁻ release associated with respiration	⇑		⇑		⇑	

FIGURE 9. Conditions found to transform mitochondrial SQ˙ from a compound involved in energy-linked respiration to an $O_2^{\cdot-}$ generator. Deviation of electrons to oxygen out of sequence is associated with an impairment of energy gain. Activation of the non-energy-linked respiration pathway is due to increased cytosolic NADH levels. Mitochondria were equilibrated either with N_2 (Contr.) or O_2. In the case of N_2, $K_3Fe(CN)_6$ as artificial electron acceptor was present. P/O, ADP to oxygen ratio; RC, respiratory control index. For details, see Nohl et al.[7,23,24]

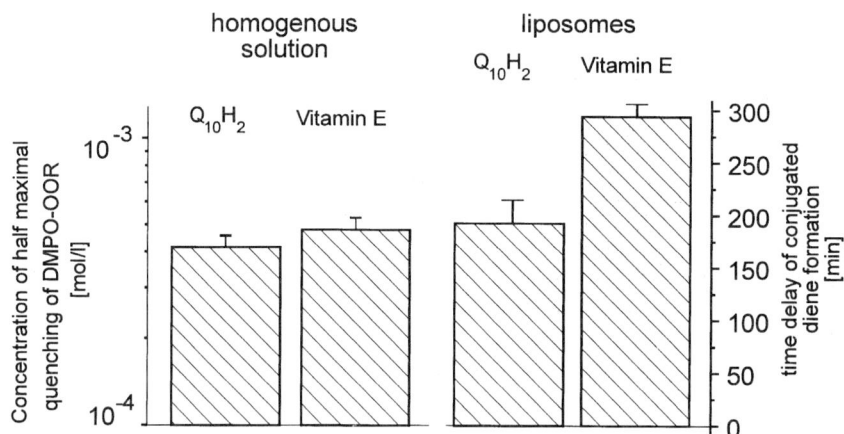

FIGURE 10. Correlation between peroxyl radical scavenging activities of reduced coenzyme Q ($Q_{10}H_2$) and vitamin E in a homogenous system (left graph). The concentration of half maximal quenching was calculated from the decay of the respective ESR signal intensity of the peroxyl radical spin adduct (DMPO-OOR) versus antioxidant concentration. Concentrations: DMPO, 0.6 mM; vitamin E, 0–3 mM; ubiquinol, 0–3 mM. Time delay of AAPH-induced (azo-bis-amidinopropane hydrochloride) conjugated diene formation in liposomes consisting of soybean phosphatidylcholine and antioxidant (ratio 16/1) (right graph). For details, see Nohl et al.[18]

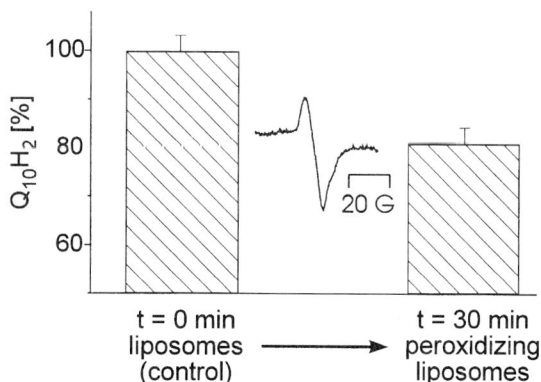

FIGURE 11. Oxidation of reduced coenzyme Q ($Q_{10}H_2$) in liposomes exposed to AAPH at 37°C for lipid peroxidation. Inset: SQ·-related ESR signal observed during lipid peroxidation (LPO). For methodological reasons LPO from AAPH was initiated by UV irradiation at 77°K to stabilize SQ· in the frozen matrix possibly formed. SQ· under these conditions were not obtained when $Q_{10}H_2$ was irradiated with UV in the absence of LPO initiator AAPH and lipids. For details, see Nohl *et al.*[18]

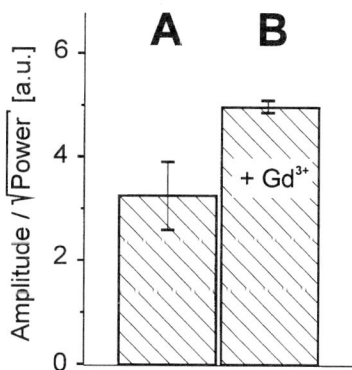

FIGURE 12. Saturation behavior of the SQ·-related ESR signal generated during LPO in liposomes from an interaction of lipid radicals with incorporated reduced coenzyme Q ($Q_{10}H_2$). In the presence of Gd^{3+} (B), the relaxation of the SQ· radical was enhanced, resulting in an increased ESR signal, which was not seen with the control samples (A) under saturation conditions (20 mW). For details, see Nohl *et al.*[18]

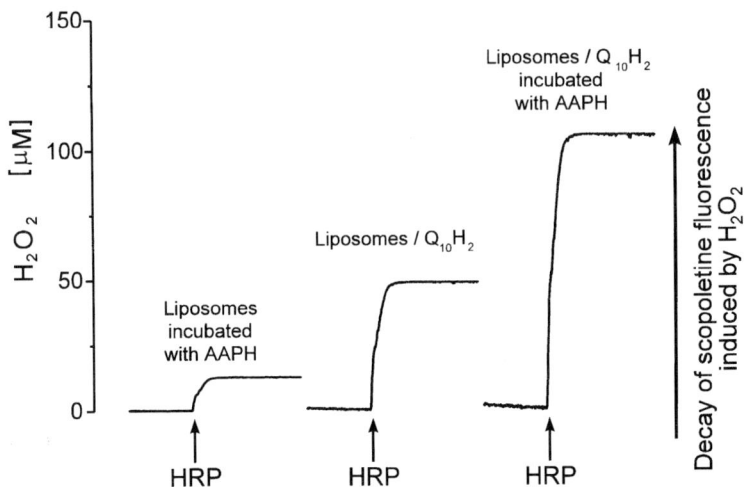

FIGURE 13. H_2O_2 formation as by-product of the antioxidant activity of reduced coenzyme Q $(Q_{10}H_2)$ in liposomes subjected to conditions of lipid peroxidation (LPO). Control samples were incubated without the LPO initiator AAPH. The amount of H_2O_2 was assessed from the catalase-sensitive fluorescence decay of scopoletine catalyzed by horseradish peroxidase (HRP). For methodological details, see Loschen et al.[25]

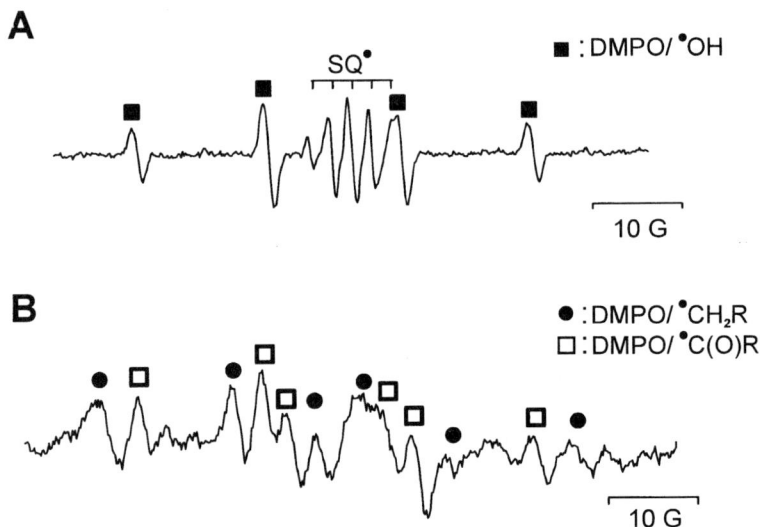

FIGURE 14. ESR spectra of DMPO spin adducts observed during reaction of SQ· with (A) H_2O_2, and (B) linoleic acid hydroperoxide (LOOH). Based on computer simulations the superimposed spectra could be assigned to (A) DMPO/·OH adduct ($a_N = 14.9$ G, $a_H = 14.9$ G) superimposed by the SQ· signal, and (B) DMPO/·CH$_2$R adduct ($a_N = 23.7$ G, $a_H = 14.8$ G) and DMPO/·C(O)R adduct ($a_N = 16.9$ G, $a_H = 13.7$ G), which result from the decomposition of the highly unstable lipid alkoxyl radical. The obtained signals of DMPO spin adducts were not sensitive to SOD and were not observed in the absence of either ubiquinone-0 or hydroperoxide.

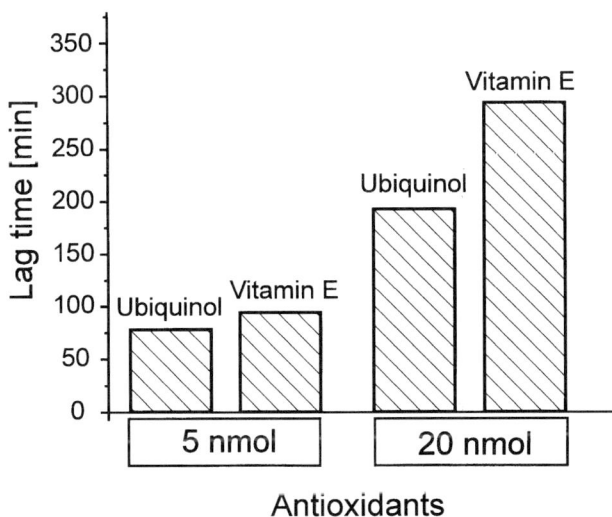

FIGURE 15. Antioxidant activities of vitamin E and ubiquinol assessed from the lag time of AAPH-induced conjugated diene formation in liposomes with different amounts of vitamin E or ubiquinol as antioxidants. The concentration of the phospholipids was kept constant for all experiments (320 nmol soybean phosphatidylcholine).

autoxidation, forming H_2O_2 from $O_2^{\bullet-}$ radicals (FIG. 13). SQ· not amenable to the water phase were found to react both with H_2O_2 giving rise to HO· radical formation (eqs. 3 and 4; FIG. 14A), and lipid peroxides (LOOH), resulting in the formation of alkoxyl radicals (LO·) (eqs. 5–7; Fig. 14B), which were detected by electron spin resonance (ESR) spin trapping.

$$SQ^{\bullet-} + H_2O_2 \rightarrow Q + HO^- + HO\cdot \tag{3}$$

$$HO\cdot + DMPO \rightarrow DMPO/\cdot OH \tag{4}$$

$$SQ^{\bullet-} + LOOH \rightarrow Q + HO^- + LO\cdot \tag{5}$$

$$LO\cdot \text{ (decomposition at r.t.)} \rightarrow \cdot CH_2R + \cdot C(O)R \tag{6}$$

$$\cdot CH_2R + \cdot C(O)R + DMPO \rightarrow DMPO/\cdot CH_2R + DMPO/\cdot C(O)R \tag{7}$$

The formation of these secondary antioxidant-derived prooxidants increases the possibility that more QH_2 was present for a reaction with split products of lipid peroxidation. This explains the concentration-dependent decline in the antioxidant capacity of QH_2 that was not seen with vitamin E (FIG. 15). The antioxidant function of QH_2 is necessarily linked to radical chain reactions, which under certain conditions may stimulate prooxidant formation. In biomembranes having electron transfer functions, SQ· can be directly reduced to QH_2, which will increase the antioxidant balance. Inhibition of radical chain reactions, which may be propagated by ubisemiquinones, can also be expected if stabilizing binding sites for this QH_2-derived reaction product are available.

CONCLUSION

At present the biological function of coenzyme Q is not fully understood. Three distinct activities are under discussion: bioenergetic activities (energy conservation, proton/electron translocation), prooxidant functions (signal transduction, oxidative stress, free radical theory of aging), and antioxidant activities (recycling of chromanoxyl radicals, balance of oxidative stress). Age-related decrease of mitochondrial Q is not responsible for the impairment of energy conservation. The decreased efficiency of energy gain from respiration is a general phenomenon compulsorily linked to $O_2^{\cdot-}$ formation via redox-cycling Q. Autoxidation of membrane-associated SQ· occurs under nonphysiological conditions: (1) increasing nonstabilized anionic SQ· species over normal, (2) membrane localization of SQ· close to the aqueous phase, and (3) shift of the midpoint potential to more negative values. Antioxidant-derived reaction products of QH_2 are prooxidants impairing termination of oxidative membrane damage with increasing concentrations.

REFERENCES

1. RAUCHOVA, H., Z. DRAHOTA & G. LENAZ. 1995. Function of coenzyme Q in the cell: Some biochemical and physiological properties. Physiol. Res. **44:** 209–216.
2. MITCHELL, P. 1975. Protonmotive redox mechanism of the cytochrome bc_1 complex in the respiratory chain: Protonmotive ubiquinone cycle. FEBS Lett. **56:** 1–6.
3. CHANCE, B., D.F. WILSON, P.L. DUTTON & M. ERECINSKA. 1970. Energy-coupling mechanisms in mitochondria: Kinetic, spectroscopic, and thermodynamic properties of an energy-transducing form of cytochrome b. Proc. Natl. Acad. Sci. USA **66:** 1175–1182.
4. BRANDT, U. 1996. Bifurcated ubihydroquinone oxidation in the cytochrome bc1 complex by proton-gated charge transfer. FEBS Lett. **387:** 1–6.
5. LENAZ, G., M. BATTINO, C. CASTELLUCCIO, R. FATO, M. CAVAZZONI, H. RAUCHOVA, C. BOVINA, G. FORMIGGINI & G.P. CASTELLI. 1990. Studies on the role of ubiquinone in the control of the mitochondrial respiratory chain. Free Radical Res. Commun. **8:** 317–327.
6. LENAZ, G., F. FATO, M. BATTINO, M. CAVAZZONI, H. RAUCHOVA, C. CASTELLUCCIO, J. SVOBODOVA & G. PARENTI-CASTELLI. 1990. Role of diffusion in mitochondrial electron transfer. *In* Highlights in Ubiquinone Research. G. Lenaz, O. Barnabei & A. Rabbi, Eds.: 52–57. Taylor & Francis. London.
7. NOHL, H., V. BREUNINGER & D. HEGNER. 1978. Influence of mitochondrial radical formation on energy-linked respiration. Eur. J. Biochem. **90:** 385–390.
8. GENOVA, M.L., C. CASTELLUCCIO, R. FATO, G.P. CASTELLI, M.M. PICH, G. FORMIGGINI, C. BOVINA, M. MARCHETTI & G. LENAZ. 1995. Major changes in complex I activity in mitochondria from aged rats may not be detected by direct assay of NADH:coenzyme Q reductase. Biochem. J. **311:** 105–109.
9. CADENAS, E., A. BOVERIS, C.I. RAGAN & A.O.M. STOPPANI. 1977. Production of superoxide radicals and hydrogen peroxide by NADH ubiquinone reductase and ubiquinol cytochrome c reductase from beef heart mitochondria. Arch. Biochem. Biophys. **180:** 248–257.
10. BOVERIS, A., E. CADENAS & A.O.M. STOPPANI. 1976. Role of ubiquinone in the mitochondrial generation of hydrogen peroxide. Biochem. J. **156:** 435–444.
11. SOHAL, R.S. 1993. Aging, cytochrome oxidase activity, and hydrogen peroxide release by mitochondria. Free Radical Biol. Med. **14:** 583–588.
12. SOHAL, R.S. & A. DUBEY. 1994. Mitochondrial oxidative damage, hydrogen peroxide release, and aging. Free Radical Biol. Med. **16:** 621–626.
13. NOHL, H. & K. STOLZE. 1992. Ubisemiquinones of the mitochondrial respiratory chain do not interact with molecular oxygen. Free Radical Res. Commun. **16:** 409–419.
14. DING, H., C.C. MOSER, D.E. ROBERTSON, M.K. TOKITO, F. DALDAL & P.L. DUTTON. 1995. Ubiquinone pair in the Qo site central to the primary energy conversion reactions of cytochrome bc1 complex. Biochemistry **34:** 15979–15996.

15. NOHL, H., L. GILLE, K. SCHOENHEIT & Y. LIU. 1996. Conditions allowing redox-cycling ubisemi-quinone in mitochondria to establish a direct redox couple with molecular oxygen. Free Radical Biol. Med. **20:** 207–213.

16. NOHL, H., K. STANIEK & L. GILLE. 1997. Imbalance of oxygen activation and energy metabolism as a consequence or mediator of aging. Exp. Gerontol. **32:** 485–500.

17. SCHÖNHEIT, K. & H. NOHL. 1996. Oxidation of cytosolic NADH via complex I of heart mito-chondria. Arch. Biochem. Biophys. **327:** 319–323.

18. NOHL, H., L. GILLE & K. STANIEK. 1997. Endogenous and exogenous regulation of redox-proper-ties of coenzyme Q. Mol. Aspects Med. **18:** S33–S40.

19. SUZUKI, H. & T.E. KING. 1983. Evidence of a ubisemiquinone radical(s) from the NADH-ubiquinone reductase of the mitochondrial respiratory chain. J. Biol. Chem. **258:** 352–358.

20. OHNISHI, T. & B.L. TRUMPOWER. 1980. Differential effects of antimycin on ubisemiquinone bound in different environments in isolated succinate cytochrome c reductase complex. J. Biol. Chem. **255:** 3278–3284.

21. DE VRIES, S., S.P.J. ALBRACHT, J.A. BERDEN & E.C. SLATER. 1981. A new species of bound ubisemiquinone anion in QH_2: cytochrome c oxidoreductase. J. Biol. Chem. **256:** 11996–11998.

22. NOHL, H. 1990. Is redox-cycling ubiquinone involved in mitochondrial oxygen activation? Free Radical Res. Commun. **8:** 307–315.

23. STOLZE, K. & H. NOHL. 1994. Effect of xenobiotics on the respiratory activity of rat heart mito-chondria and the concomitant formation of superoxide radicals. Environ. Toxicol. Chem. **13:** 499–502.

24. NOHL, H., V. KOLTOVER & K. STOLZE. 1993. Ischemia/reperfusion impairs mitochondrial energy conservation and triggers $O_2^{\bullet-}$ release as a by-product of respiration. Free Radical Res. Commun. **18:** 127–137.

25. LOSCHEN, G., L. FLOHE & B. CHANCE. 1971. Respiratory chain linked H_2O_2 production in pigeon heart mitochondria. FEBS Lett. **18:** 261–264.

Reactive Oxygen Intermediates, Molecular Damage, and Aging

Relation to Melatonin

RUSSEL J. REITER,[a,d] JUAN M. GUERRERO,[b] JOAQUIN J. GARCIA,[a] AND DARIO ACUÑA-CASTROVIEJO[c]

[a]*Department of Cellular and Structural Biology, The University of Texas Health Science Center, 7703 Floyd Curl Drive, San Antonio, Texas 78284-7762, USA*
[b]*Department of Medical Biochemistry and Molecular Biology, The University of Seville School of Medicine, Avda Sanchez Pizjuan 4, 41009-Seville, Spain*
[c]*Catadratico de Universidad, Dpto. Fisiologia, F. Medicina Avd. Madrid 11, E-18012 Granada, Spain*

ABSTRACT: Melatonin, the chief secretory product of the pineal gland, is a direct free radical scavenger and indirect antioxidant. In terms of its scavenging activity, melatonin has been shown to quench the hydroxyl radical, superoxide anion radical, singlet oxygen, peroxyl radical, and the peroxynitrite anion. Additionally, melatonin's antioxidant actions probably derive from its stimulatory effect on superoxide dismutase, glutathione peroxidase, glutathione reductase, and glucose-6-phosphate dehydrogenase and its inhibitory action on nitric oxide synthase. Finally, melatonin acts to stabilize cell membranes, thereby making them more resistant to oxidative attack. Melatonin is devoid of prooxidant actions.

In models of oxidative stress, melatonin has been shown to resist lipid peroxidation induced by paraquat, lipopolysaccharide, ischemia-reperfusion, L-cysteine, potassium cyanide, cadmium chloride, glutathione depletion, alloxan, and alcohol ingestion. Likewise, free radical damage to DNA induced by ionizing radiation, the chemical carcinogen safrole, lipopolysaccharide, and kainic acid are inhibited by melatonin. These findings illustrate that melatonin, due to its high lipid solubility and modest aqueous solubility, is able to protect macromolecules in all parts of the cell from oxidative damage. Melatonin also prevents the inhibitory action of ruthenium red at the level of the mitochondria, thereby promoting ATP production. In humans, the total antioxidative capacity of serum is related to melatonin levels. Thus, the reduction in melatonin with age may be a factor in increased oxidative damage in the elderly.

Although an age-associated loss of melatonin has been known for almost two decades,[1–3] it wasn't until 1993 that it was definitively shown that melatonin is a free radical scavenger[4] and antioxidant.[5,6] Soon thereafter, a connection between oxygen radicals, aging, and a reduction in melatonin levels was proposed.[7,8] The initial discoveries have been followed by numerous reports that have confirmed both the age-related depression in melatonin[9] as well as the antioxidant capabilities of the chief secretory product of the pineal gland.[10,11] Although interest is great in the potential causal association between the drop in melatonin in the elderly and the degenerative signs of aging, the research findings are suggestive but incomplete and therefore claims of a causal relationship are yet premature. The purpose of this report is to summarize what is known about the changes in mela-

[d]Tel: 210/567-3859; fax: 210/567-6948; e-mail: reiter@uthscsa.edu

tonin after middle age, its ability to function as an antioxidant, and the potential role of oxygen-derived radicals in aging and age-related diseases.

MELATONIN PRODUCTION AND LEVELS

Although melatonin is found in a variety of tissues in both the plant[12] and animal kingdom,[13] in vertebrates melatonin is synthesized in a limited number or organs. In mammals, the production of melatonin is known to occur in the pineal gland, retinas, gastrointestinal tract, and possibly in a small number of other tissues (FIG. 1). However, melatonin measured in the blood of mammals is derived almost exclusively from the pineal gland. In this organ, the indole is synthesized primarily at night, and, because of its rapid release, a nighttime increase in circulating concentrations of the indole parallels its production within the pineal gland. Once in the blood, melatonin is rapidly transferred to other bodily fluids, where it manifests a nocturnal rise, like that seen in the circulation, albeit of lower

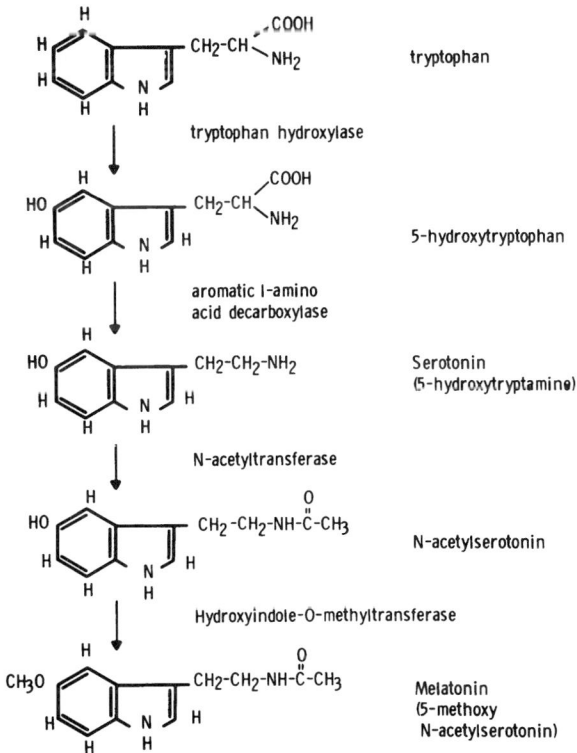

FIGURE 1. Biosynthesis of melatonin from the amino acid tryptophan, with serotonin as an intermediate. In the pineal gland the bulk of the melatonin produced is formed during the dark phase of the light:dark cycle. Once produced, pineal melatonin is quickly released into the blood, so serum levels of melatonin are considered a reliable index of pineal biosynthetic activity.

amplitude. There are no known morphophysiological barriers to melatonin, and it is generally believed that it quickly enters every cell in the organism. In some cases it interacts with membrane receptors situated on the cell surface,[14] whereas in other cases it may bind to sites in the cytoplasm, for example, to calmodulin,[15,16] and in the nucleus possibly to specific binding sites.[17]

Although well known to be highly lipid soluble,[18] melatonin also exhibits considerable solubility in aqueous media.[19] Intracellularly, the highest concentrations of melatonin, estimated by radioimmunoassay, have been detected in the nucleus,[20] but melatonin has also been immunocytochemically identified in the cytosol.[21] Except in those tissues where it is produced, intracellular levels of melatonin are generally considered to be low (no higher than the nanomolar range). After puberty and until animals are middle-aged, the circadian rhythm of melatonin seems to be highly stable, and as a result the quantity of melatonin the pineal synthesizes and releases remains uniform on a night-to-night basis. When human blood melatonin levels are measured over a 24-h period, there are clearly adult individuals who exhibit a robust nocturnal melatonin increase, whereas other subjects may have a nighttime melatonin rise only half as large. As a result, throughout a lifetime some individuals probably produce twice as much melatonin in their pineal gland as do others. Gender seems to be an insignificant factor with reference to the amount of melatonin the pineal synthesizes, with males and females producing about equal quantities.

With progressive age, the function of the pineal gland begins to wane, resulting in a reduction in nighttime melatonin synthesis (FIG. 2).[3,9,22] This leads to similarly reduced

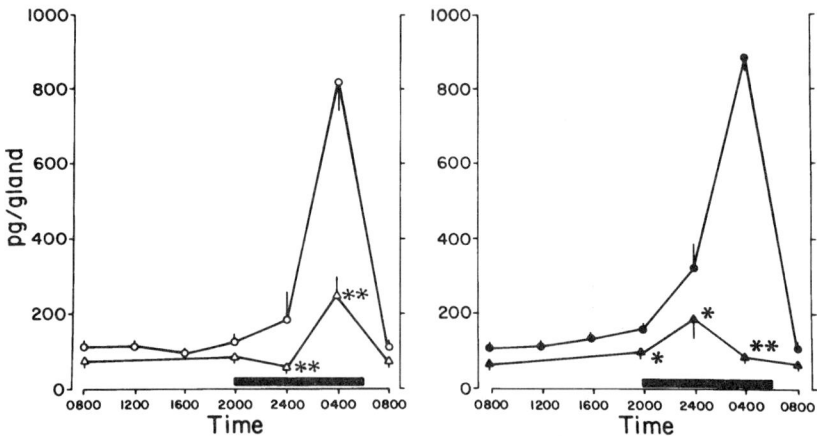

FIGURE 2. Circadian rhythm in pineal melatonin levels in young (2-mo-old) and old (18-mo-old) female and male Syrian hamsters. As in other mammals, melatonin production is low during the day and high at night. Late in life the nighttime production of melatonin becomes greatly attenuated, another feature characteristic of all mammals in which the melatonin rhythm has been measured. Left: ○—○, 2-mo-old, ♀; △—△, 18-mo-old, ♀; ▬, darkness; ** $p < 0.001$. Right: ●—●, 2-mo-old, ♂; ▲—▲, 18-mo-old, ♂; ▬, darkness; * $p < 0.05$; ** $p < 0.001$. (Reiter et al.[1] With permission from Science.)

melatonin levels in the blood and likely in tissues. Although the reduction in pineal melatonin production seems unrelated to any particular life event, for example, menopause in human females, the most marked reduction in its synthesis and secretion seems to occur between 40–60 years of age, although there are probably great individual variations. Thus, many individuals beyond 60 years of age may have no discernible nighttime rise in circulating melatonin levels. There have been no studies performed in an attempt to relate a low-amplitude melatonin rhythm or an early age-related decrease in melatonin with overt signs of aging in humans. There is one procedure, that is, food restriction, that significantly defers the signs of aging and increases life span in rats, and that also substantially delays the age-associated drop in pineal melatonin synthesis.[23] Whether the higher melatonin levels in food-deprived rats relates to the slower aging and longer survival of these animals remains unknown.

MELATONIN AS A FREE RADICAL SCAVENGER

Several criteria must be met to characterize free radical toxicity and the ability of a proposed scavenger to neutralize a radical. One important requirement is the detection of the free radical metabolite either by electron spin resonance (ESR) or by measuring a "unique" reaction product. Tan *et al.*[4] met this criterion for melatonin as a hydroxyl radical (•OH) scavenger by showing that, under *in vitro* conditions, the presence of melatonin in a reaction mixture of •OH (formed by the exposure of H_2O_2 to ultraviolet light at 254 nm) and the spin trap 5,5-dimethyl-1-pyroline *N*-oxide (DMPO) reduced the ESR signal (the 1:2:2:1 quartet) generated by the •OH-DMPO adduct. This reduction in the signal due to the presence of melatonin indicated that melatonin successfully competed for •OH, thereby reducing the number that formed •OH-DMPO adducts.

In a more thorough study of the interactions of melatonin with the •OH, Matuszak and colleagues[24] used the amplitude of the ESR signal for the •OH-DMPO adduct in the presence of increasing concentrations of melatonin to calculate the rate constant (K_r) for the reaction of melatonin with the •OH. The data used to calculate the K_r are shown in FIGURE 3. FIGURE 3A illustrates the amplitude of the ESR signal for the •OH-DMPO adduct versus the concentration of melatonin in the solution, whereas FIGURE 3B shows that the relationship between the reciprocals of the amplitudes and the melatonin concentration is linear; from these data the K_r for the reaction between melatonin and the •OH was calculated to be 2.7×10^{10} M^{-1} s^{-1}.

Although the ability of melatonin to neutralize the •OH is important because of the very high toxicity of this reactant, melatonin's scavenging activity is not confined to the •OH. According to Pieri *et al.*[25] melatonin is a better scavenger of the peroxyl radical (LOO•) than is vitamin E. This determination was based on the ability of each of these molecules to prevent the lysis of human erythrocytes exposed to an azo-initiator [2, 2'-azo-bis (2-amidinopropane) dihydrochloride] of the LOO•. The finding that melatonin is a more powerful LOO• scavenger than vitamin E is remarkable, because this vitamin is considered the premier lipid antioxidant and LOO• scavenger. Follow-up studies on melatonin as a LOO• scavenger have been somewhat less enthusiastic about how efficacious it is in this regard. According to Scaiano,[26] melatonin probably is very effective at neutralizing this radical, but this worker made no claims about its efficacy relative to that of vitamin E. Marshall *et al.*,[27] on the other hand, on the basis of *in vitro* studies, concluded that melato-

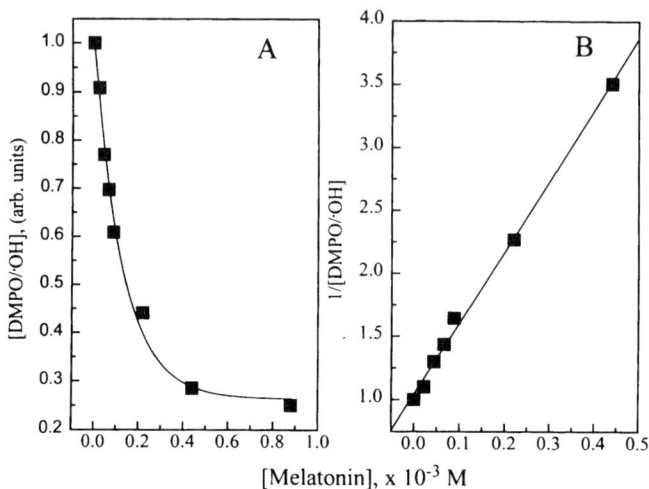

FIGURE 3. Competition between the spin trapping agent, DMPO, and melatonin for ·OH, as measured by electron spin resonance spectroscopy (ESR). A: This illustrates the change in the amplitude (second component) of the ESR signal for DMPO/·OH in a solution to which is added increasing concentrations of melatonin. B: This is a plot of the reciprocal of the amplitudes of the ESR signal versus increasing concentrations of melatonin. These findings show that melatonin successfully competes with DMPO for the ·OH. (Matuszak et al.[24] With permission from Elsevier Science Inc.)

nin does not favorably compare with vitamin E in terms of its ability to neutralize the LOO•. Thus, although melatonin's efficacy as a direct LOO• scavenger may be disputed, there is little doubt that it is highly effective as a lipid antioxidant,[10,28,29] a function that is probably mediated by several actions of melatonin.[10,29–31]

Besides its ability to detoxify the •OH and possibly the LOO• radicals, melatonin also protects against the molecular toxicity initiated by singlet oxygen ($^1O_2$) (FIG. 4). The interaction of light with photosensitive chemicals generates $^1O_2$, a molecule that may be responsible for some of the cellular damage associated with ischemia-reperfusion injury.[32] When cerebellar granule cells were exposed to the photosensitive dye, rose bengal and light, DNA fragmentation, indicative of apoptosis and cell death, were consequences.[33] A significant neuroprotective action of melatonin in this model suggested to the authors that the indole most likely scavenges $^1O_2$, thereby counteracting its potential toxicity. While the evidence clearly only provides indirect support for the idea that melatonin scavenges $^1O_2$, Reszka et al.[34] found that melatonin effectively scavenges photosensitized $^1O_2$, with a rate constant of 7.6×10^9 M^{-1} s^{-1}. Consistent with this action, King and Scaiano[35] surmised that melatonin, which is produced in the retina, reduces cytotoxicity in this organ when the light-exposed retina absorbs protoporphyrin (a photosensitizing agent); melatonin in this case was believed to function as a protective radical trap.

The production of excessive amounts of nitric oxide (NO•), a free radical generated by the inducible isoform of NO synthase (NOS), is known to cause cytotoxic changes, related at least in part to the fact that NO• reacts with $O_2^{\bullet-}$ to produce the reactive oxidant species

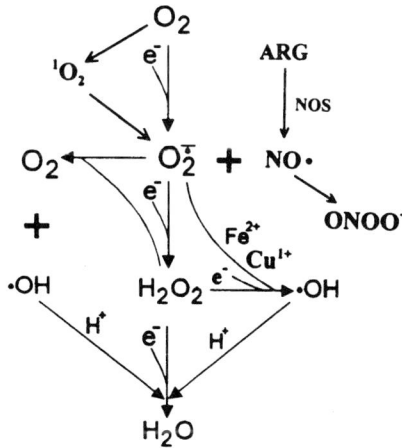

FIGURE 4. This figure illustrates the reduction of oxygen and the formation of reactive oxygen intermediates, some of which are free radicals. A one electron (e^-) reduction of oxygen (O_2) generates the superoxide anion radical ($O_2^{\cdot-}$); in the presence of the enzyme superoxide dismutase (of which there are several isoforms), $O_2^{\cdot-}$ is reduced to hydrogen peroxide (H_2O_2). H_2O_2 is either metabolized to nonharmful products, for example, H_2O, by the enzymes catalase and glutathione peroxidase or, in the presence of such transition metals as Cu^{1+} or Fe^{2+}, H_2O_2 is converted to the highly toxic hydroxyl radical ($\cdot OH$) in what is known as the Fenton reaction. Oxygen can also be converted to a high energy state (singlet oxygen or $^1O_2$) which is also toxic to tissues; $^1O_2$ degrades with the release of energy, which is measured as chemiluminescence. $O_2^{\cdot-}$ reacts with nitric oxide ($NO\cdot$) to form another toxic agent, peroxynitrite anion ($ONOO^-$). $NO\cdot$ is enzymatic generated from arginine (ARG) in the presence of the enzyme nitric oxide synthase (NOS).

peroxynitrite ($ONOO^-$) (FIG. 4).[36] Besides the inherent toxicity of the anion, $ONOO^-$ also eventually degrades into a hydroxyl-like radical species that is likewise toxic to cellular molecules. Because of this, the discovery that melatonin also scavenges $ONOO^-$ is an important aspect of the antioxidant capability of the indole.[37] In the study in question, Gilad *et al.*[37] showed that melatonin dose dependently reduced the peroxynitrite-induced oxidation of dihydrorhodamine, suggesting a possible direct interaction of melatonin with ground state $ONOO^-$, although other explanations for the findings are possible. This group further found that the acute inflammatory response associated with carrageenan administration to animals was markedly reduced when melatonin was also injected.[38] Although $ONOO^-$ is not the only cytotoxic agent induced by carrageenan, it is a major one.

The mechanisms of melatonin's scavenging actions have recently been summarized.[39] Melatonin is a highly electroreactive molecule that acts primarily as a powerful electron donor, thereby detoxifying electron-deficient reactive oxygen species. In the process of donating an election, melatonin itself becomes a radical, the melatonyl cation radical, which has very low reactivity.[40] Once generated, the cation is not recycled back to melatonin, but it reacts with the $O_2^{\cdot-}$ to produce N^1-acetyl-N^2-formyl-5-methoxykynuramine (FIG. 5). The ability of the melatonyl cation radical to scavenge the $O_2^{\cdot-}$ is important, not so much because of the toxicity of the $O_2^{\cdot-}$, but because it reduces the formation of the highly toxic $\cdot OH$ (after its conversion to H_2O_2; FIG. 4) and of $ONOO^-$.

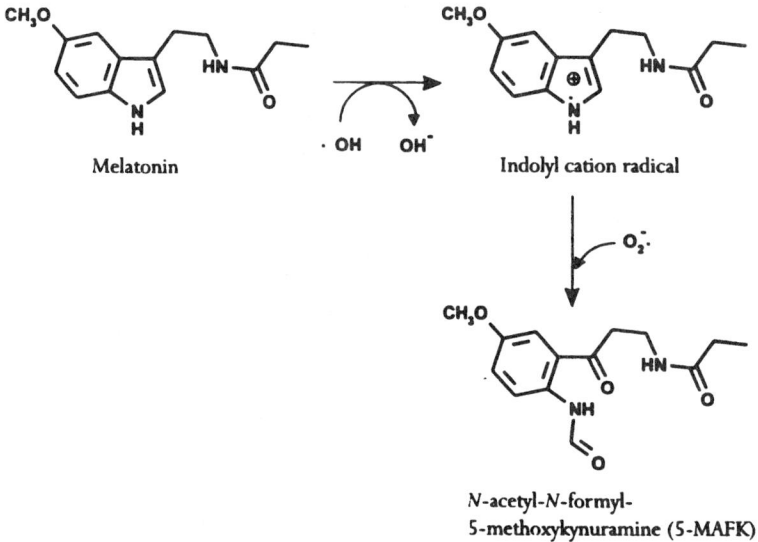

FIGURE 5. Nonenzymatic metabolism of melatonin when it reacts with a free radical such as the OH·, as shown here. Melatonin detoxifies free radicals by electron donation and thereby itself becoming a radical, the indolyl (or melatonyl) cation radical, with low toxicity. The latter molecule is then believed to scavenge the $O_2^{\bullet-}$ radical to generate a kynuramine, which is excreted in the urine.

There are two reports showing that melatonin also scavenges hypochlorous acid (HOCl).[27,41] Although it indirectly detoxifies the $O_2^{\bullet-}$ after it donates an election to more toxic species,[40] melatonin has no known direct interactions with this weakly oxidizing agent. H_2O_2, which is a reactive oxygen intermediate in the reduction of oxygen (FIG. 4), is not directly influenced by melatonin, although melatonin stimulates its metabolism to nontoxic molecular species (see below).

SPECIFICITY OF MELATONIN AS A FREE RADICAL SCAVENGER

5-Methoxytryptamine is a product of indoleamine metabolism in the pineal gland that has been proposed to have endocrine activity;[42] little, however, is known about its secretion from the pineal gland (if this occurs at all) or about blood or tissue levels of this amine. Tan et al.[4] did examine its ability to scavenge the •OH during their in vitro structure-activity studies. Under the conditions used, 5-methoxytryptamine was found to reduce •OH-DMPO adduct formation, with roughly a 50% efficiency compared to that of melatonin. A similar conclusion was reached by Matuszak and co-workers;[24] they too feel 5-methoxytryptamine exhibits •OH scavenging ability but that it is slightly less effective in this regard than is melatonin (TABLE 1).

In the biosynthesis of melatonin, its immediate precursor is N-acetylserotonin (FIG. 1). According to Tan et al.[4] it lacks any ability to detoxify the •OH. Melatonin is primarily metabolized in the liver to 6-hydroxymelatonin; this hydroxylated indole has been found

TABLE 1. Calculated Rate Constants (K_r) for the Scavenging of the •OH by Methoxylated and Hydroxylated Indoles[24]

Molecule	K_r (10^{10} M^{-1} s^{-1})
5-Hydroxytryptamine (serotonin)	1.7 ± 0.3
N-Acetyl-5-methoxytryptamine (melatonin)	2.7 ± 0.3
5-Methoxytryptamine	2.3 ± 0.1
6-Hydroxymelatonin	1.1 ± 0.3
6-Chloromelatonin	1.95 ± 0.1

to function weakly as an •OH scavenger and, additionally, to possess some prooxidative potential.[24] Another hydroxylated indole, serotonin (FIG. 1), was also found by both Tan *et al.*[4] and Matuszak and colleagues[24] to both weakly scavenge and promote •OH generation. A synthetic analogue of melatonin, 6-chloromelatonin, while also scavenging the •OH, does so less efficiently than melatonin (TABLE 1).

The methoxylated β-carboline, pinoline (6-methoxy-1,2,3,4-tetrahydro-β-carboline or 5-methoxytryptoline) has been preliminarily investigated as a free radical scavenger (FIG. 6). In the pineal gland pinoline is reportedly formed during the metabolism of melatonin with 5-methoxytryptamine as an intermediate.[43] It possesses biological activity in mammals in nanomolar concentrations, and recently interest has focused on this molecule as a potential free radical scavenger. According to Pähkla *et al.*[44] pinoline possesses free radical scavenging activity similar to that of melatonin; this conclusion was based on studies

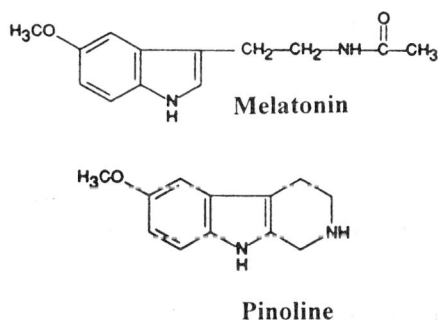

FIGURE 6. Melatonin (*N*-acetyl-5-methoxytryptamine) and its tricyclic metabolite pinoline (6-methoxy-1,2,3,4-tetrahydro-β-carboline or 5-methoxytryptoline). Melatonin has documented free radical scavenging and antioxidant activity, and preliminary investigations with pinoline suggest it may have similar actions.

conducted using three different *in vitro* systems. We also found this β-carboline to be a potential antioxidant *in vitro* (unpublished observations). It is likely that the free radical scavenging and antioxidant activity of pinoline will continue to be an area of active investigation in the foreseeable future, particularly because β-carbolines are potentially important as neuroprotective agents against oxidative damage.[45]

MELATONIN'S ACTIONS ON ENZYMES RELATED TO ANTIOXIDATIVE DEFENSE

Besides its actions in direct free radical scavenging [11,28,39,46,47] and membrane stabilization,[31] melatonin has actions on enzymes that either generate or metabolize reactive oxygen intermediates, thereby further increasing its protective activity toward free radicals. Superoxide dismutase (SOD), of which there are several isoforms, is considered a major antioxidative enzyme, because it dismutates $O_2^{\cdot-}$ to H_2O_2, thereby not only removing the anion but also reducing the formation of $ONOO^-$ (FIG. 4). Studies published by Antolin and co-workers[48] found an increase in tissue mRNA levels for both manganese and copper SOD in animals treated with melatonin, although the activities of these enzymes were not actually measured.

Once generated, H_2O_2 can be easily converted into the highly reactive and destructive •OH via the Fenton reaction (FIG. 4). To reduce this possibility, two enzymes have evolved, catalase and gluthathione peroxidase (GPx), which metabolize H_2O_2 to H_2O (FIG. 7). Pharmacologically,[49] and possibly physiologically as well,[50] melatonin stimulates the activity of GPx to remove H_2O_2 from cells; in doing so glutathione (GSH) is converted to oxidized glutathione (GSSG); GSSG is reduced back to GSH in the presence of the enzyme GSSG reductase (FIG. 7).

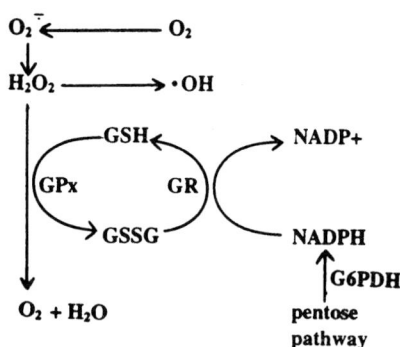

FIGURE 7. H_2O_2 is metabolized to nontoxic products by glutathione peroxidase (GPx); in the process glutathione (GSH) is oxidized to form oxidized glutathione (GSSG). GSSG is converted back to GSH by the enzyme glutathione reductase (GR); this reaction requires the cofactor NADPH, which is generated by the enzyme glucose-6-phosphate dehydrogenase (G6PDH). GPx, GR, and G6PDH are all reportedly stimulated by melatonin, thereby lowing the concentration of H_2O_2 and reducing •OH formation.

The activity of this enzyme is also stimulated by melatonin, thereby replenishing the important antioxidant GSH. GSSG reductase requires the cofactor NADPH, which is generated from NADP in a reaction catalyzed by glucose-6-phosphate dehydrogenase (FIG. 7); the activity of this enzyme is also reportedly increased in the presence of melatonin.[51] Thus, several important antioxidative enzymes seem to be stimulated by melatonin, actions that would generally protect cells from oxidative damage.

The activity of the enzyme nitric oxide synthase (NOS) determines the amount of NO· generated. As noted above, NO· reacts with the $O_2^{\bullet-}$ to form the toxic agent $ONOO^-$. Thus, NOS can be considered a proxidative enzyme, and any factor that reduces its activity would be considered an antioxidant. Recent studies have shown that in both the cerebellum[16] and hypothalamus[52] physiological levels of melatonin reduce NOS activity. Thus, inhibition of NO· production may be another means whereby melatonin reduces oxidative damage under conditions such as neural ischemia/reperfusion where NO· is felt to be important in terms of the resulting damage.[53]

OXIDATIVE DAMAGE, AGING, AND MELATONIN

The concept that accumulated free radical damage may be responsible in part for some of the degenerative signs of aging was initially proposed by Harman in 1956.[54] Since then, this theory has received considerable experimental support, to the point where many biologists with an interest in aging processes now consider free radicals and the related damage as being an integral component of aging itself as well as of a variety of age-related diseases.[55–58]

That the pineal gland, which, as demonstrated above, loses its ability to produce[9] the free radical scavenger[4,12,24] and antioxidant[10,28,39,59] melatonin, may have a role in aging is of increasing interest.[7,46,60] Besides the loss of melatonin as an antioxidant, there are other reasons why diminished pineal function may impact the processes of aging.

The rhythmic production of melatonin is known to be involved in the synchronization of circadian rhythms,[61] and reduced melatonin levels, which occur with aging,[9] may not provide a sufficiently strong signal to function as a chronobiotic and, as a result, biological rhythmicity may deteriorate and age-related changes occur.[62] In a somewhat more exotic theory, Kloeden and colleagues[63] argued for the existence of a centralized clock that coordinates the genetic switching of all cells that age. They localized the clock in the pineal gland and predicted that melatonin is the messenger of the clock. These and other theories that potentially functionally link the pineal gland and melatonin to aging have recently been summarized.[60]

The emphasis of the present report relates to the potential association between the reduction in melatonin and the consequential loss of its antioxidant capabilities with advancing age.[7,9,46,59] In the studies summarized herein, however, there is no way to discriminate between the free radical scavenging activities of melatonin and other actions of the indole that may have been beneficial in ensuring successful aging or prolonging longevity.

Relatively few studies have examined the effects of prolonged melatonin supplementation on mean and maximal life span of animals. In the ciliated, aerobic protozoan, *Paramecium tetraurelia*, the addition of melatonin (0.043 mM) to the nutrient media increased the mean clonal life span of these organisms by percentages ranging from 20.8–24.2%

over that of nonsupplemented controls.[64] Additionally, maximal clonal life span was prolonged from 14.8–24.0% over that of control paramecia. In this report, the addition of much higher doses of melatonin (0.215 and 0.430 mM) caused a decline in the survival parameters monitored. Other well-known antioxidants, that is, vitamins E and C, are known to similarly increase mean clonal life span of *Paramecium tetraurelia*, suggesting that the mechanism by which all three substances, that is, the two vitamins and melatonin, promotes longevity of this one-celled organism is via free radical scavenging.

As with most tests of molecules that may function to prolong life, the other reports in which an alteration in life span was noted after melatonin administration have used rodents as the experimental animal. According to Pierpaoli and Regelson,[65] nightly (but not daytime) melatonin supplementation of the drinking water given to either BALB/c females, New Zealand black females, or C57BL/6 male mice prolonged their life and preserved their youthful state; in these studies melatonin supplementation was begun after the animals were at least middle-aged. The work summarized in this report was accomplished prior to the discovery that melatonin was an antioxidant, so this explanation was not a consideration as a possible mechanism by which melatonin increased the life span of mice. Rather, the authors theorized that a stimulatory effect of melatonin on the immune system and pituitary-thyroid axis may have been the beneficial actions of melatonin that increased the survival of the mice. In a much less cited article, this same group of workers reported that melatonin, rather than prolonging life, reduced longevity in C3H/He female mice.[66]

Besides melatonin supplementation, Pierpaoli and Regelson[65] also reported that the transplantation of pineal glands from young mice to old mice caused the recipients to retain a youthful state and survive longer. These findings have been criticized because the pineal gland of inbred strains of mice reportedly does not produce melatonin,[67] although this is debated,[68] and so melatonin could not have been a factor in the maintenance of life span. More importantly would seem to be that transplanted pineal glands, because their sympathetic innervation is destroyed, lose their synthetic and secretory capabilities.[69]

Also in mice, Lenz and colleagues found that the treatment of NZB/W female mice with melatonin prolonged the life of these animals. In this study, the treated mice received a 100 µg melatonin injection either between 08.00–10.00 h or between 17.00–19.00 daily throughout their life; especially when given at 08.00–10.00 h, melatonin significantly enhanced the survival of the animals. The authors also noted that melatonin protected the kidney from age-related damage and delayed the onset of proteinuria. Inasmuch as it was not the goal of this study to investigate the actions of melatonin, how the indole achieved the protective effects on kidney morphology and physiology was not extensively discussed. The findings of Lenz *et al.*[70] seem to contrast somewhat with those of Pierpaoli and Regelson,[65] because in the latter report daytime administration of melatonin was without effect on the duration of survival, whereas Lenz *et al.* reported otherwise.

In the only study conducted using rats as the experimental animal, Oaknin-Bendahan and co-workers[71] found that giving melatonin in the drinking water for 16 months increased the number of rats that survived to 26–29 months of age. Although roughly 50% of the control rats survived to this age, melatonin supplementation (available throughout each 24 h period) resulted in about a 90% survival rate at the same age. Whereas these findings suggest melatonin may be beneficial in terms of preserving longevity, the impact of the data is lessened by the fact that a putative melatonin antagonist had a similar beneficial effect in terms of the percentage of animals that survived to 26–29 months.

Collectively, the findings linking supplemental melatonin to increased survival are not compelling for several reasons. First, some of the data is contradictory relative to that published in other reports. Second, the number of animals that have been used has been small, and they were usually maintained under less than ideal husbandry conditions (not barrier maintained) for studies of this type. Finally, the total number of reports is limited. Thus, although the data are suggestive of an association between melatonin and longevity, more thorough investigations must be carried out to prove or disprove this relationship.

CONCLUDING REMARKS

An interaction between free radicals and the chief secretory product of the pineal gland, melatonin, seems well established. Additionally, both *in vitro* and *in vivo* melatonin has been shown to have significant antioxidative potential, and in this capacity it greatly limits lipid peroxidation,[10,25,59,72–74] nuclear DNA damage,[75,76] and possibly the degradation of proteins.[77] Because free radical damage has frequently been implicated in aging[7,28,55,56] and age-related diseases,[57,58,59] it is not surprising that melatonin, which is greatly diminished with increasing age,[9] would be surmised to be involved in these processes.[7,11,46,51,59,60] Tests of this possibility remain incomplete, however, and considering the potential importance of these interactions, it is anticipated that reports relating to these issues will appear in the near future. These studies are essential inasmuch as melatonin is readily available as a dietary supplement.

REFERENCES

1. REITER, R.J., B.A. RICHARDSON, L.Y. JOHNSON, B.N. FERGUSON & D.T. DINH. 1980. Pineal melatonin rhythm: Reduction in aging Syrian hamsters. Science **210**: 1372–1373.
2. REITER, R.J., C.M. CRAFT, J.E. JOHNSON, Jr., T.S. KING, B.A. RICHARDSON, G.M. VAUGHAN & M.K. VAUGHAN. 1981. Age-associated reduction in nocturnal melatonin levels in female rats. Endocrinology **109**: 1295–1297.
3. TOUITOU, Y., M. FEVRE, M. LAGUGVEY, A. CARAYON, A. BOYDON & A. REINHART. 1981. Age and mental health related circadian rhythms of plasma levels of melatonin, prolactin, luteinizing hormone and follicle stimulating hormone. J. Endocrinol. **91**: 367–395.
4. TAN, D.X., L.D. CHEN, B. POEGGELER, L.C. MANCHESTER & R.J. REITER. 1993. Melatonin: A potent, endogenous hydroxyl radical scavenger. Endoc. J. **1**: 57–60.
5. TAN, D.X., B. POEGGELER, R.J. REITER, L.D. CHEN, S. CHEN, L.C. MANCHESTER & L.R. BARLOW-WALDEN. 1993. The pineal hormone melatonin inhibits DNA-adduct formation induced by the chemical carcinogen safrole *in vivo*. Cancer Lett. **70**: 65–71.
6. PIERREFICHE, G., G. TOPALL, G. COURBAIN, I. HENRIET & H. LABORIT. 1993. Antioxidant capacity of melatonin in mice. Res. Commun. Chem. Pathol. Pharmacol. **80**: 211–223.
7. POEGGELER, B., R.J. REITER, D.X. TAN, L.D. CHEN & L.C. MANCHESTER. 1993. Melatonin, hydroxyl radical-mediated oxidative damage, and aging: A hypothesis. J. Pineal Res. **14**: 151–168.
8. REITER, R.J., B. POEGGELER, D.X. TAN, L.D. CHEN, L.C. MANCHESTER & J.M. GUERRERO. 1993. Antioxidant capacity of melatonin: A novel action not requiring a receptor. Neuroendocrinol. Lett. **15**: 103–106.
9. REITER, R.J. 1992. The aging pineal gland and its physiological consequences. Bioessays **14**: 169–175.
10. REITER, R.J. 1997. Antioxidant actions of melatonin. Adv. Pharmacol. **38**: 103–117.
11. REITER, R.J., L. TANG, J.J. GARCIA & A. MUÑOZ-HOYOS. 1997. Pharmacological actions of melatonin in oxygen radical pathophysiology. Life Sci. **60**: 255–2271.

12. HARDELAND, R. & C. RODRIQUEZ. 1995. Versatile melatonin: A pervasive molecule serves various functions in signaling and protection. Chronobiol. Int. **12:** 157–165.

13. REITER, R.J. 1991. Pineal melatonin: Cell biology of its synthesis and of its physiological interactions. Endocr. Rev. **12:** 151–180.

14. GODSON, C. & S.M. REPPERT. 1997. The Mel la melatonin receptor is coupled in parallel signal transduction pathways. Endocrinology **138:** 394–404.

15. HUERTO-DELGADILLO, L., F. ANTON-TAY & G. BENITZ-KING. 1994. Effects of melatonin on microtubule assembly depend on hormone concentration: Role of melatonin as a calmodulin antagonist. J. Pineal Res. **17:** 55–62.

16. POZO, D., R.J. REITER, J.R. CALVO & J.M. GUERRERO. 1997. Inhibition of cerebellar nitric oxide synthase and cyclic AMP production by melatonin via complex formation with calmodulin. J. Cell. Biochem. **65:** 430–442.

17. CARLBERG, C. & I. WEISENBERG. 1995. The orphan receptor family RZR/ROR, melatonin and 5-lipoxygenase: An unexpected relationship. J. Pineal Res. **18:**171–178.

18. COSTA, E.J.X., R. HARZER LOPES & M.T. LAMY-FREUND. 1995. Solubility of pure bilayers to melatonin. J. Pineal Res. **19:**123–126.

19. SHIDA, C.S., A.M.L. CASTRUCCI & M.T. LAMY-FREUND. 1994. High melatonin solubility in aqueous medium. J. Pineal Res. **16:** 198–201.

20. MENENDEZ-PELAEZ, A., B. POEGGELER, R.J. REITER, L.R. BARLOW-WALDEN, M.I. PABLOS & D.X. TAN. 1993. Nuclear localization of melatonin in different mammalian tissues: Immunocytochemical and radioimmunoassay evidence. J. Cell. Biochem. **53:** 373–382.

21. MENENDEZ-PELAEZ, A. & R.J. REITER. 1993. Distribution of melatonin in mammalian tissues: The relative importance of cytosolic versus nuclear localization. J. Pineal Res. **15:** 59–69.

22. IGUCHI, H., K. KATO & H. IBAYASHI. 1982. Age-dependent reduction in serum melatonin concentration in healthy human subjects. J. Clin. Endocrinol. Metab. **55:** 27–29.

23. STOKKAN, K.A., R.J. REITER, K.O. NONAKA, A. LERCHL & D.J. JONES. 1991. Food restriction retards aging of the pineal gland. Brain Res. **545:** 66–72.

24. MATUSZAK, Z., K.J. RESZKA & C. CHIGNELL. 1997. Reaction of melatonin and related indoles with hydroxyl radicals: EPR and spin trapping investigations. Free Radical Biol. Med. **23:** 367–372.

25. PIERI, C., F. MORONI, M. MARRA, F. MARCHESELLI, & R. RECCHIONI. 1994. Melatonin is an efficient antioxidant. Arch. Gerontol. Geriatr. **20:** 159–165.

26. SCIANO, J.C. 1995. Exploratory laser flash photolysis study of free radical reactions and magnetic field effects in melatonin chemistry. J. Pineal Res. **19:** 189–195.

27. MARSHALL, K.A., R.J. REITER, B. POEGGELER, O.I. ARUOMA & B. HALLIWELL. 1996. Evaluation of the antioxidant activity of melatonin *in vitro*. Free Radical Biol. Med. **21:** 307–315.

28. REITER, R.J., D. MELCHIORRI, E. SEWERYNEK, B. POEGGELER, L.R. WALDEN-BARLOW, L. CHUANG, G.G. ORTIZ & D. ACUÑA-CASTROVIEJO. 1995. A review of the evidence supporting melatonin's role as an antioxidant. J. Pineal Res. **18:**1–11.

29. MELCHIORRI, D., R.J. REITER, E. SEWERYNEK, M. HARA, L.D. CHEN & G. NISTICO. 1996. Paraquat toxicity and oxidative damage: Reduction by melatonin. Biochem. Pharmacol. **51:** 1095–1099.

30. ACUÑA-CASTROVIEJO, D., G. ESCAMES, M. MACIAS, A. MUÑOZ-HOYOS, A. MOLINA CARBALLO, M. AROUZO, R. MONTES & F. VIVES. 1995. Cell protective role of melatonin in the brain. J. Pineal Res. **19:** 57–63.

31. GARCIA, J.J., R.J. REITER, J.M. GUERRERO, G. ESCAMES, B.P. YU, C.S. OH & A. MUÑOZ-HOYOS. 1997. Melatonin prevents changes in microsomal membrane fluidity during induced lipid peroxidation. FEBS Lett. **408:** 297–300.

32. KUKREJA, R.C., K.E. LOESSER, A.A. KEARNS, S.A. NASEEM & M.L. HESS. 1993. Protective effect of histidine during ischemia-reperfusion in isolated rat hearts. Am. J. Physiol. **264:** H1370–H1381.

33. CAGNOLI, C.M., C. ATABAY, E. KHARLAMOVA & H. MANEV. 1995. Melatonin protects neurons from singlet oxygen-induced apoptosis. J. Pineal Res. **18:** 222–228.

34. RESZKA, K.J., Z. MATUSZAK, P. BILSKI, L.J. MARTINEZ & C.F. CHIGNELL. 1996. Antioxidant properties of melatonin. Abstracts of the 18th annual Meeting of the Bioelectromagnetics Society. p. 176.

35. KING, M. & J.C. SCAIANO. 1997. The excited states of melatonin. Photochem. Photobiol. **65:** 538–542.

36. PRYOR, W.A. & G.L. SQUADRITO. 1995. The chemistry of peroxynitrite: A product from the reaction of nitric oxide with superoxide. Am. J. Physiol. **268:** L699–L722.

37. GILAD, E., S. CUZZOCREA, B. ZINGARELLI, A.L. SALZMAN & C. SZABO. 1997. Melatonin is a scavenger of peroxynitrite. Life Sci. **60:** PL169–PL174.

38. CUZZOCREA, S., B. ZINGARELLI, E. GILAD, P. HAKE, A.L. SALZMAN & C. SZABO. 1997. Protective effect of melatonin in carrageenan-induced models of local inflammation. J. Pineal Res. **23:** 231–236.

39. POEGGELER, B.H. 1997. Melatonin: Antioxidative protection by electron donation. *In* Reactive Oxygen Species in Biological Systems: An Interdisciplinary Approach. D.L. Gilbert & C.A. Colton, Eds. Plenum. New York. In press.

40. HARDELAND, R., R.J. REITER, B. POEGGELER & D.X. TAN. 1993. The significance of the metabolism of the neurohormone melatonin: Antioxidative protection and formation of bioactive substances. Neurosci. Biobehav. Rev. **17:** 347–357.

41. CHAN, T.Y. & P.L. TANG. 1996. Characterization of the antioxidant effects of melatonin and related indoleamines *in vitro*. J. Pineal Res. **20:** 187–191.

42. PEVET, P. & F. RAYNAUD. 1990. 5-Methoxytryptamine: Physiological effects and possible mechanisms of action. Adv. Pineal Res. **4:** 209–216.

43. AIRAKSINEN, M.M., J.C. CALLAWAY, P. NYKVIST, L. RÄGO, E. KARI & J. GYNTHER. 1993. Binding sites for [³H]pinoline. *In* Melatonin and the Pineal Gland. Y. Touitou, J. Arendt & P. Pevet, Eds.: 83–86. Excerpta Medica. Amsterdam

44. PÄHKLA, R., M. ZILMER, T. KULLISAAR & L. RÄGO. 1997. Antioxidative capacity of pinoline in three *in vitro* assay systems. Pharmacol. & Toxicol. **80** (Suppl. 1): abstr. 41.

45. KAWASHIMA, Y., A. HORIGUCHI, M. TAGUCHI & K. HATAYAMA. 1995. Synthesis and pharmacological evaluation of 1,2,3, 4-tetrahydro-β-carboline derivatives. Chem. Pharm. Bull. **43:** 783–787.

46. REITER, R.J., M.I. PABLOS, T.T. AGAPITO & J.M. GUERRERO. 1996. Melatonin in the context of the free radical theory of aging. Ann. N.Y. Acad. Sci. **786:** 362–378.

47. HARDELAND, R., I. BALZER, B. POEGGELER, B. FUHRBERG, H. URIA, G. BEHRMANN, R. WOLF, T.J. MEYER & R.J. REITER. 1995. On the primary functions of melatonin in evolution: Mediation of photoperiodic signals in a unicell, photoxidation, and scavenging of free radicals. J. Pineal Res. **18:** 104–111.

48. ANTOLIN, I., C. RODRIQUEZ, R.M. SAINZ, J.C. MAYO, H. URIA, M.L. KOTLER, M.J. RODRIQUEZ-COLUNGA, D. TOLIVA & A. MENENDEZ-PELAEZ. 1996. Neurohormone melatonin prevents cell damage: Effect on gene expression for antioxidative enzymes. FASEB J. **10:** 882–890.

49. BARLOW-WALDEN, L.R., R.J. REITER, M. ABE, M. PABLOS, A. MENENDEZ-PELAEZ, L.D. CHEN & B. POEGGELER. 1995. Melatonin stimulates brain glutathione peroxidase activity. Neurochem. Int. **26:** 497–502.

50. PABLOS, M.I., R.J. REITER, G.G. ORTIZ, J.M. GUERRERO, M.T. AGAPITO, J.I. CHUANG, & E. SEWERYNEK. 1998. Rhythms of glutathione peroxidase and glutathione reductase in the brain of chicks and their inhibition by light. Neurochem. Int. **32:** 69–75.

51. PIERREFICHE, G. & H. LABORIT. 1995. Oxygen radicals, melatonin and aging. Exp. Gerontol. **30:** 213–227.

52. BETTAHI, I., D. POZO, C. OSUNA, R.J. REITER, D. ACUÑA-CASTROVIEJO & J.M. GUERRERO. 1996. Physiological concentrations of melatonin inhibit nitric oxide synthase activity in the rat hypothalamus. J. Pineal Res. **20:** 205–210.

53. GUERRERO, J.M., R.J. REITER, G.G. ORTIZ, M.I. PABLOS, E. SEWERYNEK & J.I. CHUANG. 1997. Melatonin prevents increases in neural nitric oxide and cyclic GMP production after transient brain ischemia and reperfusion in the Mongolian gerbil (*Meriones unguiculatus*). J. Pineal Res. **23:** 24–31.

54. HARMAN, D. 1956. Aging: A theory based on free radical and radiation chemistry. J. Gerontol. **11:** 298–300.

55. STARKE-REED P.S. & C.N. OLIVER. 1989. Protein oxidation and proteolysis during aging and oxidative stress. Arch. Biochem. Biophys. **275:** 559–567.

56. HARMAN, D. 1992. Free radical theory of aging. Mutat. Res. **275:** 257–266.

57. HARMAN, D. 1995. Free radical theory of aging: Alzheimer's disease pathogenesis. Age **18:** 97–119.
58. WARNER, H.R. & P. STARKE-REED. 1997. Oxidative stress and aging. *In* Oxygen, Gene Expression and Cellular Function. L.B. Clerch & D.J. Massaro, Eds.: 139–167. Marcel Dekker. New York.
59. REITER, R.J. 1995. Oxidative processes and antioxidative defense mechanisms in the aging brain. FASEB J. **9:** 526–533.
60. REITER, R.J. 1995. The pineal gland and melatonin in relation to aging: A summary of the theories and of the data. Exp. Gerontol. **30:** 199–212.
61. ARMSTRONG, S.M. & J.R. REDMAN. 1993. Melatonin and circadian rhythimicity. *In* Melatonin. H.S. Yu & R.J. Reiter, Eds.: 187–224. CRC. Boca Raton.
62. ARMSTRONG, S.M., & J.R. REDMAN. 1991. Melatonin: A chronobiotic with anti-aging properties. Med. Hypotheses **34:** 300–309.
63. KLOEDEN, P.E., R. RÖSSLER & O.E. RÖSSLER. 1991. Does a centralized clock of aging exist? Gerontology **36:** 314–322.
64. THOMAS, J.N. & J. SMITH-SONNEBORN. 1997. Supplemental melatonin increases clonal life span in the protozoan, *Paramecuim tetraurelia*. J. Pineal Res. **23:** 123–130.
65. PIERPAOLI, W. & W. REGELSON. Pineal control of aging: Effect of melatonin and pineal grafting on aging mice. Proc. Natl. Acad. Sci. USA **91:** 787–791.
66. PIERPAOLI, W., A. DALL'ARA, E. PEDRINIS & W. REGELSON. 1991. The pineal control of aging. The effects of melatonin and pineal grafting on the survival of older mice. Ann. N.Y. Acad. Sci. **621:** 291–313.
67. GOTO, M., I. OSHIMA, T. TOMITA & S. EBIHARA. 1989. Melatonin content of the pineal in different mouse strains. J. Pineal Res. **7:** 195–204.
68. CONTI, A. & G.J.M. MAESTRONI. 1996. HPLC validation of a circadian melatonin rhythm in the pineal gland of inbred mice. J. Pineal Res. **20:** 138–144.
69. REITER, R.J. 1967. The effect of pinealectomy, pineal grafts, and denervation of the pineal gland on the reproductive organs of male hamsters. Neuroendocrinology **2:** 138–146.
70. LENZ, S.P., S. IZUI, H. BENEDIKTSSON & D.A. HART. 1995. Lithium chloride enhances survival of NZB/W lupus mice: Influence of melatonin and timing of treatment. Int. J. Immunopharmacol. **17:** 581–592.
71. OAKNIN-BENDAHAN, S., Y. ANIS, I. NIR & N. ZISAPEL. 1995. Effects of long-term administration of melatonin and a putative antagonist on the aging rat. Neuroreport **6:** 785–788.
72. YAMAMATO, H.A. & H.W. TANG. 1996. Melatonin attenuates L-cysteine-induced seizures and lipid peroxidation in the brain of mice. J. Pineal Res. **21:** 108–113.
73. MILLER, J.W., J. SELHUB & J.A. JOSEPH. 1996. Oxidative damage caused by free radicals produced during catecholamine autoxidation: Protective effects of *0*-methylation and melatonin. Free Radical Biol. Med. **21:** 241–249.
74. ACUÑA-CASTROVIEJO, D., A. COTO-MONTES, M.G. MONTI, G.G. ORTIZ & R.J. REITER. 1997. Melatonin is protective against MPTP-induced striatal and hippocampal lesions. Life Sci. **60:** PL23–29.
75. UZ, T., P. GIUSTI, D. FRANCESCHINI, A. KHARLAMOV & H. MANEV. 1996. Protective effect of melatonin against hippocampal DNA damage induced by intraperitoneal administration of kainate to rats. Neuroscience **73:** 631–636.
76. TANG, L., R.J. REITER, Z.R. LI, G.G. ORTIZ, B.P. YU & J.J. GARCIA. 1998. Melatonin reduces the increase in 8-hydroxy-deoxyguanosine levels in the brain and liver of kainic acid-treated rats. Mol. Cell. Biochem. **178:** 299–303.
77. LI, Z.R., R.J. REITER, O. FUJIMORI, C.S. OH & Y.P. DUAN. 1997. Cataractogenesis and lipid peroxidation in newborn rats treated with buthionine sulfoximine: Protective actions of melatonin. J. Pineal Res. **22:** 117–123.

Recycling and Redox Cycling of Phenolic Antioxidants

V.E. KAGAN[a] AND Y.Y. TYURINA

Departments of Environmental and Occupational Health and Pharmacology, University of Pittsburgh, Pittsburgh, Pennsylvania 15238

ABSTRACT: Effectiveness of phenolic antioxidants in protecting against oxidative stress depends on their reactivity towards reactive oxygen species and the reactivity of the antioxidant phenoxyl radicals towards critical biomolecules. Reduction of phenoxyl radicals by intracellular reductant (ascorbate, thiols) as well as by enzymes or intermediates of electron transport (*e.g.*, in mitochondria and the endoplasmic reticulum) recycles phenolic antioxidants, thus enhancing antioxidant protection. Several cascades may be involved in physiologically relevant recycling of vitamin E from its phenoxyl radicals. The two major ones are dihydrolipoic acid → (GSH) → ascorbate, and enzymes of electron transport → coenzyme Q. Importantly, phenoxyl radicals of vitamin E are not directly reduced by intracellular thiols. By contrast, a number of natural phenolic compounds that act as very effective scavengers of reactive oxygen species and organic radicals, may generate reactive secondary radicals of antioxidants. These secondary radicals react and modify critical intracellular targets (lipids, proteins, and DNA). As a result, the role of these phenolic compounds as biological antioxidants may be limited because of their ability to cause cyto- and genotoxic effects. Typical examples are some estrogens and phenolic drugs (*e.g.*, the antitumor drug, etoposide) that can protect lipids but oxidize GSH and protein sulfhydryls. Moreover, phenoxyl radicals produced in the course of radical scavenging by some phenolic compounds (*e.g.*, phenol) are capable of oxidizing both proteins and lipids. Hence, reactivity of phenoxyl radicals should be considered as a critical factor in the development of new antioxidant protectants.

Antioxidant enzymes (SOD, catalase, peroxidases), which are critical to protection of cells against radicals of relatively low reactivity or molecular oxygen reduction intermediates (superoxide, hydrogen peroxide), do not regulate highly reactive radicals and oxygen species, such as hydroxyl, alkoxyl, or peroxyl radicals. Protective function against the latter is accomplished in cells by antioxidants, low molecular weight molecules capable of donating an electron (hydrogen) to oxidizing radicals:

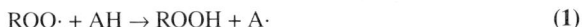

$$ROO\cdot + AH \rightarrow ROOH + A\cdot \tag{1}$$

where ROO· = peroxyl radical and AH = antioxidant in its reduced form.

As a result, reactive radicals are converted into stable molecular products; the antioxidant molecule, however, is transformed into a radical, usually one of relatively low reactivity. Given that highly reactive radicals can directly damage critical macromolecular targets in cells, their reaction with antioxidants can be viewed as a protective mechanism that slows down free radical oxidation in a radical/antioxidant exchange reaction. The fate of an antioxidant radical formed in the reaction (1) is extremely important as well.

There are three major pathways for an antioxidant radical. (1) It may recombine with another radical—antioxidant or reactive—and thus enhance or weaken the antioxidant

[a]To whom correspondence should be addressed: Dr. Valerian Kagan, Department of Environmental and Occupational Health, University of Pittsburgh, 260 Kappa Drive, Pittsburgh, PA. 15238, USA. Tel: 412/967-6516; fax: 412/624-1020; e-mail: kagan@vms.cis.pitt.edu

potency, respectively. (2) It may interact with physiologically relevant reductant and undergo "antioxidant recycling" (*i.e.*, enhancement of antioxidant protection). (3) Finally, it may directly interact with vitally important molecules and cause cytotoxic effects. These reactions represent three different metabolically important pathways that determine whether a radical scavenger will function as an efficient protective antioxidant or as a cytotoxic compound (and hence should not be used as an antioxidant). Below, these redox reactions of antioxidant radicals are considered with respect to phenolic compounds (vitamin E, its isomers and homologues) as well as some phenolic drugs and environmental agents (*e.g.*, etoposide and phenol).

RECYCLING OF VITAMIN E

In 1979, Packer *et al.* reported a very important direct observation of vitamin E phenoxyl radical reduction by ascorbate in a simple chemical system.[1] A number of subsequent studies in which more complex systems were used—liposomes and suspensions of subcellular organelles—confirmed that ascorbate (vitamin C) can, indeed, reduce the phenoxyl radical of α-tocopherol, that is, regenerate vitamin E.[2–4] Importantly, similar interactions between vitamin E and vitamin C were found *in vivo*: antioxidant function of vitamin E was shown to be enhanced by vitamin C supplementation.[5] This interaction, however, involves two vitamins, E and C, whose levels are not regulated by metabolism but depend on dietary supplementation. Is vitamin C unique in its ability to reduce vitamin E phenoxyl radicals?

Studies conducted in the late 1980s and early 1990s discovered additional pathways. Enzymes of membrane electron transport are known to act as one electron donor. Hence, it is likely that components of an electron transport chain can act as "free radical reductase," that is, catalyze one electron reduction of vitamin E phenoxyl radicals. Indeed, it has been demonstrated that mitochondrial and microsomal electron transport can reduce phenoxyl radicals of vitamin E[6,7] and other phenolic antioxidants (*e.g.*, BHT, BHA, and their homologues).[8] Liver and heart mitochondrial and microsomal enzymes recycle vitamin E by NADH-, succinate- or NADPH-dependent electron transport.

NADPH-cytochrome P450 reductase is the major contributor to vitamin E recycling activity of microsomes.[9] The role of vitamin E recycling in membrane antioxidant protection was further supported by the findings on NADPH-dependent inhibition of AMVN-induced lipid peroxidation in liver microsomes under conditions where NADPH-dependent vitamin E recycling occurred but NADPH-supported lipid peroxidation was suppressed by an iron chelator, deferoxamine. It has been demonstrated that in the presence of deferoxamine NADPH-driven electron transport in liver microsomes recycled vitamin E as evidenced by (1) sparing endogenous vitamin E and (2) quenching chromanoxyl radicals generated by the azo-initiator, AMVN, from exogenously added chromanol-α-C6.[10] Vitamin E recycling and its protective effects against microsomal lipid peroxidation are enhanced in the presence of ubiquinone Q_{10}.

Ubiquinone Q_{10} and cytochrome c were identified as possible reduction sites for vitamin E radicals in mitochondria and submitochondrial particles.[11] In proteoliposomes with reconstituted liver mitochondrial complex II, ubiquinones were found to be obligatory components of the vitamin E radical reduction by succinate-ubiquinone oxidoreductase.

Thus, electron carriers in microsomes, mitochondria, and submitochondrial particles, as well as in mitochondrial complexes integrated into liposomes, regenerate α-tocopherol from its phenoxyl radical by ubiquinone-dependent reduction. This suggests that redox interactions of coenzyme Q with α-tocopherol are important in the antioxidant protection of electron transport membranes. What are the mechanisms of this interaction between the two antioxidants? Mellors and Tappel[12,13] suggested that coenzyme Q could recycle vitamin E. The one-electron redox potential for ubiquinol [$E_{7.0}$ ($QH_2/Q\cdot$) = -0.24 V] is more negative than that for tocopherol [$E_{7.0}$ (T-OH/T-O·) = 0.48 V],[14] suggesting that ubiquinol may reduce the vitamin E phenoxyl radical (formed when vitamin E quenches a peroxyl radical), thus regenerating vitamin E:

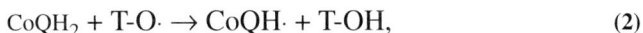

$$CoQH_2 + T\text{-}O\cdot \rightarrow CoQH\cdot + T\text{-}OH, \tag{2}$$

where $CoQH_2$ and $CoQH\cdot$ equal reduced and semireduced forms of ubiquinone (coenzyme Q), respectively; T-O· equals the vitamin E phenoxyl radical; and T-OH equals vitamin E.

In organic solvent, the interaction of ubiquinols with tocopheroxyl radicals is very efficient: the rate constant for this reaction is about 10^6 $M^{-1}s^{-1}$,[15] that is, higher than the rate constant for the reaction of ubiquinol with peroxyl radicals. Thus ubiquinols should preferentially reduce tocopheroxyl radicals rather than peroxyl radicals.

The antioxidant interaction between vitamin E and coenzyme Q is very important because it couples antioxidant function of vitamin E with the major metabolic pathways through electron transport–dependent reduction of coenzyme Q. Because electron transport is often associated with one-electron reduction of oxygen to produce superoxide, one would wonder if there is any role that superoxide may play in the reduction of coenzyme Q and recycling of vitamin E. Experiments in simple aprotic chemical systems (DMSO) demonstrated that one electron reduction of ubiquinone-10 by superoxide ion resulted in the formation of ubisemiquinone-10 radicals, which donated an electron to the α-tocopherol phenoxyl radical formed from interaction with superoxide.[16] Thus ubiquinone-dependent redox cycling of α-tocopherol from its radical may be superoxide driven. This suggests that vitamin E/coenzyme Q interactions may have a special physiological function, that is, protection against superoxide-driven oxidations.

Importantly, the antioxidant interaction of coenzyme Q with vitamin E happens not only in artificial *in vitro* systems but also protects cells against oxidative stress. Evidence for this comes from our experiments in which we used a newly developed technique to quantitate oxidative stress in different classes of phospholipids using metabolic labeling with *cis*-parinaric acid (PnA), a fluorescent fatty acid containing four conjugated double bonds.[17] This procedure creates molecular species of phospholipids in cells in which one of the fatty acid residues is substituted for an oxidation-sensitive PnA. Subsequent HPLC with fluorescence detection permits us to quantitate the loss of PnA in different phospholipids of cells exposed to low doses of oxidative stress. As shown in FIGURE 1, incubation of vitamin E-loaded HL-60 cells with a lipid-soluble azo-initiator of peroxyl radicals, 2,2′-azobis (2,4-dimethylvaleronitrile) (AMVN), caused a pronounced loss of PnA in all major classes of phospholipids—phosphatidylcholine, phosphatidylethanolamine, phosphatidylinositol, and phosphatidylserine. The oxidation of phospholipids was enhanced when both AMVN and xanthine oxidase/xanthine (a source of superoxide) were added to the incubation medium. Ubiquinone (oxidized form) exerted significant protection of phospholipids against oxidation induced by AMVN plus xanthine oxidase/xanthine. No protec-

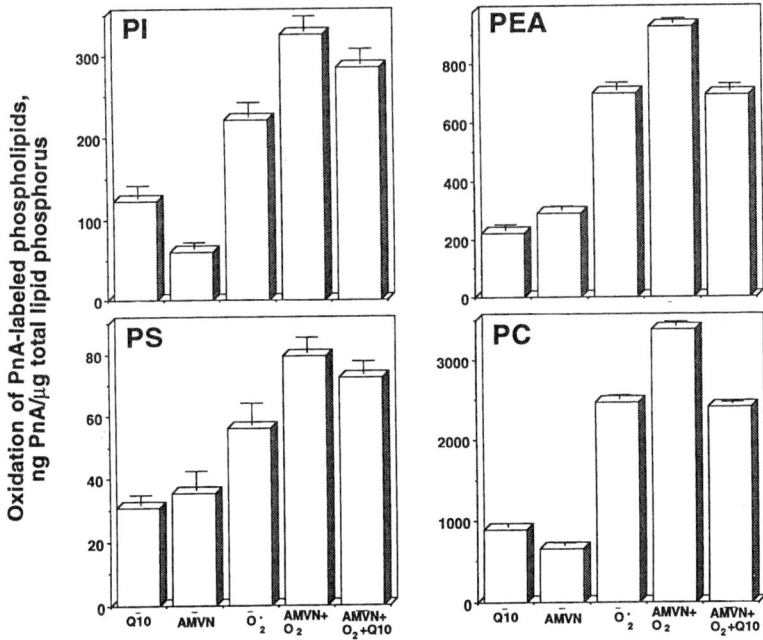

FIGURE 1. Effect of ubiquinone Q_{10} on the oxidation of *cis*-parinaric acid–labeled phospholipids induced by AMVN and xanthine oxidase/xanthine in vitamin E-loaded HL-60 cells. Cells (10^6 cells/mL) loaded with α-tocopherol (20 nmol/10^6 cells, 24 h at 37 °C) and Q_{10} (20 nmol/10^6 cells, 24 h at 37 °C) were prelabeled with PnA (2 μg/10^6 cells, 2 h at 37 °C) and incubated under aerobic conditions in the dark for 2 h at 37 °C in the presence of AMVN (500 μM) and xanthine (300 μM)/xanthine oxidase (5 mU/mL). Lipids were extracted and resolved by HPLC.[17] PI, phosphatidylinositol; PEA, phosphatidylethanolamine; PS, phosphatidylserine; PC, phosphatidylcholine. All values are means ± SEM (n = 3).

tion was afforded by ubiquinone against AMVN-induced phospholipid peroxidation (in line with numerous results on the lack of a direct antioxidant effect by oxidized ubiquinone). We conclude that electron transport/superoxide-driven reduction of ubiquinone in vitamin E–loaded cells significantly enhances their antioxidant capacity.

In addition to electron transport–supported recycling of vitamin E, there are other reducing cascades that may be involved in enhancement of antioxidant protection in cells or extracellular environments. In particular, thiols may play a prominent role in vitamin E recycling. Although direct reaction of vitamin E phenoxyl radicals with thiols is very slow,[18] thiols may participate in vitamin E recycling through their recycling of vitamin C. Dihydrolipoic acid is particularly effective in reducing dehydroascorbate[19] and maintaining high enough levels of ascorbate to provide for effective recycling of vitamin E.[20] This reductive cascade operates to recycle vitamin E by dihydrolipoate: vitamin E radicals are reduced by ascorbate, and dehydroascorbate thus formed may be reduced by dihydrolipoate. The importance of the cascade for antioxidant protection of cells is again emphasized by its coupling to metabolism through reduction of lipoate by universal intracellular sources of reducing equivalents, NADH and NADPH.[21,22]

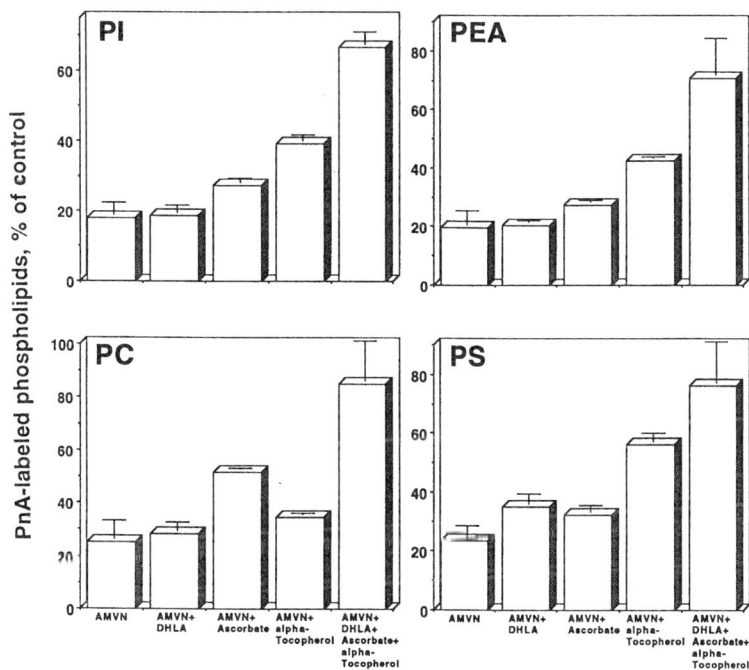

FIGURE 2. Effect of dihydrolipoic acid (DHLA), ascorbate, and α-tocopherol on the oxidation of *cis*-parinaric acid–labeled phospholipids induced by AMVN in HL-60 cells. Cells (10^6 cells/mL) loaded with α-tocopherol (30 nmol/10^6 cells, 2 h at 37 °C), ascorbate (200 μM, 2 h at 37 °C), DHLA (400 μM, 3 h at 37 °C), and prelabeled with PnA (2 μg/10^6 cells, 2 h at 37 °C) were incubated under aerobic conditions in the dark for 2 h at 37 °C in the presence of AMVN (500 μM). Lipids were extracted and resolved by HPLC.[17] PI, phosphatidylinositol; PEA, phosphatidylethanolamine; PS, phosphatidylserine; PC, phosphatidylcholine. All values are means ± SEM (n = 3).

Using our cell culture model system we were able to demonstrate directly the effectiveness of dihydrolipoate in antioxidant protection of different phospholipids against AMVN-induced oxidation (FIG. 2). Under conditions where α-tocopherol alone, ascorbate alone, and dihydrolipoic acid alone produced only slight (if any) inhibition of phospholipid peroxidation, the combination of the antioxidants, α-tocopherol plus ascorbate plus dihydrolipoic acid, caused almost complete protection against phospholipid peroxidation.

REDOX-CYCLING OF ETOPOSIDE AND PHENOL

While oxidation of ascorbate and reduced coenzyme Q by phenoxyl radicals is essential for effective recycling of antioxidants, their low reactivity towards critical intracellular targets (*e.g.*, GSH and protein thiols) is a prerequisite for noncytotoxic antioxidant potency of phenolic compounds. Conversely, oxidation of thiols by phenoxyl radicals may cause cytotoxic effects even though a given phenolic compound may protect lipids or nucleic acids against oxidative damage. We will use two examples to illustrate this statement: a phenolic antitumor drug, etoposide (VP-16), and phenol, the major metabolite of benzene by phase I biotransformation reactions.

The phenoxyl radical is an inevitable intermediate in the reaction of VP-16 with peroxyl radicals, as well as in the peroxidative activation of VP-16 by different oxidizing enzymes. Peroxyl radicals generated by endogenous or exogenous sources may be used for VP-16 activation in a form that is potentially more therapeutically effectual. AMVN-induced formation of the VP-16 phenoxyl radical can be directly observed by ESR (FIG. 3). Similar to vitamin E, ascorbate regenerates VP-16 from its phenoxyl radicals. Moreover, VP-16 acts as a potent antioxidant that protects intracellular phospholipids against AMVN-induced oxidation. In contrast to vitamin E, however, not only ascorbate but also thiols (GSH, dihydrolipoic acid, and protein sulfhydryls) are directly oxidized by the VP-16 phenoxyl radical.[23] In combinations, ascorbate plus GSH and ascorbate plus metallothionein act independently and additively in reducing the VP-16 phenoxyl radical. Ascorbate is more reactive: the VP-16–dependent oxidation of GSH or metallothionein commenced only after complete oxidation of ascorbate. In the combination of ascorbate

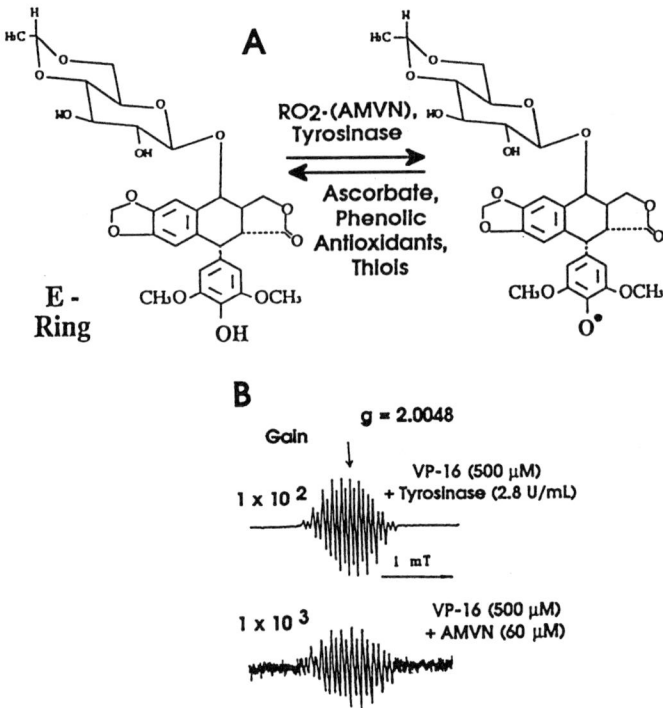

FIGURE 3. Generation and reduction of the VP-16 phenoxyl radical and ESR spectra of the VP-16 phenoxyl radical generated by tyrosinase or AMVN. A: a scheme of VP-16 redox cycling; B: ESR spectra of VP-16 phenoxyl radical.

plus dihydrolipoic acid, ascorbate is also more reactive towards the VP-16 phenoxyl radical than dihydrolipoic acid, but ascorbate concentration is maintained at the expense of its regeneration from dehydroascorbate by dihydrolipoic acid.

To elucidate the role of endogenous thiols in the reduction of VP-16 phenoxyl radicals, we used K562 human leukemia cells grown in Dulbecco's modified Eagle's medium which does not contain vitamin C (ascorbate), thus excluding the ascorbate-dependent reduction of VP-16 phenoxyl radicals.[24] We found that VP-16 phenoxyl radicals were reduced by endogenous reductant in K562 cell homogenates and that intracellular thiols were mainly responsible for the effect. Depletion of endogenous thiols resulted in almost complete inhibition of the ability of cell homogenates to reduce VP-16 phenoxyl radicals. Importantly, treatment of K562 cells with buthionine-S,R-sulfoximine (BSO, a specific inhibitor of gamma-glutamyl cysteine synthetase) led to potentiation of VP-16–induced DNA damage and to an increase in VP-16–induced growth inhibition, suggesting that, in the absence of ascorbate, modulation of endogenous thiols may be an important factor determining the oxidative metabolism and cytotoxic activity of VP-16. In conclusion, the ability of VP-16 phenoxyl radicals to oxidize thiols may be an important mechanism of its cytotoxicity.

In a similar way, phenoxyl radicals generated as intermediates in one-electron oxidation of phenol are reactive enough to oxidize directly a number of vital intracellular small reductant (e.g., NADH, NADPH, and GSH) but are also important structural and functional components of cells (e.g., lipids and proteins). Reduction of the highly reactive oxygen and organic radicals by phenols[25] is not the only source of phenoxyl radicals in cells. Generation of phenoxyl radicals can also be catalyzed by enzymes (e.g., isoforms of cytochrome P-450, tyrosinase and different peroxidases).[26,27] As a result, toxic effects of a phenolic compound (e.g., etoposide and phenol) may be mediated by an extremely potent redox-cycling mechanism with three amplifying cascades (FIG. 4): (1) Phenolic compound (P-OH) is oxidized in a reaction catalyzed by an enzyme (oxidase or oxygenase) to its phenoxyl radical, P-O·; (2) P-O· is reduced back to etoposide via oxidation of intracellular thiols (RSH) (FIG. 4), that is, sulfhydryl groups of proteins and glutathione. P-OH is thus repeatedly available as a substrate for the enzyme, whereas thiols undergo one electron oxidation to reactive thiyl radicals (RS·). These thiyl radicals can further react to generate disulfide anion radicals (RS-S$^{•-}$R) that can donate an electron to oxygen. (3) Superoxide anion radical (O2$^{•-}$) thus produced can form, in the presence of transition metal complexes, the extremely reactive hydroxyl radical (HO·) that damages DNA and other critical biomolecules, ultimately inducing cyto- and genotoxic effects. This vicious circle can be interrupted by nutritional nonphenolic antioxidants (e.g., ascorbate), that can reduce P-O· or phenolic antioxidants [e.g., vitamin E and its homologues, butylated hydroxytoluene (BHT)] that compete with P-OH for oxidase or oxygenase and produce phenoxyl radicals of low reactivity that do not react with thiols. Thus, although in model systems enzymatically catalyzed oxidative metabolism of phenolic compounds produces their quinoid derivatives, in the presence of intracellular thiols the process is transformed into a thiol-burning, redox-cycling mechanism that bursts forth several different reactive free radicals. It is the enzymatically triggered radicals, phenoxyl radicals and/or subsequently generated thiyl and oxygen radicals, that are responsible for the ultimate cyto- and genotoxic effects of phenolic compounds.

In summary, slight differences in the reactivity of phenoxyl radicals may have dramatic consequences for either protective antioxidant properties of a given phenolic compound or

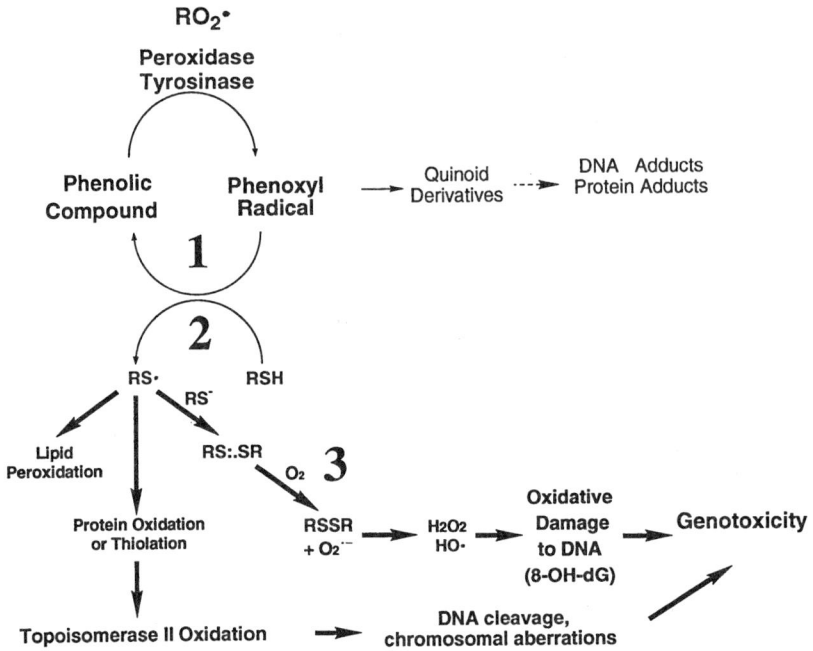

FIGURE 4. A scheme of redox cycling of phenolic compounds, resulting in their cyto- and geno-toxicity.

its damaging cyto- and genotoxic effects in cells and body fluids. Phenolic compounds whose phenoxyl radicals exert relatively low reactivity oxidize only selective intracellular constituents (*e.g.*, ascorbate and ubiquinol) and undergo recycling, which enhances anti-oxidant protection. By contrast, chemically reactive phenoxyl radicals have significantly lower selectivity and attack vitally important targets to produce nonfunctional oxidatively modified molecules. The damaging effects of these phenolic compounds may be enhanced by their enzymatically catalyzed redox cycling that triggers the production of a wide array of reactive oxygen species. Combined, these considerations imply that effective radical scavenging is only one of several substantial prerequisites for effective antioxidant protection. Another one (that may have an even greater significance) is the secondary reaction of phenoxyl radicals with biomolecules.

REFERENCES

1. PACKER, J.E., T.F. SLATER & R.L. WILLSON. 1979. Direct observation of a free radical interaction between vitamin E and vitamin C. Nature **278:** 737–738.
2. NIKI, E., J. TSUCHIYA, R. TANIMURA & Y. KAMIYA. 1982. The regeneration of vitamin E from alpha-chromanoxyl radical by glutathione and vitamin C. Chem. Lett. **6:** 789–792.
3. SCARPA, M.A., M. RIGO, M. MAIORINO, F. URSINI & C. GREGOLIN. 1984. Formation of α-toco-pherol radical and recycling of α-tocopherol by ascorbate during peroxidation of phosphati-dylcholine liposomes. Biochim. Biophys. Acta **801:** 215–219.

4. BISBY, R.H. & A.W. PARKER. 1991. Reactions of the α-tocopheroxyl radical in micellar solutions studied by nanosecond laser flash photolysis. FEBS Lett. **290:** 205–208.

5. TRABER, M.G., R. RAMAKRISHNAN & H.J. KAYDEN. 1994. Human plasma vitamin E kinetics demonstrate rapid recycling of plasma RRR-alpha-tocopherol. Proc. Natl. Acad. Sci. USA **91**(21): 10005–10008.

6. MAGUIRE, J.J., V.E. KAGAN & L. PACKER. 1992. Electron transport between cytochrome c and alpha-tocopherol Biochem. Biophys. Res. Commun. **188:** 190–197.

7. KAGAN, V.E., E.A. SERBINOVA & L. PACKER. 1990. Recycling and antioxidant activity of tocopherol homologues of differing hydrocarbon chain length in liver microsomes. Arch. Biochem. Biophys. **282:** 221–225.

8. KAGAN, V.E., E.A. SERBINOVA & L. PACKER. 1990. Generation and recycling of radicals from phenolic antioxidants. Arch. Biochem. Biophys. **280:** 33–39.

9. GOLDMAN, R., I.B. TSYRLOV, J. GROGAN & V.E. KAGAN. 1997. Reactions of phenoxyl radicals with NADPH-cytochrome P-450 reductase and NADPH: Reduction of the radicals and inhibition of the enzyme. Biochemistry **36**(11): 3186–3192.

10. KAGAN, V.E., E.A. SERBINOVA, A. SAFADI, J. CATUDIOC & L. PACKER. 1992. NADPH-dependent inhibition of lipid peroxidation in rat liver microsomes. Biochem. Biophys. Res. Commun. **86:** 74–80.

11. MAGUIRE, J.J., V.E. KAGAN, E.A. SERBINOVA, B.A. ACKRELL & L. PACKER. 1992. Succinate-ubiquinone reductase linked recycling of α-tocopherol in reconstituted systems and mitochondria: Requirement for reduced ubiquinone. Arch. Biochem. Biophys. **292:** 47–53.

12. MELLORS, A. & A.L. TAPPEL. 1966. The inhibition of mitochondrial peroxidation by ubiquinone and ubiquinol. J. Biol. Chem. **241:** 4353–4356.

13. MELLORS, A. & A.L. TAPPEL. 1966. Quinones and quinols as inhibitors of lipid peroxidation. Lipids **1:** 282–284.

14. NETA, P. & S. STEENKEN. 1982. One electron redox potentials of phenols, hydroxy and aminophenols and related compounds of biological interest. J. Phys. Chem. **93:** 7654–7659.

15. MUKAI, K., S. KIKUCHI & S. URANO. 1990. Stopped-flow kinetic study of the regeneration reaction of tocopheroxyl radical by reduced ubiquinone-10 in solution. Biochim. Biophys. Acta **1035:** 77–83.

16. STOYANOVSKY, D.A., A.N. OSIPOV, P.J. QUINN & V.E. KAGAN. 1995. Ubiquinone-dependent recycling of vitamin E radicals by superoxide. Arch. Biochem. Biophys. **323:** 343–351.

17. RITOV, V.B., S. BANNI, J.C. YALOWICH, B.W. DAY, H.G. CLAYCAMP, F.P. CORONGIU & V.E. KAGAN. 1996. Non-random peroxidation of different classes of membrane phospholipids in live cells detected by metabolically integrated *cis*-parinaric acid (PnA). Biochim. Biophys. Acta **1283:** 127–140.

18. RAO, R.D.N., V. FISCHER & R.P. MASON. 1990. Glutathione and ascorbate reduction of the acetaminophene radical formed by peroxidase. J. Biol. Chem. **265:** 844–847.

19. KAGAN, V.E., A. SHVEDOVA, E.A. SERBINOVA, S. KHAN, C. SWANSSON, R. POWELL & L. PACKER. 1992. Dihydrolipoic acid — A universal antioxidant both in the membrane and in the aqueous phase. Reduction of peroxyl, ascorbyl and chromanoxyl radicals. Biochem. Pharmacol. **44:** 1637–1649.

20. PACKER, L., E.H. WITT & H.J. TRITSCHLER. 1995. Alpha-lipoic acid as a biological antioxidant. Free Radical Biol. Med. **19**(2): 227–250.

21. HARAMAKI, N., D. HAN, G.J. HANDELMAN, H.J. TRITSCHLER & L. PACKER. 1997. Cytosolic and mitochondrial systems for NADH- and NADPH-dependent reduction of alpha-lipoic acid. Free Radical Biol. Med **22**(3): 535–542.

22. PACKER, L. & H.J. TRITSCHLER. 1996. Alpha-lipoic acid: The metabolic antioxidant. Free Radical Biol. Med. **20**(4): 625–626.

23. KAGAN, V.E., J.C. YALOWICH, B.W. DAY, R.R. GOLDMAN & D.A. STOYANOVSKY. 1994. Ascorbate is the primary reductant of the phenoxyl radical of etoposide (VP-16) in the presence of thiols both in cell homogenates and in model systems. Biochemistry **33:** 9651–9660.

24. YALOWICH, J.C., Y.Y. TYURINA, V.A. TYURIN, W.P. ALLAN & V.E. KAGAN. 1996. Reduction of phenoxyl radicals of the antitumor agent, etoposide (VP-16), by glutathione and protein sulfhydryls in human leukemia cells: Implications for cytotoxicity. Toxicol. In Vitro **10:** 59–68.

25. FOTI, M., K.U. INGOLD & J. LUSZTYK. 1994. The surprisingly high reactivity of phenoxyl radicals. J. Am. Chem. Soc. **116:** 9440–9447.

26. WHITE, R.E. & M.J. COON. 1980. Oxygen activation by cytochrome P-450. Annu. Rev. Biochem. **49:** 315–356.
27. JOB, D. & H.B. DUNFORD. 1976. Substituent effect on the oxidation of phenols and aromatic amines by horseradish peroxidase compound I. Eur. J. Biochem. **66:** 607–614.

The Chemistry and Biological Effects of Flavonoids and Phenolic Acids[a]

KEVIN D. CROFT[b]

*Department of Medicine, University of Western Australia, P.O. Box X2213,
Perth, Western Australia 6001*

ABSTRACT: Flavonoids and phenolic acids are widely distributed in higher plants
and form part of the human diet. Recent interest in these substances has been stimulated by the potential health benefits arising from the antioxidant activity of these
polyphenolic compounds. This review outlines the basic chemistry, biosynthesis, and
structure-activity relationships of these compounds with respect to their antioxidant
activity. Although there is considerable *in vitro* evidence establishing antioxidant
activity for polyphenolics found in the diet, there are few studies in humans on the
absorption and bioavailability of these compounds. The possible *in vivo* antioxidant
effects of the flavonoids is even less well understood. For example, controlled human
intervention studies with beverages, such as red wine, that are rich in polyphenolic
compounds, have yielded conflicting results. Our own work and that of others suggests that the final effects of such beverages may be a balance between the well-
described prooxidant effects of alcohol and its metabolism and the antioxidant effects
of the polyphenolic constituents. There is a need for further studies to increase our
understanding of the absorption and *in vivo* biological effects of this family of compounds.

Phenolic compounds are widely distributed in plants. One of the major groups of phenolic
compounds is the flavonoids, which are important in contributing to the flavor and color of
many fruits and vegetables and products derived from them, such as wine and tea. There is
much interest in the biological effects of flavonoids, inasmuch as evidence that diets rich
in fruit and vegetables appear to protect against cardiovascular disease and some forms of
cancer.[1–3] Because oxygen free radicals and lipid peroxidation are thought to be involved
in several conditions, such as atherosclerosis, cancer, and chronic inflammation,[4] the antioxidant activity of flavonoids has been of primary interest. There have been several excellent recent reviews on the antioxidant activity of the flavonoids.[5–7]

Despite major advances in understanding the *in vitro* antioxidant activity of flavonoids
and a number of studies on absorption in animals, there is little data available on either the
absorption or antioxidant effects of flavonoids *in vivo* in humans. This brief review will
outline our current understanding of the antioxidant activity of flavonoids and phenolic
acids, their bioavailability, and the most appropriate methods for assessing antioxidant
effects *in vivo*.

CHEMISTRY AND BIOSYNTHESIS

The term phenolic compound embraces a wide range of plant substances that possess
an aromatic ring bearing one or more hydroxyl substituents. They frequently occur
attached to sugars (glycosides) and as such tend to be water soluble. The flavonoids are the

[a]The financial support of the National Heart Foundation of Australia and the Medical Research
Foundation of Royal Perth Hospital are gratefully acknowledged.
[b]Tel. 61 8 9224 0275; fax: 61 8 9224 0246; e-mail:kcroft@cyllene.uwa.edu.au

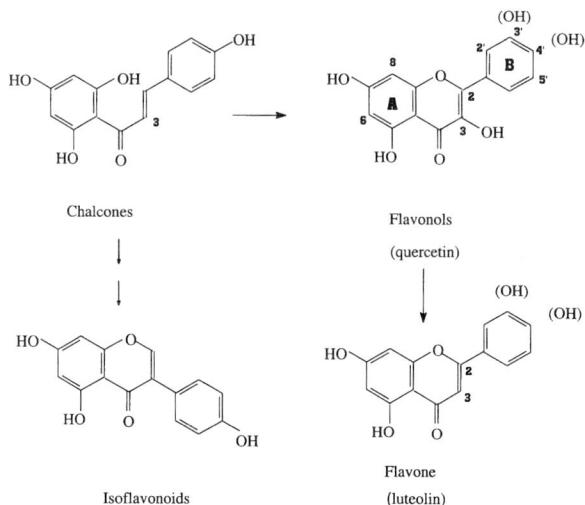

FIGURE 1. Major structural features and numbering system of the flavonoids.

largest single group of phenolic compounds. Flavonoids are C15 compounds composed of two phenolic rings connected by a three carbon unit. The flavonoids are biosynthetically derived from acetate and shikimate[8] such that the A ring has a characteristic hydroxylation pattern at the 5 and 7 position. The B ring is usually 4', 3'4', or 3'4'5'-hydroxylated. FIGURE 1 shows these major structural features with examples of the chalcone, flavonol, and flavone groups. The isoflavonoids are derived by cyclization of the chalcones, such that the B ring is located at the 3 position (FIG. 1). Other major groups of flavonoids include the catechins (often found as esters with gallic acid in tea) and the anthocyanidins, which are highly colored pigments (FIG. 2).

Of the simple phenolic acids, cinnamic acid and its derivatives are widespread in plants. They are derived from the shikimate pathway via phenylalanine or tyrosine,[8] and major examples are coumaric acid (single hydroxyl group) and caffeic acid (FIG. 3). Oxidation of the side chain can produce such derivatives of benzoic acid as protocatechuic

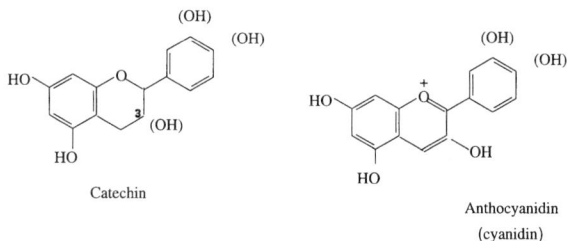

FIGURE 2.

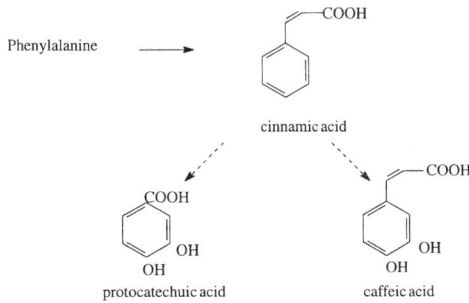

FIGURE 3. Possible biosynthetic formation of the phenolic acids.

acid. These compounds are usually found in nature as glucose ethers or in ester combination with quinic acid.[8]

Some major dietary sources of flavonoids are outlined in TABLE 1. The daily intake of flavonoids has been estimated to be between 20 mg to 1 g.[3] The flavanols, particularly catechin and catechin-gallate esters, and the flavonol quercetin are found in such beverages as green and black tea[9] and red wine.[10] Quercetin is also a predominant component of onions, apples and berries. Such flavanones as naringin are mainly found in citrus fruits.

ANTIOXIDANT ACTIVITY OF FLAVONOIDS

Free radicals are produced in the body as part of normal metabolism, for example, superoxide, $O_2^{\bullet-}$ and nitric oxide, NO·, which have important physiological functions. In general, free radicals are highly reactive and can attack membrane lipids, for example, generating a carbon radical, which in turn reacts with oxygen to produce a peroxyl radical, which may attack adjacent fatty acids to generate new carbon radicals. This process leads to a chain reaction producing lipid peroxidation products.[4] By this means a single radical may damage many molecules by initiating lipid peroxidation chain reactions. Because of the potential damaging nature of free radicals, the body has a number of antioxidant defense mechanisms that include such enzymes as superoxide dismutase and catalase; copper and iron transport, storage proteins; and both water-soluble and lipid-soluble molecular antioxidants. Oxidative stress may result when antioxidant defenses are unable

TABLE 1. Some Dietary Sources of Flavonoids and Phenolic Acids

Flavonoid	Source
Catechins	tea, red wine
Flavanones	citrus fruits
Flavonols (*e.g.*, quercetin)	onions, olives, tea, wine, apples
Anthocyanidins	cherries, strawberries, grapes, colored fruits
Caffeic acid	grapes, wine, olives, coffee, apples, tomatoes, plums, cherries

to cope with the production of free radicals, and may result from the action of certain toxins or by physiological stress.[4]

Flavonoids can act as antioxidants by a number of potential pathways. The most important is likely to be by free radical scavenging, in which the polyphenol can break the free radical chain reaction. For a compound to be defined as an antioxidant it must fulfill two conditions: first, when present at low concentrations, compared to the oxidizable substrate, it can significantly delay or prevent oxidation of the substrate;[11] second, the resulting radical formed on the polyphenol must be stable so as to prevent it from acting as a chain-propagating radical.[11] This stabilization is usually through delocalization, intramolecular hydrogen bonding, or by further oxidation by reaction with another lipid radical.[12] A number of studies have been carried out on the structure — antioxidant activity relationships of the flavonoids.[5,7,13–15] The main structural features of flavonoids required for efficient radical scavenging could be summarized as follows:[7] (1) an ortho-dihydroxy (catechol) structure in the B ring, for electron delocalization; (2) 2,3 double bond in conjugation, with a 4-keto function, provides electron delocalization from the B ring; (3) hydroxyl groups at positions 3 and 5 provide hydrogen bonding to the keto group. These structural features are illustrated in FIGURE 4.

The nonflavonoid phenolic acids may also be good antioxidants, particularly those possessing the catechol-type structure, such as caffeic acid.[16–18] Recent studies have indicated that simple cell-derived phenolic acids, such as 3-hydroxyanthranilic acid, may also be efficient coantioxidants for α-tocopherol, able to inhibit lipoprotein and plasma lipid

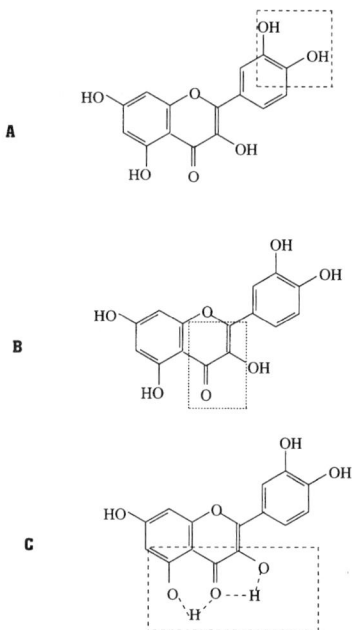

FIGURE 4. Structural groups for radical scavenging.

peroxidation in humans.[19] The possible interaction between flavonoids and phenolic acids with other physiological antioxidants, such as ascorbate or tocopherol, is another possible antioxidant pathway for these compounds. The synergistic interaction of these antioxidants may be exemplified by the enhancement of the antiproliferative effect of quercetin by ascorbic acid, possibly due to its ability to protect the polyphenol from oxidative degradation.[20]

Another pathway of apparent antioxidant action of the flavonoids, particularly in oxidation systems using such transition metal ions as copper or iron, is chelation of the metal ions. Chelations of catalytic metal ions may prevent their involvement in Fenton-type reactions, which can generate highly reactive hydroxyl radicals (reaction 1 and 2).[11]

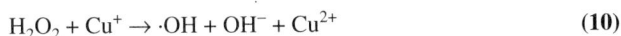

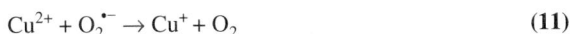

$$H_2O_2 + Cu^+ \rightarrow \cdot OH + OH^- + Cu^{2+} \tag{10}$$

$$Cu^{2+} + O_2^{\cdot -} \rightarrow Cu^+ + O_2 \tag{11}$$

The ability of polyphenolics to react with metal ions may also render them pro-oxidant. For example, in a recent study by Cao et al.,[13] using three different oxidation systems, flavonoids had potent antioxidant activity against peroxyl radicals generated from AAPH and against hydroxyl radicals but were prooxidant with Cu^{2+}. Presumably flavonoids can reduce Cu^{2+} to Cu^+ and hence allow the formation of initiating radicals.[13] Caffeic acid has also been shown to have prooxidant activity on Cu^{2+}-induced oxidation of low-density lipoprotein (LDL).[21] It should be noted that this prooxidant activity was seen only in the propagation phase of the oxidation not in the initiation phase in which caffeic acid inhibited LDL oxidation, in agreement with previous findings.[16–18]

The possible pro-oxidant effects of flavonoids may be important *in vivo* if free transition metal ions are involved in oxidation processes. In the healthy human body, metal ions appear largely sequestered in forms unable to catalyze free radical reactions.[22] However, injury to tissues may release iron or copper,[23] and catalytic metal ions have been measured in atherosclerotic lesions.[24] In these cases the potential for flavonoids to act as pro-oxidants can not be ignored.

ASSESSMENT OF OPTIMAL DIETARY INTAKE

The assessment of optimal dietary intake first requires an understanding of the absorption and bioavailability of the flavonoids and then appropriate measures of their effects on steady state oxidative damage *in vivo*. Although data on the bioavailability of flavonoids in humans is scarce, there is enough evidence to suggest that flavonoids are absorbed in significant quantities.[25] Such flavonoids as quercetin can be absorbed both as the free aglycone and glycoside, and have been detected in blood and urine.[26,27] There is some evidence that peak absorption may be two to three hours after ingestion.[28] Flavonoids that are absorbed may form such conjugates as glucuronides or sulfates in the liver. The metabolism of flavonoids is determined by the hydroxylation pattern, with compounds having 5,7 and 3′,4′ hydroxylation being susceptible to hydrolysis and heterocyclic ring cleavage by microbiological degradation in the colon.[29] It is uncertain if hydrolysis of flavonoid glycosides is necessary for absorption, however, recent methods have detected flavonoids as glycosides in human plasma.[30] Catechins is another major group of flavonoids that have been shown to be absorbed, present in plasma after one hour and also detected in a 24-hour urine test after a single oral dose.[31] Further evidence for the absorption of flavonoids comes from a number of studies with isoflavonoids. Plasma concentrations of daidzein and genistein were found to be 15–40 times higher in Japanese men compared to men eating a European diet.[32] This reflects the high soy bean (which is rich in isoflavonoids) content of the Japanese diet. Humans supplemented with a single soy bean drink containing 2 mg of isoflavone showed plasma concentrations of 2 μM after 6.5 hours.[33]

Although many aspects of flavonoid metabolism and bioavailability are still not known, there is enough evidence to suggest that some flavonoids are found in the plasma in concentrations to have biological effects. Improvements in methodologies for measuring flavonoids in plasma will continue to provide valuable information in this area.

The development of suitable methods and biomarkers for determining oxidative damage in humans is critical to the assessment of nutritional antioxidants.[34] For example, methods for the assessment of total lipid peroxidation in humans have been limited to measurements of malondialdehyde (MDA) excretion or hydrocarbon gases in exhaled air. These methods can be nonspecific and influenced by components in the diet or environment.[34] The F_2-isoprostanes formed by free radical damage to arachidonic acid in the body can be measured in plasma or urine and may be a good measure of steady state lipid peroxidation.[35] Whereas the ideal method of analysis requires negative chemical ionization, GC-MS, a commercial EIA kit, has become available, and measurements in agreement with the GC-MS method have been reported.[36] For measurement of oxidative DNA damage, the urinary excretion of 8-hydroxydeoxyguanine has been used but may suffer from several technical problems as well as a contribution from dietary intake.[34] Assessment of oxidative damage to proteins can involve measurement of specific amino acid adducts with MDA or derived by attack of reactive oxygen or nitrogen species (*e.g.*, hydroxytyrosine or nitrotyrosine).[34]

Assessment of the *in vivo* effect of dietary flavonoids can be very difficult. Recently we have studied the effects of antioxidant polyphenolics in beverages such as red wine on LDL oxidation.[18] Because oxidative damage to LDL has been linked to the development of atherosclerosis and heart disease, the rich flavonoid content of red wine has led to its appeal as possibly being beneficial against heart disease.[37] Although a number of *in vitro* studies clearly show strong antioxidant effects of red wine phenolics against LDL oxidation,[18,37,38] several human intervention trials have given conflicting results.[39–41] This may arise from the fact that alcohol itself is a pro-oxidant, and the overall effects of a beverage may be due to a balance between its pro-oxidative and antioxidant components.[37] In addition, most studies use oxidative susceptibility of isolated LDL, which may not necessarily relate to oxidative damage occurring *in vivo*.

CONCLUSION

There is good evidence that many flavonoids and phenolic acids are good antioxidants *in vitro*. The possible pro-oxidative effects of flavonoids due to interaction with metals ions may be important in some conditions where there are available or releasable transition metals. Although there is still much to be learned about the absorption and bioavailability of these compounds, there is evidence that they may occur in plasma at concentrations high enough to have biological effects. On the basis of our current scientific knowledge, it is not yet possible to determine the optimal level of dietary antioxidants to prevent disease. The application of the latest "biomarkers" of oxidative damage could help to answer questions about the potential antioxidant effects of dietary flavonoids in humans.

ACKNOWLEDGMENTS

The author would like to thank his colleagues, Ian Puddey, Lawrie Beilin, and Trevor Mori, for many helpful discussions and Ph.D. student Rima Abu-Amsha for her work on antioxidants in red wine.

REFERENCES

1. BLOCK, G. & L. LANGSETH. 1994. Antioxidant vitamins and disease prevention. Food Technol. **July:** 80–84.
2. BLOCK, G. 1992. A role for antioxidants in reducing cancer risk. Nutr. Rev. **50:** 207–213.
3. HERTOG, M.G.L., E.J.M. FESKENS, P.C.H. HOLLMAN, M.B. KATAN, & D. KROMHOUT. 1993. Dietary antioxidant flavonoids and risk of coronary heart disease. The Zutphen elderly study. Lancet **342:** 1007–1011.
4. HALLIWELL, B. 1994. Free radicals, antioxidants and human disease: Curiosity, cause or consequence? Lancet **344:** 721–724.
5. RICE-EVANS, C.A., N.J. MILLER & G. PAGANGA. 1996. Structure-antioxidant activity relationships of flavonoids and phenolic acids. Free Radical Biol. Med. **20:** 9.
6. COOK, N.C. & S. SAMMAN. 1996. Flavonoids, chemistry, metabolism, cardioprotective effects and dietary sources. J. Nutr. Biochem. **7:** 66–76.
7. BORS, W., W. HELLER, C. MICHEL & M. SARAN. 1990. Flavonoids as antioxidants: Determination of radical-scavenging efficiencies. Methods Enzymol. **186:** 343–355.
8. MANN, J. 1978. Secondary Metabolism. Oxford Chemistry Series. Clarendon Press. Oxford.
9. STAGG, G.V. & D.J. MILLIN. 1975. The nutritional and therapeutic value of tea—a review. J. Sci. Food Agric. **26:** 1439–1459.
10. FRANKEL, E.N., A.L. WATERHOUSE & P.L. TEISSEDRE. 1995. Principle phenolic phytochemicals in selected Californian wines and their antioxidant activity in inhibiting oxidation of human low-density lipoproteins. J. Agric. Food Chem. **43:** 890–894.
11. HALLIWELL, B., R. AESCHBACH, J. LOLIGER & O.I. ARUOMA. 1995. The characterization of antixoidants. Food Chem. Toxicol. **33:** 601–617.
12. SHAHIDI, F. & P.K.J.P.D. WANASUNDARA. 1992. Phenolic Antioxidants. Crit. Rev. Food Sci. Nutr. **32:** 67–103.
13. CAO, G., E. SOFIC & R.L. PRIOR. 1997. Antioxidant and pro-oxidant behaviour of flavonoids: Structure–activity relationships. Free Radical Biol. Med. **22:** 749–760.
14. VAN ACKER, S.A.B.E., D.J. VAN DEN BERG, M.N.J.L. TROMP, D.H. GRIFFIOEN, W.P. VAN BENNEKOM, W.J.F. VAN DER VIJGH & A. BAST. 1996. Structural aspects of antioxidant activity of flavonoids. Free Radical Biol. Med. **20:** 331–342.
15. CHEN, Z.Y., P.T. CHAN, K.Y. HO, K.P. FUNG & J. WANG. 1996. Antioxidant activity of natural flavonoids is governed by number and location of their aromatic hydroxyl groups. Chem. Phys. Lipids **76:** 157–163.
16. LARANJINHA, J.A.N., L.M. ALMEIDA & V.M.C. MADEIRA. 1994. Reactivity of dietary phenolic acids with peroxyl radicals. Antioxidant activity upon low density lipoprotein peroxidation . Biochem. Pharmacol. **48:** 487–494.
17. NARDINI, M., M. D'AQUINO, G. TOMASSI, V. GENTILI, N. DI FELICE & C. SCACCINI. 1995. Inhibition of human LDL oxidation by caffeic acid and other hydroxycinnamic acid derivatives. Free Radical Biol. Med. **19:** 541–552.
18. ABU-AMSHA, R., K.D. CROFT, I.B. PUDDEY, J.M. PROUDFOOT & L.J. BEILIN. 1996. Phenolic content of various beverages determines the extent of inhibition of human serum and low density lipoprotein oxidation *in vitro*: Identification and mechanism of action of some cinnamic acid derivatives from red wine. Clin. Sci. **91:** 449–458.

19. THOMAS, S.R., P.K. WITTING & R. STOCKER. 1996. 3-Hydroxyanthranilic acid is an efficient, cell-derived co-antioxidant for α-tocopherol, inhibiting human low density lipoprotein and plasma lipid peroxidation. J. Biol. Chem. **271:** 32714–32721.
20. KANDASWAMI, G., E. PERKINS, D.S. SOLONIUK, G. DRZEWIECKI & E. MIDDLETON. 1993. Ascorbic acid enhanced antiproliferative effect of flavonoids on squamous cell carcinoma *in vitro*. Anticancer Drugs **4:** 91–96.
21. YAMANAKA, N., O. ODA & S. NAGAO. 1997. Pro-oxidant activity of caffeic acid, dietary non-flavenoid phenolic acid, on Cu^{2+}-induced low density lipoprotein oxidation. FEBS Lett. **405:** 186–190.
22. HALLIWELL, B. & J.M.C. GUTTERIDGE. 1990. The antioxidants of human extracellular fluids. Arch. Biochem. Biophys. **280:** 1–8.
23. HALLIWELL, B., J.M.C. GUTTERIDGE & C.E. CROSS. 1992. Free radicals, antioxidants and human disease: Where are we now? J. Lab. Clin. Med. **119:** 598–620.
24. SMITH, C., M.J. MITCHINSON, O.I. ARUDMA & B. HALLIWELL. 1992. Stimulation of lipid peroxidation and hydroxyl radical generation by the contents of human atherosclerotic lesions. Biochem. J. **286:** 901–905.
25. HOLLMAN, P.C.H. 1997. Bioavailability of flavonoids. Eur. J. Clin. Nutr. **51:** S66–S69.
26. HOLLMAN, P.C.H. *et al.* 1995. Absorption of dietary quercetin glycosides and quercetin in healthy ileostomy volunteers. Am. J. Clin. Nutr. **62:** 1276–1282.
27. COVA, D. *et al.* 1992 . Pharmacokinetics and metabolism of oral diosmin in healthy volunteers. Int. J. Clin. Pharmacol. Ther. Toxicol. **30:** 279–286.
28. HACKETT, A.M. 1983. The metabolism and excretion of (+)- 14C-cyanidenol-3 in man following oral administration. Xenobiotica **13:** 279–286.
29. GRIFFITHS, L. 1982. Mammalian metabolism of flavonoids. *In*: The Flavonoids: Advances in Research. J. Harborne & T. Mabry, Eds.: 681–718. Chapman and Hall. London.
30. PAGANGA, G. & C.A. RICE-EVANS. 1997. The identification of flavonoids as glycosides in human plasma. FEBS Lett. **401:** 78–82.
31. LEE, M.J. *et al.* 1995. Analysis of plasma and urinary tea polyphenols in human subjects. Cancer Epidemiol. Biomarkers Prev. **4:** 393–399.
32. ADLERCREUTZ, H., H. MARKKANEN & S. WATANABE. 1993. Plasma concentrations of phytoestrogens in Japanese men and women consuming a traditional Japanese diet. Am. J. Clin. Nutr. **342:** 1209–1210.
33. XU, X. *et al.* 1994. Daiadzein is a more bioavailable soymilk isoflavone than is genistein in adult women. J. Nutr. **124:** 825–832.
34. HALLIWELL, B. 1996. Oxidative stress, nutrition and health. Experimental strategies for optimization of nutritional antioxidant intake in humans. Free Radical Res. **25:** 57–74.
35. MORROW, J.D. & L.J. ROBERTS. 1996. The isoprostanes: Current knowledge and directions for future research. Biochem. Pharmacol. **51:** 1–9.
36. WANG, Z. *et al.* 1995. Immunological characterization of urinary 8-epi-prostaglandin F2α excretion in man. J. Pharmacol. Exp. Ther. **275:** 94–100.
37. Puddey, I.B. & K.D. Croft. 1997. Alcoholic beverages and lipid peroxidation: Relevance to cardiovascular disease. Addiction Biol. **2:** 269–276.
38. FRANKEL, E.N., J. KANNER, J.B. GERMAN, E. PARKS & J.E. KINSELLA. 1993. Inhibition of oxidation of human low density lipoprotein by phenolic substances in red wine. Lancet **341:** 454–457.
39. FUHRMAN, B., A. LAVY & M. AVIRAM. 1995. Consumption of red wine with meals reduces the susceptibility of human plasma and low density lipoprotein to lipid peroxidation. Am. J. Clin. Nutr. **61:** 549–554.
40. SHARPE, P.C., L.T. MCGRATH, E. MCLEAN, I.S. YOUND & G.P. ARCHIBOLD. 1995. Effect of red wine consumption on lipoprotein (a) and other risk factors for atherosclerosis. Q. J. Med. **88:** 101–108.
41. DE RIJKE, X.B., P.N.M. DEMACKER, N.A. ASSEN *et al.* 1996. Red wine consumption does not effect oxidizability of low density lipoproteins in volunteers. Am. J. Clin. Nutr. **63:** 329–334.

The Antioxidant and Biological Properties of the Carotenoids[a]

NORMAN I. KRINSKY[b]

Department of Biochemistry, School of Medicine and USDA Jean Mayer Human Nutrition Research Center on Aging, Tufts University, Boston, Massachusetts 02111, USA

ABSTRACT: Much effort has been expended in evaluating the relative antioxidant potency of carotenoid pigments in both *in vitro* and *in vivo* experiments. It is quite clear that *in vitro*, carotenoids can inhibit the propagation of radical-initiated lipid peroxidation, and thus fulfill the definition of antioxidants.

When it comes to *in vivo* systems, it has been much more difficult to obtain solid experimental evidence that carotenoids are acting directly as biological antioxidants. In fact, under nonphysiological circumstances, carotenoids may act as prooxidants. These results can be modified by altering the oxidant stress, the cellular or subcellular system, the type of animal, and environmental conditions, such as oxygen tension. Results of this type raise the question as to whether it is still appropriate to group the carotenoids with such antioxidant vitamins as vitamin E and vitamin C. Thus, the biological properties of the carotenoids may be much more related to the *products* of the interaction of carotenoids with oxidant stress, that is, such breakdown products as apocarotenoids and retinoids.

PROPERTIES OF CAROTENOIDS

The chemical and biological properties of carotenoids have been the subject of several earlier reviews.[1-3] It appears to be quite logical that the biological actions of this interesting family of compounds must derive from their physical and chemical properties. The majority of the 600 carotenoids found in nature are 40 carbons in length and may be pure hydrocarbons, called carotenes, or possess oxygenated functional groups, in which case they are called xanthophylls. The structures of many of the carotenoids and xanthophylls found in human serum are depicted in FIGURE 1. The long-chain conjugated polyene structure accounts for the ability of these compounds to absorb visible light but also makes them quite susceptible to oxidation. This latter property is closely related to their ability to act as antioxidants.

BIOLOGICAL ACTIONS OF CAROTENOIDS

The biological actions that will be described here include antioxidation, both *in vitro* and *in vivo*, as well as data regarding the prooxidant nature of carotenoids.

[a]This study was funded by Grant RO1CA66914 from the National Institutes of Health, Bethesda, MD.

[b]Address correspondence to Dr. Norman I. Krinsky, Department of Biochemistry, Tufts University School of Medicine, 136 Harrison Avenue, Boston, MA 02111-1837, USA. Tel: 617/636-6861; fax: 617/636-2409; e-mail: nkrinsky_mna@opal.tufts.edu

443

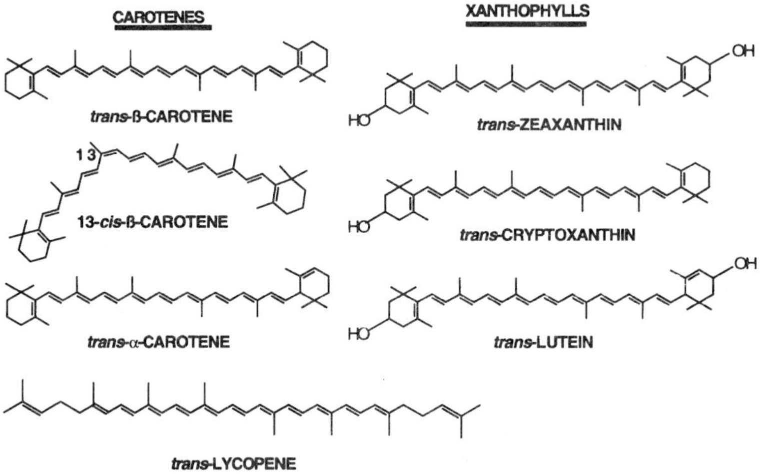

FIGURE 1. The hydrocarbon carotenes and oxygenated xanthophylls present in human serum.

Antioxidant Actions of Carotenoids—Singlet Oxygen Quenching

There are several distinct mechanisms whereby carotenoids function as antioxidants. The first mechanism that was clearly described is the ability of carotenoids to quench the highly reactive form of oxygen known as singlet oxygen ($^1O_2$).[4] $^1O_2$ is usually formed through photochemical reactions involving the absorption of light by a sensitizer molecule (S), which converts it to the singlet sensitizer (^1S). Through a process known as intersystem crossing (ISC), ^1S is then converted to a metastable triplet state (^3S). The ^3S can react with ground state oxygen, which exists in a triplet configuration ($^3O_2$), in an energy transfer reaction, to form ground state sensitizer (S) and $^1O_2$. These reactions are depicted below:

$$S + light \rightarrow {}^1S$$

$$^1S \rightarrow ISC \rightarrow {}^3S$$

$$^3S + {}^3O_2 \rightarrow S + {}^1O_2$$

It is this $^1O_2$ that is a highly reactive species, capable of oxidizing nucleic acids, various amino acids in proteins, and unsaturated fatty acids. Fortunately for plants, as well as us, carotenoids (CAR) are the most effective quenchers of $^1O_2$ found in nature, via the following reactions:

$$^1O_2 + CAR \rightarrow {}^3O_2 + {}^3CAR$$

$$^3CAR \rightarrow CAR + heat$$

Again, because of the long conjugated polyene nature of these molecules, they lose the excess energy in the excited state (^3CAR) via vibrational and rotational interactions with the solvent system, ultimately reforming the ground state CAR, ready to begin another cycle of $^1O_2$ quenching. It has been estimated that each carotenoid molecule can quench 1000 $^1O_2$ molecules before they react chemically and form products. Many of these products have been characterized and consist of carbonyls and epoxides, as shown in FIGURE 2.[5]

Antioxidant Actions of Carotenoids—Radical Reactions

In addition to quenching $^1O_2$, there have been many reports of the ability of various carotenoids to interfere with radical-initiated reactions, particularly with those that result in lipid peroxidation. These antioxidant actions have been reviewed in the past,[6,7] and newer results suggest that this protection is unlike that seen with classical dietary antioxidants such as α-tocopherol. The unique antioxidant effects of β-carotene were first pointed out by Burton and Ingold in 1984.[8] The more recent studies of the *in vitro* antioxidant activities of β-carotene and other carotenoids continue to demonstrate excellent activity, in some cases showing a more powerful antioxidant action that α-tocopherol.[9] The chemical reaction between radical species and carotenoids should result in products, and these have been characterized by several groups.[10,11] In many cases, the products are very similar to those seen with $^1O_2$ reactions, although Liebler and McClure have demonstrated recently both substitution products, as well as carotene-radical adducts.[12]

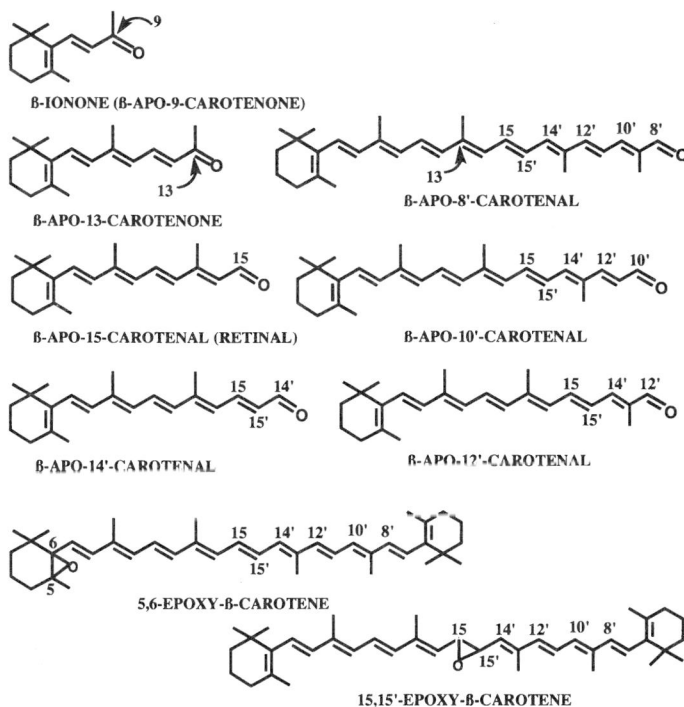

FIGURE 2. Carotenoid carbonyls and epoxides formed by singlet oxygen reactions.

Evidence for in Vivo *Antioxidant Actions of Carotenoids*

The ultimate test for an antioxidant is whether its effects can be demonstrated in animals or humans. There are now several examples of carotenoid antioxidant action being expressed in humans. Allard *et al.* supplemented smokers with β-carotene and found that the breath pentane levels (BPO) of the smokers dropped to the level of nonsmokers but found no change in the breath ethane output.[13] Gottlieb *et al.* depleted nonsmokers of carotenoids for two weeks and found a significant decrease in BPO levels after supplementation with a very high dose of β-carotene (120 mg/day) for four weeks, whereas no significant change was detected in the group supplemented with 15 mg/day β-carotene.[14] Dixon *et al.* placed women on a carotene-deficient diet for 68 days, during which time they demonstrated increased susceptibility of LDL to Cu-initiated oxidation and increased levels of TBARS in serum. All of these elevated levels were normalized upon repletion of the diets with β-carotene.[15] However, the results of β-carotene supplementation on resistance of LDL to oxidative stress is still controversial. Meydani *et al.* have demonstrated that supplementing elderly women with β-carotene at 90 mg/day for three weeks resulted in an increase in the plasma antioxidant capacity, as measured by changes in the formation of phospholipid hydroperoxides after treating plasma with the radical generator, AAPH.[16]

Even though cystic fibrosis patients are supplemented with large amounts of vitamin E, they still show significantly elevated levels of malondialdehyde (MDA) in their serum. Lepage *et al.* have treated such children with β-carotene for two months and report normalization of the MDA levels in 11/12 children.[17] Winklhofer-Roob *et al.* have also reported a decrease in MDA levels in children with cystic fibrosis after β-carotene supplementation, as well as increased resistance of LDL to oxidant stress.[18]

Prooxidant Actions of Carotenoids

Burton and Ingold were the first to present evidence that β-carotene can act as a prooxidant during radical-initiated lipid peroxidation, although this was only observed at 100% oxygen.[8,19] Since then, many reports have appeared supporting the concept that at 100% oxygen tension β-carotene acts as a prooxidant. Palozza and her associates have demonstrated that β-carotene is an effective antioxidant in protecting both rat liver microsomes[20] and mouse thymocytes[21] against radical-induced lipid peroxidation under air, but at 100% oxygen, the carotenoid acts as a prooxidant. Truscott has presented a theoretical basis for the antioxidant and prooxidant effects of β-carotene under low and high oxygen tensions.[22]

CONCLUSIONS

The experimental data are quite convincing that β-carotene, as well as other carotenoids, can act as antioxidants *in vitro*, but the evidence is much less convincing for the *in vivo* situation. With respect to the prooxidant claims for β-carotene, there had been much confusion about high oxygen tensions promoting prooxidation. Many people have interpreted the *in vitro* results as being applicable to tissues exposed to a high oxygen tension such as the lung and assuming that they would be particularly susceptible to a prooxidant effect of β-carotene. However, the data indicate that the prooxidant effect is seen at 100% oxygen and not at ambient conditions where oxygen is 21%, nor at physiological, or tissue levels, where the oxygen tension is equivalent to 1–2% oxygen. Thus, there seems to be very little support of the concept that β-carotene acts as a prooxidant in the body.

REFERENCES

1. KRINSKY, N.I. 1994. The biological properties of carotenoids. Pure Appl. Chem. **66:** 1003–1010.
2. BRITTON, G. 1995. Structure and properties of carotenoids in relation to function. FASEB J. **9:** 1551–1558.
3. KRINSKY, N.I. 1998. Carotenoid properties define primary biological actions and metabolism defines secondary biological actions. *In* Free Radicals, Oxidative Stress, and Antioxidants: Pathological and Physiological Significance. T. Özben, Ed.: 323–332. NATO Advanced Study Institute. London.
4. FOOTE, C.S. & R.W. DENNY. 1968. Chemistry of singlet oxygen. VIII. Quenching by β-carotene. J. Am. Chem. Soc. **90:** 6233–6235.
5. STRATTON, S.P., W.H. SCHAEFER & D.C. LIEBLER. 1993. Isolation and identification of singlet oxygen oxidation products of β-carotene. Chem. Res. Toxicol. **6:** 542–547.
6. KRINSKY, N.I. 1989. Antioxidant functions of carotenoids. Free Radical Biol. Med. **7:** 617–635.
7. PALOZZA, P. & N.I. KRINSKY. 1992. Antioxidant effects of carotenoids *in vitro* and *in vivo*: An overview. Methods Enzymol. **213.** 403–420.
8. BURTON, G.W. & K.U. INGOLD. 1984. β-Carotene: An unusual type of lipid antioxidant. Science **224:** 569–573.
9. MILLER, N.J., J. SAMPSON, L.P. CANDEIAS, P.M. BRAMLEY & C.A. RICE-EVANS. 1996. Antioxidant activities of carotenes and xanthophylls. FEBS Lett. **384:** 240–242.
10. HANDELMAN, G.J., F.J.G.M. VAN KUIJK, A. CHATTERJEE & N.I. KRINSKY. 1991. Characterization of products formed during the autoxidation of β-carotene. Free Radical Biol. Med. **10:** 427–437.
11. KENNEDY, T.A. & D.C. LIEBLER. 1991. Peroxyl radical oxidation of β-carotene: Formation of β-carotene epoxides. Chem. Res. Toxicol. **4:** 290–295.
12. LIEBLER, D.C. & T.D. MCCLURE. 1996. Antioxidant reactions of β-carotene: Identification of carotenoid-radical adducts. Chem. Res. Toxicol. **9:** 8–11.
13. ALLARD, J.P., D. ROYALL, R. KURIAN, R. MUGGLI & K.N. JEEJEEBHOY. 1994. Effects of beta-carotene supplementation on lipid peroxidation in humans. Am. J. Clin. Nutr. **59:** 884–890.
14. GOTTLIEB, K., E.J. ZARLING, S. MOBARHAN, P. BOWEN & S. SUGARMAN. 1993. β-Carotene decreases markers of lipid peroxidation in healthy volunteers. Nutr. Cancer **19:** 207–212.
15. DIXON, Z.R. *et al.* 1994. Effects of a carotene-deficient diet on measures of oxidative susceptibility and superoxide dismutase activity in adult women. Free Radical Biol. Med. **17:** 537–544.
16. MEYDANI, M., A. MARTIN, J.D. RIBAYA-MERCADO, J. GONG, J.B. BLUMBERG & R.M. RUSSELL. 1994. β-Carotene supplementation increases antioxidant capacity of plasma in older women. J. Nutr. **124:** 2397–2403.
17. LEPAGE, G., J. CHAMPAGNE, N. RONCO, A. LAMARRE, I. OSBERG, R.J. SOKOL & C.C. ROY. 1996. Supplementation with carotenoids corrects increased lipid peroxidation in children with cystic fibrosis. Am. J. Clin. Nutr. **64:** 87–93.
18. WINKLHOFER-ROOB, B.M., H. PUHL, G. KHOSCHSORUR, M.A. VAN'T HOF, H. ESTERBAUER & D.H. SHMERLING. 1995. Enhanced resistance to oxidation of low density lipoproteins and decreased lipid peroxide formation during β-carotene supplementation in cystic fibrosis. Free Radical Biol. Med. **18:** 849–859.
19. BURTON, G.W. 1989. Antioxidant action of carotenoids. J. Nutr. **119:** 109–111.
20. PALOZZA, P., G. CALVIELLO & G.M. BARTOLI. 1995. Prooxidant activity of β-carotene under 100% oxygen pressure in rat liver microsomes. Free Radical Biol. Med. **19:** 887–892.
21. PALOZZA, P., C. LUBERTO, G. CALVIELLO, P. RICCI & G.M. BARTOLI. 1997. Antioxidant and prooxidant role of β-carotene in murine normal and tumor thymocytes: Effects of oxygen partial pressure. Free Radical Biol. Med. **22:** 1065–1073.
22. TRUSCOTT, T.G. 1996. β-Carotene and disease: A suggested pro-oxidant and anti-oxidant mechanism and speculations concerning its role in cigarette smoking. J. Photochem. Photobiol. B: Biol. **35:** 233–235.

Structural and Functional Changes in Proteins Induced by Free Radical–mediated Oxidative Stress and Protective Action of the Antioxidants *N-tert*-Butyl-α-phenylnitrone and Vitamin E[a]

D. ALLAN BUTTERFIELD,[b,d] TANUJA KOPPAL,[b] BEVERLY HOWARD,[b] RAM SUBRAMANIAM,[b] NATHAN HALL,[b] KENNETH HENSLEY,[b] SERVET YATIN,[b] KERRY ALLEN,[b] MICHAEL AKSENOV,[c] MARINA AKSENOVA,[c] AND JOHN CARNEY[c]

[b]*Department of Chemistry and Center of Membrane Sciences and*
[c]*Department of Pharmacology, University of Kentucky, Lexington, Kentucky 40506-0055, USA*

ABSTRACT: The free radical theory of aging proposes that reactive oxygen species (ROS) cause oxidative damage over the lifetime of the subject. It is the cumulative and potentially increasing amount of accumulated damage that accounts for the dysfunctions and pathologies seen in normal aging. We have previously demonstrated that both normal rodent brain aging and normal human brain aging are associated with an increase in oxidative modification of proteins and in changes in plasma membrane lipids. Several lines of investigation indicate that one of the likely sources of ROS is the mitochondria. There is an increase in oxidative damage to the mitochondrial genome in aging and a decreased expression of mitochondrial mRNA in aging. We have used a multidisciplinary approach to the characterization of the changes that occur in aging and in the modeling of brain aging, both *in vitro* and *in vivo*. Exposure of rodents to acute normobaric hyperoxia for up to 24 h results in oxidative modifications in cytosolic proteins and loss of activity for the oxidation-sensitve enzymes glutamine synthetase and creatine kinase. Cytoskeletal protein spin labeling also reveals synaptosomal membrane protein oxidation following hyperoxia. These changes are similar to the changes seen in senescent brains, compared to young adult controls. The antioxidant spin-trapping compound *N-tert*-butyl-α-phenylnitrone (PBN) was effective in preventing all of these changes. In a related study, we characterized the changes in brain protein spin labeling and cytosolic enzyme activity in a series of phenotypically selected senescence-accelerated mice (SAMP), compared to a resistant line (SAMR1) that was derived from the same original parents. In general, the SAM mice demonstrated greater oxidative changes in brain proteins. In a sequel study, a group of mice from the SAMP8-sensitive line were compared to the SAMR1-resistant mice following 14 days of daily PBN treatment at a dose of 30mg/kg. PBN treatment resulted in an improvement in the cytoskeletal protein labeling toward that of the normal control line (SAMR1). The results of these and related studies indicate that the changes in brain function seen in several different studies may be related to the progressive oxidation of critical brain proteins and lipids. These com-

[a]This work was supported in part by NIH Grants (AG-10836 and AG-05119).
[d]Address correspondence to D. Allan Butterfield, Department of Chemistry and Center of Membrane Sciences, University of Kentucky, Lexington, KY 40506-0055. Tel: 606/257-3184; fax: 606/257-5876; e- mail: dabcns@pop.uky.edu

ponents may be critical targets for the beneficial effects of gerontotherapeutics both in normal aging and in disease of aging.

FREE RADICALS AND OXIDATIVE STRESS

The evolutionary process selected oxygen over other gases because of its ready availability, the high energy yield of oxidation, easy distribution in its gaseous state, solubility in biocomponents, and its efficient recycling using the processes of respiration and photosynthesis. Oxygen, however, is also the main source of damaging free radicals, which have been suggested to cause aging and ultimately the death of the organism.[1] There are other sources of free radicals, namely, ionizing radiations like X-rays, ultrasound, photochemical reactions, and biochemical and enzymatic processes; however, the human body is not exposed to all of these as frequently as it is to oxygen-derived radicals.

Oxidative injury is the result of an attack on cellular components by highly reactive, toxic oxygen moieties, collectively referred to as reactive oxygen species (ROS). Hydroxyl radicals, peroxyl radicals, superoxide anions, hydrogen peroxide, and nitric oxide are all a part of the ROS family. The half-lives of these free radicals generated in the cell vary from nanoseconds for the highly reactive hydroxyl radical to seconds for nitric oxide and peroxyl radicals. Also the reactivity of these radicals varies from the aqueous environment to those reacting deep within the membrane lipid bilayer. Oxy radicals, like hydroxyl radicals, have a very short life span, are extremely reactive, and hence attack the cellular components present in the vicinity of their production, whereas nitric oxide is very stable and relatively benign, except when it reacts, at diffusion-limited rates, with the superoxide anion to form peroxynitrite. Peroxynitrite is highly reactive and toxic to the cell, affecting several cellular components, leading to loss in structure and function.[2,3] Hence, the radical damage occurring in the cell is all pervasive. Intracellularly, mitochondria are a major source of free radicals.[4] Normal metabolism in healthy individuals uses the electron transport system in the mitochondria for energy production and in the process gives rise to a host of ROS. There are also various enzymatic and nonenzymatic metal-catalyzed systems capable of generating free radicals. To counteract these damaging free radical species, highly effective antioxidant systems have been developed that include enzymes like glutathione peroxidase, glutathione reductase, S-methyl transferase, superoxide dismutase (SOD), and catalase that can either act as repair agents or as antioxidant enzymes by eliminating precursors like hydrogen peroxide and superoxide from the cellular system. Also, there are proteins like hemoglobin, transferrin, and ceruloplasmin that bind ferrous and copper ions and prevent radical generation through Fenton chemistry; and proteases, ribonucleases, and lipases that preferentially degrade the modified components of proteins, DNA, and lipids, respectively. In addition, protection against free radical damage can be afforded by the inclusion of certain vitamins (vitamin C and vitamin E), carotenoids (β carotene), flavanoids, and other antioxidants in the diet, which inhibit the initiation of the free radical processes or can act as chain-breaking antioxidants.[5]

In spite of the development of various antioxidant systems to counteract the damaging effects of ROS, with age, the cell succumbs to oxidative stress, which has been defined as an imbalance that is shifted towards the prooxidant system relative to the antioxidant systems in the body, leading to cell damage and ultimately cell death.[6] Oxidative stress is known to cause lipid peroxidation, protein oxidation, DNA fragmentation, impairment of cellular energy status, and disruption of ion homeostasis.[7–9] Free radical–mediated oxidative stress has also been implicated in causing damage leading to the pathology of aging[10,11] and such age-associated disorders as stroke, amyotrophic lateral sclerosis, Parkinson's disease, and Alzheimer's disease (AD).[7,12–14] Models of aging, such as hyperoxia, have also been investigated.[15–17]

OXIDATIVE STRESS AND AGING

The free radical theory of aging[1] states that the progressive erosion of cellular components occurring due to free radical damage leads to aging and ultimately results in the death of the organism. Several studies report that the rate of metabolism is directly related to the rate of aging. Oxygen consumption and ROS production are closely related, and hence it is hypothesized that, in animals having high metabolic rates, the levels of ROS are also elevated due to increased oxidative stress. *In vivo* studies have shown that the level of oxidative stress increases during aging.[7,10,18–20] It has never been conclusively established whether this increase in oxidative stress is due to decreased antioxidant levels or due to an increase in production of prooxidant molecules in the cell. Thus oxidative stress can play an important role in aging either by affecting the efficiency with which the antioxidant defenses and/or repair mechanisms operate or by causing structural and functional changes within those molecules, or it can accelerate aging by altering the gene expression of the various cellular components.

Free radical oxidative stress with consequential protein oxidation and lipid peroxidation can lead to cell death. Our laboratory has been involved with factors associated with oxidative stress that alter the physical and chemical states of cortical synaptosomal membranes. Several *in vivo* and *in vitro* models of oxidative stress have been developed and studied for this purpose.[12,21–23] The focus of this review is a summary of work done in our laboratory on changes in protein structure and function in three *in vivo* models of free radical–induced oxidative stress, namely, hyperoxia, ischemia–reperfusion injury (IRI), and accelerated senescence, and the protection offered by the free radical scavenger, *N-tert*-butyl-α-phenylnitrone (PBN), against these damages. This review also summarizes similar protein damage seen in an *in vitro* model of oxidative stress, that is, synaptosomes exposed to amyloid β-peptide, a peptide implicated in the pathology of AD, and the protective effects of the antioxidant vitamin E against the ensuing damage.

MARKERS OF MEMBRANE PROTEIN DAMAGE

Protein Conformational Changes

Alterations in protein conformations can lead to increased aggregation, fragmentation, distortion of secondary and tertiary structure, susceptibility to proteolysis, and diminution of normal function. The technique of EPR (electron paramagnetic resonance), in conjunction with protein- and lipid-specific spin labels, is used to study membrane protein and lipid conformational changes.[24] The power of EPR spin labeling methods derives from the extreme sensitivity of EPR, the information that can be obtained about motion and polarity of the local microenvironment near the paramagnetic center of the spin label, the relatively simple spectra that need to be analyzed, that opaque samples not susceptible to light-scattering effects common to optical spectroscopy can be efficiently studied, and the fact that generally, only the spin label is paramagnetic, that is, the biological system is EPR silent and hence, does not interfere with the spectrum. The sulfhydryl-selective spin label MAL-6 (2,2,6,6-tetramethyl-4-maleimidopiperidine-1-oxyl), which covalently binds only to the -SH groups on the proteins, is the predominant spin label employed. MAL-6 is a stable paramagnetic nitroxide that generates an EPR spectrum on binding membrane proteins. Depending on whether the MAL-6 binds to -SH groups deeply within clefts of the protein or close to the protein surface, the spin label is either strongly (S) or weakly (W) immobilized, respectively. This difference in spin-label motion causes both a broad and a narrow low-field line. The ratio of the spectral amplitudes of the M_I low-field resonance line of the weakly immobilized site (W) to that of the strongly immobilized site (S), referred to as the W/S ratio, is a sensitive measure of changes in the physical state of the protein.[24] Decreased W/S ratios indicate increased protein–protein interaction and decreased segmental motion and/or conformational changes in the proteins that were labeled; the converse is also the case.[24]

Earlier studies in our laboratory using other conditions of free radical–induced oxidative stress, such as hydroxyl radical generation,[21] sepsis-associated lipopolysaccharide,[22] or menadione[23] have shown W/S ratios to be lowered in each case. Hence, the W/S ratio can be used as a valuable marker of protein alterations.

Protein Oxidation

Oxidation of proline, histidine, arginine, lysine, and other amino acid residues on proteins leads to the formation of carbonyl derivatives,[10] that is, protein carbonyls can act as a marker of protein oxidation. Protein carbonyl content has been shown to increase with age in several *in vitro* oxidative models, including the gerbil brain and human postmortem brain tissue.[20,25] The spectroscopic assay using dinitrophenylhydrazine (DNPH),[20] in which the carbonyl is reacted with DNPH to form the hydrazone, is one way of determining the levels of protein carbonyls. Other methods involving immunochemistry[26] and histofluorescence[27] have also been used to measure protein carbonyls.

Loss of Enzyme Activity

Oxidative modifications of active-site residues of proteins can lead to loss of enzyme function and activity. Glutamine synthetase (GS) and creatine kinase (CK) are two oxidatively sensitive enzymes.[10] Age-related decline in activities of GS and CK have been observed in gerbil brain and human brain tissue, with the more sensitive GS showing a greater decrease.[25,28,29] GS activity is determined using the cytosolic fraction isolated from homogenized brain neocortices following the assay as described.[30,31]

PROTECTIVE ACTION OF THE FREE RADICAL SCAVENGERS PBN AND VITAMIN E

Spin traps are molecules that have been used in EPR for trapping highly reactive, unstable radicals or radicals that are present in very low concentrations. The spin trap itself is nonparamagnetic but on reacting with the transient free radical species it forms an

EPR-active spin adduct. Spin traps are usually molecules with a nitrone moiety. In earlier studies administration of the free radical spin trap PBN was shown to reduce protein carbonyls of aged gerbils to levels comparable to those seen in young gerbils. This was also accompanied by reversal of memory loss exhibited by the aged animals.[25,32,33] PBN has also been shown to be effective in head trauma.[34]

Vitamin E, unlike PBN, does not rapidly traverse the blood brain barrier, but its efficacy as a free radical antioxidant lies in the fact that it is highly lipophilic and hence can reach the highly oxidation susceptible sites in the membrane lipid bilayer, namely the polyunsaturated fatty acids. Its phenolic head group scavenges the radicals, while the lipophilic side chain aligns with the lipid fatty acid chains to offer membrane stability. Also, molecules like vitamin C and glutathione are involved in recycling vitamin E to its reduced state, which is essential in its scavenging role. Free radical damage to the lipid bilayer can cause impairment of enzyme structure and function[35-37] and can lead to dysfunction in membrane permeability.[38] Studies conducted by other groups have shown the effectiveness of vitamin E in protecting against Aβ toxicity in PC-12 cells[39] and against Aβ-induced production and toxicity of the lipid peroxidation product, 4-hydroxynonenal (HNE), in cultured neuronal cells.[37]

PBN PROTECTION IN *IN VIVO* ANIMAL MODELS OF ISCHEMIA/REPERFUSION, HYPEROXIA, AND SENESCENCE

Stroke represents a major age-related neurodegenerative disorder and is the third leading cause of death in the United States. As a model of stroke, IRI has been shown to be associated with free radical oxidative stress.[40-42] As shown in FIGURES 1 and 2, using the protein-specific spin label MAL-6, the effects of 10-minute ischemia followed by a 1-hour and 14-hour reperfusion were investigated, demonstrating the protective action of PBN against IRI-induced protein damage.[41] For the PBN protection study the gerbils were pretreated with the nitrone 30 minutes prior to the injury. Results showed the W/S ratios comparable to those of the control values at both 1-hour and 14-hour reperfusion time points.[41] However, if instead of administering PBN preinjury, PBN were administered 6 hours postinjury, this nitrone would not attenuate the drastic change in W/S ratios, suggesting its most effective use at the initial stages of reperfusion when free radical damage can be intercepted.

Hyperoxia represents a model of accelerated oxidative stress and aging.[15,43] The hyperoxic conditions developed in our laboratory involved keeping gerbils in a $1 \times 1.5 \times 0.5$ meter clear polycarbonate hyperoxia chamber with oxygen, at 1 atm pressure, constantly monitored at 90–100% for the given time period. Control animals were left outside the hyperoxia chamber for the same time period as the study. EPR studies using MAL-6, protein carbonyl measurements, and GS activity assays were measured according to previous literature methods.[20,24,31] Using the same gerbil model for hyperoxia as mentioned here, we previously had shown that both adult and aged rodents exhibit synaptosomal protein oxidation and that this damage reaches its peak at 24 hours in adult animals.[16] After an exposure to hyperoxic conditions for 48 hours, however, the adult rodents recover, as judged by the W/S ratios, whereas the aged gerbils continue to show further damage.[16] This result is consistent with the notion that with age the ability to recover from oxidative stress-induced damages also decreases. If this protein damage observed were indeed due to free radical–mediated oxidative stress, then free radical scavengers like PBN should

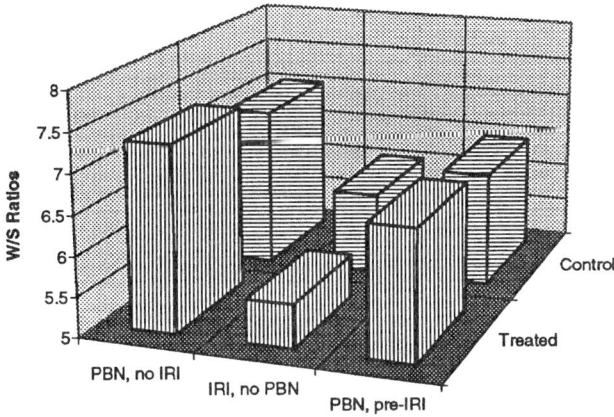

FIGURE 1. Average W/S ratios of MAL-6-labeled cortical synaptosomal membranes of adult gerbils injected with 300 mg/kg PBN, subjected to 10-minute ischemia without PBN pretreatment and 30 minutes preischemia PBN, and then given 1-hour reperfusion. Control animals were present for each group. $N = 5$–7, $p < 0.01$

attenuate the extent of damage. Hence, PBN was administered intraperitoneally into adult gerbils at 10, 20, and 40 mg/kg body weight. After a 24-hour exposure to hyperoxia, the animals were decapitated, and the isolated synaptosomes were used for all the studies.[17]

The results obtained were in accordance with our predictions of a hyperoxia-induced free radical damage and the moderating action of PBN in preventing this damage (FIGURES 3 and 4). W/S ratios of MAL-6 covalently attached to cortical synaptosomal membrane proteins from animals in hyperoxia without PBN were much lower than those injected with PBN (FIG. 3). There was no difference in W/S ratios of MAL-6 attached to membrane proteins of animals with PBN injections and those left outside the hyperoxia chamber. All

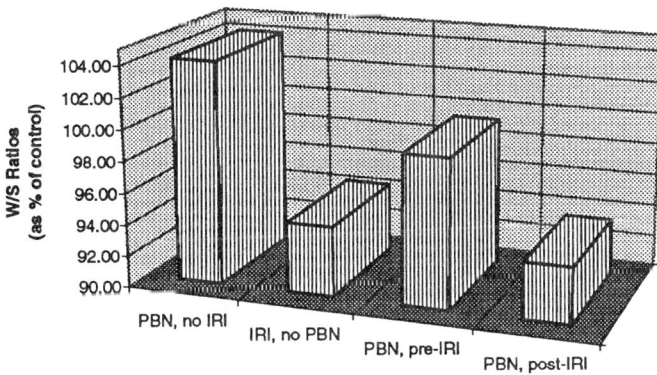

FIGURE 2. Effect of 300 mg/kg PBN given either 30-minute preischemia or 6-hour postreperfusion, on average W/S ratios of MAL-6 spin-labeled brain cortical synaptosomal membranes of adult gerbils. The animals were given 10-minute ischemia with 14 hours reperfusion, and the results are represented as a percent of the controls present for each group. $N = 5$–7, $p < 0.01$.

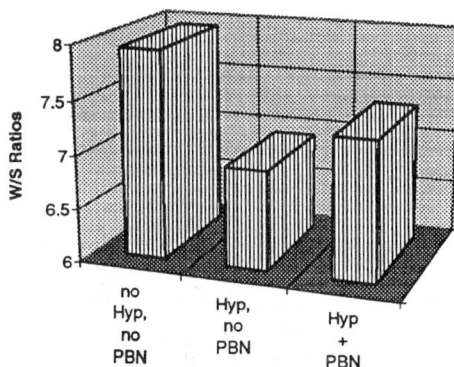

FIGURE 3. Effect of PBN (20 mg/kg) on average W/S ratios of MAL-6-labeled synaptosomal membranes of gerbils placed in 90–100% O_2 for 24 hours. The p value of the hyperoxic group versus the normoxic was found to be $p < 0.00001$, and versus the hyperoxic injected with PBN was $p < 0.0004$.

three doses of PBN provided effective prevention against protein oxidation, but 20 mg/kg body weight was found to be the optimum dose, based on the W/S ratios of MAL-6. Results seen on measuring the GS activity were similar to that obtained for the EPR study. Hyperoxia increased protein carbonyl levels and reduced GS activity and injection of animals with PBN-attenuated loss in GS activity (FIG. 4) but did not significantly lower the amounts of protein carbonyls formed. The nitroxide spin label Tempol [2,2,6,6-tetramethyl piperidine-1-oxyl-4-ol] also was effective in preventing protein oxidation.[17]

Senescence-accelerated prone mice (SAMP8) showed all signs of aging with respect to memory and behavior.[44,45] Because aging is thought to be associated with free radical oxi-

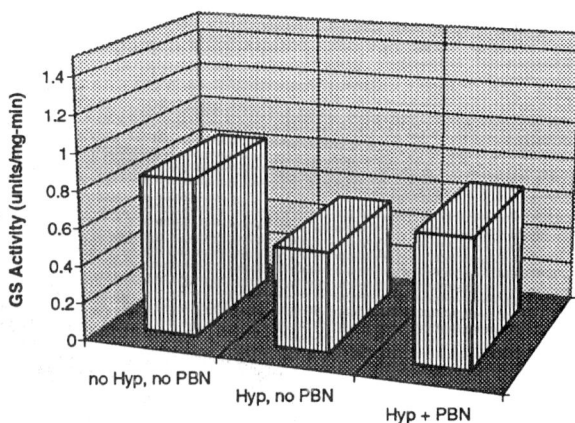

FIGURE 4. Effect of PBN treatment (20 mg/kg) on the GS activity of gerbils placed in 90–100% O_2 for 24 hours. The GS was isolated from the cytosol isolated from homogenized cortex of the gerbil brain. The p values for the hyperoxic group versus the control normoxic and the hyperoxic treated with PBN were less than 0.000005 and 0.0004, respectively.

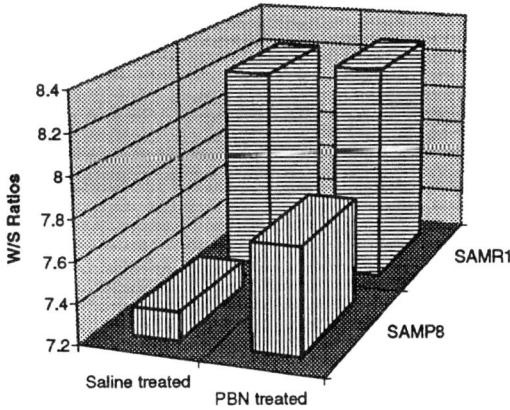

FIGURE 5. Graph represents average W/S ratios of MAL-6-labeled synaptosomes isolated from SAMP8 and SAMR1, injected i.p. with either saline or PBN (30 mg/kg), for 14 days. *p*-value for the SAMP8 saline versus PBN treated was found to be less than 0.001.

dative stress, we reasoned that PBN would modulate membrane protein oxidation in senescence-accelerated mice. The accelerated aging model was first developed by Takeda[46] as the senescence-accelerated mouse (SAM). The SAMP8 strain has a shorter life span and shows many typical signs of aging, whereas its genetic counterpart, the SAMR1 (senescence accelerated resistant), does not and lives longer.[47] PBN injected into the SAMP8 mice caused nearly a 50% increase in their life span.[48]

For our study, PBN was injected into the SAMP8 and SAMR1 mice daily, intraperitoneally, at 30 mg/kg body weight, for 14 days. Control SAMP8 and SAMR1 mice used in this study were given the same dose of saline injections for the same time period. Twenty-four hours after the final dose the animals were decapitated and examined for the various

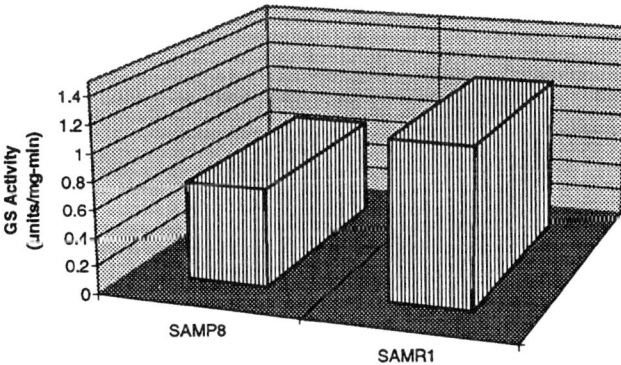

FIGURE 6. Graph represents GS activity of the cytosolic extract isolated from the cortices of the SAMP8 and and SAMR1 mice. *p* value < 0.05.

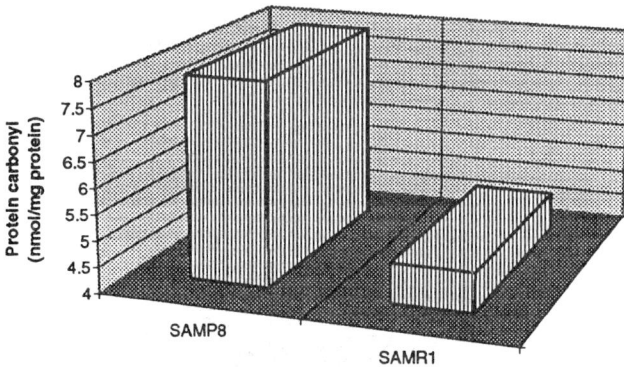

FIGURE 7. The graph shows the amounts of protein carbonyls present in the synaptosomes isolated from the cortices of the SAMP8 and SAMR1 mice.

studies, as described earlier. Consistent with oxidative damage to cortical synaptosomes with age, the W/S ratios of MAL-6 attached to membrane proteins from SAMP8 mice were significantly lower, protein carbonyls were higher, and GS activity was lower than those of the SAMR1 mice (FIGURES 5–7). By contrast, the SAMP8 mice injected with PBN showed higher W/S ratios, lower protein carbonyls, and higher GS activity when compared to those that were injected with only saline, and the values were also comparable to those obtained for SAMR1 (FIGURES 5–7), suggesting decreased oxidative stress. When these same strains of mice, treated with saline, were kept under hyperoxia for 24 hours, we observed a decrease in W/S ratios for both SAMP8 and SAMR1, indicating protein oxida-

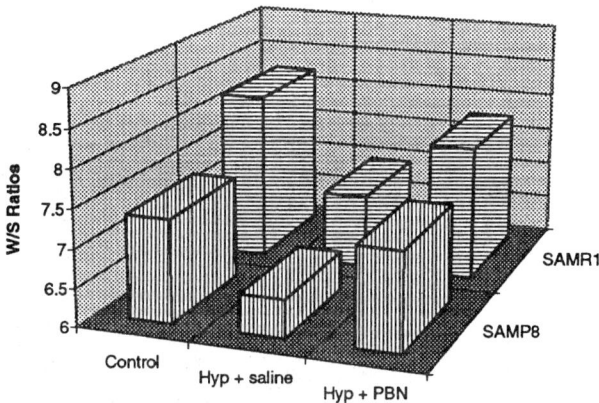

FIGURE 8. Effect of PBN treatment (30 mg/kg) on the W/S ratios of MAL-6-labeled synaptosomes isolated from SAMP8 and SAMR1 mice kept under 90–100% O_2 for 24 hours. Hyperoxic versus normoxic of SAMP8 and SAMR1 showed $p < 0.0001$ and $p < 0.001$, respectively. With respect to the hyperoxic but treated with PBN, the p values for SAMP8 and SAMR1 were less than 0.00001 and 0.002, respectively.

tion (FIG. 8). Again, treatment with PBN helped prevent lowering of the W/S ratios, to the extent that the mean values were comparable to those of the control mice (FIG. 8).

These protective results using PBN in the three *in vivo* models, IRI, hyperoxia, and accelerated senescence, suggest that damage during aging, consistent with the free radical theory of aging, is caused by free radical–mediated oxidative stress, and, hence, free radical scavengers may, if not prevent, then at least modulate or delay the onset of aging.

VITAMIN E AND Aβ (25–35)-INDUCED DAMAGE TO CORTICAL SYNAPTOSOMAL MEMBRANES

Aβ, a 39–43 amino acid–length peptide, cleaved from the transmembrane amyloid precursor protein (APP) and the major component of senile plaques (SP) in the brain of AD patients, is thought to be closely involved in the neurotoxicity of AD for a number of reasons. Genetic mutations on APP and presenilin genes, thought to be associated with APP processing, lead to AD; Down syndrome patients develop AD if they live long enough; and many of the pathological hallmarks of the AD brain are seen in APP overexpressing mice.[49] We and several other laboratories have demonstrated Aβ-associated free radical oxidative stress in neuronal systems.[7] To investigate structure and function of brain membrane systems exposed to Aβ, we have used cortical synaptosomes. Aβ (1–40) and Aβ (1–42) are the peptides found in the brain of AD patients. The region 25–35 of these peptides has been shown to be crucial for their toxicological properties.[50] Hence, for the purposes of understanding the molecular mechanics of the peptide action, we chose Aβ (25–35) for our studies.[51,52]

Aβ (25–35), obtained commercially from M. D. Enterprises, (Manhattan Beach, CA), was incubated with gerbil brain neocortical synaptosomal membranes at a final concentration of 1 mg/mL. A second aliquot was treated with Aβ (25–35) and vitamin E (final concentration 5 μM). Control samples were analyzed with each experiment. All the samples were incubated at 37°C for six hours, and all samples, except those for EPR studies, were frozen until analyzed. EPR spin labeling studies with MAL-6, protein carbonyls, cell sur-

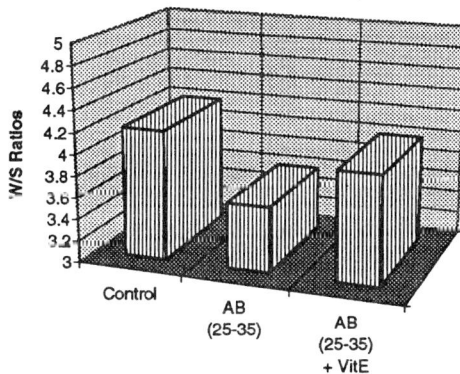

FIGURE 9. Effect of Aβ (25–35) with and without vitamin E on W/S ratio of MAL-6-labeled synaptosomes. Aβ (25–35) significantly reduces W/S ratio compared to control; $n = 3$, $p < 0.003$. Aβ (25–35) incubated with vitamin E restores the W/S ratio back to control values.

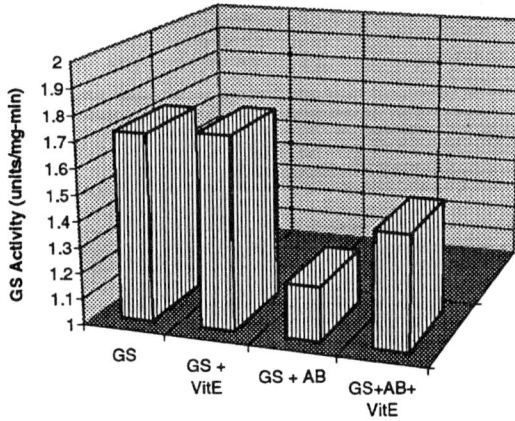

FIGURE 10. Effect of incubating GS with Aβ (25–35) in the presence and absence of vitamin E. Control values of GS activity, in the absence of any peptide, is chosen as 100%. Vitamin E alone does not alter GS activity. Aβ (25–35) significantly lowers GS activity ($n = 3$, $p < 0.005$). Vitamin E restores loss in activity caused by Aβ (25–35).

vival, and GS activity were measured according to methods mentioned earlier in this review.

The W/S ratios of MAL-6-labeled synaptosomal membranes incubated with Aβ (25–35) decreased from the control value, consistent with protein oxidation (FIG. 9). By contrast, samples having both Aβ (25–35) and vitamin E were protected against protein oxidation and showed W/S values similar to those of controls. GS activity dropped to 65% of the control on addition of Aβ (25–35) but was observed to be about 85% of control by addition of vitamin E (FIG. 10). Protein carbonyl levels, elevated by Aβ (25–35) treatment, were

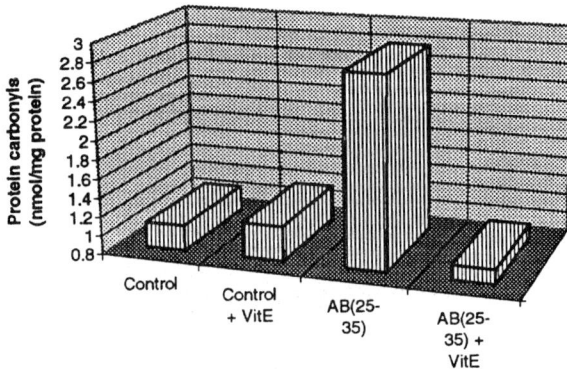

FIGURE 11. Protein carbonyl content of synaptosomes treated with Aβ (25–35) is significantly higher than untreated controls ($p < 0.01$; $n = 3$), suggesting that Aβ (25–35) caused oxidative modification of membrane proteins, resulting in a net increase in the carbonyl content. Samples treated with both the peptide and vitamin E were statistically identical to untreated controls.

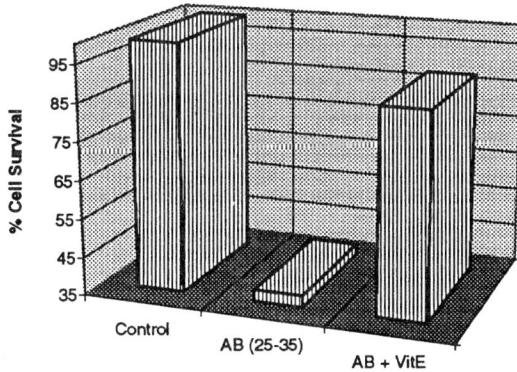

FIGURE 12. Loss in cell survival, as measured by the MTT reduction method, shows that Aβ (25–35) exposure resulted in the death of nearly 60% of the cells. Vitamin E was able to significantly protect from Aβ (25–35)-induced cell death and results in the survival of more than 90% of the cells.

comparable to control values in the presence of vitamin E (FIG. 11). In addition, toxicity to hippocampal neurons caused by Aβ (25–35) was partially prevented by vitamin E (FIG. 12).

Aβ causes lipid peroxidation,[29,53-56] and HNE, as one major lipid peroxidation product, is able to significantly modify the structure of brain membrane proteins.[35] Aβ forms HNE,[36] and both Aβ and HNE alter the function of ion-motive ATPases.[36,37] Free radical scavengers have been shown to prevent Aβ-induced formation of HNE in neuronal membranes.[36] Recently increased protein carbonyls, and decreased GS activity, in cell-free solution, following Aβ addition were demonstrated, and these changes were abrogated by PBN.[26] Taken together these vitamin E and PBN results are consistent with the notion that Aβ-associated free radicals and/or the oxidation products they produce are toxic to cells[7] and may be relevant to the known oxidative stress in AD brain.[57-59]

CONCLUSIONS

Oxidative stress in *in vivo* and *in vitro* developed animal models provide a good insight into the effects of various free radical insults that the human body has to withstand during the course of its life span. Since the free radical theory on aging was first formulated, there is a growing consensus that free radicals are largely accountable for the widespread damage seen in aging. Although the exact cause, sequence, and mechanisms underlying the damage have not yet been determined, progress has been made in developing and testing drugs and molecules that can attenuate this damage and increase the healthy life span of individuals. Tests conducted in animals have increased the awareness of people and led them to follow caloric restriction[60] and/or a diet rich in antioxidants, like vitamins and carotenoids. Studies with molecules like PBN have opened avenues for the use of nitrone-related molecules as potential therapeutic agents in the treatment of various oxidative stress disorders. The beneficial effects of such molecules have helped provide a better understanding of the action of free radical scavengers in mitigating cellular damage.

Aging is the single greatest risk factor for AD and is a significant factor in stroke. Our studies have provided insights into free radical damage and its modulation in *in vivo* and *in vitro* models of these age-related neurodegenerative disorders, but much more research needs to be performed, for example on the role of peroxynitrite-induced damage to brain membranes,[61] to completely understand the molecular basis of aging and age-related neurodegenerative disorders.

REFERENCES

1. HARMAN, D. 1992. Free radical theory of aging. Mutat. Res. **275:** 257–266.
2. RADI, R., J.S. BECKMAN, K.M. BUSH & B.A. FREEMAN. 1991. Peroxynitrite-induced membrane lipid peroxidation: The cytotoxic potential of superoxide and nitric oxide. Arch. Biochem. Biophys. **298:** 431–437.
3. SCHULTZ, J.B., R.T. MATTHEWS & M.F. BEAL. 1995. Role of nitric oxide in neurodegenerative diseases. Curr. Opin. Neurol. **8:** 480–486.
4. MIGUEL, J. & J.E. FLEMING. 1984. A two step hypothesis on the mechanism of *in vitro* cell aging: Cell differentiation followed by intrinsic mitochondrial mutagenesis. Exp. Gerontol. **19:** 31–36.
5. POULIN, J.E., C.COVER, M.R., GUSTAFSON & M.B. KAY. 1996. Vitamin E prevents oxidative modification of brain and lymphocyte band 3 during aging. Proc. Natl. Acad. Sci. USA **93:** 5600–5603.
6. SIES, H. 1985. *In* Oxidative stress. H. Sies, Ed.: 1–8. Academic Press. London.
7. BUTTERFIELD, D.A. 1997. β-Amyloid-associated free radical oxidative stress and neurotoxicity: Implications for Alzheimer's disease. Chem. Res. Toxicol. **10**(5): 495–506.
8. MECOCCI, P., U. MACGARNEY & M.F. BEAL. 1994. Oxidative damage to mitochondrial DNA is increased in Alzheimer's disease. Ann. Neurol. **36:** 747–757.
9. MATTSON, M.P., R.J. MARK, F. KATSUTOSHI & A.J. BRUCE. 1997. Disruption of brain cell ion homeostasis in Alzheimer's disease by oxyradicals, and signaling pathways that protect therefrom. Chem. Res. Toxicol. **10:** 507–517.
10. STADTMAN, E.R. 1992. Protein oxidation and aging. Science **257:** 1220–1224.
11. BUTTERFIELD, D.A. & E.R. STADTMAN. 1997. Protein oxidation processes in aging brain. Adv. Cell Aging Gerontol. **2:** 161–191.
12. COLE, P., N.C. HALL, J.M. CARNEY, O.J. PLANTE & D.A. BUTTERFIELD. 1996. Free radical oxidative stress and membrane alterations in ischemia-reperfusion injury. Recent Res. Dev. Neurochem. **1:** 99–109.
13. KITANI, K., A. AOBA & S. GOTO, EDS. 1996. Pharmacological Intervention in Aging and Age-associated Disorders. Ann. N.Y. Acad. Sci. **786:** 1–460.
14. JENNER, P. & C.W. OLANOW. 1996. Oxidative stress and the pathogenesis of Alzheimer's disease. Neurology **47:** 161–170.
15. STARKE-REED, P.E. & C.N. OLIVER. 1989. Protein oxidation and proteolysis during aging and oxidative stress. Arch. Biochem. Biophys. **275:** 559–567.
16. HENSLEY, K., B.J. HOWARD, J.M. CARNEY & D.A. BUTTERFIELD. 1995. Membrane protein alterations in rodent erythrocytes and synaptosomes due to aging and hyperoxia. Biochim. Biophys. Acta **1270:** 203–206.
17. HOWARD, B.J., S. YATIN, K. HENSLEY, K.L. ALLEN, J.P. KELLY, J. CARNEY & D.A. BUTTERFIELD. 1996. Prevention of hyperoxia-induced alterations in synaptosomal membrane-associated proteins by *N-tert*-butyl-α-phenylnitrone (PBN) and 4-hydroxy-2,2,6,6-tetramethylpiperidine-1-oxyl (Tempol). J. Neurochem. **67:** 2045–2050.
18. SOHAL, R.S., P.L. TOY & K.J. FARMER. 1987. Age-related changes in the redox status of the housefly, *Musca domestica*. Arch. Gerontol. Geriatr. **6:** 95–100.
19. SOHAL, R.S., A. MULLER, B. KOLETZKO & H. SIES. 1985. Effect of age and ambient temperature on n-pentane production in adult housefly, *Musca domestica*. Mech. Ageing Dev. **29:** 317–326.
20. OLIVER, C.N., B. AHN, E.J. MOERMAN, S. GOLDSTEIN & E.R. STADTMAN. 1987. Age-related changes in oxidized proteins. J. Biol. Chem. **262:** 5488–5491.

21. HENSLEY, K., N. HALL, W. SHAW, J.M. CARNEY & D.A. BUTTERFIELD. 1994. Electron paramagnetic resonance investigation of free radical induced alterations in neocortical synaptosomal membrane protein infrastructure. Free Radical Biol. Med. **17:** 321–331.

22. BELLARY, S.S., K.W. ANDERSON, W.A. ARDEN & D.A. BUTTERFIELD. 1995. Effect of lipopolysaccharide on the physical conformation of the erythrocyte cytoskeletal proteins. Life Sci. **56:** 91–98.

23. TRAD, C.H., W. JAMES, A. BHARDWAJ & D.A. BUTTERFIELD. 1995. Selective labeling of membrane protein sulfhydryl groups with methanethiosulfonate spin label. J. Biochem. Biophys. Methods **30:** 287–299.

24. BUTTERFIELD, D.A. 1982. Spin labeling in disease. *In* Biological Magnetic Resonance, Vol. 4. L.J., Berliner, Ed.: 1–78. Plenum Press. New York.

25. CARNEY, J.M., P.E. STARKE-REED, C.N. OLIVER, R.W. LANDRUM, M.S. CHENG & J.F. WU. 1991. Reversal of age-related increase in brain protein oxidation, decrease in enzyme activity, and loss in temporal and spatial memory by chronic administration of the spin trapping compound *N-tert*-butyl-α-phenylnitrone. Proc. Natl. Acad. Sci. USA **88:** 3633–3636.

26. AKSENOV, M.Y. M.V. AKSENOVA, J.M. CARNEY & D.A. BUTTERFIELD. 1997. Oxidative modification of glutamine synthetase by amyloid beta peptide. Free Radical Res. **27:** 267–281.

27. HARRIS, M.E., K. HENSLEY, D.A. BUTTERFIELD, R.E. LEEDLE & J.M. CARNEY. 1995. Direct evidence of oxidative injury by the Alzheimer's amyloid β-peptide in cultured hippocampal neurons. Exp. Neurol. **131:** 193–202.

28. SMITH, C.D., J.M. CARNEY, P.E. STARKE-REED, C.N. OLIVER, E.R. STADTMAN, R.A. FLOYD & W.R. MARKESBERRY. 1991. Excess brain protein oxidation and enzyme dysfunction in normal aging and in Alzheimer's disease. Proc. Natl. Acad. Sci. USA **88:** 10540–10543.

29. BUTTERFIELD, D.A., K. HENSLEY, M. HARRIS, M.P. MATTSON & J.M. CARNEY. 1994. β-Amyloid peptide free radical fragments initiate synaptosomal lipoperoxidation in a sequence-specific fashion: Implications to Alzheimer's disease. Biochem. Biophys. Res. Commun. **200:** 710–715.

30. ROWE, W.B., R.A. REMIZO, V.P. WELLNER & A. MEISTER. 1970. Glutamine synthetase. Methods Enzmol. **17:** 900.

31. MILLER, R.E., R. HADENBERG & H. GERSHAM. 1978. Regulation of glutamine synthetase in cultured 3T3-L1 cells by insulin, hydrocortisone, and dibutyryl cyclic AMP. Proc. Natl. Acad. Sci. USA **57:** 1418.

32. FLOYD, R.A. & J.M. CARNEY. 1991. Age influence on oxidative events during brain ischemia/reperfusion. Arch. Gerontol. Geriatr. **12:** 155–177.

33. CLOUGH-HELFMAN, C. & J.W. PHILLIS. 1991. The free radical trapping agent *N-tert*-butyl-α-phenylnitrone (PBN) attenuates cerebral ischemia injury in gerbils. Free Radical Res. Commun. **15:** 177–186.

34. SOUVIK, S., H. GOLDMAN, M. MOREHEAD, S. MURPHY & J.W. PHILLIS. 1994. α-Phenyl-*tert*-butyl-nitrone inhibits free radical release in brain concussion. Free Radical Biol. Med. **16**(6): 685–691.

35. SUBRAMANIAM, R., F. ROEDIGER, B. JORDAN, M.P. MATTSON, J.N. KELLER, G. WAEG & D.A. BUTTERFIELD. 1997. The lipid peroxidation product, 4-hydroxy-2-trans-nonenal, alters conformation of cortical synaptosomal membrane products. J. Neurochem. **69:** 1161–1169.

36. MARK, R.J., M.A. LOVELL, W.R. MARKESBERY, K. UCHIDA & M.P. MATTSON. 1997. Evidence that 4-hydroxynonenal mediates disruption of ion homeostasis and neuronal death by amyloid-β peptide. J. Neurochem. **68:** 255–264.

37. MARK, R.J., K. HENSLEY, D.A. BUTTERFIELD & M.P. MATTSON. 1995. Amyloid β-peptide impairs ion-motive ATPase activities: Evidence for a role in loss of neuronal Ca^{2+} homeostasis and cell death. J. Neurosci. **15:** 6239–6249.

38. ALVERADO, A., D.A. BUTTERFIELD, B.A. WATKINS, B.H. CHUNG & B. HENNIG. 1995. Lipid-induced alterations in membrane fluidity contribute to endothelial barrier dysfunction. Int. J. Biochem. Mol. Biol. **27:** 665–673.

39. BEHL, C., J.B. DAVIS, R. LESLEY & D. SCHUBERT. 1994. Hydrogen peroxide mediates amyloid β protein toxicity. Cell **77:** 817–827.

40. HALL, N.C., J.M. CARNEY, M.S. CHENG & D.A. BUTTERFIELD. 1995. Ischemia/reperfusion induced changes in membrane proteins and lipids of gerbil cortical synaptosomes. Neuroscience **64:** 81–89.

41. HALL, N.C., J.M. CARNEY, M.S. CHENG & D.A. BUTTERFIELD. 1995. Prevention of ischemia-reperfusion-induced alterations in synaptosomal membrane-associated proteins and lipids by *N-tert*-butyl-α-phenylnitrone and difluoromethylornithine. Neuroscience **69:** 591–600.

42. HALL, N.C., J.M. CARNEY, O.J. PLANTE, M. CHENG & D.A. BUTTERFIELD. 1997. Effect of 2-cyclohexene-1-one-induced glutathione depletion on ischemia/reperfusion-induced alterations in the physical state of brain synaptosomal membrane proteins and lipids. Neuroscience **77:** 283–290.

43. STADTMAN, E.R. & B.S. BERLETT. 1997. Reactive oxygen-mediated protein oxidation in aging and disease. Chem. Res. Toxicol. **10:** 485–494.

44. YAGI, H., T. KATOH, I. AGIGUTI & T. TAKEDA. 1988. Age-related deterioration of ability of acquisition in memory and learning in senescence-accelerated mouse: SAM-P/8 as an animal model of disturbances in recent memory. Brain Res. **474:** 86–93.

45. OHTA, A., T. HIRANO, H. YAGI & S. TANAKA. 1989. Behavioral characteristics of the SAM-P/8 strain in Sidman active avoidance task. Brain Res. **498:** 195–198.

46. TAKEDA, T., M. HOSOKAWA, S. TAKESHITA & M. IRINO. 1981. A new murine model of accelerated senescence. Mech. Ageing Dev. **17:** 183–194.

47. FLOOD, J. & J. MORLEY. 1992. Early onset of age-related impairment of aversive and appetite learning in SAM-P/8 mouse. J. Gerontol. **47:** 52–59.

48. EDAMATSU, R., A. MORI & L. PACKER. 1995. The spin trap tert-butyl-α-phenylnitrone prolongs the life span of the senescence mouse. Biochem. Biophys. Res. Commun. **211:** 847–849.

49. SELKOE, D.J. 1996. Amyloid β-protein and the genetics of Alzheimer's disease. J. Biol. Chem. **271:** 18295–18298.

50. PIKE, C.J., A.J. WALENCEWICZ-WASSERMAN, J. KOSMOSKI, D.H. CRIBBS, C.G. GLABE & C.W. COTMAN. 1995. Structure-activity analyses of β-amyloid peptides: Contributions of the β 25–35 region to aggregation and neurotoxicity. J. Neurochem. **64:** 253–265.

51. SUBRAMANIAM, R., T. KOPPAL, M. GREEN, S. YATIN, B. JORDAN & D.A. BUTTERFIELD. 1998. The free radical antioxidant vitamin E protects cortical synaptosomal membrane proteins from amyloid β-peptide (25-35) toxicity but not from hydroxynonenal toxicity: Relevance to the free radical hypothesis of Alzheimer's disease. Neurochem. Res. In press.

52. YATIN, S.M., M. AKSENOV & D.A. BUTTERFIELD. 1998. The antioxidant vitamin E modulates amyloid β-peptide–induced creatine kinase activity inhibition and increased protein oxidation: Implications for the free radical hypothesis of Alzheimer's disease. Neurochem. Res. In press.

53. KOPPAL, T., R. SUBRAMANIAM, J. DRAKE, M.R. PRASAD & D.A. BUTTERFIELD. 1998. Vitamin E protects against amyloid peptide (25-35)–induced changes in neocortical synaptosomal membrane lipid structure and composition. Brain Res. **786:** 270–273.

54. GRIDLEY, K.E., P.S. GREEN & J.W. SIMPKINS. 1997. Low concentrations of estradiol reduce beta-amyloid-induced toxicity, lipid peroxidation, and glucose utilization in human SK-N-SH neuroblastoma cells. Brain Res. **778:** 158–165.

55. DANIELS, W.M., S.J. VAN RENSBURY, J.M. VAN ZYL & J.J. TALJAARD. 1998. Melatonin prevents beta-amyloid-induced lipid peroxidation. J. Pineal Res. **24:** 78–82.

56. BRUCE-KELLER, A.J., J.G. BEGLEY, W. FU, D.A. BUTTERFIELD, D.E. BREDESEN, J.B. HUTCHINS, K. HENSLEY & M.P. MATTSON. 1998. Bcl-2 protects isolated plasma and mitochondrial membranes against lipid peroxidation induced by hydrogen peroxide and amyloid beta-peptide. J. Neurochem. **70:** 31–39.

57. SMITH, C.D., J.M. CARNEY, T. TATSUMO, E.R. STADTMAN, R.A. FLOYD & W.R. MARKESBERY. 1992. Protein oxidation in aging brain. Ann. N.Y. Acad. Sci. **663:** 110–119.

58. HENSLEY, K., N. HALL, R. SUBRAMANIAM, P. COLE, M. HARRIS, M. AKSENOV, M. AKSENOVA, P. GABBITA, J.F. WU, M.J. CARNEY, M. LOVELL, W.R. MARKESBERY & D.A. BUTTERFIELD. 1995. Brain regional correspondence between Alzheimer's diseaze histopathology and markers of protein oxidation. J. Neurochem. **65:** 2146–2156.

59. MARKESBERY, W.R. 1997. Oxidative stress hypothesis in Alzheimer's disease. Free Radical Biol. Med. **23**(1): 134–147.

60. GABBITA, S.P., K. HENSLEY, D.A. BUTTERFIELD & J.M. CARNEY. 1997. The effect of age and diet on mitochondrial respiration and lipid membrane status: An electron paramagnetic resonance investigation. Free Radical Biol. Med. **23:** 191–201.

61. KOPPAL, T., J. DRAKE, S. YATIN, B. JORDAN, S. VARADARAJAN, L. BETTENHAUSEN & D.A. BUTTERFIELD. 1998. Peroxynitrite-induced alterations in synaptosomal membrane proteins: Insight into oxidative stress in Alzheimer's disease. J. Neurochem. In press.

Roundtable Discussion

How Best to Ensure Daily Intake of Antioxidants (from the Diet and Supplements) That is Optimal for Life Span, Disease, and General Health

CHAIR: RUSSEL J. REITER[a]

PARTICIPANTS: THE ANTIOXIDANT SESSION SPEAKERS PLUS
IVOR DREOSTI, MOHSEN MEYDANI, AND JAMES JOSEPH

[a]Department of Cellular and Structural Biology,
The University of Texas Health Science Center,
7703 Floyd Curl Drive, San Antonio, Texas 78284-7762, USA

Free radical reactions go on continuously throughout the cells and tissues. These reactions interact with each other to variable degrees, depending on their location and the chemical reactivity of their free radical intermediates. Their adverse effects can be muted by antioxidants. Thus, many people attempt to eat diets "rich in antioxidants" and/or supplement them with one or more antioxidants. The list of antioxidant supplements keeps growing; lipoic acid, melatonin, and flavonoids are the latest additions.

Because free radical reactions play an essential role in life, it is important to consider the possibility that as the amounts and kinds of antioxidant supplements consumed increase, a point may be reached where the beneficial effects of inhibiting deleterious free radical reactions are outweighed by the adverse effects on essential reactions.

The question posed to the panel is complex. Little is known about the metabolism of antioxidants, for example, their rates of absorption, time course of the distribution of the compounds and their metabolites in the cells and tissues, and possible interactions between the compounds. Are there alternatives to the arduous task of investigating the effects of combinations of diets and supplemental antioxidants on the diseases and life spans of mice and rats, and, epidemiologically, in humans?

Denham Harman

RUSSEL J. REITER, Chair (*The University of Texas Health Science Center, San Antonio*): This discussion will consider diet or supplements that may be optimal for life span, disease, and general health. Actually, this is not the first time issues of this type have been discussed. In 1992 in Switzerland, a group of free radical biologists, some of whom I suspect are in this room today, made a proclamation in reference to what should be done about vitamin supplementation and how the information should be made available to the general public. At this point, we want some opinions as to obviously what antioxidants may be important for good health and what the expectations may be if you are taking antioxidants? What are the precautions one should consider? Are there disease states under which people perhaps should not take antioxidants? The issue, for example, of vitamin E stimulating the immune system is generally considered beneficial except perhaps in individuals with autoimmune disease. These are the types of issues, I think, that could be discussed, but this is something you as a group will determine. I don't think there are going to be any proclamations coming as a result of this, but it will give us some idea of how generally the group feels about the use of antioxidants by the lay public. I've spoken over the years with many free radical biologists and other scientists, and they seem rather enthusiastic about taking antioxidants for their own benefit. I don't know if everyone shares that feeling in this room.

PABLO ARILAR JR. (*Darwin, Australia*): I have had some training in the USA. I was convinced, with reluctance, by my wife to attend this conference. I know you have all the intentions of prolonging life, but what is life if it is not as meaningful or as useful as it should be in the long term? I'm not proposing euthanasia, mind you, but what is the point in prolonging life if one will not be much use to the community or to your wife, if you're a man. I'm confused with all these antioxidants, and the speakers say that we have to do more studies.

REITER: Thank you very much. These issues should be discussed. The speakers from the afternoon and the morning sessions are here, and they're going to give their opinions on such matters. There are several of them that have prepared statements. I'm going to start with Dr. Meydani.

MOHSEN MEYDANI (*Tufts University, Boston*): I have a slide (shown in the table below) that I want to show first. Because I am a nutritionist and also working in aging research, I want to present a slide showing the plasma levels of the antioxidant nutrients, vitamin C, vitamin E, and β-carotene, and their RDA levels.

These are nutrients with known antioxidant activities. The RDA of 8–10 mg set for vitamin E was based on the customary intake of the population in the USA. These RDAs ensure that we should get at least this amount in order to avoid deficiency and to be healthy. Most of the studies are now looking at the antioxidant function of these nutrients in bodily systems and testing their role in disease prevention and aging. One of the main goals of our Center (Jean Mayer USDA Human Nutrition Research Center on Aging at Tufts University) is to identify what the RDA of these nutrients should be for people over the age of 50 years. We would like to define the requirements for these and other nutrients for people over the age of 50, 65, or even older. We are considering the antioxidant function of these nutrients for the prevention of diseases or improving body functions in the elderly. Dependent on what system and which function, the level of antioxidant nutrients needed to prevent dysfunction and disease and to provide optimal health appears to be different. We are certain that the levels of these nutrients that are needed to be in the diet of a population in order to prevent diseases or certain conditions is much higher than what we

Antioxidant Vitamins

	Plasma Level	RDA F	RDA M	Concentration in Supplements
Vitamin C	0.5–1.0 mg/dL (30–60 μmol/L)	50 mg	60 mg	50–1,000 mg
Vitamin E	0.6–1.8 mg/dL (12–35 μmol/L)	8 mg	10 mg	30–400 IU
β-Carotene[a]	0.2–100 μg/dL (0.01–1.91 μmol/L)	4.8 mg	6 mg	1.5–15 mg

[a]Calculated as equivalent of vitamin A.

have in the RDA. It is important to keep in mind that the purpose of these studies is not to increase longevity but to improve the quality of life through diet and to avoid such situations as being bed bound in hospitals or becoming institutionalized for a long period of time. So the purpose is to improve the quality and not the quantity. Based on what we currently know about these nutrients and our own studies, the RDA level of these antioxidants does not provide enough to prevent the risk of disease in later age. I can talk very comfortably about vitamin E, but other colleagues have more authority to speak about vitamin C and β-carotene. I can talk about vitamin E, because I have quite extensive experience with this vitamin. Based on studies to improve immune function conducted at our center and other studies testing for its effect on the cardiovascular system, it appears that 200 IU of vitamin E is needed in order to improve the immune system and to prevent cardiovascular disease. So this level is quite a bit different from what we have for the RDA for this vitamin, which is not set based on certain functions. We did not hear very much about vitamin C in this conference, but I can recall the recent work of Mark Levine who conducted a very nice and well-controlled study to define the requirement for vitamin C. Basically, his study showed that 60 milligrams of vitamin C is not enough to maintain a healthy life. His recommendation based on the biochemical reaction that he used came to 200 milligrams of vitamin C per day. This is the only study that was well conducted, but probably more controlled studies are needed to confirm such a recommendation.

REITER: Thank you. I'm going to ask Dr. Norm Krinsky to comment on β-carotene with regard to the recommended dose.

NORMAN KRINSKY (*Tufts University, Boston*): There is still no official RDA or RDI for individual or total carotenoids. Several investigators have calculated the average daily intake of total or specific carotenoids and the minimum serum value associated with reduced risk of chronic diseases. Based on these values, these investigators have come forward with recommendations with respect to daily intake of carotenoids. The only known function of dietary carotenoids is limited to those pigments that serve as pro-vitamin A carotenoids and that can be metabolized to retinol, retinal, and retinoic acid. There is also evidence that individuals placed on a carotenoid-deficient diet are more susceptible to oxidative stress and that dietary supplementation with β-carotene can increase the antioxidant capacity of various groups. In children suffering from cystic fibrosis and supplemented with vitamin E, the addition of β-carotene to their diet further increases their antioxidant capacity. However, we still need more studies to see how much supplementary carotenoid is necessary to alter the antioxidant capacity and whether this effect can be observed in well-nourished individuals.

In my presentation earlier in the meeting, I emphasized the important role of metabolites of β-carotene, formed either enzymatically or via oxidative reactions, to the overall effects attributed to carotenoids. However, I believe that the area of concern with respect to dietary carotenoids is whether an adequate supply of lutein and zeaxanthin is available in our diet. These two xanthophylls are the only carotenoids found in the human retina and are located in the macula region. There is excellent epidemiological evidence showing that consumption of these carotenoids is associated with a reduced risk of developing age-related macular degeneration (AMD), a relentless, irreversible form of blindness that affects the elderly, and if we live long enough, our risk of developing this disease grows very large. Whether AMD is actually related to a deficiency of lutein and zeaxanthin has not yet been proven, but it would seem wise to insure that we ingest a diet rich in these two carotenoids. Lutein and zeaxanthin are found primarily in dark green leafy vegetables, and such food items should definitely be part of the Five-a-Day program recommended by the American Cancer Society and the USDA. One wonders why the National Eye Institute, when they initiated a program of antioxidant supplementation to prevent eye diseases, did not include these two carotenoids rather than β-carotene, which is not even present in the human retina. Furthermore, the approval of fat substitutes, which can lead to a depletion of these pigments, may increase the risk of the elderly members of our population of developing AMD.

REITER: I believe Dr. Ivor Dreosti has some comments regarding this matter.

IVOR DREOSTI (*CSIRO Division of Human Nutrition, Adelaide, South Australia*): I take a slightly different view. I certainly would agree entirely with what Dr. Meydani says about quality of life, to improve the quality of life as much as to extend life span. For this reason if one can avoid diseases like macular degeneration, ocular cataracts, or some of the more serious life-threatening diseases for 10 to 15 years, that in itself is a great benefit, even if the life span is not increased. As far as actual supplementation goes, I would like to take a more cautious view. If we look at disease risk reduction, the reduction that is obtained by people who eat a high intake of fruits and vegetables is of the same order as the risk reduction for these various diseases that you get by taking very high levels of individual supplements. So, as I said yesterday, I really do believe that we have to have an integrated approach and if we have that, the need for high levels of individual supplements is probably very much less. Also, if you have interaction on an interplay, the likelihood of doing harm by having too high a level of any one supplement, I think, is very much reduced. So I'm still in favor of supplementing at a lower level but with a wider range of them, because I believe that, at this stage, we don't have all the information about all the appropriate antioxidants. I'd still like to see the approach through dietary means. However, if we are going to consider a supplement approach, then we should have a wider range, including the carotenoids Dr. Krinski was talking about, the luteins and the xeaxanthins that appear to be concentrated in the macular, which, at the risk of making a teleological error, suggests that the macula requires those particular two carotenoids. I believe if we are going to supplement, we should try to supplement at lower levels, but with a wider range of antioxidants. As we get more information, we can bring other antioxidants into play.

REITER: Dr. Stephan Christen has a comment.

STEPHEN CHRISTEN (*University of California, Berkeley*): I'd like to comment on supplementation, too. I don't want to make any recommendations, because I'm not a medical doctor, so all I can talk about is the scientific process. I think that's probably where the problem and the confusion is, because, of course, if you do science, you have to know what's causing what. That is why we normally use a reductionist approach. Based on epidemiology, we then study alpha-tocopherol, β-carotene, or whatever and use them in intervention studies at doses higher than those normally present in the diet. What you're doing there is more like a pharmacological intervention, not preventive supplementation, the topic of this discussion. They are two totally different things. Pharmacological intervention is a pragmatical approach. I totally agree that if you can decrease heart disease by supplementing people with 400 IU of alpha-tocopherol, then you should do that. I mean, you don't really need to know more than that. That's a pragmatical approach, but that does not directly relate to what we should discuss in terms of a general health recommendation. I think we need to slightly increase the RDA; that seems to be sensible to me. But perhaps more important is that we should tell people to have a sensible diet! I think that's of primary importance. There are no magic bullets. Also, high doses of supplements may be harmful. This has not been fully addressed, in my opinion. It's been shown, for example, that under certain conditions alpha-tocopherol supplementation can increase cancer in animals. I believe that we need to do some more work before we can and should tell people to eat large amounts of vitamins in order to be healthy.

ANTHONY W. LINNANE (*Epworth Hospital, Melbourne, Victoria, Australia*): It seems to me that there are two separate questions being discussed in relation to vitamin C. One is the dosage required to prevent the development of scurvy and the need for the maintenance of collagen synthesis. Second, there is the question of vitamin C taken in much larger doses as a therapeutic intervention for the prevention of oxidating damage. However, there is yet a third point that I would like to emphasize, and that is the need for recommending a potential antioxidant therapy for particular target tissues. I think this is the biggest gap in our knowledge, and it is a question that has not been properly addressed. I would like to propose the possibility that different tissues have a need for different antioxidant compounds and, indeed, different animals have different needs.

A simple demonstration of the way in which species manage different compounds is provided by the observation of feeding sheep and cattle from the same feedlot. On slaughter the sheep fat is white, whereas the beef carcass fat is strongly colored yellow. The yellow color is due to β-carotene, which is handled quite differently by the two animal species.

If an experimenter wishes to extrapolate experimental results obtained from test tube experiments to whole animals, such extrapolations are fundamentally unsound. It is the responsibility of the experimenter to prove in whole tissue or animal experiments the physiological validity of the test tube observation. I suggest that in the future, it will emerge that various tissues require different antioxidants in order to obtain a therapeutic benefit.

KRINSKY: Responding to Dr. Linnane, I agree with Dr. Meydani that we should not limit our thinking by recomming levels of compounds that will prevent scurvy but should be thinking about population needs to achieve some level of optimum health. For this reason, the RDA values look at deficiency diseases and are not concerned with optimum health. The problem, however, is to define optimum health, and for what populations.

JAMES JOSEPH (*Tufts University, Boston*): The target tissue in which we carry out our work is the brain. If the brain shows deterioration as a function of age to the point that we lose our cognitive and motor function, then the health care costs to try to manage or restore such function are enormous. At this point the health care community is not very efficient in trying to "fix" either the behaviors or the neuronal degeneration that leads to the behavioral dysfunction. Our research suggests that perhaps the best course of action may be to prevent age-related changes in neural function through antioxidant diets and foods. The components of the foods may not be as effective in reducing oxidative stress on brain function as their interaction in the food. In other words, "the whole may be more than the sum of the parts." Thus, if we can obtain our antioxidants from foods it may be more effective than taking supplements. In the case of the aging brain, and most of you know that the brain does not age as a single unit, we are seeing that areas that control motor function, such as the striatum or cognitive function, and the hippocampus or frontal cortex, show significant change with age. In cell models that we have described in the talk and the chapter, we have shown that such antioxidants as vitamin E and the flavonoid quercetin have had significant effects on reducing the effects of oxidative stress, such as calcium disregulation and cell death in these models. In the whole animal models that used radiation or oxygen exposure to induce oxidative stress, we have been able to use food extracts, such as those from blueberries or strawberries to significantly reduce the effects of the oxidative stress. In fact, there is some indication that some of these diets given over 9 months can actually reduce the early effects of aging on the brain. So, I would like you to think in terms of the "gestalt" or the whole quality of the foods, inasmuch as there are substances that may contribute to the antioxidant effects of these foods that we have yet to explore. The important factors are the target tissue or cell and the antioxidant activity of the food.

PETER EVANS (*University of Glasgow*): Speaking as one with both a toxicological and nutritional interest presently in public health, there are a couple of points I believe are pertinent. The first is a sense of urgency. As we heard from Dr. Kitani, the demographic changes with the growth of the elderly section of the population are happening now and will occur more rapidly in the next few decades. The whole bureaucracy of deciding the nutritional RDA, involving national, European Union, and the WHO, are time consuming, and certain changes will have to be implemented before all the fine detail is worked out.

The other point arises from an early experience I had when I moved to Glasgow and was approached by a young girl asking if I had a cigarette to spare! Not only is there a large proportion of the population that smokes, but this is coupled with a low dietary intake of fresh fruits and vegetables. Unfortunately, in many areas, for adverse social and economic reasons, availability of fresh fruits and vegetables is difficult and costs can be high. Thus what is needed from the public health perspective, following the idea of suppling the polynutrients in food, are practical means to ensure that the required nutritious foods are made more readily available for actual consumption by the vulnerable people in the population.

FRANKLIN L. ROSENFELD (*Baker Medical Research Institute, Melbourne, Australia*): I'd like to draw the panel out on the question of relative deficiency. We've talked about deficiencies of antioxidants that would be a deficiency in the face of the normal requirement, but I would submit that under conditions of stress or disease the requirement may be much greater. Thus there may be a possibility of using antioxidants as a pretreatment or "preconditioning" before certain stress states, such as cardiac surgery. We have had some experience of the sort of changes in the myocardium that can take place in the heart during cardiac surgery and other stressful conditions. The requirements for nutrients and antioxidants may be quite high. So we can then perhaps focus on substances we can supplement people or patients with *before* we subject them to a deliberate planned stress. What would they be?

LARS ERNSTER (*University of Stockholm*): I would like to raise a general question in connection with this RDA. I think we have to take into consideration combination effects as well. It has been emphasized by several speakers during this meeting that you cannot take vitamin E without vitamin C, and you cannot take vitamin E without having access to a reductant, because the radicals formed from vitamins C and E in the course of their antioxidant function may act as prooxidants unless these vitamins are regenerated by an endogenous reductant, such as glutathione, lipoic acid, or ubiquinol. This is an important point, about which there is general agreement. However, one should not believe that these vitamins are harmless under all conditions. You can take huge amounts of vitamin C, but you have to think about that if you also fortify your diet with iron, because vitamin C may become a potent prooxidant. This may be balanced by vitamin E, but, again, the latter may become a prooxidant unless it is continuously regenerated. A dual effect of vitamin C, that is, acting as an anti- and prooxidant, has been demonstrated both *in vitro* and *in vivo*, in the latter case both in brain ischemia–reperfusion experiments with rats and in connection with open-heart surgery in humans. A second point I want to make is that we should also be aware of the fact that when we speak about beneficial effects of antioxidants in connection with degenerative diseases and aging we are most often—in fact almost invariably— referring to prevention or delay of progression rather than therapy. I think this is an important point to keep in mind when talking not only to the public but also to epidemiologists and to clinicians, in general.

MEYDANI: In response to Dr. Norman Krinsky's comment, the slide that I have shown lists β-carotene as pro-vitamin A, and I agree that we do not have an RDA for β-carotene. With regard to an earlier comment, I do not think that we can generalize his idea for testing the role of the antioxidant in those tissues where they concentrate. Maybe for lutein, which concentrates in macula, one can test its function in that specific tissue, but we cannot have the same approach for other nutrients. For example, vitamin E accumulates in adipose tissue, where its function is minimal. But testing a functional system in the body and examining how it is affected by a nutrient or aging would be quite useful. For example, immune function declines with age and makes the elderly become more prone to infectious disease. Can we improve that by diet? Can increases in fruit and vegetable intake or antioxidants affect this system? Blueberries may have a strong antioxidant capacity, but do they have an effect on a bodily function? We know that vitamin E supplementation improves the immune system, but how are we going to get that amount of vitamin E. Cooking oil contains a good amount of this vitamin, but in order to get 200 IU of vitamin E from food, we would have to consume a cup of vegetable oil, which we know is not advisable, or eat lots of almonds, which is not practical for the elderly. Further, the function of the digestive system also changes with aging, which may have an impact on nutrient absorption. At this time, we do not know what the requirements of the elderly for many nutrients are nor how we can achieve those requirements in order to increase the quality of life. I believe that we have to investigate such nutrients as vitamin E and vitamin C for bodily functions systematically. That is a big task and we have a long way to go.

HANS NOHL (*Veterinary University of Vienna, Vienna, Austria*): I would like to make some general comments and remarks on the topic of this roundtable discussion. The idea that elevation of antioxidants will affect life span and age-related diseases is based on the assumption that oxygen radicals are causatively responsible for these biological changes. I do not want to discuss this point again, as we all know there are still controversial standpoints concerning the rank of oxygen radicals, especially with regard to the complicated process of aging. Although an increasing number of age-related clinical diseases have been recognized to be linked to an imbalance between reactive oxygen species formation and antioxidant defense activities, supplementation by antioxidants cannot always be expected to be useful. For instance, copper and iron-storage diseases give rise to oxidative stress; however, administration of ascorbic acid as an antioxidant in this case will stimulate pathogenesis of these diseases via recycling of transition metals involved in the Fenton reaction. Another aspect are rate constants, which determine the probability of the elimination of dangerous radicals through antioxidants administered. Due to the high reactivity of radicals formed in living systems, a beneficial effect can only be expected if the antioxidants applied have higher reactivities with radicals than do biomolecules that should be protected. The successful detoxification of reactive oxygen species by antioxidants often requires relatively high concentrations to protect biomolecules. The probability of a collision between the anti- and prooxidant, which is a prerequisite of the radical-scavenging efficiency, requires the accessibility of antioxidants to the site(s) where radicals exist. High tissue levels of antioxidants may, however, exert undesired side effects, for instance, from some antioxidant-derived reaction products. This was clearly demonstrated by the formation of powerful prooxidants resulting from the antioxidant activity of ubiquinol. Supplementation of deficient antioxidants appears to be reasonable in order to reestablish the biological equilibrium without the risk of antioxidant-related side effects. This is also valid for a number of age-related health risks, such as carcinogenesis and cardiovascular diseases, as well as for metabolic disorders associated with antioxidant deficiencies. Normalization of the physiological antioxidant status in these cases may therefore prolong life expectancy. Due to the many uncertainties described, retardation of the aging process through antioxidants remains to be demonstrated.

KEVIN D. CROFT (*University of Western Australia, Perth*): On the basis of our current scientific knowledge, it is not yet possible to answer the question of what the optimal level of dietary antioxidants is to prevent disease. There is little data on absorption and bioavailability of dietary flavonoids and even less information on the ability of these compounds to inhibit oxidative damage *in vivo*. To evaluate the usefulness of various antioxidants, we need to conduct human intervention studies and assess effects on oxidative damage using the best available "biomarkers." It is only from such studies that the complex interactions of various dietary components on oxidative damage can be evaluated. My suggestion would be that before we can consider any recommendation, apart from eating a wide range of fruits and vegetables, that we need to do those sort of studies in order to characterize the absorption and antioxidant effects of components, such as flavonoids in the diet.

CARLETON J.C. HSIA (*SynZyme Technologies, Irvine, California*): The activation or priming of inflammatory white blood cells (*e.g.*, from smoking), in part, increases the risk of acute stroke and heart attack. To reduce the incidence of stroke and heart attack, I recommend supplemental dietary intake of such antioxidant vitamins as vitamin C and vitamin E. In addition, more potent oral antioxidant agents are being developed. These new antioxidant drugs may prolong and improve the quality of life in old age.

JEAN-PAUL CURTAY (*Paris*): Obviously, we're not going to save the world in the next five minutes, but I heard that the RDA is going to be reevaluated next year for the eleventh version. That is going to be revolutionary, because for the first time, it's not going to be based on the prevention of scurvy or beri-beri or pellagra but on the optimizing of cognitive function and prevention of chronic disease. Of course, everybody agrees now that we need to eat more fruits and vegetables. No pill will replace the 500 to 1000 molecules in each of those ingredients, but take the example of vitamin E. All the foods rich in vitamin E are major ones to prevent oxidation of polyunsaturated fatty acids. But the vitamin E in those foods is used to prevent rancidity of the polyunsaturated fatty acids of the food, so it's absolutely technically impossible just to get the minimum requirement, to protect your own fatty acids. You get ataxia if you're young, and you lack absorption of fat-soluble vitamins. I don't think we can escape the recommendation of supplementation when you add the fact that the aged person has a reduced absorption. Recently, it was shown in France (they created a center on nutritional research on aging) that the absorption of vitamin E in the aged is half that of a 25-year-old. That's completely new data. So how can you do it?

CARL DRANSFIELD (*Kindee, Australia*): I would point out that the population is one step ahead of you. They're already taking the supplements. Sixty-one percent of Australians and probably a larger proportion of Americans are already doing it. We're talking while the horse has bolted; the stable is already empty. In Australia, you say to me, "food will provide the nutrition." Of the top 10 materials sold as foods by supermarkets in Australia (and this is from their own organization publication), 6 of the top 10 were various sizes of Coca-Cola. The only two foods in the top 10 were canned beet root and canned beans. We're kidding ourselves unless we move now to actually lay down at least an average— not the optimum, but the best for the most across the range of known safe nutrients, for example, C, E, betacarotene, zinc, selenium, and folate. Unless we do that, the population is going to finish it themselves. I mean they're doing it. While you're experimenting with gerbils, they're experimenting with their children. We have to look at it because at the moment governments are moving to stop us. The latest regulations here in my country don't allow us to buy selenium except by prescription, and there is almost no doctor in this country that knows anything about selenium, so how do you get your hands on it? And yet, our soil in Australia is depleted of selenium. You need to look at this aspect of it, because if you don't, and don't do it soon, you will lift your heads, and the stable door will be locked because the only nutrients available will be prescribed by GPs who do six hours of nutrition in their entire career and haven't the slightest idea what you're talking about.

ROBERT J. BRADBURY (*Seattle*): As one of the three people I think at this conference who is actually from industry, I'd like to comment that I don't see how you can expect to get funding to do the additional studies, additional research, additional expansion of understanding these processes unless you, as academic researchers, make a commitment and put yourselves on the line, saying what you intend to provide to the public. There was a conference two weeks ago in Seattle on aging and age-related diseases, and at least three of the presidents of the industry companies said we are here to extend your life span, which, if you understand gerontology, is the natural consequence of extending the health span. So I'd like to know how many of you would go on record saying we are here to extend the human life span. With regard to CoQ_{10} (particularly with the doctor who is recommending when you come in with heart disease), you're recommended to decrease your meat intake, and then they give you lovastatin and pravastatin, which decreases your CoQ_{10}. So, in these conditions, what is an optimal array, a good CoQ_{10} supplementation? You can get it from 3 mg to 100 mg in the health food stores right now. The third thing that I think hasn't been addressed at this conference at all is gene regulation? The hematochromatosis gene was discovered very recently, and you could expect this to be a key critical factor in how much iron you supplement with. Richard Cutler at Genox has found that there are nonresponders with regard to betacarotene supplementation in the people he examines for betacarotene levels in the blood. There are going to be differences at the genetic level. What do you do about educating the population and intervening specifically with individuals with these types of differences in metabolism?

DR. ANDERSON: Just a brief comment on this. We have now for years been asking hospital laboratories to introduce a simple assay for determining coenzyme Q in the blood. If a patient doesn't know whether he should take 50 or 100 mg a day or 3 mg a day, he should go to the lab and find out what the maximal increase is in the blood level.

JOSEPH: Many years ago it was said that you could not do anything for anyone with a "Charley horse between the ears." In other words, if someone can look at all the publicity concerning, for example, smoking and eating a high fat diet, and still continue to do these things, we could sit up here all day and make recommendations for the RDAs for various antioxidants and vitamins and it would be meaningless. The taking of vitamins and antioxidants must be combined with a lifestyle that will maximize their effectiveness. We have the responsibility to control what we eat and to exercise. We can say it, but we cannot do it. We can recommend, but if people insist on living a lifestyle that will minimize the action of these agents and asking us to fix it when it breaks, then they have to remember that it is much harder to fix than it is to keep it from happening in the first place.

MARIA A. FIATARONE SINGH (*Tufts University, Boston*): If you look at the literature, there's an overwhelming abundance of epidemiologic evidence that good diets, high in fruits and vegetables, are protective. Yet, as scientists, we have not taken that information and done a single randomized controlled trial of a high fruit and vegetable diet for cancer protection. Instead, we have tried to reduce it to one or two isolated nutrients and then to put a lot of work into trying to see if those things are helpful. Typically, they have not been that helpful. To me, it's very much like exercise. It's as if we said, "Well, exercise works," and then somebody said, "Well, I think if I just injected catecholamines, that might work just as well." If we went down that pathway it would be equally futile. There are two things to me that seem to be driving this search for a perfect antioxidant, and one of them seems to be a belief that behavior cannot be changed. I think that, although behavior is difficult to change, it's not impossible, and if we invested millions and millions of dollars into how to change people's diets behaviorally, that might be of more benefit than developing the perfect pill. The second driving force is that a lot of this research is supported by the pharmaceutical and nutritional industries, who have a very big invested interest in developing a pill form of a diet, just as in the exercise field, growth hormone is being reported as a way to replace exercise. Liposuction might also be thought of as a high tech way to replace exercise. Those are analogies, and they are very very prominent, and I would just offer that I think behavior change is not futile. It's something that we need to work harder at as a scientific community.

KENICHI KITANI (*National Institute for Longevity Sciences, Obu, Japan*): As far as I can see, this discussion is getting more and more philosophical, not practical, and I have a suggestion before you close this session. Would you ask each of these speakers, beginning with yourself, whether or not he or she is taking any pills or other special nutrients for the specific purpose of extending his or her functional life span? It doesn't matter what kind of pill or nutrient it is. I'm not taking any supplementary medicines and, also, I am not inclined to do anything. To my knowledge, personally, I don't know any Japanese gerontologist who is enthusiastic about taking any kind of supplement to prolong life span.

MEYDANI: I agree with the comment that changing the behavior of a population is very hard. In the last ten years, between the two large studies in the United States (NHANES II and III), and despite all the publicity made by different health organizations for the reduction of fat intake, preliminary data show that intake of calories from fat sources has dropped by only one percent. Thus, it is hard to change public dietary habits, but it is possible. It requires lots of work and publicity from health organizations to make sure that our message gets out.

JOSEPH: I will volunteer. Look at the Seventh-Day Adventists in the United States. Look at their lifestyle. They don't drink, they don't smoke, and they are vegetarians. Look at the vegetarian diet. I have never read anything negative about it.

REITER: I really hate to bring this to a close, but there are liquid antioxidants waiting for us. Before we adjourn, I certainly want to thank the panel.

Comment

KEVIN D. CROFT
*Department of Medicine, University of Western Australia,
Perth, Western Australia 6001*

On the basis of our current scientific knowledge, it is not yet possible to answer the important question of what the optimal level of dietary antioxidants is to prevent disease. There is epidemiological evidence that diets rich in fruits and vegetables may help to protect us from cardiovascular disease and some forms of cancer. This may be due to the natural antioxidants they contain or to some other constituent. There is good evidence that oxidative damage to lipids, proteins, or DNA by reactive oxygen and nitrogen species in the body is linked to several disease processes. The major dietary antioxidants are as follows: (1) Vitamin E is an essential antioxidant in humans involved in protecting lipids against oxidation. There is no strong evidence that supplementation above current recommended intake is beneficial in preventing, for example, atherosclerosis. (2) Vitamin C has multiple metabolic roles and is a good water soluble antioxidant; however, it can exert prooxidant actions by interaction with iron or copper ions. (3) There is epidemiological evidence that high levels of carotenoids are associated with diminished risk of cancer or heart disease, particularly in smokers. However, the carotenoids are not well-established antioxidants and may exert effects by other actions. (4) The flavonoids are plant phenolic compounds that can protect against lipid peroxidation. Fruits, vegetables, tea, red wine, and soybean products are rich in isoflavonoids, which are phytoestrogens as well as antioxidants. Dietary intake may vary from a few milligrams to hundreds of milligrams per day. There is little data on absorption and bioavailability and even less information on the ability of these compounds to inhibit oxidation *in vivo*.

To evaluate the usefulness of various antioxidants, we need to first examine their effects as antioxidants *in vitro,* that is, their ability to protect the major target molecules (lipids, proteins, and DNA) against damage by reactive oxygen and nitrogen species. Such studies help to determine if compounds have a direct antioxidant effect, they are relatively easy to carry out, and many methods are available. On the other hand, to evaluate the effect of antioxidants on "steady state" oxidative damage *in vivo* is more difficult. The application of the latest "biomarkers" of oxidative damage could help to answer questions about the effects of dietary antioxidants in humans. There are several candidate biomarkers worthy of consideration. (1) Isoprostanes (free radical oxidation products of membrane arachidonic acid) can be measured in urine and plasma and may be a measure of whole body lipid peroxidation. (2) 8-Hydroxy-deoxyguanosine can be measured in urine as a biomarker of DNA damage and repair. (3) Specific amino acid products formed by attack of reactive nitrogen species on proteins, for example, nitrotyrosine, are excreted in urine and may be a useful index of protein damage; however, the possible confounding effects of oxidized protein in the diet must be considered.

With the continued development of methods in this area, we are now in a position to carefully examine and evaluate the effects of dietary and other antioxidants. The speculation and excitement about dietary antioxidants and some indication of their optimal levels needs to be put onto a firm scientific basis.

Comment

VALERIAN E. KAGAN
University of Pittsburgh, Pittsburgh, Pennsylvania 15238

Mother Nature created a sophisticated antioxidant system to protect us against oxidative stress and oxidative damage. This system is highly regulated and balanced. We can't understand the functioning of the system if we focus on one particular antioxidant (*e.g.*, vitamin E) without considering antioxidant interactions. During the last decade, we have learned that antioxidants are not isolated from each other but interact and "talk to each other." Vitamin C recycles vitamin E directly from its phenoxyl radical, and coenzyme Q recycles vitamin E through electron transport-dependent enzymatic reactions. Dihydrolipoate (and possibly some other thiols) recycle vitamin C. Packer's lab showed that dihydrolipoic acid is, in turn, regenerated from lipoic acid by cytosolic and mitochondrial NADH- and NADPH-dependent enzymatic reactions. In other words, there is an antioxidant network whose functioning is maintained by major metabolic pathways. An important practical implication is that antioxidant protection can be enhanced by refined tuning of the antioxidant recycling redox cascades. How do the antioxidant recycling systems change with age? We do not know that. This may be an exciting and promising field of future research.

An essential implication of this is that any attempt to bolster antioxidant protection through use of synthetic antioxidants (*e.g.*, probucol, BHT, and ebselen) or natural phytochemicals with radical scavenging properties (*e.g.*, polyphenols and flavonoids) will inevitably interfere with the endogenous antioxidant system. What if probucol- or polyphenol-derived phenoxyl radicals will be recycled by ascorbate and coenzyme Q instead of vitamin E? Will this switch of recycling mechanisms from usual endogenous pathways to new targets (*e.g.*, synthetic antioxidants) be beneficial or detrimental for the overall antioxidant protection? We do not have an answer to this question.

Another important part of the discussion is the metabolism of synthetic antioxidant molecules. Phenolic molecules may be very effective scavengers of free radicals. Yet, their effects in cells may be very harmful. One of the reasons is that these molecules can act as "biochemical parasites": they can be metabolically activated to free radical intermediates, phenoxyl radicals, that will nonproductively use the antioxidant recycling mechanisms. For example, phenolic compounds are good substrates for peroxidases and form phenoxyl radicals as one-electron oxidation intermediates. The phenoxyl radicals thus produced will be consuming ascorbate, NADH, or NADPH to get reduced back to the parent phenolic compound, which will again become a substrate of a peroxidase-catalyzed reaction. Functioning of this vicious circle will be exhausting, reducing equivalents of cells and body fluids. Moreover, phenoxyl radicals generated from some phenolic compounds will directly oxidize thiols (GSH, cysteine, homocysteine, and protein sulfhydryls) to produce disulfide anion radicals, potent reductants that can generate the superoxide via a one-electron reduction of oxygen. Thus, not only thiols will be wasted nonproductively, but oxidative stress will be enhanced due to generation of new oxygen radicals. Finally, cytochrome P-450-catalyzed biotransformation reactions produce a great variety of hydroxylated xenobiotics that may have quite outstanding radical scavenging properties. Nobody, however, will consider using hydroxylated derivatives of benz(α)pyrene as antioxidants. Similarly, I doubt anybody will try urushiol (the active component of poison ivy and a good radical scavenger) for antioxidant protection. The reason in both of these cases is clear: these compounds are known to be extremely toxic. Thus, we have to be cautious in our searches for new effective synthetic or natural antioxidants.

Finally, I agree with the comments made by Dr. Linnane. Antioxidants should be tissue specific. Phenolic antioxidants that may become substrates of peroxidases and thus activated to cytotoxic species should not be used for protection of peroxidase-rich tissues. Antioxidants that exert their effects at low oxygen tension should not be used for protection of highly aerated tissues. The latest grim example is betacarotene. A seminal paper by Ingold and Burton (1984) contained an important warning. Although vitamin E may be good for antioxidant protection of tissues where oxygen pressure is high, β-carotene, which exerts its antioxidant properties at low oxygen pressures, is not an appropriate antioxidant in pulmonary tissues. Maybe we should not have been so surprised that the results of a large-scale epidemiological CARET study were so disappointing.

Comment

MOHSEN MEYDANI
*JMUSDA Human Nutrition Research Center on Aging at Tufts University,
Boston Massachusetts, 02111*

VITAMIN E

There is evidence from epidemiological studies and clinical trials showing a beneficial effect of vitamin E in degenerative diseases. This beneficial effect is observed with the intake of doses of vitamin E above the current RDA of 8–10 IU.[1] The reduction of risk of cardiovascular disease and its morbidity and mortality appears to be associated with supplemental intake of 100–200 IU of vitamin E.[2–6] Our recent study[7] investigating a dose/response relationship between vitamin E and immune response in elderly subjects indicated that a level of vitamin E greater than currently recommended enhances certain relevant *in vivo* indexes of T cell–mediated function in healthy elderly responses. From this study, 200 IU appears to be an optimal dose for boosting the immune response in the elderly. A considerably larger dose of vitamin E (2,000 IU) has also been reported to ameliorate the progression of Alzheimer's disease.[8] However, the side effects of such a large dose over a long period of time (several years) need to be determined.

Increasing intake of this vitamin above the RDA requires selection and inclusion of special foods with high contents of this vitamin, such as wheat germ and almond and vegetable oils into the daily diet. However, increasing vitamin E intake to 100–200 IU through food sources alone is impractical. Therefore, supplemental intake of this vitamin is needed if intake of vitamin E at the level of 100–200 IU per day is desired. This should be in addition to consumption of 5–8 servings of fruits and vegetables as recommended by the U.S. dietary guidelines.

VITAMIN C

There is also epidemiological evidence linking the reduced incidence of mortality and reduced risk of cancer and heart disease and cataract to a high intake of vitamin C.[5,9–11] The current RDA sets the recommended level of vitamin C at 60 mg/day, which is the amount necessary to maintain adequate body reserves (> 300 mg) to prevent scorbutic symptoms for at least four weeks on a diet lacking vitamin C.[1] This RDA level does not take into consideration those functions of vitamin C not related to scurvy, such as catecholamine biosynthesis, xenobiotic metabolism, and antioxidant functions. A recent controlled clinical study by Levine et al.[12] found that 200 mg of vitamin C is an optimal dose to maintain steady levels of vitamin C in plasma and tissues. They have recommended that the RDA be raised to 200 mg/day, which could be obtained through diet in the form of fruits and vegetables.

REFERENCES

1. NATIONAL RESEARCH COUNCIL. 1989. Recommended Dietary Allowances. 10th ed. Washington D. C.: National Academy Press.
2. MEYDANI, M. 1995. Vitamin E. Lancet **345:** 170–175.
3. STAMPFER, M.J., C.H. HENNEKENS, J.E. MANSON, G.A. COLDITZ, B. ROSNER & W.C. WILLETT. 1993. Vitamin E consumption and the risk of coronary disease in women. N. Engl. J. Med. **328:** 1444–1449.
4. GEY, K.F. & P. PUSKA. 1989. Plasma vitamins E and A inversely correlated to mortality from ischemic heart disease in cross-cultural epidemiology. Ann. N.Y. Acad. Sci. **570:** 268–282.
5. LOSONCZY, K.G., T.B. HARRIS & R.J. HAVLIK. 1996. Vitamin E and vitamin C supplementation use and risk of all-cause and coronary heart disease mortality in older persons: The established populations for epidemiologic studies of the elderly. Am. J. Clin. Nutr. **64:** 190–196.
6. STEPHENS, N.G., A. PARSONS, P.M. SCHOFIELD et al. 1996. Randomized, controlled trial of vitamin E in patients with coronary diease: Cambridge Heart Antioxidant Study (CHAOS). Lancet **347:** 781–786.
7. MEYDANI, S.N., M. MEYDANI, J.B. BLUMBERG et al. 1997. Vitamin E supplementation enhances in vivo immune response in healthy elderly: A dose-response study. J. Am. Med. Assoc. **277:** 1380–1386.
8. SANO, M., M.S. ERNESTO, R.G. THOMAS et al. 1997. A controlled trial of selegiline, alpha-tocopherol, or both as treatment for Alzheimer's disease. N. Engl. J. Med. **336:** 1216–1222.
9. MEYDANI, M. 1995. Antioxidant vitamins. Fron. Clin. Nutr. **4:** 7–14.
10. GEY, K.F., G.B. BRUBACHER & H.B. STAHELIN. 1987. Plasma levels of antioxidant vitamins in relation to ischemic heart disease and cancer. Am. J. Clin. Nutr. **45:** 1368–1377.
11. GEY, K.F., H.B. STAHELIN & M. EICHHOLSER. 1993. Poor plasma status of carotene and vitamin C is associated with higher mortality from ischemic heart disease and stroke: Panel Prospective Study. J. Clin. Invest. **71:** 3–6.
12. LEVINE, M., C. CNRY-CANTILENA, Y. WANG et al. 1996. Vitamin C pharmacokinetics in healthy volunteers: Evidence for a recommended dietary allowance. Proc. Natl. Acad. Sci. USA **93:** 3704–3709.

Synergistic Multitherapy as a Rational Strategy in Aging: Multifactorial Process

SORIN RIGA AND DAN RIGA

Department of Psychiatric Research, "Gh. Marinescu" Hospital of Psychiatry and Neurology, 10, Berceni Road, RO-75622 Bucharest 8, Romania

Aging is a certainty and also carries on as a multifactorial process, so that the diversity of aging phenomena reflects diversity in the mechanisms involved. In this respect, the rational strategy for deceleration and retardation of the aging process is synergistic multitherapy. A practical result of this multi-intervention concept is the Antagonic-Stress® drug (RO Patent 105891/1992; PCT/WO 33486/1995 International Patent Publication), a new therapeutic and pharmaceutical system in antiaging and antioxidative stress therapy (2 papers on the 16th World Congress of Gerontology, Adelaide, 1997). The etiopathogenic and synergistic (vs. monotherapy) interventions on Antagonic-Stress® in (brain) aging processes were demonstrated in (1) preclinical patterns, multiple corrections of hypoanabolism, hypercatabolism, oxidative stress, accompanied with lipofuscinolysis and brain aging pigment diminution; and also in (2) clinical psychogeriatrics, nootropic (anti-impairment), psychotonic (antifatigue, energoactive), antidepressant, psychostabilizer, anxiolytic, adaptative (antistress) actions.[1-4] The biological (orthomolecular) nature of its substances, synergism of multiple composition and actions, etiopathogenic, and polyinterventions in the aging process explain the multiple and positive results of Antagonic-Stress® in brain senescence, and also the efficacy and superiority of this multivalent drug versus monotherapy for the aging phenomena.

REFERENCES

1. Popa, R., F. Schneider, G. Mihala, P. tef nig, I.G. Mihala, R.M. Tie & R. Mateescu. 1994. Antagonic-stress superiority versus meclofenoxate in gerontopsychiatry (Alzeimer type dementia). Arch. Gerontol. Geriatr. Suppl. **4:** 197–206.
2. Riga, S. & D. Riga. 1994. Antagonic-stress: A therapeutic composition for deceleration of aging. I. Brain lipofuscinolytic activity demonstrated by light and fluorescence microsopy. Arch. Gerontol. Geriatr. Suppl. **4:** 217–226.
3. Riga, S. & D. Riga. 1994. Antagonic-stress: A therapeutic composition for deceleration of aging. II. Brain lipofuscinolytic activity demonstrated by electron microscopy. Arch. Gerontol. Geriatr. Suppl. **4:** 227–234.
4. Predescu, V., D. Riga, S. Riga, J. Turlea, I.M. Bărbat & L. Botezat-Antonescu. 1994. Ann. N.Y. Acad. Sci. **717:** 315–331.
5. Schneider, F., R. Popa, G. Mihalas, P. Stefanigă, I.G. Mihalas, R. Măties & R. Mateescu. 1994. Ann. N.Y. Acad. Sci. **717:** 332–342.
6. Riga, S. & D. Riga. 1995. Ann. N.Y. Acad. Sci. **717:** 535–550.

Cell Signaling as a Major Factor of Intervention into the Aging Process

HANS NIEDERMÜLLER, ALOIS STRASSER, AND GERHARD HOFECKER

Department of Physiology and Ludwig Boltzman Institute of Experimental Gerontology, Veterinary Medical University, Vienna, A-1210 Wien, Austria

Aging research nowadays has come to a time where it is not enough to determine the changes of parameters, but it becomes necessary to investigate into mechanics underlying those changes and their modulation. At the cellular and molecular level cell–cell and cell–matrix interactions, mediated by multiple signal cascades, are of particular interest. Thus we determined in rats aged 9 and 30 months (1) concentrations and activities of signal molecules, such as G proteins, cAMP, diacylglycerol (DAG), inositolphosphates (IPs) and kinases (cellular) and collagens, and proteoglycans and fibronectin (extracellular) *in vivo* in the backskin. as well as in isolated fibroblasts and keratinocytes; and (2) cell proliferation. We tried to retard the aging process in the skin by topical application, or addition to cell cultures, of fetal mesenchymal cells, collagens and proteoglycans, *L. bifidus* extracts, and Soya matrix and compared the above-mentioned parameters with those obtained by stimulation of skin cells with growth factors (GFs). We found (1) no change in the amount of G_s proteins but a reduction of the binding capacity; lower concentrations of cAMP, DAG, and IPs; a reduced activity of protein kinase C *in vivo* and *in vitro*; a higher collagen cross-linking; lower proteoglycan concentration; and no change of the amount of fibronectin in the old rat skin; (2) a reduction of cell proliferation; and (3) a more or less extensive restoration of these parameters by all of the above-mentioned stimuli, to an extent sometimes exceeding the effects of GFs. So we conclude that all of the above-mentioned influences modulate the aging process of the skin and its cells by intervention into the signaling pathways, by mediating new signals to the cells, and hence by readjusting damaged feedforward systems in the cells.

Are Free Radicals a Cause of Aging?

JOSÉ REMACLE

Facultés Universitaires de Namur, 61 Rue de Bruxelles, 5000 Namur, Belgium

Free radicals have been proposed to be one of the main causes of the slow and gradual deterioration of cell functions with time. The reason for this statement is their constant production in all the cellular compartments, cytoplasm, mitochondria, peroxisomes, nucleus, ER, and their toxic effect on the main biological molecules, the peroxidation of unsaturated fatty acids, and the hydroxylation and break of DNA and of the peptidic chains.

There are, however, several lines of evidence showing that the situation is much more complex: One is that cells are very well ptotected against the free radicals by enzymatic or chemical antioxidant and repair mechanisms and are very efficient for the correction or replacement of modified molecules. On the other side, transformed cells do not show the characteristic aging pattern while they also produce large amounts of free radicals.

We have used an *in vitro* model of cultured human fibroblasts where the aging of cells can be evaluated by their gradual shift from one cell type to another. Stresses like overproduction of free radicals were able to drastically accelerate the process, which was counteracted by an increase in antioxidant level. Two other main observations were also obtained. First, the level of energy available by the cells influenced the level of the cell shift after the stress, and second, other stresses also gave the same effect.

We developed a theoretical model that incorporated the cell defense, the energy production, and the genetic adaptability of the cells in order to explain the effect of stresses like free radicals on cellular aging.

A New Genetic Conception of Aging and the Possible Mechanisms of Aging Pathologies

TEIMURAZ LEZHAVA

Department of Genetics, Tbilisi State University, Tbilisi-380028, Republic of Georgia

To reveal chromosome functional organization at the late stages of ontogenesis and to find some explanation for aging pathologies, we studied mutation level (aneuploidy and chromosome aberrations); heterochromatin regions (heat absorption of condensed chromatin; rRNA transcriptional activity; Ag-positive NORs and associations of acrocentric chromosomes); and repair intensity of unscheduled DNA synthesis and SCEs in lymphocyte cultures of individuals aged from 70 to 144 years. The obtained results show that a leading factor in aging is chromosome progressive heterochromatinization (condensation of eu- and heterochromatin). Tightly condensed (heterochromatinized) chromosome regions are genetically inert. The increase in heterochromatinization with aging inhibits repair enzymes and causes secondary increases of chromosome aberration levels. This indicates a key role of heterochromatinization in the aging process and the generation of a number of aging pathologies (Lezhava, 1984, 1991, 1996).

The Pineal Peptide Preparation, Epithalamin, Slows Aging in *Drosophila melanogaster*, Mice, and Rats

V.N ANISIMOV,[a] S.V. MYL'NIKOV,[b] AND V. KH. KHAVINSON[c]

[a]*N.N. Petrov Research Institute of Oncology,* [b]*St. Petersburg State University,*
[c]*Institute of Bioregulation and Gerontology, St.Petersburg, Russia*

In 1973 the first evidence was published that administration of the low-molecular-weight pineal peptide preparation Epithalamin® was followed by restoration of the estrus cycle in old female rats with persistent estrus syndrome and by lowering of the threshold of sensitivity of the hypothalamic-pituitary complex to feedback inhibition by estrogens in old animals (Anisimov *et al.*, 1973). Since this work the effect of Epithalamin® on the function of reproductive, neuroendocrine, and immune systems, as well as on life span, was systematically studied in our and our colleagues' experiments. Treatment with Epithalamin® increases the mean life span of all species studied (*D.melanogaster,* SHR and C3H/Sn mice, LIO rats) in ranges from 11% to 31%, p < 0.05. Ninety percent of mortality as well as maximum life span (100%) mortality were increased in flies, C3H/Sn mice, and rats. However, were not changed in SHR mice. Mortality rate was decreased by 52% in *D.melanogaster* and in rats, by 27% in C3H/Sn mice and was not changed in SHR mice exposed to Epithalamin®. Treatment with the pineal peptide increased MRDT in flies, C3H/Sn mice, and rats. It was shown that Epithalamin® increases synthesis and secrtion of melatonin in rats and inhibits free radical processes in rats and in *D.melanogaster*. Thus, exposure to Epithalamin® was followed by positive effects on parameters of life span in three species that could be related to this antioxidative potential.

Similar Gene Expression Patterns in Senescent and Hyperoxically Blocked Fibroblasts

GABRIELE SARETZKI,[a] THOMAS VON ZGLINICKI,[a] AND
BRYANT VILLEPONTEAU[b]

[a]Institute of Pathology, Charité, Humboldt University Berlin, Germany
[b]Geron Corp, Menlo Park, California, USA

We have shown recently that mild chronic hyperoxia irreversibly inhibits the proliferation of fibroblasts.[1] This proliferation block occured after one to three population doublings but was otherwise indistinguishable from proliferative senescence with regard to cell morphology, position in the cell cycle, telomere length, and mitochondrial respiration and water content.[2] To test further whether this oxidatively induced cell cycle block is, in fact, phenotypically identical to senescence, we compare here the expression of senescence-associated genes in young proliferating, young quiescent, senescent, and hyperoxically treated fibroblasts by using quantitative RT-PCR. Both inhibitors of cyclin-dependent kinases and further senescence-associated genes as identified from an enhanced differential display screen[3] were included. Although our results generally confirm the differential expression of the genes between young and senescent fibroblasts, expression levels for the tested genes were similar between senescent and hyperoxically blocked fibroblasts. This result corroborates our suggestion that the hyperoxically induced cell cycle block is a form of premature senescence.

REFERENCES

1. VON ZGLINICKI, T., G. SARETZKI, W. DÖCKE & C. LOTZE. 1995. Exp. Cell Res. **220**: 186–193.
2. VON ZGLINICKI, T. & C. SCHEWE. 1995. Cell Biochem. Funct. **13**: 181–187.
3. LINSKENS, M.H.K., J. FENG, W.H. ANDREWS, B.E. ENLOW, S.M. SAATI, L.A. TONKIN, W.D. FUNK & B. VILLEPONTEAU. 1995. Nucleic Acids Res. **23**: 3244–3251.

Altered Gene Expression in the Brain of the Senescence-accelerated Mouse

RYOYA TAKAHASHI, YUKI MAKABE, AND SATARO GOTO

Department of Biochemistry, School of Pharmaceutical Sciences, Toho University, 2-2-1 Miyama, Funabashi, Chiba 274, Japan

The senescence~accelerated mouse strain, SAMP8//Th, has a much shorter life span (approx. 50% of control strain, the accelerated senescence–resistant strain, SAMR1/Th) and exhibits early onset of learning and memory deficits. In an attempt to find molecular biological differences between SAMP8//Th and SAMR1/Th mice, we investigated age-related changes in the expression of various genes (myelin basic protein (MBP), actin, cyclophilin (Cyp), hsc70, hsp90, and glyceraldehyde-3-phosphate dehydrogenase (GAPDH)) in different brain regions (cerebrum, brain stem, and cerebellum) of both strains. The expression of certain genes (Cyp, actin, hsc70, hsp90, and actin) was promoted in all regions of the brain examined. Interestingly, the rate of age-related changes in these mRNA levels were negatively correlated with the life span of the animals. No significant difference in the profiles of change of MBP mRNA in the cerebrum and brain stem was observed between the two lines of mouse at any age. In the cerebellum, however, the MBP mRNA level was decreased significantly in the SAMP8//Th, compared to the SAMR 1/ Th. A similar tendency was found for GAPDH mRNA. Thus, the expression of certain genes was altered in the brain of the SAMP8 strain during postnatal development and aging. Such altered gene expression may be relevant to shorter life span and learning–memory impairment observed early in the life of this strain of mouse.

Prolongation of the Healthy Life Span with Early Diagnosis of T(14;18) Translocation

I. SEMSEI, Gy. SZEGEDI, M. ZEHER, I. TAKACS, AND P. SEBOK

Molecular Biological Research Laboratory, 3rd Department of Internal Medicine,
University Medical School of Debrecen, 4004 Debrecen, Hungary

The drifting away by cells from a properly differentiated state can be caused by internal factors, for example, free radicals. Such age-related diseases as the autoimmune diseases, Sjögren's syndrome, and certain lymphomas can be linked to the effect of free radicals. Genomic instability due to free radicals can result in different types of mutations, such as translocations, with increasing age.

Our aim was to test patients with Sjögren's syndrome to see if they had the early sign of follicular lymphoma: t(14;18), chromosome translocation. DNA from peripheral lymphocytes were islolated and the translocation was diagnosed using the polymerase chain reaction technique. Our results showed a high frequency of translocation that could result in the development of lymphoma.

Early diagnosis of the translocation could result in earlier treatment and, in turn, lengthening of the life span. Elimination of the translocation-bearing cells could prevent the invasion of the whole body with translocation-containing cells, preventing the outbreak of lymphoma that shortens the possible life span substantially.

Diagnostic methods, such as the one mentioned above, that can indicate the early drift away of cells from the properly differentiated state, can prove useful tools for prolonging the healthy life span, if we can reverse the effects caused by aging in time.

Genetic Dissection of Senile Hypoactivity in *Drosophila melanogaster*

CHRISTOPHER DRIVER

National Institute of Aging, Parkville, Victoria, Australia 3052

Many senile changes in behavior that occur in *Drosophila* reflect changes observed in aging humans: these include a loss of short-term memory, a failing ability to perform complex tasks, and a substantial fall in activity (hypoactivity). Genetic mutants are being used to explore the basis of this hypoactivity further.

Studies on the mutants *dunce*, *rutabaga*, *iav*, *ebony*, and *amnesiac* indicate a role for cAMP in protecting against this aging process. These studies also localize the principal cells in this activity to nuclei located within the mushroom bodies, structures analogous to the frontal lobes in humans. Neurotransmitters implicated as negative effectors include noradrenaline, the *amnesiac* peptide, and acetylcholine (nicotinic receptors).

Evidence has also been found to implicate other components in this degenerative process: mitochondria (Arking La strain), free radicals (*rosy*), and a motor neuron–derived trophic agent (*Passover*). The strains in parentheses are the strains from which such a conclusion is derived.

A new activity mutant, *akaal*, is described. It is unresponsive to the noradrenaline agonist, ephedrine, and the acetylcholine agonist, nicotine. It is less active initially, and loses activity more slowly so that in old age it is more active than control strains. It also lives longer than control strains by about 15 percent. Studies on this and similar mutants are expected to produce further data on the aging phenotype.

Structural Changes of Chromosomes Induced by Heavy Metal Salts during *in Vivo* and *in Vitro* Aging

TINA JOKHADZE

Department of Genetics, Tbilisi State University, Tbilisi-380028, Republic of Georgia

The cytogenetic effect of inorganic copper ($CuSO_4$), cadmium ($CdCl_2$), and nickel ($NiCl_2$) was studied in cells of long-term (144-hour) human cultures (*in vitro* aging model) and for the copper and cadmium in elderly individuals (80–90 years old). Copper and nickel salts increased the incidence of aberant cells during *in vitro* aging (12.20 ± 1.62% and 20.00 + 2.2%); the copper did the same in elderly individuals (14.25 ± 1.74%). In the controls we found 5.25 ± 1.10% *in vitro* and 3.94 ± 1.96% *in vivo* aging. Treatment with cadmium chloride did not induce any changes in the background index. Differences in the effect of the studied salts may be due to their different effects on chromatin modification.

Total Extent and Cellular Distribution of Mitochondrial DNA Mutations in Aging

S.A. KOVALENKO, J. KELSO, G. KOPSIDAS, AND A.W. LINNANE

Centre for Molecular Biology and Medicine, Austin and Repatriation Medical Centre, Banksia St., West Heidelberg, Victoria 3081, Australia

In 1989 we proposed a comprehensive hypothesis concerning the accumulation of mtDNA mutations with human aging.[1] Central idea to the proposal is that random mtDNA mutations occur in cells throughout life, thereby contributing to a gradual loss of cellular bioenergy capacity within tissues and organs associated with general senescence and diseases of the aging. The generality of the proposal has now been widely confirmed. Although it has been shown that mtDNA from skeletal muscle of aged subjects is extensively mutated and that mtDNA mutations progressively accumulate with age,[2] it has been questioned whether the total amount of mutation is sufficient for the observed decrease in tissue bioenergy capacity. Furthermore, the differential accumulations of mtDNA bearing 4,977 bp deletion in various cells/tissues during human aging suggest that the aging process in different cells/tissues may proceed at a different rate.[3] In this communication to investigate the age-dependent accumulation and the cellular distribution of specific mtDNA deletions as well as the distribution of multiple mtDNA deletions within the tissue (human deltoid muscle), we have employed the new sensitive chimeric *in situ* PCR hybridization technique and extralong PCR analysis (XL-PCR). The *in situ* results indicate that 4,977 bp mitochondrial mutation is not evenly distributed among all cells of a given tissue: they are located in different single cells of aged individuals with high density. The frequency of affected cells in the "common" deletion muscle fibers was significantly higher in old-age subjects compared to young. XL PCR then shows the accumulation of the extensively mutated mtDNA in muscle fibers from old-age individuals, suggesting the possible mechanism of cellular bioenergy decline with age. XL-PCR analysis of longitudinally sectioned skeletal muscle fibers also revealed that the process of accumulation of mtDNA changes is random and occus first in just a few tissue cells from young subjects, and then with age the mutations accumulate and are eventually observed in all cells of old-age (over 70 years) subjects. Each cell of aged tissue had a different unique pattern of multiple mtDNA rearrangements, indicating the different, mutational rate in the skeletal muscle cells, conceivably indicating a different rate of aging of individual cells. These findings thus parallel enzyme histochemical analyses, demonstrating the occurrence of an age-related tissue bioenergy mosaic and supporting our suggestion that multiple mtDNA mutations could be responsible for the observed decline in bioenergetic capacity of aged tissue.

REFERENCES

1. LINNANE, A.W. *et al.* 1989. Lancet, **1:** 642–645.
2. KOVALENKO, S.A. *et al.* 1997. Biochem. Biophys. Res. Commun. **232:** 147.
3. LINNANE, A.W. *et al.* 1992. Mutat. Res. **275:** 195–208.

Coenzyme Q_{10} Treatment Improves the Tolerance of the Senescent Myocardium to Pacing Stress in the Rat

FRANKLIN L. ROSENFELDT, MICHAEL A. ROWLAND, PHILIP NAGLEY, AND ANTHONY W. LINNANE

Baker Institute and Monash University, Melbourne, Australia

BACKGROUND

In the elderly the results of interventions that stress the myocardium, such as coronary bypass surgery and angioplasty are inferior to those in the young. A possible contributing factor is in age-related reduction in cellular energy production. Coenzyme Q_{10} (CoQ_{10}) is a redox carrier in the mitochondrial energy-producing electron transport chain, which may improve energy production and thus the tolerance of the senescent heart to stress. This study compared the recovery of senescent and young rat hearts after stress produced by rapid pacing and tested the effect of CoQ_{10}

METHODS

Young (4.8 ± 0.1 months) and senescent (35.3 ± 0.2 months) rats were given daily intraperitoneal injections of CoQ_{10} (4 mg/kg) or vehicle for 6 weeks. Their isolated working hearts were subjected to the aerobic stress of ventricular pacing at 510 bpm for 120 minutes.

RESULTS

In senescent hearts prepacing cardiac work was 74% and oxygen consumption (MVO_2) 66% of that in young hearts. CoQ_{10} treatment abolished these age differences. Postpacing recovery of work and MVO_2 was expressed as a percentage of prepacing levels. The untreated senescent hearts, compared to young, showed reduced recovery of work: 16.8 ± 4.3 vs. $44.5 \pm 7.4\%$ ($p < 0.01$), and MVO_2 61.3 ± 4.0 vs. $74.1 \pm 5.0\%$ ($p = 0.06$). CoQ_{10} treatment in senescent hearts improved recovery of work, (48.1 ± 4.1 vs. $16.8 \pm 4.3\%$; $p < 0.0001$), MVO_2 (82.1 ± 2.8 vs. $61.3 \pm 4.0\%$; $p < 0.01$), and efficiency (58.0 ± 3.5 vs. $25.7 \pm 6.5\%$; $p < 0.001$) in treated versus untreated hearts, respectively. Postpacing levels of these parameters in CoQ_{10}-treated senescent hearts were similar to those in young hearts where CoQ_{10} produced no benefit.

CONCLUSIONS

Senescent rat hearts show reduced tolerance to aerobic stress compared to young hearts. Pretreatment with CoQ_{10} improves the tolerance of the senescent myocardium to aerobic stress and thus may benefit aged patients undergoing cardiac interventions.

Response of the Human Myocardium to Hypoxia and Ischemia Declines with Age

Correlation with Increased Mitochondrial DNA Deletions

FRANKLIN L. ROSENFELDT, JUSTIN A. MARIANI, ROUCHONG OU, SALVATORE PEPE, PHILIP NAGLEY, AND ANTHONY W. LINNANE

Baker Institute and Monash University, Melbourne, Australia

BACKGROUND

In elderly patients, recovery of cardiac function after myocardial infarction or cardiac surgery is inferior to that in younger patients, suggesting that the senescent myocardium is more sensitive to stress. Deletions in mitochondrial DNA (mtDNA) accumulate with age, which may cause defective respiratory chain activity, reduced energy production, and decreased tolerance to stress. Our aim was to compare the response of atrial strips from senescent and younger human hearts to hypoxia or ischemia *in vitro* and to correlate decreased function with mtDNA deletions.

METHOD

Atrial strips from 58 right atrial appendages (RAA) discarded at surgery were paced in an organ bath, subjected to either 30 minutes of hypoxia (N_2) or simulated ischemia (N_2 and no perfusion), and then allowed to recover for 30 minutes. Recovery of developed tension (DT) was expressed as a percentage of prestress levels. In 45 RAA the mtDNA[4977bp] deletion was quantified by PCR. In 14 patients, both physiological and PCR results were obtained.

RESULTS

Senescent tissues demonstrated significantly less recovery after hypoxia and ischemia than younger patient tissues (TABLE 1). PCR analysis revealed age-associated increases in frequency of the deletion ($r^2 = 0.28$ $p < 0.01$). There was a significant correlation between decreased recovery of DT and increased amounts of the deletion ($r^2 = 0.36$ $p < 0.05$).

TABLE 1. Percent Recovery of DT (Mean ± SEM)

				p value	
Age (yr)	< 60 (A)	60–69 (B)	≥ 70 (C)	A vs. B	A vs. C
Hypoxia (n)	65.8 ± 5.2 (15)	43.1 ± 5.5 (7)	48.9 ± 6.8 (7)	< 0.05	< 0.05
Ischemia (n)	68.9 ± 4.6 (6)	51.4 ± 15.5 (9)	35.8 ± 7.7 (4)	< 0.05	< 0.05

CONCLUSIONS

In the senescent human myocardium, both hypoxia and ischemia lead to marked reductions in contractile function compared to the younger myocardium. Age-associated deletions of mtDNA may play a role in dysfunction of the senescent myocardium, which may explain the poorer recovery after stressful cardiac interventions.

Impairment of Mitochondrial Function in the Aging Heart

Molecular and Pharmacological Aspects

G. PARADIES, G. PETROSILLO, AND F.M. RUGGIERO

Department of Biochemistry and Molecular Biology and CNR Unit for the Study of Mitochondria and Bioenergetics, University of Bari, Bari, Italy

Aging has a profound effect on cardiac performance. The well-known age-dependent decrement in heart performance may be related to changes in the activity and in the properties of several mitochondrial proteins and enzymatic systems involved in energy metabolism. Mitochondrial anion transport proteins are responsible for the flux of metabolites that occur across the inner mitochondrial membrane. Cytochrome c oxidase is the terminal enzyme complex of the mitochondrial electron transport chain responsible for virtually all oxygen consumption in mamals. The normal functioning of these membrane-associated proteins is essential for the bioenergetics of the cardiac cell. The effect of aging on the transport activity of the pyruvate, phosphate, carnitine, and ADP carriers as well as on the cytochrome c oxidase activity in rat heart mitochondria was investigated. The activity of all these protein systems is reduced with aging. This reduced activity is not due to a lower protein content in the membrane. Cardiolipin is known to be essential for the optimal functioning of all these protein systems. The cardiolipin level was markedly reduced with aging. This decrease may be due either to alteration of cardiolipin biosynthes or to oxyradical-induced lipid peroxidation. Thus the age-dependent decrement in the activity of these protein systems may be attributed to a lower cardiolipin content in the membrane. Treatment of aged rats with acetyl-L-carnitine, a natural biomolecule that acts by stimulating cellular energy metabolism, was able to restore the mitochondrial content of cardiolipin, thereby restoring the activity of the anion carrier proteins, and that of cytochrome oxidase, to the level of young control rats. The observed age-dependent decline in these membrane-associated proteins activities may play an important role in the etiopathology of declining cardiac competence with aging.[1,2]

REFERENCES

1. Paradies, G., F.M. Ruggiero, G. Petrosillo, M.N. Gadaleta & E. Quagliariello. 1995. Mech. Ageing Dev. **84:** 103–112.
2. Paradies, G., F.M. Ruggiero, G. Petrosillo & E. Quagliariello. 1996. Ann. N.Y. Acad. Sci. **786:** 252–263.

Aluminosilicate Particulate and Beta-Amyloid *in Vitro* Interactions

A Model of Alzheimer Plaque Formation

PETER EVANS AND CHARLES HARRINGTON

Department of Public Health, University of Glasgow, 2 Lilybank Gardens, Glasgow, Scotland G12 8RZ, and Brain Bank Laboratory, MRC Centre, Cambridge, England

Alzheimer's dementia has been related to amyloid β-protein (Aβ) aggregates in the senile plaques in the brain. The aetiopathogenesis of the disease has also been linked to the presence of aluminosilicate deposits within the plaque cores. Analogous model aluminosilicate particles induce the generation of oxyradicals by activated microglial cells in culture.[1] In the present *in vitro* study, the capacity of various model aluminosilicate particulates to interact with synthetic Aβ(1-42) residue peptides has been studied using an immunochemical assay. Of the aluminosilicate samples examined, the strongest interactions were obtained with kaolin mineral clay aluminosilicates. The findings indicate that comparable *in vivo* deposition of fibrillar amyloid aggregates within the Alzheimer plaques may be exacerbated by interactions with environmental aluminosilicates, and hence enhance Aβ-stimulated and microglia-generated free radical neuro-toxicity and degeneration.

REFERENCE

1. EVANS, P.H., E. PETERHANS, T, BURGE & J. KLINOWSKI. 1992. Dementia **3**: 1–6.

Does Extracellular Ascorbate Promote Oxidative Damage to the Aging Brain?

MARIA M. ROMANAS, MITCHELL R. EMERSON, STANLEY R. NELSON,
FRED E. SAMSON, AND THOMAS L. PAZDERNIK

Smith Research Center, University of Kansas Medical Center,
Kansas City, Kansas 66160, USA

Ascorbate is the major known antioxidant in brain extracellular fluid (ECF), but ascorbate-driven Fenton reactions may generate HO· in brain ECF when metals are delocalized, as occurs locally in trauma or stroke. This may occur generally in aging. Ascorbate (Asc), metal, and H_2O_2 were incubated in Chelex-treatea phosphate buffer (pH 7.4) or in human cerebrospinal fluid (CSF). Fluorescence was used to follow salicylate hydroxylation or *cis*-parinaric acid oxidation in these systems. Salicylate hydroxylation occurred immediately in a rapid O_2-dependent phase and a slower-O_2 independent phase. *Cis*-parinaric acid oxidation, on the other hand, followed a lag period, began after the rapid phase of salicy late hydroxylation had ended, and proceeded most rapidly with low O_2. This chemistry can be understood in terms of ternary complexes (Asc-metal-O_2), previously described (Martell, 1982). Initially, electron transfer in an Asc-metal-O_2 complex occurs rapidly to form an ascorby·-metal-$O^{·-}$ complex. This latter complex reduces a second metal ion that, in turn, reacts with H_2O_2 to generate HO·. Therefore, the "ascorbate-driven Fenton reaction" is mediated by a superoxide-like complex intermediate. The ascorbate-metal-H_2O_2 complex probably mediates oxidations independent of HO• in a manner analogous to "activated bleomycin," a ternary bleomycin-metal-H_2O_2 complex. Human CSF dramatically protected against both salicylate hydroxylation and *cis*-parinaric acid oxidation. This protection appeared to be due to low molecular weight chelators in human CSF that compete with ascorbate for metal binding and hinder the formation of the Asc-metal-O_2 or Asc-metal-H_2O_2 complexes necessary for the prooxidant activity of ascorbate. We predict, therefore, that extracellular ascorbate does not promote oxidative damage even if metals are delocalized in the aging brain.

Low Levels of 4977 BP-deleted Molecules of Mitochondrial DNA in the Presence of High OH⁸DG Contents in Healthy Subjects and Alzheimer's Disease Patients

A.M.S. LEZZA, P. MECOCCI, A. CORMIO, A. CHERUBINI, M. FLINT BEAL,
P. CANTATORE, U. SENIN, AND M.N. GADALETA

Universita di Bari, 70125 Bari, Italy

The 4977 bp-deleted molecules of mitochondrial DNA (mtDNA^{4977}) and the mtDNA OH^{8}dG levels have been measured in some brain areas of healthy subjects and Alzheimer's disease (AD) patients. The percentage of mtDNA^{4977} increased with age in all examined subjects, whereas that of OH^{8}dG did so only in healthy subjects. A highly significant, positive correlation between the two damages of mtDNA has been found in the frontal area of the healthy subjects. Furthermore, the mtDNA^{4977} was always lower, and the OH^{8}dG was always higher in the AD patients than in the healthy subjects. This was observed also when the frontal and the parietal areas of the same younger healthy subjects or AD patients were compared but not when the same comparison was made in the oldest individuals of both groups and in the cerebellum. This behavior suggests that a unique process, that is, an oxidative stress, takes place both in aging and in AD, where it is only amplified, and that the opposite or the similar trends of accumulation of the two mtDNA damages depend on the onset time and on the extent of the age-related oxidative stress in each brain area or in each subject. A high level of oxidized bases on mtDNA might slow-down the mtDNA replication, so that where the mtDNA oxidative damage is earlier and/or heavier less mtDNA deleted molecules accumulate. The slowdown of the mtDNA replication could, in turn, be responsible for the death of "at risk" neurons in the regions most affected by the AD.

Correlations between Lipofuscin Accumulation and Aging Neuropathology

DAN RIGA AND SORIN RIGA

Department of Psychiatric Research, "Gh. Marinescu" Hospital of Psychiatry and Neurology, 10, Berceni Road, RO-75622 Bucharest 8, Romania

As a time-dependent phenomenon, progressive neuronal lipofuscin accumulations in the mammalian central nervous system constantly coexist and are significantly correlated with neuronal loss, decrease in the volume of neurosoma, dendritic destructions and aberrations, axonal enlargements to meganeurites, decrease of mitochondria number and area, diminution ot Nissl bodies (polyribosomes and rough endoplasmic reticulum) number and volume, decrease of RNA content (mainly ribosomal) and with amyloid precursor protein mRNA and amyloid β-protein levels, as well as glial lipofuscin storages (in perineuronal, from neuropil and pericapillary glias). Causal (direct) connections, critical lipofuscin concentrations, which generate cascades of negative subcellular events, and associative (indirect) impairment correlations determine the neuropathological profile of the aging. These direct and associate neuropathologic consequences of lipofuscin accumulations have multiple and detrimental impacts from neuronal function to central nervous system physiology: damage of cytoskeleton and consecutively of intracellular and axonal transport, decrease of protein and RNA biosynthesis, alteration of lysosomal activity, increase of catabolic processes, fundamental breakdown in neuronal homeostasis to learning and memory impairment, decrease of visual and hearing acuity, and diminution of coordination and motor performance.

Proteasome but Not Lysosomal Protease Activities Decline with Age in the Liver of Male F344 Rats

T. HAYASHI, Y. NAKANO, R. TAKAHASHI, AND S. GOTO

Department of Biochemistry, School of Pharmaceutical Sciences,
Toho University, 2-2-1 Miyama, Funabashi, Chiba 274, Japan

We found that half-lives of several proteins, including oxidatively modified ones introduced into hepatocytes from old mice, are significantly extended, when compared with those for young animals. Ward and his collaborators reported that proteasome activities in liver extracts appear to be unchanged in aging rats, except the peptidylglutamyl peptide hydrolyzing activity. We studied age-related changes of activities and the number of 20S and 26S proteasomes of male F344 rat livers separated from the other proteases on glycerol gradients. The activities for three fluorogenic substrates declined gradually with age (from 8 to 27 months old), the change in the peptidylglutamyl peptidase activity being most remarkable, that is, 53% difference between young and old animals. By contrast, activities of lysosomal cathepsins for three age groups were similar. Immunoblot analyses revealed that amounts of a proteasome subunit did not change significantly with age, suggesting the possibility that the molecular activity of the proteasome declined with age probably because of posttranslational modification(s) of proteasomal proteins. Thus, the decline of proteasome activity is likely involved in the age-related extension of half-lives of cellular proteins.

Macrophage Membrane Proteins That Recognize Carbohydrate Chains of Oxidized Erythrocytes

MASATOSHI BEPPU, SHIGETOSHI EDA, AND KIYOMI KIKUGAWA

Tokyo University of Pharmacy and Life Science,
1432-1 Horinouchi, Hachioji, Tokyo 192-03, Japan

We have previously shown that macrophages recognize oxidatively damaged erythrocytes through binding to the carbohydrate chains of the oxidized cell surface. Here, we report detection and isolation of the macrophage membrane receptors for oxidized erythrocytes. The human monocytic leukemia cell line, THP-1, was differentiated into macrophages by PMA, and the membrane was prepared and solubilized by Triton X-100. The solubilized membrane proteins were adsorbed on a PVDF membrane by dot blotting, and oxidized human erythrocytes were applied onto the dots. Oxidized but not unoxidized erythrocytes bound to the dots. Oxidized erythrocytes also bound to 50, 80, and 120 kDa protein bands of a Western blot of the solubilized macrophage proteins. The binding was prevented when the polylactosamine-type carbohydrate chains had been removed from the erythrocyte surface by endo-β-galactosidase, indicating that these types of carbohydrate chains are involved in the binding. The lectin-like proteins of the macrophage membrane were separated by affinity chromatography using polylactosamine-containing glycoproteins (band 3, human lactoferrin) and DEAE cellulose ion-exchange chromatography. The 50 kDa protein was successfully isolated after SDS-PAGE. The amino atcid sequence of the N-terminal region of the 50 kDa protein suggested that this was a new macrophage protein.

Aging and Dietary Protein: Effects of Collagen and Muscle Fiber Distribution in the Rat Diaphragm

EMILIO A. JECKEL-NETO,[a] TATIANA RIBEIRO DA SILVA,[a]
ALESSANDRA L. ROSA,[a] YOSHITAKE ITO,[b] TSUNEKO SATO,[b]
AND HISASHI TAUCHI[b]

[a]Laboratory of the Biology of Aging, PUCRS Institute of Geriatrics,
90610-000 Porto Alegre, Brazil
[b]Institute for Medical Science of Aging, Aichi Medical University, 480-11 Nagakute, Japan

To detect age changes in muscle fiber and collagen distribution of the diaphragm, 129 male Donryu rats were single-housed in a specific pathogen-free (SPF) facility and were submitted to dietary restriction. The animals were divided in three groups and received 40%, 20%, and 10% of dietary protein, respectively. Subsets of each group were sacrificed at the ages of three weeks, and 6, 12, 18, 24, and 30 months old. The diaphragms were excised and measured morphometrically to determine their thickness and volume. Fragments of the costal area were taken, and paraffin sections were cut, oriented by the muscle fibers cross-section. One section per sample was stained with sirius red-fast green to detect the collagen in the diaphragm. A second section was processed immumohistochemically using antiparvalbumin to determine the slow- and fast-twitch muscle fibers. Image analysis procedures were done to separate the muscle fibers types as well as to determine their relative distribution and size, and to measure the amount of collagen in the endomysium and perimysium. Both thickness and volume of the muscle increased with age in the three groups. The relative distribution of fiber types within the diaphragm muscle showed that there was an exchange of fast-twitch fibers for slow-twitch ones at an early age and in older animals. The diameter of fast-twitch fibers decreased with age, whereas the slow-twitch fiber diameter remained constant. In the three groups, the amount of collagen did not present significative alterations. These results show that collagen does not contribute to the changes in the thickness and volume of the muscles and that those changes would be due to modifications in the number and size of fast-twitch muscle fibers, independently of the amount of dietary protein.

Characterization of Age-dependent Multiple Yellow Fluorescent Components in Rat Kidney

KIYOMI KIKUGAWA, MASATOSHI BEPPU, AKIHIDE SATO,
AND HIROYOSHI KASAI

*Tokyo University of Pharmacy and Life Science,
1432-1 Horinouchi, Hachioji, Tokyo 192-03, Japan*

Yellow fluorescent components deposited age dependently in rat kidney were extracted in an aqueous solution and characterized after separation. The yellow fluorescence detected in the 105,000-g supernatant was fractionated by gel filtration into 5 yellow fluorescent fractions A, B (B_1, B_2, and B_3), and C showing the same fluorescence spectra at 400/620 nm. The components in fraction A were converted into the smaller molecular weight components in fraction B on treatment with 4 M urea or protease, suggesting that they were proteinous. The smallest molecular-weight fluorescent components in fraction C were adherent to solid cellulose materials. The fluorescent components in all the fractions were soluble in water and insoluble in chlorofom/methanol, indicating that they were not lipidic materials. The fluorophores in these fractions were kept stable on borohydride treatment but readily converted into nonfluorescent components on heavy-metal ion treatment. The characteristics of the yellow fluorescence were different from those of bluish fluorophores generated through lipid peroxidation.

Age-dependent Effect of Estrogen on Oxidative Stress of Ovariectomy-induced Osteoporotic Bones

SE IN OH,[a] MEE SOOK LEE,[b] CHANG-MO KANG,[c] YOUNG DO KOH,[c]
KYU HWAN KIM,[b,c] EUI-JU YEO,[c] AND SANG CHUL PARK[b,c]

[a]Department of Food and Nutrition, Seo Il Junior College,
[b]Hannam University
[c]Department of Biochemistry, Seoul National University, Seoul 110-799, South Korea

Osteoporosis is a typical disorder of the elderly, but the molecular mechanism has not been fully understood. Estrogen has been shown to play a role in mitigating the pathogenesis. Because the senile changes are deeply associated with oxidative stress, it might be assumed that the osteoporosis would be related to oxidative stress and estrogen would prevent the pathogenesis by reducing the oxidative damages. To test our hypothesis, an animal osteoporosis model was used. In the present experiment, osteoporosis was induced by ovariectomy in young (6 months of age) and old (18 months of age) female Sprague Dawley rats. Estradiol was injected to the animals intraperitoneally two times a week after ovariectomy. Eight weeks later, animals were sacrificed and the level of oxidative stress was analyzed by monitoring protein carbonylation content and the activities of glutathione using enzymes, including glutathione peroxidase, reductase, and transferase in the bone. The level of protein carbonylation increased dramatically in the bone of old ovariectomized rats compared to that of young animals. The administration of estradiol prevented the ovariectomy-induced protein carbonylation in the old animals. In addition, the activities of glutathione using enzymes (GP, GR, and GST) increased in the old ovariectomized animals, which also decreased to normal levels by estrogen treatment. Our data suggest that estrogen modifies ovariectomy-induced oxidative damage, and its effect is different according to the age of the animals.

Accelerated Aging of the Female Reproductive System in Mice following Exposure to Tobacco Smoke

C. GARY GAIROLA

University of Kentucky, Tobacco and Health Research Institute, Kentucky 40546-0236, USA

Epidemiological studies suggest that cigarette smoking is deleterious to female reproductive health. Women who smoke develop a variety of menstrual problems and reach their menopause earlier than nonsmokers. To determine if tobacco smoke were toxic to the reproductive system, we exposed female C57Bl mice daily for 60 weeks in a nose-only exposure system to mainstream cigarette smoke generated under controlled conditions. Smoke exposures were routinely monitored by measuring blood carboxyhemoglobin, particulate intake, and urinary cotinine. Increased levels of biomarkers in the exposed group confirmed effective exposure of animals to smoke. Estrual cyclicity and follicular atresia were compared in bench control, sham-treated, and smoke exposed mice. Vaginal cytology data showed that the length of the estrous cycle in the smoke-exposed group was significantly increased over that in control or sham-treated mice. Microscopic analysis of the serial sections of ovaries showed that the number of primordial follicles in the smoke-exposed group was significantly reduced in comparison to the age-matched, sham-treated, and bench control groups at 30- and 60-week exposure points. These results show accelerated loss of oocytes in smoke-exposed mice. Similar studies performed in a rat model, however, failed to produce any of the effects seen in mice, thus indicating species differences in response to cigarette smoke. Overall, the results suggest that exposure to tobacco smoke hastens aging of the female reproductive system in mice.

Endogenous Antioxidant Defense and Lipid Peroxidation in Elderly Patients with Diabetes Mellitus

L.H. CHEN, Y. OSIO, AND J.W. ANDERSON

Nutritional Sciences Program, University of Kentucky, Lexington, Kentucky 40506, USA

Endogenous enzymatic and nonenzymatic antioxidant defenses and lipid peroxidation were determined in the blood of 31 male diabetic elderly patients and 30 male elderly controls. Sixteen without cardiovascular disease (CVD) and 14 with CVD. The mean ± SD of the ages of the patients was 66 ± 5, and those of the control subjects were 69 ± 5. Serum glucose levels of diaibetic patients were 213 ± 81 mg/dL, and those of control subjects were 95 ± 14 mg/dL. Among the diabetic patients, 13 subjects were obese with BMI > 30, 26 subjects had poor control of diabetes (oxyhemoglobin > 7%), and 25 subjects had retinopathy. Blood samples were collected and analyzed for antioxidant defense parameters and lipid peroxidation. Diabetic patients had significantly lower blood-reduced glutathione levels and CuZn-superoxide dismutase activity when compared to the control subjects. There were no significant differences in plasma vitamin E levels and the activities of catalase and glutathione peroxidase in erythrocytes (RBC) between the two groups. Diabetic patients and control subjects with CVD had significantly higher malondialdehyde levels in both RBC and plasma when compared to the control subjects without CVD, indicating increased lipid peroxidation in the blood of diabetic patients and CVD patients. The results suggest that a decline of endogenous antioxidant defense capability contributes to oxidative stress in elderly diabetic patients.

The Reliability of the Mitochondrial Electron-transport Membranes and Implications for Aging

V.K. KOLTOVER

Institute of Chemical Physics, RAS, Moscow Region, 142432, Russia

The goal of this study was to show that an essential increase in production of oxyradicals may be expected when catalytic conditions for the mitochondrial enzymes turn away from the physiological optimum. The theory of reliability served as the methodological approach. The redox-cycling ubisemiquinone (SQ·) and the lipid fluidity in the rat heart mitochondria were measured by EPR. Anoxia/ischemia, substrate redundancy, and other factors, capable of turning the operation conditions apart from the physiological optimum, cause an increase in the reactivity of the mitochondrial SO· to oxygen, along with the relevant increase in superoxide production. The loss of the control of the electron flow through SQ· correlated with the increase in the lipid fluidity. The old rats' mitochondria demonstrated much higher variability in the superoxide production, as compared with that for young rats. Thus, the reliability of the mitochondrial electron-transport decreases with aging of animals, in accordance with the free radical theory of aging.[1,2]

REFERENCES

1. KOLTOVER, V.K. 1997. Theor. Biol. **184:** 157–163.
2. KOLTOVER, V.K. 1996. Chemical Physics Reports **15:** 109–115.

The Breathing Exercise, *Qigong*: A Traditional Exercise Approach for the Improvement of Symptoms in Aged Diabetics Through Modulation of Psychological State

SHAOJIN DUAN,[a] TONG XIN TIAN,[c] YING-JIAN DUAN,[b] YEN-RONG MAO,[b] AND CHUN-ZHI ZHANG[b]

[a]*The 2nd Department of Basic Medicine, Guang An Men Hospital, China Academy of Traditional Chinese Medicine, Beijing 100053 China*
[b]*Institute of High Energy Physics Academia Sinica*
[c]*Lao Yang Xianggong Society*

The breathing exercise, *qigong*, is one of three important therapeutic regimes, including the Chinese herbal medicine recipe, acupuncture, and *moxibustion* and *qigong* in traditional Chinese medicine. However, *xianggon*, as a physical exercise, belongs to the category of breathing exercises. Thirty-five aged diabetic patients (58.05 ± 7.19 years old) exercised *xianggong* for one year (and in some cases longer) to improve clinical symptoms: body weight, spirit, appetite, uptake of water and food, urination, sleep, and vision. After one year of *xianggong*, clinical symptoms in most from the aged volunteers were significantly improved, such as an increase in body weight; an improvement of spirit, appetite, and vison; and a decrease of urination. Sleep in 28 of the 35 subjects (80%) was significantly improved. The depression index and index score of anxiety in 22 out of 35 aged diabetics (62.85%) dropped, based on the evaluated scale of the Self-Rating Depression Scale, the Depression Status Inventory, and the Self-Rating Anxiety Scale, respectively. According to psychological theory and viewpoints of behavioral psychology, every action of the *xianggong* exercise, especially, smiling and relaxation, enables the whole body to be in the best situation to improve its pathological state. Devout faith in *qigong* enabled aged diabetics to be psychologically regulated and provided for an improved spleen, decreased depression and anxiety, indicating that blood sugar and glucagon came down and that insulin was raised in the aged diabetic volunteers. The free radical mechanism showing improvement of the aged diabetics has been studied in our laboratory.

Maintaining Mobility in Nursing Home Residents

ARNOLD H. GREENHOUSE AND SUSAN O'BRIEN

Reno VA Medical Center and Nevada Geriatric Education Center,
Sanford Center for Aging, University of Nevada, Reno, Nevada, USA

Although maintaining mobility is critical for health, well-being, and quality of life, loss of this function is common in nursing home residents. Volunteers in Professional Service (VIPS) was designed to train volunteers in promoting increased activity, particularly walking, in these patients. Additional objectives included enabling return to a less restrictive environment, improving resource utilization, and enhancing treatment efficacy. Since October 1994, more than 400 individuals have been mobilized by 24 volunteers who worked every day of the year. About 50% of patients participated at any given time. Interventions began 24–48 hours after admission. In less than two years, there were 1500 contacts that otherwise would not have occurred, the average number exceeding 100 during the usual inpatient stay. Improvement scaling showed that 75% of patients achieved all of their individual treatment goals. Over 75% returned to independent community living. Many additional benefits resulted from the VIPS initiative.

Double-blind Trial of Huperzine-A (HUP) on Cognitive Deterioration in 314 Cases of Benign Senescent Forgetfulness, Vascular Dementia, and Alzheimer's Disease

MA YONG-XING, ZHU YUE, GU YUE-DI, YU ZHEN-YAN, YU SAI-MEI, AND YE YONG-ZHEN

Research Division of Aging and Antiaging, Shanghai Geriatric Institute, Huadong Hospital, 221 West Yan An Road, Shanghai 200040, China

HUP, a new alkaloid extracted from *Huperzia serrata* (Thumb) Trev, by Liu, is a potent anticholinesterase with minimal toxicity. Tong has found that HUP may improve the learning and retrival function of rats, and its facilitation actions were due to an effect on the central cholinergic system.

BENIGN SENESCENT FORGETFULNESS (BSF) TREATED WITH 0.03–0.05 mg HUP im b.i.d.

The first clinical trials used the double blind method on 120 patients with cognitive deterioration, with memory quotient (MQ) (WMS) < 100. The mean values of MQ of 60 treated and 60 controls were 76.27 ± 14.08 and 77.97 ± 12.55 (p > 0.05), respectively. The dosage was 0.03 mg im B.i.d. for 14–15 days. The mean values of MQ after the treatment (the interval between pre- and posttests of WMS with A and B form, respectively, is 1–2 months) of treated and controls were 91.85 ± 13.73 (p < 0.01) and 82.24 ± 15.10 (p > 0.05), with the MQ increase of 15.82 ± 10.02 and 4.40 ± 8.72, respectively, (p < 0.01). The effective rates were 68.33 and 26.37% in the two groups. No significant side effects were observed.

The second trial included 16 patients of the HUP treatment group (0.03–0.05 mg im B.i.d for 4 wks) with IQ (WAIS) < 105 (95.0 ± 7.6). The IQ increased to 100.7 ± 12 (p < 0.01) after the treatment. The value of IQ increase is 5.7 ± 0.68 as compared with 3.0 ± 0.36 in the hyperboric oxygenation treatment group (2.5 ATA, 80′, 4 wks in 10 patients (p < 0.01)).

BSF TREATED BY 0.1 MG HUP po q.i.d.

The clinical trials used the double blind method on 88 patients with cognitive deterioration, with MQ (WMS) < 100. The mean values of MQ of 44 treated and 44 controls were 82.8 ± 14.3 and 81.5 ± 14.4 (p > 0.05), respectively. The dosage was 0.1 mg po Q.i.d. The mean values of MQ after the treatment (the interval between pre- and posttests of WMS with A and B form, respectively, is 2 months) of treated and controls were 93.5 ± 14.5 (p < 0.01) and 85.5 ± 16.5 (p < 0.01). The effective rates were 68.18 and 34.09% in the two groups. No significant side effects were observed except gastric discomfort (2), dizziness (1), insomnia (1), and mild excitement (1) in the treated group.

VASCULAR DEMENTIA AND ALZHEIMER'S DISEASE

A clinical trial of vascular dementia (25) and Alzheimer's disease (55) was conducted on 40 treated and 40 control patients with the same dosage as for BSF. The MQ of the treated group increased from 50.40 ± 18.49 to 59.74 ± 18.73 ($p < 0.05$); that of the control group increased from 53.95 ± 14.74 to 55.85 ± 16.28 ($p > 0.05$). The MQ increase of 9.37 ± 10.38 is significantly higher than that of 1.90 ± 10.36 ($p < 0.01$). The effective rate of the treated group was 60%, significantly higher than that of 35% in the control group ($p < 0.05$). No significant side effects were observed except gastric discomfort or nausea (3) and dizziness (3) in the treated group.

It is concluded that huperzine is an effective and safe drug to improve cognitive and memory function in the aged and preaged.

Double-blind Trial of Aniracetam on Cognitive Deterioration in 622 Cases of Benign Senescent Forgetfulness, Vascular Dementia, and Alzheimer's Disease

MA YONG-XING,[a] YU ZHEN-YAN,[a] JIN YONG-SHOU,[a] YU SAI-MEI,[a]
YE YONG-ZHEN,[a] GAO ZHI-XU,[b] FANG YONG-SHENG,[b] HONG ZHEN,[c]
AND DING ZHAO-LAN[c]

[a]Research Division of Aging and Antiaging, Shanghai Geriatric Institute,
Huadong Hospital, 221 West Yan An Road, Shanghai 200040, China
[b]Shanghai Mental Health Center, 600 South Wan Ping Road, Shanghai 200030, China
[c]Institute of Neurology, Shanghai Medicial University, 138 Yi Xue Yuan Road,
Shanghai 200032, China

BENIGN SENESCENT FORGETFULNESS (BSF)

The clinical trail of aniracetam, which is one of the analogues of piracetam, was performed on 448 patients with cognitive deterioration, with MQ (WMS) < 100. The mean values of MQ of treatment (mean age 63.9 ± 8.3 yrs) and controls (mean age 64.4 ± 7.7 yrs) were 79.8 ± 15.7 and 81.2 ± 14.0 (p > 0.05), respectively. The dosage was 0.2 mg t.i.d. for 1–2 months. The mean values of MQ after the treatment (the interval between pre- and posttests of WMS with A and B set, respectively, is 1–2 months) of treated and controls were 92.1 ± 15.9 and 86.8 ± 17.1, with the MQ increment of 11.89 ± 13.22 and 5.95 ± 12.04, respectively (p < 0.001). The effective rates were 75.36% and 30.8% in the two groups. The Chosen Reaction Time (CRT) significantly decreased in the treatment group. The decrement value in CRT of red light (69.04 ± 115.43 ms) and green light (64.65 ± 103.55 ms) of the treatment group was significantly greater than that of the control group. The decrement value in CRT of red light (69.40 ± 115.43 ms) and green light (64.65 ± 103.55 ms) of the treatment group is significantly greater than that of the control group (red light: 5.45 ± 66.21 ms; green light: 2.35 ± 9.31 ms) (p < 0.05) in 40 patients. This indicated that aniracetam can improve the function of the central nervous system.

VASCULAR DEMENTIA

A clinical trial of vascular dementia was conducted on 30 treated and 30 control patients with the same dosage as BSF. The total response rate of the treated group was 83%; that of the control (placebo) group was 60% (p < 0.05). The MQ of the treated group increased from 69 ± 15 to 84 ± 17 (p < 0.01); that of the control group increased from 74 ± 16 to 80 ± 20 (p > 0.05).

ALZHEIMER'S DISEASE

The clinical trail was performed on 114 patients (M, 67, F, 47, mean age: 67 ± 10). Fifty-seven patients were in the treatment group; the other 57 patients formed the control group. The MQ of the treatment group increased from 58 ± 16 to 66 ± 21 ($p < 0.01$) and that of the control increased from 59 ± 14 to 62 ± 14 ($p > 0.05$). The total effective rate of the treatment group was 68%, and that of the controls was 37% ($p < 0.05$).

No significant side effects were observed. It is concluded that aniracetam is an effective and safe drug to improve the cognitive and memory function in the aged and preaged.

The Relationship between the Cognitive Deterioration, Pulse Transit Time, Microcirculation, and Free Radical Metabolism in the Preaged and Aged

MA YONG-XING, ZHU YUE, CHEN SHU-YING, WANG ZAN-SHUN,
YU ZHEN-YAN, QUI ZHI-JUN, AND YU SAI-MEI

Research Division of Aging and Antiaging, Institute of Research of Gerontology
and Geriatrics of Shanghai, 221 West Yan An Road, Huadong Hospital,
Shanghai 200040, China

Memory quotient (MQ; WMS), reaction time (RT, shorter RT means better brain function), pulse transit time (PTT, shorter transit time means faster transit rate and rather severe arteriosclerosis), and free radical metabolism (plasma LPO and erythrocyte SOD, GSH, and catalase) were determined among 112 healthy and 446 unhealthy preaged and aged persons. The results were as follos: (1) healthy age group: PTT < 30 ms, 43 cases; MQ, 82.92 ± 18.21; PTT > 30 ms, 69 cases; MQ 94.25 ± 16.6; $p < 0.01$; (2) unhealthy age group: PTT < 25 ms, 139 cases; MQ, 85.99 ± 25.89; PTT > 25 ms, 307 cases; MQ, 89.50 ± 23.13; $p > 0.05$ ($t = 1.470$); (3) healthy and unhealthy aged: LPO < 5 nM/mL, 167 cases; MQ, 97.99 ± 17.06; LPO > 5 nM/mL; 66 cases; MQ, 89.87 ± 16.64; $p < 0.001$. (4) Simple reaction time (SRT) and choice reaction time (CRT) were determined in 49 cases with LPO < 4.35 nM/mL (lower LPO = LL) and 45 cases with LPO > 4.35 nM/mL (higher LPO = HL). (a) SRT of red light: LL group, 354.18 ± 74.96 ms; HL group, 427.96 ± 155.38 ms; $p < 0.01$. (b) SRT of green light: LL, 327.14 ± 62.31; HL, 394.73 ± 153.78; $p < 0.01$. (c) CRT of red light: LL, 549.91 ± 104.81; HL, 586.2 ± 139.47; $p > 0.05$ ($t = 1.434$). (d) CRT of green light: LL, 551.98 ± 126.97; HL, 616.43 ± 171.93; $p < 0.05$. (5) SOD and PTT were determined in 26 cases of severe coronary heart disease. Group of PTT > 30 s, 13 cases; SOD, 442.19 ± 50 mg/gHb; PTT < 30 s, 13 cases; SOD, 400.1 ± 46 mg/gHb; $p < 0.05$. (6) No significant differences were found between the subgroups in SOD, GSH, and catalase parameters. We conclude that cognitive deterioration in the aged is intimately correlated with arteriosclerosis and free radical metabolism.

The MQ (WMS) microcirculation of nail bed and SOD content of RBC were determined in 200 healthy and 195 unhealthy middle-aged and aged persons. The total weighted mean value of microcirculation was significantly higher (means bad state) in the MQ < 100 group (1.95 ± 1.27) than in the MQ > 100 group (1.25 ± 0.73) ($p < 0.01$) in the healthy. A high correlation between the mean value of the weighted mean microcirculatory flow velocity, aggregation rate of RBC, and the MQ were observed too. The mean value of total weighted mean of microcirculation in SOD > 410 and SOD < 410 mg/gHb in the healthy group was 1.90 ± 1.05 and 2.81 ± 2.04, respectively ($p < 0.05$), and that in SOD > 470 and SOD < 423 mg/gHb in the unhealthy group was 1.67 ± 1.01 and 2.19 ± 1.07, respectively ($p < 0.05$). It is suggested that the cognition function is better in the better microcirculatory state, and vice versa, and the microcirculatory state is better in the higher SOD group, and vice versa too.

Free Radicals and Arteriosclerosis

MA YONG-XING,[a] WANG ZAN-SHUN,[a] CHEN SHU-YING,[a] ZHU HAN-MIN,[a]
GUI LIANG-ZHEN,[b] AND CHEN WEI-QING[b]

[a]*Research Division of Aging and Antiaging, Institute of Research of Gerontology and
Geriatrics of Shanghai, 221 West Yan An Road, Huadong Hospital,
Shanghai 200040, China*
[b]*Department of Pathology of Shanghai Children's Hospital, 130 Feng Lin Road,
Shanghai Children's Hospital, Shanghai 200032, China*

FREE RADICALS AND ARTERIOSCLEROSIS OF MIDDLE-AGED AND AGED WISTAR RATS

High correlations were found between physiological degenerative nonatherosclerotic arterosclerotic changes, in which the parameters are parallelism, coarse branch, fine branch, reflectivity of elastic membrane of aorta; and LPO, SOD, catalase, GSH-Px, peroxidase, G-6-PD as the parameters of free radicals in 40 middle-aged (9 months) and 16 aged (24 months) rats who were not given high cholesterol foods.

In the elastic fibers of the elastic membrane, the moderately severe pathological changes in the specimens of rats with serum LPO $\geq$ 6.48 nmol/mL were observed in 100% of parallelism, 100% of reflectivity, and 85.7% of wave form, significantly higher than that in the LPO < 6.48 nmol/mL group (0%, 0%, and 14.3%, respectively). In mild and moderately severe pathological changes of parallelism, the mean levels of serum LPO were 5.07 ± 0.61 nmol/mL and 6.62 ± 2.03 nmol/mL, respectively ($p < 0.001$). It is indicated that free radicals are related to the damage of elastic fibers of the elastic membrane in the aorta because of the degree of pathological changes positively correlated with the serum level of LPO.

A higher level of LPO, SOD, catalase, POD, and G-6-PD were found in the groups with more severe changes in the elastic membrane.

It has been suggested that free radical damage may be one of the pathogenic mechanisms of physiological degenerative nonatherosclerotic arteriosclerosis. The pathological changes of the elastic membrane of the aorta would be more severe if the free radical production were much greater or if the regulation ability failed, thus resulting in decompensation.

THE CORRELATION BETWEEN FREE RADICALS AND ARTERIAL COMPLIANCE, AND PERIPHERAL RESISTANCE OF THE HEALTHY MIDDLE-AGED AND AGED

The catalase (Cat), peroxidase (POD), GSH-PX, SOD, and G-6-PD content of the erythrocyte, serum LPO, arterial compliance (AC), and peripheral vascular resistance (PVR) were determined in 81 healthly middle-aged and aged, persons. Significantly higher AC was found in the persons with higher level Cat and POD: (a) Cat > 13 U/mgHb; AC, 2.29 ± 0.79; Cat < 12.9 U/mgHb; AC, 1.60 ± 0.61 ($p < 0.01$); (b) POD > 45 U, AC, 2.58 ± 0.76; POD < 44.0, AC, 1.88 ± 0.70 ($p < 0.05$) The decreasing tendency of AC was found in persons with a higher LPO level. The tendency of the higher PVR was found in rather lower-level Cat and POD groups.

SOD and pulse transit time (PTT, shorter PTT means faster transit rate and rather severe arteriosclerosis) were determined in 26 persons with severe coronary heart disease: PTT > 30 s, 13 cases; SOD, 442.19 ± 50 mg/gHb; PTT < 30 s, 13 cases; SOD, 400.1 ± 46 mg/gHb, $p < 0.05$. It is suggested that arteriosclerosis is closely related to the free radical metabolic disturbance.

The Age-retarding Effect of Reinhardt and Sea Cucumber Capsule (RSC)

MA YONG-XING, WANG CHUAN-FU, YU ZHEN-YAN, YUAN ZHAO-HUI, CHUN SU-YING, AND XIE HUI

Research Division of Aging and Antiaging, Institute of Research of Gerontology and Geriatrics of Shanghai, 221 West Yan An Road, Huadong Hospital, Shanghai 200040, China

After treatment with 3 caps of RSC (a kind of sea animal product, 0.28 g/cap), t.i.d. for two months in 44 (NK) and 49 (LPO, SOD) healthy persons, the mean value of NK (nature killer cells) activity (method: lactic acid releasing, $14.5 \pm 7.3\% \rightarrow 18.6 \pm 7\%$) was significantly increased ($t = 2.6890$, $p < 0.01$), serum LPO significantly decreased ($4.86 \pm 0.82 \rightarrow 4.39 \pm 0.77$ nM/mL, $t = 2.9248$, $p < 0.01$), and erythrocyte SOD content significantly increased ($4857 \pm 410 \rightarrow 4929 \pm 463$ U/gHb, $t = 2.9248$, $p < 0.05$). There were no significant changes in NK, LPO, or SOD in the control group.

After the same dosage treatment of RSC in 53 cases of the aged, the mean value of SOD significantly increased ($4094 \pm 349 \rightarrow 4637 \pm 471$ U/gHb, $t = 6.7406$, $p < 0.001$), with no significant change in the control group ($4314 \pm 391 \rightarrow 4527 \pm 496$ U/gHb, $t = 1.8160$, $p > 0.05$). The increment of the RSC group (543 ± 540 U/gHb) is significantly higher than that of the control group, (213 ± 574 U/gHb), $t = 2.593$, $p < 0.01$. The LPO tendency decreased in the RSC group ($4.08 \pm 0.63 \rightarrow 3.88 \pm 0.61$, $t = 1.764$, $p > 0.05$), but slightly increased in the control group. No significant changes of GSH and catalase were found between the RSC group and the control group. The increment of the NK activity of the treatment group was significantly higher than that of the control group ($p < 0.05$).

The mean micro-blood-flow rate was significantly increased in the RSC group (80 cases, $1783.0 \pm 271.6 \rightarrow 1922.8 \pm 404.5$ µM/s, $t = 2.6215$, $p < 0.05$); no significant change in the control group (47 cases) was found.

The clinical trail of RSC was performed using the double blind method on 398 patients with hypomnesis (MQ of 294 cases < 100). The mean values of MQ (WMS) in the treated (mean age 73.7 ± 5.4 yrs.) and the control (mean age 71.7 ± 5.7 yrs.) groups were 91.3 ± 17.3 and 90.6 ± 16.5 ($p > 0.05$), respectively. The dosage was 0.28 g × 3 t.i.d. for two months. After the treatment, the mean MQ values of treated and control groups were 101.4 ± 18.8 and 94.5 ± 17.1, with the MQ increase of 10.63 ± 11.71 and 3.95 ± 8.29, respectively ($t = 6.5261$, $p < 0.001$). The total effective rates were 58.8% and 38.46% (Ridit $x^2 = 23.86$, $p < 0.01$) in the two groups.

No significant side effects were observed.

No beneficial effects of RSC on sexual hormone and cardiovascular function were found.

The cephalic SOD content of the fruit fly (40-day life) was more significantly increased in the middle concentration of the RSC treated group (TG) than that in the control group (CG): male, CG: 51.74 ± 1.22 U/mg, TG: 70.30 ± 1.79 U/mg, $p < 0.01$; female, CG: 22.85 ± 7.01 U/mg, TG: 46.85 ± 1.17 U/mg, $p < 0.05$.

The cephalic LPO content of the fruit fly (40-day life) was more significantly decreased in the middle and high concentration of the RSC TG than that in the control group (CG). In the former concentration, male, CG: 4.72 ± 0.08 mg/g; TG: 3.86 ± 0.17 mg/g, $p < 0.05$; female, CG: 4.98 ± 0.13 U/mg; TG: 4.30 ± 0.20 mg/g, $p < 0.05$. In the latter concentration, male, CG: 4.72 ± 0.08 mg/g; TG: 3.66 ± 0.11 mg/g, $p < 0.05$; female, CG: 4.98 ± 0.13 U/mg; TG: 4.29 ± 0.20 mg/g, $p < 0.05$.

The mean life span of the female fruit fly was significantly elongated among low, middle, and high concentrations of TG ($59.23 \pm 14.36 \rightarrow 64.83 \pm 12.23$, 65.06 ± 14.28, 62.40 ± 14.64 days, $p < 0.05 \sim 0.01$). The results of these experiments support the beneficial effects observed clinically. It is suggested that RSC, as a sea animal product, may provide longevity and healthy protection against aging.

Alzheimer's Disease

A Hypothesis on Pathogenesis

DENHAM HARMAN

University of Nebraska College of Medicine, Department of Medicine, 600 South 42nd Street, Omaha, Nebraska 68198-4635, USA

Senile dementia of the Alzheimer's type (SDAT) is the major cause of dementia. SDAT cases may be categorized into two groups: (1) late onset, after about age 60; 90–95% of cases; largely nonfamilial, that is, sporadic, and (2) early onset, before about age 60; 5–10% of cases; most, if not all, are familial. It is a systemic disorder whose major manifestations are in the brain. The major risk factor for SDAT is age; the prevalence increases exponentially with age.

The neurons involved in the brain lesions in both early and late-onset SDAT may be aging at a faster than normal rate, for the same lesions are seen in smaller numbers in normal older individuals. Inasmuch as free radical reactions have been implicated in aging, the foregoing suggests that the level of these reactions may be higher in the involved neurons

Thus, it is hypothesized that SDAT is caused by increased free radical reaction levels in brain neurons associated with SDAT. This may be accomplished by (1) mutations in mitochondrial (mt) DNA and/or nuclear (nuc) DNA in a somatic cell early in development that adversely affects mt function in neuronal daughter cells—late onset, sporadic SDAT; (2) mutations in maternal mtDNA and/or nucDNA that impair mitochondria in offspring— SDAT associated with chromosomes 1 and 14; (3) mutations in the gene on chromosome 21 for the amyloid precursor protein (APP); and (4) increased formation in all cells of both normal APP and superoxide dismutase—Down's syndrome.

The related disorder, dementia pugilistica, may be caused by increased neuronal free radical reaction levels secondary to old brain hemorrhages, whereas the etiology of Parkinson's disease could be similar to that of late-onset, sporadic SDAT.

The incidence of late-onset, sporadic SDAT may be decreased by efforts to minimize free radical reactions involved in initiation. Clinical decline of SDAT patients may be slowed by measures that lower the level in the involved neurons of more-or-less random deleterious free radical reactions.

Relationship between Fatty Acid Unsaturation, Sensitivity to Lipid Peroxidation, and Maximum Life Span in the Liver of Mammals

R. PAMPLONA,[a] M. PORTERO-OTIN,[a] D. RIBA,[a] M. LÓPEZ-TORRES,[b]
AND G. BARJA[b]

[a]Basic Medical Sciences, Faculty of Medicine, Lleida University, Spain
[b]Animal Biology-II, Faculty of Biology, Complutense University, Madrid, Spain

Unsaturated fatty acids are the tissue macromolecules most sensitive to oxidative damage. The free radical theory of aging thus predicts the presence of relatively low degrees of fatty acid unsaturation in the tissues of longevous animals. In agreement with this prediction, fatty acid analyses of liver lipids in six mammals ranging in maximum life span (MLSP) from 3.5 to 46 years showed that the number of double bonds, the peroxidizability index, and the sensitivity to lipid peroxidation are negatively correlated with MLSP. The low double content of longevous mammals was not due to a low polyunsaturated fatty acid content (PUFA); instead, it was mainly due to a redistribution between types of PUFAs from the highly unsaturated docosahexaenoic (n-3) and arachidonic (n-6) acids to the less unsaturated linoleic acid (n-6) in longevous animals. The same has been found in liver mitochondria from two endotherms with extraordinarily high longevity in relation to their body size and metabolic rate: pigeon and humans.[1] This redistribution suggests that the mechanism involved is the presence of low desaturase activities, specially ó-6 desaturase, in longevous animals. Correlations of fatty acids with basal metabolic rate had the opposite sense and were, in general, poorer than with MLSP. We propose that the low degree of fatty acid unsaturation of tissues from mammals with large body size could have been selected during evolution to decrease metabolic rate but also, and perhaps most importantly, to protect the tissues against oxidative damage, while maintaining, at the same time, an appropriate membrane fluidity.

REFERENCE

1. PAMPLONA, R., J. PRAT, S. CADENAS, C. ROJAS, R. PÉREZ-CAMPO, M. LÓPEZ-TORRES & G. BARJA. 1996. Mech. Ageing Dev. **86:** 53–66.

Localization in a Region of Complex I and Mechanism of the Higher Free Radical Production of Brain Nonsynaptic Mitochondria in the Short-lived Rat Than in the Longevous Pigeon

G. BARJA AND A. HERRERO

Department of Animal Biology-II (Physiology), Faculty of Biology,
Complutense University, Madrid 28040, Spain

Free radical production and leak of brain nonsynaptic mitochondria were higher with pyruvate/malate than with succinate in rats and pigeons. Rotenone, antimycin A, and myxothiazol maximally stimulated free radical production with pyruvate/malate but not with succinate. Simultaneous treatment with myxothiazol plus antimycin A did not decrease the stimulated rate of free radical production brought about independently by these two inhibitors with pyruvate/malate. Thenoyltrifluoroacetone did not increase free radical production with succinate. No free radical production was detected in complex IV. Free radical production and leak with pyruvate/malate were higher in the rat (maximum longevity 4 years) than in the pigeon (maximum longevity 35 years). These differences between species disappeared in the presence of rotenone. Both ethoxyformic anhydride and *p*-chloromercuribenzoate totally abolished the increase in free radical generation produced by addition of rotenone to pyruvate/malate-supplemented mitochondria. The results localize the main free radical production site of nonsynaptic brain mitochondria in complex I, between the sites of ferricyanide and ubiquinone reduction. This suggests that the complex I iron-sulfur centers are the free radical generators of these mitochondria. They also suggest that the low free radical production of pigeon brain mitochondria is due to a low degree of reduction of complex I in the steady state in this highly longevous species.

Heart and Brain OXO^{8}dG in the Genomic DNA of Rats and Pigeons and Maximum Longevity

A. HERRERO AND G. BARJA

Department of Animal Biology-II (Physiology), Faculty of Biology, Complutense University, Madrid 28040, Spain

Birds are unique inasmuch as they combine a high rate of oxygen consumption with a high maximum life span (MLSP). A similar situation is present in primates, including humans, which show MLSPs higher than predicted from their rates of O$_2$ consumption. Previous studies from our laboratory have shown that free radical production and free radical leak (percent of total electron flow directed to free radical production) are lower in mitochondria from various tissues (including heart and brain) of pigeons (MLSP = 35 years) than in those of rats (MLSP = 4 years); rats and pigeons are homeothermic animals with similar body size and basal metabolic rate. MLSP is around four times higher in birds than in the majority of mammals of the same size and metabolic rate when all studied species are considered. In order to clarify if the lower rate of free radical production of the pigeon could lead to lower oxidative DNA damage rates in this species than in rats, which could explain, in part, the widely different rates of aging of both animals, we measured levels of 8-oxo-7,8-dihydro-2′deoxyguanosine (oxo^{8}dG)/dG in the heart and brain genomic DNA by HPLC with simultaneous ultraviolet and coulometric electrochemical detection. To our knowledge there are no previous data about oxo^{8}dG/dG in birds. The results show no differences in brain oxo^{8}dG/dG between rats and pigeons. Nevertheless, heart oxo^{8}dG/dG was three-fold lower in the pigeon than in the rat, in agreement with the free radical theory of aging and with the lower rates of free radical production observed in pigeon heart versus rat heart mitochondria.

Cell Proliferation Restriction:
Is It the Primary Cause of Aging?

ALEXANDER N. KHOKHLOV

*Evolutionary Cytogerontology Sector, School of Biology,
Moscow State University, 119899 Moscow, Russia*

According to the hypothesis we have been elaborating for several years, cell proliferation restriction *in vivo*, due to appearance in the process of differentiation of cell populations with low or zero proliferative activity, leads to a sharp (10- or even 100-fold) decrease of the average cell replication rate for the organism. It induces an increase with time of the average amount of "senescence" macromolecular damage (most probably, DNA damage) per cell. This accumulation does not take place in the populations from fast proliferating cells, because "young" cells appearing after division do not have the defects (or have only a few of them). It leads to persistent "dilution" of existing or newly arising defects. The senescence DNA damage (it is implied that they are not mutations but physicochemical defects incompatible with DNA replication or inducing serious chromosomal aberrations, leading the cell to death) arises spontaneously in any cell under treatment from different chemical and physical factors (free radicals, ionizing radiation, and heat movement of molecules), and later can be eliminated by a DNA repair system, disappear together with dying cells, or remain in nondividing cells, increasing the defect "weight" in the cell population. The accumulation of DNA damage in tissues and organs leads to changes in function arising from different "age" diseases and, finally, increase in the probability of the organism's death, that is, senescence. It is necessary, however, to emphasize that we do not consider proliferation restriction as the primary cause of aging. Naturally, no one organism ages because of the slowdown in proliferation of its cells. It ages because of accumulation of DNA damage due to cell proliferation restriction. Thus, the primary cause of aging is exactly the accumulation of DNA damage inasmuch as it necessarily leads to functional impairment and an increase in the probability of death. One can easily imagine a situation wherein, with the help of some factor, the accumulation of DNA damage could be stopped or DNA repair activated. Therefore it is reasonable to ask why investigations on stationary cultures (the "stationary phase aging" model we elaborate) are beneficial for experimental gerontology. The answer seems obvious to us: with the help of the model, we can discover the concrete character of senescence DNA damage in cell populations and the peculiarities of their accumulation. Besides, the model gives us the possibility of express testing different physical and chemical factors, that could retard (or, ideally, fully stop) the accumulation.

Study of "Stationary Phase Aging" of Cultured Cells under Various Types of Proliferation Restriction

SERGE S. AKIMOV AND ALEXANDER N. KHOKHLOV

Evolutionary Cytogerontology Sector, School of Biology, Moscow State University, 119899 Moscow, Russia

During recent years we have intensively elaborated the "stationary phase aging" model as an alternative to the Hayflick model. According to our concept, various alterations accumulate in stationary cultured cells, and these alterations are similar to those during *in vivo* aging. In most cases, experiments were performed with the cells, which ceased to proliferate due to contact inhibition. It was presumed that it is the most "physiologic" way to inhibit cell proliferation, similar to the cessation of proliferation *in vivo*. However, according to our hypothesis, other methods of cell proliferation restriction, which do not promote immediate cell death, have to induce stationary phase aging too. In the present work we compared the stationary phase aging of murine 3T3 Swiss fibroblasts cultured in Dulbecco's modified Eagle medium (DMEM) with 10% bovine calf serum (BCS), DMEM with 0.25% BCS, and DMEM with 10% BCS and 500 units/mL heparin (structural and functional analogue of a component of intercellular matrix-heparan sulfate). The culture medium was changed every five days. In the medium with 0.25% serum, cells quickly cease to proliferate because of the absence of growth factors. Heparin inhibits cell proliferation most probably due to imitation of the presence of extracellular matrix components (like heparan sulfate in the organism). The cells were plated in 24-well plates in DMEM with 10% BCS. Twenty-four hours later the medium in some wells was changed to DMEM with 0.25% BCS or with heparin. Then, from the second day after plating, we evaluated (by radioautography) the capacity of the cells to reactvate DNA synthesis upon stimulation with fresh medium containing 10% BCS. It was found that in the process of the three-week stationary phase aging of cells cultured in low serum or heparin-containing medium, they gradually deepened in the resting phase (as well as the cells aging *in vivo* or *in vitro*). It manifested a persistent decrease with time of the DNA synthesis reactivation index (*i.e.*, the level of the cell nuclei labeling index increase the day after stimulation). We concluded that under the types of cell proliferation restriction investigated, the main peculiarities of stationary phase aging should be similar. The findings permit the use of low serum and heparin-containing medium for further investigations of the stationary phase aging phenomenon.

Index of Contributors

(Italic page numbers refer to comments made in discussion.)